T0337843

Fundamentals of Terahertz Devices and Applications

Fundamentals of Terahertz Devices and Applications

Edited by

Dimitris Pavlidis
Florida International University, USA

This edition first published 2021

Registered Offices
John Wiley & Sons, Inc., 111 River Street, Hoboken, NJ 07030, USA
John Wiley & Sons Ltd, The Atrium, Southern Gate, Chichester, West Sussex, PO19 8SQ, UK

Editorial Office
The Atrium, Southern Gate, Chichester, West Sussex, PO19 8SQ, UK

For details of our global editorial offices, customer services, and more information about Wiley products visit us at www.wiley.com.

Wiley also publishes its books in a variety of electronic formats and by print-on-demand. Some content that appears in standard print versions of this book may not be available in other formats.

Library of Congress Cataloging-in-Publication Data
Names: Pavlidis, Dimitris, editor.
Title: Fundamentals of terahertz devices and applications / edited by
Dimitris Pavlidis.
Description: First edition. | Hoboken, NJ : John Wiley & Sons, Ltd., 2021.
| Includes bibliographical references and index.
Identifiers: LCCN 2020041516 (print) | LCCN 2020041517 (ebook) | ISBN
9781119460718 (cloth) | ISBN 9781119460725 (adobe pdf) | ISBN
9781119460732 (epub)
Subjects: LCSH: Terahertz technology.
Classification: LCC TK7877 .F86 2021 (print) | LCC TK7877 (ebook) | DDC
621.381/33–dc23
LC record available at https://lccn.loc.gov/2020041516
LC ebook record available at https://lccn.loc.gov/2020041517

Cover Design: Wiley
Cover Image: © Comstock/Getty Images,
© kertlis/Getty Images

Set in 9.5/12.5pt STIXTwoText by Straive, Pondicherry, India

SKYBFE9EA28-3AA0-4F71-B082-BFFB93924E37_070621

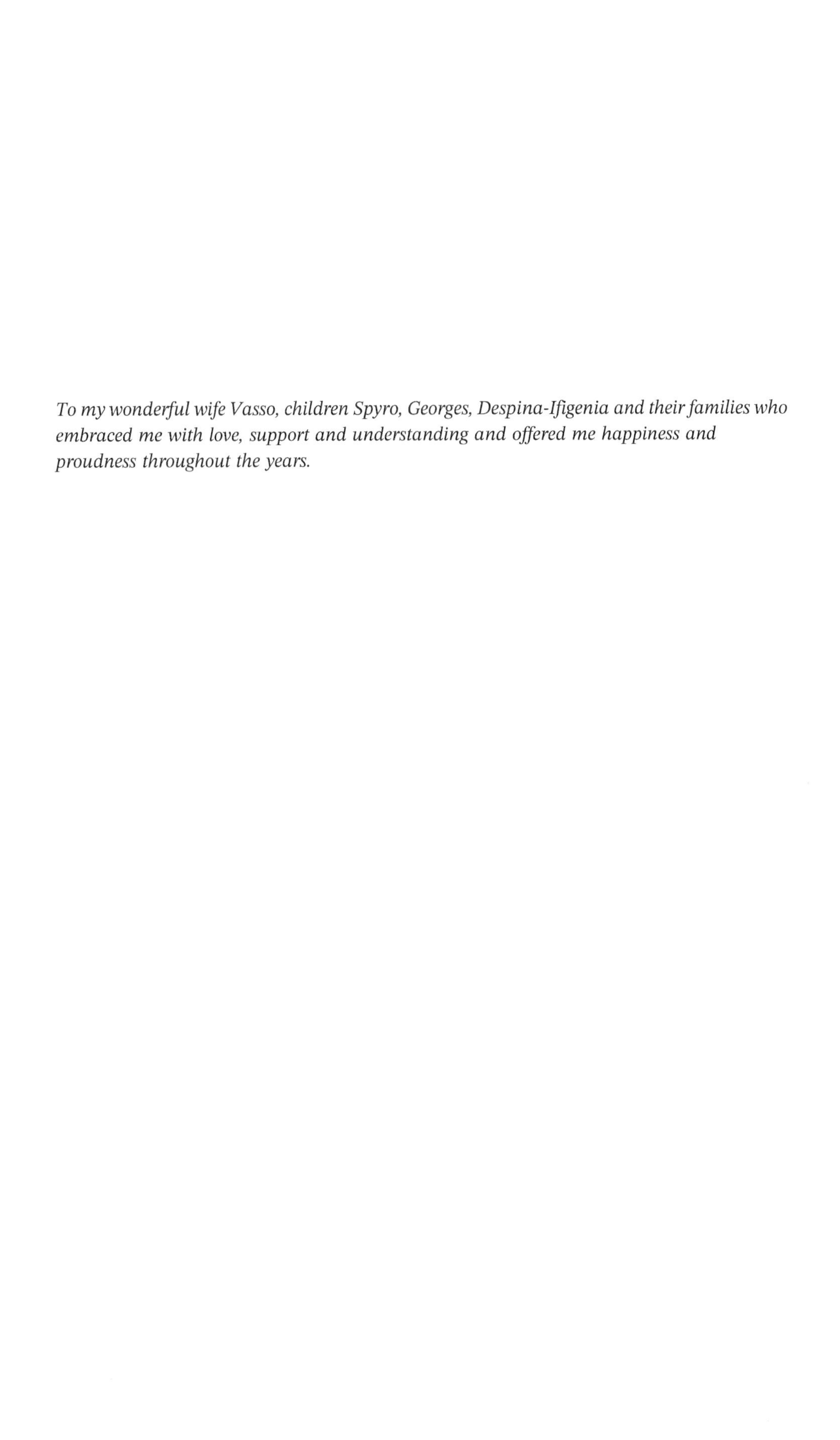

To my wonderful wife Vasso, children Spyro, Georges, Despina-Ifigenia and their families who embraced me with love, support and understanding and offered me happiness and proudness throughout the years.

Contents

About the Editor

Dimitris Pavlidis has been a Professor in Electrical Engineering and Computer Science at the University of Michigan (UofM) from 1986 to 2004 and a Founding Member of UofM's first of its kind NASA THz Center in 1988. He served as a Program Director in Electronics, Photonics, and Magnetic Devices (EPMD) at the National Science Foundation and at present is a Research Professor at Florida International University. He received the decoration of "Palmes Académiques" in the order of Chevalier by the French Ministry of Education and Distinguished Educator Award of the IEEE/MTT-S and is an IEEE Life Fellow.

List of Contributors

Maria Alonso-delPino
Jet Propulsion Laboratory, California Institute of Technology, Pasadena
CA, USA and
Department of Microelectronics Technical University of Delft
Delft, The Netherlands

Naznin Akter
Department of Electrical and Computer Engineering
Florida International University
Miami, USA

Masahiro Asada
Institute of Innovative Research
Tokyo Institute of Technology
Tokyo, Japan

Darwin Blanco
Department of Microelectronics
Technical University of Delft
Delft, The Netherlands

Elliott R. Brown
Departments of Physics and Electrical Engineering
Wright State University
Dayton, USA

Guillermo Carpintero del Barrio
Departamento Teoría de la Señal y Comunicaciones
Universidad Carlos III de Madrid
Madrid, Spain

Guillaume Ducournau
Univ. Lille, CNRS, Centrale Lille, Univ. Polytechnique Hauts-de-France, UMR 8520
Institute of Electronics, Microelectronics and Nanotechnology (IEMN)
Villeneuve d'Ascq Cedex, France

B. Globisch
Department of Physics
Technische Universität Berlin
Berlin, Germany
and
Fraunhofer Institute for Telecommunications
Heinrich Hertz Institute
Berlin, Germany

Mona Jarrahi
Department of Electrical and Computer Engineering
University of California
Los Angeles, USA

Chong Jin
ams AG,
Premstätten, Austria

Wojciech Knap
CENTERA Laboratories
IHPP-Polish Academy of Sciences
Warsaw, Poland
and Laboratoire Charles Coulomb, CNRS & Université de Montpellier
Montpellier, France

Jean-François Lampin
Univ. Lille, CNRS, Centrale Lille, Univ. Polytechnique Hauts-de-France, UMR 8520
Institute of Electronics, Microelectronics and Nanotechnology (IEMN)
Villeneuve d'Ascq Cedex, France

Nuria Llombart Juan
Department of Microelectronics
Technical University of Delft
Delft, The Netherlands

Alain Maestrini
Jet Propulsion Laboratory
California Institute of Technology
Pasadena, USA

Imran Mehdi
Jet Propulsion Laboratory (JPL)
Pasadena, USA

Tadao Nagatsuma
Graduate School of Engineering Science
Osaka University, Toyonaka
Osaka, Japan

Roberto Paiella
Department of Electrical and Computer Engineering and Photonics Center
Boston University
Boston, USA

Nezih Pala
Department of Electrical and Computer Engineering
Florida International University
Miami, USA

Dimitris Pavlidis
College of Engineering and Computing
Department of Electrical and Computer Engineering
Florida International University
Miami, USA

Emilien Peytavit
Univ. Lille, CNRS, Centrale Lille, Univ. Polytechnique Hauts-de-France
UMR 8520
Institute of Electronics, Microelectronics and Nanotechnology (IEMN)
Villeneuve d'Ascq Cedex, France

A. Rivera
Departamento Teoría de la Señal y Comunicaciones Universidad Carlos III de Madrid
Madrid, Spain

Daniel Segovia-Vargas
Departamento Teoría de la Señal y Comunicaciones
Universidad Carlos III de Madrid
Madrid, Spain

Berardi Sensale-Rodriguez
Department of Electrical and Computer Engineering
The University of Utah,
Salt Lake City, USA

Michael Shur
Electrical, Computer, and Systems Engineering Department & Physics, Applied Physics and Astronomy
Rensselaer Polytechnic Institute
New York, USA

Jose V. Siles
Jet Propulsion Laboratory
California Institute of Technology
Pasadena, USA

A. Steiger
Working Group 7.34 Terahertz Radiometry
Physikalisch-Technische Bundesanstalt
Berlin, Germany

Safumi Suzuki
Department of Electrical and Electronic Engineering,
Tokyo Institute of Technology,
Tokyo, Japan

About the Companion Website

This book is accompanied by a companion website:

https://www.wiley.com/go/Pavlidis/FundamentalsofTHz

The website includes:

- Solutions to the Exercises

1

Introduction to THz Technologies

Dimitris Pavlidis

College of Engineering and Computing, Department of Electrical and Computer Engineering, Florida International University, Miami, FL, USA

Understanding the fundamentals of Terahertz devices and applications requires thorough consideration of passive and active components together with system perspectives. In terms of passive components, antennas are a key element for signal handling, while signal generation and detection can be achieved by various means such as photoconductive (PC) devices, photomixers, plasmonic PC devices, and duantum cascade lasers (QCLs). In addition to these approaches are based on optical concepts, electronic approaches are also explored and implemented. These include advanced devices using two-dimensional (2D) layer technology, plasma field-effect transistor detectors, diode multipliers, and resonant-tunneling diodes (RTDs). THz systems combine such passive and active devices for responding to various application needs such as communication and sensing.

System operation at THz frequencies requires signal generation, emission, propagation, and reception. A key element for such systems is antennas that are discussed in Chapter 2. To fulfill the resolution or sensitivity requirements of most submillimeter-wave instruments, especially very high gain reflector-based antennas are necessary. These are illuminated by antenna feeds integrated with the transceiver/receiver front-ends and based on horns or silicon lens antennas. Horn antennas are easily connected to a waveguide-based front-ends and can be easily manufactured while presenting good radiation properties. Lenses can on the other hand be easily integrated with bolometer detectors and silicon-based font-ends. They are often used to couple to direct detectors instruments and used in planar form with superconducting-insulator-superconducting (SIS) and hot-electron bolometric (HEB) mixers as well as in PC systems on bow-tie and logarithmic spiral form.

New THz antenna arrays based on horns and lenses benefited from advances in photolithography and micromachining to respond to the needs of multi-pixel systems operating at submillimeter wave-bands. Another important point for successful THz system operation is the reduction of transmission losses which can be high in metals. To overcome this difficulty, superconducting-based microstrip lines can be used and employed in phased arrays. A disadvantage of this approach is the need for cryogenic cooling for operation which makes

Fundamentals of Terahertz Devices and Applications, First Edition. Edited by Dimitris Pavlidis.

Companion website: www.wiley.com/go/Pavlidis/FundamentalsofTHz

it impractical. Of major importance is the good understanding of the operation of integrated lens antennas and the way one can analyze them. Consideration of this type will be presented together with detailed discussions on elliptical lens and semi-hemispherical lens antennas, excitation of shallow lenses by leaky-wave/Fabry–Perot feeds, and fly-eye antenna arrays.

THz sources and receivers benefit from the availability of technologies relying on ultrafast photoconductors and PIN-based photodiodes and operate at frequencies that can exceed 300 GHz. These are analyzed and compared in detail in Chapter 3 by considering the associated optical and transport physics, but also practical effects such as contact effects, thermal stress, and circuit limits. A variety of THz PC sources are studied including PC-switches, photomixers, p-i-n photodiodes, and metal-semiconductor-metal (MSM) bulk photoconductors. The fundamental principles of THz antenna coupling are discussed and the input impedance, as well as the increase in the equivalent isotropic radiated power (EIRP) of the transmitting antenna, are reviewed for planar antennas on dielectric substrates. Resonant antennas and self-complementary antennas are also studied. Good understanding of material growth is necessary for ultrafast photoconductors and low-temperature GaAs, as well as InGaAs are considered for this purpose. To characterize with high precision THz components and in particular their power properties, a new, traceable thin-film pyroelectric detector technology is discussed. Wireless communications and spectroscopy are two major applications of THz technology. These are extensively discussed together with device as well as signal processing considerations for their better understanding.

Further information on the generation of THz continuous waves based on the optical heterodyne approach is provided in Chapter 4 by employing two-slightly detuned infrared lasers. The ultrafast photoconductors necessary for this purpose are based on subpicosecond carrier lifetime semiconductors such as low-temperature grown GaAs and InGaAs:Fe and uni-travelling-carrier (UTC) InP/InGaAs photodiodes. Electrical models are extracted for the photoconductors, PIN photodiodes, and UTCs, and their efficiency and maximum power achieved are examined. Attention is paid on the characteristics of backside illuminated and waveguide-fed UTC photodiodes. Planar and micromachined antennas are being considered for photomixing systems and attention is paid on their on wafer as well as free-space characterization.

Plasmonics based approaches can be used to enhance the performance of PC antennas and are discussed in Chapter 5. Good understanding of the photoconductor physics is necessary for this purpose together with consideration of the impact of the PC antenna and its operation as emitter and detector of pulsed and continuous-wave (CW) THz radiation. The fundamentals of plasmonics are analyzed for better performance optimization of THz devices and design considerations are made for plasmonic nanostructures. Studies are also performed on PC THz devices with plasmonic contact electrodes, large area plasmonic PC nanoantenna arrays, and plasmonic PC THz devices with optical nanocavities.

QCLs are promising devices for THz signal generation. Chapter 6 discusses their design, state-of-the-art performance and limitations, and potential for improvement based on novel materials systems. Intersubband (ISB) transitions in quantum wells (QWs) allow laser emission at THz frequencies and provide a solution to the difficulty encountered due to the lack of materials with sufficiently small bandgap energies. The basic physics involved in them is reviewed including optical absorption and emission processes and phonon-assisted nonradiative transitions. Considerations are made for the design of the QC gain medium and optical cavity, as well as the use of plasmonic waveguides to achieve strong optical confinement. Other QCL properties of interest are spectral coverage, output power, and temperature characteristics. The limitations imposed in the use of GaAs/AlGaAs QWs due to, the presence of THz-range optical phonons and thus ability to cover the entire THz spectrum emit without cryogenic cooling can be overcome through the use of GaN/AlGaN QWs, where the optical phonon frequencies are above THz range, and SiGe which has significantly weaker electron–phonon and photon–phonon interactions compared to III–V compound semiconductors.

2D layer technology can be used for various devices including those operating at THz as described in Chapter 7. Of interest is their very strong tunable electromagnetic response at THz, which can be utilized for realizing active devices such as amplitude and phase modulators as well as active filters. Beam shaping and real-time terahertz imaging can be achieved using metamaterial structures as well as large arrays. Graphene and graphene-based, as well as transition-metal dichalcogenides, offer the possibility of realizing terahertz devices. Their modeling is discussed and system applications of them are considered using modulator arrays in terahertz imaging.

To respond to the needs of THz sensing, imaging, and communication technology for detectors with high responsivity, selectivity, and large bandwidth plasma wave electronics are explored in Chapter 8. Different material systems can be investigated for this purpose and responsivities up to tens of kV/W and noise equivalent power (NEP) down to the sub-pW/Hz1/2 range have been achieved. Very high-speed communications can take advantage of their bias dependent tuning and possibility of very high modulation frequency up to 200 GHz. Devices studied for this purpose include field-effect transistors with resonant and broadband detection characteristics. Silicon and graphene materials are used for such devices Graphene and 2D-layered materials (Black Phosphorous), as well as diamond, are other possible candidates.

Details on multipliers fundamentals and their space applications are provided in Chapter 9. Their basic properties are analyzed together with a consideration of their noise characteristics. A practical approach is presented for the design of frequency multipliers and the evolution of THz frequency multiplier technology is discussed with an emphasis on the building of local oscillators. The design and fabrication of modern terahertz frequency multipliers are discussed and the case study of 2.7 THz balanced triplers is analyzed. Power combining together with integration considerations are also made. A new generation of room-temperature terahertz Schottky diode-based frequency multiplier sources presents 1 mW of output power at 1.6 THz and measured conversion efficiencies follow the theoretical limit predicted by physics-based numerical models. These yield a very significant increase in performance above 1 THz in both conversion efficiency and generated output power.

Frequency multipliers using diodes have offered the possibility of generating up to THz signals using initially hybrid approaches and later on planar and integrated design. These are discussed in Chapter 10 with main emphasis on GaN-based approaches which offer the possibility of handling the high-power levels currently possible at millimeter-wave frequencies, enabling compact size signal generation at THz. Theoretical considerations of GaN Schottky diodes using analytical and numerical approaches allow a better understanding of their non-linear properties and the way they can be best optimized. Parameters of interest to be studied are the device structure (materials, composition, geometry), breakdown voltage, I–V characteristics, as well as parameters the series resistance and C–V characteristics. They can be correlated to performance properties such as power handling capability, losses, and nonlinearity. Optical and E-Beam lithography may be used for diode fabrication. The latter opens the possibility for sub-micron anode realization, meeting the requirements of THz applications. Small but also large-signal characterization allows to extract their properties and derive models for their circuit applications. They can also assist in explaining difficulties arising in performance optimization from periphery effects, dislocation assisted reverse current. The large-signal network analyzer (LSNA) method can provide rapid evaluation of diodes which is important for rapid device development and multipliers. various multiplier device types, designs, and fabrication approaches are being considered for frequency multipliers. This includes GaN-based vertical device and heterojunction designs, i.e. InN/GaN, transistors.

RTDs is a good candidate for THz oscillators at room temperature and are discussed in Chapter 11. Promising results with oscillation frequency up to 650 GHz have been reported in the 90s for RTDs with planar antennas. Improvement in the RTD structure for short electron delay time and the antennas for low conduction loss allowed more recently demonstration of operation reaching ~2 THz and both low, i.e. GaAs and InP, and large bandgap materials, i.e. GaN have been used for their fabrication. Progress on structure optimization for high-frequency and high-output power operation, resonator and radiator type, frequency-tunable RTD oscillators, and compact THz sources allow their consideration for applications such as wireless data transmission, spectroscopy, and imaging.

Wireless communication systems at THz are described in Chapter 12. Since the electromagnetic spectrum is saturated on most already allocated frequencies, systems operating above 100 GHz, i.e., in the 200–320 GHz range draw considerable interest for very high-speed wireless transmissions. Electronic and photonic building blocks are of interest for this purpose. THz transmitters, receivers, and the basic architecture of transmission systems are discussed together with various devices suitable for T-ray communication such as photomixers and approaches suitable for the generation of modulated THz signals. Integration approaches, ways of interconnection, and antennas are key components to be investigated for the realization of THz communication systems. Communication links using both electronic- and photonic based approaches are also described.

The interest into solar system objects and the interstellar medium has led in space instrument investments and consideration of THz technologies that allow insight into solar system objects and the interstellar medium. The technology and engineering aspects of the heterodyne receiver which is the system of choice for conducting high-resolution spectroscopy for space applications I described in Chapter 13. Its critical components such as mixers (Schottky diode, SIS Mixer, and Hot-Electron Bolometric Mixers) and local oscillators (frequency multiplied chains) are also analyzed together with three distinct space science applications for THz instruments and how these applications are currently driving technology development. These include planetary science and miniaturization, astrophysics, and THz array receivers, as well as, earth science: and active THz systems.

2

Integrated Silicon Lens Antennas at Submillimeter-wave Frequencies

Maria Alonso-delPino[1,2], Darwin Blanco[2] and Nuria Llombart Juan[2]

[1] *Jet Propulsion Laboratory, California Institute of Technology, Pasadena, CA, USA*
[2] *Department of Microelectronics, Technical University of Delft, Delft, The Netherlands*

2.1 Introduction

Most of submillimeter-wave instruments, especially for space applications, make use of very high gain reflector-based antennas to fulfill the desired resolution or sensitivity requirements. This reflector based quasi-optical systems are illuminated by antenna feeds integrated with the transceiver/receiver front-end. At submillimeter-wave frequencies, these antenna feeds are mostly based on horns or silicon lens antennas.

For single-pixel heterodyne split block waveguide-based instruments such as [1], horn antennas are typically preferred due to their straightforward connection to a waveguide-based front-end architecture, their manufacturability using metal machining processes, and their good radiation properties. For example, the diagonal horn and Pickett-Potter horn achieve a relatively good performance that is compatible with a simple split-block fabrication process [2, 3]. For better performance, electroforming is a viable option to fabricate corrugated horns, which are commercially available for frequencies of 1.47 THz [4].

On the other hand, lenses are widely used for coupling into a planar antenna architecture that is often integrated with a bolometer detector or silicon-based front-ends. They have been widely used to couple to direct detectors instruments as in [5, 6]. There are also many examples of superconducting-insulator-superconducting (SIS) and hot-electron bolometric (HEB) mixers based on planar antenna architectures for heterodyne instrumentation [7, 8]. Moreover, they are also commonly used for standalone photoconductive systems using broadband antennas such as bow-ties, logarithmic spirals as in [9–11]. These hybrid antennas have multiple advantages compared to waveguide type of antennas: low loss, easy integration with receiver, and low cost of manufacture. The antenna and detector are processed using photolithographic processes on a wafer and the lenses can be fabricated using milling techniques. These lenses are usually made of silicon, which is set to match the permittivity of the substrate, are comparatively inexpensive and easy to assemble, and are air-coated

Fundamentals of Terahertz Devices and Applications, First Edition. Edited by Dimitris Pavlidis.

Companion website: www.wiley.com/go/Pavlidis/FundamentalsofTHz

with a quarter wavelength Parylene layer. They operate over large bandwidths providing good performances.

Multiple applications in the submillimeter-wave band require the use of multi-pixel systems in order to maximize the data output or reduce the image acquisition time of an imaging system. The development of antenna focal plane arrays has been challenging due to the packaging and fabrication limitations of these antennas, especially above 0.5 THz. The recent advances in photolithographic, laser micro-machining, and metal computer numerical control (CNC) machining fabrication have enabled the development and growth of new terahertz antenna arrays, which are based on horns or lenses.

There are few examples at THz frequencies of horn arrays above 300 GHz due to the complexity of fabrication and integration. Initially, the approach was to individually packed horn antennas, as the horn array used in [12]. However, this approach limits the inter-pixel spacing on the focal plane array and consequently the sampling of the observation image. In order to reduce the distance between the elements, the full array needs to be fabricated on the same metal block, which requires milling or drilling techniques. Milling techniques have been used in W-band such as the 31-pixel array at 100 GHz shown in [13]. For higher frequencies, drilling techniques with custom drill bits have been successfully employed allowing the fabrication of multi-flare angle horn arrays as in [14]. Other less conventional techniques are laser silicon micromachining, which can be employed for the milling of corrugated horns as demonstrated at 2 THz in [15], and photolithographic processes, which has been demonstrated in [16]. Both methods rely on stacking together a number of thin gold plated silicon wafers with tapered holes etched at 90°.

Transmission lines at terahertz frequencies suffer from high losses due to the metal losses, which make the development of phased arrays impracticable. However, it is not the case if the microstrip lines are made of a superconducting material. This approach was used for the BICEP2 instrument, where an array of 10 × 10 different phased arrays coupled to an array of horns was developed [17]. Each phased array was composed of 100 pixels, a transition edge sensor (TES) detector and a horizontal and vertical slot, for horizontal and vertical polarization detector. Nevertheless, superconductor materials require cryogenic cooling to operate, which makes this solution not viable for many applications.

Even though integrated lens antennas are the most suited for focusing on a planar antenna, there a very few systems implemented with large lens arrays. There have been some examples of lens arrays fabricated and assembled individually as in [18]. However, it has not been until the last years that a great development on integrated lens arrays has been made at terahertz frequencies. Advances on silicon micromachining have enabled the fabrication of large arrays of lenses on a single block piece. An array of 989 silicon pixels integrated with Kinetic Inductance Detectors (KIDs) has been developed at 1.4 and 2.8 THz [19]. The silicon lenses were fabricated on a single silicon block using laser micromachining. Another example has been the use of a photolithographic process based on deep reactive ion etching (DRIE) to fabricate shallow lenses as in [20].

In terms of performance, silicon lens antennas have been able to reach a high quasi-optical coupling efficiency for both single-pixel and focal plane array architectures. It was traditionally achieved using resonant double slot antennas as in [21] and [22], and now broadband using a leaky-wave slot as in [23]. New analysis methods and optimization techniques have been developed to improve the coupling between the lens and antenna mechanisms [24]. While books on horn and phased array design are extensive, the techniques to analyze integrated lenses antennas, especially for high directive feed antennas, i.e. leaky-wave antennas, are not that common. This chapter explains how integrated lens antennas work and can be analyzed.

Recently, a new lens feed concept has been developed in [25, 26] coupling the shallow photo-lithographic silicon lenses to a waveguide feed system with extremely low loss and high efficiency. The concept, which uses a novel leaky-wave/Fabry–Perot resonator also formed monolithically, has negligible Ohmic losses at THz frequencies while producing a very directive field inside the lens; hence producing a highly isolated and directed output beam. These monolithic leaky-wave feeds have now demonstrated the highest directivity-to-loss ratio ever reported in the THz band. Moreover, whereas traditional Fabry–Perot resonator-based antennas typically have very narrow bandwidths, this leaky-wave feed has yielded bandwidths in excess of 15%, well matched to most implemented or proposed THz heterodyne systems to date. This new lens feed concept unveils a wide range of possibilities for direct detection and heterodyne instruments, thanks to the high performance achieved for its use on highly packed focal plane arrays. This chapter will also address the design of these new lens feeds.

This book chapter is organized as follows. First, we cover the design and the analysis of elliptical lens antennas, providing analytical formulation to synthesize the lens and compute the radiation of the lens excited by a feed. Second, the semi-hemispherical lens antenna is described, from its synthesis to its radiation properties while comparing it with the elliptical lens. Third, we explain the excitation of shallow lenses by leaky-wave/Fabry–Perot feeds: starting from the analysis of the leaky-wave effect and computing its propagation constant, then analyzing its radiation into an infinite medium (primary field), and finishing up with the optimization of the shallow-lens geometry. Last, we explain how to develop a fly-eye antenna array by describing the fabrication using silicon micromachining of the full lens antenna with DRIE techniques, evaluating the surface accuracy of the lens, and providing some examples of fabricated antennas. In addition, the chapter concludes with some worked examples to make the reader consolidate and reflect on the lessons learned.

2.2 Elliptical Lens Antennas

Integrated silicon lens antennas are well suited for submillimeter-wave applications because of their focusing capabilities into a planar antenna. At submillimeter-wave frequencies, planar antennas suffer from high power loss due to the excitation of multiple surface wave modes on thick dielectric substrates. When a radiating source is placed on a dielectric surface, the rays propagate through the dielectric reaching the dielectric-air interface. At that point, the rays that reach the substrate edge with an angle above the critical angle, defined as $\theta_c = a\sin(1/n_{subs})$ being n_{subs} the refraction index of the substrate, are reflected back into the substrate generating reflections along the transversal axis of the dielectric (see Figure 2.1a). These are trapped waves that do not radiate into free space; thus they represent an efficiency loss.

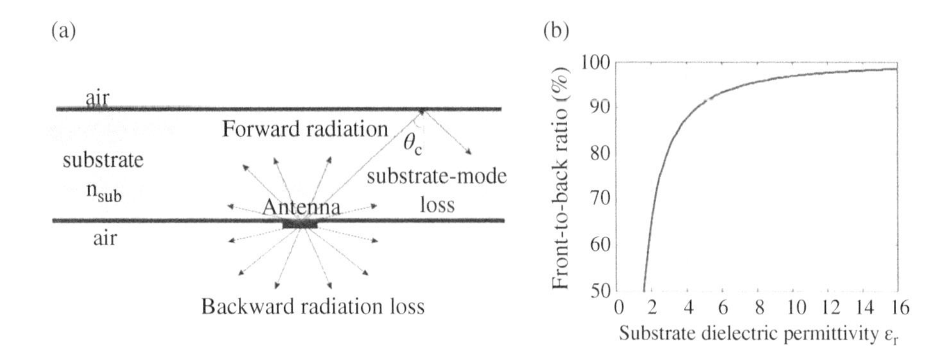

Figure 2.1 (a) Sketch of a planar antenna printed on a dielectric substrate. (b) Front-to-back ratio of an elementary dipole antenna as a function of the permittivity of the substrate [27]. *Source*: Modified from Rutledge *et al.* [27]; John Wiley & Sons.

One way to mitigate this loss is to reduce the thickness of the substrate until it is electrically thin ($\approx\lambda_{subs}/100$, being λ_{subs} the wavelength in the substrate $\lambda_{subs} = \lambda_0/n_{subs}$) as in [28, 29] but at the cost of a poor front to back radiation. The antenna is integrated on a thin dielectric membrane which is typically less than 5 μm thick and realized on silicon, SiN, or SiO_2 dielectrics. These antennas will couple weakly to the substrate and they will radiate the same amount of energy in the front and back hemispheres as if they were suspended in free space. In Ref. [30], a membrane of 1 μm of SiN was fabricated for an antenna working at 700 GHz.

Another approach is to use a thick substrate together with a lens of the same material to couple efficiently the radiation into free space as proposed by Kominami et al. [31]. The antenna is placed between two infinite mediums, one on the top and one on the bottom, and the top of the dielectric medium is curved in order to couple the radiation into a directive beam without having critical angle reflections. The front-to-back ratio of an elementary dipole planar antenna between the two mediums can be approximated as $\eta_{front\text{-}to\text{-}back} = 1 - 1/\varepsilon_r^{3/2}$ [27]. Silicon and quartz are two of the materials that are used for the fabrication of integrated lenses due to their low dielectric loss and still high front-to-back ratios. As shown in Figure 2.1b, when using a dielectric of silicon ($\varepsilon_r \approx 11.9$) the power radiated to the silicon is around 97.5% of the power radiated to the air, while if the lens is quartz ($\varepsilon_r \approx 4$), the power radiated is around 87.5%. The lens is put on the backside of the antenna on the substrate radiates most of their power into the dielectric side making the pattern unidirectional and also providing thermal and mechanical stability.

2.2.1 Elliptical Lens Synthesis

In the following example, we will derive the canonical geometry (elliptical lens) that achieves a planar wave front at the aperture, starting from a spherical wave at its focus. This demonstration is based on a geometrical optics (GO) approach, or also called ray tracing. We start by imposing that all the rays in Figure 2.2a have the same electrical length:

$$k_d r_1 + k_o r_2 = k_d h \tag{2.1}$$

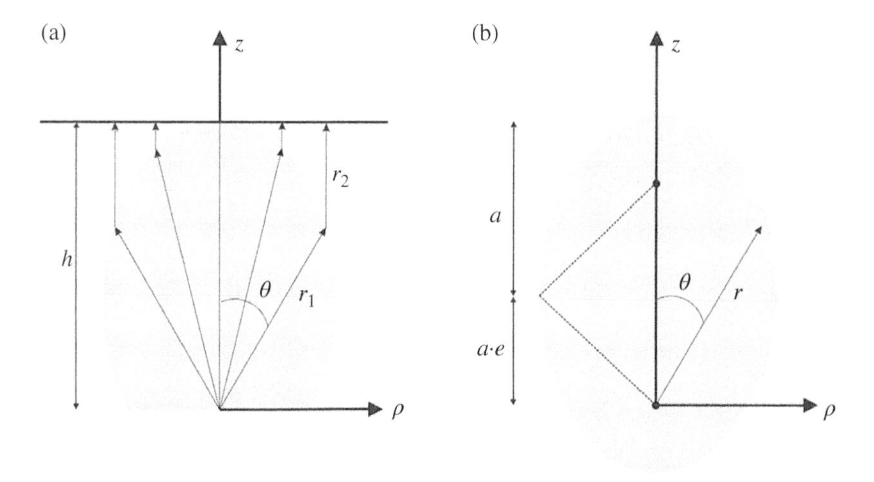

Figure 2.2 Geometrical parameters of an elliptical lens.

where k_0 and $k_d = \sqrt{\varepsilon_r} k_0$ are the wave numbers in free space and the dielectric mediums, respectively. Equation (2.1) can be rewritten as:

$$r_2 = \sqrt{\varepsilon_r}(h - r_1) \tag{2.2}$$

And we can also equate the projection in the z-axis of these rays:

$$r_1 \cos\theta + r_2 = h \tag{2.3}$$

using (2.1) and (2.2) we conclude on the following equation:

$$r_1 = h \frac{1 - \sqrt{\varepsilon_r}}{\cos\theta - \sqrt{\varepsilon_r}} \tag{2.4}$$

On the other hand, we know that an elliptical lens geometry can be described in polar coordinates using the following expression:

$$r_1 = a \frac{1 - e^2}{1 - e\cos\theta} \tag{2.5}$$

where a and e are the semi-major axis and eccentricity, respectively (see Figure 2.2b).

By recognizing in (2.5) that:

$$e = \frac{1}{\sqrt{\varepsilon_r}} \tag{2.6}$$

$$h = a(1 + e) \tag{2.7}$$

we can conclude that a spherical wave front produced by an antenna at the focus point of a lens of with an ellipsoidal shape will be transformed into a planar waveform. The antenna is placed at the second focus of an ellipse defined by the equation:

$$\left(\frac{x}{a}\right)^2 + \left(\frac{y}{b}\right)^2 = 1 \tag{2.8}$$

where b is the semi-minor axis of a ellipse. The eccentricity of the lens e, relates the geometric focus ellipse to the optical focus of the lens with the following relationship:

$$e = \frac{\sqrt{b^2 - a^2}}{b} = \frac{1}{\sqrt{\varepsilon_r}} \tag{2.9}$$

using Eqs. (2.7) and (2.8) we can derive the foci of the ellipse, defined as c as:

$$c = \sqrt{b^2 - a^2} \tag{2.10}$$

and the semi-minor axis b as:

$$b = \frac{a}{\sqrt{1 - 1/\varepsilon_r}} \tag{2.11}$$

The radiation pattern obtained by elliptical lenses is the one that reaches the highest possible directivity when illuminated with a spherical phase front generated by the feeding antenna.

2.2.2 Radiation of Elliptical Lenses

For electrically moderate and large lenses, the use of high frequency analysis techniques is typically the preferred approach. These techniques are based on Physical Optical (PO) where the field scattered by the lens is evaluated from an equivalent surface currents approximated using the incident field. These currents can be evaluated directly either at the lens surface [32] or over a planar aperture, just above the lens as shown in Figure 2.3a. In both cases, the incident field coming from a known feed is propagated until these considered surfaces using a GO approach. Then, the equivalent current densities are integrated with the free space Green's function to obtain the field radiated by the lens antenna.

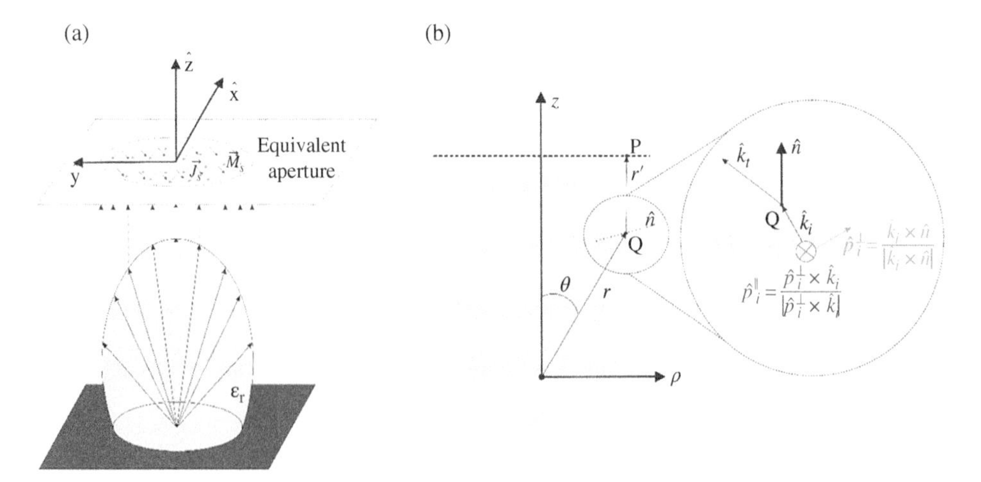

Figure 2.3 (a) Equivalent aperture on top of the elliptical lens antenna. (b) Scheme of the ray tracing of the elliptical lens antenna at a point Q in the surface.

In this section, we will derive an approximate analytical expression of these equivalent currents over a planar surface on top of the lens. This expression is valid only when the feed phase center is located in the lower focus of the ellipse and the lens surface is in the far field of the feeder. This expression can be used to perform a fast optimization of the lens and planar feeder. In order to explain this analysis method we will first evaluate the equivalent current distribution over a plane parallel to the lens aperture (i.e. that lies outside of the lens, see Figure 2.3a). The reference system used to evaluate the radiation patterns of the lens antenna is the one shown in Figure 2.3a. There are three approximations to calculate these equivalent surface currents:

First is a GO approximation of the fields. The fields over the aperture are well-approximated by the fields transmitted by the lens considering a locally flat surface to derive the transmission coefficients. The dyadic representation of the transmitted field to the incident field on the lens surface is:

$$\vec{E}_t(P) = \widetilde{T}(Q) \cdot \vec{E}_i(Q) S(Q) e^{-jk_0 r\prime} \tag{2.12}$$

$$\vec{H}_t(P) = \frac{1}{\zeta_0} \hat{z} \times \vec{E}_t(P) \tag{2.13}$$

where ζ_0 is the free space wave impedance, $S(Q)$ is the spreading factor that ensures the power density within the solid angle of the feed is equal to the power density transmitted by the lens. $\widetilde{T}(Q)$ is the transfer function calculated using the Snell's law. $\vec{E}_i(Q)$ is the radiated field by the feed placed at the lower focus of the ellipse, at an observation point Q (see Figure 2.3b). The fields in (2.12) and (2.13) are propagating in the $\hat{z}$ direction.

Second is an approximation of the equivalent currents. The equivalent electric and magnetic currents are obtained from these GO fields:

$$\vec{M}_s(\rho', \phi') \approx \vec{E}_t \times \hat{z}\Big|_S \tag{2.14}$$

$$\vec{J}_s(\rho', \phi') \approx \hat{z} \times \vec{H}_t\Big|_S \tag{2.15}$$

which are defined for $\rho < D/2$ with $D = 2b$ (twice the minor axis of the ellipse) associated with the GO field domain, and zero outside this domain.

The third approximation relies on the fact that we assume the incident field on the lens surface $\vec{E}_i$ to have a spherical wave front, which is correct when the feed phase center is in the lower focus and the lens surface in the feed far field:

$$\vec{E}_i\left(\vec{r}\,'\right) = \vec{f}\,(\theta', \phi') \frac{e^{-jkr\prime}}{r'} \tag{2.16}$$

$$\vec{H}_i\left(\vec{r}\,'\right) = \frac{1}{\zeta_d} \hat{r}' \times \vec{E}_i\left(\vec{r}\,'\right) \tag{2.17}$$

where $\zeta_d = \zeta_0 / \sqrt{\varepsilon_r}$ is the medium wave impedance. The reference system used to evaluate this field is taken at the lower focus of the elliptical lens.

Next, we will use these approximations to express the radiation in the far-field of the elliptical lens, when illuminated from this incident field $\vec{E}_i$.

2.2.2.1 Transmission Function $\widetilde{T}(Q)$

Let's now analyze an arbitrary point Q over a lens surface. The transmitted magnetic and electric field outside of the lens $\vec{E}_t(P), \vec{H}_t(P)$ are related to the incident field $\vec{E}_i(Q)$ through the transmission function $\widetilde{T}(Q)$ as given in (2.17). In general, at each point Q, the incident field can be decomposed into parallel-TM $E_i^{\|}(Q^-)$ and perpendicular-TE $E_i^{\perp}(Q^-)$ components:

$$\vec{E}_i(Q^-) = E_i^{\|}(Q^-)\hat{p}_i^{\|}(Q) + E_i^{\perp}(Q^-)\hat{p}_i^{\perp}(Q) \tag{2.18}$$

where $\hat{p}_i^{\|}(Q)$ and $\hat{p}_i^{\perp}(Q)$ are the parallel and perpendicular polarization vectors (see Figure 2.3b), and can be expressed in spherical coordinates for the on-focus feed as:

$$\hat{p}_i^{\perp} = \frac{\hat{k}_i \times \hat{n}}{\left|\hat{k}_i \times \hat{n}\right|} = \hat{\phi} \tag{2.19}$$

$$\hat{p}_i^{\|} = \frac{\hat{p}_i^{\perp} \times \hat{k}_i}{\left|\hat{p}_i^{\perp} \times \hat{k}_i\right|} = \hat{\theta} \tag{2.20}$$

which brings to:

$$E_i^{\perp}(Q^-) = E_i^{\phi}(Q^-) \tag{2.21}$$

$$E_i^{\|}(Q^-) = E_i^{\theta}(Q^-) \tag{2.22}$$

For a feed placed in the center of the coordinate system, the propagation vector see Figure 2.3b inside and outside of the lens are defined as:

$$\vec{k}_i = k_d\hat{r} = k_d(\sin\theta\hat{\rho} + \cos\theta\hat{z}) \tag{2.23}$$

$$\vec{k}_t = k_0\hat{z} \tag{2.24}$$

The normal vector to the lens surface defined in cylindrical coordinates for an elliptical lens is:

$$\hat{n} = \frac{(\cos\theta - e)\hat{z} + \sin\theta\hat{\rho}}{\sqrt{1 + e^2 - 2e\cos\theta}} \tag{2.25}$$

Then, the transmitted field can be expressed as:

$$\vec{E}_t(Q^+) = \tau^{\perp}(Q)E_i^{\perp}(Q^-)\hat{p}_t^{\perp}(Q) + \tau^{\|}(Q)E_i^{\|}(Q^-)\hat{p}_t^{\|}(Q) \tag{2.26}$$

being $\tau^{\perp}(Q)$, $\tau^{\|}(Q)$ the Fresnel transmission coefficients which are evaluated assuming a locally flat boundary (see Figure 2.4):

$$\tau^{\perp}(Q) = \tau^{\perp}(\theta_i) = \frac{2\zeta_0\cos\theta_i}{\zeta_0\cos\theta_i + \zeta_d\cos\theta_t} \tag{2.27}$$

$$\tau^{\|}(Q) = \tau^{\|}(\theta_i) = \frac{2\zeta_0\cos\theta_i}{\zeta_0\cos\theta_t + \zeta_d\cos\theta_i} \tag{2.28}$$

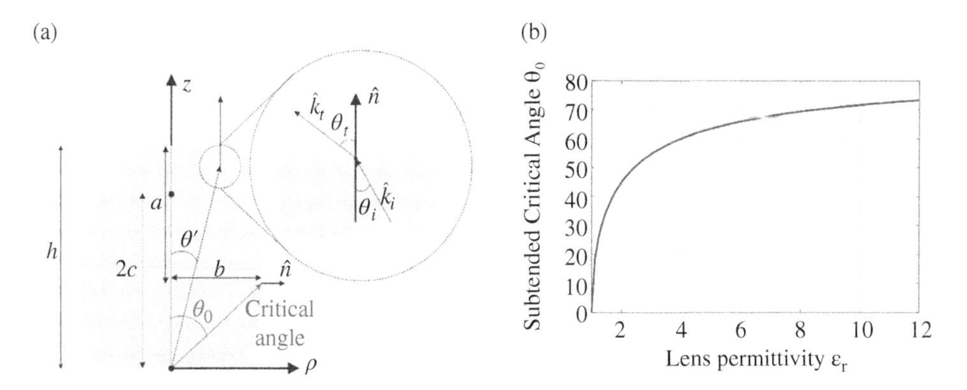

Figure 2.4 (a) Scheme of the critical angle calculation on an elliptical lens. (b) Subtended critical angle vs permittivity of the elliptical lens.

where $\cos\theta_i = \hat{k}_i \cdot \hat{n} = \frac{1 - e\cos\theta}{\sqrt{1 + e^2 - 2e\cos\theta}}$. Using (2.27) and (2.28) in the equation (2.26), and considering that $\hat{p}_t^{\perp} = \hat{\phi}$ and $\hat{p}_t^{\parallel} = \frac{\hat{p}_t^{\perp} \times \hat{k}_t}{|\hat{p}_t^{\perp} \times \hat{k}_t|} = \hat{\rho}$, we finally arrive at the expression of the transmitted fields at the equivalent aperture:

$$\vec{E}_t(P) = \left[\tau^{\perp}(\theta_i) E_i^{\phi}\left(\vec{r}\right)\hat{\phi} + \tau^{\parallel}(\theta_i) E_i^{\theta}\left(\vec{r}\right)\hat{\rho}\right) e^{-jk_0 r'} \tag{2.29}$$

$$\vec{H}_t(P) = \frac{1}{\zeta_0}\left(-\tau^{\perp}(\theta_i) E_i^{\phi}\left(\vec{r}\right)\hat{\rho} + \tau^{\parallel}(\theta_i) E_i^{\theta}\left(\vec{r}\right)\hat{\phi}\right] e^{-jk_0 r'} \tag{2.30}$$

The spatial domain of (2.29 and 2.30) is determined by the GO approximation. In case of a full angular lens as depicted in Figure 2.4a, the GO transmitted fields will be zero after the critical angle. For a flat interface between a dense medium and free space, the angle of refraction θ_t increases as the incident angle θ_i increases ($\theta_t > \theta_i$). Total internal reflection ($\Gamma = 1$) occurs when the angle of refraction is 90° and the incident angle is known as the critical angle $\theta_c = \sin^{-1}\left(1/\sqrt{\varepsilon_r}\right)$. After this angle there is no transmission, and therefore, this angle defines the domain of the equivalent aperture. As this angle lies at the center distance between the two focuses, the radius of the equivalent aperture is therefore the minor axis b. Thus, the fields in (2.29 and 2.30) are zero for a region outside the lens antenna aperture, i.e. $\rho > D/2$ with $D = 2b$ in the case of a full angular lens. In the case of a smaller angular lens such as in [33], the spatial domain of the fields will be given by edge angle of the lens instead. Using the definition of the critical angle and that of the normal vector, $\cos\theta_i = \frac{1 - e\cos\theta}{\sqrt{1 + e^2 - 2e\cos\theta}}$, we can evaluate the subtended critical angle, θ_0, seen from the lens feeder in Figure 2.4a at the critical angle:

$$\tan\theta_0 = \frac{b}{ea} \tag{2.31}$$

As an example, Figure 2.4b shows θ_0 as a function of the relative permittivity of the elliptical lens. It can be observed how the lower the permittivity, the smaller the angle is, and therefore, a more directive feed will be required to efficiently illuminate the lens above the critical angle.

2.2.2.2 Spreading Factor *S*(*Q*)

Let's now analyze the power balance in an elliptical lens antenna. We can evaluate the incident power crossing an area A (see Figure 2.5a) by considering each ray incident on a point Q as a plane wave. The orthogonal and parallel components of the incident power are defined as follows:

$$P_i^{\perp}(Q) = S_i^{\perp}(Q)A = \frac{|E_i^{\perp}(Q)|^2}{2\zeta_d} A\cos\theta_i \tag{2.32}$$

$$P_i^{\parallel}(Q) = \frac{\left|E_i^{\parallel}(Q)\right|^2}{2\zeta_d} A\cos\theta_i \tag{2.33}$$

where S_i represents the amplitude of the Poynting vector. The same can be done for the reflected power towards inside of the lens:

$$P_r^{\perp}(Q) = \frac{|\Gamma^{\perp}(Q)E_i^{\perp}(Q)|^2}{2\zeta_d} A\cos\theta_i \tag{2.34}$$

$$P_r^{\parallel}(Q) = \frac{\left|\Gamma^{\parallel}(Q)E_i^{\parallel}(Q)\right|^2}{2\zeta_d} A\cos\theta_i \tag{2.35}$$

and for the transmitted power:

$$P_t^{\perp}(Q) = \frac{|\tau^{\perp}E_i^{\perp}(Q)|^2}{2\zeta_0} A\cos\theta_t \tag{2.36}$$

$$P_t^{\parallel}(Q) = \frac{\left|\tau E_i^{\parallel}(Q)\right|^2}{2\zeta_0} A\cos\theta_t \tag{2.37}$$

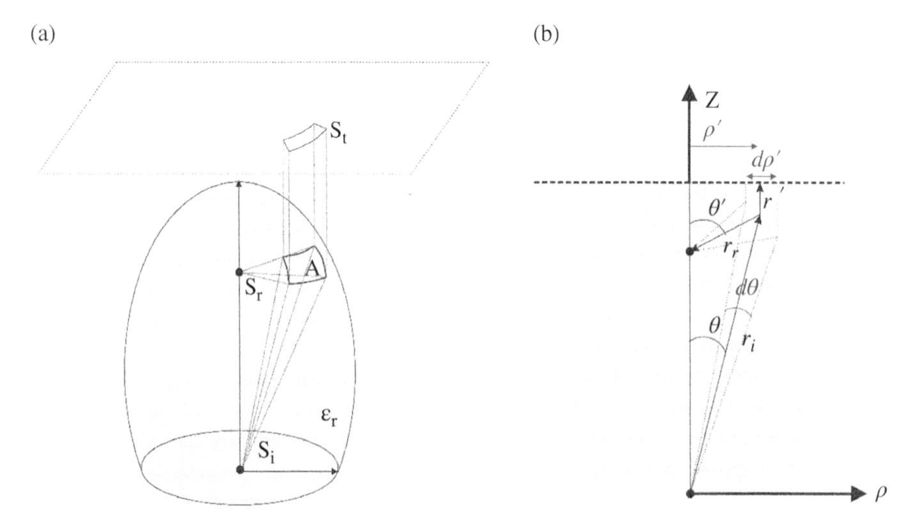

Figure 2.5 (a) Incident, reflected, and transmitted power density in a dielectric lens. (b) Referenced geometrical parameters.

The relation between the incident, reflected and the transmitted power density is represented in Figure 2.5a. The power transmitted by the ellipse *Pa* and propagating in parallel to the z-axis through the aperture area of $dA_a = \rho' d\rho' d\phi$, is equal to the transmitted power, P_t, by the ellipse through the area $dA_i = r_i d\Omega_i = r_i^2 \sin\theta d\theta d\phi$; where r_i is the distances between the first focus and the lens surface (see Figure 2.5b). Moreover, the transmitted power is related to the power incident, on the lens surface, P_i (i.e. neglecting the reflected power):

$$P_a = P_t = P_i |\tau|^2 \frac{\zeta_d}{\zeta_0} \frac{\cos\theta_t}{\cos\theta_i} \tag{2.38}$$

where $P_a = S_a dA_a$ and $P_i = S_i dA_i$; S_a and S_i are the amplitude of the Poynting vectors of the aperture, and incident fields, respectively. Moreover, τ is the Fresnel coefficient in transmission, θ_i and θ_t are the incident and transmitted angles on the lens surface, respectively. By using the relations $\rho' = r_i \sin\theta$, (2.38) can be simplified as:

$$S_a = S_i |\tau|^2 \frac{\zeta_d}{\zeta_0} \frac{\cos\theta_t}{\cos\theta_i} r_i \frac{d\theta}{d\rho'} \tag{2.39}$$

The amplitude of the aperture and incident Poynting vectors can be expressed as:

$$S_a = \frac{|E_a|^2}{2\zeta_0} \tag{2.40}$$

$$S_i = \frac{|E_i|^2}{2\zeta_d} \frac{1}{r_i^2} \tag{2.41}$$

where $|E_a|$ is the amplitude of the electric field on the equivalent aperture of the lens, and $|E_i|$ is the amplitude of the far-field pattern of the incident electric field. Moreover, the term $\frac{d\theta}{d\rho'}$ can be simplified as:

$$\frac{d\rho'}{d\theta} = \frac{d[r_i(\theta)\sin\theta]}{d\theta} = \frac{dr_i(\theta)}{d\theta}\sin\theta + r_i(\theta)\cos\theta \tag{2.42}$$

where $r_i(\theta) = a\frac{1-e^2}{1-e\cos\theta}$. The term $\frac{dr_i(\theta)}{d\theta}$ can be calculated as:

$$\frac{dr_i(\theta)}{d\theta} = \frac{-a(1-e^2)}{(1-e\cos\theta)^2}(e\sin\theta) = \frac{r_i(\theta)e\sin\theta}{e\cos\theta - 1} \tag{2.43}$$

by substituting (2.43) in (2.42):

$$\frac{d\rho'}{d\theta} = \frac{d[r_i(\theta)\sin\theta]}{d\theta} = r_i(\theta)\left[\frac{e - \cos\theta}{e\cos\theta - 1}\right] \tag{2.44}$$

by substituting (2.40), (2.41), and (2.44) in (2.39):

$$\frac{|E_a|^2}{2\zeta_0} = \frac{|E_i|^2}{2\zeta_d}\frac{1}{r_i^2}|\tau|^2 \frac{\zeta_d}{\zeta_0}\frac{\cos\theta_t}{\cos\theta_i}\frac{e\cos\theta - 1}{e - \cos\theta} \rightarrow |E_a| = |\tau E_i|\frac{1}{r_i}\sqrt{\frac{e\cos\theta - 1}{e - \cos\theta}}\sqrt{\frac{\cos\theta_t}{\cos\theta_i}} \tag{2.45}$$

For an elliptical lens with $e = \frac{1}{\sqrt{\varepsilon_r}}$, one can represent $\cos\theta_i$ and $\cos\theta_t$ as:

$$\cos\theta_i = \frac{1 - e\cos\theta}{\sqrt{1 + e^2 - 2e\cos\theta}} \tag{2.46}$$

$$\cos\theta_t = \frac{\cos\theta - e}{\sqrt{1 + e^2 - 2e\cos\theta}} \tag{2.47}$$

by substituting (2.46) and (2.47) in (2.45), the amplitude of the field at equivalent aperture of the lens is related to the amplitude of the incident field as:

$$|E_a| = |\tau E_i|\frac{1}{r_i}\sqrt{\frac{e\cos\theta - 1}{e - \cos\theta}}\sqrt{\frac{\cos\theta - e}{1 - e\cos\theta}} \rightarrow |E_a| = |\tau E_i|\frac{S(r0)}{r_i} \tag{2.48}$$

Therefore, the spreading factor can be defined as:

$$S(r) = \sqrt{\frac{\cos\theta_t}{\cos\theta_i}\frac{e\cos\theta - 1}{e - \cos\theta}} = 1 \tag{2.49}$$

This is because, in the considered geometry (feed at the focus), the output GO phase front is planar and consequently has no power spreading.

2.2.2.3 Equivalent Current Distribution and Far-field Calculation

The equivalent current distributions on the aperture, using the incident field is defined as in (2.21 and 2.22) and the phase relation derived in (2.1), can then be written as follows:

$$\begin{aligned}\vec{\boldsymbol{M}}_s(\rho',\phi') \approx \vec{E}_t \times \hat{z}|_S &= \left[\tau^{\perp}(\theta_i)f^{\phi}(\theta',\phi')\hat{\rho}' - \tau^{\|}(\theta_i)f^{\theta}(\theta',\phi')\hat{\phi}'\right]\frac{e^{-jk_d a(1+e)}}{r'(\theta')} \\ &= M_S^{\rho'}(\rho',\phi')\hat{\rho}' + M_S^{\phi'}(\rho',\phi')\hat{\phi}'\end{aligned} \tag{2.50}$$

$$\begin{aligned}\vec{\boldsymbol{J}}_s(\rho',\phi') \approx \hat{z} \times \vec{H}_{t|S} &= \frac{1}{\zeta_0}\left[-\tau^{\perp}(\theta_i)f^{\phi}(\theta',\phi')\hat{\phi}' - \tau^{\|}(\theta_i)f^{\theta}(\theta',\phi')\hat{\rho}'\right]\frac{e^{-jk_d a(1+e)}}{r'(\theta')} \\ &= J_S^{\rho'}(\rho',\phi')\hat{\rho}' + J_S^{\phi'}(\rho',\phi')\hat{\phi}'\end{aligned} \tag{2.51}$$

where $r'(\theta')$ is given by (2.5) and $\tan\theta' = \rho'/z'$ where $z' = a\sqrt{1 - \left(\frac{\rho'}{b}\right)^2} + c$. The vectorial components of these equivalent current distributions can be also expressed in Cartesian coordinates using $J/M_S^{x'}(\rho',\phi') = J/M_S^{\rho'}(\rho',\phi')\cos\phi' - J/M_S^{\phi'}(\rho',\phi')\sin\phi'$ and $J/M_S^{y'}(\rho',\phi') = J/M_S^{\rho'}(\rho',\phi')\sin\phi' + J/M_S^{\phi'}(\rho',\phi')\cos\phi'$.

With these equations, we have determined the surface currents on the lens aperture, and thus, the far field patterns in the reference system used in Figure 2.6 can be obtained using the following expressions:

$$\vec{E}_\theta\left(\vec{r}\right) = -j\frac{k_0 e^{-jk_0 r}}{4\pi r}\int_0^{\frac{D}{2}}\int_0^{2\pi}\left(M_S^{\phi} + \eta_0 J_S^{\theta}\right)e^{jk_0\rho'\sin\theta\cos(\phi - \phi')}\rho' d\rho' d\phi' \tag{2.52}$$

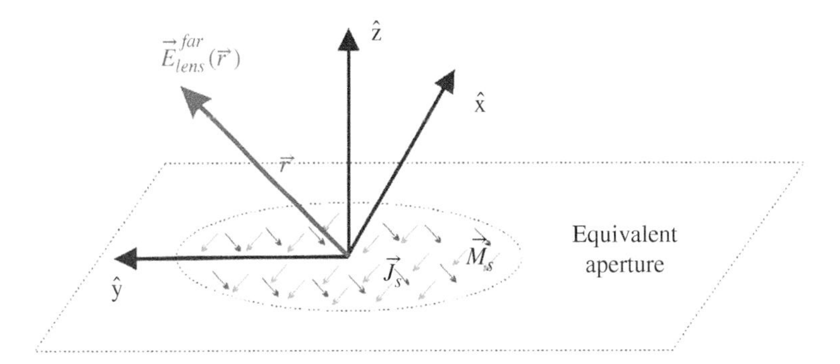

Figure 2.6 Reference system for the evaluation of the far fields radiated by the elliptical lens antenna.

$$\vec{E}_\phi\left(\vec{r}\right) = -j\frac{k_0 e^{-jk_0 r}}{4\pi r}\int_0^{\frac{D}{2}}\int_0^{2\pi}(M_s^\theta + \eta_0 J_s^\phi)e^{jk_0\rho' \sin\theta\cos(\phi-\phi')}\rho' d\rho' d\phi' \tag{2.53}$$

where $J/M_s^\theta = J/M_s^{x'}\cos\theta\cos\phi + J/M_s^{y'}\cos\theta\sin\phi$ and $J/M_s^\phi = -J/M_s^{x'}\sin\phi + J/M_s^{y'}\cos\phi$; and η_0 is the intrinsic impedance of free space.

2.2.2.4 Lens Reflection Efficiency

One of the main differences of lenses w.r.t. reflectors is the fact that in a reflector all the incident field is transformed into a transmitted field as the surface can be modeled as a perfect electric conductor (PEC). However, in the case of lenses, the radiation principle is based on the refraction law; thus some incident energy is transmitted to the air but some part is reflected inside of the lens toward the top focus (see Figure 2.7a) [34]. GO/PO field approximation only includes the effect in the far field of the first transmitted rays. Part of the energy is actually reflected in the lens interface. This energy will eventually be radiated to the far field via multiple reflections or it will be lost in the lens material. In the GO/PO field analysis,

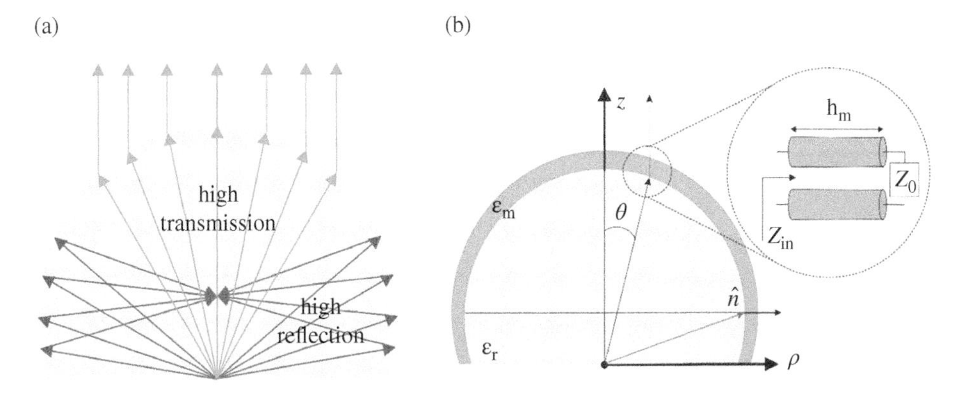

Figure 2.7 (a) High transmission and high reflection region of a silicon made elliptical lens. (b) Elliptical lens with a matching layer and its equivalent transmission line representation.

this reflected energy is simply considered a loss in efficiency. That is typically a good assumption since the multiple reflections do not usually contribute to the radiated field main beam, as they have a random phase, but to the far side-lobes. It means that the broadside gain of the antenna can be related as the directivity multiply by the radiation efficiency:

$$G(\theta,\phi) = D(\theta,\phi)\eta_r \tag{2.54}$$

where the radiation efficiency, η_r, is the ratio between the power radiated by the lens into the air and the power radiated by the primary source inside the lens $\eta_r = P_{rad}^{lens}/P_{rad}^{feed}$.

The amount of energy reflected inside the lens depends on how the lens feed illuminates the lens surface (see Figure 2.7b). We can identify two distinct zones as shown in Figure 2.7a. The top part of the lens is where the energy transmission is the highest, and therefore the most efficient part, whereas the lateral part leads to high reflected energy (even total at the critical angle) and therefore is the least efficient one [34]. That is why the lens feed should be designed to illuminate only the top part of the lens.

In order to improve the reflection efficiency of lens antennas, we can use a quarter wavelength impedance transformers. This impedance transformer is typically designed for broadside radiation (top part of the lens). It can be easily analyzed using a transmission line model for the *TE* and *TM* polarizations (corresponding to orthogonal and parallel polarizations, respectively), with the characteristic impedances of the lines representing the different mediums indicated with i defined as $Z_{0i}^{TE} = \zeta_i k_i/k_{zi}$ and $Z_{0i}^{TM} = \zeta_i k_{zi}/k_i$ with $k_{zi} = \sqrt{k_i^2 - k_d^2 \cos^2\theta_i}$ as shown in Figure 2.7b. Using this model, we can derive the condition for no reflection when $Z_{in} = \zeta_d$ which implies that the permittivity of the antireflective layer material is $\varepsilon_{rm} = \sqrt{\varepsilon_r}$ and its thickness $h_m = \frac{\lambda_0}{4\sqrt{\varepsilon_{rm}}}$.

The reflections coefficients of a silicon elliptical lens with and without anti-reflection layer are shown in Figure 2.8 as a function of the lens feeder angle (as in Figure 2.4a). Total internal reflection ($\Gamma = 1$) occurs when the incident angle onto the lens surface is 90° corresponding to about 72° for the feed angle. Before that angle, there is a total transmission angle for the parallel polarization at the called Brewster's angle. The same figure shows the

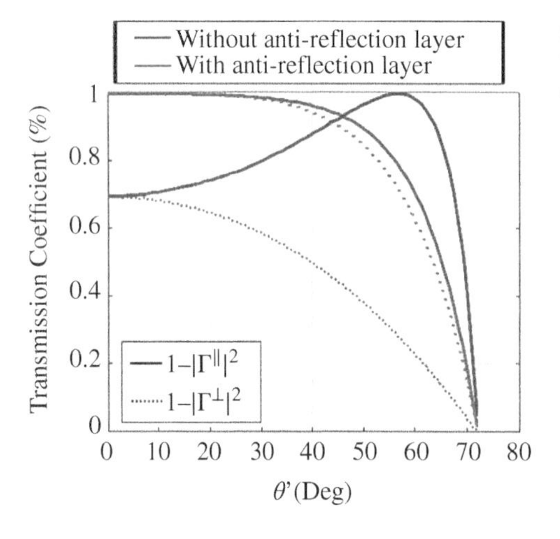

Figure 2.8 Parallel and perpendicular transmission coefficient for a lens made in silicon with and without matching layer for a silicon lens ($\varepsilon_r \approx 11.9$).

improved transmission coefficient of a lens made in silicon with an antireflective layer compared to the same lens without the antireflective layer. As it is shown, the transmission coefficient remains high until up 40° approximately with the use of an antireflective layer, and the parallel and perpendicular coefficients have now comparable values and therefore they will not introduce asymmetries in the azimuthal angle. It is important to mention that the introduction of an antireflective layer in lenses does not remove the problem of the critical angle. In both cases, the energy transmitted goes to zero at about 72°.

Silicon is the dielectric material employed in integrated lenses and circuit substrates. It is an inherently strong and excellent material in terms of losses, due to its high resistivity ($\geq$10 kΩ cm), which lowers the loss tangent to $\tan\delta \leq 10^{-4}$. It has also a large thermal conductivity, meaning that for cryogenic applications it will absorb little of the incoming heat load and conduct it efficiently to the edge where the heat sinks are mounted. Moreover, thanks to its uniformity and homogeneity, it is a material well suited for polarization measurements. The high permittivity of silicon ($\varepsilon_r \approx 11.9$) allows designs with strong focusing and smaller thicknesses; however, it presents a major drawback in terms of reflection losses. At submillimeter-wave frequencies, the plastic Parylene-C is extensively used as an antireflective layer thanks to the close match to the ideal index and its deposition through pyrolysis [35]. Other plastics such as Cirlex have suitable permittivity, but they are glued to the silicon with lossy epoxy adhesive [36]. Other methods include the synthesis of artificial dielectric coatings through the patterning of subwavelength grooves through micro-machining processes [37]. This artificial dielectric coating can synthesize the desired permittivity and behave as a continuous medium providing the closest match to the ideal antireflective coating.

2.3 Extended Semi-hemispherical Lens Antennas

While elliptical lenses reach the highest possible directivity, extended hemispherical lenses are often employed to ease their fabrication. This alternative configuration can have a close performance to the elliptical when the primary feed is designed in such a way to excite only the upper part of the dielectric lens and a dense material is used. The dimensions of the lens are chosen to approximate the focusing properties of the elliptical lens when the feed is placed at the foci. The synthesis of the ellipse is achieved using a semi-hemisphere of unit radius defined as:

$$x^2 + y^2 = 1 \text{ for } y \geq 0 \tag{2.55}$$

and an extension defined by L (see Fig. 2.9). The distance $b + c$ should be equal to $L + 1$ so both lenses are overlaid. Thus, the extension height L can be defined as:

$$L = b + c - 1 \tag{2.56}$$

It can also be synthesized the other way around, from an extended semi-hemispherical lens defined with the R and L, we can obtain the equivalent ellipse with the following expressions:

$$b = \frac{R(L+1)}{1 + \frac{1}{\sqrt{\varepsilon_r}}} \tag{2.57}$$

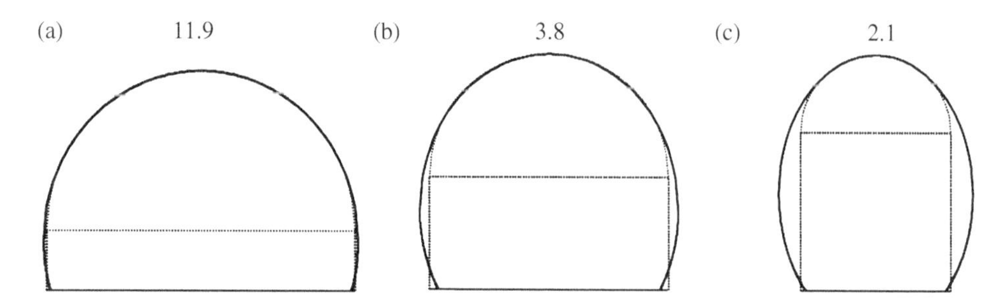

Figure 2.9 Synthesis of an elliptical lens from an extended hemispherical lens for silicon, fused silica and PTFE. Note that the extended hemisphere is a good approximation of the elliptical lens at high dielectric constants.

$$c = R(L + 1) - b \tag{2.58}$$

$$a = b\sqrt{1 - \frac{1}{\sqrt{\varepsilon_r}}} \tag{2.59}$$

Depending on the dielectric constant, the fitting of the elliptical lens and the extended hemispherical lens varies. Three examples are shown in Figure 2.9 where both lenses of silicon ($\varepsilon_r \approx 11.9$), quartz ($\varepsilon_r \approx 4.5$) and polytetrafluoroethylene (PTFE) ($\varepsilon_r \approx 2$). The rays from a broadside plane wave incident onto the extended semi-hemispherical lens, as opposed to the elliptical lens, do not come to a point focus, thus, which means that the lens introduces an aberration to the radiated fields. But as we will see in the following sections, even if it does not couple well to a planar equi-phase front, the extended hemispherical lens will couple well to a Gaussian-beam system.

2.3.1 Radiation of Extended Semi-hemispherical Lenses

In Section 2.2, we computed the radiation pattern from an elliptical lens antenna by deriving the currents on a planar aperture above the lens and, from them, we obtained the radiated fields from the lens antenna. In this section, we will explain how to compute the radiated fields of the extended hemispherical lens antenna from the currents evaluated on the lens surface. A PO method provides an approximation of these surface currents over a lens of several wavelengths. Using again the Love's Equivalence Principle, the radiated fields from the antenna feed obtained previously are used to compute the equivalent magnetic and electric sheet currents outside of the lens surface:

$$\vec{M}_l(\theta, \phi) = -\hat{n} \times \vec{E}^t \tag{2.60}$$

$$\vec{J}_l(\theta, \phi) = \hat{n} \times \vec{H}^t \tag{2.61}$$

where $\hat{n} = \hat{\imath}_r = \hat{\imath}_x \sin\theta \cos\phi + \hat{\imath}_y \sin\theta \sin\phi + \hat{\imath}_z \cos\theta$ is the normal vector to the surface of the lens (Figure 2.10) and $\vec{E}^t$ and $\vec{H}^t$ are the radiated fields from the antenna feed transferred outside the boundary of the lens. The transmitted fields can be calculated similarly that how it was done for the elliptical lens. The PO assumes that the equivalent currents can

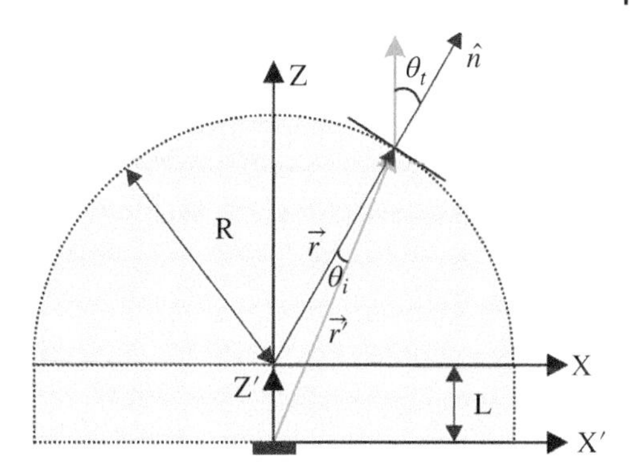

Figure 2.10 Sketch of the extended semi-hemispherical lens antenna parameters.

be formulated directly using the transmitted fields. Those are computed by approximating the lens surface locally by a flat dielectric-air interface and using the Fresnel transmission coefficients for plane waves as explained previously. Thus, the radiated field from the feed is decomposed into their parallel and perpendicular components to the incidence plane (see Figure 2.10) and multiplied by the corresponding transmission coefficient. The total field is defined by $\vec{E}^t = \vec{E}^t_{|} + \vec{E}^t_{|}$, and the total magnetic field can be equivalently calculated as $\vec{H}^t = \hat{i}^t_r \times \vec{E}^t/\eta$. Each of the components is given by:

$$\vec{E}^t_{|} = \vec{E}^f_{|}\tau_{|}\hat{i}^t_{|} \tag{2.62}$$

$$\vec{E}^t_{\perp} = \vec{E}^f_{\perp}\tau_{\perp}\hat{i}^t_{\perp} \tag{2.63}$$

where the $\tau_{||}$ and $\tau_{\perp}$ are Fresnel transmission coefficients for a dielectric lens of permittivity ε_r (2.27 and 2.28). This time the incident angle is evaluated using the normal vector corresponding to the hemispherical lens. The propagation vectors of the incident and transmitted fields are defined as follows:

$$\hat{i}_{r'} = \hat{i}_x \sin\theta' \cos\phi' + \hat{i}_y \sin\theta' \sin\phi' + \hat{i}_z \cos\theta' \tag{2.64}$$

$$\hat{i}^t_r = \hat{i}_{r'}\sqrt{\varepsilon_r} + \hat{i}_r\sqrt{\varepsilon_r}\cos\theta_i\sqrt{1 + \frac{1-\varepsilon_r}{\varepsilon_r \cos^2\theta_i}} \tag{2.65}$$

$$\hat{i}^t_{|} = \hat{i}_{\perp} \times \hat{i}^t_r \tag{2.66}$$

Once the PO surface currents are evaluated via the transmitted fields, one can obtain the far-field patterns. Those patterns can be obtained using the reference system shown in Figure 2.8, and integrating the PO surface currents over the lens hemispherical surface, as follows:

$$\vec{E}_{\theta}\left(\vec{r}\right) = -j\frac{k_0 e^{-jk_0 r}}{4\pi r}\int_0^{\pi/2}\int_0^{2\pi}\left(M^{\phi}_l + \eta J^{\theta}_l\right)e^{\,jk_0 R\cos\varphi}R^2 \sin\theta' d\theta' d\phi' \tag{2.67}$$

$$\vec{E}_{\phi}\left(\vec{r}\right) = -j\frac{k_0 e^{-jk_0 r}}{4\pi r}\int_0^{\pi/2}\int_0^{2\pi}\left(M^{\theta}_l + \eta J^{\phi}_l\right)e^{\,jk_0 R\cos\varphi}R^2 \sin\theta' d\theta' d\phi' \tag{2.68}$$

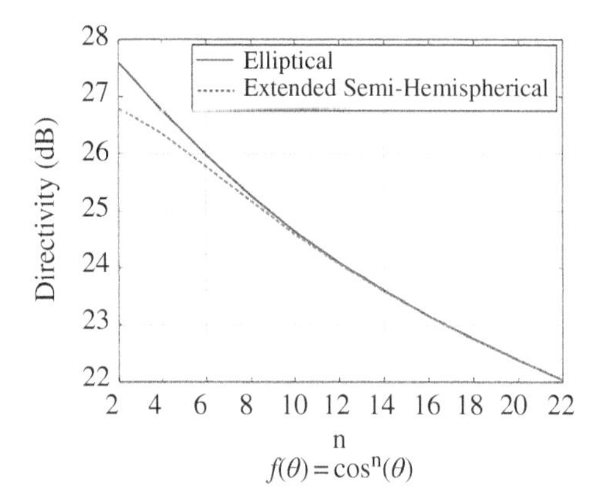

Figure 2.11 Directivity of an elliptical lens and an extended hemispherical lens as a function of the feed illumination. A feed illumination of a $f(\theta) = \cos^n\theta$ is used in the example to illuminate a silicon lens of diameter 7.65 λ and an extended hemispherical lens of $L = 0.375$.

where R is the radius of the hemisphere of the lens, $\cos\varphi = \hat{i}_r \cdot \hat{i}^t_r, \eta$. The equivalent magnetic and electric currents are defined as $\vec{M}_l = M^\theta_l \hat{i}_\theta + M^\phi_l \hat{i}_\phi$, $\vec{J}_l = J^\theta_l \hat{i}_\theta + J^\phi_l \hat{i}_\phi$.

Figure 2.11 shows the directivity of a silicon elliptical lens and a hemispherical lens (i.e. synthesized from the elliptic geometry) as a function of the subtended angle by the feed. As it is shown in the figure, the elliptical lens provides the highest directivity compared to the hemispherical lens, however, the difference is only noticeable when using a feed with low directivity (low n). As the directivity increases, the performance of the elliptical and extended semi-hemispherical is equivalent.

2.4 Shallow Lenses Excited by Leaky Wave/Fabry–Perot Feeds

As mentioned in Section 2.3, the amount of energy reflected inside of the lens depends on how the lens feed illuminates the lens surface. The top part of the lens is the most efficient part of the lens since it is where the transmitted energy is the highest, and the least efficient area is the lateral part, leading to high reflected energy. Thus, in order to have a highly efficient lens antenna, we need a feed that illuminates only the top part of the lens, a.k.a. a shallow dielectric lens.

The use of shallow dielectric lenses, as the lens antenna is shown in Figure 2.12, presents great advantages in terms of fabrication and electrical performances at submillimeter-wave frequencies. Lower cost and better surface accuracies can be achieved when the lens presents a shallow curvature for both silicon micromachining techniques (laser and DRIE photolithography). On the other hand, since the top part of the lens is the most efficient in terms of reflection, we need a feed radiating in an infinite dielectric medium with very large directivity in order to illuminate properly this lens. For example, a lens with a solid angle of 20° will need a $\cos^n\theta$ feed with 19 dB directivity. Generating this type of radiation with low loss is challenging at high frequencies. In Ref. [25], a very directive lens feed, radiating most of the energy well below the air-dielectric critical angle, with wide bandwidth and a very low-loss at THz frequencies was proposed using a Fabry–Perot/leaky-wave concept.

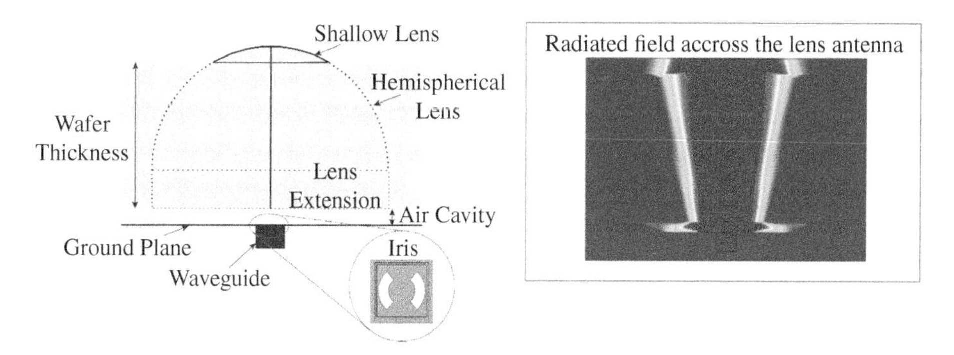

Figure 2.12 Sketch of the silicon lens antenna fed by a leaky-wave feed geometry and its radiated field across the antenna.

Leaky wave antennas (LWAs) [38], also referred as electromagnetic band-gap (EBG) antennas [39], Fabry–Perot Antennas (FPA) [40] or resonant cavity antennas [41], use a partially transmissive resonant structure that can be made of a thin dielectric superstrate [38] or by using frequency selective surfaces (FSS) [42] to increase the effective area of a small antenna. These antennas are used to achieve high directivity from a point source by the excitation of a pair of nearly degenerated TE_1/ TM_1 leaky-wave modes. These modes propagate in the resonant region by means of multiple reflections between the ground plane and the superstrate, while partially leaking energy into the free space. The amount of energy radiated at each reflection is related to the LW attenuation constant and can be controlled by the FSS sheet-impedance or the dielectric constant. At the resonant frequency, where the real part and imaginary part of the complex leaky-wave wavenumber are similar, these antennas radiate a pencil beam. For the air cavity of thickness h_0 and dielectric super-layer of thickness h_s, the maximum directivity at broadside is achieved at the resonant condition, i.e. the thickness of the resonant air cavity is $h_0 = \lambda_0/2$, and that of the super-layer is $h_s = \lambda_0/4\sqrt{\epsilon_r}$, [38]. Under this condition, the couple of TE/TM leaky-wave modes can propagate with same phase velocity, creating a nearly uniform phase distribution in the aperture. It has been seen that the generated aperture field is also very well polarized, due to a compensation effect between the TE and TM modal tangential field components [39]. However, this type of LWA also generates an undesired spurious TM_0 leaky-wave mode, conceptually associated with the transverse electromagnetic (TEM) mode of the perfectly conducting walls parallel plate waveguide [43]. This mode radiates near the Brewster angle creating spurious lobes in the E-plane reducing the beam efficiency. An iris containing a double slot for single-polarization or two double slots for dual-polarization will suppress the spurious TM_0 mode and provide the matching between the waveguide's fundamental modes and the silicon interface [44].

In a LWA, the maximum directivity is directly proportional to the super-layer permittivity but inversely proportional to the relative bandwidth. This resonant behavior is well known as the main drawback of LWA and limits their use to narrowband applications. However, this drawback can be partially solved when the leaky-wave antenna is radiating in a semi-infinity medium [25]. Indeed the use of a resonant air gap cavity for small antennas radiating into a dense medium ensures a highly directive beam with most of the energy being radiated for angles smaller than the air-dielectric critical angle over bandwidths ranging from 10% to 40% depending on the lens material [45].

2.4.1 Analysis of the Leaky-wave Propagation Constant

In this section, we will analyze the propagation constant, $k_{\rho lw}$, of the leaky waves propagating in the air cavity. For this evaluation the propagation constant, we will approximate the aperture field of Fabry–Perot leaky-wave antennas as $E_{ap} \propto \frac{e^{-jk_{\rho lw}\rho}}{\sqrt{\rho}}$ and use the approximate analytical equations provided in [46]. The value of this propagation constant depends on the impedance, Z_L, seen from the top of the half-wavelength cavity in the equivalent transmission line model. For a dielectric quartz (being $\varepsilon_{r_{quartz}} \approx 4.45$) super-layer with a quarter wavelength thickness, this impedance is $Z_L = \zeta_0/\varepsilon_{r_{quartz}}$. The pointing direction of the leaky-wave modes is in this case 17°. If we need a more directive antenna, a lower Z_L is required in order to achieve less attenuated mode, implying stronger multiple-reflections (i.e. more resonant), and thus, less bandwidth.

If instead of a dielectric super-layer stratification we consider an infinite layer of silicon, the impedance seen on top of the cavity is $Z_l = \zeta_0/\sqrt{\varepsilon_{r_{silicon}}}$. Since $\sqrt{\varepsilon_{r_{silicon}}} \approx \varepsilon_{r_{quartz}}$, the impedances associated with the quartz super-layer and the infinite silicon layer are comparable and consequently, the leaky-wave propagation constants. Figure 2.13 shows the real and imaginary parts of the propagation constant for the three possible propagating modes: two *TM* and one *TE*. On the left axis of the figure, the propagation constants are normalized to the free space propagation constant k_0. The main mode TM_1 and TE_1 points towards approximately 18°, whereas the non-desired mode TM_0 radiates towards larger angles, i.e. 81°. However, if we calculate the pointing angles inside the dielectric, from the right

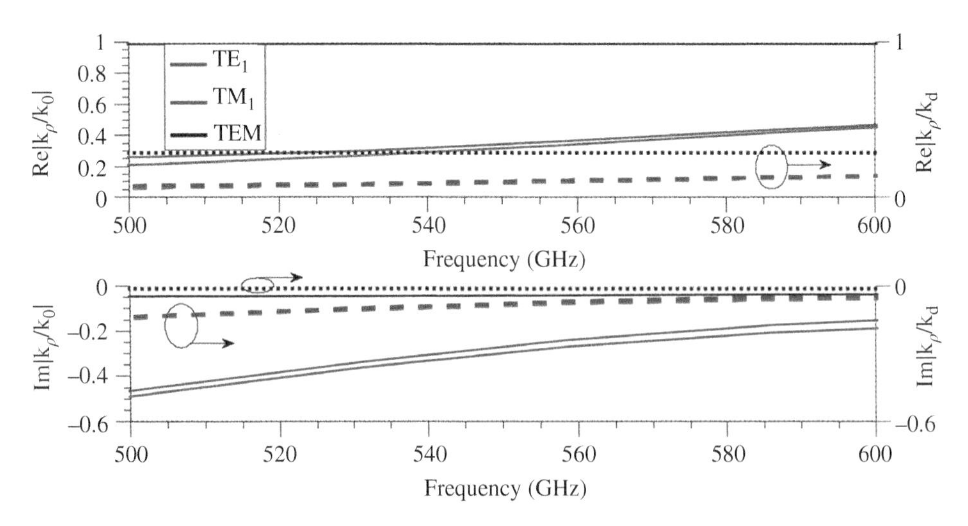

Figure 2.13 Real and imaginary parts of the propagation constants k_{lw} of the leaky-wave modes present in an air cavity (h = 275 μm) and infinite silicon dielectric medium. On the left axis, k_{lw} is normalized to the free space propagation constant, k_0, whereas k_{lw} (shown in the right axis) is normalized to the propagation constant in the dielectric, $k_d = k_0\sqrt{\varepsilon_r}$.

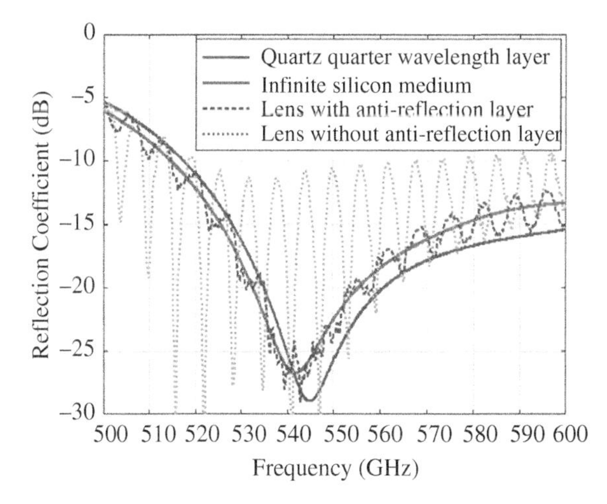

Figure 2.14 Input reflection coefficient of a waveguide loaded with a double-slot iris in the presence of an air cavity (h = 275 μm) shown in [25]. The feed dimensions are: R_{in} = 109.7 μm, α = 50° and wg = 367.6 μm. Several cases are considered on top of the cavity: a quartz quarter wavelength super-layer ($\varepsilon_{rq} = \sqrt{\varepsilon_{rs}}$); an infinite silicon medium (ε_{rs} = 11.9); and an extended; hemispherical lens with R = 6 mm and L_t/R = 0.2 with and without a coating. *Source*: Modified from Llombart et al. [25]; John Wiley & Sons.

axis of the figure, where the propagation constant is normalized to $k_d = k_0\sqrt{\varepsilon_r}$ instead of k_0, we obtain values of about 6° for the main poles, which will translate into ε_r times more directive pattern inside the dielectric medium without compromising the bandwidth.

Another advantage of this configuration is that the leaky-wave mode shows the same frequency behavior as the quartz super-layer, even if a more directive feed is radiated inside the dielectric, and therefore will have similar antenna impedances and bandwidth. Figure 2.14 shows the simulated reflection coefficient using a full-wave simulator of a square waveguide loaded by a double-slot iris in the presence of the resonant air-cavity. Several cases are shown in the figure, all of them with the same cavity and feed dimensions: an infinite silicon dielectric medium, a quartz quarter wavelength super-layer, and a silicon lens on top of the cavity with and without coating. The infinite silicon medium and quartz quarter wavelength super-layer present very similar reflection coefficients since they both present very similar cavity load impedances Z_l. A comparison in the reflection coefficient with and without an anti-reflection layer of $\varepsilon_r = 3.45$ is shown in Figure 2.14. Note the strong effect of the multiple reflections on the reflection coefficient when an antireflective coating is not used.

All in all, when employing LWA, one should consider different dielectric combinations, e.g. infinite quartz layer, quartz cavity, and silicon lens, depending on the trade-offs between the bandwidth and the directivity [45].

2.4.2 Primary Fields Radiated by a Leaky-wave Antenna Feed on an Infinite Medium

This part will explain how to calculate the radiated fields into an infinite medium of a certain permittivity starting from a known aperture distribution. These fields can be then used

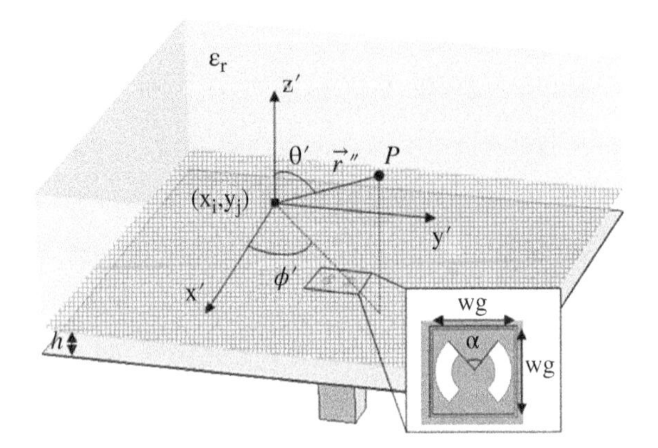

Figure 2.15 Sketch of the leaky-wave feed with its main parameters. The red grid represents the field aperture grid.

as the incident fields in the previously explained lens analysis. We will use the example of a leaky-wave feed over silicon but it can be applied to any source or lens dielectric material.

The basic geometry of the feed is shown in Figure 2.15. A waveguide exciting the fundamental mode mounted under an infinite ground plane with an iris. Above the iris, there is a resonant air cavity and a silicon slab that extends infinitely in z and the xy plane.

Using Love's Equivalence Principle, the electric fields computed at the aperture plane ($z = h$) can be used to compute the equivalent magnetic and electric currents outside of the surface where $\hat{n}$ is the normal vector to the aperture surface, in this case $\hat{n} = \hat{z}$, and $\vec{E}_a$ and $\vec{H}_a$ are the electric fields computed at the aperture plane using the full-wave simulator. The use of a full-wave simulator will provide the accuracy to obtain the fields on an aperture grid, which will allow the computation of the currents using the free space Green's function and the equivalence principle. The infinite silicon medium can be simulated by setting absorbing boundaries in the full-wave simulator. An electric field probe is set on the air-silicon interface and will be the origin of our primary field origin of coordinates xyz (the plane is marked in red in Figure 2.15).

The lens can be in the radiative near field or far field of the feed. In the case of the leaky-wave feed, due to its high directivity, the lens will be in most cases in the radiative near field. In the radiative region of the near-field, the field can still be represented by a local spherical wave with an angular distribution that depends on the distance, and therefore the same PO explained previously can still be used. The near and far-field effect is illustrated in Figure 2.16, where the fields are computed over a sphere of two certain radius $\rho = 4.5\lambda$ and $\rho = 20\lambda$. The far field is calculated independently of the lens geometry, as the dependency in r is eliminated (i.e. the term e^{-jkr}/r). This dependency is added when the equivalent currents are computed over the lens. From the phase shown in the radiation patterns, we can see that the phase center is not in the plane of the iris ground plane, as shown in Figure 2.15, it is below and it varies with frequency. Thus, in the contrary to the design equations on the previous sections, the lens extension height L and radius R will need adjustment to maximize the aperture efficiency of the overall antenna. This aperture efficiency will be

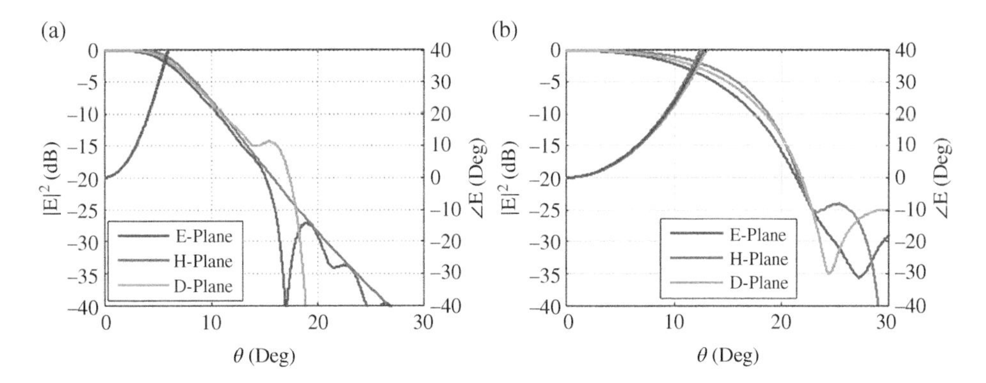

Figure 2.16 Amplitude and phase of the electric centered at a central frequency 550 GHz in the (a) far-field and (b) near-field at $\rho = 4.5\ \lambda$ of the leaky-wave feed radiating over an infinite silicon medium.

computed over an aperture of diameter D (see Fig. 2.17a), which is smaller than the lens radius R because of the high directivity of the leaky-wave feed.

2.4.3 Shallow-lens Geometry Optimization

This part will cover the design of a leaky-wave feed with an integrated silicon lens in the order of $4-20\lambda$ as described in [33]. We will start by choosing a leaky-wave feed that is matched over 15% bandwidth and provides the highest directivity by using a resonant Fabry–Perot air-cavity. For a desired aperture diameter, the procedure to design the overall integrated silicon lens dimensions, shown in Figure 2.17a is, as follows:

1) The waveguide, iris, and air cavity are optimized at the central frequency. After setting the air cavity thickness to $\lambda_0/2$,the other dimensions of the waveguide and iris can be

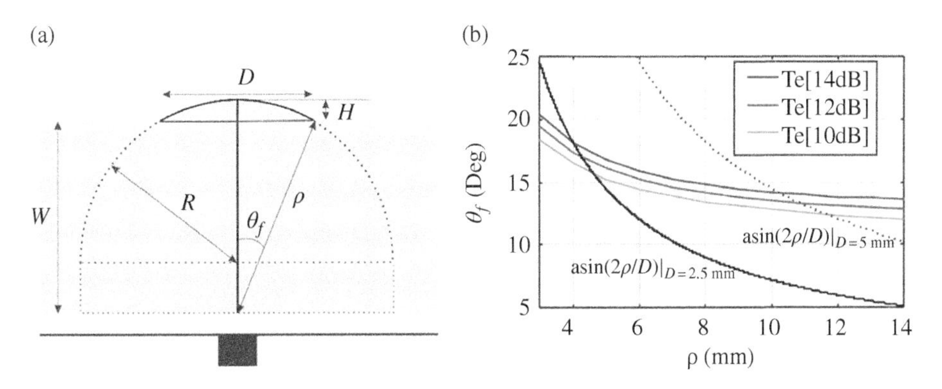

Figure 2.17 (a) Drawing of the basic parameters of the shallow lens antenna geometry. (b) Taper angle θ_f as a function of ρ for a shallow lens of diameter $D = 2.5$ mm and $D = 5$ mm, calculated at the central frequency of 550 GHz.

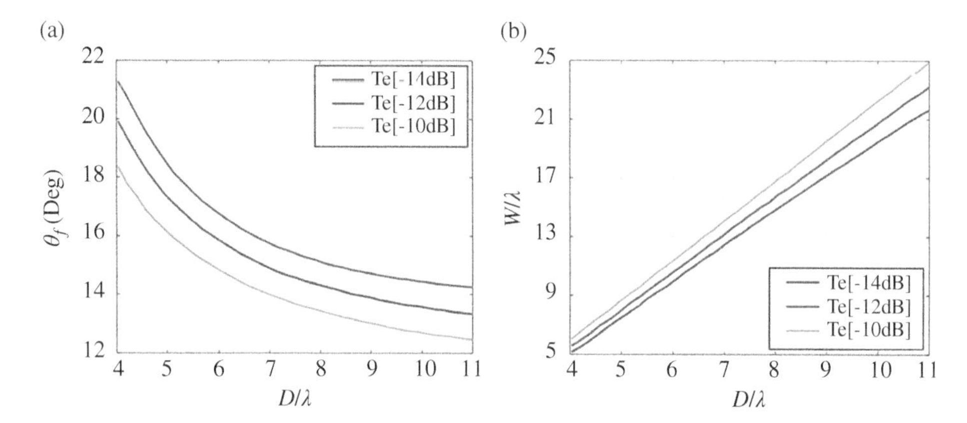

Figure 2.18 Optimum (a) taper angle θ_f and (b) lens thickness W as a function of each diameter D that maximize the lens antenna performance using the procedure described.

optimized with a full-wave simulator for maximum radiation efficiency. The dimensions and reflection coefficient are shown in Figure 2.14 for a central frequency of 550 GHz. Next, the electric field components in the aperture plane are exported from the 3D simulator in order to perform the optimization of the lens geometry using the formulation previously explained.

2) Obtain the basic dimensions W and taper edge angle θ_f. From the exported fields we will compute the radiated field from the leaky-wave feed inside the infinite silicon for different ρ. After this, we will calculate the taper edge angle θ_f for a certain taper field value in the 10–14 dB region. The field taper is chosen according to the tradeoff between the spillover and the taper efficiency, equivalent to the tradeoff in reflector antennas. This computation of the field at different ρ is necessary because, for the dimension of the lens aperture, the feed is placed on the near field of the antenna. This dependency can be seen in Figure 2.17b, when we move toward the far-field region this dependency fades, and the curves in Figure 2.17b get flatten (θ_f becomes a constant with ρ). The optimal ρ and θ_f value will be given by the intersection of this curve with $\rho = {}^{D}\!/_{2\,\sin\theta_f}$. Figure 2.18a shows the optimal ρ and θ_f values for different diameters ranging from $4 - 11\lambda_0$. And Figure 2.18b shows the optimum distance W defined as $W = \sqrt{\rho_{opt}^2 - \left(\frac{D}{2}\right)^2}$ where the shallow lens surface should be placed.
3) The shallow lens surface is optimized to maximize the antenna directivity. The lens curvature is defined by the radius R and the height H, or equivalently by the extension height L (see Figure 2.19a). For planar antennas, the optimum L and R are known, see Section 2.3. However, as stated previously, because the phase center of the leaky wave is below the waveguide aperture, this makes the optimum lens surface differ from the standard cases [33]. Then, we will vary the height and radius of the lens until the optimum is achieved. The variation of the directivity as a function of the extension height L is shown in Figure 2.18b. To perform this optimization, full-wave simulators are not recommended as the computational time and complexity of the global structure is considerably. Instead, the PO techniques described previously provide a good compromise between quality of

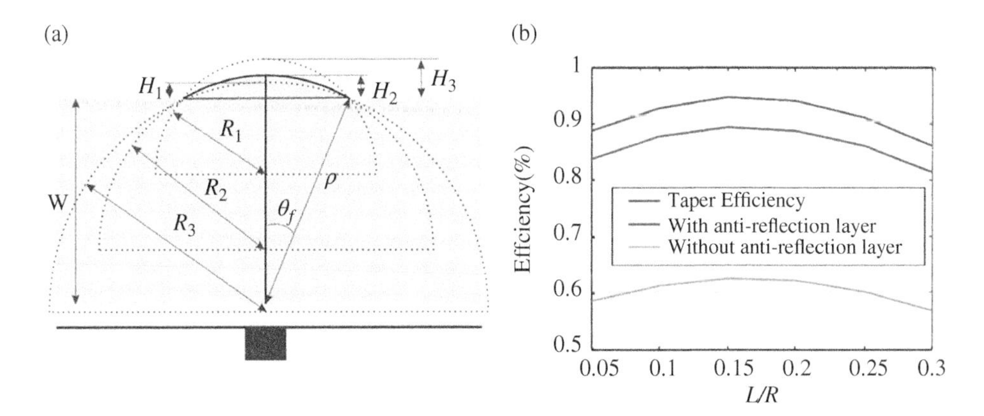

Figure 2.19 (a) The shallow lens of diameter D is defined by a corresponding H and R. (b) Taper and aperture efficiency as a function of the lens extension height L.

the results and computational time which makes it fundamental for integration into optimization of lenses. The optimum at different field tapers for each diameter are summarized in Figure 2.20a and b.

4) Following these two steps, we can obtain the highest directivity for a desired aperture diameter D for the designed leaky-wave feed. The directivity attainable with this architecture is shown in Figure 2.21a. Note that for the chosen field tapers of -10 dB to -14 dB the spillover remains below 0.25 dB. The Gaussicity, calculated with the expressions on Section II is shown in Figure 2.21b. The beamwaist and the phase center of the lens have been adjusted to maximized the Gaussicity for each diameter. As it is shown very high Gaussicity values can be achieved across all the diameters due to the high symmetry and shape of the leaky-wave field.

2.5 Fly-eye Antenna Array

The main advantage of having a lens antenna with a waveguide feed is its seamless integration in an array configuration for waveguide-based receivers at submillimeter-wave-frequencies. As explained in previous sections, the leaky-wave feed illuminates a small aperture of diameter D of the lens, making the resulting effective lens surface a very shallow lens. These shallow lenses can be packed together forming an array and it can be fabricated using silicon micromachining techniques. The feed and lens array can be entirely fabricated over a few silicon wafers and stacked to the rest of the instrument (see Figure 2.22).

The micromachining techniques used for the fabrication of heterodyne receivers and sources at submillimeter-wave frequencies are based on CNC metal block machining, as well as wafer-level processing using photolithographic techniques. Compared with metal block machining, wafer-level processing allows a higher integration of the whole receiver, which reduces the volume, mass, and losses. The different waver level processing solutions in literature, from the use of SU-8 [36] machining to lithography, electroplating, and

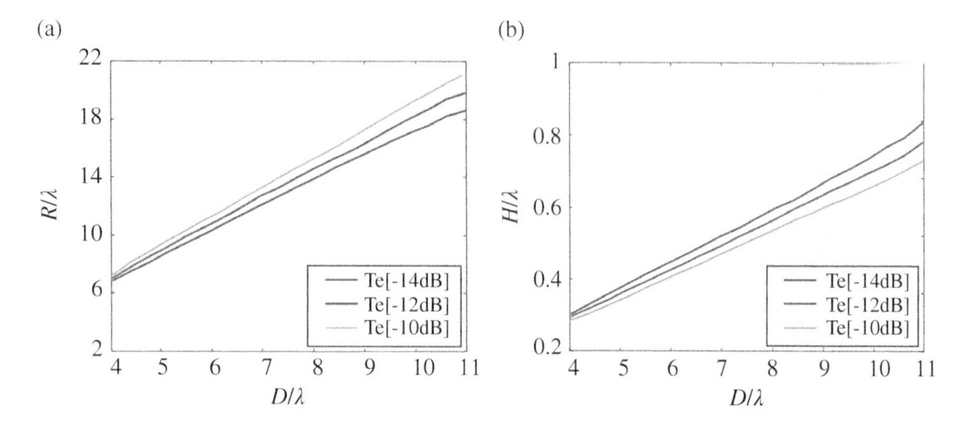

Figure 2.20 (a) Radius R and (b) height H of the lens as a function of the diameter D for different field tapers.

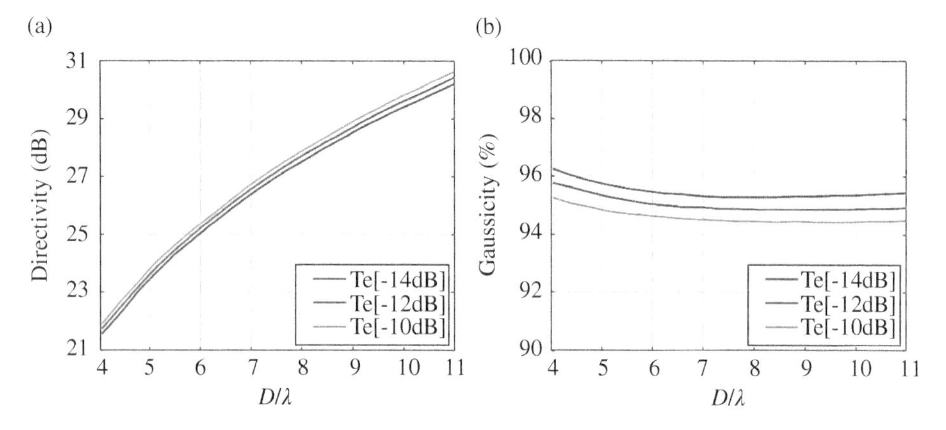

Figure 2.21 (a) Directivity and (b) Gaussicity achieved for the shallow lens antenna Reflection coefficient centered at a central frequency f for the dimensions shown in the table.

molding (LIGA)-based processes [37] that use thick resist and electroplating processes to form the waveguide walls, suffer from limitations in terms of multi-etching capabilities and non-uniformity problems. Silicon micromachining based on DRIE allows the fabrication of high aspect ratio features while maintaining straight sidewalls and smooth multi-depth surfaces. This technology is very attractive for the development of heterodyne receivers and sources as the technology allows the fabrication of complex circuit features that require multiple waveguides of various depths, especially as the wavelength decreases, when machining tolerances become a concern.

As previously stated, the presented antenna composed by a leaky-wave feed and the shallow lens can be fabricated in four silicon wafers using DRIE processes. The first wafer consists of the iris and the waveguide feature. The second wafer contains the air cavity. The third wafer contains dummy silicon wafer to achieve the correct lens

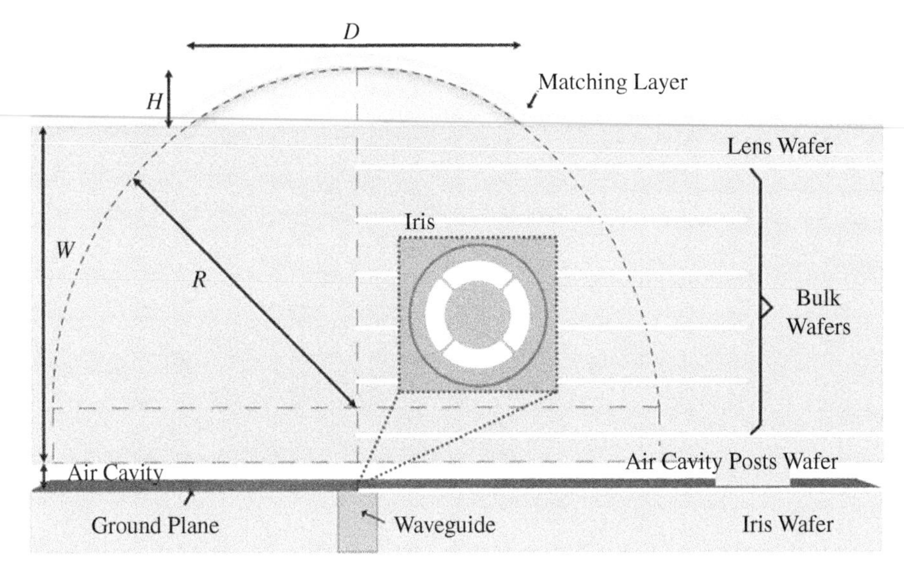

Figure 2.22 Reflection coefficient centered at a central frequency f for the dimensions shown in the table.

thickness. The fourth wafer contains the lens surface. The four different wafers (i.e. iris air gap, bulk wafers, and lens) will be assembled together on a single wafer stack. Note that high resistivity silicon wafers, i.e. 10 kΩ cm, are required to reduce the dielectric absorption loss. And, the wafers need to be double-sided polished in order to have good surface contact between all the wafers in the stack, avoiding air gaps. The alignment, which is a challenge when working at such high frequencies, has been solved by using a silicon compressive pin that is slightly larger than the pocket and can be compressed slightly when put into place [26].

2.5.1 Silicon DRIE Micromachining Process at Submillimeter-wave Frequencies

In order to achieve the depth accuracy and smoothness of the features on silicon, Silicon-on-insulator (SOI) wafers are commonly employed, since thin layers of SiO_2 act as etch stops of the silicon. For example, [26] presented an iris of the leaky-wave antenna at 1.9 THz fabricated on a 2.5 μm silicon membrane. Compared with a thermal oxide membrane, the crystalline silicon provides more robustness to compressive stress. In that case, a SOI wafer with a 2 μm device layer was used because, together with the metallization, the membrane would reach the desired 2.5 μm. And a handling wafer of 400 μm defined the waveguide on the back of the membrane.

The step by step fabrication of the slot iris and waveguide process is shown in Figure 2.23. Thermally grown SiO_2 is used as a hard mask in the front and back of the SOI wafer. Since the small size of the double slot iris, an i-line Canon stepper is used to illuminate the SiO_2. And for the rest of the waveguide and alignment features, regular UV photolithography can be employed. The pre-etching of the features, i.e. the slot iris and the waveguide, on the front and back of the SiO_2 layer are performed using photoresist and an inductively coupled

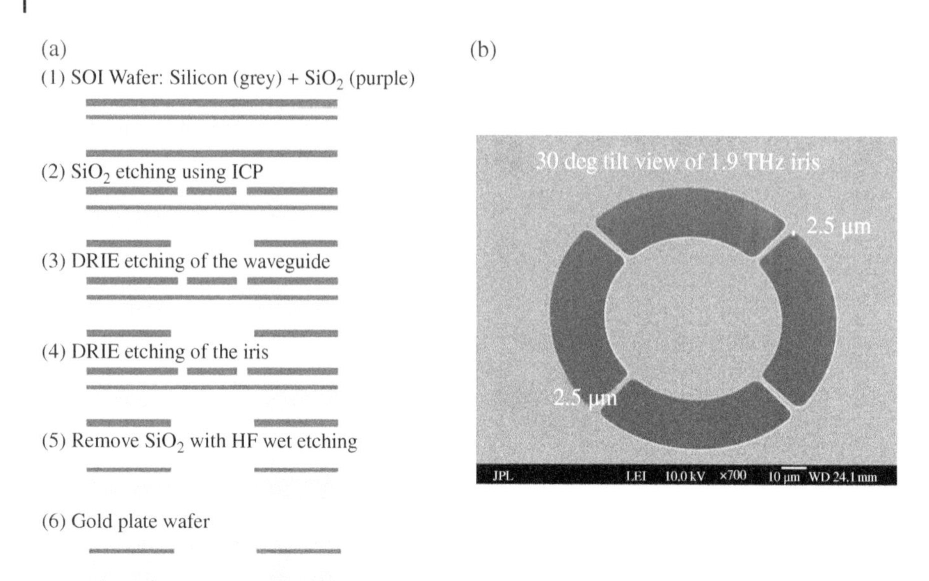

Figure 2.23 (a) Sketch of the membrane fabrication process that contains the iris and waveguide of the leaky-wave feed. (b) SEM of the iris developed at 1.9 THz in [26] using the explained process. *Source*: Alonso-delPino et al. [26]; IEEE.

plasma (ICP) reactive ion etcher. Then, the silicon on the front backside is etched using a PlasmaTherm DRIE system. While the selectivity of the silicon to photoresist is around 70 : 1, the selectivity of the silicon to SiO_2 can be optimized to reach 130 : 1, which allows to etch high depth features with high control. The SiO_2 was removed using a hydrogen fluoride HF solution.

At the end of the process, the overall wafer was sputtered with gold, used due to its high conductivity, immunity to oxidation, and ease of deposition. The overall results of this process can be observed in the scanning electron microscope (SEM) image in Figure 2.23b, showing a very clean and well-defined pair of double slots.

The rest of the wafers that define the lens are processed similarly as the procedure described but, because the radiation is going through the wafers, it is necessary the use of high resistivity silicon wafers (the resistivity around 10 kΩ cm) to avoid the introduction of absorption losses.

2.5.1.1 Fabrication of Silicon Lenses Using DRIE

Silicon shallow lens arrays have been fabricated either by laser micro-machining process or using photolithographic processes based on DRIE. Laser micro-machining allows the fabrication of 3-D geometries with accuracy as presented in [33], however, it is a linear process where the cost depends on the laser time, which might not be the most cost-efficient method for large lens arrays. A novel DRIE silicon process presented in [47] allows the fabrication of arrays of lenses on a single wafer and in parallel. This section will provide an overview of the process and a fast method to estimate the overall fabrication accuracy without needing to test the lens antenna.

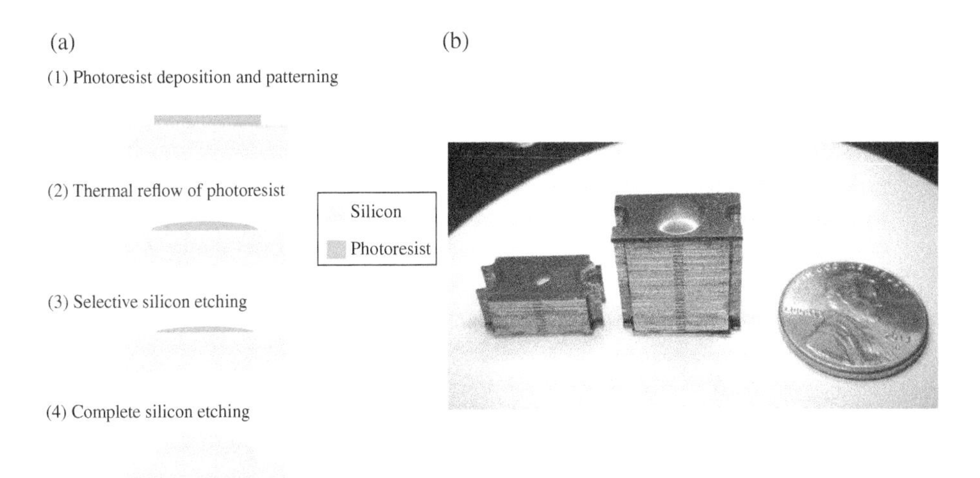

Figure 2.24 (a) Sketch of the fabrication process of the shallow silicon lens. (b) Photograph of shallow lens antennas at 1.9 THz of diameter 2.6 and 6.3 mm presented in [26]. *Source*: Alonso-delPino et al. [26]; IEEE.

The process to fabricate the silicon shallow lens consists of four steps, illustrated in Figure 2.24a. The first step consists of the patterning of the photoresist on a high resistivity wafer with the desired lens aperture diameter. The thickness and aperture diameter of the photoresist applied onto the silicon wafer defines the thickness and curvature of the lens surface. Multiple photoresist coatings can be applied to achieve the desired thickness. Next, the photoresist reflows by applying heat, i.e. around 110 °C on a hot plate, to the wafer. The surface tension applied to the photoresist by its coating above the glass transition temperature, makes the surface reflow into a spherical shape. Last, the photoresist shape is transferred into the silicon wafer using a DRIE process. The photoresist and silicon etching selectivity, controlled by the CF_4 and O_2 gas ratio and the DC and RF power applied in the process, defines the curvature of the overall lens. The examples shown below a selectivity of 1 : 1.3 was used to achieve a total height of 475 μm for a 360 μm of photoresist. The surface roughness achieved with the process in the order of hundreds of nanometers.

The last step of the process consists of applying an antireflective coating to the lens which is essential to reduce the high reflection losses that occur by using a dielectric with high permittivity. A coating with the polymer Parylene is usually employed as matching layer at submillimeter-wave frequencies.It has an index of refraction around 1.64, which is not the ideal for silicon, but it is close enough to considerably reduce the reflections. The coating conformal is deposited using vapor deposition which allows high control of the thickness and uniformity.

2.5.1.2 Surface Accuracy

The fabricated surface of the lens can be analyzed independently of the rest of the antenna by the accurate characterization of the actual fabricated surface. Surface profilometers are used to map the 3D profile of the lens surface with high resolution. The basic parameters of this surface in terms of radius R, diameter of the aperture D, height H are computed with an

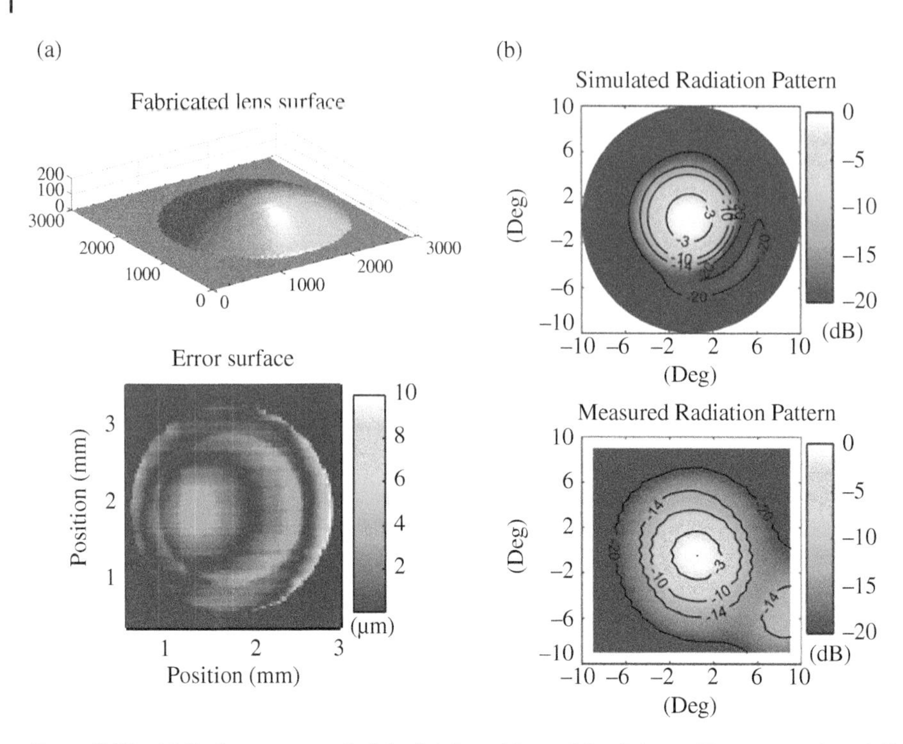

Figure 2.25 (a) Surface measured of the fabricated lens of $D = 2.6$ mm. The error surface defined as the difference between the measured surface and a perfect sphere is shown underneath. (b) Computed radiation pattern of the measured lens surface and the measured radiation pattern of the whole lens antenna at 1.9 THz from [26].

optimization procedure using, for example, the optimization toolbox in Matlab and a basic cost function to minimize. The cost function compares the fabricated surface and an ideal spherical surface until it finds the best fitting sphere. By evaluating the error between both surfaces we can have a sense of how spherical the fabricated lens is, and identify flaws that can improve the fabrication process. For example, Figure 2.25a shows the profile of a fabricated shallow lens of $D = 2.6$ mm measured with a profilometer and its error surface obtained when compared with a perfect sphere. In this case, since the edges of the lens did not provide a good fit with the sphere, an improvement of the performance could be achieved by adjusting the illumination to minimize the lens area illuminated. This adjustment can be achieved by decreasing the lens thickness W, which will decrease the illumination of the lens edges, i.e. increase the edge field taper.

One can have an estimation of the effects of the aberrations in the radiation pattern by translating the error surface into a phase error surface. We can assume that the field distribution on top of the aperture has a Gaussian field distribution of a certain taper, for example −14 dB, with a phase of $e^{jk_0\left(\sqrt{\varepsilon_{Si}}\,\vec{r}(\theta,\varphi)+\vec{r}\,'(\theta,\varphi)\right)}$, where $\vec{r}(\theta,\varphi)$ is the distance from the phase center of the waveguide feed to the fabricated lens surface and $\vec{r}\,'(\theta,\varphi)$ is the distance from the fabricated lens surface to the lens aperture plane. The resulting radiation pattern is obtained by calculating the Fourier transform of this resulting field distribution. From these

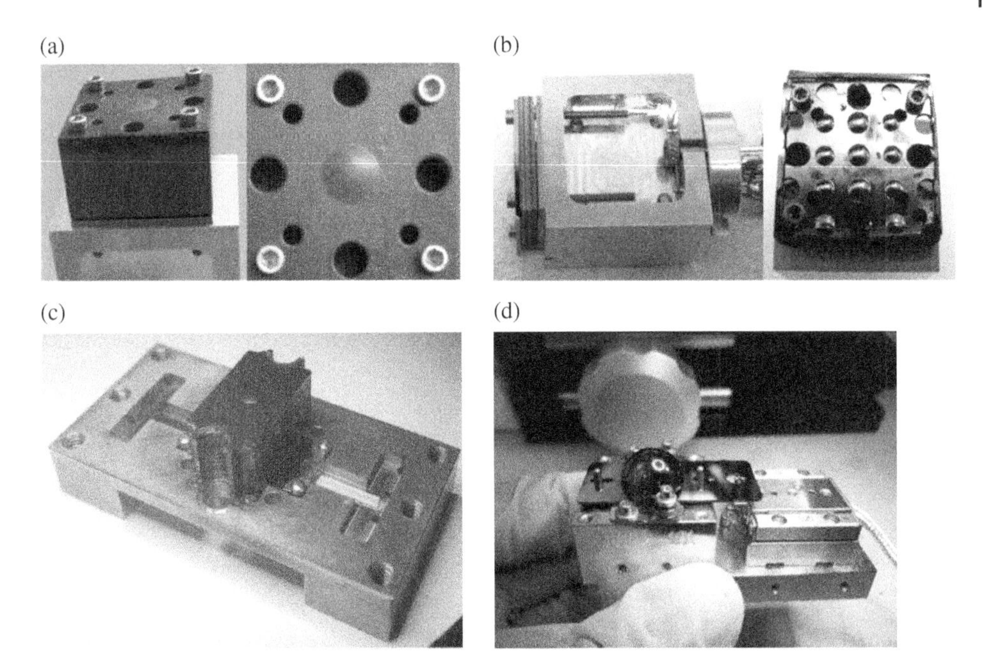

Figure 2.26 Photographs of different lens antenna prototypes fed by leaky-wave feeds (a) at 550 GHz with lens laser micro-machined [33]. *Source*: Alonso-DelPino et al. [33]; IEEE. (b) At 550 GHz with lens fabricated using DRIE silicon micromachining [25]. *Source*: Llombart et al. [25]; IEEE. (c) At 1.9 THz integrated with tripler all in silicon micromachining package [26]. *Source*: Alonso-delPino et al. [26]; IEEE. (d) At 550 GHz integrated with a piezo-electric motor in order to perform beam-scanning [48]. *Source*: Alonso-delPino et al. [48]; IEEE.

fields, the directivity and Gaussicity loss can be computed to have a sense of how our antenna would perform. In this example, the small lens of 2.6 mm aperture would result in a directivity loss of 0.2 dB and a Gaussicity loss close to 2%. Figure 2.25b shows the radiation patterns of the fabricated lens at 1.9 THz for a 2.6 mm diameter lens using the method explained compared with the measured radiation pattern.

2.5.2 Examples of Fabricated Antennas

In this section, we will show different example implementations of leaky-wave antennas feeding silicon lenses. Figure 2.18 shows the photograph of the different prototype examples we will comment on.

Figure 2.26a and b show two prototypes of leaky-wave microlens antennas fabricated at 550 GHz. The first lens was fabricated using laser micromachining, while the second lens was fabricated using DRIE silicon micromachining. Even though both antennas obtained a good agreement with the expected results, the DRIE silicon micromachined lens portrayed some aberrations that were visible in in the radiation pattern measurements showing an increase in the secondary lobes and a small tilt.

Figure 2.26c shows a microlens antennas at 1.9 THz integrated with a 1.9 THz Schottky-based tripler all in the same wafer. The antenna was synthesized on 12 resistivity silicon wafers and the tripler waveguide circuit was synthesized on three wafers. All the wafers, including the lenses, were fabricated using DRIE silicon micromachining process, aligned using the silicon compressive pin technique, and finally glued together. An alignment better than 2 μm was achieved across the wafer stack [26]. Even though in this case, the lenses suffered from aberrations caused by the lack of the surface control in the fabrication process, well-focused beam patters very similar to the design ones were achieved.

On another note, new efforts are now being investigated to enable lens beam-scanning capabilities in the system front end. Figure 2.26d shows a highly integrated beam scanning lens-antenna using a piezo-electric motor demonstrated operating at 550 GHz presented in [48]. A hemispherical lens was glued on top of a silicon wafer containing alignment marks processed using DRIE silicon micromachining. The piezoelectric motor displaced the lens around ±1 mm from the center position of the lens, providing a beam scan of ±25°. Not only this method can be employed for improving the alignment of the lens with the feed, but it also has the potential to enable beam-scanning capabilities on the system front end for future terahertz imaging systems.

The results achieved so far show a great potential to use these dielectric lens antennas in the development of future focal plane arrays at terahertz frequencies. By using the leaky-wave waveguide feed, we only need a small part of the surface of the lens, which reduces the reflection losses and phase errors that these type of lenses suffer. But most of all, it allows the use of photolithographic process when fabricating the lens. The fabrication of the lenses using photolithographic process reduces the cost, with the same performance achieved with other fabrication methods, such as laser micromachining.

Exercises

E2.1 Derivation of the Transmission Coefficients and Lens Critical Angle

In Section 2.2, we defined the Fresnel reflection coefficients $\tau^{\perp}(Q)$, $\tau^{\parallel}(Q)$ as:

$$\tau^{\perp}(Q) = \tau^{\perp}(\theta_i) = \frac{2\zeta_0 \cos\theta_i}{\zeta_0 \cos\theta_i + \zeta_d \cos\theta_t} \tag{2.69}$$

$$\tau^{\parallel}(Q) = \tau^{\parallel}(\theta_i) = \frac{2\zeta_0 \cos\theta_i}{\zeta_0 \cos\theta_t + \zeta_d \cos\theta_i} \tag{2.70}$$

where $\cos\theta_i = \hat{k}_i \cdot \hat{n} = \frac{1 - e\cos\theta}{\sqrt{1 + e^2 - 2e\cos\theta}}$. Similarly, the transmission coefficients can be obtained from the reflection coefficients $\tau^{\perp} = 1 + \Gamma^{\perp}$ and $\tau^{\parallel} = \left(1 + \Gamma^{\parallel}\right)\frac{\cos\theta_i}{\cos\theta_t}$. Thus, they are represented as:

$$\Gamma^{\perp} = \frac{\zeta_0 \cos\theta_i - \zeta_d \cos\theta_t}{\zeta_0 \cos\theta_i + \zeta_d \cos\theta_t}$$

$$\Gamma^{\parallel} = \frac{\zeta_0 \cos\theta_t - \zeta_d \cos\theta_i}{\zeta_0 \cos\theta_t + \zeta_d \cos\theta_i}$$

Note that the angle of refraction θ_t increases as the incident angle θ_i inside of the lens increases ($\theta_t > \theta_i$). Total internal reflection ($\Gamma = 1$) occurs when the angle of refraction is

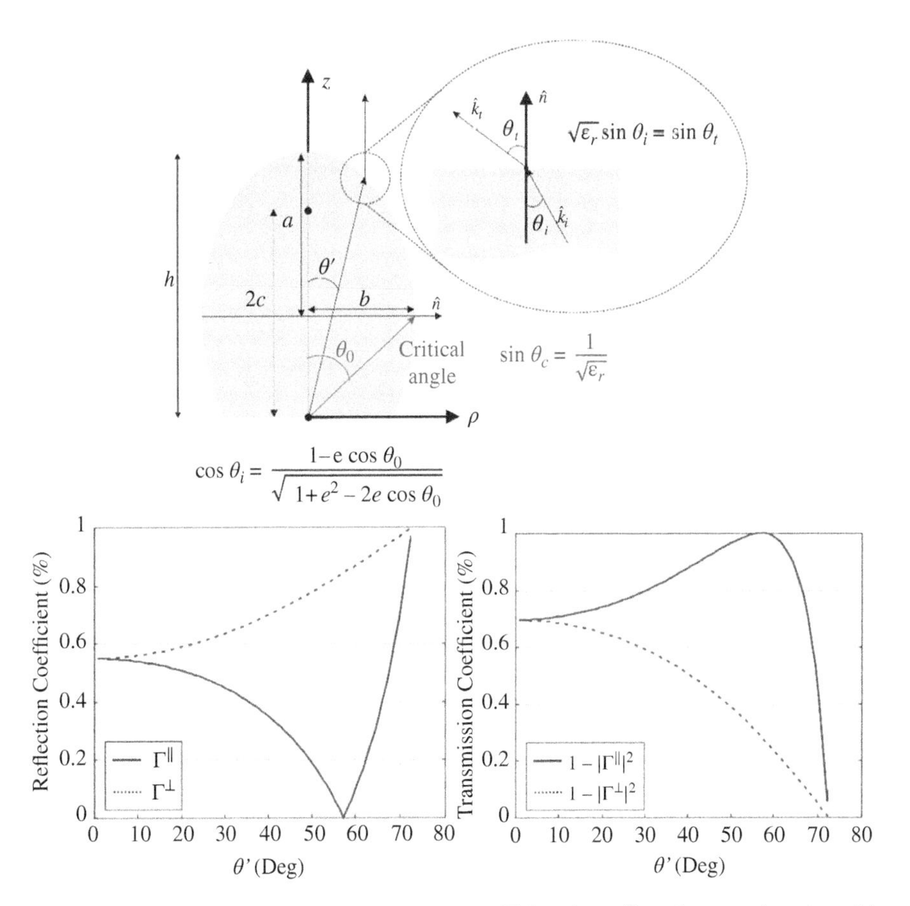

Figure 2.27 (a) Reflection and (b) transmission coefficient for a silicon lens as a function of the angle. The reflection coefficient becomes 1 for angles larger or equal than the critical angle. The transmission coefficient becomes one for the Brewster angle.

90° and an incident angle known as the critical angle $\theta_c = \sin^{-1}\left(1/\sqrt{\varepsilon_r}\right)$. After this angle there is no transmission, therefore, this angle defines the domain of the equivalent aperture. As this angle lies at the center distance between the two focuses, the radius of the equivalent aperture is therefore the minor axis b. The incident angle that creates zero reflection in the parallel reflection coefficient is called Brewster's angle. An example of the reflection and transmission coefficients for a lens made in silicon are shown in Figure 2.27a and b, illustrating the critical and Brewster angles.

E2.2

It is common to use an elliptical silicon lens antenna to be integrated with a high frequency detector. The detector, which operates at 350 GHz, it is placed on silicon ($\varepsilon_r = 11.9$). In order to couple the radiation into the detector an antenna consisting of two slots is used.

1) Give the length and separation of the slots in order to generate a field radiated inside an infinite silicon medium that could be used to illuminate efficiently the lens.
2) Design the lens dimensions such that its far field is characterized by a directivity of 30 dBs.
3) What are the sidelobe level, directivity, and reflection efficiency of lens antenna?
4) Assuming we aimed at the same directivity of the lens antenna, how would the two slot antenna and lens dimensions change if the dielectric material was $\varepsilon_r = 4$? What about the reflection efficiency?

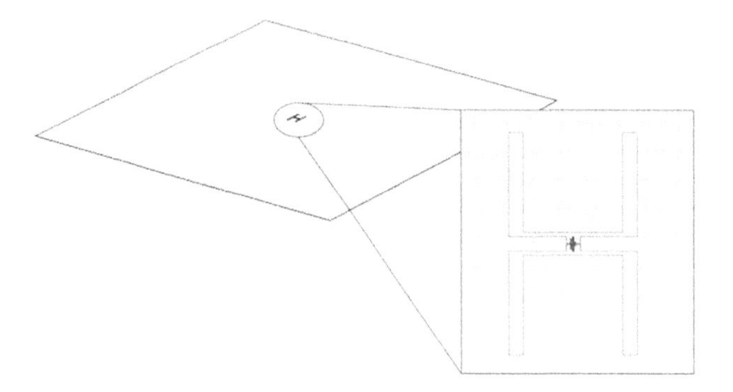

Justify your answers with figures and try to give a physical insight into the problem.

E2.3

For automotive radar applications at 70 GHz, we need to design an elliptical lens antenna with a directivity of 30 dB fed by a rectangular waveguide as shown in the figure. The waveguide can be modeled considering the field in the aperture is approximated by the TE10 mode.

1) Make a study of the optimal dimensions of the waveguide and the lens to reach the specifications for the following different lens materials: materials: silicon ($\varepsilon_r = 11.9$), quartz ($\varepsilon_r = 4$) and plastic ($\varepsilon_r = 2$)
2) How does the change physical dimensions with the lens dielectric material?
3) How does the critical angle change with the lens dielectric material? And the directivity inside the dielectric and the reflection efficiency of the optimized waveguide feed?
4) How does the radiated patterns by the several lens antenna compared to each other?

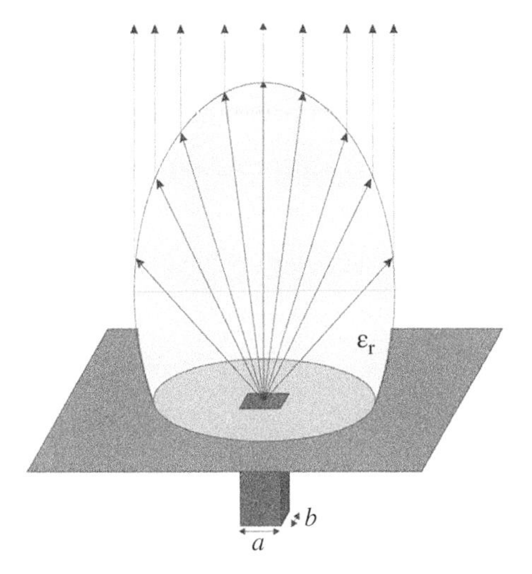

Justify your answers with figures and try to give a physical insight into the problem.

References

1 Gulkis, S., Allen, M., Backus, C. et al. (2007). Remote sensing of a comet at millimeter and submillimeter wavelengths from an orbiting spacecraft. *Planetary and Space Science* **55** (9): 1050–1057.

2 Love, W. (1962). The diagonal horn antenna. *Microwave Journal* **V**: 117–122.

3 Johansson, J.F. and Whyborn, N.D. (1992). The diagonal horn as a sub-millimeter wave antenna. *IEEE Transactions on Microwave Theory and Techniques* **40** (5): 795–800.

4 RPG Radiometer Physics GmbH https://radiometer-physics.de

5 Day, P., Leduc, H., Goldin, A. et al. (2006). Antenna-coupled microwave kinetic inductance detectors. *Nuclear Instruments and Methods in Physics Research Section A: Accelerators, Spectrometers, Detectors and Associated Equipment* **559** (2): 561–563.

6 Sarmah, N., Grzyb, J., Statnikov, K. et al. (2016). A fully integrated 240-GHz direct-conversion quadrature transmitter and receiver chipset in SiGe technology. *IEEE Transactions on Microwave Theory and Techniques* **64** (2): 562–574.

7 Skalare, A., van de Stadt, H., de Graauw, Th. et al. (1992). Double-dipole antenna SIS receivers at 100 and 400 Gilz. *Proceedings of the 3rd Mt. Conference Space Terahertz Technology*, Ann Arbor, MI, 222–233.

8 Semenov, A.D., Hubers, H.-W., Richter, H. et al. (2003). Superconducting hot-electron bolometer mixer for terahertz heterodyne receivers. *IEEE Transactions on Applied Superconductivity* **13** (2): 168–171.

9 Kormanyos, B.K., Ostdiek, P.H., Bishop, W.L. et al. (1993). A planar wideband 80–200 GHz subharmonic receiver. *IEEE Transactions on Microwave Theory and Techniques* **41** (10): 1730–1737.

10 Garufo, A., Carluccio, G., Freeman, J.R. et al. (2018). Norton equivalent circuit for pulsed photoconductive antennas—part II: experimental validation. *IEEE Transactions on Antennas and Propagation* **66** (4): 1646–1659.

11 Garufo, A., Llombart, N., and Neto, A. (2016). Radiation of logarithmic spiral antennas in the presence of dense dielectric lenses. *IEEE Transactions on Antennas and Propagation* **64** (10): 4168–4177.

12 Chattopadhyay, G., Glenn, J., Bock, J.J. et al. (2003). Feed horn coupled bolometer arrays for SPIRE – design, simulations, and measurements. *IEEE Transactions on Microwave Theory and Techniques* **51** (10): 2139–2146.

13 Kangas, M.M., Ansmann, M., Horgan, B. et al. (2005). A 31 pixel flared 100-GHz high-gain scalar corrugated nonbonded platelet antenna array. *IEEE Antennas and Wireless Propagation Letters* **4**: 245–248.

14 Leech, J., Tan, B.K., Yassin, G. et al. (2011). Multiple flare-angle horn feeds for sub-mm astronomy and cosmic microwave background experiments. *Astronomy and Astrophysics* **532**: A61.

15 Lubecke, V.M., Mizuno, K., and Rebeiz, G.M. (1998). Micromachining for terahertz applications. *IEEE Transactions on Microwave Theory and Techniques* **46** (11): 1821–1831.

16 Lee, C., Chattopadhyay, G., Decrossas, E. et al. (2015). Terahertz antenna arrays with silicon micromachined-based microlens antenna and corrugated horns. International Workshop on Antenna Technology (iWAT), 70–73.

17 Chattopadhyay, G., Kuo, C.-L., Day, P. et al. (2007). Planar antenna arrays for cmb polarization detection. Infrared and Millimeter Waves, 2007 and the 2007 15th International Conference on Terahertz Electronics. IRMMW-THz. Joint 32nd International Conference on, 2007, 184–185.

18 Iacono, A., Freni, A., Neto, A., and Gerini, G. (2011). In-line x-slot element focal plane array of kinetic inductance detectors. Antennas and Propagation (EUCAP), Proceedings of the 5th European Conference on, 2011, 3316–3320.

19 Bueno, J., Murugesan, V., Karatsu, K. et al. (2018). Ultrasensitive kilo-pixel imaging array of photon noise-limited kinetic inductance detectors over an octave of bandwidth for THz astronomy. *Low Temperature Physics* **193**: 96. https://doi.org/10.1007/s10909-018-1962-8.

20 Lee, S.-K., Kim, M.-G., Jo, K.-W. et al. (2008). A glass reflowed microlens array on a Si substrate with rectangular through-holes. *Journal of Optics A: Pure and Applied Optics* **10** (4): 044003.

21 Zmuidzinas, J. and LeDuc, H.G. (1992). Quasi-optical slot antenna SIS mixers. *IEEE Transactions on Microwave Theory and Techniques* **40**: 1797–1804.

22 Filipovic, S., Gearhart, S., and Rebeiz, G.M. (1993). Double-slot antennas on extended hemispherical and elliptical silicon dielectric lenses. *IEEE Transactions on Microwave Theory and Techniques* **41** (10): 1738–1749.

23 Neto, A., Bruni, S., Gerini, G., and Sabbadini, M. (2005). The leaky lens: a broad-band fixed-beam leaky-wave antenna. *IEEE Transactions on Antennas and Propagation* **53** (10): 3240–3246.

24 Llombart, N., Dabironezare, S.O., Carluccio, G. et al. (2018). Reception power pattern of distributed absorbers in focal plane arrays: a fourier optics analysis. *IEEE Transactions on Antennas and Propagation* **66** (11): 5990–6002.

25 Llombart, N., Chattopadhyay, G., Skalare, A., and Mehdi, I. (2011). Novel terahertz antenna based on a silicon lens fed by a leaky wave enhanced waveguide. *IEEE Transactions on Antennas and Propagation* **59** (6): 2160–2168.

26 Alonso-delPino, M., Reck, T., Jung-Kubiak, C. et al. (2017). Development of silicon micromachined microlens antennas at 1.9 THz. *IEEE Transactions on Terahertz Science and Technology* **7** (2): 191–198.

27 Rutledge, D.B., Neikirk, D.P., and Kasilingam, D.P. (1983). Integrated circuit antennas. In: *Infrared and Millimeter-Waves*, vol. **10** (ed. K.J. Button), 1–90. New York: Academic Press.

28 Rebeiz, G.M. (1992). Millimeter-wave and terahertz integrated circuit antenna. *Proceedings of the IEEE* **80** (11): 1748–1770.

29 Moussessian, A., Wanke, M.C., Li, Y. et al. (1998). A terahertz grid frequency doubler. *IEEE Transactions on Microwave Theory and Techniques* **46** (11): 1976–1981.

30 Rebeiz, G., Regehr, W., Rutledge, D. et al. (1987). Submillimeter-wave antennas an thin membranes. Antennas and Propagation Society International Symposium, 1987, 1194–1197.

31 Kominami, M., Pozar, D., and Schaubert, D. (1985). Dipole and slot elements and arrays on semi-infinite substrates. *IEEE Transactions on Antennas and Propagation* **33** (6): 600–607.

32 Carluccio, G., Albani, M., and Neto, A. (2012). An iterative physical optics algorithm for the analysis and design of dielectric lens antennas. *2012 IEEE International Symposium on Antennas and Propagation and USNC-URSI National Radio Science Meeting*, Chicago, Illinois, USA (8–14 July 2012).

33 Alonso-DelPino, M., Llombart, N., Chattopadhyay, G. et al. (2013). Design guidelines for a terahertz silicon micro-lens antenna. *IEEE Antennas and Wireless Propagation Letters* **12**: 84–87.

34 Neto, A., Maci, S., and De Maagt, P.J.I. (1998). Reflections inside an elliptical dielectric lens antenna. *IEE Proceedings – Microwaves, Antennas and Propagation* **145** (3): 243–247.

35 Gatesman, A.J., Waldman, J., Ji, M. et al. (2000). An anti-reflection coating for silicon optics at terahertz frequencies. *IEEE Microwave and Guided Wave Letters* **10** (7): 264–266.

36 Nitta, T., Sekiguchi, S., Sekimoto, Y. et al. (2014). Anti-reflection coating for cryogenic silicon and alumina lenses in millimeter-wave bands. *Journal of Low Temperature Physics* **176** (5): 677–683. https://doi.org/10.1007/s10909-013-1059-3.

37 Busse, L.E., Florea, C.M., Frantz, J.A. et al. (2014). Anti-reflective surface structures for spinel ceramics and fused silica windows, lenses and optical fibers. *Optical Materials Express* **4** (12): 2504–2515. https://doi.org/10.1364/OME.4.002504.

38 Jackson, D.R., Oliner, A.A., and Ip, A. (1993). Leaky-wave propagation and radiation for a narrow-beam multiple-layer dielectric structure. *IEEE Transactions on Antennas and Propagation* **41** (3): 344–348.

39 Lee, Y., Yeo, J., Mittra, R., and Park, W. (2005). Application of electromagnetic bandgap (EBG) superstrates with controllable defects for a class of patch antennas as spatial angular filters. *IEEE Transactions on Antennas and Propagation* **53** (1): 224–235.

40 Guerin, N., Enoch, S., Tayeb, G. et al. (2006). A metallic Fabry-Perot directive antenna. *IEEE Transactions on Antennas and Propagation* **54** (1): 220–224.

41 Foroozesh, A. and Shafai, L. (2010). Investigation into the effects of the patch-type FSS superstrate on the high-gain cavity resonance antenna design. *IEEE Transactions on Antennas and Propagation* **58** (2): 258–270.

42 Sarabandi, K. and Behdad, N. (2007). A frequency selective surface with miniaturized elements. *IEEE Transactions on Antennas and Propagation* **55** (5): 1239–1245.

43 Llombart, N., Neto, A., Gerini, G. et al. (2008). Impact of mutual coupling in leaky wave enhanced imaging arrays. *IEEE Transactions on Antennas and Propagation* **56** (4): 1201–1206.

44 Neto, A., Llombart, N., Baselmans, J.J.A. et al. (2014). Demonstration of the leaky lens antenna at submillimeter wavelengths. *IEEE Transactions on Terahertz Science and Technology* **4** (1): 26–32.

45 Campo, M.A., Blanco, D., Carluccio, G. et al. (2018). Circularly polarized lens antenna for Tbps wireless communications. *2018 48th European Microwave Conference (EuMC)*, Madrid, 1147–1150.

46 Neto, A. and Llombart, N. (2006). Wideband localization of the dominant leaky wave poles in dielectric covered antennas. *IEEE Antennas and Wireless Propagation Letters* **5**: 549–551.

47 Llombart, N., Lee, C., Alonso-delPino, M. et al. (2013). Silicon micromachined lens antenna for THz integrated heterodyne arrays. *IEEE Transactions on Terahertz Science and Technology* **3** (5): 515–523.

48 Alonso-delPino, M., Jung-Kubiak, C., Reck, T. et al. (2019). Beam scanning of silicon lens antennas using integrated piezomotors at submillimeter wavelengths. *IEEE Transactions on Terahertz Science and Technology* **9** (1): 47–54.

3

Photoconductive THz Sources Driven at 1550 nm

Elliott R. Brown[1], *B. Globisch*[2,3], *Guillermo Carpintero del Barrio*[4], *A. Rivera*[4], *Daniel Segovia-Vargas*[4], *and A. Steiger*[5]

[1] *Department of Physics and Electrical Engineering, Wright State University, Dayton, OH, USA*
[2] *Department of Physics, Technische Universität Berlin, Berlin, Germany*
[3] *Fraunhofer Institute for Telecommunications, Heinrich Hertz Institute, Berlin, Germany*
[4] *Departamento Teoría de la Señal y Comunicaciones, Universidad Carlos III de Madrid, Madrid, Spain*
[5] *Working Group 7.34 Terahertz Radiometry, Physikalisch-Technische Bundesanstalt, Berlin, Germany*

3.1 Introduction

3.1.1 Overview of THz Photoconductive Sources

The THz region is at minimum the decade-wide band of the electromagnetic spectrum between 300 GHz and 3.0 THz[1], with the millimeter-wave region below and the far-infrared region above. The first experiments were conducted around 1900 by H. Rubens, M. Czerny, and B.W. Snow [1, 2]. At that time, the so-called residual-radiation method was applied in order to extract radiation with a wavelength between 15 and 100 μm (20–3 THz) from the spectrum of a blackbody [2]. However, it took another 80 or so years until the development of the first femtosecond pulse laser in the 1970s allowed for the generation and detection of broadband THz radiation [3–5]. The reason for the long time delay is because the generation and detection of THz radiation are relatively difficult. Thus, the first successful broadband THz radiation relied on an optoelectronic technique: a femtosecond laser source being applied to induce a transient current in a light-sensitive, biased semiconductor having short photocarrier relaxation time. Soon thereafter, THz optoelectronic generation of continuous-wave (CW) radiation was demonstrated by the mixing of two frequency-offset lasers on a similar semiconductor [6]. These so-called photoconductive (PC) sources have served as the workhorses for the development of coherent terahertz systems in the following decades.

Amongst many emerging technologies, these THz PC sources have arguably created the biggest impact on the THz field over the past 30 years. They are used in a wide variety of instruments and systems ranging from time- and frequency-domain spectrometers (FDSs), to impulse–reflection biomedical imagers and free-space-communications systems. In all of

Fundamentals of Terahertz Devices and Applications, First Edition. Edited by Dimitris Pavlidis.

Companion website: www.wiley.com/go/Pavlidis/FundamentalsofTHz

these applications, an inherent advantage of THz PC sources is *useful bandwidth*—instantaneous bandwidth in time-domain sources, and tuning bandwidth for frequency-domain sources. In most cases, the *useful bandwidth* is ~1.0 THz or greater which far exceeds the capability of competing solid-state electronic sources such as semiconductor fundamental-frequencyamplifiers and frequency multiplier chains. Competitive solid-state, nonlinear-opticalsources based on difference-frequency-generation (DFG) have comparable bandwidth, but great difficulty producing significant power much below 1.0 THz where PC sources are strongest. And photonic sources such as quantum cascade lasers (QCLs) cannot operate at room temperature and at ~1.0 THz or below where $h\nu << k_BT$ (for $T = 300$ K, $h\nu = k_BT$ at $\nu = 6.25$ THz) [7].

As the name suggests, PC devices are based on the internal photoelectric effect in solids, usually some type of semiconductor composite or heterostructure. As such, usually, two free carriers (an electron and a hole) are produced per incident photon (intrinsic photoconductivity), but sometimes just a free electron (extrinsic photoconductivity). In the presence of a large internal electric field, the photocarriers are accelerated towards electrodes which, if "ohmic" allow induced electrical current in the external circuit. The most popular external circuit by far is the planar antenna since in principle it can create free-space THz radiation with low loss and, if designed properly, can cover a very large bandwidth itself, especially with the self-complementary type antennas (see Section 3.4).

The fundamental reason for the high bandwidth of PC devices is the ultrafast nature of the impulse response function. The leading edge of the impulse response (i.e. rise-time) is determined by the photon excitation process and photocarrier acceleration in an electric field E. Photon excitation is known from quantum mechanics to be very fast (~10 fs or less). And although not as fast as the photon excitation, photocarrier acceleration, $a = eE/m^*$, can also be very fast in semiconductors where $m^* \ll m_0$–the photocarrier mass in vacuum. This is quite distinct from all electronic devices, such as transistors, where the rise-time always has device-limited delay, usually by device capacitance. The lagging edge (i.e., fall-time) is the photocarrier relaxation process, which is relatively slow (~1 ns or longer) in normal semiconductor devices in which relaxation occurs by radiative recombination (the inverse process of the excitation). However, it can be reduced greatly to ~1 ps or less in specially designed "ultrafast" semiconductor materials by nonradiative recombination, or in certain devices (e.g. p–i–n photodiodes (PIN-PDs)) by ultrafast transit time across the active region.

As ultrafast photoconductors have evolved over the decades, this reduction in fall time has been accomplished while maintaining the excellent electron-transport properties (e.g. the mobility) that semiconductors naturally provide. Hence, the only fundamental limit on the impulse response, and therefore the frequency response (i.e. transfer function) is the photocarrier relaxation time, which entails a single time constant τ. As such, the fundamental frequency roll-off is just 6 dB/octave, comparable to the best electronic sources [8], but with much more *useful bandwidth*. This assumes, of course, that the capacitance C of the PC device is tolerably small such that there is not a comparable RC time constant (with R usually representing the differential resistance of the THz load circuit). However, this is not difficult to satisfy through prudent electrode and antenna engineering.

Some related benefits of ultrafast photoconductors are spectral purity and *isolation*. Provided that the laser intensities are not too high, the instantaneous photocurrent is perfectly quadratic in the optical electric fields, i.e. $i(t) = \eta eP(t)/h\nu = c\,[E(t)]^2$, where η is the external quantum efficiency, $h\nu$ is the photon energy, E the optical electric field, and c is a constant. Unlike nonlinear electronic devices such as mixers and harmonic multipliers, there is no cubic or higher-order terms, meaning that there is no intermodulation between the frequency-offset lasers driving a photomixer, or between different spectral components of an ultrashort laser pulse. And since so few of the photocarriers relax by radiative recombination, there is negligible photonic feedback to the laser(s). Hence, optoelectronic isolation is very high except for the small dielectric reflection at air–semiconductor interfaces, which can be made negligible by antireflection coatings.

In spite of these inherent advantages, THz PC devices have several drawbacks which we will also address in this chapter. The primary one is low average output power P_{THz} and the associated optical-to-THz conversion efficiency ($\varepsilon = P_{THz}/P_{opt}$). As will be discussed later, both of these have improved dramatically in recent years for devices driven at 1550 nm such that P_{THz} can readily exceed 10 μW (−20 dBm) in CW photomixers, and 100 μW (−10 dBm) in pulsed, time-domain devices. However, what continues to frustrate THz researchers, and impede the migration of systems engineers to the THz field, is that these power levels are still well below those typically produced by microwave sources in RF systems (e.g. local oscillators for down-converting mixers in radio and radar receivers). The situation is not unlike that in the mid- and long-wave infrared where active systems, such as lidar, have not become popular because of lacking solid-state laser power.

Ours is not the first review of PC sources in the 1550-nm region. We follow at least two review articles that include this topic with focus on CW photomixers [9, 10]. However, ours may be the first that contrasts the two most common types of devices—p–i–n-based and MSM-based photodiodes—on an equitable quantitative basis and in the context of the two common implementations as PC switches and photomixers. We also broach the subtle differences between the two device types and modalities whether used as a source or detector. This is a topic that is not well understood, or at least not well explained by the vast majority of THz literature, but of utmost importance to many scientists and engineers just entering the field.

We also emphasize THz metrology—one of the most problematic and daunting topics in the THz field, especially the measurement of THz power. Through collaboration with the PTB in Berlin, Germany, we describe perhaps the most accurate THz power sensor in the world today, and then demonstrate it an exemplary fashion. Our demonstration was made on a commercial THz PIN-PD source similar, or identical, to that which many researchers around the world are already using. For pedagogical completeness, we also address many pitfalls in THz power measurements, such as the leakage of 1550-nm drive laser power, and far-infrared thermal re-emission from the substrate (usually InP) on which the device is fabricated.

3.1.2 Lasers and Fiber Optics

A fundamental advantage of 1550 compared to 780-nm lasers is higher relative photon flux, $f = P/h\nu = P\lambda/hc$ [photons/s]. The ratio of the fluxes at 1550 and 780 nm is simply $\lambda_{1550}/\lambda_{780} \approx 2.0$. Although seemingly trivial compared to the other differences, it is generally true that ultrafast PC devices, be they CW photomixers or PC switches, generally produce THz power proportional to the optical power squared, so that 2.0 difference in photon flux can

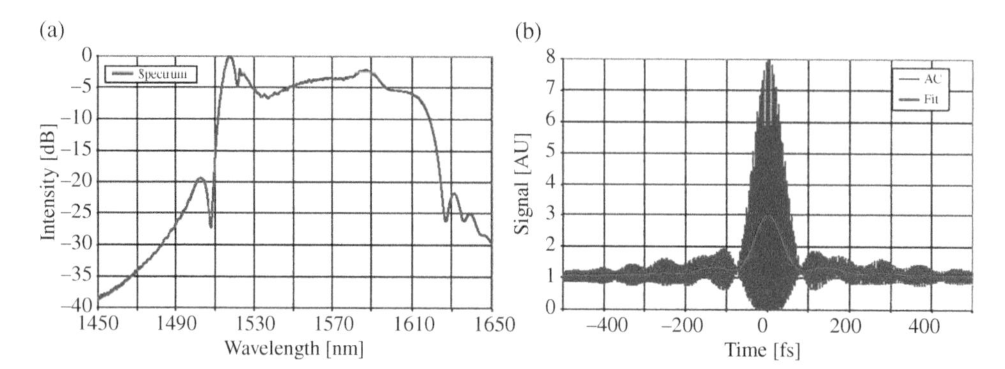

Figure 3.1 (a) Power spectrum from commercial EDFA mode-locked laser. (b) Time-domain single-pulse characteristics based on measurement of autocorrelation function.

produce up to ~4× difference in THz power, per mW of optical power. This is often underemphasized in the THz literature.

The primary shortcoming of the cross-gap GaAs-based PC sources is that they must be pumped by lasers (pulsed or CW) operating in the spectral range ~ 780–850 nm. While available, such lasers are more expensive and difficult to integrate with fiber-optic components than lasers in the longer-wavelength 1550-nm region (common fiber Telecom band). This is largely because of the existence of the erbium-doped fiber amplifier (EDFA), first demonstrated in 1986, and so successful that it changed the history of fiber-optic communications almost overnight. And not only can it be fabricated for CW amplification but also for pulsed operation too as a mode-locked laser (MLL). Figure 3.1a shows the output spectrum and single pulse characteristics of a 1550-nm fiber MLL routinely used by the authors [11]. Its spectrum is relatively flat across the entire range from 1515 to 1615 nm, which corresponds to a frequency bandwidth of $\Delta\nu = 12.3$ THz. Figure 3.1b shows the time-domain autocorrelation function and a fit using a csch function, from which the FWHM of the time domain pulse is found to be ≈90 fs. And in the 780–850 nm region, there is no such fiber amplifier.

A related advantage of 1550-nm operation comes from semiconductor technology. This was the development of InGaAsP-based quantum-well distributed-feedback (DFB) lasers for CW operation, and ultimately for pulsed operation too. Again, materials science played a key role here as the ternary compound $In_{0.53}Ga_{0.47}As$ is lattice matched to InP substrates and has excellent semiconductor properties. Quaternary compounds of $In_XGa_{1-X}As_YP_{1-Y}$ have similar properties. Such DFB lasers are used widely in fiber-optic telecommunications and are key to the success of THz CW photomixers today.

Ultimately, optical-fiber coupling of THz PC devices is highly desirable for many reasons, including compactness of system integration, mechanical and thermal stability, and cost. Again, 1550-nm operation is advantageous compared to 780 nm for reasons pertaining to optical-fiber technology. For example, single-mode fiber displays much lower loss at 1550 nm, as shown in Figure 3.2[2]. Historically, the optical fibers operating between ~780 and 850 nm were developed first [12]. But then with the advent of EDFA amplifiers, the interest shifted to the 1300- and 1550-nm bands because of the surprisingly low loss in these

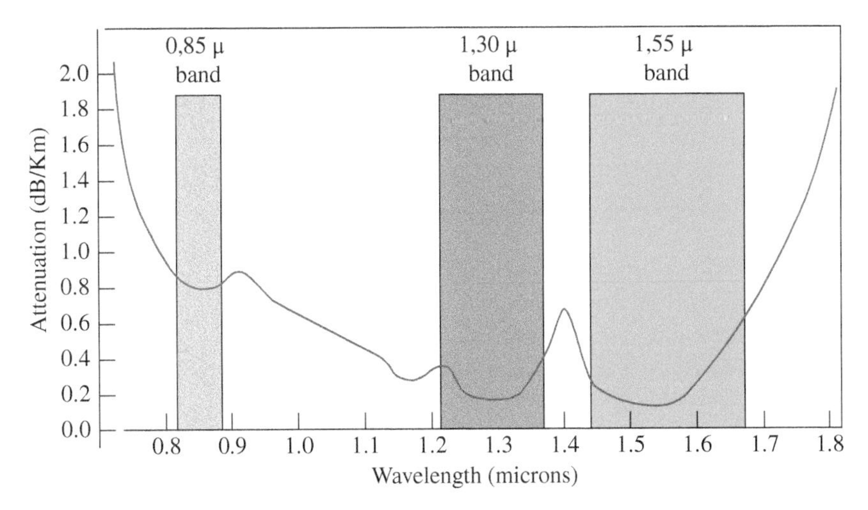

Figure 3.2 Optical fiber absorption vs wavelength. *Source*: Modified from http://photonicswiki.org/index.php?title=Dispersion_and_Attenuation_Phenomena..

bands with a minimum attenuation of ~0.2 dB/km exhibited in the 1550-nm region. A related issue is that 1550-nm fiber has lower dispersion[3] which is mostly a problem with ultrafast pulsed lasers (pulse widths <<1 ps) because of their very high spectral bandwidth, as shown in Figure 3.1a.

But there are other advantages of 1550-nm fibers as well. For example, the core diameter of a single-mode fiber is ≈9 μm compared to ≈6 μm for a 780 nm fiber[4]. For given amount of power, the intensity (P/area) is approximately twice that in 780-nm fiber. This higher intensity can induce Brillouin scattering and other possible nonlinear processes much more readily in a 780-nm fiber[5]. This is mostly an issue with pulsed lasers, which have orders-of-magnitude higher peak power than in CW lasers. Finally, a practical advantage stemming from the larger fiber core diameter is the coupling of fibers to devices. This is always tricky, requiring special equipment such as a multi-axis (usually five) goniometer. But the larger core diameter alleviates the procedure significantly and tends to make both active and passive fiber-optic components lower cost at 1550 nm than at 780 nm. These advantages of 1550-nm optical fiber have already been utilized in at least one system application, which is local-oscillator distribution for arrays of mm-wave coherent receivers, such as those in the Atacama Large Millimeter Array (ALMA) radio observatory in Chile [13].

3.2 1550-nm THz Photoconductive Sources

3.2.1 Epitaxial Materials

3.2.1.1 Bandgap Engineering

Until recent years, the majority of the successful THz instruments and systems have been constructed with PC devices made from some type of epitaxial GaAs designed for ultrafast response. Perhaps the most common is low-temperature-grown (LTG) GaAs, which is now about 30 years old but still used worldwide. However, like any form of GaAs, to display strong intrinsic photoconductivity it must be driven by photons having an energy at or above its energy bandgap, which is 1.42 eV at 300 K, corresponding to a wavelength of 874 nm. A logical drive wavelength is around 780 nm (1.59 eV) because of the wide selection of lasers there, both CW and pulsed. The best CW lasers are generally GaAs-based semiconductor lasers, and the most tunable is the solid-state ($TiAl_2O_3$) laser, which is also the best choice for ultrashort pulse applications since it is relatively easy to mode lock.

In practice, LTG–GaAs and most other ultrafast photoconductors are grown by molecular beam epitaxy—an elegant growth technique capable of producing many exotic materials, including heterostructures, quantum wells, nanocomposites, nanoparticle arrays, quantum dots, and nonstoichiometric materials. As such, it is used to fabricate a large variety of materials in high-speed electronic and optoelectronic devices, and ultrafast photoconductors are no exception. One problem with MBE is that it can only support a slow growth rate (<1 μm/h), so the epitaxial layers are generally kept thin—1–2 μm being typical. And to produce a high crystalline quality, MBE can only be done for epitaxial materials that are lattice-matched (or nearly) to an extremely smooth substrate. For LTG–GaAs, the simple choice has always been semi-insulating, "epi-ready" GaAs substrates, which are readily available and affordable.

For 1550-ultrafast PC devices, the material must have a bandgap less than 0.80 eV and be lattice-matched to some "epi-ready" substrate. The obvious choice for nearly 30 years has been $In_XGa_{1-X}As$ epilayers on SI, "epi-ready" InP substrates, the lattice-matching fraction being $X = 0.53$. And the band-gap for this fraction is ≈0.75 eV at room-temperature so just below the 1550-nm photon energy. The most successful 1550-nm-driven THz PC devices have been designed with the $In_{0.53}Ga_{0.47}As$ as their "absorbing" layer where the photocarriers are generated and where the electron- and hole transport properties must be good to achieve excellent overall device performance. In undoped, highly crystalline $In_{0.53}Ga_{0.47}As$ for example, the electron and hole mobilities are $\approx 12 \times 10^3$ and 300 cm^2/V-s, respectively. These are roughly 50% higher than the corresponding values in GaAs of similar quality, largely because of the lower electron and hole effective masses in the $In_{0.53}Ga_{0.47}As$. And although InP substrates are roughly 2× more expensive than GaAs ones, this pales in comparison to the actual MBE growth cost, be it on GaAs or InP.

The $In_{0.53}Ga_{0.47}As$-on-InP strategy offers other important materials features that are utilized depending on the device type, as discussed in the next section. One is the use of barrier layers, such as epitaxial $In_{0.52}Al_{0.48}As$ (also latticed matched to InP) on InP. Another is precise doping (a generic benefit of MBE) of materials like Be—a common acceptor in $In_{0.53}Ga_{0.47}As$ that can compensate for its natural n-type behavior, reducing the total free-carrier concentration and thereby increasing the resistivity. A third is doping with a preferred atomic species, such as Fe, that creates a high concentration of deep electron levels, roughly at the middle of the $In_{0.53}Ga_{0.47}As$ bandgap. This not only compensates the n-type nature but also creates ultrafast (<1 ps) electron–hole recombination centers, which is desirable in some device types. Another interesting option is atomic erbium which when used at high enough concentration incorporates as crystalline ErAs nanoparticles which create ultrafast electron–hole recombination and also the possibility of "extrinsic" photoconductivity as described later.

With thin epitaxial layers, it is also prudent to drive ultrafast photoconductors far enough above the band edge to get significant photon coupling into the "absorbing" epilayer. Since the cross-gap absorption in all semiconductors generally follows the Beer–Lambert law, this

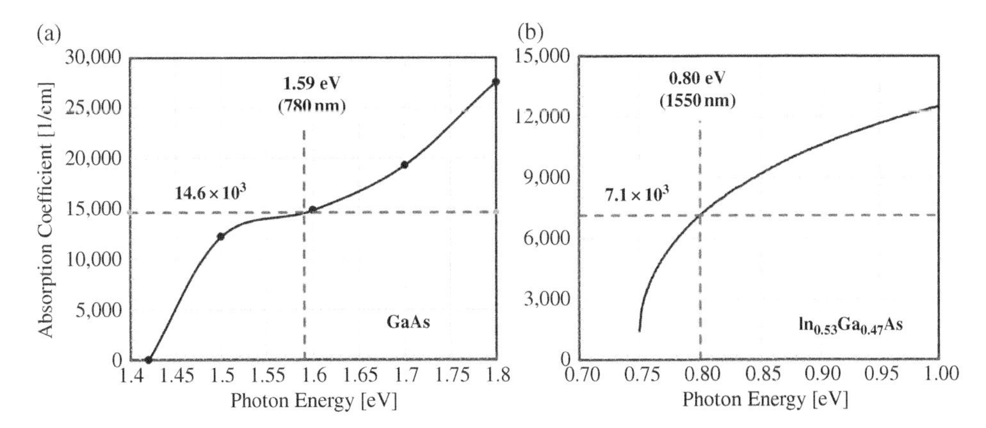

Figure 3.3 (a) Near-band-edge absorption coefficient of GaAs at room temperature. (b) Same for $In_{0.53}Ga_{0.47}As$.

can be measured by the fractional absorption, $\eta_A = [1 - \exp(-\alpha T)]$ where α is the absorption coefficient and T is the epilayer thickness. The curves of α vs photon energy are shown in Figures 3.3a,b for GaAs and $In_{0.53}Ga_{0.47}As$, respectively. The curve for GaAs is taken from a curve fit to highly accurate experimental data and [14] predicts an α of 14.6×10^{-4} at 780 m. The curve for $In_{0.53}Ga_{0.47}As$ is taken from a sophisticated analytic expression developed for limited experimental results [15] and predicts an α of $7.1 \times 10^{-3}/cm^1$ at 1550 nm. Clearly, the greater energy separation between a 780 nm photon and the GaAs bandgap ($\Delta E = 0.17$ eV), compared to a 1550 nm photon and the $In_{0.53}Ga_{0.47}As$ bandgap ($\Delta E = 0.05$ eV), is the main reason for the difference in absorption coefficient. The corresponding fractional absorption is plotted in Figure 3.4a. Because of its higher absorption coefficient at 780 nm, the η_A of GaAs approaches unity much faster vs epitaxial thickness than $In_{0.53}Ga_{0.47}As$ does at 1550 nm. Notice however that the η_A for a 1-μm-thick GaAs layer (≈0.77) is only slightly higher than that of a 2-μm-thick InGaAs layer (≈0.75). Hence, depending on the device type, the use of a thicker InGaAs epitaxial layer can compensate for its weaker α.

A disadvantage of 1550 operation compared to 780-nm operation is laser-diode tuning in CW operation of photomixers, for example. Figure 3.4b shows the difference frequency

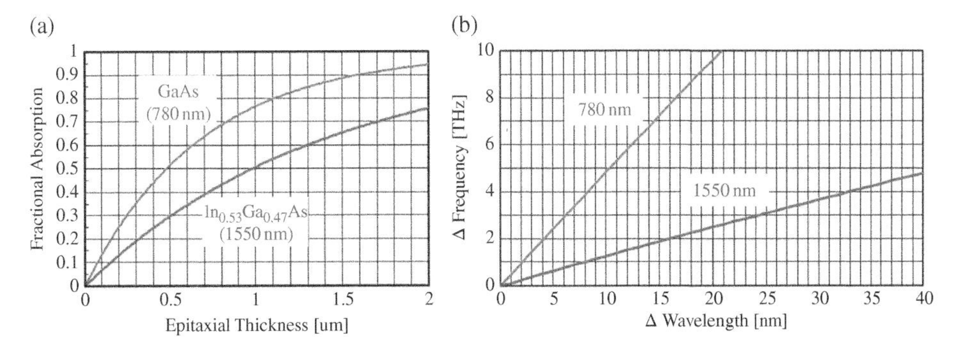

Figure 3.4 (a) Fractional absorption vs epitaxial thickness for a GaAs layer at 780 nm, and an $In_{0.53}Ga_{0.47}As$ layer at 1550-nm. (b) Difference frequency between two single-frequency lasers vs wavelength difference, one fixed at the nominal wavelength and the other tuned away from it.

between two single-frequency DFB lasers assuming one is fixed at the nominal wavelength, and the other is tuned to longer wavelengths. But from Figure 3.4b, a much larger change in difference wavelength is required at 1550 nm to obtain a given THz difference frequency. This leads to a larger required temperature control range for the 1550-nm DFB lasers, which is not available from standard thermoelectric coolers, for example.

3.2.1.2 Low-Temperature Growth

From a materials standpoint, LTG-GaAs represents one of the biggest breakthroughs in THz PC-device history. As such, it has served as a touchstone for the development of LTG $In_{0.53}Ga_{0.47}As$ for ~25 years. Of particular interest are the interplay of growth temperature, post-growth annealing, and doping with beryllium as first studied extensively in LTG-GaAs [16, 17]. The main finding for LTG GaAs was that at growth temperatures below 300 °C, the GaAs is nonstoichiometric due to the incorporation of As atoms on Ga lattice sites [18, 19]. These defects are called arsenic antisites (As_{Ga}) and their concentration increases for decreasing growth temperatures. At $T_G = 200$ °C, As_{Ga} concentrations above $10^{20}/cm^3$ were reported [18, 19]. The energy level of the As_{Ga} defects is approx. 0.75 eV below the conduction band (CB) minimum and, therefore, almost mid-bandgap in GaAs [20]. The conductivity of as-grown LTG-GaAs is determined by hopping conductivity in this deep-defect band [20, 21].

When LTG-GaAs is annealed after growth at temperatures above 300 °C the excess arsenic precipitates and forms arsenic clusters, whereas the size and the spacing of these clusters depend on the duration and the temperature of the annealing process [21–23]. Due to the precipitation of As_{Ga}, the resistivity of LTG-GaAs increases as the probability of hopping conductivity decreases exponentially with the separation of the defect sites [23, 24]. In addition, the arsenic precipitates form Schottky-barriers in the material, which further increases the resistivity [25]. During the annealing process, As_{Ga} defects diffuse via gallium vacancies (V_{Ga}), which are incorporated during the growth process in concentrations of up to $10^{18}/cm^3$ [26, 27]. The precipitation of arsenic antisites could be significantly reduced when LTG-GaAs was *p*-doped with beryllium during the growth process [28]. The Be atoms incorporate substitutionally, and their small size helps reduce the lattice strain caused by the As_{Ga} defects, which is the principal driving force of the As-precipitation [29].

In addition to these effects on the electrical properties, low-temperature growth had a great impact on the carrier lifetime of GaAs. Positively charged arsenic antisites ($As_{Ga}{}^{+}$) are effective recombination centers with relatively large capture cross sections for electrons $\sigma_e = 7 \times 10^{-15}$ cm^2 and holes $\sigma_e = 6 \times 10^{-17}$ cm^2, respectively [16, 30]. The ionization of As_{Ga} defects is caused by gallium vacancies in as-grown LTG-GaAs [31]. Electron lifetimes as short as 100 fs were obtained in as-grown LTG-GaAs, whereas the electron lifetime increases for annealed samples due to the precipitation of (ionized) antisite defects [23]. The optimization and tuning of the growth parameters thus allowed for the fabrication of PC devices suitable for the excitation with femtosecond pulses from a Ti-sapphire laser centered at 780 nm [32].

When low-temperature growth is applied to $In_{0.53}Ga_{0.47}As$ on InP substrates, As_{Ga} defects are incorporated in LTG-InGaAs for growth temperatures below 300 °C [33] (in the remainder of this work we use the abbreviation InGaAs for $In_{0.53}Ga_{0.47}As$ in some instances). In analogy to LTG-GaAs, the concentration of As_{Ga} increases with decreasing growth

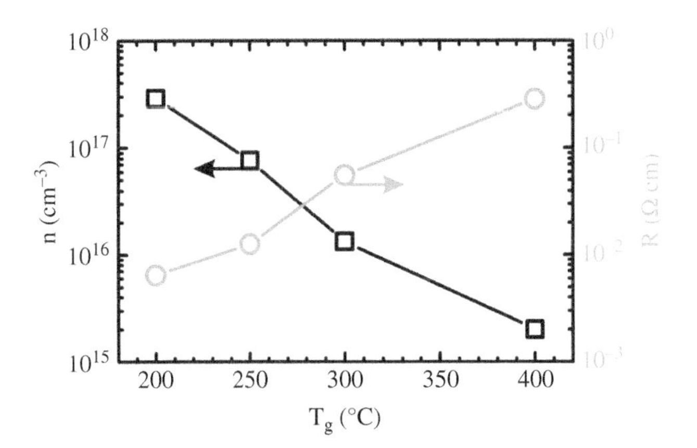

Figure 3.5 Residual electron concentration (black squares) and resistivity (blue circles) of 1 μm LTG-InGaAs grown on top of an InP substrate as a function of the growth temperature T_g in the MBE determined by ambient Hall measurements. The increase of n for lower growth temperatures is caused by an augmented incorporation of AsGa defects, which are partially ionized at room temperature.

temperatures and can reach a fraction of approx. 1% of the total number of As Atoms in the lattice, which corresponds to $10^{20}/cm^3$ [33, 34]. The fundamental difference between As_{Ga} defects in LTG-GaAs and LTG-InGaAs is the energy level of the defect. Whereas it is almost mid-bandgap in LTG-GaAs the ionization energy in uncompensated LTG-InGaAs is 30–40 meV [33, 34]. The important consequence is that a considerable amount of the As_{Ga} defects is thermally ionized at room temperature, i.e. As_{Ga} defects act as donors in LTG-InGaAs, which results in n-conductive material with carrier concentrations in the $10^{17}/cm^3$ range for growth temperatures around 200 °C [33]. Figure 3.5 shows the residual electron concentration n and the resistivity R of as-grown InGaAs with a thickness of 1 μm determined by ambient Hall measurements as a function of the growth temperature in the MBE. The increase of the carrier concentration for lower growth temperatures due to the incorporation of As_{Ga} defects can be clearly seen. As a consequence, the resistivity of the material is lower than 0.01 Ω cm. Due to these unfavorable electrical properties, as-grown LTG-InGaAs is not directly applicable as THz photoconductor.

In contrast to LTG-GaAs, post-growth annealing of LTG-InGaAs cannot increase the resistivity nor decrease the residual electron concentration considerably, although arsenic precipitates are formed [34]. The reason is the fundamentally different origin of the conductivity of as-grown LTG-InGaAs compared to as-grown LTG-GaAs. In the latter case, the dominant mechanism is hopping conductivity between mid-bandgap As_{Ga} defects, whereas thermally ionized As_{Ga} defects cause the conductivity of LTG-InGaAs. Since the probability of hopping conductivity decreases exponentially with the distance between the defect sites, annealing has a great impact on the electrical properties of LT-GaAs. In LTG-InGaAs, this effect is much smaller, since the probability of thermal ionization is directly proportional to the defect concentration.

In order to reduce the residual electron concentration in LTG-InGaAs, the material is commonly p-doped with beryllium [35]. Thereby, the resistivity can be increased to 10–100 Ω cm and the residual electron concentration decreases to $10^{14}/cm^3$ [36, 37]. In addition to the compensating effect, Be has a great impact on the carrier dynamics after the optical excitation. When the doping concentration exceeds the residual electron concentration of LTG-InGaAs, the conductivity remains electron-like—even for Be doping concentrations of $2 \times 10^{18}/\text{cm}^3$. The reason is that Be dopants tend to ionize additional antisite defects, which increases the concentration of fast trapping centers in LTG-InGaAs [34]. The carrier dynamics as well as the performance as PC THz emitter and detector of Be-doped LTG-InGaAs/InAlAs (MQWs) were studied by several groups. The main findings are that the electron lifetime can be reduced to below 1 ps for Be doping concentrations between $8 \times 10^{17}/\text{cm}^3$ and $12 \times 10^{18}/\text{cm}^3$ [38, 39]. The shortest electron lifetime reported to date is 140 fs [40].

3.2.2 Device Types and Modes of Operation

As for 780-nm-driven operation, the two most common device types for 1550-nm operation are the p–i–n-photodiode of Figure 3.6a and the bulk-PC of Figure 3.6b, both shown in cross-sectional view. In the bulk-PC device, the $In_{0.53}Ga_{0.47}As$ along with any doping features or internal heterobarriers is connected between two top-side ohmic contacts. The material in the gap between the electrodes defines the absorbing region. Electron–hole pairs are created by the 1550-nm drive in the gap, and when a bias voltage is applied across it, photoelectrons drift towards the positive-biased electrode and photoholes toward the negative electrode. According to the famous Shockley–Ramo theorem of device physics, the instantaneous photocarrier flow induces an electrical current in the external circuit having DC and possibly AC components. And if the external circuit includes a resistive load, such as a planar antenna radiating into free space, there can be a net AC power delivered by the device, even up to THz frequencies.

In the PIN-PD structure, the absorbing $In_{0.53}Ga_{0.47}As$ layer is sandwiched between two wider bandgap materials (to avoid ~1550 nm optical absorption), one doped n-type and the other p-type. Ohmic contacts are defined on the top and bottom of the structure, which is generally easier to fabricate if the n-layer is on the bottom and the p-layer on top. The absorbing layer can be undoped or compensation-doped such that its optical behavior

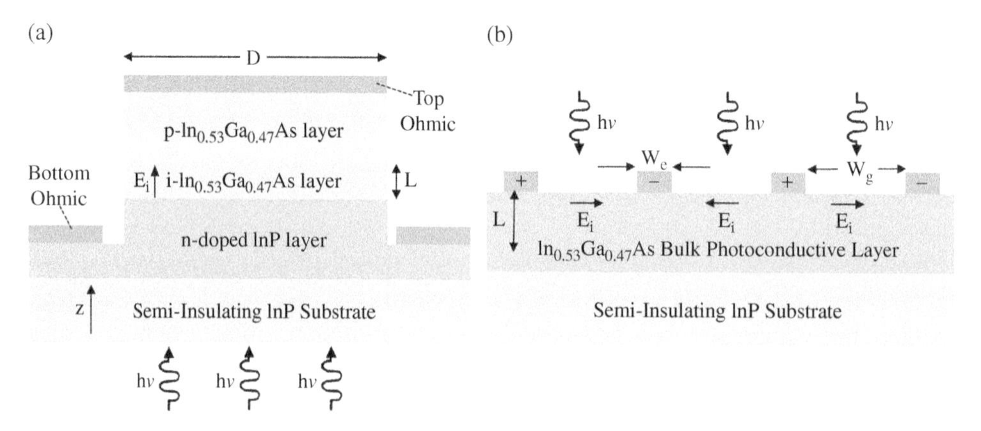

Figure 3.6 (a) Conventional 1550-nm p–i–n photodiode. (b) Conventional MSM bulk photoconductor.

and photocarrier transport properties are "intrinsic," i.e., like that of pure, crystalline $In_{0.53}Ga_{0.47}As$. In contrast, the n- and p-type regions are doped heavily enough to create significant band bending, and therefore internal electric field, across the intrinsic layer, even with zero bias on the electrodes. Hence, photocarriers generated in the intrinsic layer by incident 1550-nm photons (assumed incident from the bottom side in Figure 3.6a) drift vertically toward the appropriate electrode creating electrical current in the external circuit, and the possibility of AC power delivery by the same reasoning as above.

For each of the two device types, there are two common modes of operation: (i) ultrafast pulsed mode, also known as PC switching; and, (ii) continuous-wave mode, also known as photomixing. As described earlier, the pulsed mode was developed first in the mid-1980s followed by photomixing in the early 1990s. Both were demonstrated successfully up to THz output frequencies using GaAs driven at 780 nm. And both are quite distinct from a device physics standpoint. The PC switch is inherently a large-signal device subject to very high peak optical drive power, high instantaneous photocarrier densities, and large instantaneous electrical current in the external circuit. As such, it is limited in output power by photocarrier screening, electrical saturation, and electrical breakdown if the bias voltage is high [41]. This is particularly true when the PC switch is a type of PIN-PD, in which photocarrier densities are very high compared to those in THz bulk-PC devices.

By contrast, the photomixer is inherently a small-signal device with modest photocarrier densities and electrical currents compared to the peak values in the PC switch. As such, it is limited in output power more by inherent inefficiency, especially low PC "gain." And because the optical drive comes from two lasers, each under nominal continuous operation, the THz output is often limited by thermal overload. This favors p–i–n type photodiodes over bulk-PC devices because they tend to be more efficient. These issues will be discussed further in the analysis below.

The most successful 1550-nm-driven THz PC devices have relied on this material strategy with just a few exceptions described later. And there have been several different approaches, depending on whether the device is a bulk photoconductor or a photodiode most of them utilizing either LTG-$In_{0.53}Ga_{0.47}As$, $In_{0.53}Ga_{0.47}As$ doped with Er, $In_{0.53}Ga_{0.47}As$ compensation-doped with Be, or InGaAs/InAlAs superlattice structures. While able to achieve the sub-picosecond response required for useful THz operation, these InGaAs-based devices have generally suffered from low external responsivity, low resistivity, or low breakdown voltage. As such, the THz output power from these devices has typically been 10× lower than that from the GaAs-based devices.

3.2.3 Analysis of THz Photoconductive Sources

In this section, we provide a more detailed analysis of the two device types (bulk-PC and PIN-PD) and their two operating modes (PC switch and photomixing) when driven at 1550 nm. We start with the PC-switch analysis since it was the first THz PC device and its behavior tends to be more generic, although more complicated by nonlinear effects. Several excellent and extensive reviews have been published on this topic [42–46] so this section is restricted to those aspects which are most relevant to THz generation and detection.

3.2.3.1 PC-Switch Analysis

In PC THz time-domain spectroscopy (TDS) the broad spectral bandwidth of a femtosecond laser pulse is translated into a broadband electromagnetic pulse by applying a PC switch as the optical-to-electrical converter. When an external bias field is applied to the PC emitter, the electrons and holes excited by the femtosecond laser pulse generate a transient current in the photoconductor, which is radiated into free space [47]. One of the simplest models, which describes many of the most relevant mechanism of THz generation, is the Drude model of carrier transport. In this framework, the current density can be expressed as [46]

$$j(t) = -e\, n(t) v(t) \tag{3.1}$$

Here, $n(t)$ and $v(t)$ are the density of electrons in the CB of the photoconductor and their velocity, respectively. The constant e denotes the elementary charge. Since the effective mass of CB-electrons is significantly smaller than the effective mass of valence band (VB) holes in common III–V photoconductors, the main contribution to the transient current originates from CB-electrons. Thus, this analysis is restricted to electrons only. The dynamic equations describing the time dependence of n and v can be expressed as (see for example [42]):

$$\frac{dn}{dt} = G(t) - \frac{n(t)}{\tau_e} \tag{3.2}$$

$$\frac{dv}{dt} = -\frac{v(t)}{\tau_s} + \frac{e}{m_e^*}\left(E_{bias} - \frac{P(t)}{\eta e}\right) \tag{3.3}$$

$$\frac{dP}{dt} = -\frac{P(t)}{\tau_r} + j(t) \tag{3.4}$$

In Eq. (3.2) the term $G(t)$ represents the generation of electrons by the femtosecond laser pulse and the second term denotes the electron trapping with the time constant τ_e. This is a rather simple description of the dynamic trapping process in photoconductors, which has to be adapted to the defect structure of the PC material used [48, 39]. In Eq. (3.3) the electron velocity increases with the external bias field E_{bias}, whereas scattering processes, represented by the electron scattering time τ_s, and the (dynamic) screening of the bias field, decrease the velocity $v(t)$. The screening itself is modeled via the time-dependent polarization $P(t)$, which builds up proportional to the current density $j(t)$ and decays due to carrier recombination with time constant τ_r. In Eq. (3.3), the effective electron mass is represented by m_e^* and η is a geometrical factor [49, 50]. By solving the set of equations, the transient current density $j(t)$ can be determined. According to Maxwell's electromagnetism, the far field of an electric dipole is proportional to the time derivative of the transient current [7].

$$E_{THz}(t) \propto \frac{dj(t)}{dt} \tag{3.5}$$

Hence, the carrier dynamics inside the PC material determines the properties of the radiated THz pulse.

The electric field of the emitted THz pulse can be detected by PC sampling with an antenna structure similar to the emitter. Thereby, the PC receiver is illuminated with a portion of the same femtosecond pulse train that was used for the illumination of the PC emitter. The incoming THz pulse serves as the bias field of the PC receiver. With the help of an optical delay line, the electric field of the THz pulse is sampled sequentially in time by the receiver. Hence, the current induced in the PC receiver can be described by the convolution of the incoming THz pulse $E_{THz}(t)$ with the response function $g(t)$ of the receiver antenna [46].

$$J(t) \propto E_{THz}(t) * g(t) \tag{3.6}$$

Here, the carrier dynamics in the PC receiver can be described in analogy to the dynamical processes in the emitter. In order to point out the need of an ultra-short electron lifetime in PC THz-receivers, we analyze Eq. (3.6) in two limiting cases.

In the first limit, the response function of the PC receiver $g(t)$ is regarded as a delta-function, which means that the duration of the exciting laser pulse and the lifetime of the photocarriers are much smaller than the duration of the incoming THz pulse. In that case, the current induced inside the PC receiver is directly proportional to the electric field of the THz pulse, such that the frequency response of the receiver can be written as [46].

$$J(\omega) \propto E_{THz}(\omega) \tag{3.7}$$

Hence, this regime describes the optimal PC receiver, which resembles every frequency component of the incoming THz pulse.

In the second limit, the receiver is assumed to be an integrating detector, in which the lifetime of the photoexcited carriers exceeds the duration of the incoming THz pulse by far. In this case, the frequency response of the PC receiver can be written as [46],

$$J(\omega) \propto E_{THz}(\omega)/\omega \tag{3.8}$$

The detected current $J(\omega)$ is inversely proportional to the frequency of the incoming THz signal. Thus, the receiver becomes more insensitive with higher THz frequencies. Since high THz frequencies are especially important for applications in material science and broadband THz spectroscopy, this analysis underlines the importance of an ultrafast current response in PC THz receivers for obtaining undistorted and broadband THz pulses.

The desired properties of broadband PC switches can be summarized as follows:

- High absorption coefficient for the femtosecond laser pulse, in order to obtain high concentrations of exited carriers in the photoconductor.
- High carrier mobility, for the efficient acceleration and deceleration of the optically excited carriers.
- High breakdown fields, especially in the PC emitter, for applying high external bias fields.
- Ultrashort carrier lifetime, especially in the PC receiver, for the detection of broadband THz pulses (see Eqs. (3.6) and (3.7)).

In general, high mobility and ultrashort carrier lifetime are opposing quantities. The carrier lifetime is commonly reduced by the incorporation of (ionized) point defects acting as traps and recombination centers; however, this reduces the carrier mobility. Hence, one of the principal challenges for the design of efficient THz photoconductors is the precise adjustment of these material properties. And the aim of much current research is to understand the influence of material parameters like carrier mobility, carrier lifetime, and resistivity on the performance of the THz PC switches, aiming for an optimum design methodology.

3.2.3.2 Photomixer Analysis

Compared to PC switches, photomixers generally have much smaller peak optical power and instantaneous photocarrier density, which makes them amenable to a small-signal analysis commonly done with electronic and optoelectronic devices. Of greatest interest is the response to an impulse of incident light excitation, whose Fourier transform defines the photocurrent spectral response function $i(\omega)$. High-speed photodiodes (HF-PD) and MSM bulk photoconductors have long been of interest as detectors to support the very high data rates made possible by optical telecommunications. As such, they have both been analyzed in small-signal operation, and here we provide a short review of this analysis put on the same footing to better understand the trade-offs between the two device types. Then, large-signal operation will be analyzed and differences between the two devices inferred.

3.2.3.2.1 p-i-n Photodiode

Although over 30 years old, the seminal paper by Bowers [51] is a good starting point for the analysis of PIN-PDs operating to THz frequencies, especially as photomixers where small-signal behavior is maintained. The first quantity of interest is the *normalized* current transfer function (Fourier transform of current impulse response),[6]

$$S(\omega) = \frac{1}{1 - \exp(-\alpha L)} \times \left[\frac{1 - e^{-j\omega t_n - \alpha L}}{\alpha L + j\omega t_n} + e^{-\alpha L} \cdot \frac{e^{-j\omega t_n} - 1}{j\omega t_n} + \frac{1 - e^{-j\omega t_p}}{j\omega t_p} + e^{-\alpha L} \cdot \frac{1 - e^{-j\omega t_p + \alpha L}}{\alpha L - j\omega t_p} \right] \tag{3.9}$$

where α is the absorption coefficient, L is the thickness of the intrinsic region, and τ_n and τ_p are the transit times across the intrinsic region for electrons and holes, respectively. This expression assumes the light is incident from the bottom (n-doped) side of the device, as shown in Figure 3.6a, and assumes the absorption and bias-voltage drop occur entirely in the i-region. It accounts for the nonuniform absorption and hence photocarrier generation along the vertical (z) axis proportional to $\exp(-\alpha z)$. It does not account for other important effects such as absorption and diffusion of photocarriers generated in the top p-type InGaAs region.

Each photocarrier created by absorption in the intrinsic region accelerates in the internal electric field E toward the same-polarity region (electrons toward n-region, holes toward p-region) where they are annihilated. E is assumed large enough that the acceleration time is short compared to the transit time, and hence the transit time is given approximately by $t_n = L/v_n$ and $t_p = L/v_p$, where v_n and v_p are the saturation velocities for electrons and holes,

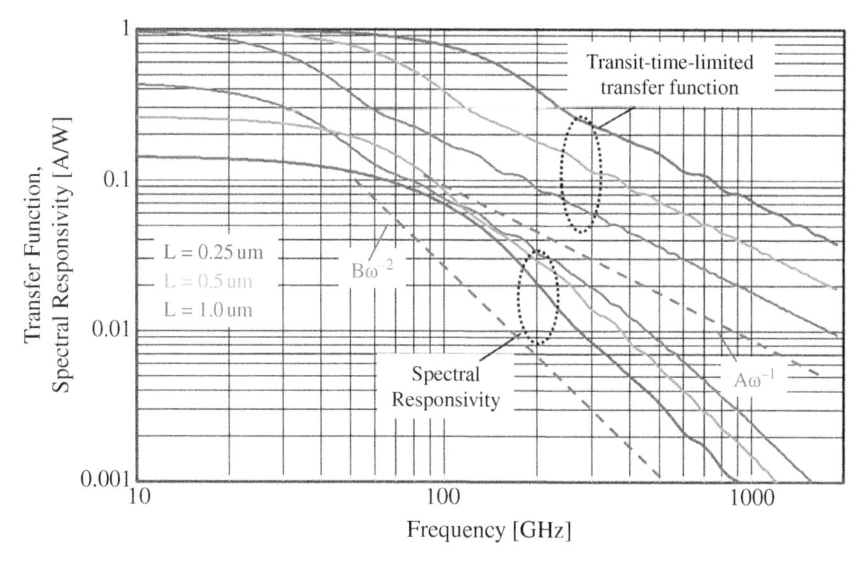

Figure 3.7 Simulation results for p–i–n InGaAs/InP photodiode, showing transit-time-limited transfer function, and spectral responsivity, both for different depletion lengths.

respectively, in the intrinsic region. Each photoelectron–hole pair can induce at most one electron in the external circuit, independent of where photoexcitation occurs. Furthermore, like most solid-state diodes, there is an internal capacitance C that diverts induced current away from the load resistance R_L, creating a characteristic RC roll-off. And thus the spectral responsivity (current delivered to R_L circuit per incident optical power) is given by

$$R(\omega) = \frac{e}{h\nu}\left[1 - R_{pin}\right]\left[1 - e^{-\alpha L}\right]S(\omega) \equiv \frac{\eta e}{h\nu}\cdot S(0)\cdot\left[1 + (\omega R_L C)^2\right]^{-1/2} \tag{3.10}$$

The middle two terms represent the fraction of incident photons (from free space) usefully absorbed, so this product is also the external quantum efficiency η.

$S(\omega)$ and $R(\omega)$ are plotted in Figure 3.7 parametrically vs L since that is the only critical design parameter according to this simple model. Listed in Table 3.1 are the essential parameters as well as $R(\omega)$ values at three important frequencies for assessing THz performance:

Table 3.1 Simulation parameters for InGaAs p–i–n photodiode.

								Resp (mA/W)@ f (THz)		
L (μm)	1 − exp(−αL)	t_n (ps)	t_p (ps)	t_{eff}	f_{tt} (GHz)	C (fF)	f_{RC} (GHz)	0.1	0.3	1
0.25	0.16	3.8	5.2	4.5	111	40	80	70	8.5	0.82
0.5	0.30	7.7	10.4	9.1	55	20	159	86	14	1.5
1	0.51	15.4	20.8	18.1	28	10	318	75	19	2.4

100, 300, and 1000 GHz. Three different L values are considered: 0.25, 0.50, and 1.0 μm. The other essential parameters are: $\alpha = 7130/\text{cm}$, $v_n = 6.5 \times 10^6$ cm/s, and $v_p = 4.8 \times 10^6$ cm/s, yielding the transit times in Table 3.1. The optical reflectance is assumed to be $R = 0.3$, consistent with the experimental value of $In_{0.53}Ga_{0.47}As$ at 1550 nm and normal incidence. The p–i–n mesa is assumed to be round with a diameter of 10.2 μm, creating an area of just over 81 μm^2. This area is chosen to be the same as that of the MSM device analyzed below, and is also convenient from a practical standpoint, being large enough to couple from free space or from optical fiber. The capacitance is estimated by $C = \varepsilon_r\varepsilon_0 \cdot A/L$, where $\varepsilon_r = 13.9$ (the DC permittivity of $In_{0.53}Ga_{0.47}As$)[7] consistent with the assumption that the internal electric field is nonzero only in the intrinsic region.

The resulting $S(\omega)$ curves in Figure 3.7 show two interesting effects. First, as L increases, $S(\omega)$ begins to roll off at successively lower frequencies. As in electrical transit-time devices [52], the transfer function is expected to roll-off quickly when the phase angle ωt_{tt} of the slowest photocarriers reaches the anti-phase condition, $\omega t_{tt} = \pi$, or $f_{tt} = 1/2t_{tt}$. Assuming $t_{tt} = (t_n + t_p)/2$ we get the values listed in Table 3.1 of $f_{tt} = 111$, 55, and 28 GHz for $L = 0.25$, 0.5, and 1.0 μm, respectively. At these frequencies, $S(\omega) \approx S(0)/(2)^{1/2}$, and at higher frequencies each curve approaches a A/ω where A is a constant. The other interesting effect in Figure 3.7 are the ripples in $S(\omega)$ that become evident at $f > f_{tt}$. These are a natural consequence of the variation of phase angle with frequency: peaks are created when $\omega t_{tt} = 2m\pi$ and troughs when $\omega t_{tt} = (2\,m - 1)\pi$, with m being any positive integer. Note that these ripples are relatively weak compared to those in electronic transit-time devices since traditional PIN-PDs create a distribution of transit times for electrons and holes alike, each transit time depending on the location (along z-axis) of photocarrier generation.

The corresponding curves of $R(\omega)$ are also plotted in Figure 3.7. The first obvious feature is that $R(\omega)$ does not approach 0 as $\omega \to 0$ because of the sub-unity external quantum efficiency, which increases with L as expected (increasing fractional absorption in i-region). And each curve begins to roll-off quickly at lower frequencies than f_{tt}, approaching a B/ω^2 dependence at the high end. This is consistent with a second roll-off mechanism, the R_LC circuit effect where C is the device capacitance listed in Table 3.1, and the load resistance R_L is assumed to be 50 Ω. At frequencies well above both f_{tt} and $f_{RC} = (2\pi R_LC)^{-1}$, the $1/\omega^2$ dependence becomes evident. Interestingly, the three curves in Figure 3.7 converge toward a common value of $R(\omega)$ at around $f = 100$ GHz and then separate again at higher frequencies. This is caused by the opposing transit-time and RC effects vs L. Unfortunately for THz operation, the $1/\omega^2$ behavior begins just above 100 GHz so that $R(\omega)$ at the THz specification frequencies in Table 3.1 decay quickly.

3.2.3.2.2 MSM Bulk Photoconductor

The MSM photoconductor has been a popular alternative to the PIN-PD for many decades, partly because of its relative simplicity in terms of fabrication and operation, but also because of its compatibility with ultrafast PC materials. It has two key differences with respect to the p–i–n diode in terms of the device physics: (i) the electron transport is primarily in-plane perpendicular to the incident photon flux, and (ii) if the active layer is made from an ultrafast PC material like LTG-InGaAs, not every photocarrier will reach an electrode before annihilation, so the PIN-PD condition of one electron induced in the external circuit per photon absorbed, does not apply. Assuming that both the photoelectron and photohole have the same lifetime, τ, the spectral transfer function for electrons and holes has the simple behavior

$$S(\omega) = \frac{1}{[1 + \omega^2\tau^2]^{1/2}} \tag{3.11}$$

and the external responsivity can be written

$$\begin{aligned} R(\omega) &= (e/h\nu)[1 - R_{msm}]\left[1 - e^{-\alpha L}\right]\left[\frac{\tau}{t_n} + \frac{\tau}{t_p}\right]\frac{S(\omega)}{\left[1 + (\omega R_L C)^2\right]^{1/2}} \\ &\equiv \eta(e/h\nu)G\frac{S(\omega)}{\left[1 + (\omega R_L C)^2\right]^{1/2}} \end{aligned} \tag{3.12}$$

Here, t_n and t_p are the photoelectron and photohole transit times, respectively, between adjacent electrodes, and G is the PC "gain." We note that the reflectance R_{msm} is generally larger than for the PIN-PD because of electrode blocking of the incident beam. Unlike the PIN-PD, both the electron and hole can reach an electrode without being annihilated in the high-field region. And if they do, they are supplanted by the same carrier type at the opposite electrode, satisfying space–charge neutrality. Hence, if $\tau > t_n$ and/or $\tau > t_p$, the PC gain can, in principle, exceed unity. Although an attractive feature in the early days of THz bulk PC sources, it has clearly never been achieved for reasons pertaining to the behavior of Eq. (3.12).

The other interesting aspect of MSM photoconductors vs PIN-PDs is design latitude. There are two parameters that can be changed rather freely in MSM: the PC gain through the photocarrier lifetime, and the device capacitance C through the MSM electrode geometry. These will have the same opposing behavior as the transit-time and capacitance of PIN-PDs, but with more design freedom. The capacitance of the MSM structure depends primarily on the interdigital electrode geometry through the well-known expression [53],

$$C = \frac{K(k)}{K(k')}\frac{\varepsilon_0(1 + \varepsilon_r)}{(W_e + W_g)}A \tag{3.13}$$

where $k \equiv \tan^2[\pi\, W_e/4(W_e + W_g)]$, K is the complete elliptic function of the first type, $k' = (1 - k^2)^{1/2}$, W_e (W_g) is the width of the MSM electrode (gap), and A is the physical area. In the present analysis and for ease of fabrication, we will assume the MSM structure is a perfect square and that the number of electrodes exceeds the number of gaps by one, so that $A \approx [N_e W_e (N_e - 1) W_g]^2$ where N_e is the number of electrodes. To reduce the reflective loss from the electrodes, we also assume $W_e < W_g$ for which a useful simplification is $k << 1/2^{1/2}$, which leads to [54]

$$\frac{K(k)}{K(k')} \approx \frac{\pi}{\ln\left[2\left(1+k'^{\frac{1}{2}}\right)/\left(1-k'^{\frac{1}{2}}\right)\right]} \tag{3.14}$$

$S(\omega)$ is plotted in Figure 3.8 for τ values of 0.3, 1.0, and 3.0 ps—all three feasible by modern ultrafast materials technology of $In_{0.53}Ga_{0.47}As$. Well above the characteristic frequency $f_\tau = (2\pi\tau)^{-1}$ each falls at a rate $1/\omega$. The f_τ values listed in Table 3.2 are 53, 159, and 531 GHz. The other design parameter is the capacitance through the MSM electrode configuration. We consider the three cases listed in Table 3.2, each having an assumed electrode width of W_e = 0.2 μm, which is readily fabricated by e-beam lithography and N_e/W_g values of 5/2.0 μm (case 1), 8/1.1 μm (case 2), and 13/0.53 μm (case 3). These values are chosen to keep the active area = 81 μm^2—the same as for the PIN-PD simulations. The calculated capacitance value for the three cases is all well below that of even the lowest PIN-PD, demonstrating an inherent advantage of MSM interdigital electrodes compared to vertical PIN-PDs. Before plotting $R(\omega)$ curves, we calculate the spectral responsivity at the three lifetimes and the three characteristic frequencies for each case, as in Table 3.2. The best overall performance by far is displayed for Case#3 having N_e = 13, W_g = 0.53 μm, C = 5.2 fF, and f_{RC} = 610 GHz.

The corresponding set of three $R(\omega)$ curves is also plotted in Figure 3.8. Because of the low device capacitance, the spectral responsivity does not approach the $1/\omega^2$ behavior till above 1.0 THz. And all three $R(\omega)$ curves asymptotically approach the same value at high frequencies, consistent with the well-known behavior of bulk photoconductors displaying a PC "gain-bandwidth" product. In the present analysis, it is actually a responsivity-intrinsic-bandwidth product, given by $R(\omega = 0)B$, where $B = (2\pi\tau)^{-1}$. From Eq. (3.12), we can compute this product as

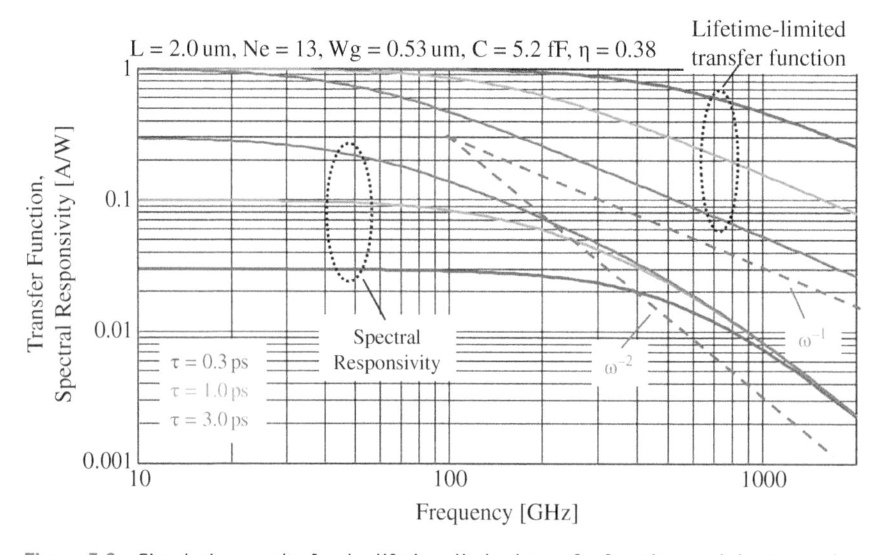

Figure 3.8 Simulation results for the lifetime-limited transfer function, and the spectral responsivity for an InGaAs MSM photodiodes of different photocarrier lifetimes on an InP substrate.

Table 3.2 Simulation parameters for InGaAs MSM photoconductor.

					Resp (mA/W)@f (THz)		
τ (ps)	$f\tau$ (GHz)	τ/t_n	τ/t_p	PC gain	0.1	0.3	1
Case 1: $N_e = 5$, $W_g = 2.0\,\mu m$, $C = 1.15\,fF$, $\eta = 0.47$, $f_{RC} = 2.77\,THz$							
0.3	531	0.0098	0.0072	0.017	9.8	8.7	4.4
1	159	0.032	0.024	0.056	28	15	4.9
3	53	0.098	0.072	0.17	47	17	5.3
Case 2: $N_e = 8$, $W_g = 1.1\,\mu m$, $C = 2.4\,fF$, $\eta = 0.44$, $f_{RC} = 1.33\,THz$							
0.3	531	0.018	0.014	0.032	17	15	6.5
1.0	159	0.06	0.045	0.105	49	27	7.3
3.0	53	0.18	0.14	0.32	82	30	7.4
Case 3: $N_e = 13$, $W_g = 0.53\,\mu m$, $C = 5.2\,fF$, $\eta = 0.38$, $f_{RC} = 0.61\,THz$							
0.3	531	0.037	0.027	0.064	29	23.5	7.4
1	159	0.12	0.09	0.21	84	42	8.3
3	53	0.37	0.27	0.64	139	47	8.4

$$R(\omega = 0)B = (e/h\nu)[1 - R_{msm}]\left[1 - e^{-\alpha L}\right]\left[\frac{1}{t_n} + \frac{1}{t_p}\right]\frac{1}{2\pi} \tag{3.15}$$

The strongest device-dependent term in this expression is the net transit *rate* $(t_n)^{-1} + (t_p)^{-1} = (t_p + t_n)/(t_p t_n)$. Since the electron and hole velocities are saturated, the only control of this parameter exists through fabrication of the gap width W_g. As displayed in Table 3.2, the MSM optical reflectance should be weakly dependent on the electrode configuration provided $W_e << W_g$.

One goal of the above analysis is to objectively compare the generic PIN-PD to the MSM photoconductor. Figure 3.9 shows a direct comparison between the PIN-PD simulations of Figure 3.7 and the best-case (#3) MSM photoconductor simulations of Figure 3.8. At frequencies below roughly 50 GHz, the PIN-PD offers the best performance with the $L = 1.0\ \mu m$ design option. But at the THz characteristic frequencies of 100 GHz and above, the MSM photoconductor provides the best performance. As described above, this arises for two reasons: (i) the MSM photoconductor provides a much lower specific capacitance per unit area than the PIN-PD, and (ii) unlike PIN-PDs, the external quantum efficiency in MSM photoconductors is rather independent of the capacitance, provided that ratio $W_e/W_g \ll 1.0$.

3.2.4 Practical Issues

There are other issues too that are usually under-emphasized in THz device analyses, and therefore deserve attention in this chapter. The three we consider here are (i) ohmic contact effects, (ii) thermal effects, and (iii) circuit power limitations. Again, we are emphasizing practical, engineering issues to help advance THz PC devices to the system-engineering level.

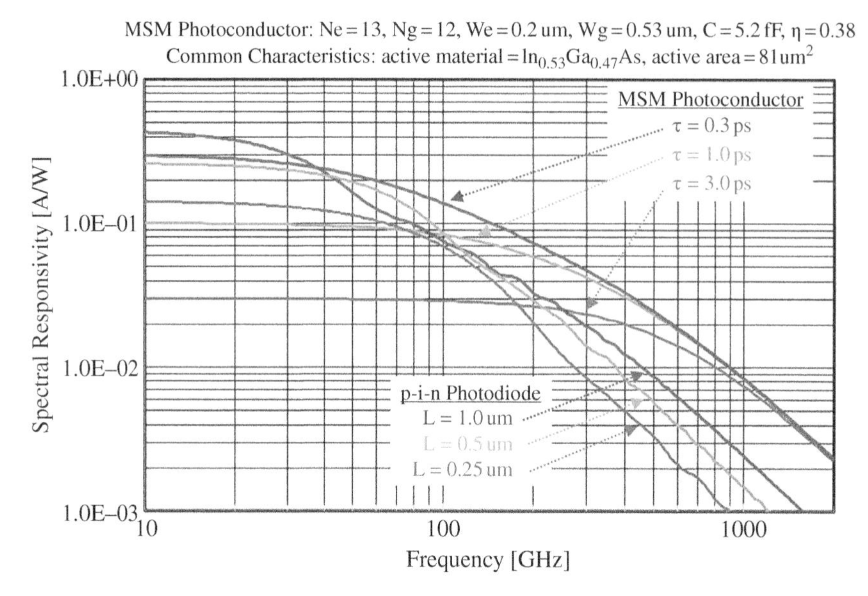

Figure 3.9 Comparison between spectral responsivities of 1550-nm InGaAs MSM-bulk photoconductor and InGaAs/InP p–i–n photodiodes.

3.2.4.1 Contact Effects

One of the least discussed issues in THz optoelectronic devices is the role of metal–semiconductor contacts. This is particularly true in MSM PC devices where for practical reasons (i.e. metal adhesion), the interdigital electrodes are often fabricated with a metallization consisting of thick Au on top of a thin adhesion metal, such as Ti. This helps prevent the undesirable outcome of the submicron-wide interdigital electrodes "peeling off" during lift-off processing, which is already an expensive process because of the required e-beam writing. The problem is that Ti on GaAs or InGaAs, even after high-temperature annealing, naturally forms a Schottky contact, not an ohmic contact. The barrier height φ_B is typically half of the semiconductor bandgap, based on the general principle that the Fermi level U_F in the metal contact tends to "pin" at the middle of the corresponding semiconductor bandgap. The implications of this at low bias voltage ($V_B \leq 2\varphi_B/e$) are shown in the Figure 3.10a. The band-bending occurs between two adjacent electrodes of the MSM structure still incurs a barrier at the anode electrode. This barrier impedes the flow of photoelectrons, which then accumulate in the potential-energy "well" and contribute to screening of the internal bias field.

However, if the MSM structure is biased with a strong enough voltage between adjacent electrodes, the Schottky barrier of the anode electrode can be overcome, meaning that the photoelectrons "see" no barrier. Clearly, this must be a bias of $V_B >> 2\varphi_B/e$, where $\varphi_B \approx 0.4$ eV for metals on InGaAs. But the precise value depends on W_g (gap-width) between the electrodes and the fixed ionized impurity or defect density N_I, assuming they collectively are positively charged. In the "depletion approximation" of semiconductor physics, the length of this depletion layer is found from Poisson's equation to be

$$X_D \approx [\{2/N_I\}(V_B + \varphi_B)(\varepsilon_0\varepsilon_r)]^{1/2} \tag{3.16}$$

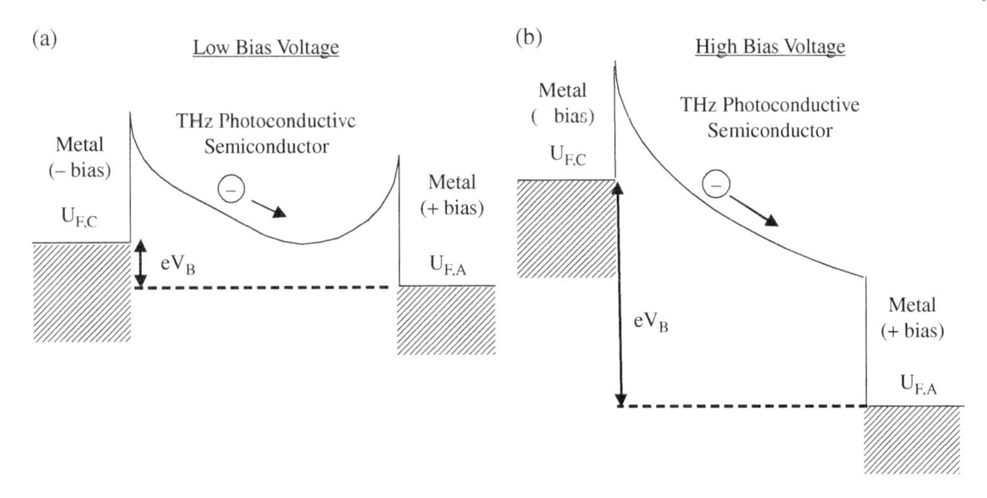

Figure 3.10 (a) Energy diagram of an MSM electrode structure with low bias applied between them. (b) Energy diagram with a high-enough bias so that the barrier to electron flow at the anode contact disappears. By convention, this is called the "punch through" condition.

So the condition that the barrier to electron transport be eliminated at the positively biased (anode) contact is that (cathode) depletion length equal the gap separation between interdigital electrodes. This is called the "punch-through" condition in MSM photodetectors, as shown graphically in Figure 3.10b. Setting $X_D = W_G = 1\ \mu\text{m}$, $\varepsilon_r = 13.9$, and $N_I = 1 \times 10^{16}/\text{cm}^3$ in Eq. (3.16), we find $V_B = 6.1$ V—a value that is routinely exceeded in MSM photoconductors. Since N_I is a difficult quantity to determine, researchers usually take an empirical approach and apply as much bias voltage as is "safe." The safety criterion is usually based on the behavior of the dark current, which at high bias levels starts to rapidly increase with V_B—a tell-tale sign of impact ionization. Fortunately, in many THz photoconductors such as LTG-InGaAs, the impact ionization is "soft," meaning that there is no critical bias voltage and no avalanche effect.

Electrical contacts are generally not an issue with PIN-PDs since the outermost p- and n-regions can be heavily doped and low-resistance ohmic contacts can be fabricated by standard methods.

3.2.4.2 Thermal Effects

Both devices considered in Figure 3.9 display a high DC spectral responsivity, i.e. $R(\omega = 0)$, >0.1 A/W. And for both device types, the DC current is given by $I_0 \approx R(\omega = 0) \cdot P_0$, where P_0 is the average incident optical power. In the case of the PIN-PD, such responsivity only occurs at a high negative bias to create junction electric fields of $>10^5$ V/cm. This is because the built-in electric field of the p–i–n structure is usually ~10^4 V/cm which is too weak to accelerate the photocarriers up to the saturation velocity on a ps time-scale. Hence, the DC power dissipation, $I_0 \cdot V_B$ is significant, of order 10 s of mW. Similarly, the MSM photoconductor must be biased externally $V_B \sim 10$ V or higher, creating a similar DC power dissipation as for the p–i–n device, if not higher. Being that the total DC power dissipation is the sum of the absorbed laser power plus the DC bias power, a longstanding trend in the MSM-bulk photoconductor field has been to reduce the DC power dissipation without sacrificing the

THz performance. From the curves in Figure 3.9, this is best done by reducing the lifetime of the MSM PC material, down to 1.0 ps or less. This exemplifies a common misconception that the reduction in photocarrier lifetime is necessary to get good performance up to ~1 THz and above. As Figure 3.9 shows, and because of the responsivity-bandwidth constant effect, a 1.0-ps-lifetime MSM material should perform approximately the same as a 0.3-ps-lifetime material at 1.0 THz, but the former will incur more DC-photocurrent power dissipation, leading to possible thermal instability. This does not consider dark current, which can be relatively large in InGaAs devices for 1550-nm operation, as discussed for MSM bulk photoconductors previously in Section 3.2.

The thermal effect of absorbed laser power is an interesting topic in its own right. It is not the electron–hole (or just electron) photo-generation per se that creates heat. On the contrary and especially in direct-bandgap semiconductors, cross-gap photogeneration is generally a very efficient process not requiring phonon interaction and thus no heat effects. However, in bulk PC (usually MSM) devices, the relaxation process of photocarriers to a deep trap or recombination center necessarily requires phonon emission to conserve energy and crystal momentum. In normal photoconductors used as infrared detectors, for example, this is of little concern since the light levels are usually very low and the phonon flux does not significantly impact the thermal equilibrium in the material. By contrast, in THz PC devices, the radiation levels are necessarily very high by comparison, and the phonons generated by photocarrier relaxation constitute enough heat to raise the temperature significantly. Note that if the relaxation time was to approach the cross-gap "natural" lifetime (typically of order 1 ns), then the relaxation would start to become radiative recombination so would not require phonon interaction at all. This is unlikely in most MSM ultrafast photoconductors which typically are loaded with fast trapping levels or midgap recombination centers, or both. It is a fair assumption, therefore, that all the absorbed laser power in MSM-type THz PC devices contributes to heating. The underlying assumption can be re-phrased by stating that the relaxation time be much less than the transit time between electrodes, or equivalently the PC gain $<<1$.

It is a different story in PIN-PDs where traps and midgap levels in the depletion layer are avoided because they reduce the collection efficiency of photo-electrons and holes to the neutral n- and p-regions, respectively. Depending on where in the depletion layer they are created, photocarriers drift in the built-in electric field (augmented in the presence of external back bias). But for reasons described above, the depletion layer is generally much longer than a mean-free-path, so the kinetic energy gained by drift is rapidly lost by phonon emission. But the transit time is generally much shorter than the radiative lifetime, at least for electrons. Hence, the heat generated is in proportion to the DC photocurrent I_0, with the proportionality constant being the potential energy $e(V_{bi} + V_B)$ lost by an electron–hole pair, where V_{bi} is the built-in potential drop, and V_B is the external bias voltage (positive for back bias). Hence, the power dissipation is $P \approx I_0\ (V_{bi} + V_B)$, consistent with Joule heating. So what happened to the photonic power initially absorbed? Once the photocarriers reach their respective neutral regions, they are majority carriers and therefore not subject to recombination. Instead, they alter the local (quasi-) Fermi level and induce current to flow in the external circuit.

We thus have two different extremes exhibited by MSM bulk photoconductors and PIN-PDs. The former is heated primarily by nonradiative photocarrier relaxation, and

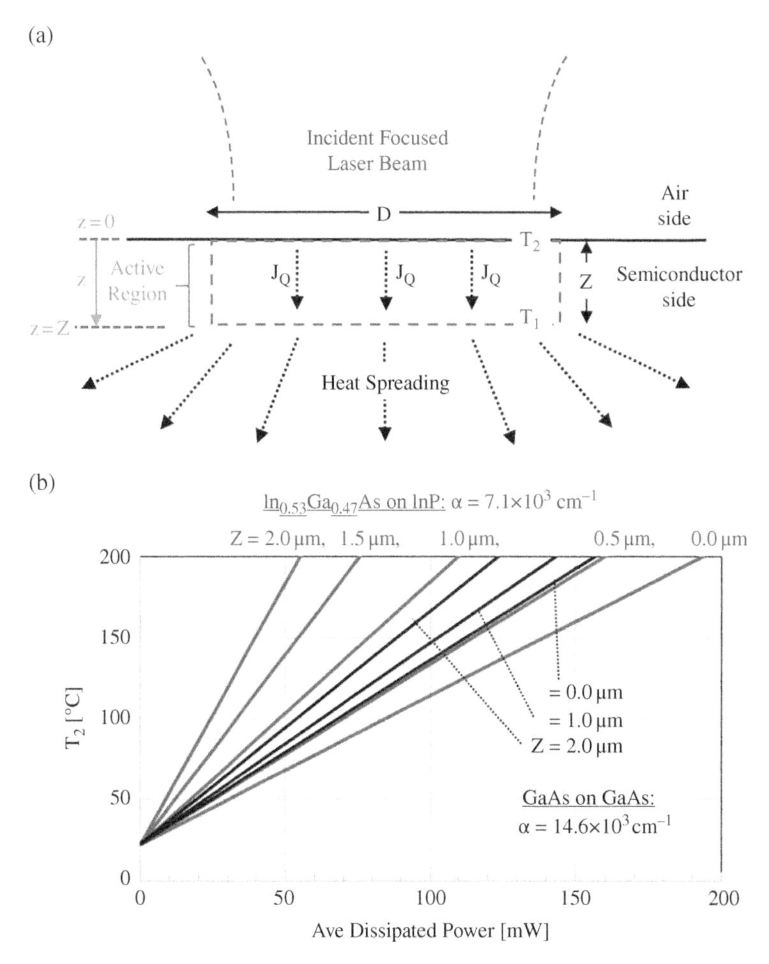

Figure 3.11 (a) Cross-sectional view of photoconductor illuminated from the top with heat flux vertical in the active layer and then spreading into the substrate. (b) Top interface temperature ($z = 0$) vs total dissipated power for an InGaAs active region on an InP substrate, and a GaAs active region on a GaAs substrate.

the latter by Joule (I–V) heating. In either case, we can estimate the thermal effects by a simple heat-transfer analysis. Most of the 1550-nm devices have active regions consisting of $In_{0.53}Ga_{0.47}As$ or a combination with $In_{0.52}Al_{0.48}As$ barriers, both of which have a notoriously low thermal conductivity (≈0.05 W/cm-K for the InGaAs alloy [55]) compared to GaAs (0.55 W/cm-K) or InP (0.68 W/cm-K). This is caused by the random-alloy nature of the InGaAs or any ternary alloy for that matter. So in addition to accounting for the heat "spreading" into the InP substrate, we also account for the vertical heat flow in the epitaxial layer according to the model shown in Figure 3.11a. The heat flux is assumed to be vertical and has an exponentially decaying heat generation density $Q = \alpha I_0 \cdot \exp(-\alpha z)$, where $z = 0$ is the semiconductor–air interface, and $z = Z$ is the epilayer–substrate interface. When laser power absorption dominates, α is then the absorption coefficient in accordance with the Beer–Lambert law. I_0 is the average laser intensity across the active region, and for simplification, we assume $I_0 = P_0/A$ where A is the active area. The steady-state temperature at $z = 0$ is denoted as T_2, and at $z = Z$ is T_1. As can be derived with elementary heat transfer theory[8], this leads to an analytic solution

$$T_2 = T_1 + \frac{I_0}{K_E}\left[Z + \frac{\exp(-\alpha Z)}{\alpha}\right] - \frac{I_0}{K_E\alpha} \tag{3.17}$$

where K_E is the thermal conductivity of the epitaxial layer. Hence, we can write the thermal resistance for the epilayer as

$$R_{TH,2} = \frac{1}{AK_E}\left[Z + \frac{\exp(-\alpha Z)}{\alpha}\right] - \frac{1}{AK_E\alpha} \tag{3.18}$$

This calculates to 701 °C/W for a 1-μm-thick-InGaAs PC device having 81-μm^2 active area, a thermal conductivity of 0.05 W/cm-K, and a 1550-nm absorption coefficient of 7100 cm^{-1}. We note that if the epilayer contains internal heterobarriers as PC devices often do[9], there will be an increase in $R_{TH,2}$ which can be estimated using phonon "effective-media" theory. However, this is a complicated topic and beyond the scope of the chapter.

We combine this epilayer vertical heat-transfer model with the traditional heat "spreading" model from a thin slab into a half-space. This is justified by the fact that in 1550-nm-driven THz devices, the epilayer (or a mesa containing the epilayer) is generally much thinner than the lateral dimension of the active device. Hence, we can assume that the heat from the thin epitaxial slab "spreads" isotropically into the substrate, as shown in Figure 3.11a. This is more justifiable in 1550-nm InGaAs-on-InP PC devices than in 780-nm GaAs based devices since the former only absorb laser power and create Joule heating in the epilayer, not in the substrate. The spreading process can be represented by another thermal resistance, $R_{TH,1}$, which leads to an estimate of the interface temperature T_1:

$$T_1 = T_0 + P_{tot}R_{TH,1} \approx T_0 + \frac{P_{tot}}{K_S\pi R_{eq}}, \text{ so } R_{TH,1} \approx \frac{1}{K_S\pi R_{eq}} \tag{3.19}$$

where T_0 is the ambient temperature (at the bottom of the substrate), P_{tot} is the total dissipated power (optical + electrical), K_S is the thermal conductivity of the substrate, and R_{eq} is the effective radius of the active region. For a circular slab (disk), R_{eq} is the disk radius, and for a square slab, $R_{eq} \approx (A/\pi)^{1/2}$. Then for the same 81-sq-μm device as described above and for $K_S = 0.68$ W/cm-K, we get $R_{TH,1} = 922$ W/cm-K, which is close to the value reported for an LTG-GaAs photomixer by a more detailed study [56].

Equations (3.17) and (3.19) can be combined to provide an estimate for the temperature T_2 at the top of the THz PC device structure. We plot this vs total average dissipated power in Figure 3.11b for two cases, (i) a GaAs epitaxial layer on a GaAs substrate, and (ii) an $In_{0.53}Ga_{0.47}As$ epilayer on an InP. In both cases, the active area is assumed to be 81 μm^2. Each curve is parametrized vs epitaxial layer thickness from 0 to 2.0 μm. In both cases, the $Z = 0$ curves correspond to the T_1 solution of Eq. (3.18) (i.e. the epilayer–substrate interface becomes the air–epilayer interface). Interestingly, the T_1 curve for the $In_{0.53}Ga_{0.47}As$ epilayer lies below that for GaAs since the InP substrate is thermally superior to GaAs, and the same is true for T_2 for a 0.5-μm-thick InGaAs epilayer. However, for a 1.0-μm InGaAs epilayer or thicker, the T_2 curves rise very quickly compared to the GaAs T_2 curves, meaning that the thermal resistance $R_{TH,2}$ of the epilayer is becoming important. For example, a PC device having a 1-μm-thick InGaAs epilayer and dissipating 50 mW of total power will rise to a temperature of just over $T_2 = 100\,°C$ at the top surface, whereas a GaAs PC device with the same epilayer thickness will have $T_2 \approx 84\,°C$. And the same InGaAs device would have a T_2 of ≈185 °C if the total power dissipation is increased to 100 mW—likely above a temperature of safe operation.

T_2 values of 150 °C or above should be considered potentially unstable in GaAs- and InP-based devices, and subject to failure by thermal runaway. Unfortunately, the experimental studies on thermal stability are sparse, perhaps the best one being conducted on LTG-GaAs devices and determining a maximum safety $\Delta T \approx 120\,°C$ [57], or $T_2 \approx 145\,°C$ according to the model presented here. But there is some agreement on the thermal runaway mechanisms, the primary one being that the bandgap of InGaAs (and all common semiconductors) shrinks monotonically with increasing temperature with a figure-of-merit of $dE_G/dT \approx -3.5 \times 10^{-4}$ eV/K [58]. The shrinking bandgap increases α, which in turn increases $R_{TH,2}$ and T_2 according to Eqs. (3.17) and (3.18). This is a regenerative feedback mechanism that is moderated by the fact that the absorbed laser power has a maximum, namely P_0. However, the shrinking bandgap also promotes the generation of free carriers, both by cross-gap thermal generation, and thermal ionization of shallow levels that are often present in ultrafast photoconductors. This represents a second regenerative feedback mechanism that can increase the dark current rapidly and catastrophically. A third mechanism is that the thermal conductivity decreases with rising temperature, usually as in most semiconductors, because of reduction in the phonon mean-free-path. Above 300 K, this dependence is generally close to $T^{-3/2}$, at least in the compound semiconductors like GaAs [59] and InP. Because of all these effects, the majority of THz researchers determine the maximum temperature indirectly by "trial and failure," and back off in operating conditions accordingly. Given the complexity of the thermal physics, this is the pragmatic approach.

The message of this exercise is that one must be careful in using InGaAs epitaxial layers in PC devices, even on the thermally robust InP substrates. Although there is a temptation to increase the InGaAs epilayer thickness to 2.0 μm to improve the external quantum efficiency and responsivity, the poor thermal conductivity of the InGaAs can result in a dangerously high temperature (T_2) at the top interface in top-side illuminated photoconductors. This ignores the possible heat transfer in the interdigital electrodes usually fabricated at this interface. However, these electrodes are generally narrow (<1 μm wide), thin (100 nm typically), and necessarily a low fill factor (~10%) of metal on the interface to minimize reflection of the incident laser beam. So although unproven, their impact on transferring heat away from the interface is thought to be negligible.

The situation changes dramatically for bottom-side illuminated devices, such as PIN-PDs coupled optically through dielectric waveguides, as shown in Figure 3.6a. In this case, the top-side metal has 100% fill factor and can be thickened up to 500 nm or more, therefore capable of transporting significant heat away from the top interface and reducing the temperature there. In this case, the maximum temperature in the device occurs somewhere deeper in the epilayer, likely near the depletion region of a PIN-PD or near the middle of a photoconductor epilayer. The analysis is considerably more complicated so not presented

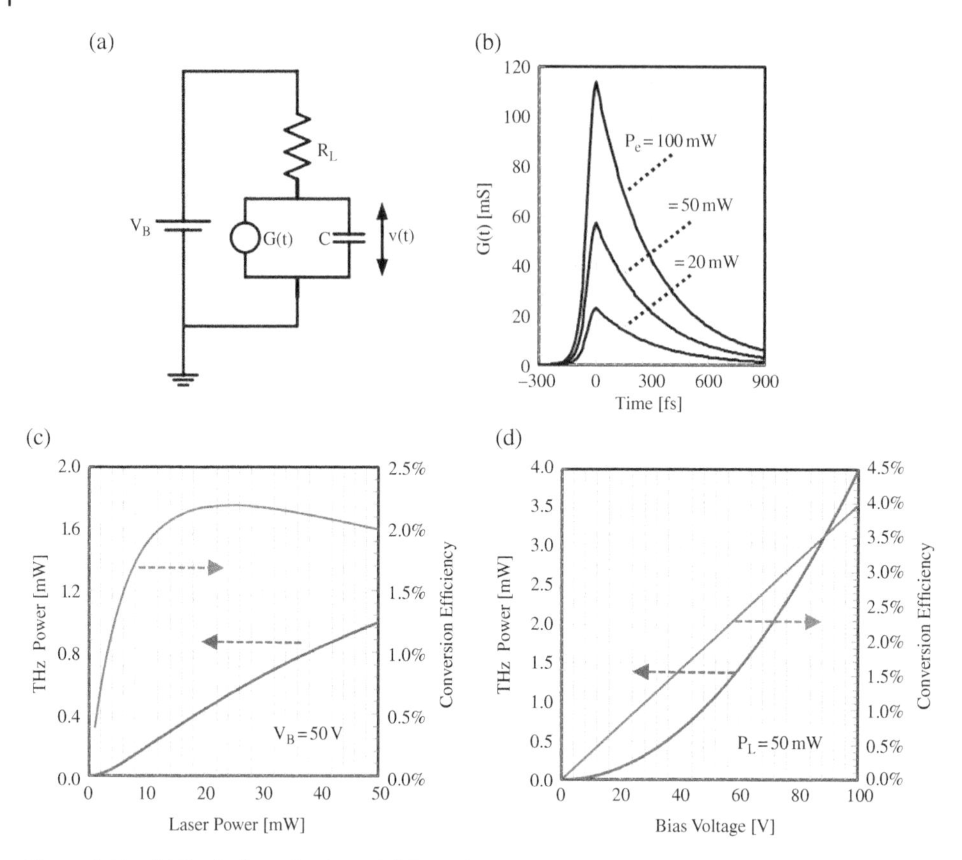

Figure 3.12 (a) Equivalent circuit model for a photoconductive switch. (b) Photoconductance vs time for the model discussed in the text at different average pulsed-laser power. (c) Average THz power vs average laser power at a DC bias of 50 V, and optical-to-THz conversion efficiency. (d) Same quantities as (c) except plotted vs bias voltage.

here, but suffice it to say that bottom-illuminated devices are much more thermally stable. This is yet another reason p–i–n (especially the uni-traveling-carrier photodiodes (UTC-PDs)) have become the favorite as THz transmitters in most system applications, as will be discussed more later in the chapter.

3.2.4.3 Circuit Limitations

A second practical limit pertains to the maximum deliverable THz power to a load. Be it a photomixer or PC switch, a very simple equivalent circuit is as shown in Figure 3.12a, where $G(t)$ is the time-dependent photoconductance, C is the capacitance, and R_L is the load resistance. In the presence of laser illumination, $G(t)$ increases and the current $i(t)$ through R_L increases as well. If we ignore C and assume that R_L is frequency independent, the instantaneous power across the load can be written,

$$P(t) = [i(t)]^2 R_L = \{V_B/[R(t) + R_L]\}^2 R_L \tag{3.20}$$

where $R(t) = 1/G(t)$. To carry the analysis further, we note that for CW photomixers, thermal effects generally limit the optical power to a level such that $R(t) \gg R_L$ at all times, and Eq. (3.20) becomes

$$P(t) = \{V_B[G(t)]\}^2 R_L \tag{3.21}$$

In photomixers, $G(t)$ is usually sinusoidal $G_0 \sin(\omega t)$ where ω is the laser difference angular frequency. Hence, we get the well-known result for the maximum time-averaged THz power at any frequency,

$$P_{THz} = \langle P(t) \rangle = (1/2)\,(V_B G_0)^2 R_L \tag{3.22}$$

A useful estimate for G_0 comes from the total DC photocurrent I_0 assuming V_B is applied through a low-impedance voltage source, $G_0 \approx I_0/V_B$ [60], which leads to the intuitive result,

$$P_{THz} \approx (1/2)(I_0)^2 R_L \tag{3.23}$$

This is a good rule-of-thumb in all photomixers that the authors have worked with, whether photodiodes or photoconductors.

The circuit effects of PC switches are more interesting because they are inherently large-signal in nature and generally not impacted as much by thermal limits. And because the peak power of MLL pulses is so high, it is possible (and desirable) for $R(t)$ to be comparable or possibly less than R_L around the peak, so that Eq. (3.20) cannot be simplified. However, there is one fundamental limit on the power delivered to the load which is worth mentioning since PC switches continue to improve in output power and efficiency, particularly at 1550-nm operation. If we represent them as square pulses of duration Δt and peak power P_p and assume that the peak power is effective in driving the instantaneous photoconductance to a high enough level that $G(t) > 1/R_L$, then the instantaneous current through R_L is $i_L \approx V_B/R_L$, and the instantaneous power is $P_L(t) = (V_B)^2/R_L$. The time average is just

$$\langle P_L(t) \rangle = \left[(V_B)^2/R_L\right]\Delta t \cdot f_P \tag{3.24}$$

where f_P is the pulse repetition frequency, assumed implicitly to satisfy $f_P\,\Delta t$ (the "duty factor") $\ll 1$. Typical fiber MLL laser pulses have FWHM of $t_p \approx 100$ fs, and $f_P = 100$ MHz. However, as described further below, another contribution to Δt should be the photocarrier lifetime τ_0, which is usually substantially longer than t_p. So a better approximation is $\Delta t = t_p + \tau_0$, or $f_P\,\Delta t = 4 \times 10^{-5}$. For R_L, we assume a self-complementary planar antenna for which $R_L = R_A = 60\pi/(\varepsilon_{eff})^{1/2}$, where ε_{eff} is the effective dielectric constant given by $(1 + \varepsilon_r)/2$, and R_A is the antenna impedance. This is the real, frequency-independent driving-gap impedance on a dielectric half-space, as derived later in Section 3.4. For InP, $\varepsilon_r \approx 12.5$ (the DC value)[10], so that $\varepsilon_{eff} = 6.75$ and $R_A = 73\,\Omega$. Finally, for the typical InGaAs-based, 1550-nm PC switch, $V_B = 50$ V and we get $\langle P_L(t) \rangle = 1.3$ mW. As will be described later in Section 3.5, this is remarkably close to the highest broadband THz power levels from 1550-nm PC switches. So like Eq. (3.23) for photomixers, Eq. (3.24) serves as a useful rule-of-thumb for PC switches. However, the former is based on a measurable DC parameter (I_0) and makes no assumption about the device impedance, whereas the latter assumes $G(t)R_L \gg 1$ during the pulse, which is difficult to justify.

To pursue the PC-switch performance further, we seek a model behavior for $G(t)$ based on the common hyperbolic secant form of MLL pulses for the rising edge of $G(t)$ and exponential decay for the trailing edge:

$$G(t) = AP_P\left[\text{sech}^2(t/t_0)\Theta(-t) + \exp(-t/\tau_0)\Theta(t)\right] \tag{3.25}$$

where P_P is the peak laser power, A is a constant in units of S/W, t_0 is related to the FWHM of the laser pulse Δt by $\Delta t = 1.763\, t_0$ [61], τ_0 is the free-carrier relaxation time, and Θ is the unit step function. For the sech^2 pulses, P_P is related to the average laser power P_{ave} by the well-known relation $P_{ave} = 2P_P\, t_0\, f_p$ (derived in problem). For example, for $\Delta t = 100$ fs and $f_p = 100$ MHz, we get $P_p = 5.7 \times 10^3$ W. The quantity A is the specific photoconductance in units of S/W. Ideally, it would be the external DC optical responsivity in A/W, divided by the bias voltage. But this ignores photocarrier screening of the bias field, the relative immobility of holes on the time scales of the laser pulse, and other possible deleterious effects (e.g. laser beam bleaching). So as a rough estimate, we might start with the typical DC optical responsivity of an InGaAs photoconductor (see Figure 3.34), $R(\omega = 0) \sim 10$ mA/W at a bias of 10 V, corresponding to $A \approx 1$ mS/W. However, to get reasonable agreement with experiment, we have to downgrade this roughly by two orders-of-magnitude to 0.01 mS/W. This should not be surprising considering the fact that at the peak of the MLL pulse, the instantaneous photocarrier density can be of order $10^{18}/\text{cm}^3$ or higher as limited by the conduction-band density-of-states. The resulting $G(t)$ is plotted in Figure 3.12b at three different average power levels: 20, 50, and 100 mW. We see that at just 20 mW of average power, the peak instantaneous conductance is $\approx$20 mS, corresponding to a gap resistance of 50 Ω.

The instantaneous power delivered to the load resistor, $P_L(t)$, is calculated from Eq. (3.20) using the circuit of Figure 3.12a, the frequency-independent value of $R_L = R_A = 73\ \Omega$, and ignoring the effect of the device capacitance. The broadband, time-averaged load power, $\langle P_L(t)\rangle \equiv P_{THz}$, is then computed by numerical integration. The results are plotted in Figure 3.12c for P_{THz} vs average laser power at $V_B = 50$ V, and in Figure 3.12d for P_{THz} vs V_B at $P_0 = 50$ mW. Both plots also show the photonic-to-THz conversion efficiency $\varepsilon = P_{THz}/P_0$. Figure 3.12c shows that at low values of P_0, P_{THz} increases superlinearly (i.e. concave up) with P_0 as is observed experimentally in most PC switches. However, at P_0 values above $\approx$20 mW, a cusp is reached above which P_{THz} becomes concave down vs P_0. This leads to a maximum in the conversion efficiency of ~2.2%, although P_{THz} continues to rise toward 1.0 mW. Physically, this is consistent with a *saturation* behavior of the circuit. Once the laser peak power begins to drive the PC-switch $G(t)$ to values $>1/R_L$, the instantaneous load current in R_L approaches a circuit-limited maximum of V_B/R_L as described above for the rule-of-thumb [62]. By contrast, the THz power vs bias voltage plotted in Figure 3.12d shows no such saturation effect P_{THz} rising roughly as $(V_B)^2$ and the conversion efficiency linear vs V_B. In both cases, however, the maximum P_{THz} approaches the rule-of-thumb value of 1.3 mW at 50 V, as expected. We emphasize that the realization of this circuit saturation requires a large enough instantaneous photoconductance according to Eq. (3.25), without limitation imposed by the photocarrier density-of-states, screening, or other effects.

3.3 THz Metrology

3.3.1 Power Measurements

3.3.1.1 A Traceable Power Sensor

Metrology is defined as the science of measurement by the International Bureau of Weights and Measures (BIPM)[11]. In this context, an absolute measurement i.e. the determination of a quantity with a physical unit has to be traced back to the International System of units (SI)[12]. The confidence interval of the result is quantified by specifying its uncertainty[13]. Such quantity is the radiant power P given in the unit Watt [W]. Measurements of the THz power which is emitted by PC devices have not been traceable to the SI in the past [63]. Manufacturers as well as users simply specify relative data such as the signal-to-noise ratio (SNR) to characterize their devices. The main difficulties are the large spectral coverage of pulsed PC and their low output power. This leads to a lack of capabilities to trace back radiant power in the THz spectral range. A suitable THz detector should fulfill the following requirements:

- Its aperture must be comparably large that the whole THz beam can be measured. Due to the long wavelength of the THz radiation, it is difficult to focus it to a small spot size.
- The spatial homogeneity should be reasonable, i.e. the variation of the power responsivity across the large aperture should be small to ensure a uniform response independent of the spot size or the location of the focused beam inside the aperture.
- The spectral power responsivity must be constant (spectrally flat) within the broad spectrum of pulsed PC devices spanning more than two orders of magnitude.
- It must be sensitive enough to measure radiant powers as low as a few microwatts.

These conditions are difficult to meet because there is no opaque material which absorbs all radiation in the entire THz range [64].

All known optically black coatings with carbon-based black paints or carbon nanotubes [65], gold-black [66], or other metal blacks such as platinum-black become more and more transparent at longer wavelengths increasing from THz to GHz frequencies. To maintain sufficient absorption at longer wavelength, the thickness of these coatings must be increased. The more material that is used to absorb the radiation, the less its temperature is changed. Consequently, the responsivity is decreased and the response time to reach thermal equilibrium is delayed. A solution of these drawbacks in the THz range of any thermal detector with a so-called "black" absorption coating is an alternative method, which avoids this additional "black" layer for complete (100%) absorption.

Already in 1934, Wilhelm Woltersdorff published an article on the "optical constants of thin metal layers in the ultra-red" [67] as the far-infrared spectral range was named more than 80 years ago. As shown in Figure 3.13a, a maximum absorption of 50% can be achieved by a thin metal layer if its sheet resistance Z is matched to half the impedance of free space ½ Z_0. This result is "independent of the wavelength" if the wavelength of the absorbed radiation is much larger than the thickness of the layer. Such wavelength-independent absorption of 50% means that the absorption is spectrally flat and independent of the angle of incidence. The power responsivity of such detectors is constant for GHz and THz radiation. The second half of the radiation is transmitted (25%) and reflected (25%) in equal amounts.

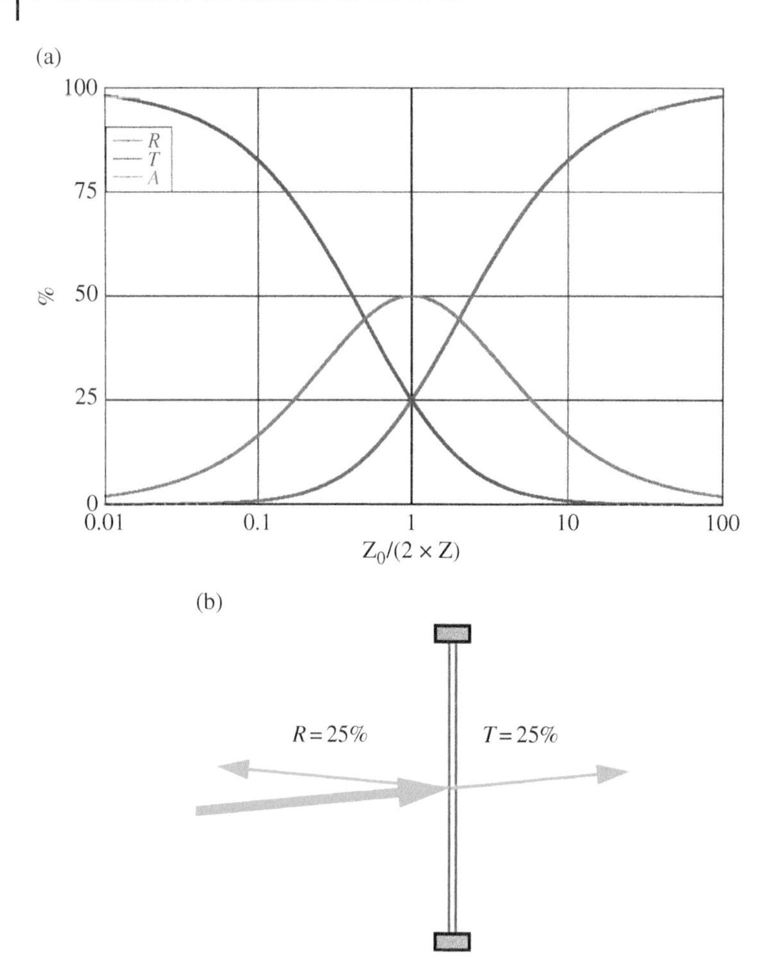

Figure 3.13 (a) Reflectance (*R*), transmittance (*T*), and absorbance (*A*) of a thin conducting layer with sheet resistance *Z*. (b) Schematic of the *R* and *T* for the SPD detector: the pyroelectric foil (yellow) is coated by semitransparent metal electrodes (red) on both sides. This results in a 50% absorbance of the incoming THz radiation (bold green arrow) and 25% transmittance and reflectance, independent of the THz frequency.

For absolute power measurements, it is important that both un-absorbed parts do not reach the absorbing metal layer again.

The first realization of the 50% absorption principle to perform absolute THz-power and energy measurements for millimeter and submillimeter waves is a photo-acoustic detector [68]. It uses thick windows under Brewster angle at the entrance and at the exit side of the partially absorbing thin metal foil. An alignment under Brewster angle is necessary to prevent standing waves inside the plane-parallel windows and between the windows and the foil. This requires linear polarized THz-radiation. Millimeter thick windows are necessary to measure the pressure change of the confined gas in between by a microphone if the gas is periodically heated by the embedded thin metal foil when chopped THz radiation is absorbed. The same periodic thermal heating mechanism can be reached also by a modulated electrical current through the conducting foil. As the resistivity of the foil is known to be ½ Z_0 = 178 Ω a built-in known current source is utilized during the calibration procedure to determine the power responsivity by electrical substitution [69]. Due to the large aperture for measuring a THz laser beam up to 30 mm in diameter without clipping, there is a considerable volume of gas to be heated by the embedded metal foil. The resulting large time constant for a complete heat balance limits the chop frequency to a few tens of Hertz, and the noise equivalent THz power (NEP) is in the order of 20 μW. The thick entrance window results in a frequency-dependent correction factor for its transmittance in the THz range[14]. These disadvantage properties, the high NEP, and the frequency-dependent transmission, prevent this photo-acoustic THz power meter of measuring a broadband source such as pulsed PC devices. Instead, it was designed to measure the monochromatic THz power of optically pumped molecular gas lasers with a typical power range spanning from a few milliwatts to several hundreds of milliwatts. The molecules of such gas lasers emit narrow-band line radiation at a low pressure of about a permille of normal pressure if they are optically pumped in continuous wave mode by a CO_2-Laser. Such powerful THz-lasers were invented already in the 1970s, and since then many types of gases and hundreds of lasing lines have been demonstrated [70].

A solution of a windowless and more sensitive THz-detector are special pyroelectric detectors (SPDs) [71]. Their favorite 50% absorption is realized by specific conducting electrodes which are used to read out the pyroelectric signal. There is no need of any additional absorbing coating because the sheet resistance of both electrodes together matches half the impedance of free space for frequency-independent absorption at THz wavelengths which are larger than the thickness of the pyroelectric sensor. It is a thin foil of polyvinylidene difluoride (PVDF)[15] material with a thickness of a few μm only. Therefore, the SPD is suited for all frequencies below 10 THz because the wavelength is then larger than 30 μm. Accurate 50% absorption is achieved because the pyroelectric PVDF material is transparent and only the conducting electrodes absorb the THz radiation. Moreover, its refraction index is around 1.5. A refraction index much smaller than 3 is a prerequisite because the surface reflection R at normal incidence is

$$R = \left(\frac{n-1}{n+1}\right)^2 \tag{3.26}$$

which amounts already to 25% for $n = 3$ even without any electrode coating. Many other pyroelectric materials such as $LiNbO_3$ or $LiTaO_3$ have a higher refractive index in the THz range and there is no electrode coating, that can realize a spectrally flat absorption.

For absolute THz-power measurements, the radiant power responsivity must be determined traceable to the International System of Units (SI). This is possible at the THz detector calibration facility of PTB [72]. Up to now, PTB is the only metrology institute that can offer this service to customers in the THz range. As the spectrally flat responsivity of the SPDs was proven by an international THz comparison [73], a calibration at single THz frequency is valid for all other THz frequencies. Due to the μm-thin pyroelectric foil and the lack of an additional absorbing coating, there is not much material to be heated by the absorbed radiation. The resulting fast response time enables chop frequencies up to a few kHz. Furthermore, the noise equivalent power of a SPD is less than 1 μW in a quiet environment. Therefore, this THz-detector can be applied to measure the total THz power of a pulsed PC emitter with known low uncertainty [74]. The result is the time average of the pulsed THz emission because the repetition rate of pulsed PC emitters is typically in the megahertz and much higher than the on–off chop frequency which is needed for the pyroelectric detection. Besides the use to determine the average THz power of TDS systems these SPDs are suited as a test piece for THz transmittance (T) and reflectance (R) measurements, as well. The importance of such test piece has been disclosed by an international comparison of TDS transmission measurements, which involved 18 academic and industrial participants and demonstrated a high variability of measurement outcomes from a set of identical samples [75]. The nonabsorbed radiation of a SPD, i.e. its 25% transmitted and reflected part can be used as a test piece because the thin PVDF sensor has a negligible optical thickness and R and T are constant for all GHz and THz frequencies (Figure 3.13b).

In summary, the calibrated windowless, spectrally and spatially flat SPDs enable manufacturers of PC emitters and complete TDS systems to specify the quality of their products and their long-term performance in a comprehensible way for customers and competitors. This is a big step forward to industrial applications of pulsed TDS systems for nondestructive testing and quality control in a noncontact modality. The business requirements, the technological maturity, and the market pull seem to converge to such TDS sensor that can probe, e.g. wet coating layers on car bodies [76], not equaled by any other technology so far.

3.3.1.2 Exemplary THz Power Measurement Exercise

To demonstrate the utility of the SPD vs an alternative, commercial pyroelectric detector, we carried out comparative measurements of a known-good-source—a CW p-i-n photomixer[16] [95]. The SPD has a 20-mm-diam aperture and a transimpedance amplifier (TIA) set to a gain of 10^8 V/A. The commercial pyroelectric detector was acquired from QMC Instruments[17] and was known to have outstanding sensitivity in the THz region [NEP < 1 nW/(Hz)$^{1/2}$], partly because of efficient coupling of a small pyroelectric element to free space via an $f/3$ Winston cone. The photomixer was electrically chopped at 11 Hz and the TIA outputs of each detector were measured with a lock-in amplifier having an integration time $\tau = 300$ ms. The RMS signal and noise vs photomixer frequency of the SPD and QMC pyroelectric detectors are plotted in Figure 3.14. The QMCI yields ~10× lower noise floor and ~3× higher signal output than the SPD, but also displays considerably more variation of the signal vs frequency, particularly below ~400 GHz. At first glance, this might appear to electrical noise. However, the voltage SNR is $>10^3$ in this region, so that the variation vs frequency is attributed to nonuniform optical coupling, as described further below.

Although not having as good of SNR, the SPD allows absolute calibration of the incident power knowing its optical responsivity. In our measurement, this was the product of the pyroelectric transduction factor, the TIA gain, and the fractional power absorbed (0.5) as described above. Adjustment by this factor leads to the signal and noise curves plotted in Figure 3.15. The remarkable fact is that these are *absolute, PTB traceable* values. At the low-frequency end (100 GHz), the power reaches ~160 μW—a level comparable to that claimed in many publications over the past several years, but generally not correlated to a traceable standard. Also notice in Figure 3.15 the optical NEP $\approx 7 \times 10^{-8}$ W, calculated as the average of the noise floor over photomixer frequency. This is a low NEP compared to competitive, traceable sensors in the THz region, such as the free-space-coupled

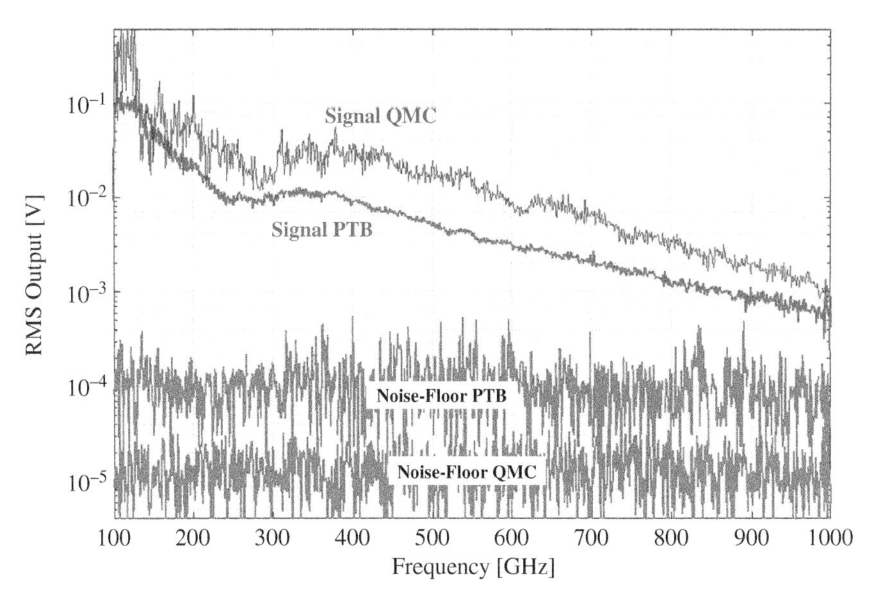

Figure 3.14 Comparison between calibrated SPD detector and a commercial pyroelectric detector made by QMCI.

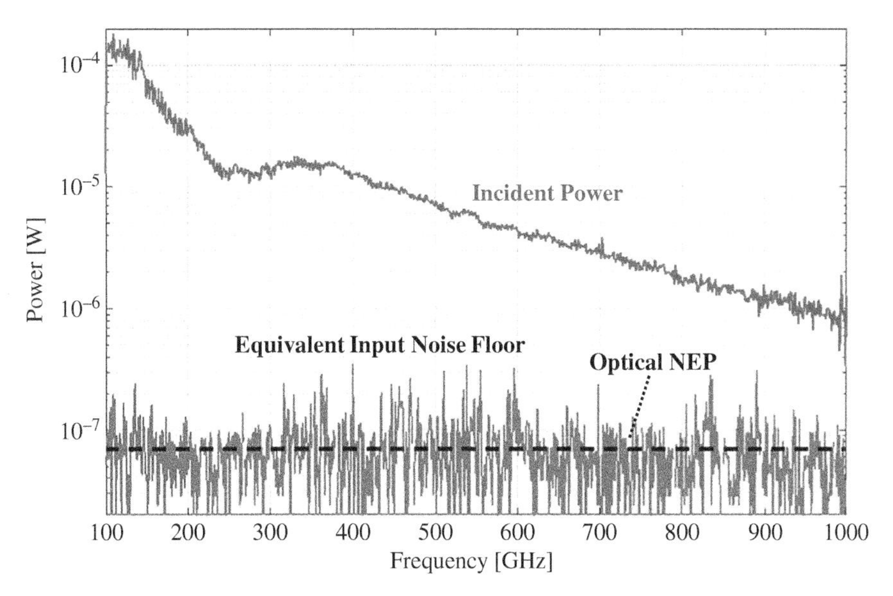

Figure 3.15 Equivalent THz input power for the output signal of Figure 3.14 obtained by dividing the signal by the optical responsivity. Also shown is the noise floor and the optical noise-equivalent power (NEP).

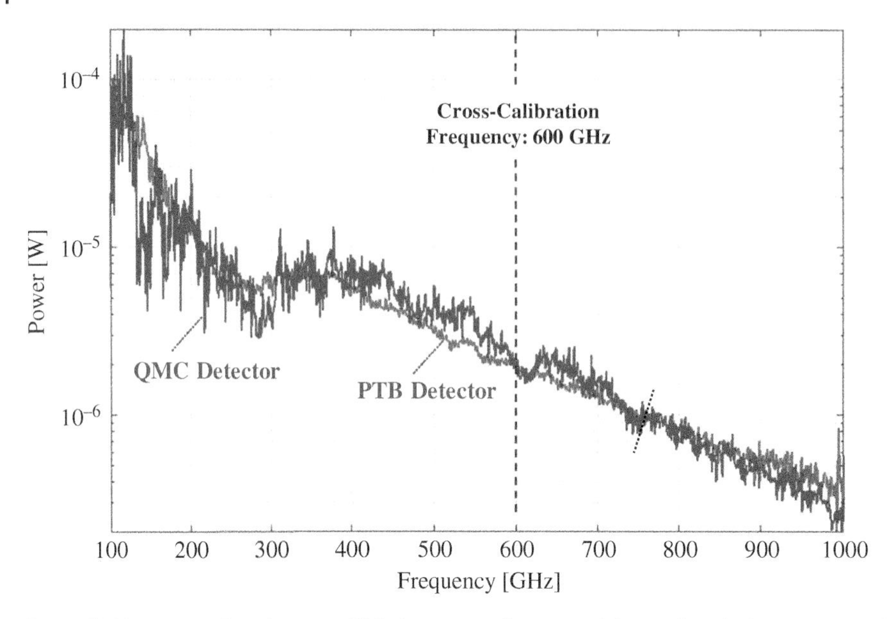

Figure 3.16 Comparison between SPD detector and commercial pyroelectric detector made by QMCI with the QMC detector cross calibrated at 600 GHz.

photo-acoustic[18] or waveguide-coupled thermistor-based sensors[19]. And an increase of the integration time would reduce this further since NEP $\propto (\tau)^{-1/2}$.

A recommended practice in the THz field is to cross-calibrate broadband detectors such as pyroelectrics and Golay cells against a traceable sensor. This is done in Figure 3.16 for the QMC pyroelectric by adjusting its output spectrum for the photomixer source to be identical to the PTB spectrum at a chosen frequency, 600 GHz. The error incurred at other frequencies is significant, especially at the low-frequency end where the estimated power goes from being ~2× too high at around 120 GHz, to >5× too low at 140 GHz. However, the error diminishes at frequencies >600 GHz. This is consistent with the fact that at the lower frequencies, the cavity in which the pyroelectric is mounted and the Winston-cone coupler begin to display electromagnetic modal effects (i.e. cavity resonance, cutoff frequencies, etc.). In fact, these effects are partly responsible for the excellent SNR of the QMC pyroelectric seen in Figure 3.14 compared to the PTB standard. The same is true for many other commercial pyroelectrics, and Golay cells too.

Once a calibrated power spectrum is obtained as shown by the PTB spectrum in Figure 3.16, one can also calibrate the optical responsivity. The result is shown for the QMC pyroelectric in Figure 3.17 in comparison to its far-infrared optical responsivity (1.2×10^4 V/W, which includes the TIA factor) measured with a filtered blackbody[20]. Notice that the QMC responsivity is 1/3–1/2 of its optical responsivity above ~500 GHz. In contrast, at the low-frequency end, the responsivity nearly reaches the optical value around 120 GHz, but then falls to ~1/10th the optical value around 140 GHz. Again, this is likely a ramification of the modal effects described above and represents a common pitfall in using uncalibrated power detectors in the THz region.

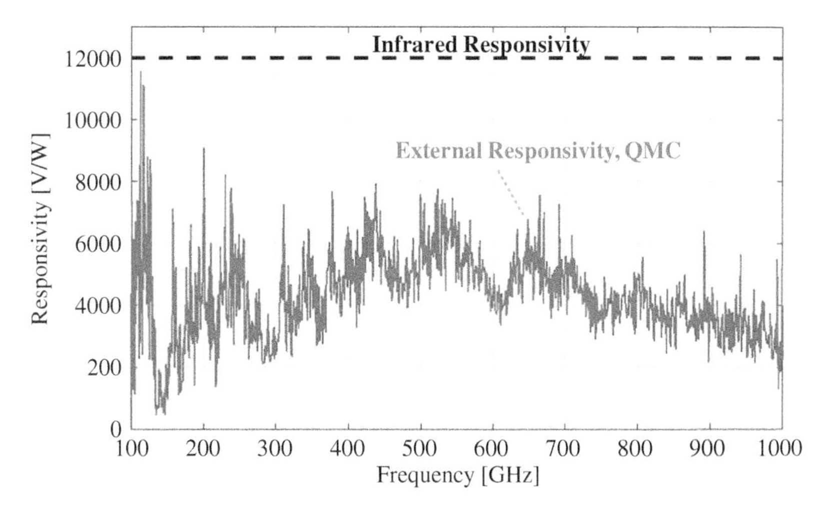

Figure 3.17 THz optical responsivity of QMC detector after cross-calibration against the SPD detector at 600 GHz. Also shown is the LWIR optical responsivity measured with a blackbody source and an optical bandpass filter.

3.3.1.3 Other Sources of Error

Whether fabricated on an InP substrate (the most common), GaAs, or Si, 1550-nm-driven PC sources generally "leak" a certain fraction of the incident laser power to free space since these substrates have very low absorption at this wavelength. Furthermore, the common hemispherical lens-coupling to free space (described below) is usually done with high-resistivity Si, which also has negligible absorption at 1550 nm. So depending on the thickness of the active epitaxial layer, the "leaked" 1550-nm power can be ~1 mW or more, and the common THz power sensors like pyroelectrics are sensitive to it, often even more sensitive than they are to the THz power. This requires the use of a filter located somewhere between the PC source and the power sensor. One good choice is a thick black-polyethylene (PE) sheet. Black PE is known to absorb visible and near-IR radiation strongly but transmit far-infrared and THz radiation rather effectively [78]. This results from the presence of small carbon particles distributed uniformly throughout the PE matrix. In our experience, *a* ~ 1-mm-thick black PE sheet is needed to make THz power measurements accurate to the 1-μW level. If it is much thinner (e.g. the ~0.1 mm black PE commonly used for trash bags), the leaked 1550-nm radiation will taint the power measurements. If it is much thicker, significant attenuation (>50%) of the THz power will occur. And this rule-of-thumb applies to a specific type of black PE with high concentration of carbon[21]. It is interesting how much optical physics is occurring in such a simple material!

An alternative filter material is expanded polystyrene (PS) foam[22]. Although bulk PS does not absorb 1550-nm very strongly, its randomized nature in the foam state is effective at scattering the 1550 radiation so that a negligible amount reaches the THz detector. And because the foam is highly rarefied (mostly air) and bulk PE has low absorption (and refractive index) at THz frequencies, the impact on the THz power is usually negligible. But again, the foam must be thick, 1 cm being suitable based on our experience.

A second source of error in THz power measurements of PC sources, shared by 1550-nm and 780-nm driven sources alike, is thermal self-radiation. As described above, PC devices (especially photomixers) operate at "junction" temperatures elevated well above ambient, often $T_J > 100$ C. And all the common PC epitaxial materials and substrates (i.e. compound semiconductors) have a significant radiative emissivity in the far infrared roughly between $\lambda = 10$ and 100 μm. This is caused by polar optical phonons. While the Si lens material has optical phonons too, they are not polar and therefore have weak optical activity. So the heating of the PC device, which usually extends much further into the substrate than the epitaxial thickness, can create significant blackbody-like radiation that passes through the Si lens with little or no absorption.

Although difficult to estimate, the thermal self-radiation is relatively easy to alleviate. In most high-quality PC devices, the majority of the dissipated power (P_0 defined above) is absorbed laser power. So the self-radiation will not affect the THz power measurement greatly if the PC device is electrically chopped instead of optically chopped. Although the thermal radiation is still present, it is not modulated and so not measured using synchronous (lock-in amplifier) demodulation. This technique works best on MSM-photoconductor devices which generally produce no coherent THz at zero bias, unlike photodiode devices which can produce significant THz even at zero bias. Therefore, a better method is to use an low-pass (in frequency) optical filter, such as a capacitive mesh. Such filters can readily be designed with their cut-on wavelength around $\lambda = 100$ μm (or shorter), and good transmission at longer λ, such that the impact on the THz power is minimal. Unfortunately, such filters do not exist in many THz labs, although they are commercially available[23].

3.3.2 Frequency Metrology

THz frequency measurements are not as demanding as power measurements because techniques developed long ago for free-space optical and infrared radiation translate well down to the THz region, and can be done under ambient conditions. For CW single-frequency sources such as photomixers, the Fabry-Perot interferometer works well at practically any frequency except those coincident with a strong water-vapor line. However, photomixing of two frequency-offset near-infrared lasers offers an alternative and generally more convenient technique which is the use of an optical spectrum analyzer (OSA). With the explosion of 1550-nm telecommunications, fiber-coupled OSAs are readily available and affordable too. An example of the optical spectrum of two offset DFB lasers around 1550 nm is plotted in Figure 3.18a. The accuracy of measuring the difference frequency is $\ll$1 GHz, which is good enough for applications like solid-state spectroscopy.

For pulsed, broadband sources such as PC-switches, a simple way to measure the relative power spectrum is by autocorrelation. A typical set-up is based on the Michelson interferometer and a broadband power detector. The source-under-test is first collimated to the highest degree practical and then divided into the two arms of the interferometer by a broadband beamsplitter. Our preferred beamsplitter is based on the same principle described above for absolute power detection, consisting of a metal film on an ultra-thin dielectric membrane[24]. An exemplary autocorrelation is shown in Figure 3.18b with the source a 1550-nm PC switch and the detector a cross-calibrated pyroelectric [79]. The resulting power spectrum, obtained by FFT, is plotted in Figure 3.18c. The problem with this

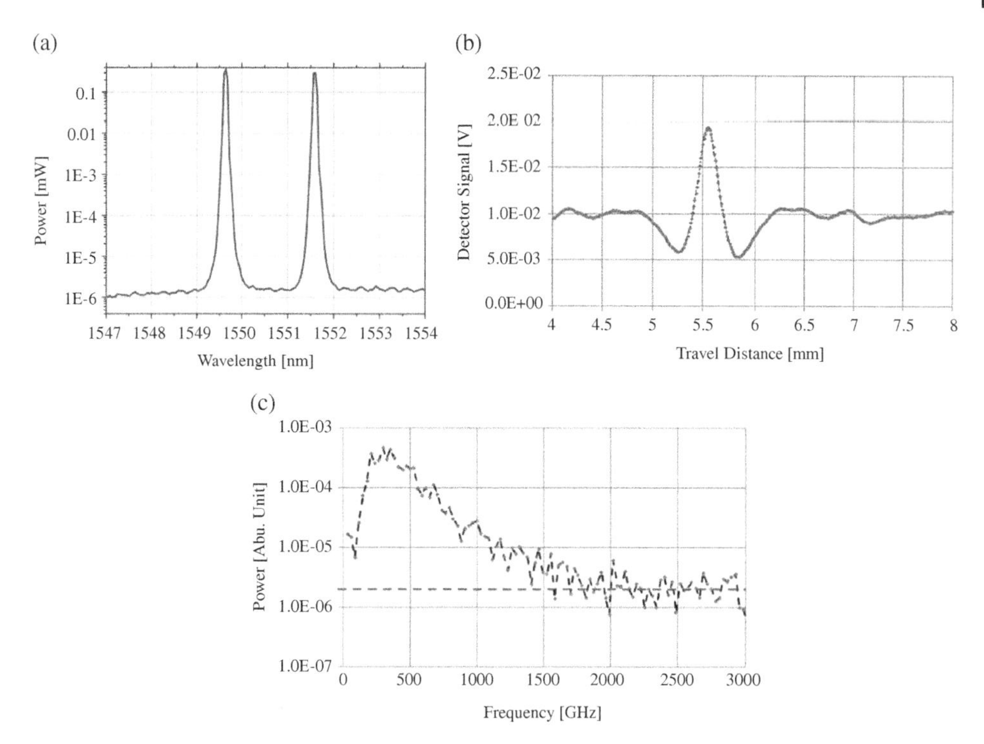

Figure 3.18 (a) Spectrum of two frequency-offset DFB lasers measured with OSA. (b) Autocorrelation function of PC switch obtained with Michelson interferometer. (c) Power spectrum of PC switch in (b) obtained by FFT.

technique is that any nonuniformity in the spectral response of the power detector is convolved with the source spectrum, creating amplitude distortion in the result. Of course, the same problem occurs in any TDS system where the total power spectrum is always a convolution between the transmit and receive PC-switch spectral characteristics.

3.4 THz Antenna Coupling

3.4.1 Fundamental Principles

THz PC devices are generally designed to radiate into free space, so must be coupled carefully to an antenna, be it a planar substrate-based antenna, or waveguide-based (i.e. horn) antenna. In order to get efficient coupling between the two, several issues have to be jointly considered. Since the PC devices are usually much smaller than the electromagnetic wavelength, an "embedded" design approach is usually favored with the device located in the driving gap of a planar antenna. This usually precludes impedance-matching circuits, which are relatively large and lossy at THz frequencies. Hence, the antenna structure has to be designed with the PC device characteristics in mind. This can be easily understood with the help of Figure 3.19a which shows an MSM photomixer integrated within two arms of a planar antenna (it could be a dipole, a bow tie, a log-spiral, whatever).

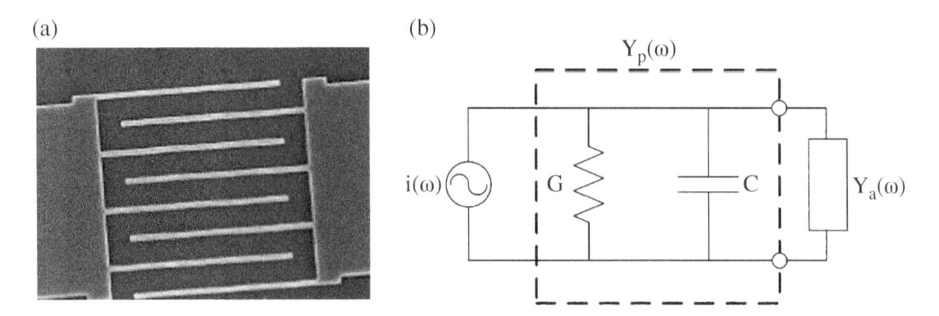

Figure 3.19 (a) SEM photograph of MSM photomixer is located in the gap between two arms of a planar antenna. (b) Equivalent circuit of the structure in (a).

Figure 3.19b shows the equivalent circuit of the integrated device-antenna combination. The circuit is shown in transmit mode with a current generator $i(\omega)$ representing the difference-frequency photocurrent of a photomixer, an admittance $(G + j\omega C)$ representing the MSM structure, and the equivalent admittance Y_a of the antenna itself. Thus, when aiming to maximize the radiated power by the THz PC source, three issues have to be considered: optimizing the PC device (as explained in previous sections), optimizing the antenna itself and, third, jointly optimizing of the antenna fully integrated with the device. This concept of jointly designing the active device fully integrated with the antenna has been referred to in the microwave region as an active integrated antenna (AIA), and in the THz region as an antenna emitters (AE). At this point, it must be emphasized that although the antenna itself would behave as a reciprocal circuit when working alone, this is not the case when integrated with an active device. This is because the THz PC devices are inherently nonlinear, as discussed previously. Hence, different metrics or parameters have to be taken into account depending on whether the devices are operated in transmit or receive mode.

Nonetheless, a universal metric can be adopted by considering the amount of power coupled to or from free space. Some degree of impedance mismatch and loss inevitably occurs, so the useful metric is the external power coupling factor, η, which is analogous to the external quantum efficiency in photonic devices. In transmit mode, it is defined by

$$P_R = \eta\, P_A \tag{3.27}$$

where P_R is the average power radiated into free space, and P_A is the "available" power from the PC device. In principle, $\eta = 1$ occurs only when a complex-conjugate impedance match occurs space between the device and the antenna, and there is no other source of loss. Two common sources of loss at THz frequencies are ohmic losses in the antenna metallization and surface-mode excitation in the semiconductor substrate.

3.4.2 Planar Antennas on Dielectric Substrates

The analysis of antennas lying on the interface between two electrically distinct media has been carried out over 30 years ago [80]. Antennas lying on a dielectric medium have some special features that distinguish them from antennas in free-space. In free-space, the power radiated from a planar antenna divides equally above and below the plane by symmetry,

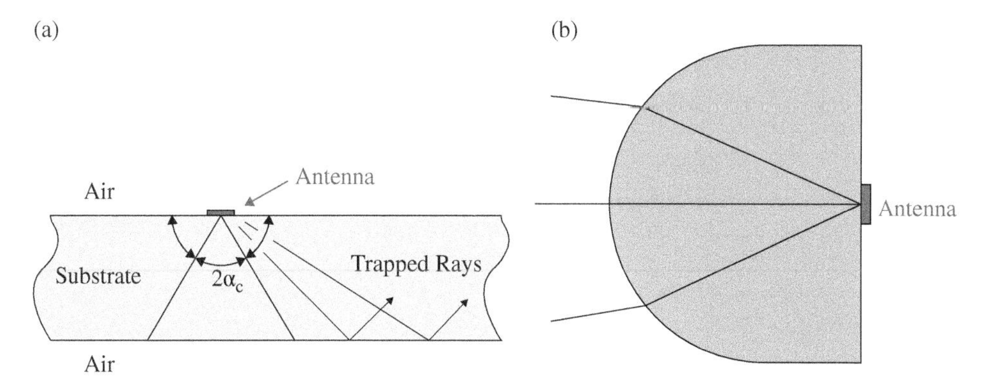

Figure 3.20 (a) Antenna on a dielectric substrate showing the rays trapped by total internal reflection, and exciting substrate modes. The critical angle is represented as α_c (b) Coupling to THz planar antenna via hyper-hemispherical dielectric lens, which eliminates the substrate modes.

while an antenna on a dielectric substrate mainly radiates primarily into the dielectric. This can be explained by the wave impedance of both media. Let us assume that a dipole is placed on the interface between air and a dielectric. If the dipole is used as a receiving antenna it measures the electric field collinear to it. This electric field is the field transmitted through the interface. In the transmission-line model of wave propagation, a wave from a high impedance material (air) is incident on a low-impedance material (dielectric). The transmitted electric field is small, so the dipole response is small. The effect is the same as looking at the voltage across a coaxial cable terminated with a low resistance. When the wave is incident from the dielectric side, the transmitted electric field is large and the dipole response is large, since the material has lower impedance than the air. The analogous situation in coaxial cable is the voltage seen at a high-impedance load. The result is that the response is larger when the wave is incident from the dielectric side. By reciprocity, the power transmitted from the dipole is larger in the dielectric.

Another issue is the thickness of the substrates at THz frequencies that is on the order of a wavelength. Under this condition, the substrate is relatively thick and its effect has to be considered. From the ray point of view shown in Figure 3.20a, the rays incident with an angle larger than the critical angle are completely reflected and trapped inside. From an electromagnetic wave standpoint, this trapped power can be explained in terms of substrate modes, which reduce the antenna radiation efficiency and cannot be avoided by changing to lower-permittivity substrate materials. This is because of fabrication restrictions: 1550-nm THz PC materials such as InGaAs PIN-PDs, or LTG InGaAs/InGaAs multilayer stacks, require lattice matching to InP substrates. And the THz dielectric constant of InP is close to its DC permittivity, i.e. $\varepsilon_r = 12.5$[25].

3.4.2.1 Input Impedance

The input impedance of an antenna is the relation between the AC voltage phasor and the AC current phasor at the input port of the antenna. It is normally a complex number that depends only on the frequency:

The substrate modes can be avoided by the inclusion of a lens on the backside of the substrate, as shown in Figure 3.20b. The lens needs to have a dielectric constant close to the InP, so a good choice is high-resistivity silicon, whose THz $\varepsilon_r \approx 11.7$[26]. The rays are then incident nearly normal to the surface and do not suffer total internal reflection (TIR). In addition, if the lens is "hyper-hemispherical" rather than hemispherical relative to the location of the antenna, the overall radiation directivity is improved. Among the possible "hyper-hemispherical" designs, one is attractive because it is "aplanatic," i.e. it adds no spherical aberration or coma. This is the hyper-hemispherical lens with an extension of $d = r/n$. The disadvantages of the substrate lens are those of any system using refractive optics: absorption and reflection loss. However, high-resistivity silicon ($\rho \geq 10^4\ \Omega\text{-cm}^2$) has acceptably low losses in the THz region.

$$Z_{in}(\omega) = R(\omega) + jX(\omega) \tag{3.28}$$

Any mismatch between the antenna impedance and the complex conjugate of the device impedance, or vice versa, implies that power coupling is not optimum. It must be emphasized that unlike transistor-based integrated circuits, no control between the antenna and the PC device is available so great care has to be taken into the design procedure to obtain a satisfactory impedance match. At microwave frequencies, $R(\omega)$ is usually accurately described as the "radiation resistance," R_r, i.e. proportional to the power radiated to free space via the relation $P_{rad} = I^2 R_r$. However, at THz frequencies there is always a series resistance R_S because of skin-effect losses in the antenna metallization, so that

$$Z_{in}(\omega) = R_s(\omega) + R_r(\omega) + jX(\omega) \tag{3.29}$$

In our experience, skin-effect losses are the most neglected effect in THz antennas, particularly in the broadband, physically long planar antennas like the spirals described below.

3.4.2.2 ΔEIRP (Increase in the EIRP of the Transmitting Antenna)

There are several antenna metrics fundamental to THz PC antennas but considered as background to the present chapter [81][27]. For example, there is antenna directivity D, antenna gain G_{ant}, effective aperture A_{eff}, and equivalent isotropic radiated power (EIRP). However, all of these fundamental metrics ignore the fact that a nonlinear (THz PC) device is integrated with the antenna, and that the currents induced in the antenna depend on this nonlinearity. For this reason, modified metrics need to be considered. For example, in transmit mode the increase in the EIRP, ΔEIRP is a parameter that includes the effect of jointly integrating the PC device and the antenna

$$\Delta EIRP(\theta,\varphi) = G(\theta,\varphi)\cdot G_{PHO}\cdot\frac{1-|\rho|^2}{L} = G(\theta,\varphi)\cdot\Delta G \tag{3.30}$$

where $G(\vartheta,\varphi)$ is related to the conventional (small signal) antenna gain, G_{PHO} is related to the gain provided by the active device and $1-|\rho|^2$ are the mismatching losses between the active device and the antenna; finally, L is the ohmic losses usually dominated by the skin-effect resistance described above. It must be emphasized that all the terms in (3.20) but ΔEIRP cannot be easily measured or estimated by numerical simulation. For that reason, the only parameter to be measured and optimized is the increase in the radiated power.

3.4.2.3 G/T or A_{eff}/T

In receive mode, a useful metric for all types of sensors is the power SNR. Unfortunately, the noise figure (or factor) concept commonly applied in RF systems is not applicable or particularly useful to THz PC devices since it presupposes linearity. THz direct detectors are inherently nonlinear elements, which is what allows them to extract information from incoming radiation in the form of a DC or low-frequency "rectified" signal. In this case, it is the SNR *at the point of detection or decision* that is generally the most important quantity in system performance,

$$\frac{S}{N} = \frac{\langle P \rangle}{\sqrt{\langle (\Delta P)^2 \rangle}} = \frac{\langle P \rangle}{S_P \cdot B_N} \propto \frac{G_{ant}}{T_{whole_{receiver}}} \propto \frac{A_{eff}}{T_{whole_{receiver}}} \tag{3.31}$$

where S_P is the power spectral density and B_N is the equivalent noise bandwidth at that point in the sensor, and $T_{whole\_receiver}$ is the noise temperature of the overall sensor, comprising the antenna and the active integrated device. Similar to above, none of the parameters in this expression but the joint G/T or A_{eff}/T can be directly measured.

3.4.3 Estimation of Power Coupling Factor

The estimation of the external power coupling factor η is undertaken in a two-fold way. On one hand, the large-signal matching between the active device and the antenna has to be considered. On the other hand, the coupling between different media, guided and free-space, is important. From the antenna point-of-view, three parameters are essential: the radiation propagation efficiency (ε_{rad}), the impedance-matching efficiency (M), and the polarization matching efficiency (ε_{pol}):

$$\eta = \varepsilon_{rad} M \varepsilon_{pol} \tag{3.32}$$

Maximizing this efficiency is crucial to THz PC device design. For Si-lens-coupled planar antennas, the propagation efficiency is limited primarily by reflection from the hemispherical lens-air interface. Since most of this radiation is at or near-normal incidence, the power reflectivity is given by $|\Gamma|^2 = [(n-1)/(n+1)]^2$, and $\varepsilon_{rad} \approx 1 - |\Gamma|^2$, where n is the THz refractive index of silicon ($n = 11.7$), which leads to $|\Gamma|^2 \approx 0.30$, and $\varepsilon_{rad} \approx 0.70$.

In receive mode, the impedance matching efficiency at the THz signal frequency can be analyzed as the ratio between the power delivered to the device acting as a passive load (P_L) and the power available from the antenna terminals acting as a generator (P_{avs}):

$$M = \frac{P_L}{P_{avs}} = \frac{4R_D R_A}{(R_D + R_A)^2 + (X_D + X_A)^2}, \tag{3.33}$$

where $Z_D = R_D + jX_D$ is the device impedance and $Z_A = R_A + jX_A$ is the antenna impedance. When $M = 1$, the maximum available power from the source is delivered to the load according to the well-known conjugated matching condition which states that the maximum power transferred to the load is obtained when $Z_A = Z_D^*$ (conjugate match, i.e. $\Gamma_A = \Gamma_D^*$). In the next section, some examples will be given for the achieved values of the mismatching factor depending on the values of the impedances and also on the non-linear conversion process. We emphasize that these mismatches tend to make M low at THz frequencies, $M \sim 1\%$ being typical.

3.4.4 Exemplary THz Planar Antennas

In most scenarios, the generated THz power must be either radiated into free space or coupled to some circuit, such as metal waveguides or resonators. In this section, the most common antenna topologies used in the THz gap are presented. Resonant antennas such as dipoles or slot antennas, as well as spirals, log-spirals, log-periodic, and bow-ties, are also presented and explained as self-complementary antennas.

3.4.4.1 Resonant Antennas

The most widely used planar antennas in the THz region are dipoles or topologies related to them. In either receive or transmit mode, planar antennas present a capacitive input impedance that has a strong influence at high frequencies. Traditionally, this capacitive part has been compensated by a series inductance (i.e. a "choke") [82]. Figure 3.21a shows an example of a dipole antenna with a series inductance, specifically a stepped-impedance low-pass filter. Once the capacitance is compensated, an antenna having the same real part as the device is desired to maximize the power delivered to or from the antenna. For instance, photomixers and back-biased PIN-PDs have a high differential input resistance [83], so an antenna with high input resistance is the best option. On the other hand, devices such as zero-bias Schottky diodes or FET plasma-wave detectors have a lower real part of the input impedance depending on how they are biased, so different antennas may be used. In [83], a detailed numerical study on some of these resonant antennas at THz frequencies is presented.

Two types of resonant dipoles are common: $\lambda/2$ and λ full length. The main difference between them is the input impedance they can provide. $\lambda/2$ dipoles have a minimum of

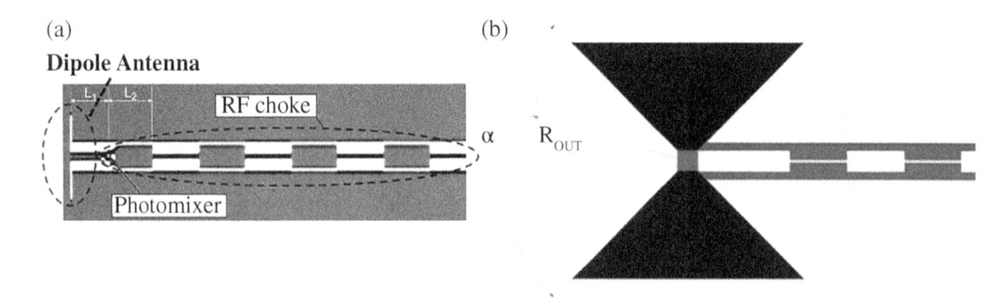

Figure 3.21 (a) Sketch of a dipole antenna with a stepped impedance low pass filter acting as RF choke. (b) Sketch of a planar bow-tie, including the antenna metallization (black), the RF choke (blue), and the photomixer (red).

voltage and a maximum of current at the driving gap, thus a minimum in the input resistance is obtained. Changing the gap-size and the width of the dipole can modify this, but for an InP substrate typically around 50 Ω is obtained at its resonance. λ dipoles have a voltage maximum and a minimum of current at the feeding point, thus a maximum in the driving-gap impedance is obtained, typically around 300 Ω on an InP substrate. A useful rule-of-thumb to calculate the physical length of the dipole as a function of frequency is the mean dielectric-constant rule, whereby the velocity of the THz drive current on a dipole at the air-dielectric interface is $v \approx c/[\varepsilon_{\text{eff}}]^{1/2}$, where $\varepsilon_{\text{eff}} = (1 + \varepsilon_r)/2$. For example, a full-wave dipole antenna on InP ($\varepsilon_r = 12.5$) has a physical length of 385 μm for 300 GHz operation and half this length for 600 GHz operation.

3.4.4.2 Quick Survey of Self-complementary Antennas

In resonant-dipole antennas, the radiation parameters depend on its electrical dimensions: length, width, etc. In this way, if a dimensional scale in the antenna is made, such as doubling the electrical dimensions, the antenna performance will be the same but at half the original frequency[28]. From that reasoning Rumsey [84] made the following profound statement: an antenna whose geometry can be described only in terms of *angles* will have a performance independent of frequency since its geometry does not change with differing dimensional scale. This is the concept of a "self-scaled" antenna, and a closely related concept is the so-called "self-complementary" antenna. Self-complementary antennas are antennas where the metallic and nonmetallic areas have the same shape and can be superimposed on each other by rotation. An antenna with a self-complementary structure has constant input impedance, independent of the frequency or shape of the antenna structure. These are broadband antennas that can achieve a bandwidth at least one octave [85], and if physically large enough, at least one decade.

According to [85], the input impedance of self-complementary antennas is equal to:

$$Z_{in} = \frac{Z_0}{2} \Rightarrow Z_{in} \cdot Z_{in}^{sc} = \frac{Z_0^2}{4} \tag{3.34}$$

where Z_0 is the intrinsic impedance of the medium, and Z_{in}, Z_{in}^{sc} are the impedances of the antenna and of its self-complement. In the case of dielectric half-space coupling, $Z_0 = \eta/(\varepsilon_{eff})^{1/2} = 120\pi//(\varepsilon_{eff})^{1/2}$, where $\varepsilon_{eff} \approx (1 + \varepsilon_r)/2$, and ε_r is the relative dielectric constant of the dielectric half-space. Hence, we get $Z_{\text{in}} = 60\pi/(\varepsilon_{eff})^{1/2}$ Ω, independent of frequency. Such self-complementary antennas are very attractive because they have excellent wideband radiation characteristics and avoid high-frequency resonant behavior, which is difficult to estimate in the THz region.

Contrary to the resonant antennas described above, the self-complementary antennas do not require a complex conjugate matching with the device. However, they show, more or less, a constant antenna impedance over a broad bandwidth. This means that optimum matching will never be achieved with a reactive source. However, depending on the variation of the device impedance, a relatively flat coupling factor can be achieved. Among the available self-complementary topologies, three are widely used: (i) the bow-tie antenna, (ii) the log-spiral, and (iii) the log-periodic antenna.

3.4.4.2.1 Bow-Tie Antennas

The bow-tie antenna has the shape depicted in Figure 3.21b. It is similar to a dipole but its arm width increases linearly away from the driving gap. A self-complementary bow-tie has an angle $\alpha = 90°$. This self-complementary topology provides the most constant input impedance of the three analyzed cases. However, its radiation pattern deteriorates with frequency, becoming strongly asymmetric in the E and H planes.

3.4.4.2.2 Log-Spiral Antennas

Log-spiral antennas are self-complementary antennas when designed with specific parameters such as the proper arm length and width. The most important characteristic of these antennas is the constant input impedance (theoretical) not depending on frequency, when both the width of the arms and spacing between them are equal. Obviously, a log-spiral should be of infinite extent to be rigorously self-complementary. Nevertheless, due to practical reasons, they are always truncated and the input impedance cannot be constant over an infinite bandwidth, both in the high and the low band. The truncation has effects on the low-frequency limit, where the size of the element is comparable to the wavelength and the energy spreads throughout all the structure. At high frequencies, the limitation is due to the source feed itself because of the roll-off and also to the truncation of the spiral on the inner side where the feed is placed.

The log-spiral design parameters are based on radii and angles:

$$r_1 = ke^{a\varphi}, r_2 = ke^{a(\varphi - \delta)} \tag{3.35}$$

where r_1 and r_2 represent the inner and external radius of the spiral, respectively. The angle δ determines the arm width, and a and k are constants, which control the growth rate of the spiral and the size of the terminal region, respectively. In Figure 3.22a, the log-spiral geometry is depicted. Its lower frequency limit is determined approximately by D_{out}, while its upper-frequency limit is determined by the size of the gap in the center. Log-spiral antennas

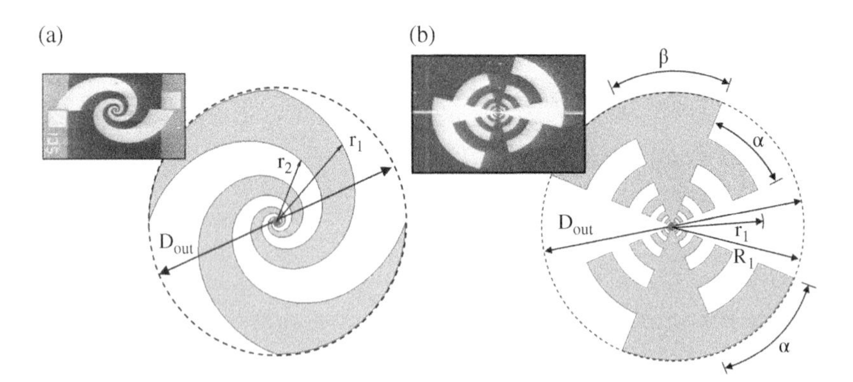

Figure 3.22 (a) Sketch of a log-spiral AE. The antenna (grey) and the photomixer (red) are shown. The inset shows a fabricated log-spiral of diameter D_{out} = 1.30 mm. Biasing pads are marked in the inset (blue). (b) Sketch of a log-periodic AE. The antenna (grey), and the photomixer (red). The inset shows a SEM picture of a log-spiral of diameter D_{out} = 0.74 mm. Biasing lines are marked in the inset (blue).

provide a relatively constant input impedance and circular polarization. Their radiation efficiency is high and they provide a symmetric and almost constant radiation pattern with frequency.

3.4.4.2.3 Log-Periodic Antennas

The log-periodic antenna has a log-periodic circular-toothed structure with tooth and bow angles of $\alpha = \beta = 45°$ (for the self-complementary case). The ratio of the successive teeth (R_{n+1}/R_n) is 0.5 while the size ratio of the tooth and anti-tooth (r_n/R_n) is equal to $\sqrt{0.5}$. Figure 3.22b shows the geometry of this antenna. Its lower working frequency is approximately determined by D_{out}, whereas its higher frequency is determined by the size of the gap. Log-periodic antennas show a weakly varying impedance with frequency, and their polarization changes too. The bigger problem is that their radiation efficiency decays with frequency, which is one reason they are not popular above ~1.0 THz. On the other hand, they have a good radiation pattern and large directivities can be achieved over a wide bandwidth. Hence, this topology is one of the most used at sub-THz frequencies.

3.5 State of the Art in 1550-nm Photoconductive Sources

3.5.1 1550-nm MSM Photoconductive Switches

3.5.1.1 Material and Device Design

As discussed above, LTG-InGaAs is commonly p-doped with beryllium to reduce the residual electron concentration [35]. Thereby, the resistivity can be increased to 10–100 Ω cm and the residual electron concentration decreases to $10^{14}/cm^3$ [36, 37]. In order to increase the resistivity of the material even more for use in bulk-PC devices, multiple quantum wells (MQWs) consisting of up to 100 periods of Be-doped $In_{0.53}Ga_{0.47}As/In_{0.52}Al_{0.48}As$ can be grown. Due to the high bandgap energy $E_G = 1.45$ eV of InAlAs, the material is transparent for 1550 nm radiation. In addition, the resistivity of as-grown InAlAs is several orders of magnitudes higher than the resistivity of InGaAs. Therefore, the effect of these InAlAs barriers is purely compensatory as the LT-defects in these barriers trap residual electrons. For Be doped LTG-InGaAs/InAlAs heterostructures a resistivity of several 100 Ω cm was reported [39, 40].

A key aspect of the InGaAs/InAlAs periodic design is that when fabricated with the MSM electrodes described above, the InAlAs barriers do not seriously impair the PC gain as might first be expected. This is because the majority of photocarriers are created near the center of the gap where the electric-field lines-of-force are nearly parallel to the plane of the InGaAs quantum wells. So the photocarriers can drift roughly the same distance between the electrodes as if the InAlAs barriers were not present, and by the Shockley-Ramo theorem, the current induced in the external circuit is near the same too. However, there is a rapid drop in the DC electric field with depth into the semiconductor as addressed in Ref. [86] and plotted for a specific IDE geometry in Figure 3.23a.

Although found by solving Laplace's equation in GaAs, the solution in InP or InGaAs/InAlAs heterostructures should look very similar. A clever fix to this problem is to fabricate

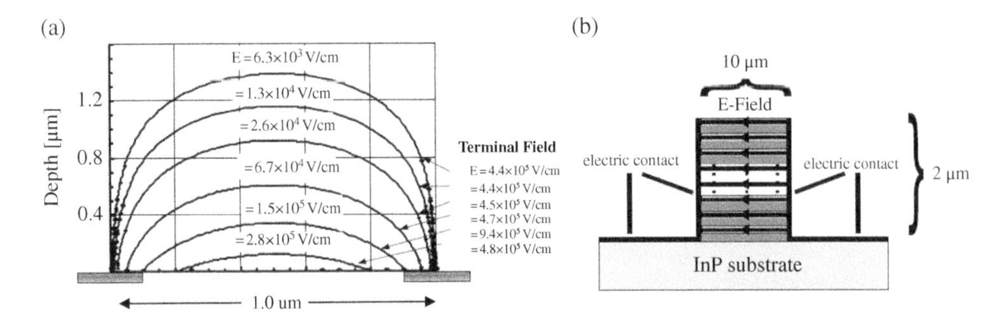

Figure 3.23 (a) Electric lines of force between two coplanar electrodes on a GaAs substrate. The electrodes are 0.2-μm wide and separated by a gap of 0.8 μm. The bias voltage is assumed to be 40 V. (b) Electric lines of force in a mesa structure with the two electrodes fabricated on the sidewalls of the mesa.

a mesa structure and place the electrodes on the sidewalls as shown in Figure 3.23b [87]. This creates a much more uniform bias field across structure, although it complicates the fabrication significantly.

3.5.1.2 THz Performance

The THz performance of the InGaAs/InAlAs PC switches have long been characterized in the TDS set-up of Figure 3.24. The Be-doped LTG InGaAs PC switch is used in the receiver, but a more generic InGaAs switch is used in the transmitter, which is DC biased, and in the receiver, which is unbiased. As in all other TDS systems, the other required components are a femtosecond laser source, an optical delay line, and THz optics to guide the emitted THz pulse from the emitter to the sample and back to the receiver. For THz detection, a time-delayed portion of the femtosecond pulse, which excited the PC emitter, is used to switch

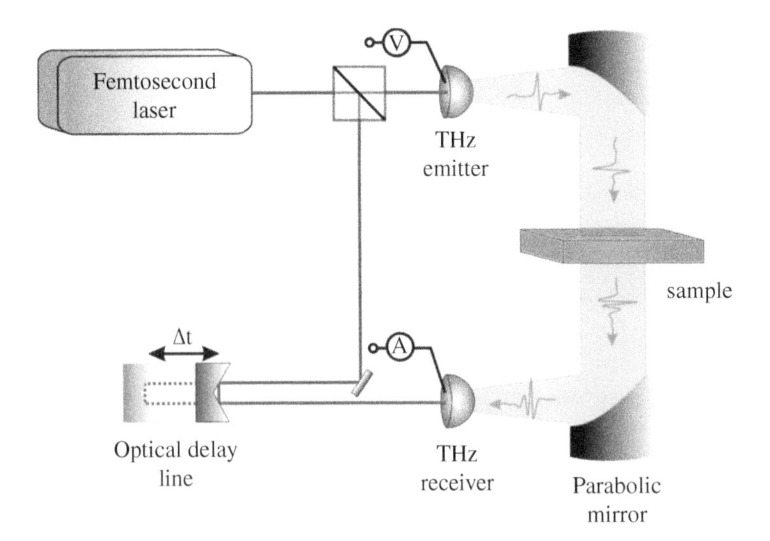

Figure 3.24 Schematic of a THz time-domain spectroscopy (TDS) setup in transmission geometry based on ultrafast bulk photoconductors.

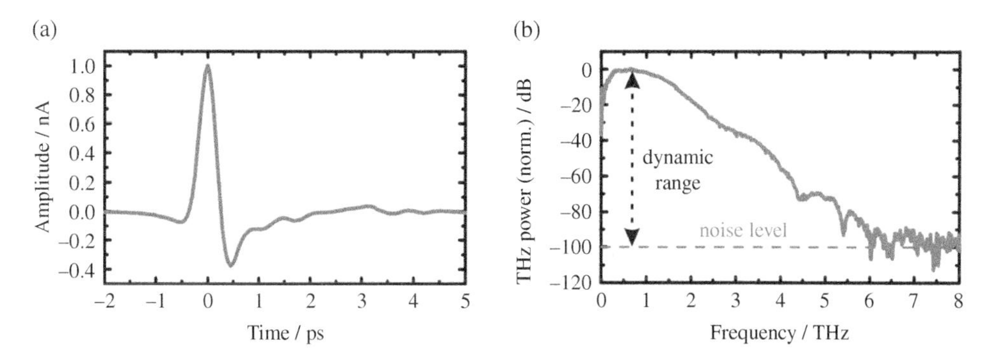

Figure 3.25 (a) Typical normalized pulse trace and corresponding normalized power spectrum recorded with THz-TDS under nitrogen purge. The noise level of the spectrum is highlighted by the horizontal dashed line. (b) The dynamic range indicated as the separation between the spectral amplitude of and the noise level.

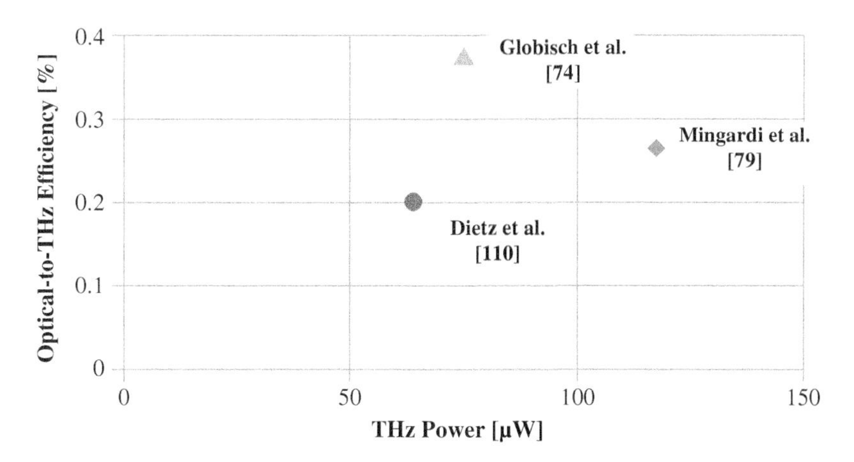

Figure 3.26 THz power and optical-to-THz conversion efficiency for state-of-the-art 1550-nm-drived PC switches.

the conductivity of the PC receiver. When THz pulse and optical sampling pulse have a temporal overlap on the receiver, a direct current can be measured in the receiver. By sampling the incoming THz pulse with the femtosecond optical pulse, the amplitude and phase of the THz pulse are recovered in the receiver. The resulting THz time-domain waveform and power spectrum are plotted in Figure 3.25a,b, respectively. The PC emitter and detector have a bandwidth of up to 6 THz [88], and a peak dynamic range of more than 90 dB is provided by THz-TDS [89, 90].

The THz average power levels corresponding to this type of PC switch are shown in Figure 3.26. As measured by the PTB standard or a cross calibration, these are integrated across the entire spectrum of Figure 3.25b, and they lie in the range 50–100 μW that just 10 years ago would have been thought by many THz researchers to be unreachable with 1550-nm PC switches. Also impressive is that the conversion efficiency $\varepsilon = P_{THz}/P_{opt}$ is just under 0.4%. While small compared to the quantum efficiencies of photonic devices or the electrical efficiencies of microwave devices, the PC switch is operating in a spectral range where solid-state photonic devices do not work (at least at room temperature) and electronic devices (e.g. pHEMTs) work but with comparably low efficiency. And the power level and efficiency are within an order-of-magnitude of the circuit-limited values plotted in Figure 3.12c,d.

3.5.2 1550-nm Photodiode CW (Photomixer) Sources

3.5.2.1 Material and Device Design

The PIN-PD analyzed above was quite generic and subject to improvements made possible by materials engineering and device design. To examine some possibilities, we show the band bending and photocarrier transport diagram in Figure 3.27a assuming the light is incident from the bottom (n) side. The THz performance is limited mostly by a trade-off in the intrinsic region: a thicker i-region absorbs more of the 1550-nm photons and thus increases the quantum efficiency. And it reduces the device capacitance. However, it also increases the transit time, which becomes particularly important at frequencies >100 GHz. An interesting feature is the band offset between the $In_{0.53}Ga_{0.47}As$ i-region and the InP n-region. The majority of the band-gap difference occurs as a VB offset, which is rather unusual for semiconductor heterostructures.

An important factor in the $In_{0.53}Ga_{0.47}As$ p–i–n structure is the relatively poor transport properties of the holes. Compared to the electron mobility of ~1×10^4 cm^2/V-s, the hole mobility is only ~250 cm^2/V-s. This means that there is a significant delay in the acceleration of photo-holes, which are naturally photo-generated at zero velocity. A simple calculation predicts that the acceleration time across an $x = 0.5$-μm i-region is given by $t = x/uE$, which is 2.0 ps for $E = 1.0 \times 10^5$ V/cm. This is an aspect of THz photoconductivity that is sometimes ignored in the THz literature but is addressed succinctly in the following section. Even more insinuating is that photo-holes have the same unity charge as photo-electrons, so when produced at the high spatial density typical of THz PC devices, they tend to screen out the internal bias electric field shown in Figure 3.27a. In other words, the photo-holes can be more deleterious than good to the THz performance.

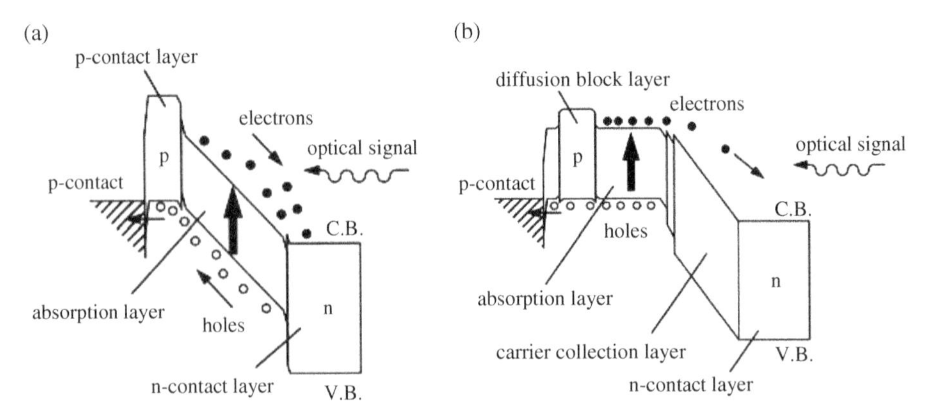

Figure 3.27 (a) conventional p–i–n photodiode with the i-region made of InGaAs and the n-region from InP. (b) UTC-photodiode with the absorbing layer of InGaAs separate from the InP collector layer.

The first and arguably most effective variant of the basic PIN-PD design was introduced approximately 20 years ago and has been in continuous development ever since. It is the design of Figure 3.27b which is essentially "unipolar" designed to collect the photo-electrons, not the photo-holes. It was dubbed the "uni-traveling-carrier photodiode (UTC-PD)" [91, 92], a novel photodiode design that utilizes only electrons as active photocarriers. As shown in Figure 3.27b, the active part of the UTC-PD consists of a p-type doped $In_{0.53}Ga_{0.47}As$ light absorption layer and an undoped (or a lightly n-type doped) InP carrier-collection layer. In the UTC-PD, the photo-generated minority electrons in the neutral absorption layer diffuse into the depleted collection layer and then drift to the cathode layer. Because the absorption layer is nominally undoped, photo-generated majority holes respond very fast within the dielectric relaxation time by their collective motion. Therefore, the photoresponse of a UTC-PD is determined primarily by the electron transport (drift) across the depleted InP layer. This is an essential difference from the conventional p-i-n-PD (Figure 3.27a), in which both electrons and holes contribute to the response, and the low-velocity hole-transport can limit the total performance [93]. There are then three features of the UTC-photodiode that allow it to perform better at THz frequencies than the conventional p-i-n device.

The first and most obvious feature is separate design of the absorbing InGaAs and depleted InP layers, taking advantage of the small conduction-band barrier presented by InP to electrons in the InGaAs. The transit time of electrons across the depletion layer is much shorter than for holes, and the current spectral response is closer to the ideal $[\sin(\omega t_{tt})/\omega t_{tt}]^2$ as shown in Figure 3.28 since all of the electrons are created outside the depletion layer and thus incur the same transit time. However, as the internal bias field increases, the impulse response improves and the transit-time ripples diminish because of velocity overshoot by a significant fraction of the photoelectrons. Hence, the transit-time and RC effects, which both limit the THz performance in the PIN-PD, are much less limiting in the UTC-PD structure. Instead, there is another speed factor, which is the diffusion time of electrons over a distance Δz within the InGaAs absorber layer, to the edge of the depletion

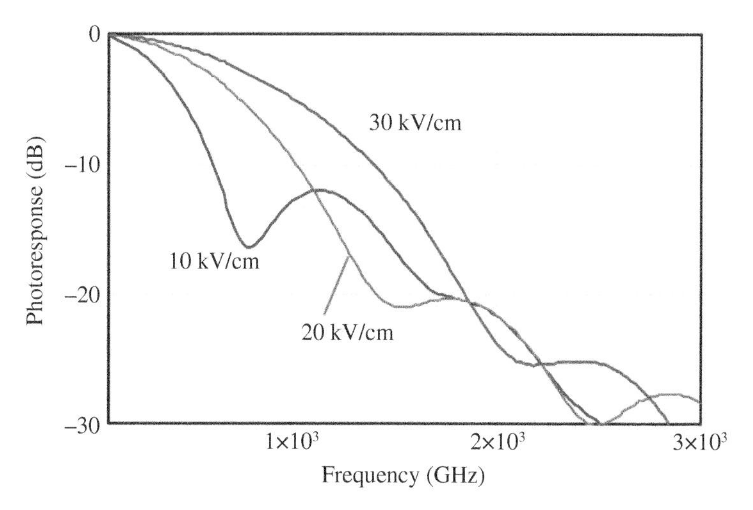

Figure 3.28 Photocurrent transfer function for UTC-PD device structure parameterized by the bias electric field across the collector region.

layer. From elementary semiconductor transport theory, this time is given approximately by $t_d = (\Delta z)^2/D$, where D is the diffusivity of electrons, estimated by Einstein's relation as $D = k_B T\mu/e = 0.026\ \mathrm{m^2/s}$. This yields $t_d = 1.5$ ps for $\Delta z = 200$ nm. Clearly, the thickness of the absorbing layer must be managed carefully in the UTC design.

A second and less obvious feature of the UTC design is a significant increase in the maximum photocurrent (before saturation) because of weaker screening by the photocarriers of the internal bias field in the depletion layer. This is categorically a nonlinear effect so not covered by the small-signal analysis made above, but it causes saturation of the photocurrent spectrum. While not so important in the use of PIN-PDs as high-speed detectors of low light intensity, it is quite important in THz photomixers where the incident light is generally at the 10–100 mW level, and the THz power tends to vary as $|i(\omega)|^2$. Estimates and experiments of the UTC-PD (electrons only) saturation current indicate an order-of-magnitude increase over standard p–i–n design [94].

3.5.2.2 THz Performance

Like MSM photodiodes, UTC-p–i–n devices are readily fabricated and packaged for THz testing with the one difference that the incident 1550-nm radiation is usually coupled to the p–i–n devices from the bottom side through the InP bottom n-type contact and depleted InP collection layer. Then, to produce THz output with minimal circuit coupling losses, the UTC-PD is fabricated in or close to the driving gap of a THz planar antenna, such as the bow-tie shown in Figure 3.29a. As described in the antenna section above, the majority of the radiation from the planar antenna on InP-based materials radiates into the substrate and is coupled to free space through a hyper-hemispherical lens, such as the high-resistivity Si lens shown in Figure 3.29b. Once into free space, the UTC-PD power has been measured by a number of metrological, although not necessarily SI-traceable techniques. The results reported by the UTC-PD inventors are plotted in Figure 3.29c showing output power levels of $\approx$100 μW at 150 GHz, and slightly over 1 μW at 1.0 THz. Notice also from Figure 3.29c that the experimental output spectrum is best fit by an effective electron drift velocity across the InP depletion layer of 4×10^7 cm/s. This is strong experimental evidence that the UTC-PD structure does support velocity overshoot in the electron transport across the InP depletion layer—a very interesting aspect of device physics and engineering both.

More recently, the UTC-PD has been characterized carefully using the THz SPD metrology described earlier in Sec. III. The experiments were carried out at HHI in Berlin in cross-comparison with advanced waveguide-integrated p–i–n technology developed independently [95]. The results are plotted in Figure 3.30 and are remarkable because the THz performance of UTC- and PIN-PDs seem to converge at frequencies above ~400 GHz. Although more study is required, the primary difference in performance between the two devices occurs between 100 and 300 GHz, which has been tentatively attributed to a difference in the antenna coupling[29].

3.6 Alternative 1550-nm THz Photoconductive Sources

As in any emerging technological field, there are many alternative THz 1550-nm PC sources although not as developed as the MSM-bulk photoconductor and p–i–n (or UTC) photodiode described above. Here, we address only 1550-nm PC sources which we have had

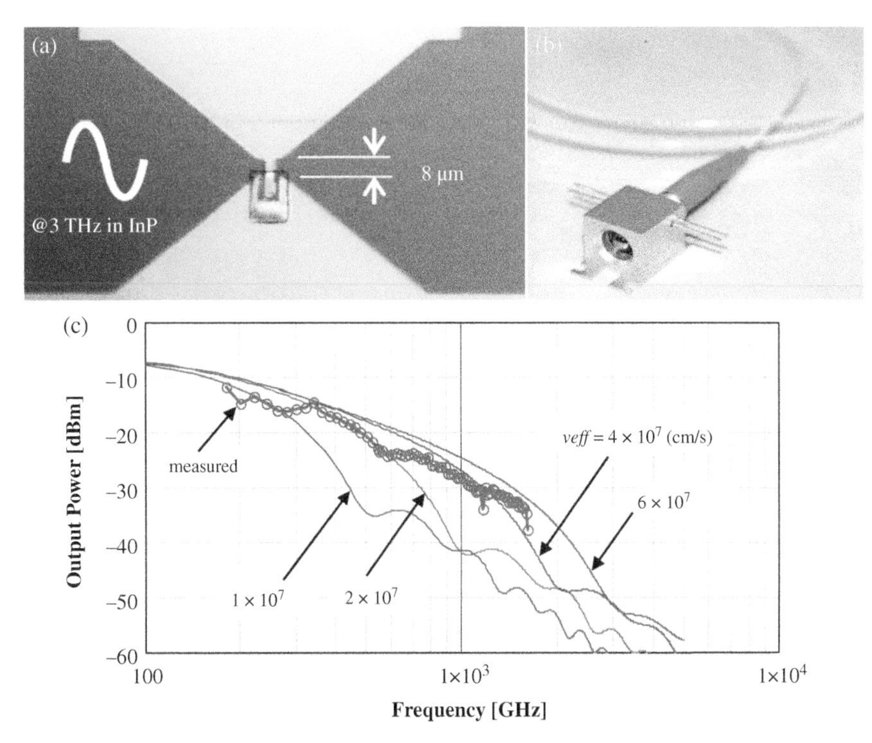

Figure 3.29 (a) UTC-PD fabricated at the gap of a bow-tie antenna. (b) Fully packaged UTC-PD including fiber-optic coupling. (c) Experimental power spectrum vs theoretical simulation parameterized by the effective drift velocity of electrons across the collector region.

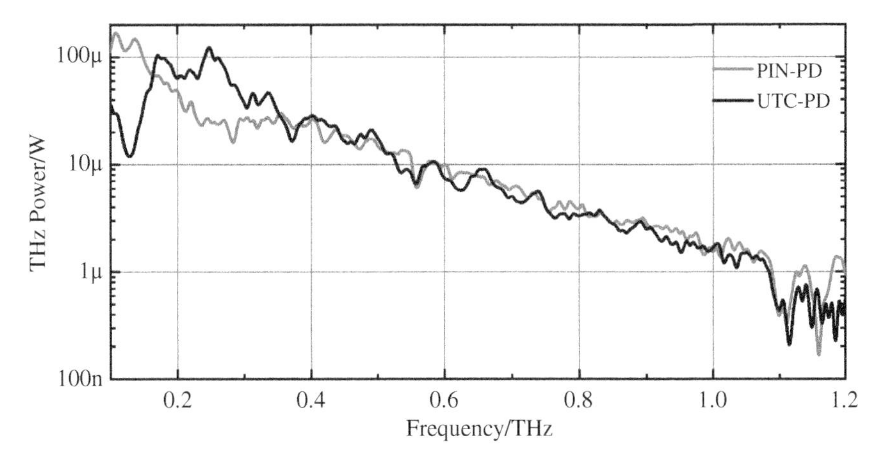

Figure 3.30 Comparison of UTC-PD (NTT) and HHI p–i–n photodiode THz power results both measured with the SPD, SI-traceable technology at PTB in Berlin, Germany.

personal experience with and are confident in the metrological results. The reader is referred to an excellent review by Preu et al. [9] for a detailed description of other 1550-nm-devices such as the *n–i–p–n–i–p* superlattice photodiodes, ErAs:InGaAs photoconductors, and large-area emitters. Other chapters of this book also address novel devices and concepts.

3.6.1 Fe-Doped InGaAs

The general goal for the design of THz photoconductors is the combination of sub-picosecond electron lifetime with high resistivity and high mobility in the same material. These properties are especially important for the development of integrated THz devices. For example, a terahertz transceiver, containing emitter and receiver in close proximity to each other on the same chip would enable compact THz reflection heads for applications in material inspection and in-line process monitoring. For this purpose, a photoconductor that works equally well as an emitter and a receiver is essential. Therefore, new THz PC materials suitable for the excitation with 1550 nm radiation, are under constant investigation.

One promising candidate is iron (Fe) doped InGaAs. The doping of III–V semiconductors with transition metals like iron, rhodium, iridium, or ruthenium has been studied extensively in the last decades [96]. The unique properties of transition metals stem from their partially filled d-orbitals, which determine their electrical and optical properties. As iron is incorporated as a deep acceptor in the middle of the bandgap of InGaAs it acts as a fast recombination center [97, 98]. In 2005, first results on photoswitches made of Fe-implanted InGaAs were demonstrated [99]. However, the bandwidth of these devices was limited to approximately 3 THz, which was significantly lower than LT-GaAs photoswitches designed for 800 nm optical excitation. In addition, the incorporation of high iron concentrations into the PC material, which is needed for sub-picosecond lifetimes, appeared to be difficult. This is mainly due to the fact that iron tends to form clusters during epitaxial growth. Nevertheless, additional promising results of photoswitches and photomixers made of InGaAs:Fe have been demonstrated [100, 101]. Recently, Fe-doped InGaAs grown by MBE was investigated as a PC material. Compared to previous approaches, MBE growth enabled doping concentrations higher than $10^{20}/\text{cm}^3$, which led to high resistivity, high mobility, and sub-picosecond lifetime in the material. When these devices were used as photoswitches in a fiber-coupled THz-TDS system, an average output power of >70 μW a peak dynamic range of 95 dB and a bandwidth higher than 5.5 THz could be demonstrated [102]. This renders InGaAs:Fe a promising material for next-generation photomixers, photoswitches, and integrated, on-chip THz devices.

3.6.2 ErAs Nanoparticles in GaAs: Extrinsic Photoconductivity

As described previously in the evaluation of transport in InGaAs photodiodes and photoconductors, the effect of holes on THz performance is dubious. And in PC switches, the effect of holes is thought to be detrimental because of the large instantaneous photohole density created by the pulsed laser, which tends to screen the internal bias field and thus impede the transport of the photoelectrons. The UTC-PD photodiode is cleverly engineered to prevent holes from being generated-in or entering the high-field depletion region, but this is not possible in the photoconductor devices. So the question naturally arises, is it possible to engineer a PC device in which holes are not photogenerated at all, just electrons? One intriguing possibility is *extrinsic photoconductivity* as shown graphically in Figure 3.31a. It entails the creation of a photoelectron in the CB of a semiconductor from an energy level lying in its band gap [103]. It has been utilized for many decades to create sensitive infrared

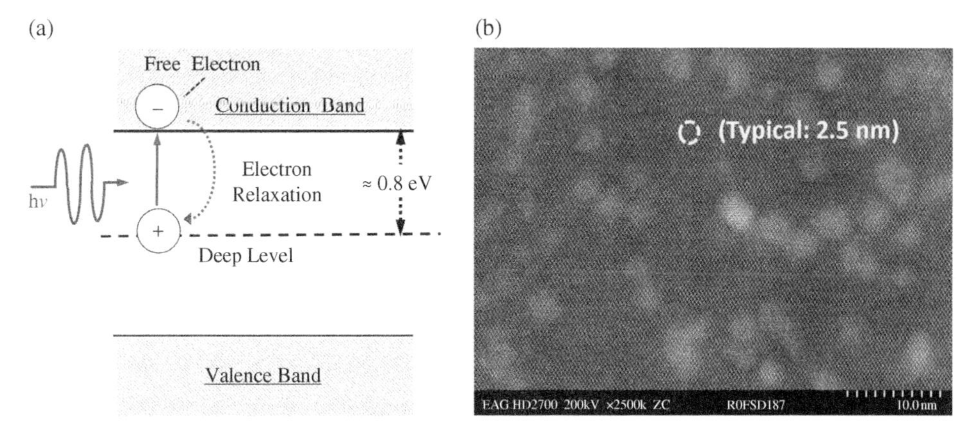

Figure 3.31 (a) Schematic diagram of photoexcitation and recombination in an extrinsic photoconductor. (b) Cross-sectional TEM image of the GaAs:Er epilayer showing the quasi-spherical ErAs quantum dots.

detectors operating at wavelengths much longer than the band-gap wavelength, and in robust semiconductors like Si, Ge, and GaAs. However, cryogenic operation is generally required when the operating wavelength is so long that $hc/\lambda < k_B T$ at room temperature.

The first challenge in utilizing extrinsic photoconductivity in THz photoconductors is absorption strength. The absorption in traditional extrinsic photoconductors occurs from point impurities, such as hydrogenic donors, which have a relatively low absorption cross section and therefore a low overall absorption coefficient compared to the cross gap value, unless very heavy doping is used. Unfortunately, heavy doping creates "impurity banding" effects which generally smear out the absorption cross section and reduce the resistivity of the material to impractical levels. However, the physics can change dramatically if the impurities cluster together during growth (or a post-growth anneal) to form nanoparticles or quantum dots. Such is the case when GaAs is heavily doped with Er during MBE growth, which spontaneously forms ErAs nanoparticles if the doping concentration is $>7 \times 10^{17}/\text{cm}^3$ [104] As shown in the cross-sectional TEMs of Figure 3.31b, the nanoparticles are spherical (or nearly so) in shape and have diameters around 2.5 nm. The Er doping concentration used for this sample was $8 \times 10^{20}/\text{cm}^3$, but the condensation process makes the particle density much lower, $\sim 2 \times 10^{18}/\text{cm}^3$. And, the material retains a high electrical resistivity at room temperature $\rho > 10^6$ Ω-cm, meaning that the nanoparticles are probably not thermally ionized.

The second challenge with extrinsic photoconductivity is getting the strong absorption at or near 1550 nm. Shown in Figure 3.32 is the absorption coefficient for the GaAs:Er sample of Figure 3.31b [105] Remarkably, the absorption coefficient is not only strong, but it is resonant and peaked at 1660 nm with α reaching almost 10^4 cm^{-1}. And the width of the resonance is broad enough that the value at 1550 nm is $\approx 7.5 \times 10^3$ cm^{-1}—comparable to the cross gap absorption in InGaAs at 1550 nm. This was a fortuitous discovery but very similar to an earlier result on similar GaAs:Er material [106]. The next task was to see if the absorption was photoconductive at 1550 nm (i.e. photoionization of the nanoparticle), since the nanoparticles are small enough to behave like quantum dots, which are known to support

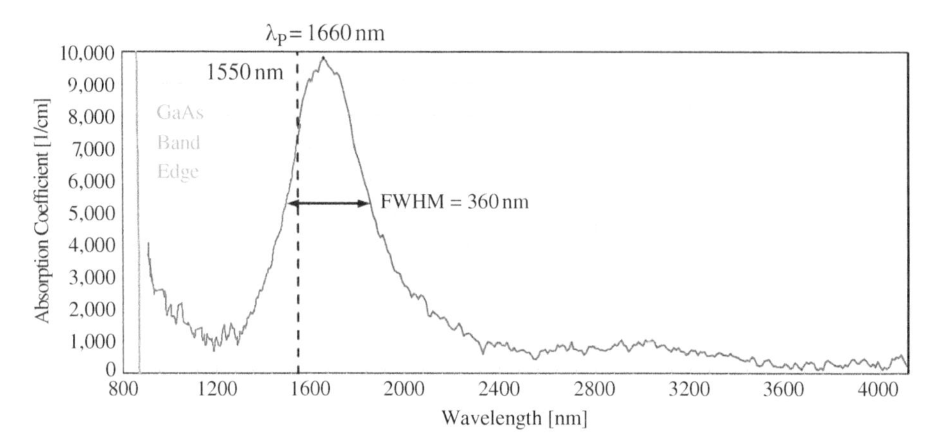

Figure 3.32 Near-infrared absorption spectrum of the GaAs with embedded ErAs quantum dots.

photonic transitions between internal quantum levels. Through a series of experiments starting with photo-Hall measurements, it was found that free carriers were produced by 1550-nm excitation, and they were electrons, not holes. Then PC switches were fabricated as $9 \times 9\,\mu$m gaps at the center of square spiral antennas [107], and the responsivity was found to be ~5 mA/W, a value somewhat lower than for LTG-InGaAs and LTG-GaAs PC switches of similar design. But the LTG materials are known to generate photo-electron–hole pairs, the holes contributing to the DC external responsivity even if they are detrimental to the THz performance.

Finally, the broadband THz power was measured using a fiber MLL having the characteristics in Figure 3.1. The power vs bias voltage and laser power are plotted in Figure 3.33a, b, respectively. The maximum power achieved was 117 μW at a bias voltage of 145 V and a laser average power P_0 = 83 mW. This corresponds to the optical-to-THz conversion

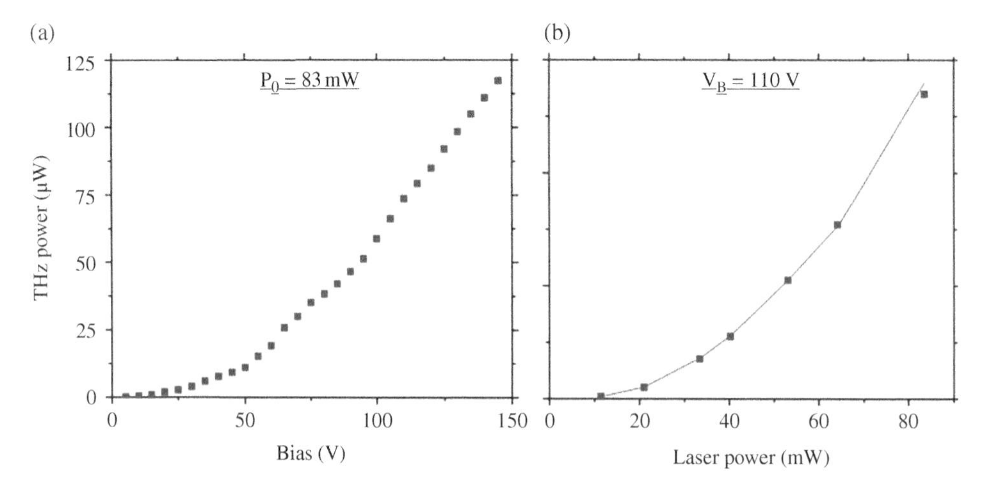

Figure 3.33 (a) Average THz power vs bias voltage for a photoconductive switch made from the ErAs: GaAs material. The average laser power was 83 mW and the pulsewidth was 90 fs. (b) Average THz power vs laser power for the same PC switch as in (a) but at a bias voltage of 110 V.

efficiency of 0.18% entered in Figure 3.26. And both curves are characteristic of PC behavior, the THz power increasing superlinearly with bias voltage and approximately quadratic vs laser power. The power measurements were made with a Golay cell that was cross-calibrated not against the PTB standard described in detail earlier, but rather a traceable waveguide detector (thermistor design) at a frequency of 110 GHz.[30] The autocorrelation function and power spectrum of the THz pulses are what were plotted in Figure 3.18b,c, respectively. The new extrinsic photoconductor was also tested as a photomixer but the THz power levels were sub-μW and the external responsivity was much lower than in pulsed mode [108]. This discrepancy is not yet understood and still under investigation.

3.7 System Applications

3.7.1 Comparison Between Pulsed and CW THz Systems

3.7.1.1 Device Aspects

The fundamental difference between optoelectronic pulsed and continuous-wave terahertz generation and detection is the duty cycle of the laser and the density of carriers after the excitation. Pulsed THz systems operating at 1550 nm wavelength are commonly driven with a femtosecond fiber laser emitting pulses with a length of approximately 100 fs and a repetition rate of 80–250 MHz. Hence, after the illumination of the photoswitch with a femtosecond laser pulse the next pulse will hit the device 12.5–4 ns later. Therefore, the photoswitch can recover from the optical excitation within this period of time, i.e. compared to the duration of the optical excitation the recombination of excited or trapped carriers can occur on a much longer time scale. In contrast, the excitation of a photomixer with an optical beat note of 0.1–3 THz modulates the excited carriers with the frequency of the beat note itself. Hence, photomixers have to recover from the excitation within 0.3–10 ps, which is more than three orders of magnitudes faster than in photoswitches [109]. These fundamentally different demands on photomixers and photoswitches stimulated the development of optimized devices for pulsed and continuous-wave terahertz generation and detection.

For photoswitching emitters, the carrier lifetime of the photoconductor is of minor importance compared to high breakdown fields and high carrier mobility [49, 110, 111]. The only constraint on the lifetime is that the material has to recover from the optical excitation within the repetition rate of the exciting femtosecond laser. Therefore, photoconductors with comparably long lifetimes of 10–100 ps can be used as THz emitters. In these devices, the longer lifetimes are achieved by reducing the density of ionized scattering centers, which increases the mobility of excited carriers and thus increases the emitted THz power [112]. With this approach, an average THz power > 0.1 mW and a spectral bandwidth >5 THz has been demonstrated [74]. The situation is fundamentally different for photoswitching receivers. Here, the electron lifetime is of great importance for an accurate sampling of the incoming THz pulse leading to broadband detection. Hitherto, different approaches have been investigated such as low-temperature grown [37], Fe-doped [100, 102], or ion-implanted InGaAs [99] as well ErAs:In(Al)GaAs [113] or Be-doped LTG-InGaAs/InAlAs heterostructures [88, 114].

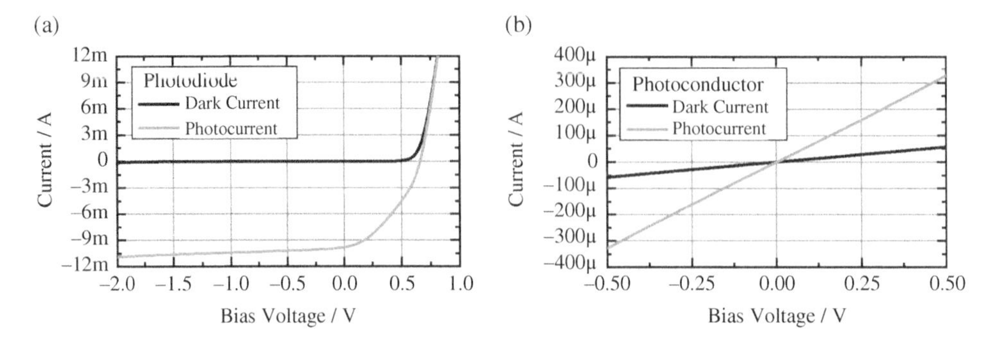

Figure 3.34 (a) I–V characteristics of a waveguide integrated photodiode and a top illuminated photoconductor with interdigitated finger electrodes based on Be-doped LTG InGaAs/InAlAs. The laser power for both is ≈30 mW, the −3 dB bandwidth of the photodiode is ≈60 GHz, and the lifetime of the photoconductor is ≈0.3 ps. More detailed information about the devices can be found in [95].

For photomixing terahertz systems, two different combinations of emitter and receiver devices are commonly used: either photoconductors as emitter and receiver or photodiode emitters combined with photoconducting receivers. These devices have been described in detail in the preceding sections. The main difference between photodiodes and photoconductors is the fact that the switching speed in PDs is determined by the transit time whereas the trapping and recombination of excited carriers determine the switching speed in photoconductors. This sets different constraints on the design parameters of PDs compared to photoconductors. Until today, photomixing systems with a bandwidth of up to 3.5 THz and a peak dynamic range of more than 100 dB have been demonstrated [115, 116].

One important remaining question is whether or not photodiodes can also be used for coherent detection. To answer this, we compare the photocurrent as a function of the applied bias voltage for a waveguide integrated PIN photodiode and a photoconductor made of Be-doped LTG InGaAs/InAlAs in Figure 3.34a and b, respectively [95]. Due to the built-in electric field between n and p contact illuminated PDs generate a photocurrent of a few mA even in the unbiased case. In contrast, photoconductors need an external bias field to generate a photocurrent. This behavior is exploited in using photoconductors as coherent CW detectors: The incoming terahertz signal is applied as a bias voltage to the photoconducting receiver and the detected photocurrent is proportional to the product of the THz field and the optical power. Hence, photoconductors combine the two important properties for a mixing process, i.e. high sensitivity to variable bias voltage and optical power. In case of a PD detector, the comparably weak electric field of the incoming THz signals adds linearly to the build-in electric field of the PD. Hence, the modulation of the build-in electric field of the PD caused by the THz field is comparably small. Therefore, PDs are not as efficient as photoconductors for the detection of coherent CW THz signals [95].

3.7.1.2 Systems Aspects

In the last 10 years, the workhorse of the THz technology was TDS. The combination of femtosecond fiber-lasers emitting at a central wavelength of 1550 nm and PC antennas sensitive to this radiation paved the way for compact, reliable, and fully fiber-coupled THz spectrometers [90]. These systems have been used in first industrial applications, for example for

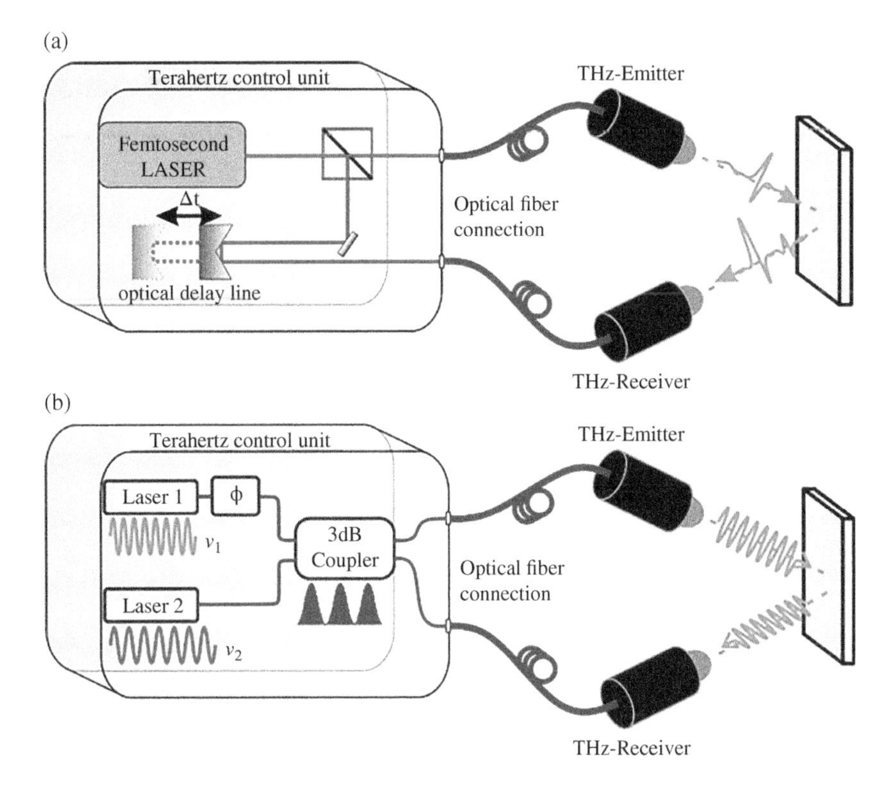

Figure 3.35 (a) Schematic of a fiber-coupled THz-TDS system and (b) a continuous-wave (frequency-domain) system in reflection geometry.

contact-free paint thickness measurements on car bodies [76, 117]. In contrast, coherent CW terahertz systems using photomixers as emitters and receivers have mainly been employed for high-resolution spectroscopy [118]. In this paragraph, we will briefly compare CW and time-domain THz systems suitable for 1550 nm excitation and give an outlook on future developments.

The building blocks of a fiber-coupled THz-TDS and a CW terahertz system are displayed schematically in Figure 3.35a and b, respectively. In THz-TDS the femtosecond laser and an optical delay line are the key optical components of the terahertz control unit whereas CW THz systems comprise two (tunable) lasers, a phase modulator or phase section as well as an optical 3 dB coupler. The current limitations of today's THz TDS systems are mainly set by the optical components: on the one hand, the measurement speed is limited to a few 10 measurements per second due to the mechanical optical delay line used. On the other hand, femtosecond fiber lasers are complex devices and femtosecond pulses require careful dispersion engineering. These drawbacks have prevented TDS systems from being used in a broader field of potential applications since the high complexity of the required components results in high system costs [119]. Therefore, alternatives to the standard THz-TDS-configuration shown in Figure 3.35a are currently an attractive research topic. Apart from concepts that try to replace the mechanical delay line by two phase-locked femtosecond lasers [120, 121], CW terahertz systems can potentially overcome the shortcomings of

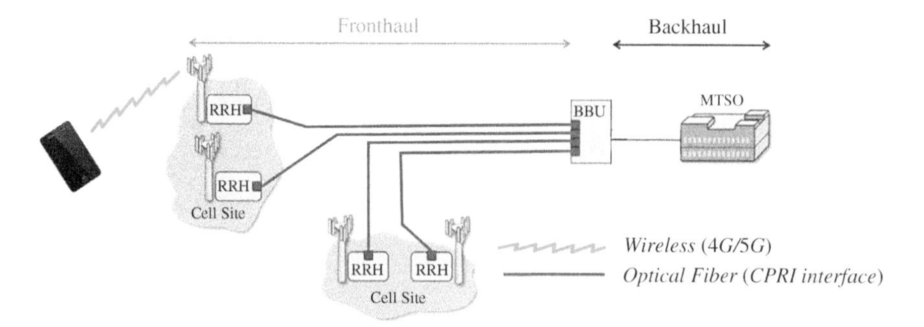

Figure 3.36 Mobile communications network architecture.

traditional THz-TDS. CW THz systems do not require any moving parts nor femtosecond lasers. In addition, they are fully compatible with photonic and electronic integration technology. The CW THz system displayed in Figure 3.35b can be integrated on a single photonic chip, including fixed frequency and fast tunable lasers, optical phase modulators, waveguide couplers as well as THz emitters and receivers. These properties render the CW THz technology extremely promising for low cost and miniaturized optoelectronic terahertz systems. In recent publications, some of these promising aspects of CW THz systems have already been demonstrated. The photonic integration of the optical components on a single chip [122, 123] as well as real-time CW THz systems with acquisition rates comparable to THz-TDS [124].

3.7.2 Wireless Communications

Photonic-based signal generation within the millimeter-wave (30–300 GHz) and terahertz (300–3000 GHz) frequency ranges have one key advantage in their application to wireless communications. Photonic techniques enable the seamless convergence of fiber optic (wired) and wireless segments of the communication networks [125], enabling the direct conversion of data signals transmitted through an optical fiber to a wirelessly radiated signal without requiring signal processing overheads. This is especially critical at the mobile fronthaul, where a natural evolution has replaced traditional coaxial-based systems on cell towers around the world, shifting to a fiber-optic-based approach as shown in Figure 3.36. The bulky, expensive and power-hungry copper cabling has been replaced by fiber for more capacity and longer-reach distances. The substitution of copper by optical fiber included the development of a new protocol running through the fiber between the baseband unit and the remote radio head, called common public radio interface (CPRI). It was established in 2003 by base station vendors such as Ericsson, Nokia, Siemens Networks, Alcatel-Lucent, NEC, and Huawei Technologies to define a publicly available specification that standardizes the protocol interface between baseband unit and remote radio head [126].

In our ever more interconnected world, the amount of data being transmitted through the existing infrastructure has exhibited an exponential growth rate, demanding higher bandwidths at each level of the network. From 2002 to 2015 internet traffic grew by a factor of more than 200[31], resulting in a compound annual growth rate (CAGR) of almost 50% mainly driven by mobile internet usage. Forecasts for the future, which include machine-to-machine connections within the "internet-of-things" and Smart Car scenarios, predict even higher increases that pose a serious challenge to the networks, especially at its fronthaul. For these applications, the bandwidth is mainly limited by the currently bandwidth of current CPRI standard. A competition has started toward defining future ultra-fast wireless communications systems that meet these demands for bandwidth, with different alternative technologies that provide access to larger bandwidths are being investigated including 60-GHz radio [127], free-space optical communications (FSO) [128], and millimeter (MMW) and Terahertz-waves (THz) [129]. To date, several ultra-broad bandwidth communication wireless links have reported operating at carrier wave frequencies above 100 GHz. The highest data rates that have been reported used carrier wave frequencies in the range of 300 GHz, generated either by electronic and photonic techniques.

Electronic techniques commonly use multiplier MMICs to generate the carrier wave signal from a low frequency (below 10 GHz) local oscillator. It has been recently demonstrated a full MMIC chipset, composed by a quadrature transmit and receive RF frontends, with data transmission speed up to 64 Gbit/s [130]. The 300 GHz carrier wave is generated by a 100–300 GHz tripler, feed with 100 GHz generated from an 8.333 GHz input in X-band using a cascade of three multiplication stages (doubler, tripler, doubler). The maximum transmitted power is −7 dBm per sideband, with a total RF transmitted power of −4 dBm on a 64 GHz bandwidth. The link is established using 24 dBi transmitter and receiver antennas, with a total antenna gain of 48 dBi, reaching 2.4 m maximum distance.

Photonic techniques have pioneered the access to the MMW and THz frequency ranges, enabling the generation of carrier waves within these ranges fully by photonics. Currently, photonic technologies lead the way in real-time wireless data transmission. The wide modulation bandwidth of photonic components allows using amplitude-shift keying (ASK) modulation format with data rates up to 50 Gbit/s to demonstrate real-time transmission of HDMI video signals [131]. A 330 GHz carrier wave signal, generated by optical heterodyning using two free-running single wavelength tunable lasers, shining onto a uni-traveling-carrier photodiode (UTC-PD). In this case, the total antenna gain is 40 dBi, and the link is demonstrated over 1 m distance.

As shown in Figure 3.37, there are two main photonic-based signal generation techniques commonly used in wireless data transmission systems. Optical heterodyning is a technique where two optical wavelengths, $\lambda 1$ and $\lambda 2$, are combined on a high-speed photodiode (which is a square-law detector). The combined optical signal, as shown in Figure 3.37, can be represented by the sum ($f1 + f2$) and difference ($f1 - f2$) optical frequencies in the time domain. Typically, the bandwidth limit on the photodetector allows observing only the difference frequency, also known as beat frequency. There are different arrangements to implement an optical heterodyne source. An optical heterodyne signal is usually achieved combining the output of two free-running CW laser diodes through a 50/50 optical coupler as shown in Figure 3.37. An electrical beat note signal or radio-frequency signal is then generated at the output of the photodetector with a frequency corresponding to the wavelength spacing of the two optical waves. This approach provides an excellent frequency tuning range, from a few tenths of MHz to 10 THz [132]. The main problem is that due to the fact that both lasing modes are uncorrelated, its frequency stability is generally poor. Phase noise of −75 dBc/Hz at an offset frequency of 100 MHz and the frequency drift more than 10 MHz/hour have been reported [133].

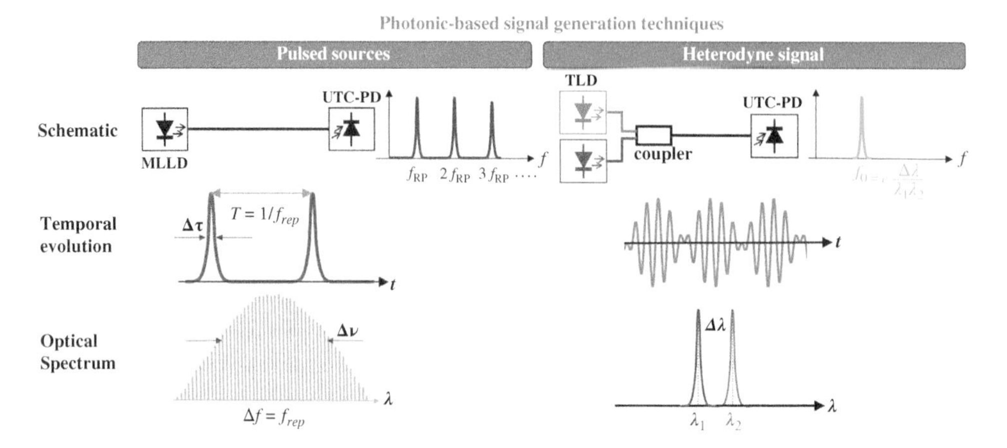

Figure 3.37 Main photonic-based signal generation concept techniques.

When high stability is required, the optical heterodyne technique can be implemented using a continuous-wave single wavelength laser source followed by an external Mach-Zehnder modulator. When properly biased (in the minimum transmission point of the modulator), injecting an RF signal from a CW RF signal source produces optical sidebands spaced twice the injected RF frequency. The current state of the art limits the maximum bandwidth to approximately 110 GHz [134]. On the other hand, it has excellent phase noise or stability of the generated signal as it depends on the RF synthesizer.

Pulsed sources have also been used to generate signals in the millimeter and terahertz frequency range, usually by means of mode-locked laser diodes (MLLDs). The optical emission is in the form of optical pulses, generated by phase-locking multiple optical longitudinal modes. The optical pulses are spaced by a time T, which is the inverse of the pulse repetition rate frequency, f_{rep}. The repetition rate is inversely proportional to the cavity length and directly proportional to the number of pulses circulating in the cavity. For monolithic laser diodes, which have cavity lengths around 400 μm, repetition frequencies in the GHz range were easily achieved. However, it has a reduced tunability, typically from 100 MHz to 1 GHz [135, 136]. The phase noise of passively MLLs based on Fabry-Perot lasers is relatively high more than −70 dBc/Hz at an offset frequency of 100 Hz. The case of active mode locking, in which the laser is driven with an electronic oscillator, the phase noise is much lower (<−75 dBc/Hz at an offset frequency of 100 Hz), presenting an excellent stability. In communication applications, it has been demonstrated that pulsed sources have the capability of generating greater RF output power than the heterodyning sources [137].

Table 3.3 summarizes the characteristics of the different photonic-based signal generation techniques in terms of three key parameters: (i) The maximum frequency refers to the highest frequency that can be achieved with a given technique, (ii) the tuning range is the ability to change the generated frequency within a certain range, and (iii) the stability of the signal, related to the random fluctuations of the frequency due to noise, which leads to a finite linewidth of the signal. This noise is dominated by the phase noise.

Table 3.3 Comparison among photonic signal generation techniques.

Photonic signal	Method	Maximum frequency	Tunability	Stability/phase noise
Heterodyne source	Two LDs (with short lasing cavity)	Excellent from 0.1 to 10 THz	Excellent from 0.1 to 10 THz	Poor: bad frequency drift, large linewidth
	CW LD + external modulator	Good around 100 GHz	Good, ≈110 GHz	Excellent: determined by electronics
Pulsed source	Mode-locked laser diode (passive/ active)	Good: Passive >1 THz Active >200 GHz	Bad from 0.1 to 1 GHz	Excellent for active; acceptable for passive

A wireless communication link using photonic-based carrier frequency generation is shown in Figure 3.38. On the transmitter module, we can always identify three main building blocks: the carrier generation module, the optical data modulation module, and the wireless transmitter module. On the receiver side, one can choose between direct detection or coherent detection (as shown), where the carrier wave frequency is down-converted to

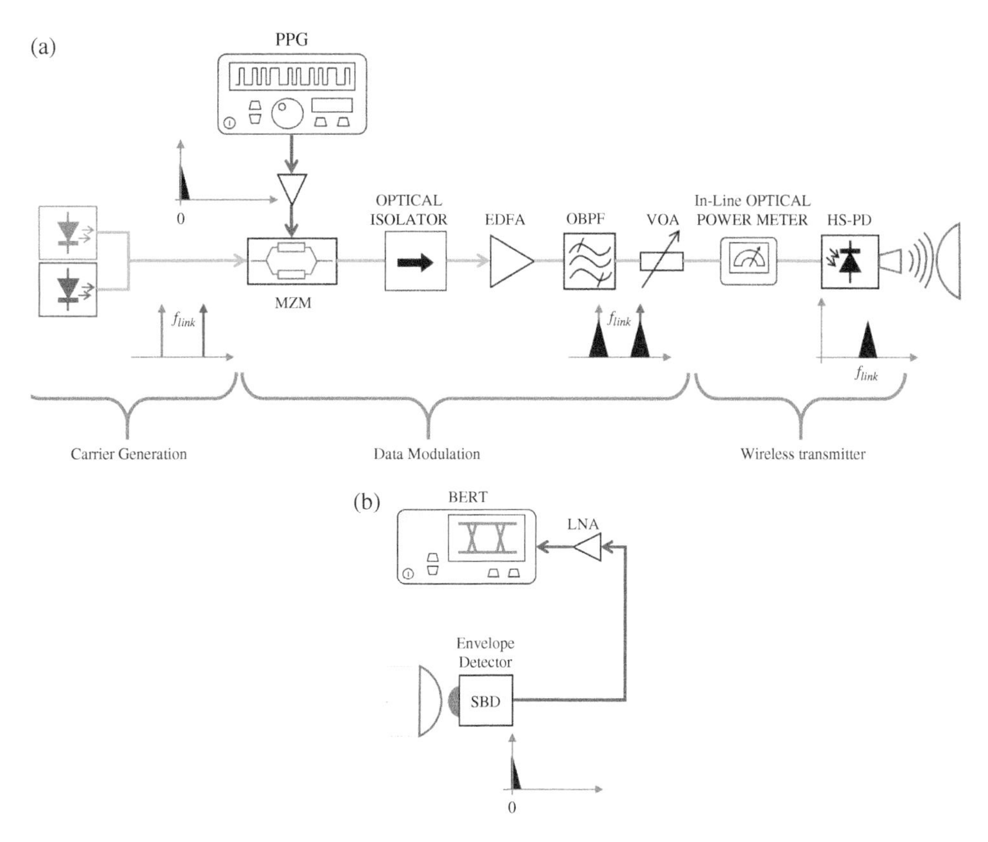

Figure 3.38 Ultra-broadband wireless link with photonic-based carrier wave generation at the transmitter and direct detection on the receiver.

baseband with a sub-harmonic mixer, which needs a local oscillator (LO) frequency, so that the intermediate frequency at the output is $f_{link} - f_{LO}$. The photonic carrier generator module was implemented by using a variety of sources. Optical heterodyne carrier generation was generated from a pair of external cavity tunable single-frequency lasers (Sacher Lion) as well as integrated dual DFB lasers. Pulsed sources were implemented through on-chip colliding pulse passive mode-locked laser diodes with repetition rates at 70 GHz and 100 GHz with on-chip multiple colliding pulse passive MLLs. Usually, a 99/1 splitter is connected to divide the optical signal into two branches, using the 1% split to monitor the output optical power and the 99% split to feed the optical data modulator.

The data modulator module comprises an electro-optical amplitude modulator (EOAM), which receives 0 dBm optical power through a polarization controller. The EOAM is biased at one of its quadrature points (with −6 V DC voltage). The data signal is generated from the Pulse Pattern Generator (PPG) source of a bit error rate tester (BERT), and Anritsu BERT-Wave MP2100. We inject a 1 Gbps nonreturn-to-zero (NRZ) On–Off Keying (OOK) data signal with a pseudorandom bit sequence (PRBS) of 223–1 word length. The maximum output voltage delivered by the PPG is 2 Vpp. The modulated optical signal is led to a second optical splitter, with 90/10 splitting ratio. The 10% split goes to an OSA to observe the optical signal at all times, and the 90% split is connected to an EDFA that boosts the optical power to compensate for the losses at the modulator (of the order of 8 dB). A variable optical attenuator (VOA) and in-line optical power meter allow to control the optical signal launched to the high-speed photodiode. The wireless transmitter converts the optical modulated signal into an electrical radio-frequency signal on the high-speed photodiode. The radio-frequency signal is then radiated into the air by means of a horn antenna. Two different HF-PD are available, a U2T XPDV4120 PIN-PD), with responsivity 0.62 A/W, and a NTT Electronics UTC-PD, with responsivity 0.4 A/W. The horn antennas have a gain of about 22 dBi.

In the wireless link shown in Figure 3.38, the receiver is based on a direct detection scheme. A zero-bias Schottky barrier diode (SBD) envelope detector is used, which rectifies the carrier wave frequency down-converting the modulated signal back to baseband. To benefit from the broadband nature of the photonic-based emitter, a quasi-optic approach is preferred. In this approach, the SBD is mounted on a broadband planar antenna that responds over a wide range of carrier frequencies and assembled on a silicon lens to increase the directivity of the module by 15 dBi of equivalent antenna gain. The responsivity of the antenna-integrated in the SBD detector module with the silicon lens at carrier frequency 70 GHz and 100 GHz is 100 V/W and 500 V/W, respectively. The SBD module is then connected by a chain of amplifiers composed by a low-noise amplifier (LNA) with 34 dB gain followed by a limiting amplifier (LA). The carrier frequency detection band is set by the SBD antenna, which ranges from 40 to 160 GHz, having a peak responsivity of 7129 V/W at 75.7 GHz. The data bandwidth limit is determined by the LNA bandwidth, which is 9 kHz to 3 GHz.

We have compared two optical signal generation schemes, pulse source, and optical heterodyne, generating a carrier wave in the F-band, around 100 GHz. For the pulse source, we used a novel on-chip multiple colliding pulse passive mode-locked laser diode structure operating at 100 GHz [138]. In this case, the photonic transmitter used the UTC-PD photodiode with a WR8 conical horn antenna. A -pseudo-random bit sequence PRBS = 2^{23}–1 with a 1

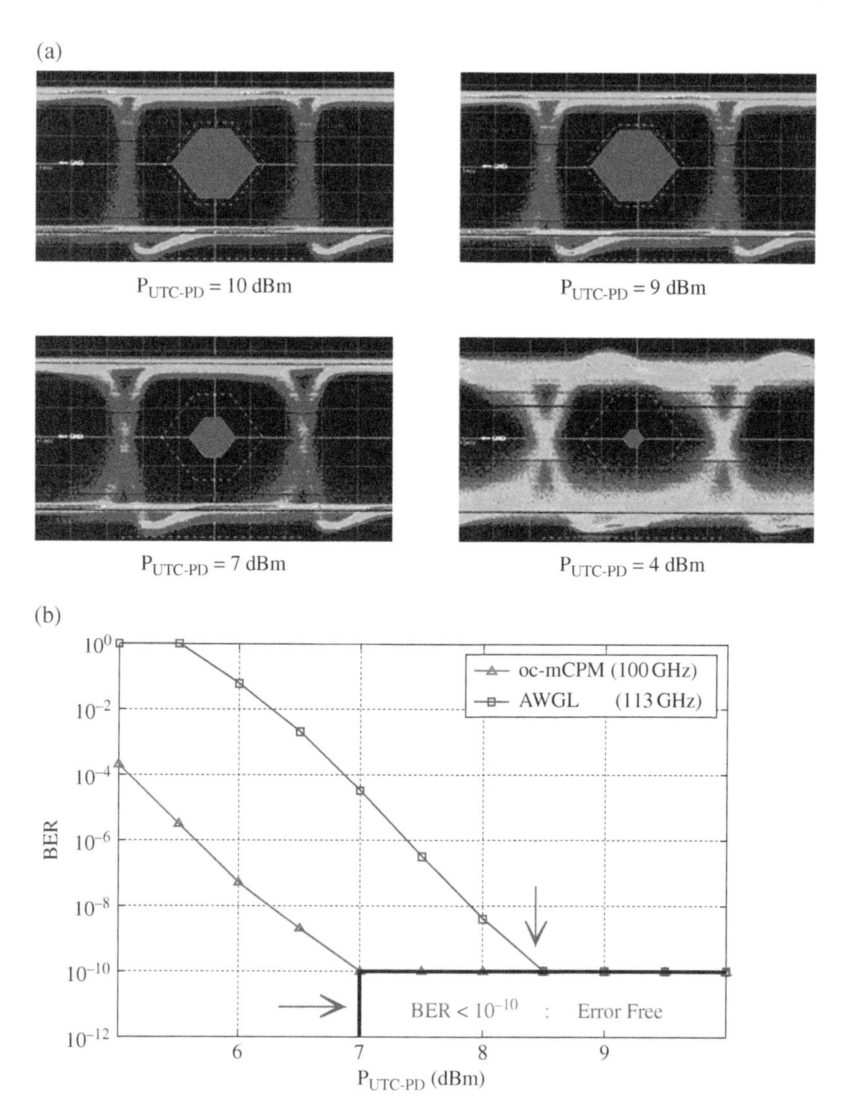

Figure 3.39 (a) Eye-patterns of the received PRBS = $2^{23}-1$ at bit rate BR = 1 Gbps in the F-band wireless link, voltage level 500 mv/Div and the span 200 ps/Div and (b) comparison of Bit Error Rate levels with pulsed source and optical heterodyne techniques versus the optical power launched into the UTC-PD.

Gbps bit rate is used to trace the eye-patterns at different optical power levels launched into the UTC-PD. The obtained eye patterns are shown in Figure 3.39a with voltage level 500 mv/div and the span 200 ps/div. The received eye-patterns are clearly open at the power launched values above $P_{UTC\text{-}PD} = 7$ dBm (photocurrent, $I_{UTC\text{-}PD} = 2$ mA), in which the time variation at zero crossing is below 27 ps. The eye-patterns start to close at lower power launched values than $P_{UTC\text{-}PD} = 7$ dBm. The time variation at zero crossing increases from 27 to 81 ps and the BER increases severely.

The evolution of the BER is evaluated versus the optical power launched into the UTC-PD. The bit error rate characteristics at 1 Gbps is shown in Figure 3.39b. As expected, the higher the optical power launched into the photodiode the lower the BER. We consider error-free transmission when the BER reaches 10^{-10}. This was achieved with the pulse source when the optical power into the UTC-PD is above $P_{\text{UTC-PD}} = 7\,\text{dBm}$, condition for which we have shown the eye diagram is wide open. The bit error rate characteristics at BR = 1 Gbps of the F-band wireless transmission link using the heterodyning method. In this case, a carrier frequency at 113 GHz was used. From the comparison of the BER measurements between the pulsed source and heterodyning source, the main improvement of the pulsed source method using oc-mCPM is that lower power than the heterodyning source is required to be launched into the UTC-PD in order to obtain error-free transmission. The pulse source only requires 7 dBm power while the heterodyning source requires around 9 dBm for achieving error-free transmission.

3.7.3 THz Spectroscopy

3.7.3.1 Time vs Frequency Domain Systems

Since the early days of THz PC devices, it has been recognized that the time-domain spectrometers (TDSs) and photomixing FDS both offer a unique capability to operate over a huge bandwidth from ~0.1 to several THz and with just one system. Here we define spectroscopy in a broad sense as being the process of determining the optical properties of a material—be it solid, liquid, or gas phase—by measurement of its optical transmittance or reflectance. The most common optical properties of interest are the broadband complex dielectric function and the absorbance spectrum. For example, the former is of primary interest in solid-state materials (e.g. graphene) and in materials having a variable parameter (e.g. free-carrier concentration in semiconductors). And the latter is of primary interest in materials displaying THz absorption resonances which are common in gases (e.g. water vapor, polar molecules) and in some solids (e.g. many sugars). Not only is bandwidth important in the former, but also measurement of the complex transmittance (amplitude and phase) so that the complex dielectric function can be derived. Resolution is most important in the latter such that the center frequency and lineshape of absorption resonances can be determined. As in the more traditional field of microwave molecular spectroscopy [139], the resonance frequency and lineshape are both useful in determining the underlying physics in any state of matter.

This was first demonstrated with 780-nm TDS systems in the 1990 timeframe [140], and then with FDS systems a few years later [141]. Although not as broadband as far-infrared Fourier transform spectrometers, the THz PC-source spectrometers provide vastly superior dynamic range and with no cryogenics required in the receiver, which far-IR FT spectrometers usually require since their source power—often from some type of far-infrared blackbody—is so weak. Like FTIR, a significant issue for THz-PC spectroscopy was, and still is, spectral resolution. TDS systems have a resolution limited by the travel distance of the optical delay line in Figure 3.24, such that $\Delta f \approx c/\Delta L$. And the highest possible resolution is determined by the pulse repetition frequency PRF of the MLLs used to drive them. So for example, a travel distance of 10 cm yields a resolution of 3 GHz for an MLL having a PRF of 100 MHz. FDS systems can provide far superior resolution, but it is not easy. The first demonstrations of high-resolution FDS spectroscopy were carried out with solid-state or dye lasers having external optical stabilization, allowing for very low drift and jitter of the difference frequency. However, for cost and portability reasons, FDS systems were quickly forced to adopt single-frequency semiconductor (diode) lasers which were just maturing at that time in a form having internal distributed Bragg reflector (DBR) structures for frequency stabilization. Unfortunately, the DBR laser diodes were superseded by the simpler distributed feedback (DFB) diode lasers which are simpler and lower cost, but more problematic for THz spectroscopy because of temperature drift, longitudinal mode-hopping, RIN noise, and other issues. Nevertheless, the THz resolution of DFB photomixing spectrometers rather routinely reaches the frequency jitter limit of ~100 MHz. And with the continual evolution of fiber-optic telecommunications, the 1550 nm DFB lasers are readily available, fiber coupled, and quite powerful up to the level of ~100 mW. A downside of 1550-nm compared to 780-nm DFBs is the more gradual THz tuning vs wavelength offset, as evidenced in Figure 3.4b. This requires greater temperature tuning between a given pair of DFBs—a problem that has been addressed by the clever use of three 1550-nm DFBs instead of two [142]. Often, in resonant "line" spectroscopy, phase information of the transmission coefficient is unnecessary; only center frequency and lineshape are required.

An exemplary block diagram of a 1550-nm TDS system was shown in Figure 3.35a. The system is entirely fiber coupled, including the delay line between the fiber laser and the THz receiver. Such 1550-nm systems are now sold commercially by several companies such as Toptica in Germany[32], Menlo Systems in Germany[33], and TeTechS in Canada[34]. An example of the TeTechS system performance is plotted in Figure 3.40a, showing a background dynamic range exceeding 70 dB at low frequencies, and a useful dynamic range out to beyond 3.5 THz. Both the transmitter and receiver in the TeTechS system are photoconductors, emphasizing the fact that they are generally superior to photodiodes in ultrafast pulsed applications like TDS systems. Plotted in Figure 3.40b are exemplary spectra from granular materials known to display strong THz resonant signatures, such as lactose monohydrate and tartaric acid. Shown in Figure 3.41 are the performance curves for the Menlo TDS system, which provides ~90 dB of low-frequency dynamic range and useful performance to at least 5.5 THz.

As demonstrated in Figure 3.40b TDS systems measure the entire available THz spectrum whether it is needed or not. And as in FTIR spectrometry, this is advantageous when one is measuring broadband dielectric properties, or searching for resonant absorption "lines" in an unfamiliar material. However, if one is interrogating a small portion of the available spectrum where, for example, absorption resonances are already known to exist, the FDS is advantageous because it can easily be adjusted to scan over a limited frequency range without sacrificing optimal performance. A promising THz system application has always been the identification of illicit substances, such as drugs and explosives, by identifying their signature "lines" in transmission measurements of solid samples (typically powders). This is similar to the widespread use of FTIR to identify molecules by virtue of their signature vibrational resonances that typically occur in the SWIR to LWIR spectral region. Unfortunately, THz vibrational signatures are usually much more complicated than IR signatures, especially in solids. They generally involve the motion of entire molecules or portions of a molecule, rather than the simpler interatomic vibrations that characterize the IR signatures. So while THz spectroscopy has been maligned by many people over the years, it is not for the want of good instruments, but rather the complex phenomenology that limits this promising application, at least in the opinion of the present authors.

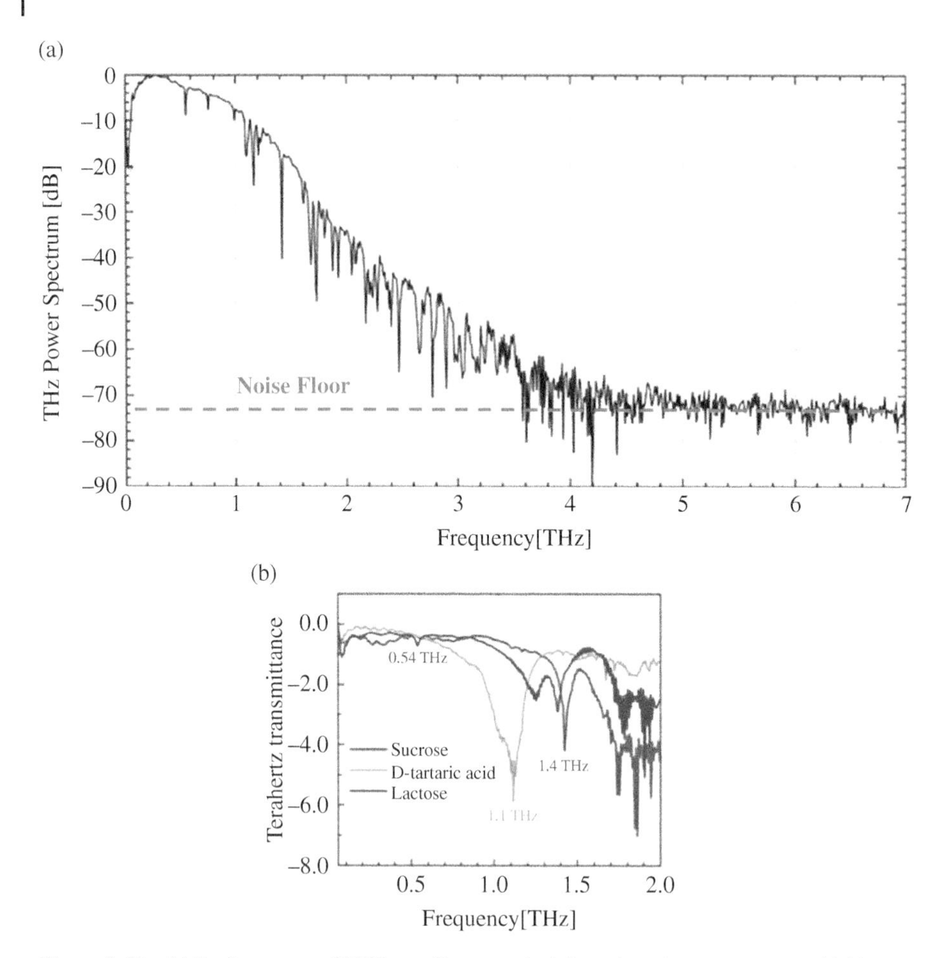

Figure 3.40 (a) Performance of 1550-nm fiber-coupled time-domain spectrometer. (b) Measurement of transmittance through powders having known-strong resonant signatures[34].

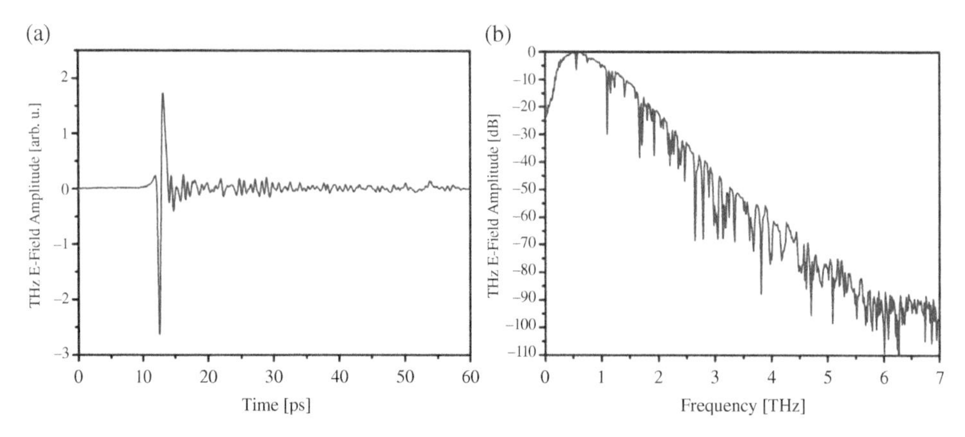

Figure 3.41 (a) THz pulse and (b) associated spectrum, for state-of-the-art fiber-laser-based 1550-nm TDS system, displaying a dynamic range of ~90 dB, and useful bandwidth >5 THz[33]. *Source*: Based on TERA K15 All fiber-coupled Terahertz Spectrometer, https://www.menlosystems.com/products/thz-time-domain-solutions/all-fiber-coupled-terahertz-spectrometer/.

An exemplary block diagram of a 1550-nm FDS system was shown in Figure 3.35b. The system is also entirely fiber coupled, including the phase shifter (or delay line) in the path of one of the single-frequency lasers before it is combined with the other. The dynamic range of this system is shown in Figure 3.42a and is comparable to the 1550-nm TDS system in Figure 3.39a at low frequencies being just over 70 dB [109]. However, the bandwidth of the FDS system is not as great as the TDS, reaching the noise floor just over 2.1 THz. This is typical of FDS systems. Because of the much lower average power in FDS than the peak power in TDS systems, higher responsivity devices are needed in the transmitter and receiver both, such as PIN-PDs in the transmitter and MSM photoconductors in the receivers. In TDS systems, simpler gap-based PC devices work well, and generally provide greater bandwidth because of their lower capacitance.

However, a big advantage of FDS systems is demonstrated in Figure 3.42b showing a "zoom-in" on one of the many water vapor lines seen in the broad scan of Figure 3.42a centered around 557 GHz [109]. The frequency step on the zoom-in is 200 MHz—a resolution that is impractically narrow in TDS systems. In fact, there are over 50 resolution elements on this one line, so the broad scan of Figure 3.42a would have easily found this line with a much coarser step, say 1 or 2 GHz. In other words, the FDS systems allow a mode whereby a broadband, coarse-step scan is done first to find resonant features. Then a zoom-in scan is done with much higher resolution. The zoom-in scan would be much faster than the broadband one because the start and stop frequencies of the second scan would only have to occur just below and above the resonant feature. This is quite unlike a TDS system where to zoom-in on any fine features of a first, coarse scan, the resolution would have to be increased by increasing the path of the delay line in Figure 3.35a. This in turn will also require a much slower second scan.

3.7.3.2 Analysis of Frequency Domain Systems: Amplitude and Phase Modulation

Independent of the system type, it is good to understand how the high *useful bandwidth* of THz spectrometers is utilized in their various applications. Because TDS systems are the oldest and most studied [7], we focus here on FDS systems and some of the details of their optical behavior and signal processing. We start with the most general case whereby one is interested in measuring the complex transmission coefficient T through a material sample, or just the optical transmittance $|T|^2$. It can be shown that the receive photomixer in any FDS system provides an output photocurrent proportional to the THz E field times a phase factor [143]:

$$i(\Delta\nu) \propto V_{THz}(\Delta\nu)G_P(\Delta\nu) = D \cdot E_{THz,0}(\Delta\nu)G_P \cos(\varphi_1 - \varphi_2) \tag{3.36}$$

Here $\Delta\nu$ is the laser difference frequency $|\nu_1 - \nu_2|$, $E_{THz,0}$ ($V_{THz,0}$) is the magnitude of the THz electric field (voltage) induced in the receive photomixer planar antenna by the THz beam from the emitter photomixer, D is a constant, and G_P is the photoconductance induced by the two laser beams in the same photomixer gap at $\Delta\nu$, assuming there is no significant attenuation or dispersion in the laser or THz paths. Also in this expression φ_1 is the difference-frequency phase change of the laser beams from their point of optical combining to the THz transmit photomixer ($\equiv\varphi_T$), plus the additional phase change of the THz beam from the transmit to the receive photomixers ($\equiv\varphi_{THz}$). φ_2 is the difference-frequency phase change of the laser beams from their point of combining to the receive photomixer

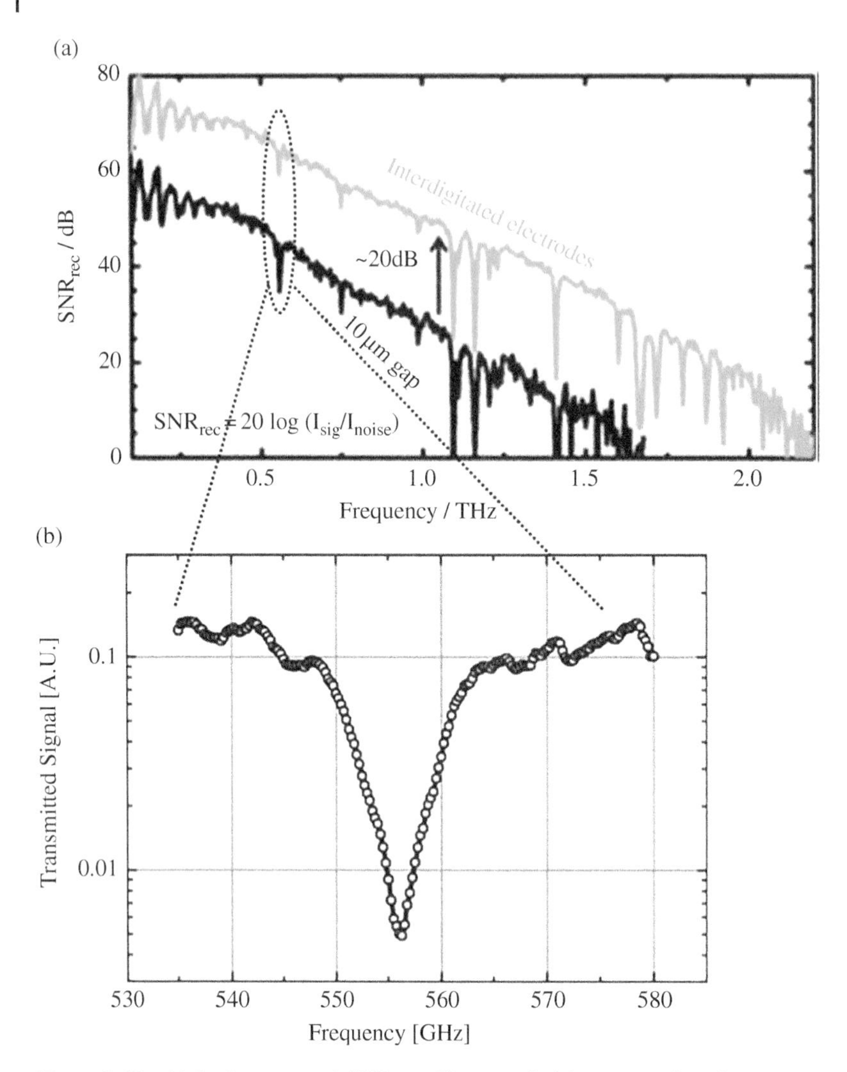

Figure 3.42 (a) Performance of 1550-nm fiber-coupled frequency-domain spectrometer having performance similar to that in Figure 3.40a. (b) Zoom-in on the water-vapor line centered around 557 GHz.

($\equiv\varphi_R$). To avoid measuring DC signals at the output and their issues with drift and $1/f$ noise, we assume the THz path is (AM) modulated (usually by bias voltage modulation of the transmit photomixer) with waveform $M(t)$ so we can write

$$i_B(\Delta\nu, t) = DM(t)E_{THz,0}(\Delta\nu)G_P \cos\left(\varphi_T + \varphi_{THz} - \varphi_R\right) \tag{3.37}$$

Since the optical frequencies and path lengths are assumed fixed, we can re-write (3.37) as

$$i_B(\Delta\nu) = DM(t)\, E_{THz,0}(f) G_P \cos\left[\Delta k(L_T + L_{THz} - L_R)\right] \equiv DM(t) E_{THz,0}(f) G_P \cos\left[\Delta k \Delta L\right] \quad (3.38)$$

where L_T and L_R, are the *optical* path lengths from point of combining to the transmit photomixer and to receive photomixer, respectively; L_R is the *optical* length from the point of combining to the receive photomixer, and the subscript B stands for "background" (i.e. no sample in THz path). For brevity, we have re-defined $f \equiv \Delta\nu$, so that $\Delta k = 2\pi f/c$ where c is the speed of light in vacuum[35], and also assume implicitly that $\Delta L > 0$, which is generally true in practical FDS systems. But if not, an essentially similar derivation could be carried out. If $M(t)$ is periodic in time, synchronous demodulation (e.g. lock-in amplifier) is effective at reducing noise and the output of the lock-in is

$$i_{B,out}(f) = FE_{THz,0}(f) G_P \cos\left[\Delta k(L_T + L_{THz} - L_R)\right] \equiv FE_{THz,0}(f) G_P \cos\left[\Delta k \Delta L\right] \quad (3.39)$$

where F is another constant.

The addition of a sample of material to the THz path generally changes the amplitude and phase of the receiver output. The sample is assumed to not distort or refract the THz beam, which leads most researchers to test samples formed into disks or slabs having parallel input and output facets. The addition of the sample then leads to a lock-in amp output

$$i_{S,out}(f) = FE_{THz,S}(f) G_P \cos\left\{\Delta k[\Delta L + L_S(n-1)]\right\} \quad (3.40)$$

where L_S is the physical thickness of the sample, and n is the refractive index, generally a function of f but often weakly so. The ratio of Eqs. (3.40) is the transmission coefficient T, which cancels F and G_P but leaves a dependence on $E_{THz,S}(f)/E_{THz,0}(f)$ times the ratio of phase terms.

The analysis of T is facilitated by the fact that most FDS systems can vary f and therefore Δk over a huge tuning range, $\Delta\nu_{max} - \Delta\nu_{min} = f_{max} - f_{min} \equiv \Delta f$, typically one decade or more. Furthermore, ΔL in Eq. (3.38) is often large enough such that

$$(\Delta k_{max} - \Delta k_{min})\Delta L \gg 2\pi \quad (3.41)$$

meaning that i_B varies sinusoidally over many cycles in the "background" scan (i.e. no sample). The period of these oscillations is determined by setting $\Delta k\, \Delta L = (2\pi f/c)\cdot\Delta L = 2\pi$, for which we get a background-spectrum oscillation frequency

$$\Delta f_B = c/\Delta L \quad (3.42)$$

If the difference frequency is scanned in discrete steps of f_S, as is usually the case, then these oscillations will be fully resolved (according to the Nyquist condition) if $\Delta f_B/f_S \geq 2$. If ΔL is too small to satisfy Eq. (3.41), it can always be increased by changing L_T or L_R, whichever is easier in the given FDS system.

Knowing ΔL from Eq. (3.42), one can substitute into Eq. (3.39) to write

$$i_{B,out}(f) = FE_{THz,0}(f) G_P \cos\left[2\pi f/\Delta f_B\right] \quad (3.43)$$

and into Eq. (3.40) to write

$$i_{S,out}(f) = F \cdot E_{THz,S}(f) \cos\left[2\pi f[(1/\Delta f_B) + L_S(n-1)/c]\right] \tag{3.44}$$

Equation (3.44) displays a different oscillation Δ_s in the f domain, found again by setting the argument to 2π, which yields

$$n = 1 + (c/L_S)\left[(\Delta f_S)^{-1} - (\Delta f_B)^{-1}\right] \tag{3.45}$$

This is physically reasonable when Δf_S is less than Δf_B, meaning that $n > 1$. In general, n is a function of f across such a wide tuning range, but if it is slowly varying compared to Δf_B, then Δf_S vs f can be obtained from the experimental data and used to estimate $n(f)$ using Eq. (3.45). Equation (3.44) also simplifies to

$$i_{S,out}(f) = F \cdot E_{THz,S}(f) \cos\left[2\pi f/\Delta f_S)\right]] \tag{3.46}$$

Upon dividing Eqs. (3.46) by (3.43), we can write the transmittance as

$$\begin{aligned} |T|^2 &= [i_{S,out}(f)/i_{B,out}(f)]^2 \\ &= [E_{THz,S}(f)/E_{THz,B}(f)]^2 \{\cos\left[2\pi\Delta f/f_S\right]\}^2/\{\cos\left[2\pi f/\Delta f_B)\right]\}^2 \end{aligned} \tag{3.47}$$

From the square of the fields we have the ratio of intensities I_S/I_B, which if the THz path is lossless except for the sample, can be written as $\exp(-\alpha L_S)$. So in principle, the absorption coefficient α can be calculated from the ratio of the two separate phase factors and the known thickness of the sample. And knowing n, we can obtain the complex index $= n + j\kappa$, where $\kappa = \alpha/2$ is the extinction coefficient.

The problem with this approach is that it is very sensitive to spurious effects in the spectrometer that are inconsistent with the model expression Eq. (3.36). A few examples are Fabry–Perot standing waves in the sample caused its parallel facets and feedback from the receiver to transmitter caused by impedance mismatch between the receive photomixer and its antenna. The latter is a problem seldom discussed in the THz literature for either photodiode or PC devices. But it is one that deserves more attention since un-wanted feedback is difficult to model and can lead to errors in the signal processing. So a less precarious method is to consider only correlated maxima (or zero crossings) of each cos in Eq. (3.47) [144]. A maxima in the denominator is given by the condition, $f_B = m{\cdot}\Delta f_B$ where m is an integer. So for a given m, the numerator is evaluated not at f_B, but rather $f_S = m\Delta f_s$. The accuracy of this procedure is incumbent on having the phase change induced by the sample being much less the background phase difference, i.e. according to Eq. (3.40), $L_S(n-1) \ll \Delta L$, which from Eq. (3.42) is equivalent to $L_S(n-1) \ll c/\Delta f_B$. Hence, the method works best on thin samples having low-to-moderate refractive index, which is often the case in FDS measurements. This technique can also be used to measure the n and κ values in samples containing resonant absorption features, such as the lactose monohydrate feature around 530 GHz [143]. However, this requires that the oscillation frequency be significantly small so that many peaks in Eq. (3.46) occur over the resonance linewidth, δf, i.e. $\Delta f_S/\ \delta f \ll 1$.

There are several other techniques used to process the signals from FDS systems depending on what one is interested in measuring, For example, if one just wants to identify the center frequency and lineshape of absorption resonances without knowing the refractive index, and the $f_S/\delta f \ll 1$ condition is satisfied, then a centered running-average filter can be carried out on the $i_B(f)$ and $i_S(f)$ signals separately to effectively filter out the rapidly varying $\cos(2\pi f/f_B)$ and $\cos(2\pi f/f_S)$ terms but pass the more slowly varying $E_{THz,B}(f)$ and $E_{THz,S}(f_S)$ components. The transmittance will then be proportional to $\exp(-\alpha \cdot L_S)$, in principle, but running average filters are not high-quality low pass filters and errors and inaccuracies can occur. Hence, a higher quality, digital low-pass filter is advisable.

When a 1550-nm TDS system has its lasers coupled through optical fiber, new techniques are enabled given the addition of a hardware component or two. For example, fiber-coupled mechanical delay lines can be implemented to vary ΔL in the above analysis, and even drive it to zero (depending on accuracy). This has the attractive feature of practically eliminating the oscillations in the background spectrum of Eq. (3.39). Then after adding the sample, the delay line is adjusted a distance Δz to eliminate the oscillations in the sample spectrum of Eq. (3.40). Because delay lines generally operate by propagating through air, we can write $\Delta z = L_S(n-1)$. The transmittance then follows from the ratio $[i_s(f)/i_B(f)]^2$ but without any dominant phase-dependent terms. The downside of this approach is measurement speed, since the attainment of $\Delta L = 0$ and then the correct Δz is a "hunt-and-peck" operation, requiring scans over the entire range of the FDS to confirm the absence of oscillations.

A more elegant development in photomixing systems is the addition of fiber-based, phase modulation. This started in 780-nm systems over 10 years ago [144–146] and since has advanced to 1550-nm and been integrated into working FDS systems. Arguably the two most advanced approaches are: (1) an optoelectronic phase modulator [147]; and (2) a fiber stretcher [148]. Both have the effect of yielding an output from the receive photomixer that has a phase equal to φ_{THz}, and an amplitude proportional to E_{THz}. To see how this works, Figure 3.43a shows a block diagram of an FDS system containing a phase modulator that operates on only one laser (ν_1) and only the transmit photomixer. To analyze this, we return

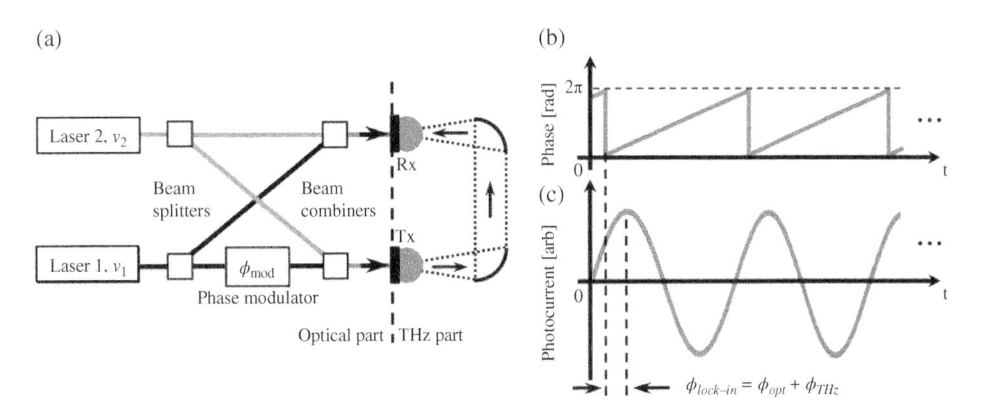

Figure 3.43 (a) Block diagram of 780-nm FDS spectrometer having a single-wavelength phase modulator in the transmit branch of the fiber-optic network. (b) Ideal sawtooth waveform applied as the electrical bias of the phase modulator. (c) Ideal receive photomixer output current when the amplitude is set to special condition described in the text (after Refs. [146, 149]).

to Eq. (3.37) but instead of amplitude modulation, we consider phase modulation of just one of the two lasers, say k_1.

$$i_B(\Delta\nu, t) = DE_{THz,0}(\Delta\nu)G_P \cos\left[\varphi_T + \varphi_{THz} - \varphi_R + \Phi(t)\right] \tag{3.48}$$

Without loss of generality, we can write $\varphi(t) = k_1 L(t)$ where L is the effective modulator optical length.

Now suppose that $L(t)$ has a periodic sawtooth waveform such $L(t) = L_M(-1 + 2t/T_m)$ where T_m is the modulation period, i.e. the inverse of the modulation frequency. If L_M is adjusted so that $k_1 L_M = \varphi_T - \varphi_R = \Delta k(L_T - L_R)$, Eq. (3.48) becomes

$$i(\Delta\nu, t) = DE_{THz,0}(\Delta\nu)G_P \cos\left[\varphi_{THz} + k_1(2L_M/T_M)t\right] \tag{3.49}$$

where $\varphi_{THz} = \Delta k \cdot L_{THz}$. Eq. (3.49) is just a pure cosine wave of frequency $f_m = k_1(2L_M/T_M) = (\nu_1/c)\,(2L_M/T_M)$, and phase φ_{THz}. Figure 3.43b,c shows the ideal sawtooth optical phase modulation waveform, and the corresponding sinusoidal output from the receive photomixer, respectively. Provided f_m is known or can be measured in some way (e.g. electrical spectrum analyzer), simple phase-sensitive lock-in demodulation at f_m will show a phase exactly equal to the THz phase. This remarkable technique is usually credited to [149] and called the *single sample point* (SSP) method.

The SSP method does pose significant practical challenges. First, even if k_1 is from the fixed-frequency laser and k_2 is tuned, φ_{THz} depends on Δk so must be adjusted at each THz frequency over the scan. Second, there is great sensitivity of the THz output to phase and frequency fluctuations in the modulator, in addition to any in the fibers. These arise, for example, from changes in ambient conditions, especially temperature. However, at least one clever scheme has been developed whereby both laser tones can be propagated through the same fiber and modulator with only one tone being modulated, so there is common-mode cancellation of such fluctuations [146, 147]. The basic idea is to couple both wavelengths with orthogonal polarization into the same polarization-maintaining (PM) fiber. Thereby, the output of the first laser is coupled to the slow axis of the PM fiber whereas the output of the second laser is coupled to the fast axis of the same fiber. Afterward, a polarization sensitive, fiber-coupled phase modulator, which is a standard device in optical telecommunications, can be used to selectively modulate one wavelength only. The advantage of this concept is that both wavelengths are guided through the same fiber at all times, such that phase fluctuations due to temperature variations affect both wavelength simultaneously. Hence they have no influence on the overall phase of the optical beat note. The layout of this phase modulation scheme is shown in Figure 3.44a.

In a more recent publication it was demonstrated that this phase modulation scheme in combination with a fast tunable y-branch laser (Modulated Grating Y-branch, Finisar WaveSource™) enabled the acquisition of coherent CW spectra with a bandwidth of more than 2 THz, a peak dynamic range of 60 dB and a frequency resolution of 1 GHz in only 0.04 seconds, as shown in Figure 3.44b [150]. High-resolution scans of water vapor lines between 1.1 and 1.2 THz are shown in Figure 3.44c. To the best of our knowledge, this is the fastest FDS system demonstrated to 2019, the year of composition for this chapter. In addition, it shows that CW THz systems can reach the same measurement rate as their time-domain counterparts without the need of a femtosecond laser and a free space optical delay line.

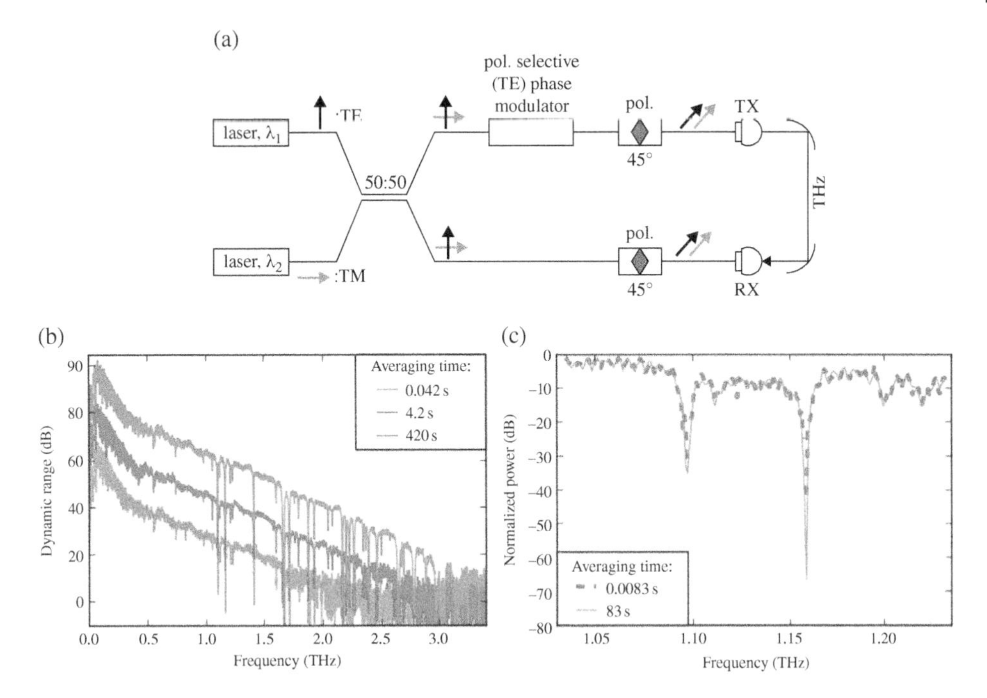

Figure 3.44 (a) Block diagram of 1550-nm fiber-based wavelength-selective phase modulation for THz phase control (after Ref. [147]). (b) Typical THz spectra obtained at a continuous update rate of 24 Hz showing single shot (acquisition time 0.042 seconds), an average of 100 spectra and an average of 10 000 spectra, taking 4.2 seconds and about 7 min, respectively. (c) Normalized spectral power showing the prominent water absorption lines at 1.1 and 1.16 THz obtained at a continuous update rate of 120 Hz. The plot compares a single-shot measurement and averaging of 100 and 10 000 spectra (after Ref. [150]).

Exercises

E3.1 Planck's law of radiation is a triumph of quantum physics and very useful in many fields of science and engineering. (a) Given a 300 K blackbody of unity emissivity, find the frequency in THz (accurate to 100 GHz) and the wavelength in micron where the brightness form of Planck's law is a maximum (can use numerical or analytic techniques). (b) What is this maximum brightness in MKS units? (c) What is the brightness of this same blackbody at 1.0 THz and how much less is this than the maximum?

E3.2 A useful function for practically any electromagnetic sensor is the *integrated* brightness (i.e. integrated over frequency). (a) Derive an expression for the integrated brightness IB in terms of the temperature and the Stefan–Boltzman constant, $\sigma = (2\pi h/15c^2)(\pi k_B/h)^4$. (b) Evaluate IB at $T = 300$ K. (c) Now consider the fractions of IB that lies above and below a certain "cutoff" frequency, $IB(>\nu c) \equiv \int_{\nu c}^{\infty} B(\nu)d\nu$ and $IB(<\nu c) \equiv \int_{0}^{\nu c} B(\nu)d\nu$ (both are very useful in evaluating the "background" radiation for photon and thermal detectors). In the THz region and for terrestrial sources, one can generally apply the Rayleigh–Jeans approximation, $h\nu c/k_B T \ll 1$. In this case, derive an expression for $IB(<\nu c)$, and then the general expression for $IB(>\nu c)$. (d) Evaluate $IB(<\nu c)$ for $\nu c = 1$ THz and $T = 300$ K (suggestion: consult Eisberg and Resnick for the genesis of Planck's law and the Stefan–Boltzman constant).

E3.3 THz radiation filter: (a) Suppose one is using a thermal detector (e.g. Golay cell) that is approximately equally sensitive to IR and THz radiation. Clearly, a low-pass filter is desirable to attenuate (i.e. "stop") the IR but "pass" the THz. For the first approximation, one can assume that the low-pass filter has a power transmission function, $T = |S_{21}|^2 = \tau\theta(\nu - \nu_C) + \theta(\nu_C - \nu)$, where θ is the unit step function and τ is the "average" transmission above ν_C. (a) Assuming that the filter is operated at room temperature but does not emit any radiation toward the detector, and that ν_C satisfies the Rayleigh–Jeans limit, write an expression for the value of τ that makes the incident power on the detector from $\nu > \nu c$ equal to that from $\nu < \nu c$. (b) Evaluate this τ for $\nu c = 1$ THz and $T = 300$, and state the answer as an insertion loss in decibel units. (c) In the stopband region, what is a better type of attenuation for the filter to have: absorptive or scattering? State your answer qualitatively assuming that the filter can be fabricated as a uniform film that completely intercepts the incoming radiation to the detector? (Clue: there are two reasons for the correct answer, one of which involves Kirchoff's law of radiative transfer.)

E3.4 Fluctuations of radiation: the quantum picture. Given the quantum statistics for radiation based on Boltzman and Planck, it is simple to derive the fluctuations from general principles of probability theory.

A) Derive the variance or mean-square fluctuations, $<(\Delta n_K)^2>$ for the radiation modeled as photons having the energy function $U_K = (n_K + ½)\hbar\omega_K$, where n_K is the number of photons in mode K. (Clue: start with the Boltzman (exponential) PDF and utilize the general result for random variables, $<(\Delta n_K)^2> = <(n - <n_K>)^2> = <(n_K)^2> - <n_K>^2$.)

B) Use the above expression to calculate the root mean square fluctuations of incident power in a single spatial mode from the sun within a spectral bandwidth of 1 GHz at the following two center wavelengths: (i) $\lambda = 0.5$ mm (THz region), and (ii) $\lambda = 0.5$ μm (visible region). Express your answer in MKSA and in dBm (decibels relative to 1 mW). (Clues: make the narrow passband approximation in both cases; approximate the sun as having a brightness temperature of 5800 K.)

Exercise: THz Interaction with Matter

E3.5 Vapor state: Molecular rotational transitions. Along with water, the Earth's atmosphere can hold many trace molecules of interest for global warming, and homeland security, many of which have measurable resonances in the THz or millimeter-wave regions.

A) A molecule of interest for the "greenhouse effect" and global warming is nitrous oxide, N_2O. Using the rigid rotor model to estimate the resonance frequencies between 100 GHz and 1.0 THz

B) A molecule of interest for homeland security is phosgene, $COCl_2$ which is a planar molecule having the structure shown in the figure below. Find a formula for the moment of inertia for a planar molecule with respect to the axis perpendicular to the plane, and use it to predict the lowest vibrational resonance frequency.

O
C
174 pm
118 pm
Cl
111.8°
Cl

E3.6 Liquid State: Debye Model. Water, the "liquid of life", has very interesting dielectric properties in the THz region. Assuming freshwater, a temperature of 25 °C, the "single" Debye model, and five frequencies of interest of 10, 100, 300, 1000, and 3000 GHz.

A) Find or calculate the Debye relaxation time and calculate the real and imaginary parts of the dielectric function.

B) Calculate the complex electric-field transmission and reflection coefficients.

C) Calculate the power transmission and reflection coefficient.

D) Calculate the power absorption coefficient, the attenuation length, and the emissivity.

E3.7 Solid state: Although a century old, the Drude model is still the first best estimate for the complex dielectric function of solid-state conductors at RF and THz frequencies.

A) Boron-doped *p*-silicon is probably the most common single-crystal semiconductor substrate in the world, being used pervasively in CMOS fabrication. The resistivity of these substrates typically falls in the range 0.1–10 Ω-cm, and the permittivity is always $\varepsilon_0 = 12$. Is this resistivity high enough for low-loss THz applications? To address this, estimate the complex conductivity at 300 GHz assuming a momentum relaxation time of 0.3 ps for resistivity values of 0.1, 1.0, and 10.0 Ω-cm.

B) Use the result of (a) to estimate the complex dielectric function at 300 GHz for the same three resistivity values.

C) Use the result of (c) to estimate the absorption coefficient [cm^{-1}] and penetration depth for the same three resistivity values.

E3.8 Scattering effects: THz propagation through fog (plus good exercise in probability).

It is well known that microwaves and millimeter waves have far better transmission through clouds and fog than infrared or visible light. But what about THz transmission? Provided that the water particles in the fog are all spherical and much smaller than a wavelength so that the scattering is in the Rayleigh limit, the radar cross section of an individual water droplet can be estimated by the expression derived from effective-medium theory of optics: $\frac{\sigma}{\pi R^2} = 4\left|\frac{\varepsilon_c - 1}{\varepsilon_c + 1}\right|^2 \left(\frac{2\pi R}{\lambda}\right)^4$ where R is the radius and ε_c is the complex dielectric function.

Assume the distribution function of droplets can be approximated as a Rayleigh pdf with the most likely radius being 10 μm and the concentration of fog particles being $50/\text{cm}^3$.

A) Find the variance of the Rayleigh-distributed fog (or cloud).

B) Find the mean value of the radar cross section at 10 GHz, 100 GHz, 300 GHz, and 1.0 THz using the "single-Debye" model for the dielectric function of water.

C) Assuming single-particle scattering only and validity of the Beer law of transmission, estimate the attenuation constant (units of cm^{-1}) and the transmission loss in dB/km at 100 GHz, 300 GHz, and 1.0 THz.

Exercise: Antennas, Links, and Beams

E3.9 Antennas: Metallic horns have long been popular as antennas at mm-wave and THz frequencies. The simplest designs transform a single-mode waveguide to free space through a length of linear taper, as shown in Figure 3.1. The waveguide end of the feedhorn is called the *throat*, while the free-space end is called the *mouth*. If the waveguide is circular, the feedhorn is often conical. The standard types of both pyramidal and conical feedhorns have a smooth taper, and the far-field radiation pattern has sidelobes. If grooves are cut in the taper section of the horn, it is possible to excite a higher-order spatial mode in such a way that the sidelobes are canceled. The far-field pattern can then be represented by a symmetric function of elevation angle θ with a pattern $F(\theta) = \text{sech}\,(a\theta)$, θ in units of rad. Calculate the following for $a = 10$:

A) The FWHM beamwidth β.

B) The pattern solid angle Ω_B exactly (by numerical integration), pencil-beam approximation, and the "ice-cream cone" approximation.

C) The exact antenna directivity D and effective aperture at $f = 300$ GHz.

D) S_{max} at a range of 1 m if the total radiated power is 1 W.

E3.10 Friis' formula for radar: Some missiles rely on monostatic radar for guidance toward specified targets. As such they are often called missile "seekers," and if they include both a transmitter and receiver are described as "fire and forget." The radar is usually located in the nose cone of the missile which, depending on the system, can provide up to ~1 ft diam of physical aperture. For the sake of estimation, let's take antenna and system specs: $f = 94$ GHz (W-band); antenna type = parabolic reflector, 25 cm diam; average transmit power = 1 W; receiver noise power $P_N = -100$ dBm (dB relative to 1 mW); linear transmit polarization.

A) Estimate D and G using $4\pi A/\lambda^2$ and given 4 dB of total loss (reflective and absorptive) between the antenna and the *Tx* and *Rx* electronics.

B) Missile seekers should be able to "acquire" targets up to a range of at least one mile. At this range, what is the smallest radar cross section that such a missile can acquire given a minimum SNR of 10 dB and assuming isotropic and randomly polarized scattering by the target?

C) By fitting the beam to an equivalent Gaussian $\exp(-a\theta^2)$, determine the detection range as a function of angle from 0° to 90° for the minimum RCS in (b).

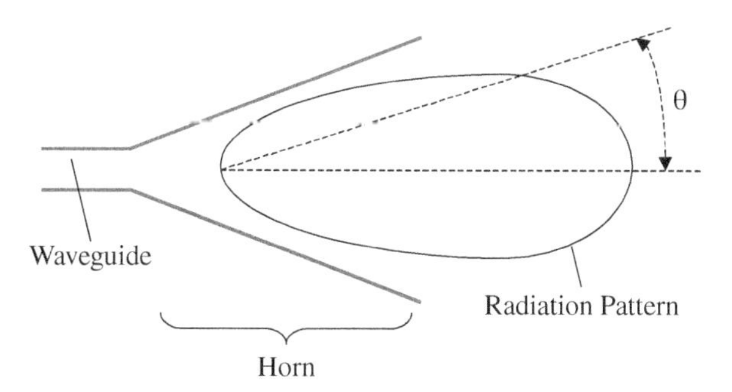

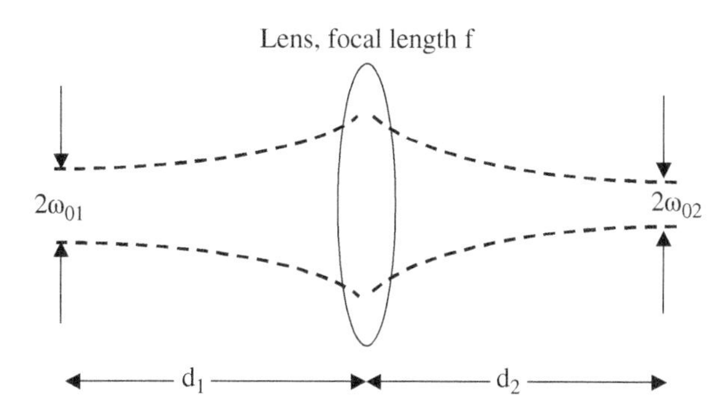

E3.11 Radiometric detection: The sun is sometimes used as a source of radiation in RF sensors because it behaves like a blackbody of brightness temperature 5800 K, which is much hotter than common incandescent thermal sources. Suppose you are designing a single-mode 94-GHz radiometric receiver to utilize the sun as a calibrated source in the upper mm-wave region where hot thermal sources are not available.

A) Derive an expression and evaluate the solid angle subtended by the photosphere of the sun with respect to the earth.

B) Start with the radiometric approximation given in the notes:
where Ω_B is the beam solid angle of the receive antenna, Ω_T is the solid angle subtended by the sun with respect to the earth, and T is the brightness temperature of the sun. Make plots of the radiometric power spectral density [in units of W/GHz] and antenna temperature (power spectral density divided by k_B) vs antenna directivity [in units of dB] assuming perfect pointing between the antenna and the sun. What is the directivity beyond which the spectral density and antenna temperature saturate according to this model?

C) Assume the cost of your receive antenna scales linearly with the aperture diameter as $1 K/m. You are tasked to find the area that maximizes the antenna-temperature-to-cost ratio. Plot this ratio vs directivity from 10 to 60 dB. What directivity maximizes this ratio?

E3.12 Gaussian beams: One of the useful applications of the TEM_{00} Gaussian-beam formalism is in predicting the propagation of radiation through optical components, such as lenses. A common scenario is the lens-focusing problem shown in Figure 3.2. A TEM_{00} Gaussian beam having beam waist ω_{01} is transformed by a thin dielectric lens of focal length f. The distance between the waist location and the lens location is $z = d_1$.

A) Find the condition on f in terms of λ, ω_{01}, and d_1, which guarantees that a second beam waist ω_{02} occurs somewhere to the right side of the thin lens.
B) Using your favorite graphical tool, plot the distance d_2 vs f for $\nu = 1$ THz, $\omega_{01} = 2$ mm, and $d_1 = 20$ cm. How many values of f can create a given d_2?
C) Solve for the possible values of f when $d_2 = 10$ cm? Of the possible values, which is the most practical one, and why?
D) A practical question, often faced in mm-wave and THz experiments and systems, is how to form the smallest possible ω_{02} when ω_{01} is given (i.e. you already have a source), f is given (i.e. you already have a lens) but d_1 and d_2 are variable. Start by finding an expression for ω_{02} in terms of ω_{01}, d_1, d_2, and f. Qualitatively, what does this solution suggest for the best values of d_1 and d_2 relative to f? According to this solution, how small can ω_{01} be?
E) In practice, d_1 cannot be made arbitrarily large because of "spillover" of the source Gaussian beam at the lens. And d_2 cannot be made arbitrarily large because of limits on the size of the experiment or instrument. This leads to the common "constrained" problem whereby $d_1 + d_2$ is limited to a maximum value of D. Fortunately, the relatively long wavelengths in the THz region (compared to the visible or infrared) allow one to fabricate lenses rather easily, so the choice of f is great. Using elementary calculus, determine the values of d_1 and d_2 in terms of f and D that minimize ω_{02}. And write the expression for the minimum ω_{02}.

Exercise: Planar Antennas

E3.13 Design of a bow-tie antenna emitter.

A 10 x 10 μm photomixing source is placed in the gap of a bow-tie antenna. Both are fabricated in the center of a rectangular electrically thick InP slab ($\varepsilon_r = 9.6$) of dimensions 1 x 1 mm. The bow-tie antenna has a width and a length of 227 μm. By running a full-wave simulation its impedance is found to be $Z_{in} = 62.56 - j9.08\ \Omega$ at $f_0 = 500$ GHz. The photomixer has a parasitic capacitance of $C = 2$ fF, and a conductance $G = (10\ \text{k}\Omega)^{-1}$. The following figure shows the antenna emitter and the RF choke:

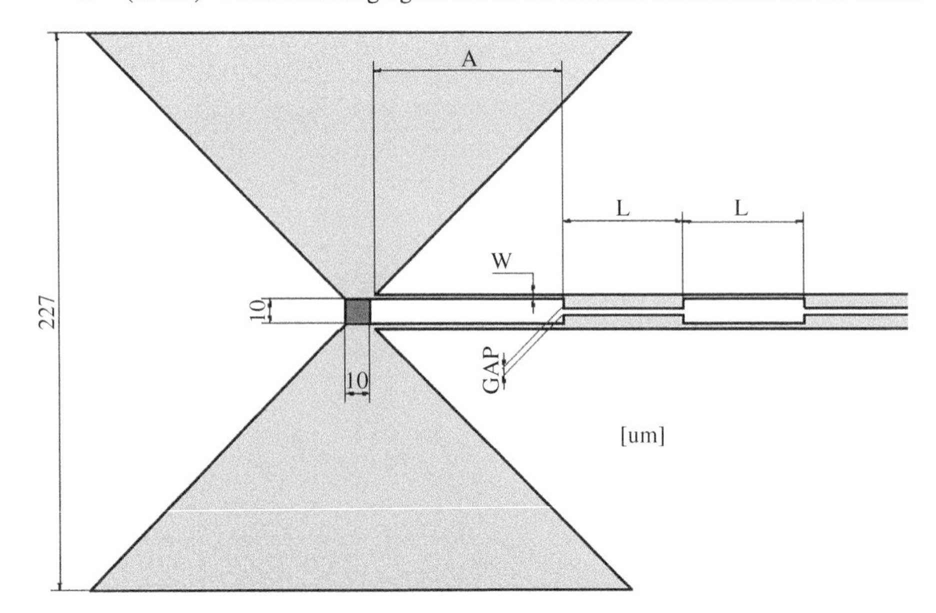

A) In order to maximize the RF rejection at f_0, what is the optimal electrical length L? How many sections would have to be used in the stepped-impedance low pass filter?

B) Calculate A for compensating the imaginary parts of both antenna and photomixer. If required, assume that high-Z sections have a $Z_0 = 70\,\Omega$ and low-Z sections a $Z_0 = 55\,\Omega$.

C) According to the Babinet's principle, is the simulated input impedance of the antenna a reasonable value?

E3.14 Radiation efficiency of a terahertz antenna lying on a high permittivity substrate (I).

A printed dipole is fed with a continuous-wave n–i–p–n–i–p photomixer source on a InP wafer of relative permittivity $\varepsilon_r = 9.6$. Since only the power radiated into the air for $Z > 0$ (see the sketch below) is useful for the intended application, the following measurement of efficiency η is defined by: $\eta = P_{Z>0}/P_{RAD}$, where P_{RAD} is the total radiated power and $P_{Z>0}$ is the useable power. The following picture illustrates the topology of the problem, on which the dipole is printed along the X-axis:

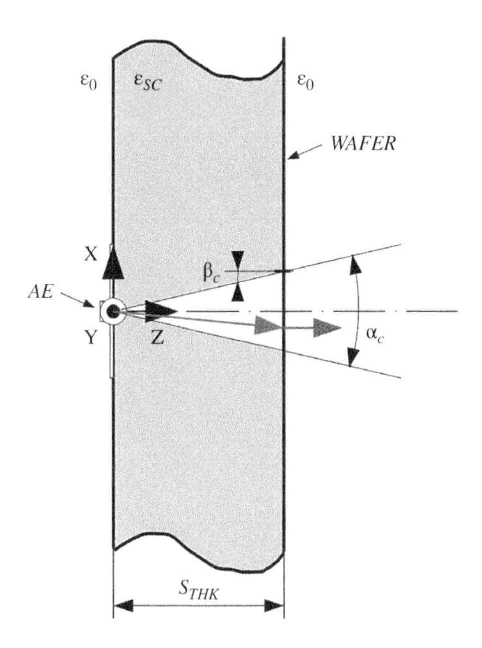

The radiation intensity $U(\theta, \Phi)$ can be calculated when embedded in an infinite medium as follows:

$$U(\theta,\Phi) = U_p(\theta,\Phi) + U_s(\theta,\Phi)$$

$$U_p(\theta,\Phi) = U_0 \cdot [\sin(\alpha)\cos(\theta) + \sin(\alpha)\cos(\theta)\cos(\Phi)]^2$$

$$U_s(\theta,\Phi) = U_0 \cdot \sin^2(\alpha)\sin^2(\Phi)$$

Where U_0 is the radiation intensity amplitude, α is the angle between the dipole and the z-axis, and $U_0(\theta, \Phi)$ and $U_s(\theta, \Phi)$ are the s and p polarization components.

A) Estimate an upper bound on the efficiency η by only taking into account the TIR in the dielectric slab.

E3.15 Radiation efficiency of a terahertz antenna lying in a high permittivity substrate (II).

In Ref. [151] W. Lukosz gives an analytic description of the power emission of dipoles embedded in infinite dielectric slabs. As shown in the following figure, part of the emitted power is totally internal reflected (TIR, red) and has no contribution to the radiated fields. The rest of the power contributes to the radiation pattern by the transmission through the semiconductor–air interface and successive internal reflections between the two media interfaces.

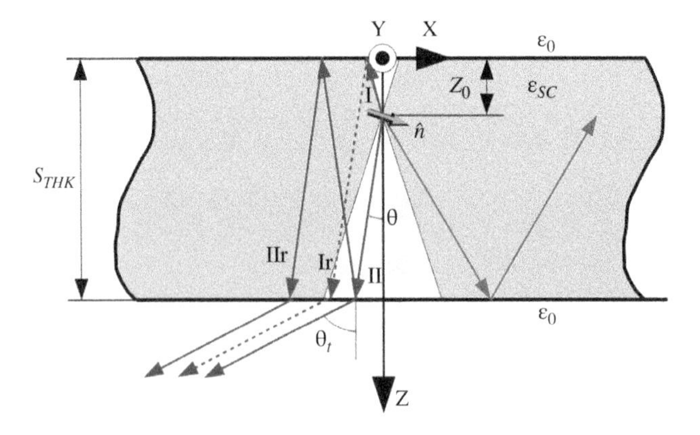

S_{THK} is the thickness of the slab, and $Z_0 = 0$ for a printed dipole parallel to x-axis and lying in the interface at $Z = 0$. According to https://en.wikipedia.org/wiki/Terahertz_radiation, the radiated intensity $U(\theta, \Phi)$ in both s- and p-polarization components can be calculated as follows:

$$U(\theta, \Phi) = U_p(\theta, \Phi) + U_s(\theta, \Phi)$$

$$U_{Z>0,p}(\theta_t, \Phi) = Q_p(\theta, \Phi) M_p(\theta) \hat{T}_p(\theta_t)$$

$$U_{Z>0,s}(\theta_t, \Phi) = Q_s(\theta, \Phi) M_s(\theta) \hat{T}_s(\theta_t)$$

Where $\theta_t = \arcsin\left(\sqrt{\varepsilon_r} \sin(\theta)\right)$ according to Snell's law, $Q_s(\theta, \Phi)$ and $Q_p(\theta, \Phi)$ are the antenna radiation pattern in the substrate when considering both slab interfaces:

$$Q_p(\theta, \Phi) = U_0 \cdot \left[\cos^2(\theta) \cos^2(\Phi) W_{-,p}(\theta)\right]$$

$$Q_s(\theta, \Phi) = U_0 \cdot \left[\sin^2(\theta) \sin^2(\Phi) W_{+,s}(\theta)\right]$$

$W_{-,p}(\theta)$ and $W_{+,s}(\theta)$ model the effect of the $Z = 0$ interface in the dipole radiation pattern:

$$W_{-,p}(\theta) = 1 + |r_p(\theta)|^2 - 2|r_p(\theta)| \cdot \cos(\arg\{r_p(\theta)\})$$
$$W_{+,s}(\theta) = 1 + |r_s(\theta)|^2 - 2|r_s(\theta)| \cdot \cos(\arg\{r_s(\theta)\})$$

And $r_p(\theta)$ and $r_s(\theta)$ is the reflection coefficient for both *s*- and *p*-components defined by:

$$r_p(\theta) = \frac{\cos(\theta) - \sqrt{\varepsilon_{SC}} \cdot \cos(\theta_t)}{\cos(\theta) + \sqrt{\varepsilon_{SC}} \cdot \cos(\theta_t)}$$

$$r_s(\theta) = \frac{\sqrt{\varepsilon_{SC}} \cdot \cos(\theta) - \cos(\theta_t)}{\sqrt{\varepsilon_{SC}} \cdot \cos(\theta) + \cos(\theta_t)}$$

θ_t and θ are related by the Snell's law:

$$\theta_t = a\sin\left(\sqrt{\varepsilon_{SC}} \cdot \sin(\theta)\right)$$

$M_s(\theta)$ and $M_p(\theta)$ describe the influence of successive reflections between the slab interfaces (solid blue ray II_r in the figure):

$$M_{p,s}(\theta) = \left|1 - r_{p,s}^2(\theta) \cdot e^{2\pi i \frac{\sqrt{\varepsilon_{SC}}}{\lambda_0} S_{THK} \cdot \cos(\theta)}\right|^{-2}$$

And $\hat{T}_s(\theta_t)$ and $\hat{T}_p(\theta_t)$ describe the transmittances in the boundary between the semiconductor and the air.

$$\hat{T}_{p,s}(\theta_t) = t_{p,s}(\theta_t) \cdot \frac{\cos(\theta_t)}{\sqrt{\varepsilon_{SC}} \cdot \cos(\theta)}$$

With $t_p(\theta_t)$ and $t_s(\theta_t)$ the transmission coefficient for both *s*- and *p*-components:

$$t_p(\theta) = \frac{4 \cdot \cos(\theta)\cos(\theta_t)}{\left(\cos(\theta) + \sqrt{\varepsilon_{SC}} \cdot \cos(\theta_t)\right)^2}$$

$$t_s(\theta) = \frac{4 \cdot \cos(\theta)\cos(\theta_t)}{\left(\sqrt{\varepsilon_{SC}} \cdot \cos(\theta) + \cos(\theta_t)\right)^2}$$

A) Assuming the same conditions as in Problem 14 (InP wafer of relative permittivity $\varepsilon_r = 9.6$), write a program (e.g. MATLAB®) that determines the maximum and the minimum expected efficiency η, defined as the total radiated power P_{RAD} divided by the power radiated into the free space for $Z > 0$, $P_{z>0}$ when considering substrates having thickness of 0.5 to $5 \cdot \lambda_{SC}$.

Exercise: Device Noise, System Noise, and Dynamic Range

E3.16 (Device noise) So many devices or parts of devices in RF and THz electronics display the canonical junction rectification I–V curve, $I = I_0$ [exp(eV/k_BT) – 1]. Suppose the device is a one-port that displays generalized Johnson–Nyquist noise and full shot noise and operates at 300 K into a load resistance R_L.

A) For $R_L = 0$ (short-circuit case) and zero bias, what is the RMS current fluctuation through the device in terms of I_0? Interpret the answer in terms of the necessary device physics behind I_0

B) If $R_L = 0$ (short-circuit case), at what bias voltage does the RMS shot-noise current equal the RMS thermal noise current?

C) Suppose this device is connected to an R_L = 50-Ω load. At what bias voltage does the noise power generated in the load from the rectifier equal the noise power from the load itself?

D) Express the noise power in (c) as a power spectral density of current fluctuations.

E3.17 (System noise figure) Every linear component in a receiver chain can be represented by a noise figure, $F_i = (SNR_{in}/SNR_{out})i = 1 + (N_{add})_i/(G_i\ k_BT_{300}\ B_i)$, where $(N_{add})_i$ is the noise added by the ith component, G_i is its gain, and B_i is its equivalent noise bandwidth.

Use these definitions to derive Friis' formula for total noise figure, F_T.

A) Suppose a receiver chain consists of a bandpass filter (BPF) having $G = -3.0$ dB and $F = 0.5$ dB, an LNA having $G = Y$ dB and $F = 2.0$ dB, a mixer having $G = -5$ dB and $NF = 6$ dB, and a power amp having $G = 30$ dB and $F = 10$ dB. What is the total noise figure if $Y = 30$ dB? (Clue: easy and fun with an Excel spreadsheet.)

B) In the limit of very large Y, the noise figure approaches asymptotically a "front-end" limited value. What is this value? And to what value can Y be reduced so that the actual F is within 1 dB of the "front-end" limit?

C) Suppose the BPF and LNA are interchanged. What is the new F_T if $Y = 30$ dB?

E3.18 (Linear dynamic range) This is an important metric in all RF sensors, defined as the difference in input power between the Rx noise power and the level that makes the receiver output compress by 1 dB (from nonlinearities in the electronics). Suppose a receiver chain consists of a BPF having $G = -3.0$ dB and $F = 0.5$ dB, an LNA having $G = 30$ dB and $F = 2.0$ dB, a mixer having $G = -5$ dB and $NF = 6$ dB, and a power amp having $G = 30$ dB and $F = 10$ dB.

A) If the LNA has the transfer function $P_{out} = G_0P_{in}/[1 + (G_0P_{in}/P_{sat})]$ where G_0 is the small-signal gain and all other components are assumed to be perfectly linear, what values of P_{in} set the 3- and 1-dB compression points of the receiver when expressed in terms of P_{sat} and G_0? (Note this is a useful gain-compression characteristic in many first-pass system calculations.) Evaluate these P_{in} values for $P_{sat} = +20$ dBm and $G_0 = 30$ dB.

B) Make a plot of P_{out} vs P_{in} for the receiver on a log–log axis showing clearly the noise floor and 1-dB compression level if the bandwidth is assumed to be 1 MHz.

C) From this plot, or from the calculations leading to the plot, evaluate the linear dynamic range of this receiver.

E3.19 (Power-law detection) A common detector type used in sensors, particularly at low signal levels, is the square-law device represented by $X_{out}(t) = A[X_{in}(t)]^2 \equiv \Re P_{in}$.

A) Of all the possible power-law detectors $X_{out} = A(X_{in})^n$, show that the square-law is the most effective at maximizing the ratio of baseband (dc) power to input (THz) power.

B) Assume a nonlinear diode is biased into a region where the I–V curve can be approximated to have the form $I = BV^2$. Use Taylor's theorem to write a general expression for the rectified current $\delta I = I – I_0$ in terms of the voltage difference δV induced at some bias point V_0. Re-express this result as $\delta I = \Re P$, where $P = (\delta V)^2/R$ and $\Re$ is the short-circuit responsivity. Write an expression for $\Re$ in terms of known derivatives.

C) One of the most useful devices through the history of THz sensors has been the metal–semiconductor, or Schottky, diode. A convenient model I–V curve for Schottky diodes is $I = I_S\{\exp(eV/nk_BT) - 1\}$, where n ("ideality" factor) > 1. Formulate the square-law short-circuit $\Re$ for a Schottky diode. Evaluate this responsivity at 300 K assuming $n = 1.0$—the best possible case.

Exercise: Ultrafast Photoconductivity and Photodiodes

E3.20 Photoconductivity in normal GaAs:

A) Use the Shockley–Hall–Read (SHR) formalism to find the steady-state electron and hole concentrations in a GaAs sample having $10^{16}/\text{cm}^3$ (fully ionized donors), $10^{15}/\text{cm}^3$ defect levels at $U_t = U_i$ with $\sigma_n = \sigma_p = 10^{-14}$ cm^2 and in which an outside laser source is generating $10^{21}/\text{cm}^3$ electron–hole pairs uniformly in the sample.

B) Suppose under the conditions in (a) that the laser is turned off. Find an expression for the hole density vs time and evaluate the excess carrier lifetime. (Clues: assume that $v_{th} = 1 \times 10^7$ cm/s, J and $E = 0$, and carrier concentrations are uniform; $n_i = 9 \times 10^6/\text{cm}^3$ in GaAs at 300 K.)

E3.21 Ultrafast GaAs photoconductivity. One of the very useful applications of the SHR formalism is in predicting the effect of deep levels on the recombination lifetime in ultrafast photoconductors. Breakthroughs in THz PC devices, for example, have occurred because of ErAs:GaAs: a composite semiconductor in which nanoparticles of semi-metallic ErAs permeate a GaAs matrix.

A) Suppose the ErAs is deposited 1% by volume into the GaAs lattice and all of it occurs as spherical nanoparticles having a mean diameter of 5 nm. What is the mean concentration of nanoparticles per unit volume, N_{ErAs}? (Although not true, the spherical particle assumption is very useful in illustrating some of the key physical issues surrounding the SHR formalism.) What is the mean geometric cross section of the nanoparticles? What is the thermal velocity $v_{th} = (3k_BT/m^*)^{1/2}$ at $T = 300$ K? What is the associated kinetic capture time, $t_c \equiv (Nv_{th}\sigma_c)^{-1}$ where N is the mean concentration and σ_e is the cross section?

B) Now suppose the deposition fraction of ErAs remains the same but the mean diameter is decreased to 2 nm. What are the new concentration of nanoparticles, the capture cross section, and the kinetic capture times? Contrast to the result in (a).

C) Now we will apply the Shockley–Read–Hall formalism to ErAs:GaAs. Start by deriving the famous expression for the net recombination rate U $(=R_n - G_n = R_p - G_p)$ in terms of the density of nanoparticles, the thermal velocity v_{th}, the density of electrons n, the density of holes p, the intrinsic carrier concentration n_i, the electron and hole capture cross-sections (σ_n and σ_p), the nanoparticle defect level (U_t), and the intrinsic energy level (U_i).

D) Now suppose the ErAs:GaAs photoconductor at 300 K is pumped by a laser having photon energy just exceeding the band-gap energy in GaAs and intensity low enough to preserve space-charge neutrality and not disturb the background free-carrier concentration. Use this SHR formalism to find an expression for the net recombination rate if the GaAs is weakly n-type with $10^{15}/\text{cm}^3$ fully ionized shallow donors, and assuming $U_t = U_i$.

E) Now suppose the laser is generating $10^{21}/\text{cm}^3$ electron–hole pairs uniformly in the sample (again, a simplification, but one that leads to the key insights of this problem with far less work). Find the steady-state electron and hole densities assuming a capture time equal to that calculated in part (a) above.

F) Now assume the same n-type material but suppose the ErAs nanoparticle level can rise above $U_t = U_i$ toward the conduction-band edge. According to the SHR formalism, how much above U_i must the nanoparticle level rise to reduce the net recombination rate by a factor of two? [Clue: to make this one easier, assume U_i lies exactly at the middle of the GaAs band gap.] Where should U_t lie to make the effective recombination go down 1000 times, corresponding to an increase in the recombination time to the order of 1 ns?

E3.22 Femtosecond MLLs.

A) A useful model for the individual pulses of modern solid-state MLLs, including fiber lasers, is the instantaneous pulse power $P(t) = P_p \operatorname{sech}^2(t/t_0)$, where P_p is the peak power and t_0 is a characteristic time. Find the full width at half-maximum Δt in terms of t_0.

B) Use the $P(t)$ expression and integration tables to find an expression for the single pulse energy E_p as a function of P_p and Δt.

C) Use the above results to find an expression for the average optical power P_{ave} in terms of E_P and the pulse repetition frequency f_P, assuming $f_P \cdot \Delta t \ll 1$.

Explanatory Notes (see superscripts in text)

1 https://en.wikipedia.org/wiki/Terahertz_radiation
2 https://networkel.com/network-fiber-optic-cable-fundamentals/
3 http://photonicswiki.org/index.php?title=Dispersion_and_Attenuation_Phenomena
4 https://www.ozoptics.com/ALLNEW_PDF/DTS0079.pdf
5 https://www.rp-photonics.com/spotlight_2007_09_01.html
6 There is a sign correction in this expression for the third term published in Bowers and Burrus.
7 http://www.ioffe.ru/SVA/NSM/Semicond/GaInAs/basic.html
8 E.R. Brown, unpublished.
9 Heterobarriers increase the thermal resistance since they generally have different phonon, and therefore, thermal transport behavior than the surrounding semiconductor material.
10 http://www.ioffe.ru/SVA/NSM/Semicond/InP/basic.html
11 https://www.bipm.org/
12 https://en.wikipedia.org/wiki/International_System_of_Units
13 https://www.bipm.org/en/publications/guides/gum.html
14 http://www.terahertz.co.uk/tk-instruments/products/absolute-thz-power-energy-meters
15 https://en.wikipedia.org/wiki/Polyvinylidene_fluoride
16 Fabricated at the Heinrich Hertz Institute, Berlin, Germany.
17 www.terahertz.co.uk
18 http://www.terahertz.co.uk/tk-instruments/products/absolute-thz-power-energy-meters
19 http://vadiodes.com/en/products-6/power-meters-erickson
20 Measured by QMCI, Cardiff, UK.
21 Sometimes called "agricultural grade" BPE
22 https://insulationcorp.com/eps/
23 http://www.terahertz.co.uk/qmc-instruments-ltd/thz-optical-components/multi-mesh-filters
24 http://www.radiabeam.com/products/THz_Products.html
25 http://www.ioffe.ru/SVA/NSM/Semicond/InP/basic.html
26 http://www.ioffe.ru/SVA/NSM/Semicond/Si/basic.html
27 Popovic and Popovic, or one of the many books specializing in Antenna Engineering
28 This assumes, of course, that there is no significant losses of any type.
29 B. Globisch, private communication.
30 Unfortunately, the PTB sensor and cross-calibrated QMC pyroelectric were not available at the time of these measurements.
31 Cisco Systems Inc. 'The Zettabyte Era: Trends and Analysis' White paper (San Jose, 2016).
32 www.toptica.com
33 www.menlosystems.com
34 www.tetechs.com
35 Like *electrical* length commonly used in microwave theory, *optical* length at a given frequency is defined by $L_O = L \cdot (c/v)$, where L is the physical length and v is the wave velocity in the medium.

References

1 Rubens, H. and Snow, B.W. (1893). On the refraction of rays of great wavelength in rock salt, sylvine, and fluorite. *Philosophical Magazine* **35**: 35.

2 Rubens, H. and Kurlbaum, F. (1903). On the heat radiation of long wave-length emitted by black bodies at different temperatures. *Astrophysical Journal* **14**: 335–347.

3 Auston, D.H. (1975). Picosecond optoelectronic switching and gating in silicon. *Applied Physics Letters* **26** (3): 101–103.

4 Mourou, G., Stancampiano, C.V., Antonetti, A., and Orszag, A. (1981). Picosecond microwave pulses generated with a subpicosecond laser-driven semiconductor switch. *Applied Physics Letters* **39** (4): 295–296.

5 Auston, D.H., Cheung, K.P., and Smith, P.R. (1984). Picosecond photoconducting Hertzian dipoles. *Applied Physics Letters* **45** (3): 284–286.

6 Brown, E.R., McIntosh, K.A., Nichols, K.B., and Dennis, C.L. (1995). Photomixing up to 3.8 THz in low-temperature-grown GaAs. *Applied Physics Letters* **66**: 285.

7 Mittleman, D.M. and Cheville, R.A. (2008). Terahertz generation and applications. *Ultrafast Optical Textbox*: 1–58.

8 Armstrong, C.M. (2012). *The Truth about THz*. IEEE Spectrum.

9 Preu, S., Dohler, G.H., Malzar, S. et al. (2011). Tunable, continuous-wave terahertz photomixer sources and applications. *Journal of Applied Physics* **109**: 061301.

10 Sartorius, B., Stanze, D., Gobel, T. et al. (2012). Continuous wave terahertz systems based on 1.5 μm telecom technologies. *Journal of Infrared Millimeter and Terazhertz Waves* **33**: 405–417.

11 Test report for MenloSystems T-Light EDFA MLL laser, Serial# T168, purchased by Wright State University.

12 Keck, F.P.K.D.B. and Maurer, R.D. (1970). Radiation losses in glass optical waveguides. *Applied Physics Letter* **17**: 423–425.

13 Payne, J.M. and Shillue, W. (2002). Photonic techniques for local oscillator generation and distribution in millimeter-wave radio astronomy. 2002 International Topical Meeting on Microwave Photonics, Awaji, Japan, 9–12.

14 Aspnes, D.E. and Studna, A.A. (1983). Dielectric functions and optical parameters of Si, Ge, GaP, GaAs, GaSb, InP, InAs, and InSb from 1.5 to 6.0 eV. *Physical Review B* **27**: 985–1009.

15 Adachi, S. (1989). Optical dispersion relations for GaP, GaAs, GaSb, InP, InAs, InSb, AlXGa1-XAs, and In1-XGaXAsYP1-Y. *Journal of Applied Physics* **66**: 6030–6040.

16 Gupta, S., Whitaker, J.F., and Mourou, G.A. (1992). Ultrafast carrier dynamics in III-V semiconductors grown by molecular-beam epitaxy at very low substrate temperatures. *IEEE Journal of Quantum Electronics* **28** (10): 2464–2472.

17 Melloch, M.R., Woodall, J.M., Harmon, E.S. et al. (1995). Low-temperature grown III-V materials. *Annual Review of Materials Science* **25**: 547–600.

18 Kaminska, M., Weber, E.R., Liliental-Weber, Z., and Leon, R. (1989). Stoichiometry-related defects in GaAs grown by molecular-beam epitaxy at low temperatures. *Journal of Vacuum Science & Technology B: Microelectronics and Nanometer Structures* **7** (4): 710.

19 Feenstra, R.M., Woodall, J.M., and Pettit, G.D. (1993). Observation of bulk defects by scanning tunneling microscopy and spectroscopy: arsenic antisite defects in GaAs. *Physical Review Letters* **71** (8): 1176–1179.

20 Look, D.C., Walters, D.C., Manasreh, M.O. et al. (1990). Anomalous Hall-effect results in low-temperature molecular-beam-epitaxial GaAs: hopping in a dense EL2-like band. *Physical Review B* **42** (6): 3578–3581.

21 Bliss, D.E., Walukiewicz, W., Ager, J.W. et al. (1992). Annealing studies of low-temperature-grown GaAs:Be. *Journal of Applied Physics* **71** (4): 1699–1707.

22 Staab, T.E.M., Nieminen, R.M., Luysberg, M., and Frauenheim, T. (2005). Agglomeration of As antisites in as-rich low-temperature GaAs: nucleation without a critical nucleus size. *Physical Review Letters* **95**: 125502.

23 Gregory, S., Tey, C.M., Cullis, A.G. et al. (2006). Two-trap model for carrier lifetime and resistivity behavior in partially annealed GaAs grown at low temperature. *Physical Review B* **73**: 195201.

24 Shklovskii, B.I. and Efros, A.L. (1984). Electronic properties of doped semiconductors. In: *Electronic Properties of Doped Semiconductors*, vol. **45** (ed. M. Cardona), 72–93. Berlin: Springer.

25 Warren, C., Woodall, J.M., Kirchner, P.D. et al. (1992). Role of excess As in Low-temperature-grown GaAs. *Physical Review B* **46** (8): 4617–4620.

26 Bliss, D.E., Walukiewicz, W., and Haller, E.E. (1993). Annealing of AsGa-related defects in LT-GaAs: the role of gallium vacancies. *Journal of Electronic Materials* **22** (12): 1401–1404.

27 Bourgoin, J.C., Khirouni, K., and Stellmacher, M. (1998). The behavior of As precipitates in low-temperature-grown GaAs. *Applied Physics Letters* **72** (4): 442–444.

28 Krotkus, A., Bertulis, K., Dapkus, L. et al. (1999). Ultrafast carrier trapping in Be-doped low-temperature-grown GaAs. *Applied Physics Letters* **75** (21): 3336–3338.

29 Gebauer, J., Zhao, R., Specht, P., and Weber, E.R. (2001). Does beryllium doping suppress the formation of Ga vacancies in nonstoichiometric GaAs layers grown at low temperatures? *Applied Physics Letters* **79**: 4313–4315.

30 Lochtefeld, J., Melloch, M.R., Chang, J.C.P., and Harmon, E.S. (1996). The role of point defects and arsenic precipitates in carrier trapping and recombination in low-temperature grown GaAs. *Applied Physics Letters* **69**: 1465–1467.

31 Luysberg, M., Sohn, H., Prasad, A. et al. (1998). Effects of the growth temperature and As/Ga flux ratio on the incorporation of excess As into low temperature grown GaAs. *Journal of Applied Physics* **83**: 561–566.

32 Krotkus, A. (2010). Semiconductors for terahertz photonics applications. *Journal of Physics D: Applied Physics* **43**: 273001.

33 Künzel, H., Böttcher, J., Gibis, R., and Urmann, G. (1992). Material properties of Ga0.47In0.53As grown on InP by low-temperature molecular beam epitaxy. *Applied Physics Letters* **61**: 1347–1349.

34 Grandidier, B., Chen, H., Feenstra, R.M. et al. (1999). Scanning tunneling microscopy and spectroscopy of arsenic antisites in low temperature grown InGaAs. *Applied Physics Letters* **74** (10): 1439–1441.

35 Kuenzel, H., Biermann, K., Nickel, D., and Elsaesser, T. (2001). Low-temperature MBE growth and characteristics of InP-based AlInAs/GaInAs MQW structures. *Journal of Crystal Growth* **227–228**: 284–288.

36 Sartorius, B., Roehle, H., Künzel, H. et al. (2008). All-fiber terahertz time-domain spectrometer operating at 1.5 microm telecom wavelengths. *Optics Express* **16** (13): 9565–9570.

37 Takazato, A., Kamakura, M., Matsui, T. et al. (2007). Detection of terahertz waves using low-temperature-grown InGaAs with 1.56 μm pulse excitation. *Applied Physics Letters* **90**: 101119.
38 Takahashi, R., Kawamura, Y., Kagawa, T., and Iwamura, H. (1994). Ultrafast 1.55-μm photoresponses in low-temperature-grown InGaAs/InAlAs quantum wells. *Applied Physics Letters* **65** (14): 1790–1792.
39 Globisch, B., Dietz, R.J.B., Stanze, D. et al. (2014). Carrier dynamics in Beryllium doped low-temperature-grown InGaAs/InAlAs. *Applied Physics Letters* **104**: 172103.
40 Globisch, B., Dietz, R.J.B., Nellen, S. et al. (2016). Terahertz detectors from Be-doped low-temperature grown InGaAs/InAlAs: interplay of annealing and terahertz performance. *AIP Advances* **6**: 125011.
41 Loata, G.C., Thomson, M.D., Löffler, T., and Roskos, H.G. (2007). Radiation field screening in photoconductive antennae studied via pulsed terahertz emission spectroscopy. *Applied Physics Letters* **91** (23).
42 Jepsen, P.U., Jacobsen, R.H., and Keiding, S.R. (1996). Generation and detection of terahertz pulses from biased semiconductor antennas. *Journal of the Optical Society of America B: Optical Physics* **13** (11): 2424–2436.
43 Tani, M., Herrmann, M., and Sakai, K. (2002). Generation and detection of THz pulsed radiation with photoconductive antennas. *Measurement Science and Technology* **13**: 1739–1745.
44 Shan, J. and Heinz, T.F. (2004). Terahertz radiation from semiconductors. In: *Ultrafast Dynamical Processes in Semiconductors*, Topics Appl. Phys., vol. **92** (ed. K.-T. Tsen), 1–59. Berlin, Heidelberg: Springer.
45 Sakai, K. (ed.) (2005). *Terahertz Optoelectronics*. Berlin, Heidelberg: Springer.
46 Jepsen, P.U., Cooke, D.G., and Koch, M. (2011). Terahertz spectroscopy and imaging – modern techniques and applications. *Laser & Photonics Reviews* **5** (1): 124–166.
47 Jepsen, P.U. and Keiding, S.R. (1995). Radiation patterns from lens-coupled terahertz antennas. *Optics Letters* **20** (8): 807–809.
48 Benjamin, S.D., Loka, H.S., Othonos, A., and Smith, P.W.E. (1996). Ultrafast dynamics of nonlinear absorption in low-temperature-grown GaAs. *Applied Physics Letters* **68**: 2544–2546.
49 Tani, M., Matsuura, S., Sakai, K., and Nakashima, S. (1997). Emission characteristics of photoconductive antennas based on low-temperature-grown GaAs and semi-insulating GaAs. *Applied Optics* **36** (30): 7853–7859.
50 Jacobsen, R.H., Birkelund, K., Holst, T. et al. (1996). Interpretation of photocurrent correlation measurements used for ultrafast photoconductive switch characterization. *Journal of Applied Physics* **79** (5): 2649–2657.
51 Bowers, J.E. and Burrus, C.A. Jr. (1987). Ultrawide-band long-wavelength p-i-n photodetectors. *Journal of Lightwave Technology* **LT-5**: 1339–1350.
52 Sze, S. (1981). *Physics of Semiconductor Devices*, 2e. New York, NY: Wiley Chapter 10.
53 Soole, J.B.D. and Schumacher, H. (1990). Transit-time limited frequency response of InGaAs MSM photodetectors. *IEEE Transactions on Electron Devices* **37** (11): 2285.
54 Gupta, K.C., Garg, R., and Bahl, I.J. (1979). *Microstrip Lines and Slotlines*. Dedham, MA: Artech House Sec. 7.3.1.
55 Adachi, S. (1983). *Journal of Applied Physics* **54** (4): 1844–1848.

56 Göbel, T., Schoenherr, D., Feiginov, M. et al. (2011). Reliability investigation of photoconductive continuous-wave terahertz emitters. *IEEE Transactions on Microwave Theory and Techniques* **59**: 2001–2007.
57 Verghese, S., McIntosh, K.A., and Brown, E.R. (1997). Optical and terahertz power limits in the low-temperature-grown GaAs photomixers. *Applied Physics Letters* **71**: 2743–2745.
58 Zielinski, E., Schweizer, H., Streubel, K. et al. (1986). Excitonic transitions and exciton damping processes in InGaAs/InP. *Journal of Applied Physics* **59**: 2196.
59 Blakemore, J.S. (1982). Semiconducting and other major properties of gallium arsenide. *Journal of Applied Physics* **53** (10): R123–R181.
60 Brown, E.R. (2003). Terahertz generation by photomixing in ultrafast photoconductors. In: *Electronic Devices and Advanced Systems Technology*, Terahertz Sensing Technology, vol. **1** (eds. D.L. Woolard, W.R. Loerop and M.S. Shur), 147–195. Singapore: World Scientific.
61 Paschotta, R. (2008). Sech2-Shaped Pulses. In Encyclopedia of Laser Physics and Technology, 1st Ed., October 2008. Wiley-VCH.
62 Suen, J.Y., Li, W., Taylor, Z.D., and Brown, E.R. (2010). Characterization and modeling of a THz photoconductive switch. *Applied Physics Letters* **96**: 141103.
63 Naftaly, M. (2015). *Terahertz Metrology*. Artech House.
64 Popović, Z. and Grossman, E.N. (2011). Terahertz Metrology and Instrumentation. *IEEE Transactions on THz Science and Technology* **1**: 133.
65 Lehman, J.H., Hurst, K.E., Radojevic, A.M. et al. (2007). Multiwall carbon nanotube absorber on a thin-film lithium niobate pyroelectric detector. *Optics Letters* **32**: 772–774. https://en.wikipedia.org/wiki/Optical_properties_of_carbon_nanotubes.
66 Lehman, J.H., Theocharous, E., Eppeldauer, G., and Pannell, C. (2003). Gold-black coatings for freestanding pyroelectric detectors. *Measurement Science and Technology* **14**: 916–922.
67 Woltersdorff, W. (1934). Über die optischen Konstanten dünner Metallschichten im langwelligen Ultrarot. *Zeitschrift für Physik* **91** (3–4): 230–252.
68 Martin, D.H. (1991). *Proceedings of SPIE* **1576**: 15762Z. https://doi.org/10.1117/12.2297820.
69 Hengstberger, F. (1989). *Absolute Radiometry*. Academic Press.
70 Douglas, N.G. (1989). Millimetre and Sumillimetre wavelength lasers. In: *Springer Series in Optical Sciences*, vol. **61**. Springer.
71 Steiger, A., Bohmeyer, W., Lange, K., and Müller, R. (2016). Novel pyroelectric detectors for accurate THz power measurements. *tm- Technisches Messen* **83**: 386.
72 Steiger, A., Kehrt, M., Monte, C., and Müller, R. (2013). Traceable THz power measurement from 1 THz to 5 THz. *Optics Express* **21**: 14466.
73 Steiger, A., Müller, R., Oliva, A.R. et al. (2016). Terahertz laser power measurement comparison. *IEEE Transactions on THz Science and Technology* **6**: 664.
74 Globisch, R., Dietz, T., Göbel, M. et al. (2015). Absolute THz power measurement of a time-domain spectroscopy system. *Optics Letters* **40**: 3544.
75 Naftaly, M. (2016). An international intercomparison of THz-TDS. Proceedings of the 41st International Conference on Infrared, Millimeter, and Terahertz Waves, https://doi.org/10.1109/IRMMW-THz.2016.7758763
76 van Mechelen, J.L.M. (2018). Predicting the dry thickness of a wet paint layer. Proceedings of the 43rd International Conference on Infrared, Millimeter, and Terahertz Waves, https://doi.org/10.1109/IRMMW-THz.2018.8509847

77 Nellen, S., Globisch, B., Kohlhaas, R.B. et al. (2018). Recent progress of continuous-wave terahertz systems for spectroscopy, non-destructive testing, and telecommunication. In: *Proc of SPIE OPTO* no. 10531.

78 Blea, J.M., Parks, W.F., Ade, P.A.R., and Bell, R.J. (1970). *Journal of the Optical Society of America* **60**: 603.

79 Mingardi, A., Zhang, W-D., Brown, E.R. et al. (2018). High power generation of THz from 1550-nm photoconductive emitters. *Optics Express* **26** (11): 14472–14478.

80 Rutledge, D., Neikirk, D., and Kasilingam, D. (1983). Integrated circuit antennas in Infrared and Millimeter Wave, Vol. 10. Academic Press, Inc.

81 Ulaby, F.T. (1997). *Fundamentals of Applied Electromagnetics*. Upper Saddle River, NJ: Prentice-Hall, Inc.

82 Duffy, S.M., Verghese, S., McIntosh, A. et al. (2001). Accurate modeling of dual dipole and slot elements used with photomixers for coherent terahertz output power. *IEEE Transactions on Microwave Theory and Techniques* **49** (6): 1032–1038.

83 Nguyen, T.K. and Park, I. (2012). Resonant antennas on semi-infinite and lens substrates at terahertz frequency. In: *Convergence of Terahertz Sciences in Biomedical Systems*, 181–193. Springer.

84 Rumsey, V.H. (1957). Frequency independent antennas. *IRE International Convention Record* Part 1 **5**: 114–118.

85 Mushiake, Y. (1992). Self-complementary antennas. *IEEE Antennas and Propagation Magazine* **34** (6): 23–29.

86 Brown, E.R. (1999). A photoconductive model for superior low-temperature-grown GaAs THz photomixers. *Applied Physics Letters* **75** (6): 769–771.

87 Roehle, H. et al. (2010). Next generation 1.5μm terahertz antennas: mesa-structuring of InGaAs/InAlAs photoconductive layers. *Optics Express* **18** (3): 2296–2301.

88 Dietz, R.J.B., Globisch, B., Roehle, H. et al. (2014). Influence and adjustment of carrier lifetimes in InGaAs/InAlAs photoconductive pulsed terahertz detectors : 6 THz bandwidth and 90dB dynamic range. *Optics Express* **22** (16): 19411–19422.

89 Kostakis, D.S. and Missous, M. (2012). Terahertz generation and detection using low temperature grown InGaAs-InAlAs photoconductive antennas at 1.55 μm pulse excitation. *IEEE Transactions on Terahertz Science and Technology* **2** (6): 617–622.

90 Vieweg, N. et al. (2014). Terahertz-time domain spectrometer with 90 dB peak dynamic range. *Journal of Infrared, Millimeter, and Terahertz Waves* **35** (10): 823–832.

91 Ishibashi, T.; Shimizu, N.; Kodama, S.; Ito, H.; Nagatsuma, T. & Furuta, T. (1997). Uni-traveling-carrier photodiodes, Tech. Dig. Ultrafast Electronics and Optoelectronics (1997 OSA Spring Topical Meeting), Incline Village, Nevada, 166–168.

92 Ito, H., Kodama, S., Muramoto, Y. et al. (2004a). High- speed and high-output InP/InGaAs uni-traveling-carrier photodiodes, IEEE J. *Selected Topics in Quantum Electronics* **10** (4): 709–727.

93 Furuta, F., Ito, H., and Ishibashi, T. (2000). Photocurrent dynamics of uni-traveling-carrier and conventional pin-photodiodes. *Institute of Physics Conference Series* **166**: 419–422.

94 Ishibashi, T. (1994). *High speed heterostructure devices Semiconductors and Semimetals*, vol. **41**, 333. San Diego: Academic Press, Chap. 5.

95 Nellen, S., Ishibashi, T., Deninger, A. et al. (2020). Experimental Comparison of UTC- and PIN-Photodiodes for Continuous-Wave Terahertz Generation. *Journal of Infrared, Millimeter, and Terahertz Waves* **41**: 343–354.

96 Zunger, A. (1986). Electronic structure of 3d transition-atom impurities in semiconductors. *Solid State Physics* **39**: 276–462.
97 Guillot, G. et al. (1990). Identification of the Fe acceptor level In Ga0.47In0.53As. *Semiconductor Science and Technology* **5**: 391–394.
98 Srocka, H.S. and Bimberg, D. (1994). Fe2+-Fe3+ level as a recombination center in In0.53Ga0.47As. *Physical Review B* **49** (15): 10259–10268.
99 Suzuki, M. and Tonouchi, M. (2005). Fe-implanted InGaAs terahertz emitters for 1.56 μm wavelength excitation. *Applied Physics Letters* **86** (5): 051104.
100 Hatem, O. et al. (2011). Terahertz-frequency photoconductive detectors fabricated from metal-organic chemical vapor deposition-grown Fe-doped InGaAs. *Applied Physics Letters* **98**: 121107.
101 Mohandas, R.A. et al. (2016). Generation of continuous wave terahertz frequency radiation from metal-organic chemical vapour deposition grown Fe-doped InGaAs and InGaAsP. *Applied Physics Letters* **119**: 153103.
102 Globisch, B. et al. (2017). Iron doped InGaAs: Competitive THz emitters and detectors fabricated from the same photoconductor. *Journal of Applied Physics* **121** (5): 053102.
103 Bube, R.H. (1992). *Photoelectronic Properties of Semiconductors*. Press: Cambridge University.
104 Poole, K., Singer, E., Peaker, A.R., and Wright, A.C. (1992). Growth and structural characterization of molecular beam epitaxial erbium-doped GaAs. *Journal of Crystal Growth* **121**: 121–131.
105 Brown, R., Mingardi, A., Zhang, W.-D. et al. (2017). Abrupt dependence of ultrafast extrinsic photoconductivity on Er fraction in GaAs:Er. *Applied Physics Letters* **111**: 031104.
106 Scarpulla, M.A., Zide, J.M.O., LeBeau, J.M. et al. (2008). Near-infrared absorption and semimetal-semiconductor transitions in 2 nm ErAs nanoparticles embedded in GaAs and AlAs. *Applied Physics Letters* **92**: 173116.
107 Middendorf, J.R. and Brown, E.R. (2012). THz generation using extrinsic photoconductivity at 1550 nm. *Optics Express* **20**: 16504–16509.
108 Zhang, W.-D., Middendorf, J.R., and Brown, E.R. (2015). Demonstration of a GaAs-based 1550-nm continuous wave photomixer. *Applied Physics Letters* **106**: 021119.
109 Sartorius, B., Stanze, D., Göbel, T. et al. (2012). Continuous wave terahertz systems based on 1.5 μm telecom technologies. *Journal of Infrared, Millimeter, and Terahertz Waves* **33**: 405–417.
110 Dietz, R.J.B. et al. (2013). 64 μw pulsed terahertz emission from growth optimized InGaAs/InAlAs heterostructures with separated photoconductive and trapping regions. *Applied Physics Letters* **103**: 061103.
111 Lloyd-Hughes, J., Castro-Camus, E., and Johnston, M.B. (2005). Simulation and optimisation of terahertz emission from InGaAs and InP photoconductive switches. *Solid State Communications* **136**: 595–600.
112 Dietz, R.J.B., Gerhard, M., Stanze, D. et al. (2011). THz generation at 1.55 μm excitation: six-fold increase in THz conversion efficiency by separated photoconductive and trapping regions. *Optical Express* **19** (27): 25911.
113 Nandi, U., Norman, J.C., Gossard, A.C. et al. (2018). 1550-nm Driven ErAs:In(Al)GaAs photoconductor-based terahertz time domain system with 6.5 THz bandwidth. *Journal of Infrared, Millimeter, and Terahertz Waves* **39** (4): 340–348.

114 Kohlhaas, R.B. et al. (2018). Improving the dynamic range of InGaAs-based THz detectors by localized beryllium doping: up to 70 dB at 3 THz. *Optics Letters* **43** (21): 5423–5426.

115 Göbel, T., Stanze, D., Globisch, B. et al. (2013). Telecom technology based continuous wave terahertz photomixing system with 105 decibel signal-to-noise ratio and 3.5 terahertz bandwidth. *Optics Letters* **38** (20): 4197–4199.

116 D. J. F. Olvera, H. Lu, A. C. Gossard, and S. Preu, "Continuous-wave 1550 nm operated terahertz system using ErAs:In(Al)GaAs photo-conductors with 52 dB dynamic range at 1 THz," Opt. Express, vol. 25, no. 23, p. 29492, 2017.

117 S. Gregory, R. K. May, K. Su, and J. A. Zeitler, "Terahertz car paint thickness sensor: Out of the lab and into the factory," Int. Conf. Infrared, Millimeter, Terahertz Waves, IRMMW-THz, no. Award 101262, p. 101262, 2014.

118 Vogt, W., Jones, A.H., Schwefel, H.G.L., and Leonhardt, R. (2018). Prism coupling of high-Q terahertz whispering-gallery-modes over two octaves from 0.2 THz to 1.1 THz. *Optics Express* **26** (24): 31190–31198.

119 Hochrein, T. (2014). Markets, Availability, Notice, and Technical Performance of Terahertz Systems: Historic Development, Present, and Trends. *J. Infrared, Millimeter, Terahertz Waves* **36** (3): 235–254.

120 Dietz, R.J.B., Vieweg, N., Puppe, T. et al. (2014). All fiber-coupled THz-TDS system with kHz measurement rate based on electronically controlled optical sampling. *Optics Letters* **39** (22): 6482–6485.

121 Bartels, A., Cerna, R., Kistner, C. et al. (2007). Ultrafast time-domain spectroscopy based on high-speed asynchronous optical sampling. *Review of Scientific Instruments* **78** (3).

122 Carpintero, G., Balakier, K., Yang, Z. et al. (2014). Photonic Integrated Circuits for Millimeter-Wave Wireless Communications. *Journal of Lightwave Technology* **8724** (20): 3495–3501.

123 Theurer, M., Göbel, T., Stanze, D. et al. (2013). Photonic-integrated circuit for continuous-wave THz generation. *Optics Letters* **38** (19): 3724–3726.

124 Liebermeister, L., Nellen, S., Kohlhaas, R.B. et al. (2018). Real-time continuous wave terahertz spectroscopy with 2 THz bandwidth. International Conference on Infrared, Millimeter, and Terahertz Waves, IRMMW-THz. https://doi.org/10.1109/IRMMW-THz.2018.8510414.

125 Nagatsuma, T., Horiguchi, S., Minamikata, Y. et al. (2013). Terahertz wireless communications based on photonics technologies. *Optical Express* **21**: 23736–23747.

126 Macknofsky, G. (2015). Understanding the Basics of CPRI Fronthaul Technology. EXFO App Note 310.

127 Stöhr, A., Akrout, A., Buß, R. et al. (2009). 60 GHz radio-over-fiber technologies for broadband wireless services. *Journal of Optical Networking* **8** (5): 471–486.

128 Cvijetic, N., Qian, D., and Wang, T. (2008). 10 Gb/s free-space optical transmission using OFDM. OFC/NFOEC 2008.

129 Song, H.J. and Nagatsuma, T. (2011). Present and future of terahertz communications. *IEEE Transactions on Terahertz Science and Technology* **1** (1): 256–263.

130 Kalfass, I., Dan, I., Rey, S. et al. (2015). Towards MMIC-based 300GHz indoor wireless communication systems, special section on terahertz waves coming to the real world. *IEICE Transactions on Electronics* **E98.C** (12): 1081–1090.

131 Nagatsuma, T. and Carpintero, G. (2015). Recent progress and future prospect of photonics-enabled terahertz communications research, special section on terahertz waves coming to the real world. *IEICE Transactions on Electronics* **E98.C** (12): 1060–1070.
132 Fan, Z. and Dagenais, M. (1997). Optical generation of a mHz-linewidth microwave signal using semiconductor lasers and a discriminator-aided phase-locked loop. *IEEE Transactions on Microwave Theory and Techniques* **45** (8): 1296–1300.
133 Hirata, M., Harada, K.S., and Nagatsuma, T. (2003). Low-cost millimeter-wave photonic techniques for gigabit/s wireless link. *IEICE Transactions on Electronics* **E86-C** (7): 1123–1128.
134 Chen, D., Fetterman, H., Chen, A. et al. (1997). Demonstration of 110 GHz electro-optic polymer modulators. *Applied Physics Letters* **70** (25): 3335–3337.
135 Sato, K. (2001). 100 GHz optical pulse generation using Fabry-Perot laser under continuous wave operation. *Electronics Letters* **37** (12): 763–764.
136 Sato, K., Kotaka, I., Kondo, Y., and Yamamoto, M. (1996). Active mode locking at 50 GHz repetition frequency by half- frequency modulation of monolithic semiconductor lasers integrated with electro absorption modulators. *Applied Physics Letters* **69** (18): 2626–2628.
137 Moeller, L., Shen, A., Caillaud, C., and Achouche, M. (2013). Enhanced THz generation for wireless communications using short optical pulses. Infrared, Millimeter, and Terahertz Waves (IRMMW-THz), 1–3.
138 Gordón, C., Guzmán, R., Corral, V. et al. (2016). On-chip multiple colliding pulse mode-locked semiconductor laser. *J. Lightwave Technol. vol.* **34**: 4722–4728.
139 Townes, C.H. and Schawlow, A.L. (1955). *Microwave Spectroscopy*. New York, NY: McGraw-Hill.
140 van Exter, M., Fattinger, C., and Grischkowsky, D. (1989). Terahertz time-domain spectroscopy of water vapor. *Optics Letters* **14** (20): 1128–1130.
141 Pine, A.S., Suenram, R.D., Brown, E.R., and McIntosh, K.A. (1995). A THz photomixing spectrometer: application to SO_2 self broadening. *Journal of Molecular Spectroscopy* **175**: 37.
142 Deninger, A.J., Roggenbuck, A., Schindler, S., and Preu, S. (2015). 2.75 THz tuning with a triple-DFB laser system at 1550 nm and InGaAs photomixers. *Journal of Infrared, Millimeter, and Terahertz Waves* **36**: 269–277.
143 Roggenbuck, A. et al. (2010). Coherent broadband continuous-wave terahertz spectroscopy on solid-state samples. *New Journal of Physics* **12**: 043017.
144 Gregory, I.S., Evans, M.J., Page, H. et al. (2007). Analysis of photomixer receivers for continuous-wave terahertz radiation. *Applied Physics Letters* **91** (15): 154103–154105.
145 Sinyukov, A.M., Liu, Z., Hor, Y.L. et al. (2008). Rapid-phase modulation of terahertz radiation for high-speed terahertz imaging and spectroscopy. *Optics Letters* **33** (14): 1593–1595.
146 Göbel, T., Schoenherr, D., Sydlo, C. et al. (2008). Continuous-wave terahertz system with electrooptical terahertz phase control. *Electronics Letters* **44** (14): 863–864.
147 Stanze, D., Gobel, T., Dietz, R.J.B. et al. (2011). High-speed coherent CW terahertz spectrometer. *Electronics Letters* **47** (23): 1255.
148 Roggenbuck, A. et al. (2012). Using a fiber stretcher as a fast phase modulator in a continuous wave terahertz spectrometer. *Journal of the Optical Society of America B* **29** (4): 614.

149 Göbel, T., Schoenherr, D., Sydlo, C. et al. (2009). Single-sampling-point coherent detection in continuous wave photomixing terahertz systems. *Electronics Letters* **45** (11): 65–66.

150 Liebermeister, L., Nellen, S., Kohlhaas, R. et al. (2019). Ultra-fast, high-bandwidth coherent cw THz spectrometer for non-destructive testing. *Journal of Infrared, Millimeter, and Terahertz Waves* **40**: 288–296.

151 Lukosz, W. (1981). Light emission by multipole sources in thin layers. I. Radiation patterns of electric and magnetic dipoles. *JOSA* **71** (6): 744–754.

4

THz Photomixers

Emilien Peytavit, Guillaume Ducournau, and Jean-François Lampin

Univ. Lille, CNRS, Centrale Lille, Univ. Polytechnique Hauts-de-France, UMR 8520
Institute of Electronics, Microelectronics and Nanotechnology (IEMN) Villeneuve d'Ascq
Cedex, France

4.1 Introduction

This chapter is mainly focused on the generation of continuous waves in the THz frequency range by using the optical heterodyne, also called photomixing of two slightly detuned, spatially overlapped, infrared laser beams in an ultrafast photodetector. We are not able to present exhaustively all the works which have been carried out in this field these last 20 years and we will focus on two standard photomixers devices: ultrafast photoconductors based on sub-picosecond carrier lifetime semiconductors (e.g. low-temperature grown GaAs, InGaAs:Fe, etc.) and uni-travelling-carrier InP/InGaAs photodiodes.

4.2 Photomixing Basics

4.2.1 Photomixing Principle

Let's assume two continuous-wave (cw) laser beams spatially overlapped of angular frequencies $\omega_1 = \omega_0 + \frac{1}{2}\omega_b$ and $\omega_2 = \omega_0 - \frac{1}{2}\omega_b$ and wave vectors having a same polarization (parallel to $\vec{e}_y$) and magnitude. In a plane wave approximation and assuming that the direction of propagation of the lasers is along the z-axis, the electric field of one laser can be written as:

$$\vec{E}_i(z,t) = E_0 \cos(\omega_i t - k_i z)\,\vec{e}_y = E_0 \cos(\omega_i t + \varphi_i(z))\,\vec{e}_y \tag{4.1}$$

Fundamentals of Terahertz Devices and Applications, First Edition. Edited by Dimitris Pavlidis.

Companion website: www.wiley.com/go/Pavlidis/FundamentalsofTHz

The total electric field is then:

$$\begin{aligned}\vec{E}(z,t) &= E_0[\cos(\omega_1 t + \varphi_1) + \cos(\omega_2 t + \varphi_2)]\vec{e}_y \\ &= 2E_0\left[\cos\left(\omega_0 t + \frac{1}{2}(\varphi_1 + \varphi_2)\right)\cos\left(\frac{1}{2}\omega_b t + \frac{1}{2}(\varphi_1 - \varphi_2)\right)\right]\vec{e}_y\end{aligned} \tag{4.2}$$

We recognize here a carrier at angular frequency ω_0 modulated by an envelope at angular frequency $\frac{1}{2}\omega_b$

The energy flux density of the superposition of the two fields is then

$$\begin{aligned}\vec{S}(z,t) &= \vec{E}(z,t) \wedge \vec{H}(z,t) = y_0\left|\vec{E}(z,t)\right|^2 \vec{e_z} \\ \left|\vec{S}(z,t)\right| &= y_0\left|\vec{E}(z,t)\right|^2 = 4y_0 E_0^2 \cos^2\left(\omega_0 t + \frac{1}{2}(\varphi_1 + \varphi_2)\right)\cos^2\left(\frac{1}{2}\omega_b t + \frac{1}{2}(\varphi_1 - \varphi_2)\right) \\ &= y_0 E_0^2(1 + \cos(2\omega_0 t + (\varphi_1 + \varphi_2))(1 + \cos(\omega_b t + (\varphi_1 - \varphi_2))\end{aligned} \tag{4.3}$$

where $y_0 = \frac{1}{Z_0} = (\varepsilon_0/\mu_0)^{1/2}$ is the admittance of free space.

In a photomixing experiment, the laser frequencies are far above the cut-off frequency of the photomixer, and only the time average (equal to zero) of its contribution is detected. The difference frequency $\omega_b = \omega_2 - \omega_1$ is chosen small enough to be followed by the photodetector. The energy flux density detected by the photomixer is, therefore

$$\left|\vec{S}(z,t)\right| = y_0 E_0^2[1 + \cos(\omega_b t + (\varphi_1 - \varphi_2)] = S_0[1 + \cos(\omega_b t + (\varphi_1 - \varphi_2)] \tag{4.4}$$

The detected energy flux density is composed of a dc part, which is the sum of the laser energy flux densities taken separately $\frac{1}{2}S_0 = \frac{1}{2}y_0 E_0^2$, and a modulated part. The frequency ω_b of the modulation of the energy flux density is the frequency difference of the two lasers. ω_b can be easily tuned by tuning the frequency of at least one of both lasers. As the laser frequencies are much larger than ω_b, only a small relative tuning range of the laser is sufficient for the photomixer's output signal to be widely tunable. In practice, the tuning range is limited by the frequency limitation of the photomixer.

4.2.2 Historical Background

The first experimental demonstration of the mixing of two optical sources is shown in 1955 by T.A Forrester et al. with a photocathode illuminated by two Zeeman components of an atomic spectral line separated by 10 GHz [1]. However, this technique has received a great deal of attention after the appearance of the laser in 1960 providing sufficiently coherent and easy to use optical sources. The first optical mixing of laser sources is done in 1962 by Javan et al. [2] with two single-mode He–Ne lasers mixed in the photocathode of a photomultiplier. The beating frequency is 5 MHz and is detected by a spectrum analyzer. The main goal of the experiment is actually the measurement of the spectral purity of the laser and not the generation of microwaves. In the same year, several reports of photomixing with beating frequencies lying in the GHz range follow this work, in which the photomixer is firstly a transparent cathode of a standard microwave traveling wave tube (TWT) [3]. Solid

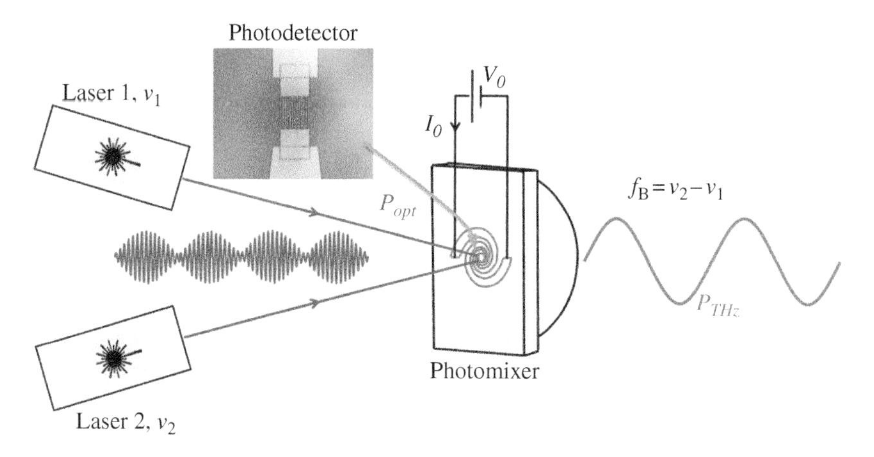

Figure 4.1 THz generation by photomixing.

state photomixers appear in a second time, with the use of a *p–i–n* junction Germanium photodiode [4, 5] and a biased photoconductive CdSe crystal [6]. In these pioneering works, the optical sources consist of near longitudinal modes of a ruby laser or a He–Ne laser. Two years later, the physical and electrical model of the microwave generation by photomixing done by Coleman et al. [7] give the key parameters for the optimization of the generated power. In the 1990s, with the progress of solid-state lasers and high-speed photodetectors, the interest in the microwave generation by the optoelectronic route is revived. The first demonstration of THz generation by photomixing is reported in 1995 by Brown and McIntosh [8] thanks to an ultrafast photoconductor, consisting of an interdigitated electrode capacitance on a low-temperature-grown GaAs (LT-GaAs) layer and pumped by two Ti: Al_2O_3 lasers working around $\lambda = 0.8\ \mu m$. The photomixer is coupled to a spiral antenna and the radiated THz waves are detected by a He-cooled bolometer a schematic of this first THz photomixing experiment is shown in Figure 4.1. LT-GaAs, discovered some years before [9], seems actually, to be a perfect semiconductor for ultrafast optoelectronics, due to a subpicosecond carrier lifetime, a high electric-field breakdown (~300 kV/cm) [10] combined with a relatively high electron mobility (~150 cm^2/Vs) [11]. The output power is quite low, i.e. 4 μW at 300 GHz and 1 μW at 800 GHz, but the noise level is reached only at 3.8 THz. The second breakthrough is achieved in 2003 by Ito et al. [12] with the report of ~80 μW at 300 GHz and 2.6 μW at 1 THz with InP-based uni-traveling carrier photodiodes (UTC-PD) initially developed for telecom applications and pumped by 1.55-μm-wavelength lasers. As far as concern the output power, no significant improvements have been achieved since these two pioneer works.

4.3 Modeling THz Photomixers

The purpose of this section is the presentation of an electrical model of a THz source based on photomixing in photodetectors. This will be followed by an estimation of the level of power that can be generated by this method.

4.3.1 Photoconductors

4.3.1.1 Photocurrent Generation

As seen in the previous part, the incident energy flux density (or intensity) in a photomixing experiment can be written as:

$$S(z,t) = \left|\vec{S}(z,t)\right| = S_0[1 + \cos(\omega_b t + (\varphi_1(z) - \varphi_2(z))] \tag{4.5}$$

If we assume an illumination on the surface area of the photoconductor being at $z = 0$, the intensity of the light impinging the photoconductor can be written as:

$$S_{z_0}(t) = S(z=0,t) = S_0[1 + \cos(\omega_b t + \varphi)] \text{ where } \varphi = (\varphi_1(0) - \varphi_2(0)) \tag{4.6}$$

A schematic of a photoconductor illuminated by an optical beatnote is shown in Figure 4.2. Here, we consider an ideal photoconductor of capacitance C, having a single type of carrier of mobility μ, a lifetime τ, and a free carrier density n_{dark} assumed to be zero. When the energy flux density impinges the photoconductor, a fraction $(1-R)$ enters the semiconductor, where R is the reflectance. In the semiconductor, light is absorbed according to the Beer–Lambert law, and the intensity of the light beam is given by $S(t,z) = S_{z_0}(t)(1-R)e^{-\alpha z}$, where α is the absorption coefficient. If we assume a homogeneous illumination onto the photoconductor of area $A = l \times w$, we have a direct relation between the time average pumping optical power P_{opt} and the time average intensity: $P_{opt} = l \times w \times S_0$. The density of electron-hole pairs created by light absorption is:

$$\mathcal{G}(z,t) = -\frac{1}{h\nu}\frac{\partial S}{\partial z} = \frac{1}{h\nu}\alpha S_{z_0}(t)(1-R)e^{-\alpha z} = \frac{1}{h\nu}\alpha(1-R)e^{-\alpha z}S_0(1 + \cos(\omega_b t + \varphi)) \tag{4.7}$$

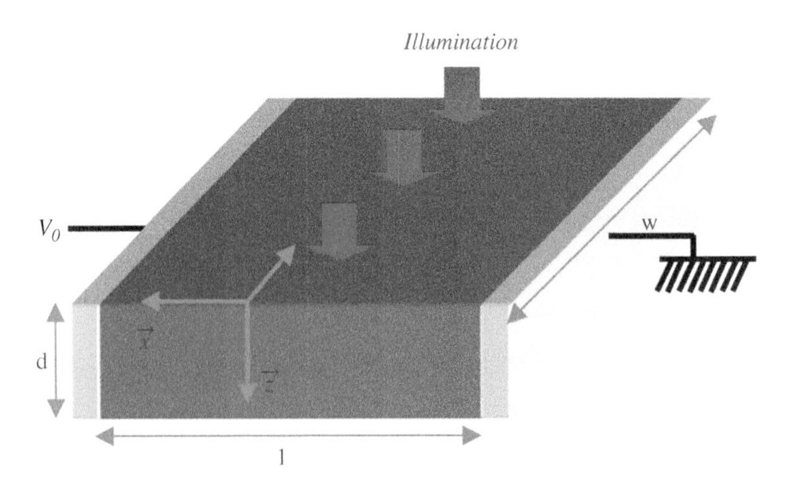

Figure 4.2 Photomixing in a photoconductor.

where $h\nu$ is the energy of one photon, $\nu = \frac{\omega}{2\pi}$ being the average frequency of the laser light ($\omega \approx \omega_1 \approx \omega_2$). The electron density in the semiconductor $n\left(\vec{r}, t\right)$ can be calculated by using the continuity equation:

$$\frac{\partial n\left(\vec{r}, t\right)}{\partial t} = \mathcal{G}\left(\vec{r}, t\right) - \frac{n}{\tau} + \frac{1}{e}\nabla\vec{J}\left(\vec{r}, t\right) \tag{4.8}$$

Where n/τ correspond to the recombination/trapping rate of the free carriers, with τ the carrier lifetime of the material. $\vec{J} = \vec{J}_{diff} + \vec{J}_{drift}$ is the current density, composed of diffusion and drift currents: $\vec{J}_{diff} = -eD\vec{\nabla}n$and $\vec{J}_{drift} = -ne\vec{v} = ne\mu\vec{E}$, where $\vec{E}$ is the electric field. Here we assume that the charge neutrality is preserved everywhere in the device. The diffusion current is negligible compared to the carrier trapping due to the short life-time of the photoconductive material, in the order of $\tau \approx 1$ ps. In a standard planar photomixer, absorption of light follows the Beer–Lambert-law, and the free electron density can be approximated by $n(z) = n_0 e^{-\alpha z}$. The diffusion current becomes $J_{diff} = -\frac{D\partial^2 n}{\partial^2 z} = D\alpha^2 n$. The typical absorption coefficient in semiconductors such as LT-GaAs is $\alpha \approx 10^4$/cm. The diffusion constant is $D = k_B T\mu/e \approx 30$ cm^2/s at room temperature, assuming a typical mobility of $\mu \approx 1000$ cm^2/V/s in low carrier lifetime materials. The diffusion contribution to $\partial n/\partial t$ is then $J_{diff} \approx n \times 3.10^9 s^{-1}$ and can be neglected since it is much smaller than the trapping rate $n/\tau \approx n \times 10^{12} s^{-1}$.

In the absence of currents or at small electric fields, where the contribution of the drift current J_{drift} to $\partial n/\partial t$ can be neglected, the continuity equation takes the form:

$$\frac{\partial n\left(\vec{r}, t\right)}{\partial t} = \mathcal{G}\left(\vec{r}, t\right) - \frac{n\left(\vec{r}, t\right)}{\tau} \tag{4.9}$$

with $\mathcal{G}\left(\vec{r}, t\right) = \frac{1}{h\nu}\alpha(1 - R)e^{-\alpha z}I_0(1 + \cos(\omega_b t + \varphi))$ and without loss of generality we can choose the time origin such as $\varphi = 0$.

The electron concentration can be then written as:

$$n(z, t) = n_0(z)\left[1 + \frac{\cos(\omega_b t + \psi)}{\sqrt{1 + (\omega_b \tau)^2}}\right] \tag{4.10}$$

with $\psi = \tan^{-1}(\omega_b\, \tau)$ and $n_0(z) = \frac{\tau\alpha(1 - R)e^{-\alpha z}S_0}{h\nu}$

The photogenerated charge carriers move toward the electrodes along the electric field lines. In general, the transit time τ_{tr} of a charge carrier moving from one electrode to the other depends on the position at which the latter is created due to different path lengths and field strengths. In the basic model analyzed here, we assume at low electric field E that $\tau_{tr} = l/v = l/\mu E$. The ratio of lifetime over transit time is also known as the photoconductive gain, $\gamma = \tau/\tau_{tr}$. It can be thought as the number of electrons delivered to the electrode for each electron-hole pair created.

The current generated by the photoconductor is given by the flux of charge carriers at the electrodes. For the simplest model, the current density is $j = j_c + j_d$ with $j_c = nev$ and $j_d = \varepsilon \frac{\partial E}{\partial t}$. By integration, the conduction photocurrent becomes:

$$\begin{aligned} I_c(t) &= \int_0^w dy \int_0^d j_c dz = evw \int_0^d n(z,t) dz \\ &= \frac{e}{h\nu} \mu E w S_0 (1-R)(1-e^{-\alpha d}) \tau \left[1 + \frac{\cos(\omega_b t + \psi)}{\sqrt{1+(\omega_b \tau)^2}} \right] \end{aligned} \tag{4.11}$$

$$I_c(t) = \frac{e}{h\nu} \gamma \eta_{opt} P_{opt} \left[1 + \frac{\cos(\omega_b t + \psi)}{\sqrt{1+(\omega_b \tau)^2}} \right] \tag{4.12}$$

where $\eta_{opt} = (1-R)(1-e^{-\alpha d})$ is sometimes called the optical quantum efficiency and is in our case the ratio of the photon flux absorbed in the active layer to the incident photon flux. As far as concerned the displacement current, if we suppose the charge density in the semiconductor negligible in comparison with the charge density on the metallic electrodes we can consider that $E = V/l = cste$ and:

$$I_d(t) = \int_0^w dy \int_0^d j_d dz = \frac{dw}{l} \varepsilon \frac{dV}{dt} = C \frac{dV}{dt} \tag{4.13}$$

with C, the electrical capacitance of the photoconductor. In a first time, we suppose that there is only a perfect dc voltage source in parallel with the photoconductor (see Figure 4.2), $V = cste$ and $I_d = 0$.

The photoresponse of the photomixer is defined as $\mathcal{R} = \bar{I}/P_{opt}$. From the last equation, we can notice that $\mathcal{R} \leq e/h\nu$. The term $e/h\nu$ is approximately 1.25 A/W for a laser wavelength of 1.55 μm and 0.63 A/W at 780 nm. Photomixers at longer wavelengths can ideally be more efficient due to the higher number of photons impinging on the photomixer at the same optical power.

4.3.1.2 Electrical Model

The current generated by a photoconductor without a load resistance has been calculated in the previous section. However, in practice, the photoconductor can be coupled to different loads depending on the cases: to an antenna to radiate the THz currents in free space, to a waveguide followed by a power meter or a spectrum analyzer, etc. In the simplest case, the impedance seen by the photoconductor is real and is in the order of 50 Ω, as it will be seen later that the radiation resistance of broadband antennas is in the same order of magnitude as the standard impedance of microwave instruments. The THz power dissipated in the load then depends also on the capacitance of the photodetector. For the analysis of the electric circuit, it is helpful to calculate the conductance of the photoconductor:

$$G(t) = \frac{I_c}{V} = \frac{e}{h\nu V} \frac{\tau}{\tau_{tr}} \eta_{opt} P_{opt} \left[1 + \frac{\cos(\omega_b t + \psi)}{\sqrt{1+(\omega_b \tau)^2}} \right] \tag{4.14}$$

At low voltage, in constant mobility regime, $v = \mu E = \mu V/l$, $\tau_{tr} = \frac{l}{v} = l^2/(\mu V)$, and $\frac{\partial G}{\partial V} = 0$ In this case, G can be written as:

$$G(t) = G_0 + G_1 \cos(\omega_b t + \psi) \tag{4.15}$$

where

$$G_0 = \frac{e}{h\nu}\frac{\mu\tau}{l^2}\eta_{opt}P_{opt} = \frac{e\mu\overline{n_0}A}{l} \tag{4.16}$$

with $\overline{n_0}$ the spatial average value of $n_0(z)$.

$$G_1 = \frac{G_0}{\sqrt{1 + (\omega_b\tau)^2}} \tag{4.17}$$

To simplify the calculation, we assume that the photoconductor is coupled to a real load admittance G_L and a band-pass filter consisting of an LC circuit, having a resonance frequency at ω_b, is added in parallel with the load in order to short-circuit the load at every frequency except the beating frequency (as shown in Figure 4.3). Voltages and currents in the circuit are time periodic (period $T = 2\pi/\omega_b$). They can be expressed in Fourier series and the band-pass filter removes all the upper harmonics. By using this simplification and by set $\psi = 0$ to lighten the notations, the voltage across the photoconductor takes the simple form:

$$V(t) = V_{dc} + V_{ac}\cos(\omega_b t + \varphi) \tag{4.18}$$

and the current:

$$I(t) = I_{dc} + I_{ac}\cos(\omega_b t + \delta) \tag{4.19}$$

It remains to calculate V_{dc}, V_{ac}, I_{dc}, I_{ac}, φ, δ. To do that, we use:

1) The relationship between the voltage and the current across the photoconductor:

   $$I(t) = G(t)V(t)$$

   where we use the complex notation, i.e. $I_{ac}\cos(\omega_b t + \delta) = Re\,(i_{ac}(\omega_b)e^{j\omega_b t})$, $V_{ac}\cos(\omega_b t + \delta) = Re\,(v_{ac}(\omega_b)e^{j\omega_b t})$, etc. we find from this latter:

$$I_{dc} = V_{dc}G_0 + \frac{1}{2}V_{ac}G_1\cos\varphi \qquad i_{ac} = G_0 v_{ac} + G_1 V_{dc} \tag{4.20}$$

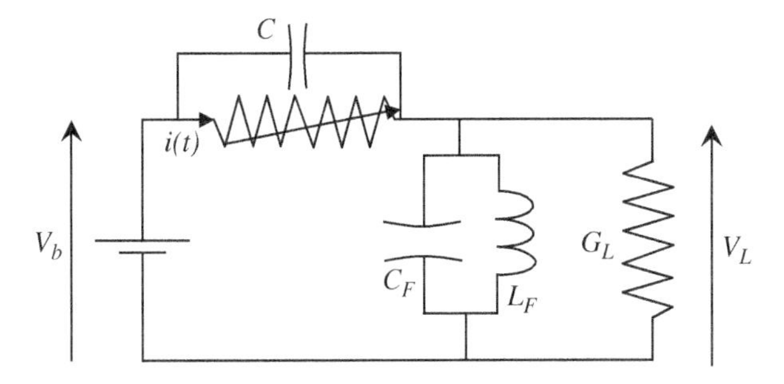

Figure 4.3 Electrical model of a photoconductor coupled to a load admittance G_L.

2) The Kirchhoff's current law at $\omega = \omega_b$ which gives (see Figure 4.4a): $-G_L V_L = I_{ac} + C\frac{dV_{ac}}{dt}$ which can be expressed in complex notation by:

$$-G_L v_{ac} = i_{ac} + jC\omega_b v_{ac} \tag{4.21}$$

Thereby, V_{ac} and φ can be calculated:

$$V_{ac} = V_{dc}\frac{G_1}{\sqrt{(G_0 + G_L)^2 + C^2\omega_b^2}} \text{ and } \varphi = \pi - a\tan\left(\frac{C\omega_b}{G_0 + G_L}\right) \tag{4.22}$$

In addition, the band-pass filter short-circuits G_L at $\omega = 0$ and then: $V_{dc} = V_b$ (see Figure 4.4b).

The LC circuit filters all the frequencies except for $\omega = \omega_b$. It is obvious that this is a calculation "trick" to obtain easily the voltage and current in the circuit, which is very difficult to use in practice since it limits dramatically the frequency band of the photomixing source. In practice, there is always a circuit which decouples dc and ac parts of the generated currents in the circuits. It is done straight when an antenna is used to radiate the ac currents in free space. A bias-T is inserted, when waveguide measurements are performed (see Figure 4.5), resulting in the same ac/dc decoupling but without canceling the signals at harmonics frequencies ($2\omega_b$, $3\omega_b$, etc.). The theoretical results obtained here can thus be used also in this case but only when the higher harmonics are negligible (which is often true ...). From the equations above, we can deduce directly the small-signal equivalent circuit at $\omega = \omega_b$ which consists of a current source $I_{ac} = |i_{ac}| = G_1 V_b$ having an internal conductance

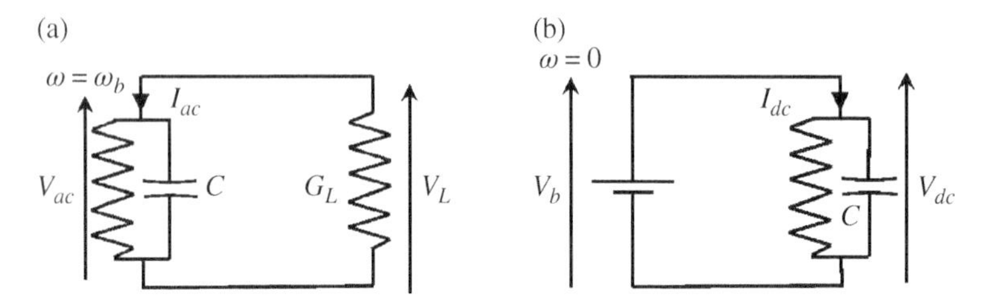

Figure 4.4 (a) Electrical circuit at $\omega = \omega_b$ and (b) $\omega = 0$.

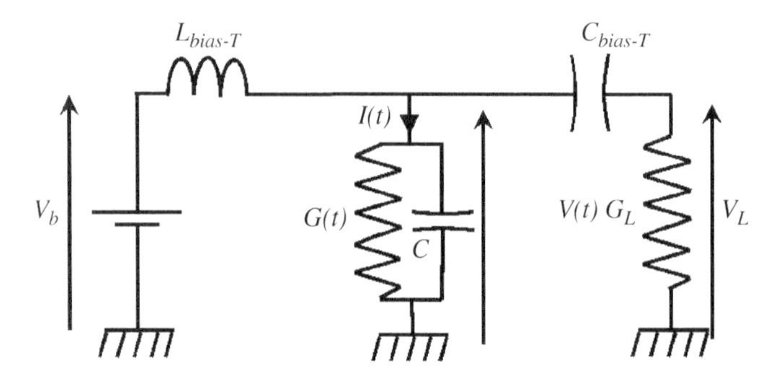

Figure 4.5 Photomixing experiment AC/DC decoupling using a Bias-T.

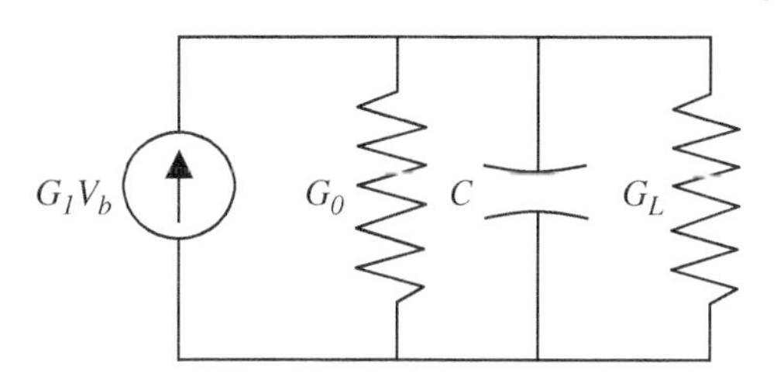

Figure 4.6 Small signal Equivalent circuit at ω_b.

G_0 in parallel with the electrical capacitance of the photodetector and loaded by a conductance G_L (see Figure 4.6).

4.3.1.3 Efficiency and Maximum Power

4.3.1.3.1 Case $G_1 = G_0$ and $C = 0$

It is rather obvious that the electrical capacitance of the photoconductor limits the performance at very high frequencies. It is however interesting to deal with the ideal case where $C\omega_b \ll (G_0 + G_L)$

From the calculations above, the power provided by the dc generator is:

$$P_{dc} = I_{dc}V_b = V_b^2\left(G_0 - \frac{1}{2}\frac{G_1^2}{G_0 + G_L}\right) \tag{4.23}$$

The average RF power dissipated in the load G_L is

$$P_{THz} = \frac{1}{2}G_L V_1^2 = \frac{1}{2}V_b^2\frac{G_L G_1^2}{(G_0 + G_L)^2} = \frac{1}{2}V_b^2\frac{G_L G_1^2}{(G_0 + G_L)^2} = \frac{1}{2}I_{ac}^2\frac{G_L}{(G_0 + G_L)^2} \tag{4.24}$$

The maximum power dissipated in the load is obtained at the impedance matching condition when $G_0 = G_L$. It is also known as the "maximum available power" of a source by microwave engineers. In the most favorable case, when the carrier lifetime is much smaller than the beating period $G_0 \approx G_1$, it can be written as:

$$P_{max} = \frac{1}{8}G_0 V_b^2 \tag{4.25}$$

The conversion efficiency is then (when $G_0 \approx G_1$) equal to:

$$\eta_{G_L = G_0} = \frac{P_{max}}{P_{dc}} \approx 0.167 \tag{4.26}$$

It is worth noting that it does not correspond exactly to the maximum efficiency which is reached when $G_L = \frac{G_0}{\sqrt{2}}$:

$$\eta_{max} = \frac{1}{\left(\sqrt{2} + 1\right)^2} \approx 0.172 \tag{4.27}$$

4.3.1.3.2 General Case

We consider here the more general case where the electrical capacitance is not negligible and the load admittance is not a pure conductance and has the form $Y_L = G_L + jB_L$. The dissipated power in Y_L is now:

$$P_{THz} = \frac{1}{2} I_{ac}^2 \frac{G_L}{(G_0 + G_L)^2 + (\omega_b C + B_L)^2} \tag{4.28}$$

which takes the form when the expression of G_1 and I_{ac} are put in Eq. (4.28).

$$P_{THz} = \frac{V_b^2 G_0^2}{2} \frac{G_L}{\left[(G_0 + G_L)^2 + (\omega_b C + B_L)^2\right]\left(1 + \omega_b^2 \tau^2\right)} \tag{4.29}$$

The maximum power dissipated in the load is obtained when $Y_L^* = Y_S$, i.e. when $G_0 = G_L$ and $B_L = -C\omega_b$ and is equal to the previous case. If the load is a pure conductance ($B_L = 0$) and if $G_L \gg G_0$, Eq. (4.29) takes the form popularized by Brown and coworkers [8]:

$$P_{THz} = \frac{I_{dc}^2}{2} \frac{R_L}{\left[(1 + (\omega_b R_L C)^2\right]\left(1 + \omega_b^2 \tau^2\right)} \tag{4.30}$$

Here, we have defined $R_L = 1/G_L$. It can be noted that $I_{dc} = \mathcal{R} \cdot P_{opt}$. We recognized the two cut-off frequencies related to the RC time constant (f_{RC}) and to the carrier lifetime in the photoconductive material (f_τ).

4.3.2 Photodiode

4.3.2.1 PIN photodiodes

As previously mentioned, another kind of semiconductor device that can be used as photomixer is the *pn* junction photodiode. Electrons and holes photogenerated in the depletion region of a *pn* junction are accelerated toward the electrodes by the built-in potential, giving rise to a photocurrent. However, *p–n* homojunctions with constant doping in the *p* and *n* regions are not very efficient photodiodes. At high doping concentrations, the depletion region in which carrier transport is efficient (drift) becomes narrow, and only a small portion of the incoming light can be absorbed in or close to the depletion region. Lower doping concentrations increase the depletion region width but also make the formation of low resistance ohmic contacts more difficult.

The main problems of *p–n* junctions for photodetectors can be overcome by adding an intrinsic layer between the *p* and *n* regions. A schematic band diagram of such a *p–i–n* diode is shown in Figure 4.7. *p–i–n* diodes allow high doping concentrations for low resistance ohmic contacts, independent from the width *w* of the intrinsic layer. This latter is an important design parameter: it should be large enough to absorb most of the impinging radiation and to reduce the capacitance of the device, but it should be as small as possible to reduce the transit time of the charge carriers. Minority carriers that are generated in the doped regions are subject to diffusion. Only after diffusing into the intrinsic region, they are accelerated efficiently by the electric field. The doped regions of *p–i–n* diodes are, therefore, often fabricated from a material with larger bandgap than the photon energy at the operation wavelength. Light absorption and photogeneration of charge carriers then only takes place in the intrinsic region, where the charge carriers are separated efficiently by the electric field. The frequency response of *p–i–n* diodes has a RC-roll-off in a similar manner to that of photoconductors. Recombination of photogenerated charge carriers is negligible, and the transit time determines the frequency response instead of carrier the lifetime as in the case of the photoconductors. The *p–i–n* photodiodes are not suited for THz photomixing because

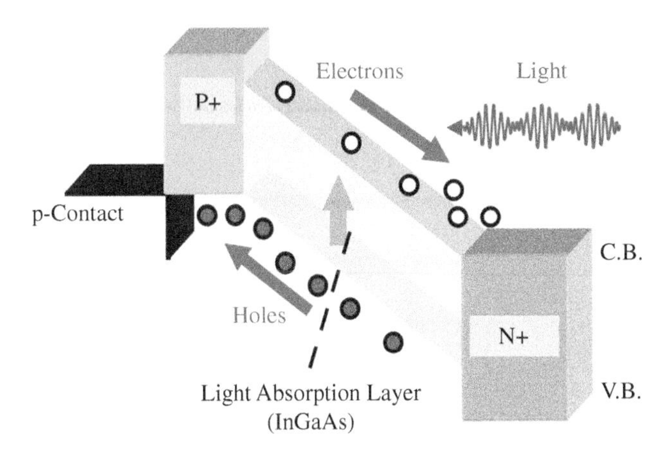

Figure 4.7 Schematic band diagram of a *p–i–n* photodiode. *Source*: From Nagatsuma et al. [13]; Copyright Wiley.

the low drift velocity of holes in most of the III–V semiconductors materials results in an intrinsic low-frequency bandwidth and a low saturation current because of the space charge accumulation in the intrinsic layer screening the bias field.

4.3.2.2 Uni-Traveling-Carrier Photodiodes

UTC-PD have been proposed in 1997 by researchers of the NTT laboratory [14, 15] in order to overcome the limitations of the InP/InGaAs *p–i–n* photodiodes in terms of frequency response and photocurrent saturation. As seen in Figure 4.8, UTC-PD's have a *p–i–n* diode-like structure in which the absorption of light and the carrier collection are achieved in two disjoint regions. The GaInAs absorption layer is heavily p-doped leading to a frequency response which is not limited anymore by the transit time of low-saturation velocity

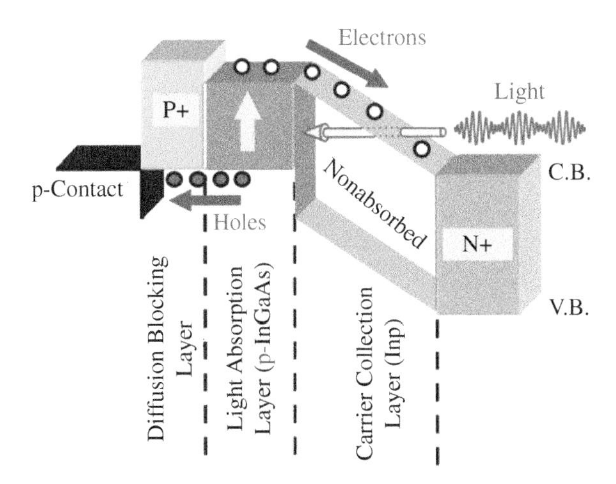

Figure 4.8 Schematic band diagram of a UTC photodiode. *Source*: From Nagatsuma et al. [13]; Copyright Wiley.

holes in GaInAs ($v_h = 5 \times 10^6$ cm/s) but by the electron motion in the InGaAs absorption and InP collection layer ($v_e = 4 \times 10^7$ cm/s in InP). The hole response time is indeed very fast since it is related to the dielectric relaxation time $\tau_R = \varepsilon/\sigma = \varepsilon/e\mu_h p_0$ where p_0 is the doping concentration. For example, if $p_0 = 1 \times 10^{18}$ cm^{-3}, $1/(2\pi\tau_R) \approx 30$ THz. Thanks to a blocking layer, electrons only diffuse/drift toward the InP collection layer, which they drift across at their overshoot drift velocity. Furthermore, the photocurrent saturation level is improved because of the reduction of the space charge effect in the collection layer induced mainly by the slow holes in the pin photodiodes.

4.3.2.3 Photocurrent Generation

In order to model the photocurrent generation in an UTC photodiode, we use the works of Ishibashi et al. [16] and Feiginov [17] and we assume, as previously, a homogeneous illumination onto a diode of area A. The current density flowing through the device consists of the sum of electron and holes conduction current densities and of the displacement current. Since it is constant along the device, it is equal to its average on the diode thickness (see Figure 4.9):

$$I(t) = \frac{A}{W}\int_0^W \left(J_e(t) + J_h(t) + \varepsilon\frac{\partial E}{\partial t}\right) dz \tag{4.31}$$

In a photomixing experiment, as already seen the current in the device consists of a dc part and an alternating component at the beating frequency ε: $I(t) = I_{dc} + I_{ac}\cos(\omega t + \varphi)$. In the following, we are interested only in the oscillating component and we use the complex notation where $I_{ac}\cos(\omega t + \varphi) = Re\,(i_{ac}(\omega)e^{j\omega t})$.

Equation (4.31) takes thus the form:

$$i_{ac}(\omega) = \frac{A}{W}\int_0^W (\,j_e(\omega) + j_h(\omega) + j\omega\varepsilon E)\, dz \tag{4.32}$$

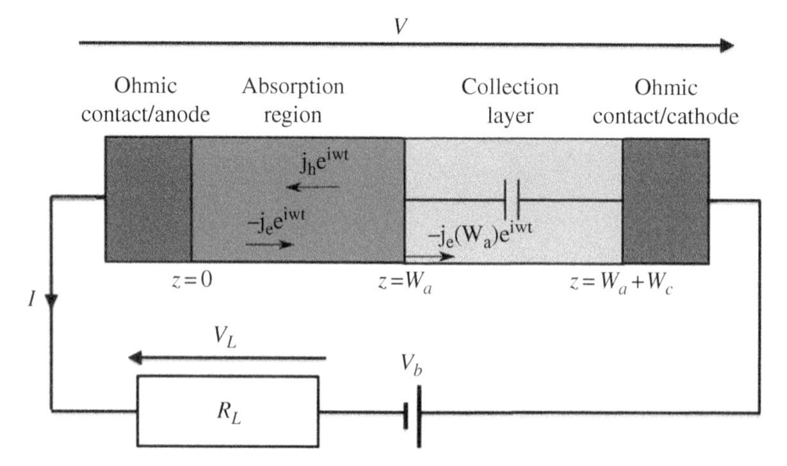

Figure 4.9 Electric model of a UTC photodiode. *Source*: Derived from Ishibashi et al. [16]; John Wiley & Sons.

And the Kirchhoff's voltage law for the oscillating terms (since $V_b = cste$) gives

$$0 = v_{ac} + R_L i_{ac} \tag{4.33}$$

where $v_{ac} = \int_0^W Edz$. By injecting Eq. (4.33) in Eq. (4.32), we obtain

$$i_{ac}(\omega) = \frac{i_c(\omega)}{1 + j\omega R_L C} \tag{4.34}$$

with

$$i_c(\omega) = \frac{A}{W}\int_0^W (j_e(\omega) + j_h(\omega))\, dz \tag{4.35}$$

with $C = \varepsilon A/W$, the electrical capacitance of the device. As a matter of fact, it can be shown that this capacitance is composed of the p-doped absorption layer capacitance in series with the nid InP collector layer $\frac{1}{C} = \frac{1}{C_a} + \frac{1}{C_c}$ with $C_a \gg C_c$, that is why we will consider from now that:

$$C \approx \varepsilon A/W_c \tag{4.36}$$

$i_c(\omega)$ can be divided into two parts:

$$\frac{A}{W}\int_0^W (j_e(\omega) + j_h(\omega))dz = \frac{A}{W}\int_0^{W_a} (j_e(\omega) + j_h(\omega))dz + \frac{A}{W}\int_{W_a}^W j_e(\omega)dz \tag{4.37}$$

The hole current in the collector is neglected since the holes are blocked by the potential barrier at the interface between the absorption layer and the collector layer. The current densities are calculated in the drift-diffusion approximation by using the poisson equation and the continuity equations in 1D:

$$\frac{\partial n}{\partial t} = \mathcal{G} - \frac{n}{\tau} + \frac{1}{e}\frac{\partial j_e}{\partial z} = \mathcal{G} - \frac{n}{\tau} - \frac{\partial}{\partial z}\left(\mu_e n(E_0 + E) - D_e\frac{\partial n}{\partial z}\right) \tag{4.38}$$

$$\frac{\partial p}{\partial t} = \mathcal{G} - \frac{p}{\tau} - \frac{1}{e}\frac{\partial j_h}{\partial z} = \mathcal{G} - \frac{p}{\tau} - \frac{\partial}{\partial z}\left(\mu_h(p + p_0)E - D_h\frac{\partial p}{\partial z}\right) \tag{4.39}$$

where p and n are the photogenerated charge carriers, p_0 is the background hole density in the absorption layer, $\mathcal{G}$ is the carrier generation rate (see the previous section related to photoconductor), and τ is the recombination lifetime in the absorption layer, E is the electric field induced by the photogenerated carriers. E_0 is a quasi field acting on the electrons in the absorption layer, which can be added by a proper design of the epitaxial layer stack in order to shorten the diffusion time of the electron out of this latter.

It can be shown according to this model [16] and by neglecting the contact resistances that:

$$i_c(\omega) = -\frac{e}{h\nu}P_{opt}\left(1 - e^{-\frac{W_a}{\delta}}\right)\frac{1}{1 + j\omega\tau_a}\frac{\sin\left(\frac{\omega\tau_{tr}}{2}\right)}{\frac{\omega\tau_{tr}}{2}}e^{-j\frac{\omega\tau_{tr}}{2}} \tag{4.40}$$

Where P_{opt} is average optical power illuminating the UTC photodiode. In this model, we neglect the reflections and losses in the structure. $\delta = \frac{1}{\alpha}$ is the absorption depth in the InGaAs layer.

Without a quasi-field in the absorption layer, the electron dynamics is related to diffusion. In that case:

$$\tau_a \approx \frac{W_A}{v_{th}} + \frac{W_A^2}{2D_e} \tag{4.41}$$

with v_{th} is the thermionic emission velocity, approximately equal to $v_{th} \approx 1 \times 10^7$cm/s in $In_{0.53}Ga_{0.47}As$ [16, 17]. For a reasonable absorption layer thickness of around 150 nm and assuming a mobility $\mu_e \approx 5000$ cm^2/V/s, it gives $\tau_a \approx 2.3$ ps and response speed is limited by the electron diffusion with a 3 dB cutoff frequency of 67 GHz. A quasi field in the absorption layer increases the frequency response by adding a drift to the electron diffusion. It is included during the epitaxial growth either by a composition gradient or a doping gradient of the InGaAs layer. It has been shown [17] that when the drift velocity $v_d > v_{th}$ the diffusion related time constant can be approximated by $\tau_a \approx \frac{W_A}{v_d}$ [17]. The drift velocity can exceed 2×10^7 cm/s, in which case the 3 dB cutoff frequency is increased to 200 GHz for the same 150-nm-thick absorption region. The faster electron transport in the absorption layer due to the built-in quasi-field is also important to reduce the carrier density and improve the response of the device in the high injection regime [18]. The electron transit time across the collector layer is given by τ_{tr}. An advantage of the UTC diodes over *p–i–n* diodes is that only electrons flow across this layer and therefore by $\tau_{tr} \approx \frac{W_C}{v_d}$ is shorter than for these latter at identical layer thickness. Moreover, electrons reach their saturation velocity at relatively low electric field strengths (<100 kV/cm) because of their high mobility. A reverse bias in the order of −1 V is sufficient for efficient operation of UTC photodiodes with ~100 nm thick collector layers. In addition, it is worth noting that velocity overshoot could play an important role in the collector layer, further decreasing τ_{tr}. The overshoot electron velocity in InP can reach 4×10^7cm/s [18]. In this case, the 3 dB frequency related to τ_{tr} is around 900 GHz for a 200-nm-thick collector layer which is clearly well beyond the cutoff frequency related to the diffusion time τ_a.

4.3.2.4 Electrical Model and Output Power

We can note from Eq. (4.40) that, in this model, the photocurrent does not depend on the voltage, in other words, the internal photoconductance of the photodiode is zero (i.e. $G = \frac{\partial I}{\partial V} = 0$). The equivalent electrical circuit of the photodiode (as shown in Figure 4.10) is thus only an AC photocurrent source ($i_c(\omega)$ defined in Eq. (4.40)) in parallel with a capacitance *C*. Obviously, in a full model of the frequency response of the photodiode, there is a small internal conductance since it is well known [19] that at high photocurrent, the frequency response of the photodiode was improved by adding an external bias voltage in order to balance the space charge screening in the device. In addition, it is clear that in a full model, the contact resistance should be added.

Concerning the power dissipated in load of admittance $Y_L = G_L + jB_L$, we can take the expression obtained in the case of a photoconductor with $G_0 = 0$. We obtain

$$P_L = \frac{1}{2}\frac{|i_c|^2 G_L}{(G_L)^2 + (\omega C + B_L)^2} \tag{4.42}$$

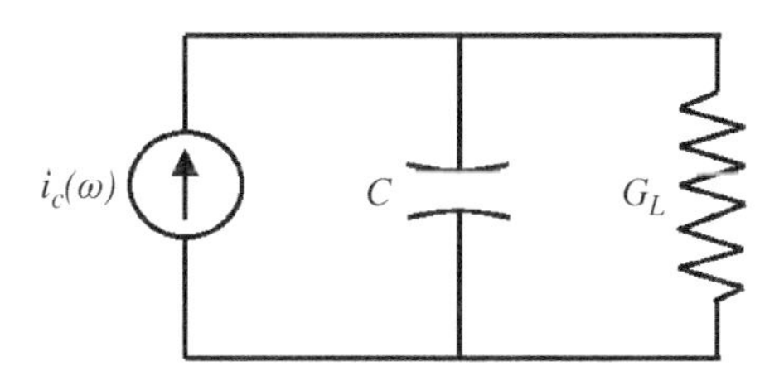

Figure 4.10 Equivalent circuit of an UTC photodiode at ω ($G_L = 1/R_L$).

which becomes when $B_L = 0$ and if we use the expression of i_c:

$$P_L = \frac{1}{2} R_L i_{dc}^2 \frac{1}{1 + (\omega R_L C)^2} \frac{1}{1 + (\omega \tau_a)^2} \left| \frac{\sin\left(\frac{\omega \tau_{tr}}{2}\right)}{\frac{\omega \tau_{tr}}{2}} \right|^2 \tag{4.43}$$

where $i_{dc} = -\frac{e}{h\nu} P_{opt}\left(1 - e^{-\frac{W_a}{\delta}}\right)$.

It is worth noting we have here three time constants instead of only two in the case of ultrafast photoconductors which can explain the highest power measured with LT-GaAs photomixers in comparison with UTC photodiodes at frequency above 2 THz (see for example [20]).

4.3.3 Frequency Down-conversion Using Photomixers

The demonstration of homodyne (or heterodyne) detection with identical photomixers performed by Verghese and coworkers [21] paved the way to practical photomixing systems by avoiding helium cooled bolometer detection. However, the photomixer used as detector has attracted little attention, although the requirements are quite different. First of all, as detectors, the photoconductor is biased only by the terahertz field which lies typically in the mV range. Therefore high breakdown voltage is not required. In counterpart, high low field mobility is a mandatory feature. As for the emitter, a low photocarrier lifetime and low electrical capacitance will be fixed the high frequency performances. In addition of a high sensitivity, a low dark current, obtained with a high resistivity photoconductive material, minimizes the Johnson noise even if the noise floor is mainly due to the shot noise, as it has been shown in time-domain spectroscopy systems [22]. With regards to 0.8-μm-laser systems, only a few improvement has been achieved since the development of the standard planar photoconductor on LT-GaAs [23]. Homodyne detection systems have been also reported using alternative photoconductive material, such as ion-implanted GaAs [24] with no improvement until the development of optical cavity LT-GaAs photoconductors that we will present in the following part [25]. Things are very different with respect to the 1.5-μm-wavelength laser systems. The development of highly sensitive, high-dark-resistivity photoconductors working at this pump wavelength is a very challenging task which opens the way to low-cost and highly reliable systems thanks to the availability of standard telecom-fiber components. Photodiodes are efficient emitters but poor detectors as we will see below. Their built-in internal electric field indeed results in a photoresponse depending slightly of the applied bias voltage unlike the photoconductors even if nonlinear effects are still possible allowing down conversion with low efficiency [26, 27]. Photoconductive materials developed for THz generation with 1.5-μm-wavelength lasers, such as high energy ion-implanted InGaAs [28], Fe:InGaAs [29], Rd:InGaAs [30], or beryllium doped low temperature grown InGaAs present a carrier lifetime reaching the sub-picosecond range, close to the values of the LT-GaAs but exhibit low dark resistivity (~1 KΩ cm), increasing the Johnson noise and thus increasing the noise equivalent power of the detector. The best results have been obtained with a 2-μm-thick multi nanolayer stack of low Beryllium doped LT-InGaAs absorbing layer, presenting a low carrier lifetime and InAlAs layer used as electron trapping layers to increase the dark resistivity of the LT-InGaAs layers [31, 32].

4.3.3.1 Electrical Model: Conversion Loss

In the general case of heterodyne detection by means of a photoconductor pumped by a beating note of two lasers beams, the dc voltage source is replaced by a RF/THz wave of frequency f which will be down-converted to an intermediate frequency f_{IF} such as f_b. The beatnote frequency f_b is then chosen such as $|f_{RF} - f_{IF}| = f_b$. Once again, the wide tunability provided by the use of a laser beatnote give the possibility to obtain a wideband heterodyne receiver only limited by the frequency response of the photoconductor. The main figure of merit of a heterodyne mixer is the conversion loss between the power of the incident THz wave (P_{THz}) and that of the down-converted signal (P_{IF}):

$$L = \frac{P_{THz}}{P_{IF}} \tag{4.44}$$

In Figure 4.11 is shown the equivalent electrical circuit in which the photoconductance biased by the incident THz wave modeled by an alternating source of internal admittance Y_s and electromotive force:

$$V_{THz}\cos(\omega_{THz}t)$$

In addition, it is assumed that the IF signal is filtered on the THz side by a coil of inductance L_d and the THz waves are filtered on the IF side by a capacitor of capacitance C_d, chosen to obtain a short-circuit at ω_{THz} and an open circuit at ω_{IF}. Conversely, L_d is chosen such that an open circuit at ω_{THz} and a short-circuit at ω_{IF} are obtained. Furthermore, as previously mentioned, the photoconductor illuminated by the optical beatnote is modeled by a time-dependent photoconductance:

$$G(t) = G_0 + G_1 \cos(\omega_b t)$$

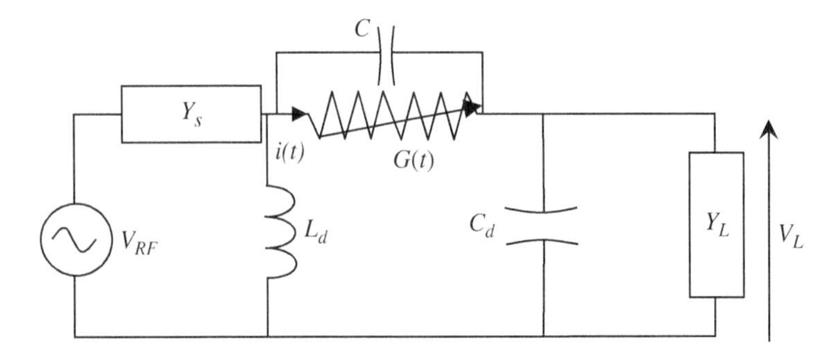

Figure 4.11 Heterodyne mixing in a photoconductor illuminated by an optical beatnote.

By neglecting the harmonics generated in the circuits, the current and the voltage across the photoconductor have the following form:

$$V(t) = V_0 \cos(\omega_{THz} + \varphi) + V_{IF} \cos(\omega_{IF} t + \delta)$$
$$i(t) = i_0 \cos(\omega_{THz} + \varphi_i) + i_{IF} \cos(\omega_{IF} t + \delta_i)$$

with $\omega_{IF} = |\omega_{THz} - \omega_b|$

By using again the relation $i(t) = G(t)V(t)$, the Kirchhoff laws at both frequencies (ω_{THz} and ω_{THz}) and by assuming that $V_{IF} \ll V_{THz}$, it is found in the simplest case where $Y_s = Y_L$ and $C = 0$ [25]:

$$L = \frac{(G_0 Z_s + 1)^4}{G_1{}^2 Z_s^2} \tag{4.45}$$

In the ideal case $G_1 = G_0$ and L has a minimum $L_{\min} = 16$ when $Ys = Y_L = G_0$

It gives in dB, $L_{dB} = 12$ dB

We have shown once again that the photoconductance G_0 must have a value close to that of load admittance or internal admittance of the THz source to reach the minimum conversion loss. The values obtained are in that case comparable to those of electronics mixers based on Schottky diodes.

4.4 Standard Photomixing Devices

4.4.1 Planar Photoconductors

Following the first demonstration of THz generation by photomixing by Brown and McIntosh [8], ultrafast photoconductors used for the generation of THz waves by photomixing have been designed on the model of the metal-semiconductor-metal photoconductors using interdigitated metallic electrodes patterned on a subpicosecond-carrier-lifetime photoconductive material. This first demonstration has been performed by using an LT-GaAs layer compatible with 800-nm-wavelength light but since then, various ultrafast photoconductive materials working at 800 or 1550 nm have been tested with always the same electrodes topology [20, 33–37]. It is shown in Figure 4.12 a scanning electron microscope (SEM) picture of this standard topology. It consists of five 200-nm-wide and 6-μm-long contact electrodes spaced from 1.8 μm apart. The active area of the photoconductor is 64 μm^2.

A THz bandwidth is obtained by using a sub-picosecond lifetime material such as LT-GaAs, LT-InGaAs, InGaAs:Fe, etc. and by limiting the electrical capacitance of the set of interdigitated electrodes at values in the order of few femtofarads. Moreover, the photodetector size should be lower than 10 microns in order to attenuate the effects due to wave interferences occurring when the wavelength of the electromagnetic wave is comparable with the dimensions of the device. The load impedance (antenna radiation impedance, waveguide characteristic impedance, power meter impedance, etc.) is in the range of 50 to 100 Ω, the $R_L C$ time constant is then also sub-picosecond. To be more specific, the planar photodetectors based on LT-GaAs shown in Figures 4.12 and 4.13 have the properties below:

- 200 fs $< \tau <$ 1 ps $\rightarrow$ 160 GHz $< f_\tau <$ 800 GHz

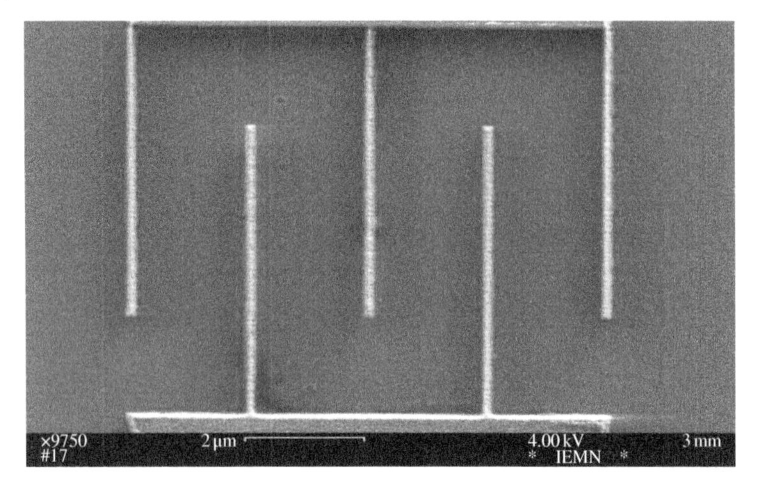

Figure 4.12 SEM picture of an ultrafast photoconductor based on interdigitated electrodes patterned on a low-carrier lifetime photoconductive material.

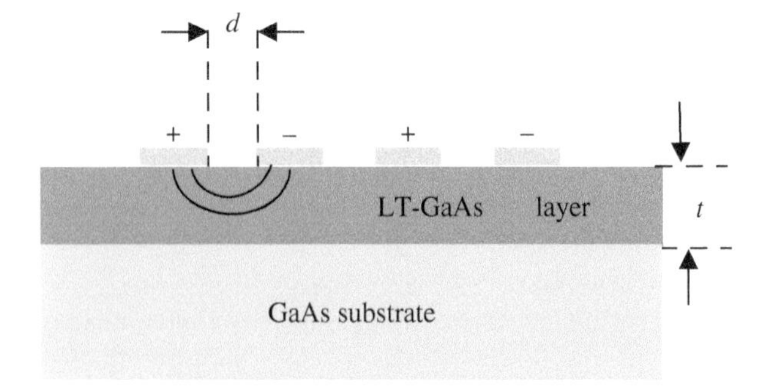

Figure 4.13 LT-GaAs planar photoconductor.

- $0.5\ \text{fF} < C < 3\ \text{fF}$ and $R_L \approx 50\ \Omega \rightarrow 25\ \text{fs} < R_LC < 150\ \text{fs}$ and $6\ \text{THz} > f_{R_LC} > 1\ \text{THz}$
- $d \approx 2\ \mu\text{m} \rightarrow$ Photoresponse $R \approx 4 \cdot 10^{-2}\text{A/W}$ with an antireflection coating and $t \approx 2\ \mu\text{m}$ (see further for the justification of the need of $t \approx d$)
- $P_{opt} \approx 100\ \text{mW}\ \mu\text{m}$ at thermal failure, $I_{dc} \approx 2\ \text{mA}$ and $V_b \approx 20\ \text{V}$
- $G_0 \approx 10\ \text{k}\Omega$

By using Eq. (4.30) with these values, we obtain an output power of $P \approx 10\ \mu\text{W}$ at 300 GHz and $P \approx 1\ \mu\text{W}$ at 1 THz.

4.4.1.1 Intrinsic Limitation

As a matter of fact, the powers emitted by the planar photomixers hardly reach 10 μW at 300 GHz and 1 μW at 1 THz [33–37], which is still largely insufficient. In fact, the only broadband detectors sensitive enough to detect this power level with a good signal-to-noise ratio (>1000) are the bolometric detectors operating at 4 K. The THz power is indeed limited by thermal effects [38]. For a given dc bias voltage, when the optical power is gradually increased, thermal failure occurs before saturation effects are observed in the THz power or dc photocurrent. Increasing the THz power can only be achieved by optimizing:

1) The ratio between the power emitted and the thermal power generated in the photodetector: i.e. the electrical and optical efficiency. This amounts to increasing the photoconductor Photoresponse ($\mathcal{R}$) or the ratio G_0/P_{opt} so that the photoconductance is closer to $G_0 = 1/50\ \Omega^{-1}$.
2) The thermal power management in close proximity of the device to hamper the temperature rise during operation.

At low electric field, when the carrier mobility is constant, it has already been shown (Eq. (4.16)) that: $G_0/P_{opt} \propto 1/l^2$ (with l the inter-electrode spacing). Furthermore, when the saturation velocity is reached at higher electric field, it is easily deduced from Eq. (4.11) that the maximum photoresponse ($\mathcal{R}_m$) is proportional to $1/l$. *The inter-electrode spacing should be decreased in order to increase* $\mathbf{G}_0/\boldsymbol{P}_{\mathbf{opt}}$ *and* $\mathcal{R}_{\boldsymbol{m}}$.

We have also to take care of the thermal properties of the device. The total thermal resistance R_{th} between the photoconductor and the substrate (assumed to be at the thermal mass) can be seen as the sum of two thermal resistances in series ($R_{th_1} + R_{th_2}$), with R_{th_1}, the contribution of the photoconductive active layer and R_{th_2} the contribution of the substrate. Once again the planar structure is penalized by the thickness of the low carrier lifetime photoconductive material required ($t \approx 2$ μm). The latter has indeed a low thermal conductivity because of the high defects density (i.e. 20 Wm^{-1}/K for LT GaAs, half as much as standard GaAs [39], and only 5 Wm^{-1}/K for standard $In_{0.53}Ga_{0.47}As$) required to obtain a subpicosecond lifetime. One should therefore reduce its thickness (t) to decrease R_{th}. On the other hand, the thermal conductivity of the standard growth substrate (InP or GaAs) is relatively low (55 Wm^{-1}/K for GaAs, 68 Wm^{-1}/K for InP), so it would seem wise to replace the semi-insulating growth substrate with a better thermal conductivity. *It is, therefore, necessary to reduce the inter-electrode spacing (l) and the thickness of the ultrafast photoconductive material* (**t**).

In the case of a planar photodetector, t and l depend on the optical absorption depth ($\delta_\alpha = 1/\alpha$) in the photoconductive material which is around 1 μm (0.7 μm at $\lambda = 780$ nm in GaAs, 1 μm at $\lambda = 1550$ nm in $In_{0.53}Ga_{0.47}As$)

1) Thus $t \approx 2$ μm is needed to optimize the optical quantum efficiency but also and especially to limit the absorption in the growth substrate in which the carrier lifetime is greater than that in the ultrafast photoconductive material by many orders of magnitude.
2) $l \approx 2$ μm is required so that the static electric field magnitude is sufficient in the whole absorption zone. Indeed, on a planar photodetector, the application of an electrical potential difference between two conductive electrodes deposited on the surface of a semiconductor considered as a dielectric produces a non-negligible static electric field down to a depth almost equal to the inter-electrode spacing. So, to have a significant electric field to a depth of 2 microns, the inter-electrode spacing must be close to this value. We will see in the next parts how this intrinsic limitation can be overcome by the use of optical cavities.

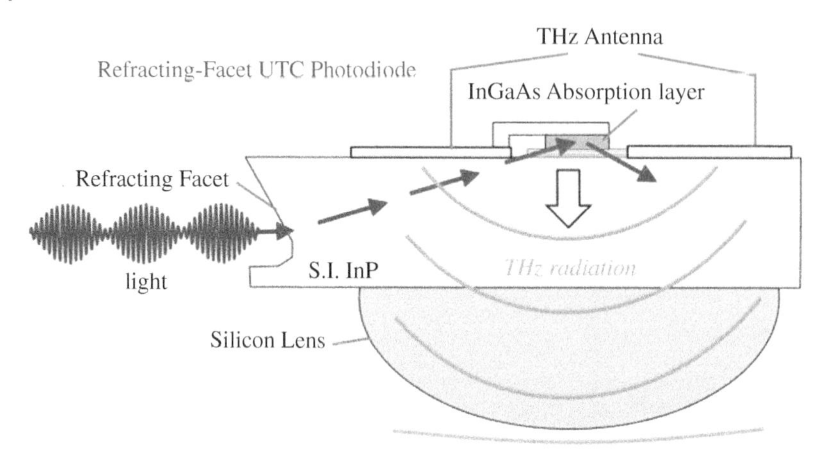

Figure 4.14 Refracting facet UTC photodiode. *Source*: Based on Nagatsuma et al. [13]; John Wiley & Sons.

4.4.2 UTC Photodiodes

4.4.2.1 Backside Illuminated UTC Photodiodes

The experimental demonstration of a cut-off frequency reaching 300 GHz [39] and a power of 20 mW generated by photomixing at 100 GHz [40] with a photocurrent reaching 25 mA has shown the potential of the uni-traveling carrier photodiode, initially developed for telecom applications, as THz photomixer. One of the main differences between both applications comes from the need of a THz antenna to radiate in free space the THz photocurrents generated in the device. As it will be seen in the next section, THz antennas monolithically integrated with photomixers are planar antennas coupled to a silicon lens, which is required to obtain an efficient extraction outside the semiconductor substrate. Below 1 THz, the most powerful photomixers are based on backside vertically illuminated UTC-PD. This structure uses a metallic opaque top p-contact electrode essential to get good thermal properties and a low contact resistance. The backside illumination is not possible when a silicon lens is added at the backside of the substrate. As shown in Figure 4.14, one possibility consists in the use of a refracting facet etched in the InP substrate used for epitaxial growth. Around 1 THz, a power of 2.6 μW has been measured when integrated with a log-periodic antenna [12] and 10.9 μW with a double-dipole resonant antenna [41]. Three-dimensional-shaped THz antennas have also been proposed to remove the need for a silicon lens. By using a transverse electromagnetic (TEM) horn antenna monolithically integrated to a UTC photodiode illuminated through the InP substrate (see Figure 4.15), a power of 1.1 μW at 1 THz has been measured [42]. At lower frequencies, the waveguide-integrated photomixers associated with a matching circuit give the best results. For example, up to 0.5 mW at 350 GHz have been obtained when an UTC PD is integrated into J-Band module [43].

4.4.2.2 Waveguide-fed UTC Photodiodes

Nevertheless, the high cut-off frequencies of the vertically illuminated UTC-PD are achieved by lowering the thickness of the absorption layer, but at the price of a low quantum

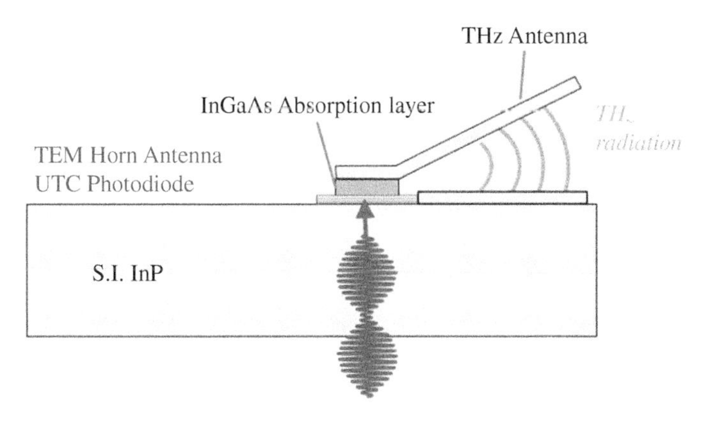

Figure 4.15 TEM horn UTC photodiode.

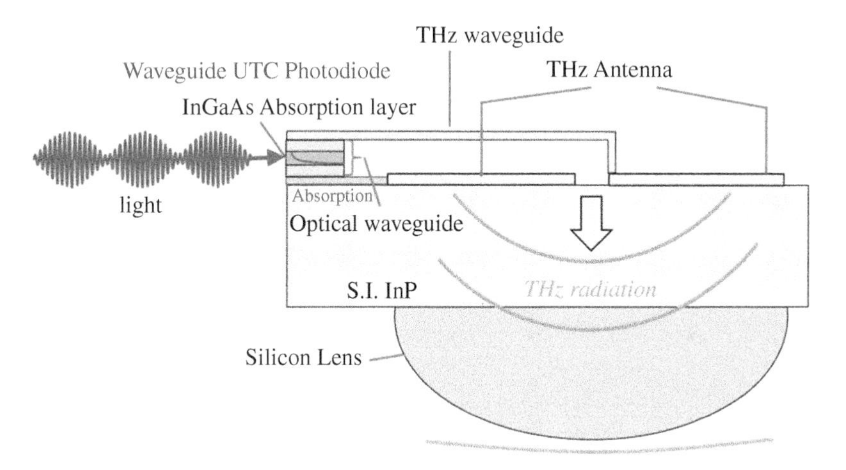

Figure 4.16 Waveguide UTC photodiode coupled to a planar antenna.

efficiency and photoresponse ($\mathcal{R} < 100$ mA/W). Using the solutions already developed for *p–i–n* PD, edge-coupled UTC-PD have been proposed to enhance photoresponse while maintaining a high frequency bandwidth and a high photocurrent saturation. For example, a waveguide-fed UTC-PD (WG-UTC-PD) is a promising structure to achieve higher output powers (see Figure 4.16). This structure is illuminated by coupling the light into the facet of a semiconductor optical waveguide obtained by thinning and cleaving the InP substrate. At 200 GHz, a maximum output power of 1 mW at a dc photocurrent of 23 mA has been measured with a WG-UTC-PD integrated to a coplanar waveguide and mounted on a Peltier cooler [44]. The integration with a silicon lensed planar antenna is somewhat easier to make than in the case of the backside illumination even if some difficulties arise from the fact that the active device is positioned at the edge of the chip. For example, an additional waveguide can be needed to feed the wideband antenna at its center. These structures integrated with a resonant antenna (dipole) have delivered 124 μW at 457 GHz and 24 μW at 914 GHz [45]. However, better results, i.e. 10 μW at 1 THz at a photocurrent of 12 mA ($\mathcal{R} = 310$ mA/W), have been achieved with an edge-coupled composite p–i–n photodiode (C-PIN) integrated with a broadband antenna [46]. A C-PIN PD is a quasi-unipolar traveling carrier photodiode. The GaInAs layer is indeed not heavily p-doped on its whole thickness. This supplementary parameter allows to both balance and reduce the response time of electrons and holes [47].

4.5 Optical Cavity Based Photomixers

4.5.1 LT-GaAs Photoconductors

We have seen that the two main geometrical parameters of a fast photodetector, namely the thickness of the photoconductive layer (t) and the inter-electrode spacing (d) are fixed by the optical absorption depth in the photoconductive material to about 2 μm. Such a limitation can be overcome by using an optical cavity. This idea has already been applied to improve the photoresponse of infrared photodiodes (see for example Refs [48–50]). We proposed some years ago to apply this concept to ultrafast photoconductors [51, 52] by using a buried metal layer as optical reflector. This metallic reflector is more suited to the development of powerful photomixers in comparison with a bragg reflector because of its better thermal properties.

4.5.1.1 Optical Modeling

In this structure (see Figure 4.17), the LT-GaAs layer (3) (or another semi-insulating photoconductor), is sandwiched between two noble metal layers (2 and 4) with low infrared loss, such as gold, silver, or copper. These two metal layers have a dual role. They serve at the same time as bias electrodes of the photoconductor and as optical mirror of the optical cavity. The bottom layer (4) acts as a near-perfect metal mirror and the top layer (2) is semi-transparent and is used to tune the transmission of the optical power in the cavity. The thickness of the layer (t) and the inter-electrode spacing (d) are now combined and can be chosen such that $t = d < \delta_\alpha$. Indeed, it is possible by using an optical Fabry–Pérot resonance, to obtain an optical quantum efficiency close to unity despite a submicron layer thickness. An infrared transparent dielectric layer (1) may be added at the surface to

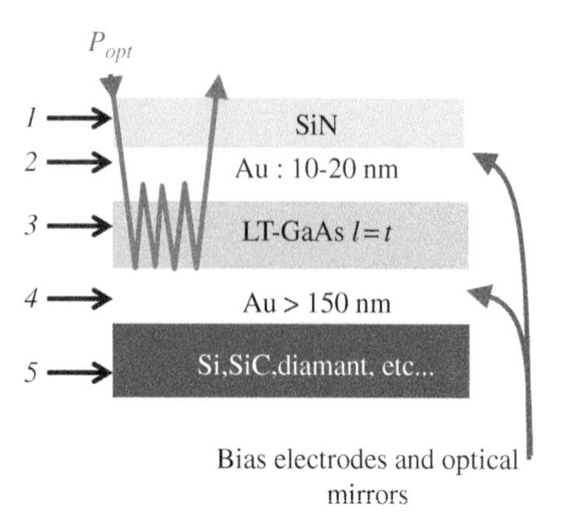

Figure 4.17 LT GaAs optical cavity photoconductor.

encapsulate the component and/or adjust the properties of the resonant cavity. This layer can consist of silicon nitride (Si_3N_4) but also silicon oxide (SiO_2) synthesized by plasma-enhanced chemical vapor deposition (PECVD). On the other hand, this structure opens the possibility to transfer the epitaxial layer of the photoconductive material on another substrate (5) by Au–Au thermocompression bonding or by soldering with an Au-In or Au-Sn eutectic. The buried metal electrode can therefore also serve as a "bonding" layer between the epitaxial layer and a host substrate chosen for its thermal and/or electromagnetic properties.

It can be shown that the resonant absorption peaks t_i of the cavity of complex refractive index ($n = Re(n) + jIm(n)$) illuminated by an optical wave of wavelength λ_0 are given by [53]:

$$t_i = t_0 + i\frac{\lambda_0}{2\,Re\,(n)} \quad \text{with } i \text{ a positive integer} \tag{4.46}$$

The first peak occurs for a thickness t_0 which depends on the phase shifts induced by the reflections on the two mirrors of the cavity. On the other hand, if it is assumed that the lower mirror is perfectly reflective, it is possible to demonstrate that the optical absorption is maximum in the semiconductor if [48]:

$$R_1 = (1 - e^{-\alpha t})^2 \tag{4.47}$$

Where R_1 is the reflectance of the top mirror assumed to be lossless (composed here of several layers that are modeled by a single mirror), α is the absorption coefficient in the semiconductor. (i.e. $\delta_\alpha = \frac{1}{\alpha} \approx 0.68$ µm in GaAs at $\lambda_0 = 780$ nm and $\alpha = 4\pi Im(n)/\lambda_0$).

In Figure 4.18 is shown the optical efficiency calculated numerically by the transfer matrix method [51] as a function of t in the case where the thicknesses of layer (1), (2), and (3) are 10, 10, and 700 nm respectively. The substrate has no influence on the optical response since a gold layer of thickness greater than 150 nm is totally opaque at these wavelengths. Numerically, we find $t_0 = 47$ nm and $t_1 - t_0 = t_{i+1} - t_i = \frac{\lambda_0}{2\,Re\,(n)} = 105$ nm as expected ($Re(n) = 3.7$).

The absorption in the photoconductor is greater than 70% of the incident optical power for $t_{0<i<8}$ (see Figure 4.18). It is worth noting that even if the mirrors are not perfect, we find, as expected, a maximum absorption ($\eta_{opt} \approx 80\%$) for peaks obtained at $t_1 = 152$ nm and

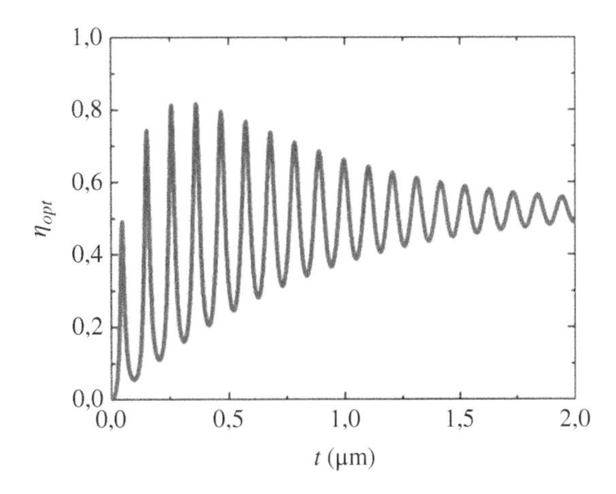

Figure 4.18 Calculated optical quantum efficiency versus active layer thickness (LT-GaAs parameters: $n = 3.7 + j0.09$ at $\lambda_0 = 780$ nm [54]).

$t_2 = 257$ nm, the closest values to $t_{max} = 200$ nm obtained using Eq. (4.47) in the case where $R_1 \approx 50\,\%$) (reflectance calculated for the layer stack: 10 nm thick gold/infinitely thick GaAs)

Thus, compared to the planar structure in which $d = 2000$ nm, *we obtain for* $d_1 = t_1 = 152$ nm *a theoretical improvement of a factor* $\frac{2000}{152} \approx 170$ *for the ratio* G_0/P_{opt} *while there is a degradation of* η_{opt} *of a factor* 0.8 *with respect to this same planar structure considering that the latter at an optical quantum efficiency* $\eta_{opt} = 100\%$.

In addition, the spectral width of the absorption resonance is compatible with the targeted application since we calculated that the absorption varies only from 1% to 2% when the optical wavelength is increased from $\lambda_0 = 770$ nm to $\lambda_0 = 790$ nm which corresponds to a beating frequency $f \approx 10$ THz.

4.5.1.2 Experimental Validation

4.5.1.2.1 Photoresponse

In order to carry out the experimental validation of the simplified model previously developed, we measured the photoresponse of an optical cavity photoconductor as a function of the inter-electrode spacing t on the same layer of LT-GaAs. We made this variation of thickness on the same sample by making a "bevel" etching of a 2.2-μm-thick LT-GaAs layer. By this method, we have obtained thicknesses ranging from 2.2 to 0.3 μm. We present Figure 4.19 the structure tested. The upper and lower metal electrodes are made of gold and serve as mirrors and bias electrodes.

In Figure 4.20 are shown the theoretical (solid line) and experimental (in squares) thickness dependences of the photoresponse measured for a constant mean electric field $\frac{V_b}{t} = 120\,\text{kV/cm}$ and an optical power $P_{opt} = 9$ mW at $\lambda_0 = 820$ nm. The $1/t$ dependence expected is convolved by the absorption peaks due to the optical resonance in the cavity. Experimental and theoretical details are further developed in Ref. [52]. It may be noted that for the peak located around $t = 250$ nm, a photoresponse of the order of 0.15 A/W has been measured, which is almost a 10 times improvement with respect to the photoresponse of a planar photoconductor. In addition, a time-resolved photoreflectance experiment has been performed on the LT-GaAs layer after epitaxial transfer (shown in the insert of Figure 4.20), which shows that the photoresponse is high despite a lifetime of less than 1 ps.

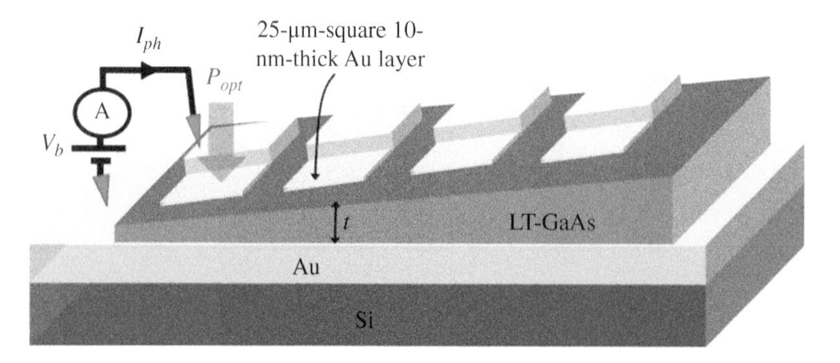

Figure 4.19 Experimental set-up aimed at photoresponse measurement.

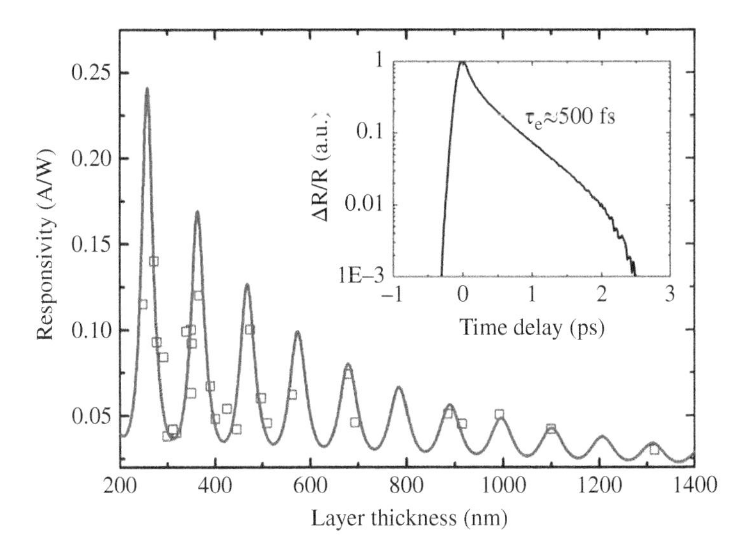

Figure 4.20 Theoretical (solid line) and experimental (in squares) photoresponses as a function of LT-GaAs layer thickness. In the inset is shown the transient differential photoreflectivity measured at $\lambda = 820$ nm on the LT-GaAs layer [52].

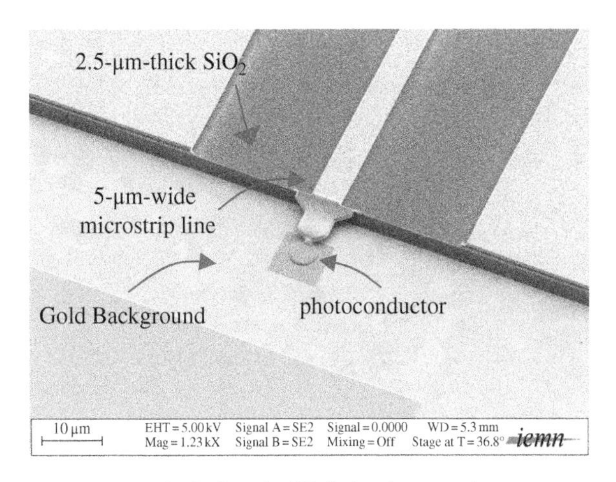

Figure 4.21 Optical cavity LT-GaAs photoconductor.

4.5.1.2.2 Photomixing Experiment in the 220–325 GHz Frequency Band

4.5.1.2.2.1 Photomixer Characteristics In Figure 4.21 is shown a SEM picture of an optical cavity LT-GaAs photoconductor using the second absorption peak. It consists of a 160-nm-thick LT-GaAs layer sandwiched between two gold layers, which serve at the same time as bias electrodes and optical mirrors of the optical cavity. The 0.4-μm-thick buried gold layer is obtained thanks to the transfer of the LT-GaAs epitaxial layer onto a 2-in.-diameter silicon wafer. The upper bias electrode consists of a 20-nm-thick semi-transparent gold layer and is

Table 4.1 Devices characteristics.

Diameter (μm)	Electrical capacitance (fF)	f_{RC} (GHz) if R_L = 50 Ω	Measured lifetime (f_s)
4	11	290	400

linked by an air bridge to a 50-Ω thin-film microstrip line patterned on a 2.55-μm-thick-SiO_2 layer. A 100-nm-thick silicon nitride layer is added to lower the top mirror reflectivity to encapsulate the device. The fabrication process of the LT-GaAs optical cavity photoconductor is detailed in [52, 55]. In the Table 4.1 are shown the main characteristics of the device.

4.5.1.2.2.2 Experimental Set-up A schematic overview of the experimental set-up is shown in Figure 4.22. A combination of half-wave plates and polarizers is used to ensure that the polarization of the two extended cavity laser diodes (ECLDs) are parallel and have equal powers. The optical beatnote is generated by spatially overlapping the two ECLDs (λ_0 = 780 nm) and used to seed a tapered semiconductor optical amplifier, the output of which is fiber-coupled. The optical wave is then focused into a 5 μm-wide gaussian spot by a lensed fiber. The wave generated in the photoconductor at the beat frequency is collected by a waveguide 220–325 GHz coplanar probe and sent to a sub-harmonic mixer driven by the spectrum analyzer or to a power meter (Erickson PM4).

4.5.1.2.2.3 Results In Figures 4.23 and 4.24 are presented the results obtained in photomixing with this device [56]. As expected, we obtained a very low photo resistance, $\frac{1}{G_0}$ = 180 Ω instead of 10 kΩ with standard planar photoconductors (and 240 Ω with a

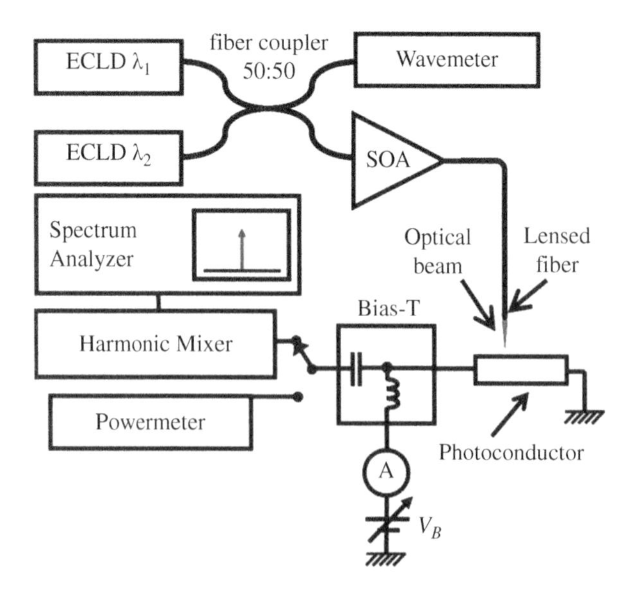

Figure 4.22 Experimental set-up. ECLD, external cavity laser diode; SOA, semiconductor optical amplifier. Lensed fiber: MFD = 5 μm.

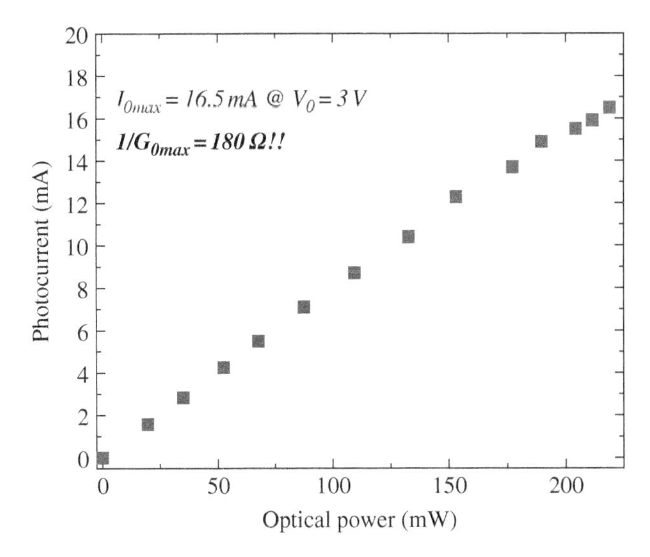

Figure 4.23 Photocurrent as a function of optical power $V_b = 3$ V.

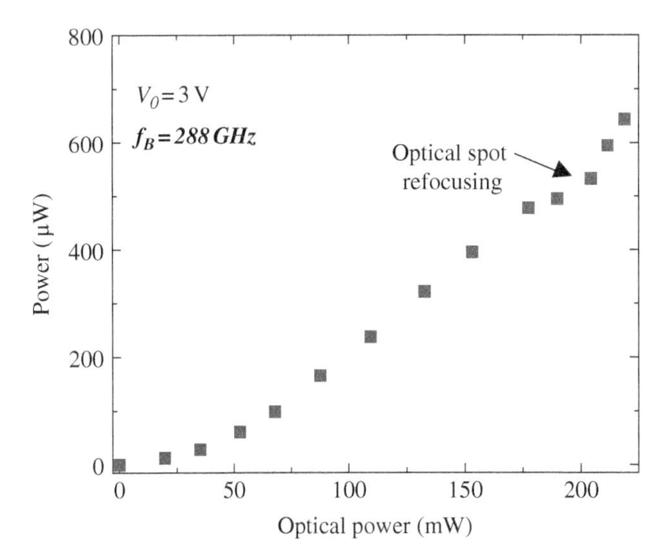

Figure 4.24 Output power at $f_B = \frac{\omega_b}{2\pi}$ = **288 GHz** and $V_b = 3$ V.

previous generation using the third optical resonance peak achieved at a thickness t_3 = 280 nm [57]). The maximum current of 16.5 mA has generated more than 600 μW at 288 GHz which once again validates the interest of reducing the thickness of the LT-GaAs layer at least for application in the lower part of the THz spectrum. It is worth noting that by using a matching circuit consisting of a ¼ wavelength transformer (see Figure 4.25) up to 1.8 mW of output power has been measured at 252 GHz with a 6-μm-diameter device delivering a photocurrent of 25 mA [58].

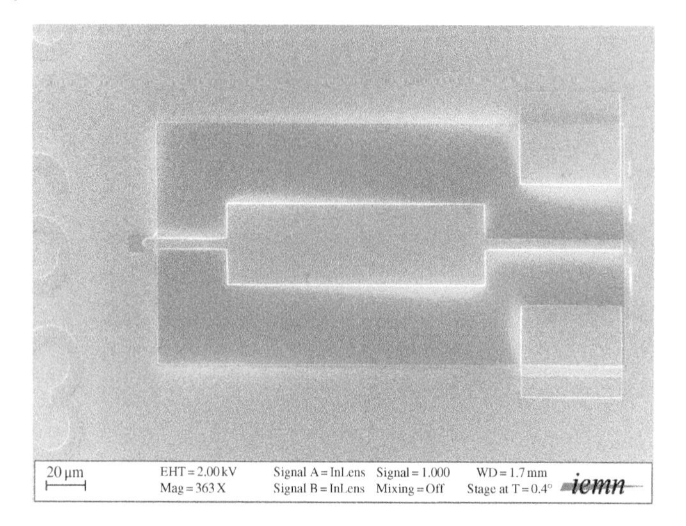

Figure 4.25 Top view of a 6-μm-diameter photoconductor coupled to the impedance matching circuit.

Table 4.2 Measured output power at 50 GHz and estimation at 1 THz as a function of device diameters.

Diameter (μm)	C (fF)	f_{RC} (GHz) if $R_L = 50\ \Omega$	Photocurrent max (mA)	Current density (mA/μm^2)	Output power (mW) at 50 GHz	Estimated power at 1 THz (μW)
2	2.7	1175	5.2	1.7	0.6	50
3	6.1	522	10.2	1.4	2.1	63
4	10.8	294	14.2	1.1	4.1	45
6	24.4	131	15.3	0.54	4.2	10
8	43.3	73	26	0.51	8.3	6

Current density is calculated by taking the maximum current density achieved experimentally.

4.5.1.2.3 Output Power Estimation at 1 THz

In Table 4.2 are shown experimental output powers obtained at 50 GHz with different device diameters as well as an estimation (using the model developed in the previous part) of the achievable power at 1 THz when taking into account the two cutoff frequencies f_{RC} and $f_\tau = 400$ GHz (measured lifetime $\tau = 400$ fs). But it should be noted that, without any adaptation circuit, we can expect more than 50 μW of output power in a 50 Ω load at 1 THz with an optical cavity LT-GaAs photoconductors of diameter $d = 3$ microns and $d = 2$ microns. It can also be seen that the current densities obtained (>160 kA/cm^2) are extremely high and it will probably be difficult to go much further. The largest diameters ($d = 8$ μm) photoconductors generating generate up to 8 mW at 50 GHz are logically less efficient at higher frequencies.

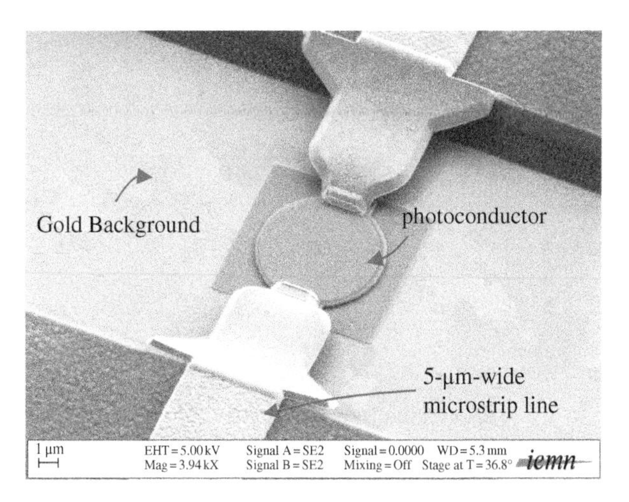

Figure 4.26 SEM micrograph of an optical cavity LT-GaAs photoconductor linked to two 50-Ω thin-film microstrip lines.

4.5.1.2.4 *Frequency Down Conversion Using An Optical Cavity LT-GaAs Photoconductor*

As already presented, it is also possible to use a photoconductor to down-convert a signal of frequency f_{THz} to a lower frequency f_{IF} compatible with the digital acquisition and signal processing systems. This is achieved by setting the optical beatnote frequency f_b such that $|f_{THz} - f_b| = f_{IF}$. It can be seen as a heterodyne mixer using a photonic oscillator. This mixing between a THz wave that polarizes the photoconductor and the optical beatnote that modulates the carrier density is possible because a photoconductor has no internal electric field unlike a photodiode. In Figure 4.26 is shown a photoconductor with two microwave accesses fabricated in order to assess the potential of the optical cavity photoconductor when used as mixer. This study has been performed by using a photoconductor of thickness $t = 160$ nm and of diameter $d = 4$ µm (see Table 4.1). Down-conversion of waves up to 325 GHz has been carried out by using multiplier chains. At each frequency point, we adjust the emission wavelength of one of the two laser diodes to have an optical beat frequency f_b such that $|f_{THz} - f_b| = 400$ MHz. A waveguide coplanar probe was used to inject the THz signal while the IF signal was collected by a 0–67 GHz coplanar probe and sent to a spectrum analyzer (Figure 4.27).

The conversion losses defined by $L = \frac{P_{THz}}{P_{IF}}$ of the order of $L = 22$ dB at 100 GHz and $L = 27$ dB in the 220–325 GHz band have been measured with an optical pump power of 100 mW. As expected, the conversion losses decrease when the optical power and therefore the photoconductance increases (see Figure 4.28). The electrical model of the experimental device shows that in the present case, with a load impedance (Z_L) and an internal source impedance such as $Z_s = Z_L$ and without THz and IF filters, (unlike the calculation made previously) the conversion loss is [25]:

$$L(dB) = 20 \log \left(\frac{(2 + Z_L G_0)^2}{Z_L G_0} \right)$$

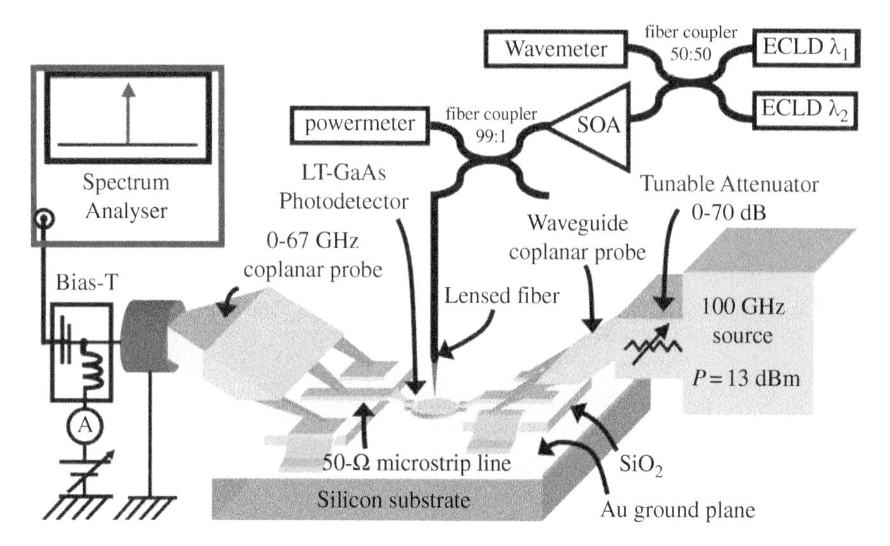

Figure 4.27 Down conversion experimental-set-up.

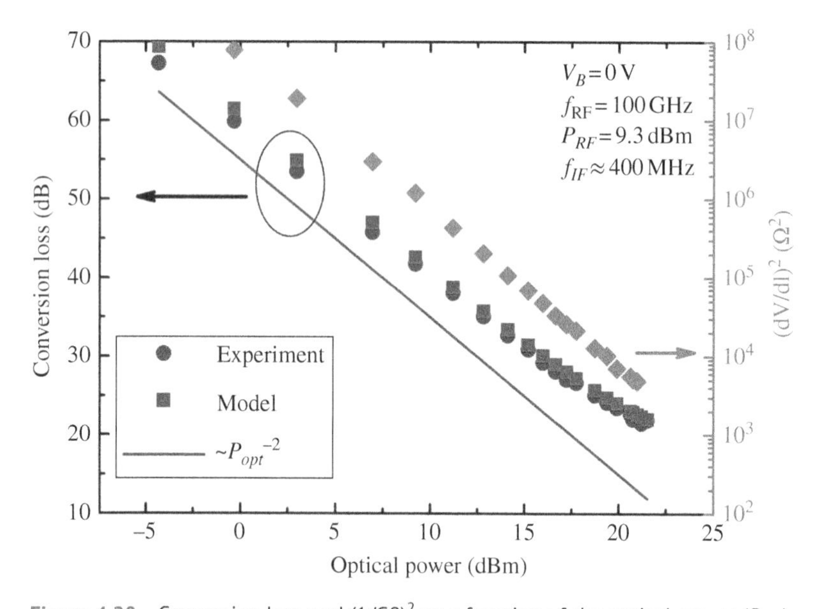

Figure 4.28 Conversion loss and $(1/G0)^2$ as a function of the optical power (P_{opt}).

which has a minimum $L_{min} = 18$ dB when $G_0 = 2/Z_L = Z_s//Z_L = (25\ \Omega)^{-1}$. This corresponds to the impedance matching between the photoconductor and its load impedance. From the above relationship, $L \propto 1/G_0^2 \propto 1/P_{opt}^2$ when $Z_L \ll 1/G_0$ and saturates when $1/G_0 \approx Z_L$. Experimental results shown in Figure 4.28 are in good agreement with this theoretical calculation. The losses related to the electrical capacitance and the carrier lifetime are of course neglected, assumption which is no longer valid at 300 GHz, which explains the higher conversion loss obtained ($L = 27$ dB).

4.5.2 UTC Photodiodes

UTC photodiodes for THz applications have active area of the order of 10 μm^2 to limit their electrical capacitance. At this scale, the contact resistance between metal and semiconductor may become a major issue. Vertically illuminated PDs using a metallic ring-based contact for mm-waves cannot be exploited efficiently anymore at THz frequencies due to the high contact resistance [59]. Most current photodiode designs for applications above 100 GHz, therefore, use a fully metalized top contact and backside illumination through the substrate [18], back-side illumination through a refracting facet [42, 60], or traveling wave structures using an edge-facet or waveguide illumination [45, 46] These illumination schemes are cumbersome and make on-wafer measurements or integration with lens-coupled THz planar antennas very challenging, especially as compared with front-side illumination.

In this context, has been proposed recently [19], a semi-transparent top contact based on a metallic layer with sub-wavelength periodic apertures covering a large fraction of the diode surface, resulting in a small contact resistance, while still maintaining a high transmittance at 1.55-μm-wavelength polarized light. This approach paves the way to efficient and powerful photomixers using front-side illumination. The proposed contact can be used with the same epilayer and mesa topology than a standard UTC-PD. Moreover, this structure is well suited to be integrated into an optical cavity by integrating the semiconductor layers between this semitransparent contact as top mirror and a buried metallic layer as bottom mirror to form a resonant cavity enhanced UTC (RCEUTC).

4.5.2.1 Nano Grid Top Contact Electrodes

The topology of the device is presented in Figure 4.29. The semi-transparent contact consists of parallel metallic strips of width $w = p - a$, with period p, aperture a, and metallization height h (Figure 4.29a). The transmittance of an infinite surface of such strips can be calculated using quasi-analytical techniques such as the coupled-modes method [61]. Figure 4.29c shows the transmittance of a p-polarized plane wave of wavelength $\lambda =$ 1.55 μm through such a surface of strips on $In_{0.53}Ga_{0.47}As$. For a choice of the semi-transparent contact parameters (a, p, h) lying in the high transmittance regions (red regions) it is possible to achieve values higher than 90%, even with 50% of the surface covered by metal, for example at $a = 0.5\,\mu m$, $h = 0.3\,\mu m$, and $p = 1\,\mu m$.

4.5.2.2 UTC Photodiodes Using Nano-Grid Top Contact Electrodes

Two parameter sets are selected for device fabrication (Figure 4.29b): (i) $a = 700$ nm, $p = 1$ μ m, $h = 300$ nm *A.3.5* (~80% transmittance) and (ii) $a = 500$ nm, $p = 1$ μm, $h = 300$ nm A.3.5 (~90% transmittance). In addition, diodes without strips, but with (iii) the contact metallization on the border of the mesa are fabricated for comparison. In that case, a 63% transmittance is obtained, corresponding to the air/semi-conductor interface ($n \sim 3.2$). Moreover, the semi-transparent contact allows the fabrication of optical cavity UTC Photodiodes by inclusion of a mirror below the active photodiode layers, improving the photoresponse of the devices [50, 58]. Schematic cross-sections of the devices are shown in Figure 4.30. All diodes are fabricated from the same epitaxy with 150 nm absorption layer and 100 nm collector layer, and a square geometry of $3 \times 3\ \mu m^2$, $4 \times 4\ \mu m^2$, and $6 \times 6\ \mu m^2$ are chosen for the mesa. SEM images of the two device types are shown in the Figure 4.31.

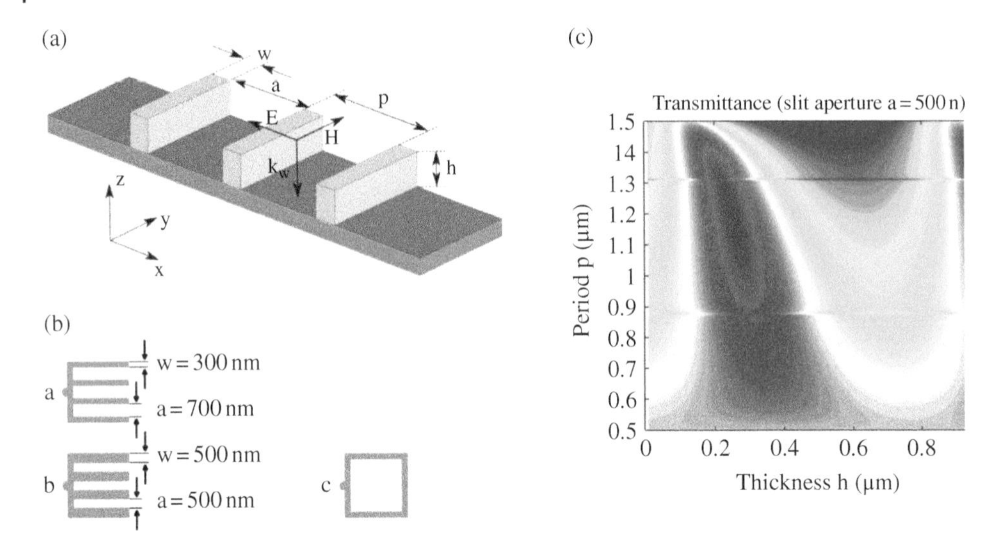

Figure 4.29 Design of the nanostructured contact. (a) Geometry of the metallic semitransparent nanostructured contact. (b) Schematic of the fabricated semitransparent contact geometries. The c-type contact without nanostructure serves as comparison. Calculated transmittance of an infinite surface of parallel strips of a perfect electrical conductor on InGaAs (n = 3.55), illuminated by a p-polarized plane wave with 1.55 μm wavelength. The calculations are performed using the coupled-mode-method and the single-mode approximation.

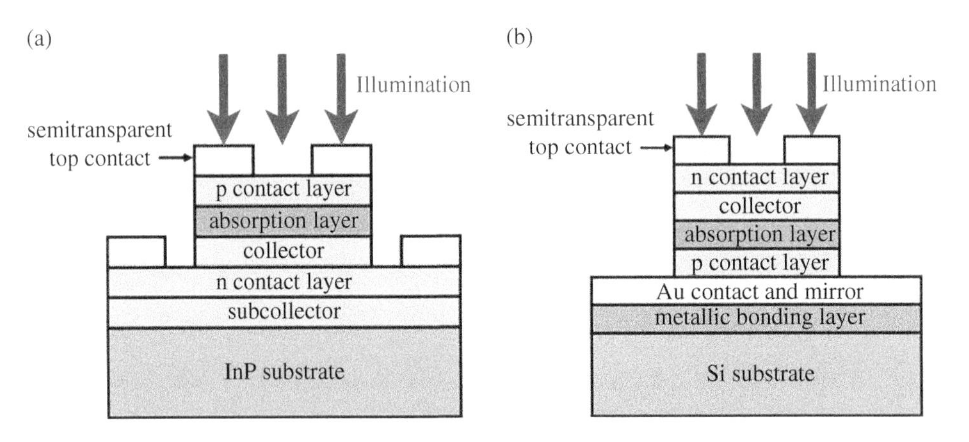

Figure 4.30 Fabrication of the PD. (a, b) Schematic cross section of the fabricated devices, (a): UTC-PD, (b): RCEUTC-PD, fabricated after wafer-bonding of the epitaxial layers to Si substrate and removal of the original InP substrate by wet-etching.

4.5.2.3 Photoresponse Measurement

In order to illustrate the enhancement of dc responsivity and THz output power of the RCEUTC, the PD's were characterized by on-wafer measurements in which the devices are illuminated by using a lensed fiber providing an optical spot of 3 μm (see Ref. [19] for further details). In the following, the RCEUTC devices are indicated by capital letters.

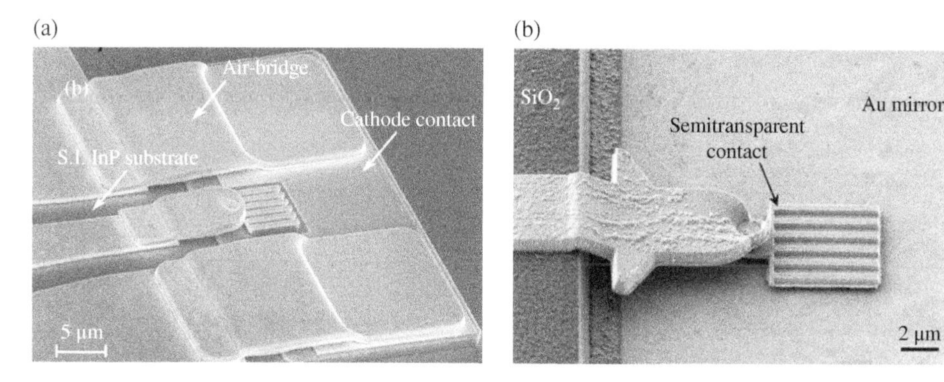

Figure 4.31 (a)SEM image of UTC-PD integrated with CPW, and (b) SEM image of RCEUTC-PD, integrated with conductor-backed-CPW lines on SiO_2 for on-wafer measurements.

Table 4.3 Dc photoresponse at 1550 nm of $6 \times 6\,\mu m^2$ UTC-PD and RCEUTC-PD devices under illumination of a single laser at low power of 320 µW.

Diode	a6	b6	c6	A6	B6
R (mA/W)	58	65	49	167	255

The dc photoresponse $\mathcal{R}$ of the larger devices ($6 \times 6\,\mu m^2$) is shown in Table 4.3. The lowest photoresponse corresponds to the device without finger contact (c6, see Figure 4.29b) and improves by nearly 30% adding the semi-transparent contact with 300 nm wide fingers (a6) and 40% by the 500 nm wide fingers (b6). By including the metallic mirror through wafer bonding, a resonant cavity is created, as evidenced by the photoresponse of the A6 and B6 devices (see Figure 4.32). The photoresponse is in both cases improved by a factor larger than three compared to the c6 device.

4.5.2.4 THz Power Generation by Photomixing

Finally, the photodiodes have been tested up to the saturation regime. In Figure 4.33, the linear and saturation regimes of the output power can be observed at high photocurrents in the devices with $3 \times 3\,\mu m^2$. The saturation is due to space charge screening at high current densities in the collector region [62] and temperature dependence of the electron transport properties. The larger device ($6 \times 6\,\mu m^2$) does not reach saturation at the measured photocurrents due to the smaller current densities, and the available optical power was not high enough to reach the saturation regime. Also, we compared two metal stacks for top-contact B3, Pt/Au and Au. The B3 Pt/Au device generates lower RF power than the B3 Au device at the same photocurrent due to the contact resistance which is larger for Pt than Au on n-InGaAs. The saturation effects can be reduced by applying a higher bias voltage, and by slightly increasing the distance between lensed fiber and diode, which reduces the maximum current density in the center of the PD at the cost of a lower photoresponse. Upon defocalisation, reducing the photoresponse by 0.5 dB, an RF power just above 750 µW at 300 GHz has been measured from a B3 device at −1.2 V bias, 83 mW optical power, and

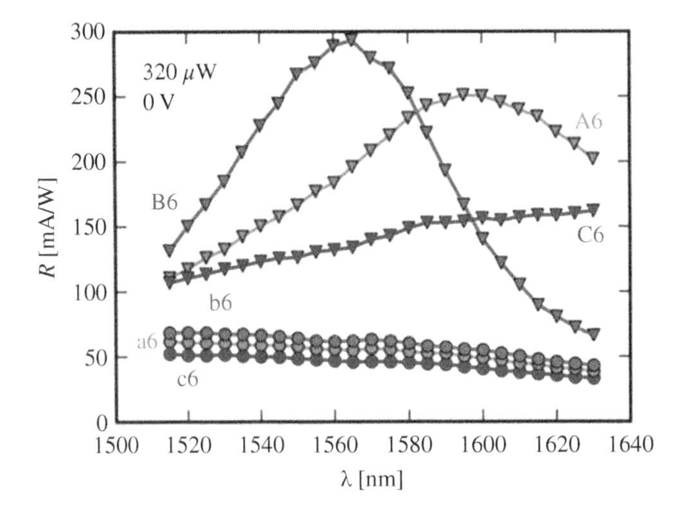

Figure 4.32 Experimental comparison of photoresponse of UTC-PD devices and RCEUTC-PD over 1510-1630 nm range.

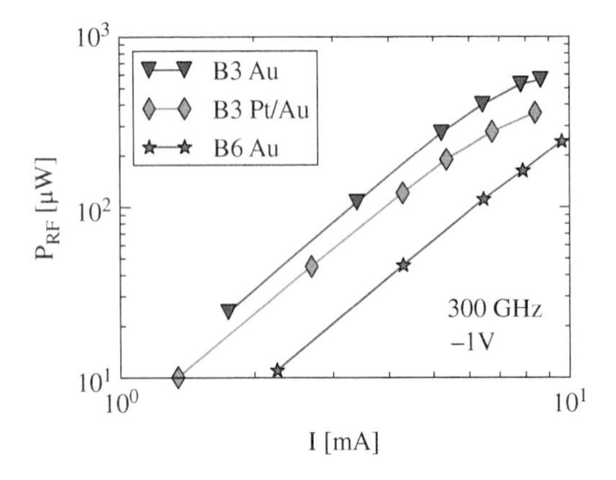

Figure 4.33 RF power generated at 300 GHz by various B-type photodiodes for $3 \times 3\ \mu m^2$ (B3) and $6 \times 6\ \mu m^2$ area (B6).

9.8 mA photocurrent. It shows that, to the best of our knowledge, the RCEUTC presented here generates the highest power for a single device at 300 GHz. These results show the real advantage of using an optical cavity consisting of a buried metallic layer (instead of a bragg reflector for example) in the case of photodiodes by improving also the electrical properties (contact resistance) and the thermal properties of the device.

4.6 THz Antennas

In the THz range, the standard rectangular metallic waveguides are lossy and the integration with a photodetector can be very challenging. The use of an integrated antenna in order to radiate directly a THz wave in free space using the generated THz currents has allowed for overcoming this difficulty. We will first focus on standard planar antennas which were the first used and then on micromachined antennas requiring more technological development.

4.6.1 Planar Antennas

The most popular terahertz antennas consist of planar conductors patterned directly on the semiconductor on which the photoconductor is made. Most of the common planar antenna geometries have been tested yet, equiangular spiral [8], logarithm-periodic [12, 63], or bow tie [64] as broadband antennas and dipole [64, 65], slot [65], dual-dipole, and dual-slot [20, 41] as resonant antennas.

Broadband antennas are generally chosen in order to take advantage of the very large intrinsic tunability of the photomixing source. In theory, they can have a purely resistive impedance over a wide frequency range. Two main principles can be used to obtain a frequency-independent antenna. Firstly, the Rumsey principle stipulates that the antenna has to be described entirely by angles [66, 67], as the bow tie or the equiangular spiral antenna. If an infinite antenna verifies the Rumsey principle the shape of the antenna remains the same regardless of the scale at which it is viewed (in the case of the spiral a rotation might be needed). Another principle that is useful to estimate the impedance of an antenna is the Babinet's principle. It states that: $Z_{Ant}Z_{Comp} = Z_s^2/4$ where Z_{Ant} is the impedance of an antenna, Z_{Comp} is the impedance of its complementary, Z_s is the impedance of free space of the surrounding medium [68]. A particular case is interesting when an antenna is equivalent to its complementary: it is self-complementary (a classical example is the infinite bow-tie antenna with an angle of 90°), in this case, the impedance of the antenna is purely resistive, independent of frequency and equal to $R_A = Z_0/2 \approx 188.4\ \Omega$ in vacuum. When a planar self-complementary antenna is patterned at the boundary between two different media of relative permittivity ε_{r_1} and ε_{r_2} the impedance becomes $R_A = Z_0/(4\varepsilon_{eff})^{1/2}$, where ε_{eff} is the effective dielectric constant: $\varepsilon_{eff} = (\varepsilon_{r_1} + \varepsilon_{r_2})/2$. Let us focus on the two most common broadband antennas for photomixers: the equiangular spiral antenna [69] and the logarithm-periodic antenna [70].

The equiangular spiral antenna patterned on a GaAs substrate (see Figure 4.34) used by Brown and McIntosh [8], is self-complementary. Contrary to what was supposed in the previous paragraph practical antennas are finite. They cannot be infinitely large then a low-frequency cut-off appears when the overall length of the spiral arm is approximately equal to the radiation wavelength. They cannot be also infinitely small close to the center. The smallest inner radius in the center of the spiral defines the high frequency cut-off. Between these two cut-off frequencies, the previous calculation can be used to approximate the radiation resistance in the working bandwidth. For GaAs, $\varepsilon_r = 12.9$ and then $R_A \approx 71\ \Omega$. In this bandwidth, the radiated field is circularly polarized, the type of polarization depending on the direction of winding of the spiral. The size of the antenna can be chosen to have a frequency cut-off around 0.1 THz. The high frequency cut-off is limited by the photodetector dimension. For instance, it has been shown that for a 8×8-μm^2 LT-GaAs planar photoconductor at the center of a spiral antenna, the radiated field is no more circularly but linearly polarized parallel to the fingers of the interdigitated electrodes above $f = 0.6$ THz [71, 72]. It shows that the standard design of the spiral antenna is rapidly inefficient at high frequency and that the terahertz wave is preferentially radiated by other conductive parts even if the

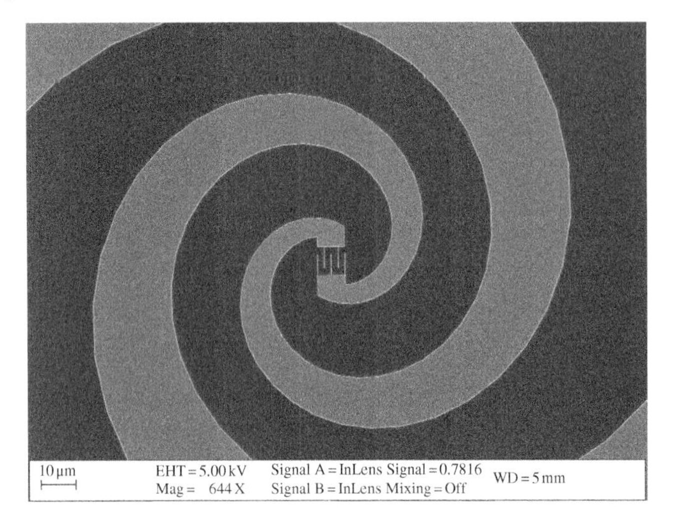

Figure 4.34 SEM picture of an equiangular spiral antenna.

guided wavelength $\lambda_g = c/(f \times \varepsilon_{eff}^{1/2}) \approx 190$ μm is much larger than the inner radius. This observation is consistent with the steeper decrease than expected of the power radiated by the spiral antenna-coupled-photomixers [73].

The log-periodic antenna can be seen as an array of tooth-shaped resonant dipole, attached perpendicularly to the bow-tie arms. This geometry does not fulfill the Rumsey principle, but it damps the resonance brought by the finite length of a real bow-tie antenna [70]. The operation bandwidth is defined by the smallest and the largest teeth. At each resonant frequency, the radiated field is linearly polarized parallel to the resonant tooth. The frequency dependency of the radiation impedance is clearly not as smooth as that of the spiral antenna. It impacts also the frequency dependency of the output power of log-periodic antenna-coupled-photomixers [63].

The radiation impedance of the planar broadband antennas on semiconductors is generally below 100 Ω. As seen before, higher antenna impedances can be a way to improve the output power and the efficiency of the photomixing process. Record output power has been achieved, whatever the type of photodetector, with resonant antenna whose impedance can reach some hundreds ohm at the resonance [20, 41, 45, 74]. However, it is worth noting, that improvements are achieved thanks to the higher antenna impedance, but also thanks to a matching condition that can be obtained near the resonance in a narrow band: the capacitance of the photodetector is compensated by an inductive behavior of the antenna. A matching circuit can be also added in order to increase the bandwidth.

The radiation pattern of a planar antenna on a high permittivity substrate is very different of the same antenna in vacuum. In this case, the most part of the electromagnetic energy is radiated into the semiconductor because of its high permittivity. If a standard substrate is used (a few hundreds of μm thick), guided electromagnetic modes are excited and trapped into the wafer (surface waves) [75]. The impedance of the antenna and its radiation pattern depend on the exact dimensions and shape of the substrate and tend to be very resonant. In practice, the characteristics of the antenna are difficult to control. In order to avoid this behavior, the standard solution consists in applying a "substrate lens" to the backside of the substrate [76, 77]. It is the analog of the solid immersion lens used in microscopy; ideally, the lens should have the same permittivity as the substrate. For a GaAs substrate ($n_{GaAs} \approx$ 3.6), high-resistivity silicon ($n_{HRSi} \approx 3.42$) spherical lenses are generally used because they are easier to manufacture and handle. Following the concepts of geometrical optics, at high frequencies, the planar antenna can be approximated as a point source. In this case, perfect stigmatism is obtained for the Weierstrass case: the height of the truncated sphere is $(1 + 1/n)R$ where R is the radius of the spherical surface of the lens. For HR Si the height should be 1.29 times the radius of the sphere, they are called hyperhemispherical lenses. Thanks to these lenses the divergence of the beam is largely reduced but a better collimation can be achieved using an extra lens or parabolic mirror. The refractive index of semiconductors being high, hemispherical lenses (height equal to the radius) can also be used with a slightly reduced efficiency. In all cases, some waves are still trapped in the substrate + lens system because of the Fresnel reflections at the silicon/air boundary, and broadband anti-reflection coatings are not available. In practice, the substrate lens is the simplest way to reduce the trapped waves of planar antennas but its spatial alignment is critical, particularly at high frequency, and it requires good translation stages with micrometer screws. The substrate lens can be bonded after active alignment but precautions must be taken to avoid gaps between the lens and the substrate that can induce parasitic reflections. Another constraint of the substrate lens is that the photomixer cannot be illuminated from the backside of the substrate.

The exact calculation of the radiation pattern of planar antennas coupled to a high permittivity substrate lens is still a difficult problem and only a few reports give simulations and practical measurements [78–80]. The following section deals with the few attempts of alternative antennas or waveguide structures which have been done in order to overcome the limits of the lens-coupled planar antenna.

4.6.2 Micromachined Antennas

Antenna on semiconductor membranes have been proposed [81], but results have been disappointing because of the low thermal conductance of the membrane, which limits the maximum optical power that can pump the photomixer. More interesting results have been obtained with a transverse electromagnetic horn antenna (TEM-HA) integrated with both a LT-GaAs photoconductor [37] and with a UTC-Photodiode [42].

A TEM-HA (Figure 4.35) can be seen as a flat version of a truncated section of a TEM biconical antenna [68]. It is an end-fire traveling wave antenna consisting of two triangular conductive sheets of angular width 2α separated by an angle 2β. Entirely defined by angles, the infinite antenna respects the Rumsey principle. In practice, the low-frequency cut-off is defined by the length of the triangle. For the particular case $\beta = 90°$ the TEM-HA is equivalent to a bow-tie antenna. The monopole configuration of the TEM-HA, with a ground plane at the symmetry plane, is shown in Figure 4.35. The 3D shape of this antenna is simpler and the conductive plane is used to prevent the radiation from spreading in the substrate (shielding).

In Figure 4.36 is shown a SEM picture of a TEM-HA Horn monolithically integrated with a LT-GaAs photoconductor [37]. The metallic triangle has been patterned on a sacrificial

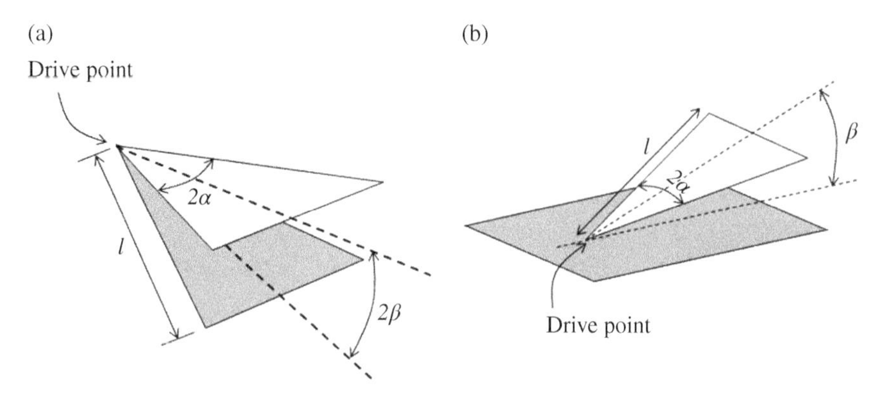

Figure 4.35 Geometry of the TEM-HA (a). Geometry of the monopole configuration of the TEM-HA (b).

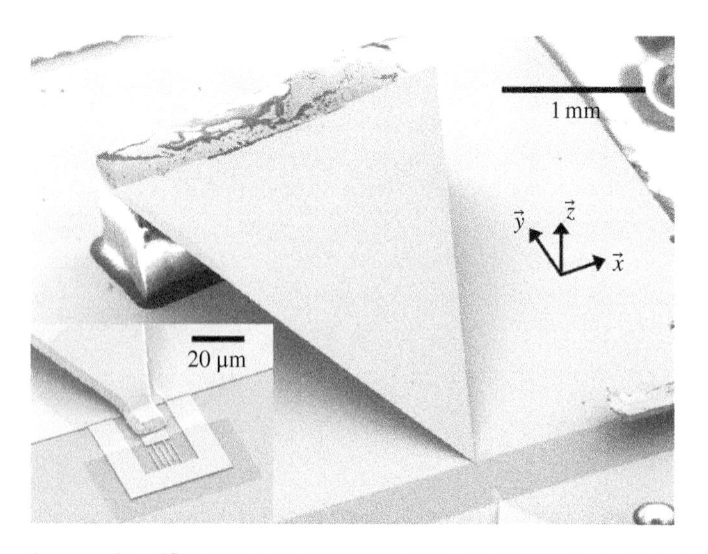

Figure 4.36 SEM picture of a THz Horn antenna. *Source*: Peytavit et al. [37]; AIP Publishing LLC.

layer and then lift up to the desired height (here $h = 800\ \mu m$) and brazed to a metalized polytetrafluoroethylene (PTFE) block.

Radiation patterns have been calculated [82] and a good agreement with measurements has been found [83] (Figure 4.37). The TEM-HA radiates linearly polarized waves directly in free space and thus resolves most of the problem of the lens-coupled planar antenna, but its fabrication needs a post-process packaging step, which is likely to limit its diffusion.

To conclude this section about antennas for photomixers, it must be noted that even if the integrated antennas are necessary for most of spectroscopy applications they have a limited efficiency especially at the highest frequencies. Improvements are still needed in this domain. Another aspect is that the evaluation of the absolute power produced by a photomixer integrated with an antenna is a difficult task: impedance variations, collection efficiency, and back reflections limits generally the power and the precision to which it can be measured (as mentioned in the last part of this chapter). The record output powers, which

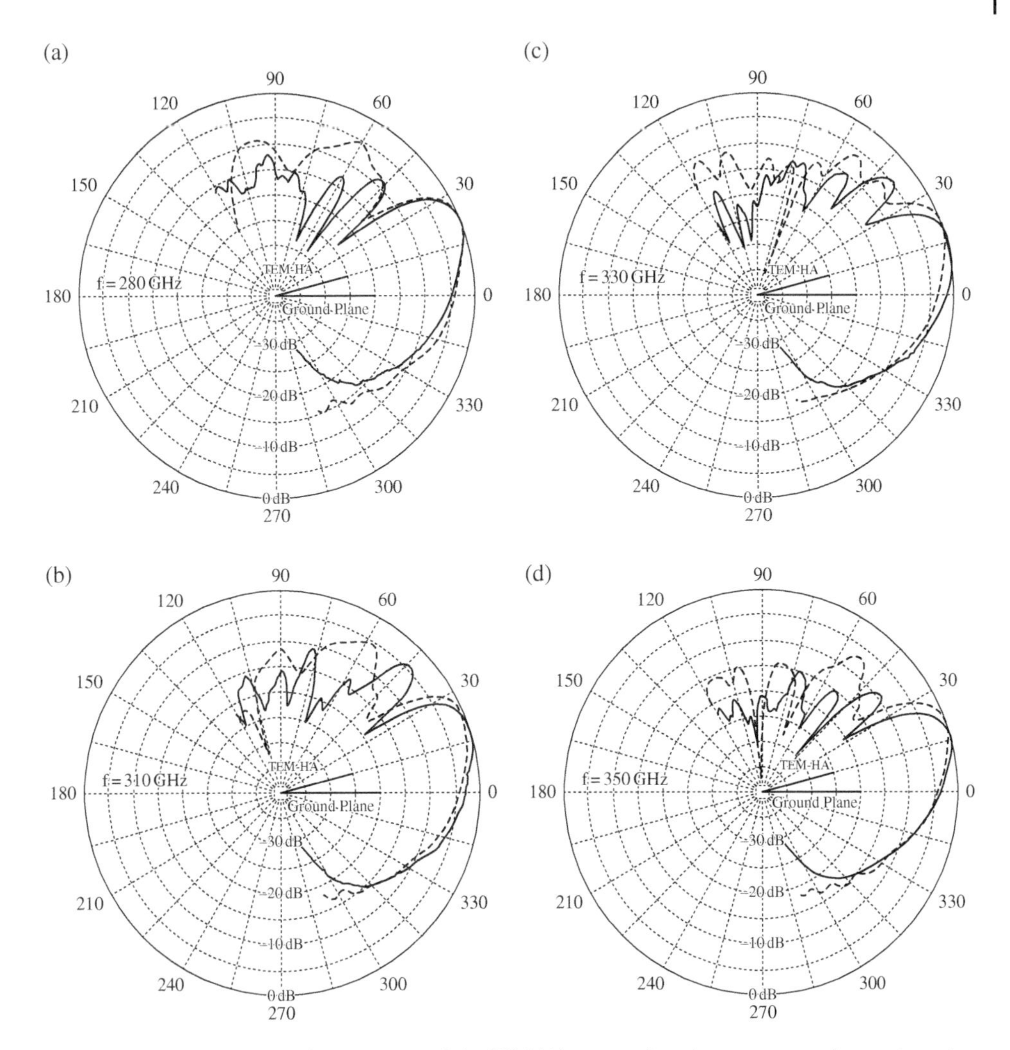

Figure 4.37 E-plane radiation patterns of the TEM-HA: comparison between experimental results (full line) and simulations (dashed line) at 280, 310, 330, and 350 GHz.

reach the milliwatt level, at least in the low-frequency range, have been reported with waveguide-coupled-photomixers [18, 43] or by means of waveguide-coplanar probes coupled to waveguide power meters [44, 57].

4.7 Characterization of Photomixing Devices

4.7.1 On Wafer Characterization

Since the early applications of THz waves, many of the photoconductive devices and photodiodes have been used and characterized in free space, driven by applications or just by the fact that wafer-level characterization has an overall quite high cost. However, accurate

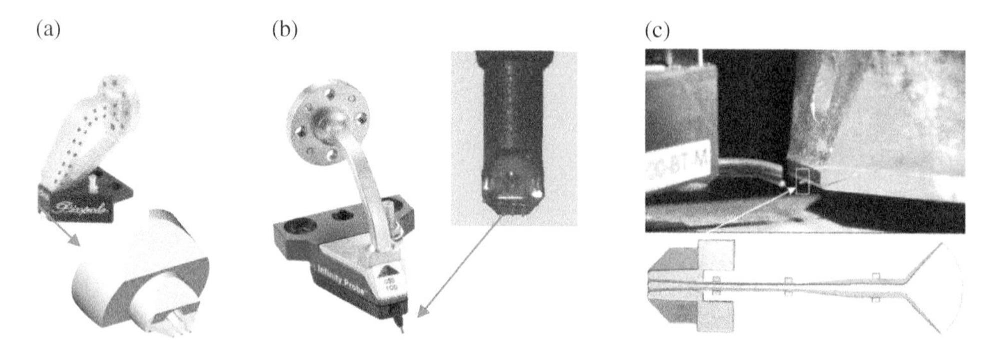

Figure 4.38 Different probes technologies: GGB probes (Picoprobes for wafer-level testing. http://www.ggb.com/) (a). *Source*: "Picoprobes for wafer-level testing." [Online]. Available: http://www.ggb.com/; GGB Industries. Infinity-micro-coax (b). *Source*: https://www.formfactor.com/product/probes/infinity/infinity-waveguide-probe/; FormFactor. up to 500 GHz (https://www.formfactor.com/product/probes/infinity/infinity-waveguide-probe/) and micromachined probes (c) [85]. *Source*: Reck et al. [85]; IEEE.

characterization of devices for modeling generally requires on-wafer testing. This can be conducted at several frequency ranges, however, depending on the probe used the band can be reduced. On-wafer measurements and device modeling is generally done using coaxial probes up to 110 GHz that can provide four frequency decades usually when combining with coaxial vector network analyzers (VNAs). Recently, VNA coaxial extensions have now reached up to 220 GHz in a single sweep (https://www.anritsu.com/en-us/test-measurement/news/news-releases/2019/2019-05-31-us01). However up to now, beyond 110 or 140 GHz, measurements are conducted mostly in waveguide configuration. In that case, the bandwidth is 1 : 1.5 related to the waveguide range. Definitions of usual waveguide bands can be found here (https://vadiodes.com/VDI/pdf/waveguidechart200908.pdf). The probe technology is frequency band depending: looking on high frequency measurements up to 500 GHz, probes integrated with micro-coaxial tips are available (https://www.formfactor.com/product/probes/infinity/infinity-waveguide-probe/). Beyond 500 GHz, Picoprobes now also enables (Picoprobes for wafer-level testing. http://www.ggb.com/) measurements up to 1.1 THz, with micro-coax connection at probe tip. Recently, new probe technology was introduced [84] using micromachined silicon circuit at probe tip, and these devices enable wafer-probing up to 1.1 THz (T-waves probes, Form Factor company) (see Figure 4.38). This technology, thanks to the microfabrication techniques is scalable in frequency and enables up to now probing up to 1.1 THz. All of these probes have a footprint corresponding to a coplanar contact (Ground Signal Ground). To ensure a good signal coupling to the device as the wavelength is reduced, the "pitch" size, i.e. the physical length between signal and ground contacts as to be reduced accordingly. In the 140–220 GHz range a 100 or 50 μm GS distance is required. Beyond 220 GHz a pitch of 50 μm becomes mandatory and the distance is further reduced down to 25 μm for 325–500 or 500–750 GHz. The highest frequency band, commercially available for on-wafer testing is 1.1 THz, corresponding to WR1.0 waveguide (750–1100 GHz). Looking at photomixer on-wafer testing, the Figure 4.39a gives an overview of a typical characterization setup for on-wafer measurements [19].

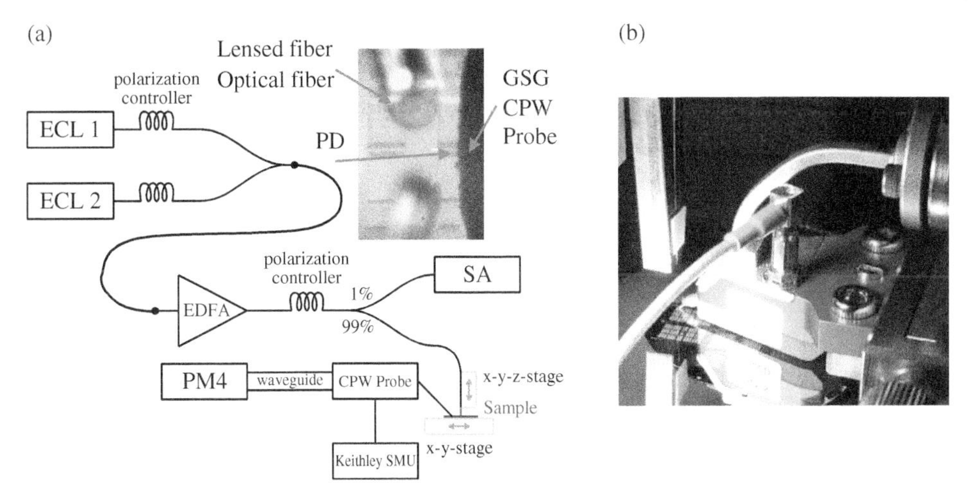

Figure 4.39 (a) Experimental setup for photomixing at wafer-level. (b) Photograph of the probe station showing the wafer-level photomixers, the fiber optic injection, and the waveguide-probe.

In this kind of photomixing setup, two external cavity lasers are used as sources for cw photomixing. A fiber coupler is used to couple the two detuned laser wavelengths and a polarization controller to adjust the polarization of the optical wave injected inside de photomixers. Indeed, depending of the photomixer structure used the sensitivity can be polarization dependent and, in that case, the polarization of the two beams has to be carefully adjusted. Generally, the laser powers, considering the extra losses of fiber components (polarization controllers, couplers, ...) are not reaching the required power to properly feed the photomixers, then optical dual-frequency signal is often injected inside an erbium-doped fiber amplifier (EDFA).

As the EDFA gain is generally depending of the wavelength, a fraction of the optical power is injected into optical spectrum analyzer (OSA) to monitor the spectrum. In this spectrum, the difference frequency of the lasers has to be monitored during the during RF power measurements. At the end, the optical signal is injected inside the photomixing device with a lensed fiber. The spot size is also depending of the photomixer structure, for the example given here, it is around 3 μm, linked to the UTC-PD diameter. The Figure 4.39b gives a photo of InGaAs/InP photodiodes during wafer-level testing. This example of measurement uses a waveguide coplanar probe in WR3.4, with integrated bias-Tee (Cascade Microtech infinity i325). Usually, the DC output of the bias-Tee is connected to a source meter to bias the device while measuring the photocurrent, meanwhile RF output (ground–signal–ground, 50-μm pitch) of the probes is feeding a waveguide-based power-meter, in that case, this is a PM5 from Virginia Diodes Inc (VDI). This power meter has the advantage to give an absolute power level measurement, and thanks to the waveguide coupling, the signal is properly coupled in the detector. However, there are two limitations: (i) the signal frequency range is limited to the probe/waveguide bandwidth and (ii) the electrical response of the probe (amplitude and phase) has to be accurately known to evaluate the photomixer performances. A first example of the frequency dependence of the output power in the 220–325-GHz band is shown in Figure 4.40. While on the range 220–270 GHz

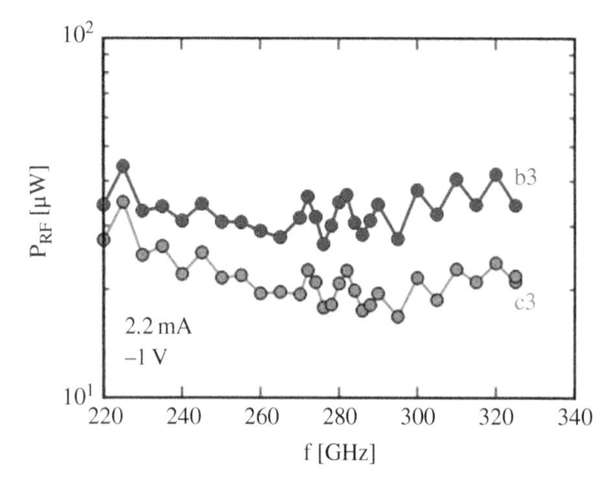

Figure 4.40 Frequency response example of a UTC-PD [19].

the photodiode roll-off is appearing, beyond 270 GHz some ripples are affecting the measurements. These oscillations are due to the frequency dependence of the load impedance presented by the coplanar probe. This roll-off is due to an inductive part of the load impedance, which increases with frequency and partially matches the load of the device and access pad capacitance.

A second example of UTC-photodiode (same UTC size) output power measurement is given by the frequency evolution (Figure 4.41a) and linearity (Figure 4.41b) of the device at 300 GHz. this measurement is done this time using a micro-machined probe, and the spectral evolution of the output power is not exactly the same, partially linked to the loading probe effect. On-wafer testing thus requires a good knowledge of the probe effect on the device as absolute power measurement is concerned.

4.7.2 Free Space Characterization

Even if on-wafer measurement is generally well suited for device modeling and cut-off frequency analysis, characterization up to the THz range is enabled using photomixer + antenna structure. In that case, the THz signal is radiated directly into the free space and different power-meters can be used.

A large part of the THz photomixing devices has been integrated with different antenna types. This can be planar antennas (bow-tie, spiral, dipoles) as well as integrated antennas used as transitions toward waveguide structures. In that case, it is generally difficult to accurately measure the output power levels because of the radiation in free space of the THz signal. In this case, the measured performance is the mixing of the photomixer + antenna structure, not the photomixer itself. Using planar antennas, generally associated with a silicon lens, the signal power level is mostly dependent of the silicon lens alignment. It can also be affected by the presence of an air gap between substrate and lens, leading to deleterious frequency response and echos in time-domain. Looking on detector side, several solutions exist today: pyroelectric detectors, Schottky diodes integrated with silicon lenses,

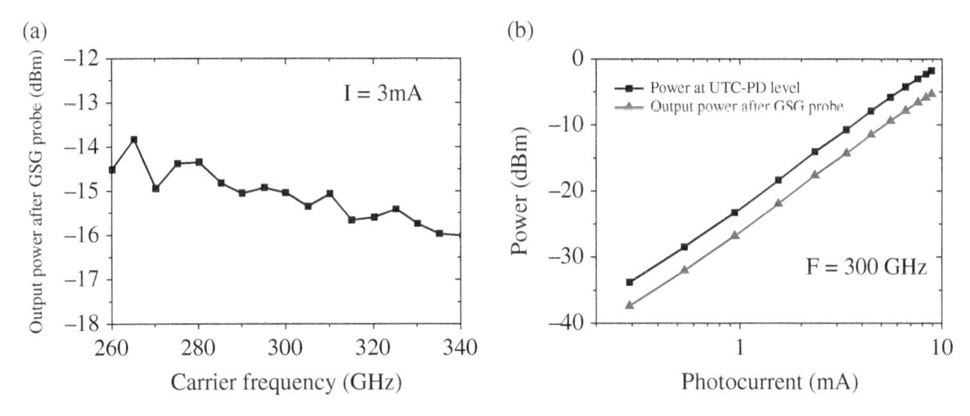

Figure 4.41 Frequency response example of and device linearity.

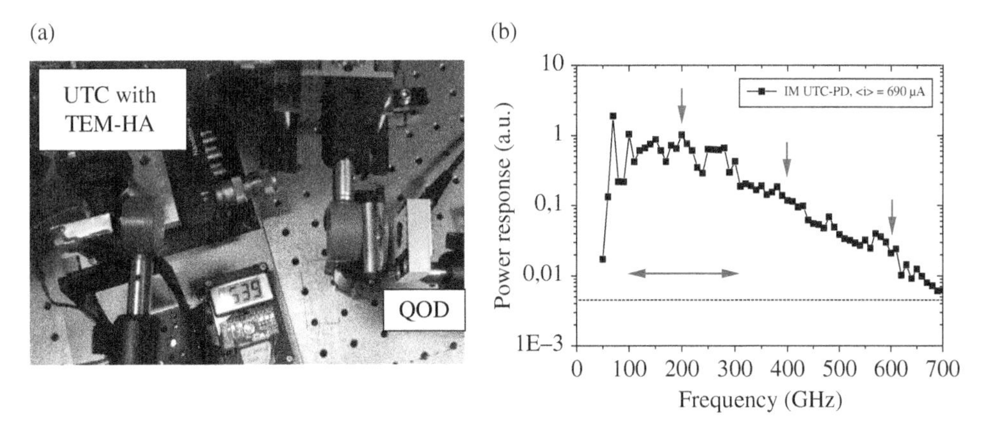

Figure 4.42 Example of free space UTC-PD measurement using a Schottky quasi-optic detector (Quasi-optic detector from VDI Inc. https://www.vadiodes.com/VDI/pdf/2009AugustNews.pdf). Some ripples on the frequency response are observed due to the stationary waves between THz source and receiver. *Source*: Modified from "Quasi-optic detector from VDI Inc." [Online]. Available: https://www.vadiodes.com/VDI/pdf/2009AugustNews.pdf; John Wiley & Sons.

as well as free-space detectors with traceable calibration from the german national metrology institute (PTB) (https://www.ptb.de/cms/en/presseaktuelles/journals-magazines/ptb-news/ptb-news-ausgaben/archivederptb-news/ptb-news-2014-2/detector-standard-for-terahertz-radiation.html), as well as Golay cells or Thomas Keating detector (https://www.ptb.de/cms/en/presseaktuelles/journals-magazines/ptb-news/ptb-news-ausgaben/archivederptb-news/ptb-news-2014-2/detector-standard-for-terahertz-radiation.html). Additional information can be found in [86].

The Figure 4.42 hereafter gives an example of the free-space characterization of a broadband nonresonant UTC-PD integrated with a TEM-horn antenna [42], this UTC is then associated to a set of lenses to feed a quasi-optic Schottky module for THz reception. As can be seen on the frequency response, even if the spectral response of the device is broadband, some ripples in the curve can easily reach 3 dB variations at specific frequencies, coming from stationary wave (S.W.) effects between the THz source and receiver. This experimental limitation is often encountered in such kind of measurements, and in that case, the use of isolators could be helpful. In addition, as soon as the power-meter or detector is designed to reduce the level of back-reflection at the detector input, these S.W. can be greatly reduced. Generally speaking, the power measurement in free-space configuration is still a challenging issue in the sub-mm or THz range and intensively investigated by the THz researchers.

Exercises

Exercise A. Photodetector Theory

E4.A1 By assuming that each photon incident on a photodetector results in an electron-hole pair reaching the contact electrodes, give the maximum photoresponse achievable as a function of the photon wavelength.

E4.A2 Demonstrate that the Photoresponse × bandwidth product of a THz photoconductor based on a low-carrier lifetime semiconductor is independent of the carrier lifetime if we neglect the electrical capacitance.

E4.A3 Calculate the photoresponse at $\lambda = 780$ nm of an ultrafast photoconductor based on a 1 μm-thick LT-GaAs layer (lifetime $\tau = 500$ fs) with contact electrodes separated from $s = 1$ μm by assuming that:

- There is no reflection of the incident light at the top and bottom interfaces of the layer.
- Absorption depth in GaAs at $\lambda = 780$ nm:$\delta \approx 0.7$ μm.
- Photocurrent results only from the electron drift at a velocity $v_s = 10^7$ cm/s.

E4.A4 Explain why a uni-traveling carrier photodiode has better current saturation bandwidth properties.

E4.A5 Give the physical origin of the response time of a UTC photodiode.

E4.A6 Plot the ratio conduction current (i_c)/average optical power (P_{opt}) of an UTC PD of surface 10 μm^2 illuminated by an optical beatnote as a function of the beatnote frequency:

UTC layer properties:

- A 150-nm-thick absorbing InGaAs layer with a gradient composition.
- A 300-nm-thick InP collector layer in which the electron is assumed to drift at their overshoot velocity.

The absorption depth in InGaAs is $\delta \approx 1$ μm.

Exercise B. Photomixing Model

1. Ultrafast Photoconductor

We study a photomixing source based on a planar photoconductor of capacitance $C = 5$ fF, photoresponse 0.02 A/W at $V_b = 5$ V pumped by an incident optical beatnote of average power $P_{opt} = 50$ mW. The carrier lifetime of the photoconductive material is $\tau = 400$ fs. It is assumed that it behaves as a perfect photoconductance up to V_b. The impedance of the antenna coupled to the photoconductor is equal to 70 Ω (purely radiative).

E4.B1.1 By using the equivalent electrical model, give the expression of the output power as a function of the frequency.

E4.B1.2 Plot the output power as a function of the frequency from 10 GHz to 1 THz.

E4.B1.3 Calculate the optical power required to reach the impedance matching between the photomixing source and the antenna. Is it realistic?

2. UTC Photodiode

The UTC PD of the question A6 is pumped by an optical beatnote at $\lambda = 1550$ nm of average power $P_{opt} = 30$ mW which is assumed to reach without loss the absorption layer. The impedance of the antenna coupled to the photoconductor is equal to 70 Ω (purely radiative).

E4.B2.1 By using the equivalent electrical model, give the expression of the output power as a function of the frequency.

E4.B2.2 Plot the output power as a function of the frequency from 10 GHz to 1 THz.

E4.B2.3 Is it possible to achieve the impedance matching condition.

Exercise C. Antennas

We want to integrate a photoconductor with a spiral antenna.

E4.C1 Calculate the impedance of an infinite self-complementary THz spiral antenna fabricated on a semi-infinite quartz substrate (refractive index of quartz in the THz range: 2.0).

E4.C2 Is the spiral antenna on quartz more favorable than the same antenna on GaAs from the point of view of impedance matching? (you can repeat exercise B1 with this new antenna impedance).

References

1 Forrester, A.T., Gudmundsen, R.A., and Johnson, P.O. (1955). Photoelectric mixing of incoherent light. *Physics Review* **99** (6): 1691–1700.

2 Javan, A., Ballik, E.A., and Bond, W.L. (1962). Frequency characteristics of a continuous-wave He-Ne optical maser. *Journal of the Optical Society of America* **52** (1): 96–98.

3 Mcmurtry, B.J. and Siegman, A.E. (1962). Photomixing experiments with a ruby optical maser and a traveling-wave microwave phototube. *Applied Optics* **51** (1): 133–135.

4 Inaba, H. and Siegman, A.E. (1962). Microwave photomixing of optical maser outputs with a PIN-junction photodiode. *Proceedings of the IRE* **50** (8): 1823.

5 Riesz, R.P. (1962). High speed semiconductor photodiodes. *Reviews of Scientific Instruments* **33** (9): 994.
6 DiDomenico, M., Pantell, R.H., Svelto, O., and Weaver, J.N. (1962). Optical frequency mixing in bulk semiconductors. *Applied Physics Letters* **1** (4): 77.
7 Coleman, P.D., Eden, R.C., and Weaver, J.N. (1964). Mixing and detection of coherent light. *IEEE Transactions on Electron Devices* **ED-11** (11): 488–497.
8 Brown, E. and McIntosh, K. (1995). Photomixing up to 3.8 THz in low-temperature-grown GaAs. *Applied Physics Letters* **66** (3): 285–287.
9 Smith, F.W., Calawa, A.R., Chen, C.-L. et al. (1988). New MBE buffer used to eliminate backgating in GaAs MESFETs. *IEEE Electron Device Letters* **9** (2): 77–80.
10 Smith, F.W., Le, H.Q., Diadiuk, V. et al. (1989). Picosecond GaAs-based photoconductive optoelectronic detectors. *Applied Physics Letters* **54** (10): 890.
11 Gupta, S., Frankel, M.Y., Valdmanis, J.A. et al. (1991). Subpicosecond carrier lifetime in GaAs grown by molecular beam epitaxy at low temperatures. *Applied Physics Letters* **59** (25): 3276.
12 Ito, H., Nakajima, F., Furuta, T., and Yoshino, K. (2003). Photonic terahertz-wave generation using antenna-integrated uni-travelling-carrier photodiode. *Electronics Letters* **39** (25): 24–25.
13 Nagatsuma, T., Ito, H., and Ishibashi, T. (2009). High-power RF photodiodes and their applications. *Laser and Photonics Reviews* **3** (1–2): 123–137.
14 Furuta, T., Ishibashi, T., Matsuoka, Y. et al. (1998). Pin photodiode with improved frequency response and saturation output. United States patent US 5,818,096.
15 Ishibashi, T., Shimizu, N., Kodama, S. et al. (1997). Uni-traveling-carrier photodiodes – OSA trends in optics and photonics series. *Ultrafast Electronics and Optoelectronics* **13**: UC3.
16 Ishibashi, T., Kodama, S., Shimizu, N., and Furuta, T. (1997). High-speed response of uni-traveling-carrier photodiodes. *Japanese Journal of Applied Physics* **36** (10): 6263–6268.
17 Feiginov, M.N. (2007). Analysis of limitations of tetrahertz p-i-n uni-traveling-carrier photodiodes. *Journal of Applied Physics* **102** (8): 084510.
18 Ito, H., Kodama, S., Muramoto, Y. et al. (2004). High-speed and high-output InP–InGaAs unitraveling-carrier photodiodes. *IEEE Journal of Selected Topics in Quantum Electronics* **10** (4): 709–727.
19 Latzel, P., Pavanello, F., Billet, M. et al. (2017). Generation of mW level in the 300-GHz band using resonant-cavity-enhanced unitraveling carrier photodiodes. *IEEE Transactions on Terahertz Science and Technology* **7** (6): 800–807.
20 Duffy, S.M., Verghese, S., McIntosh, A. et al. (2001). Accurate modeling of dual dipole and slot elements used with photomixers for coherent terahertz output power. *IEEE Transactions on Microwave Theory and Techniques* **49** (6): 1032–1038.
21 Verghese, S., McIntosh, K.A., Calawa, S. et al. (1998). Generation and detection of coherent terahertz waves using two photomixers. *Applied Physics Letters* **73** (26): 3824.
22 van Exter, M. and Grischkowsky, D.R. (1990). Characterization of an optoelectronic terahertz beam system. *IEEE Transactions on Microwave Theory and Techniques* **38** (11): 1684–1691.
23 Gregory, I.S., Tribe, W.R., Baker, C. et al. (2005). Continuous-wave terahertz system with a 60 dB dynamic range. *Applied Physics Letters* **86** (20): 204104.
24 Roggenbuck, A., Schmitz, H., Deninger, A. et al. (2010). Coherent broadband continuous-wave terahertz spectroscopy on solid-state samples. *New Journal of Physics* **12** (4): 043017.

25 Peytavit, E., Pavanello, F., Ducournau, G., and Lampin, J.-F. (2013). Highly efficient terahertz detection by optical mixing in a GaAs photoconductor. *Applied Physics Letters* **103** (20): 201107.

26 Rouvalis, E., Fice, M., Renaud, C., and Seeds, A. (2012). Millimeter-wave optoelectronic mixers based on uni-traveling carrier photodiodes. *IEEE Transactions on Microwave Theory and Techniques* **60** (3): 686–691.

27 Rouvalis, E., Fice, M.J., Renaud, C.C., and Seeds, A.J. (2011). Optoelectronic detection of millimetre-wave signals with travelling-wave uni-travelling carrier photodiodes. *Optics Express* **19** (3): 2079.

28 Chimot, N., Mangeney, J., Joulaud, L. et al. (2005). Terahertz radiation from heavy-ion-irradiated In[sub 0.53]Ga[sub 0.47]As photoconductive antenna excited at 1.55 μm. *Applied Physics Letters* **87** (19): 193510.

29 Globisch, B., Dietz, R.J., Kohlhaas, R.B. et al. (2017). Iron doped InGaAs: competitive THz emitters and detectors fabricated from the same photoconductor. *Journal of Applied Physics* **121** (5): 053102.

30 Kohlhaas, R.B., Globisch, B., Nellen, S. et al. (2018). Rhodium doped InGaAs: a superior ultrafast photoconductor. *Applied Physics Letters* **112** (10): 102101.

31 Göbel, T., Stanze, D., Globisch, B. et al. (2013). Telecom technology based continuous wave terahertz photomixing system with 105 decibel signal-to-noise ratio and 35 terahertz bandwidth. *Optics Letters* **38** (20): 4197.

32 Globisch, B., Dietz, R.J.B., Stanze, D. et al. (2014). Carrier dynamics in Beryllium doped low-temperature-grown InGaAs/InAlAs. *Applied Physics Letters* **104** (17): 1–5.

33 Mangeney, J., Meng, F., Gacemi, D. et al. (2010). CW THz generation by In 0.53 Ga 0.47 As photomixer with TEM-Horn antenna driven at 1.55 μm wavelengths. IRMMW-THz 2010 – 35th International Conference on Infrared, Millimeter, and Terahertz Waves, Conference Guide.

34 Mangeney, J., Merigault, A., Zerounian, N. et al. (2007). Continuous wave terahertz generation up to 2 THz by photomixing on ion-irradiated In 0.53 Ga 0.47 As at 1.55 μm wavelengths. *Applied Physics Letters* **91** (24): 241102.

35 Mohandas, R.A., Freeman, J.R., Rosamond, M.C. et al. (2016). Generation of continuous wave terahertz frequency radiation from metal-organic chemical vapour deposition grown Fe-doped InGaAs and InGaAsP. *Journal of Applied Physics* **119** (15): 153103.

36 Bjarnason, J.E., Chan, T.L., Lee, A.W. et al. (2004). ErAs:GaAs photomixer with two-decade tunability and 12 μW peak output power. *Applied Physics Letters* **85** (18): 3983–3985.

37 Peytavit, E., Beck, A., Akalin, T. et al. (2008). Continuous terahertz-wave generation using a monolithically integrated horn antenna. *Applied Physics Letters* **93** (11): 111108.

38 Verghese, S., Mcintosh, K.A., and Brown, E.R. (1997). Optical and terahertz power limits in the low-temperature-grown GaAs photomixers. *Applied Physics Letters* **71**: 2743–2745.

39 Ito, H., Furuta, T., Kodama, S., and Ishibashi, T. (2000). InP/InGaAs uni-travelling-carrier photodiode with 310 GHz bandwidth. *Electronics Letters* **36** (21): 1809.

40 Ito, H., Nagatsuma, T., Hirata, A. et al. (2003). High-power photonic millimetre wave generation at 100 GHz using matching-circuit-integrated uni-travelling-carrier photodiodes. *IEE Proceedings-Optoelectronics* **150** (2): 138.

41 Nakajima, F., Furuta, T., and Ito, H. (2004). High-power continuous-terahertz-wave generation using resonant-antenna-integrated uni-travelling-carrier photodiode. *Electronics Letters* **40** (20): 3–4.
42 Beck, A., Ducournau, G., Zaknoune, M. et al. (2008). High-efficiency uni-travelling-carrier photomixer at 1.55 μm and spectroscopy application up to 1.4 THz. *Electronics Letters* **44** (22): 1320.
43 Wakatsuki, A., Furuta, T., Muramoto, Y. et al. (2008). High-power and broadband sub-terahertz wave generation using a J-band photomixer module with rectangular-waveguide output port. 2008 33rd International Conference on Infrared, Millimeter and Terahertz Waves, 1–2.
44 Rouvalis, E., Renaud, C.C., Moodie, D.G. et al. (2012). Continuous wave terahertz generation from ultra-fast InP-based photodiodes. *IEEE Transactions on Microwave Theory and Techniques* **60** (3): 509–517.
45 Rouvalis, E., Renaud, C.C., Moodie, D.G. et al. (2010). Traveling-wave uni-traveling carrier photodiodes for continuous wave THz generation. *Optics Express* **18** (11): 11105–11110.
46 Henning, I.D., Adams, M.J., Sun, Y. et al. (2010). Broadband antenna-integrated, edge-coupled photomixers for tuneable terahertz sources. *IEEE Journal of Quantum Electronics* **46** (10): 1498–1505.
47 Yoder, P.D. and Flynn, E.J. (2007). Quasi-unipolar InGaAs/InP photodetection for enhanced optical saturation power and maximal bandwidth. *Applied Physics Letters* **91** (6): 062114.
48 Farhoomand, J. and McMurray, R.E. (1991). Design parameters of a resonant infrared photoconductor with unity quantum efficiency. *Applied Physics Letters* **58** (6): 622.
49 Kishino, K., Unlu, M.S., Chyi, J.-I. et al. (1991). Resonant cavity-enhanced (RCE) photodetectors. *IEEE Journal of Quantum Electronics* **27** (8): 2025–2034.
50 Li, N., Chen, H., Duan, N. et al. (2006). High power photodiode wafer bonded to Si using Au with improved responsivity and output power. *IEEE Photonics Technology Letters* **18** (23): 2526–2528.
51 Peytavit, E. and Lampin, J.-F. (2010). Photomelangeur pour la generation de rayonnement terahertz. WO2011030011 A3.
52 Peytavit, E., Coinon, C., and Lampin, J.-F. (2011). A metal-metal Fabry–Pérot cavity photoconductor for efficient GaAs terahertz photomixers. *Journal of Applied Physics* **109** (1): 016101.
53 Yariv, A. and Yeh, P. (2006). *Photonics: Optical Electronics in Modern Communications*. Oxford University Press.
54 Palik, E.D. (1998). *Handbook of Optical Constants of Solids*, vol. **3**. Academic Press.
55 Peytavit, E., Coinon, C., and Lampin, J.-F. (2011). Low-temperature-grown GaAs photoconductor with high dynamic responsivity in the millimeter wave range. *Applied Physics Express* **4** (10): 104101.
56 Peytavit, E., Latzel, P., Pavanello, F. et al. (2013). Milliwatt output power generated in the J-Band by a GaAs photomixer. Proceedings of 38th International Conference on Infrared, Millimeter and Terahertz Waves, IRMMW-THz, 1–3.
57 Peytavit, E., Lepilliet, S., Hindle, F. et al. (2011). Milliwatt-level output power in the sub-terahertz range generated by photomixing in a GaAs photoconductor. *Applied Physics Letters* **99** (22): 223508.

58 Peytavit, E., Latzel, P., Pavanello, F. et al. (2013). CW source based on photomixing with output power reaching 1.8 mW at 250 GHz. *IEEE Electron Device Letters* **34** (10): 1277–1279.

59 Demiguel, S., Campbell, J.C., Tulchinsky, D., and Williams, K.J. (2004). A comparison of front- and backside-illuminated high-saturation power partially depleted absorber photodetectors. *IEEE Journal of Quantum Electronics* **40** (9): 1321–1325.

60 Ishibashi, T., Muramoto, Y., Yoshimatsu, T., and Ito, H. (2014). Unitraveling-carrier photodiodes for terahertz applications. *IEEE Journal of Selected Topics in Quantum Electronics* **20** (6): 79–88.

61 Garcia-Vidal, F.J., Martin-Moreno, L., Ebbesen, T.W., and Kuipers, L. (2010). Light passing through subwavelength apertures. *Reviews of Modern Physics* **82** (1): 729–787.

62 Williams, K.J. and Esman, R.D. (1999). Design considerations for high-current photodetectors. *Journal of Lightwave Technology* **17** (8): 1443.

63 Mendis, R., Sydlo, C., Sigmund, J. et al. (2005). Spectral characterization of broadband THz antennas by photoconductive mixing: toward optimal antenna design. *IEEE Antennas and Wireless Propagation Letters* **4** (1): 85–88.

64 Matsuura, S., Tani, M., and Sakai, K. (1997). Generation of coherent terahertz radiation by photomixing in dipole photoconductive antennas. *Applied Physics Letters* **70** (5): 559.

65 McIntosh, K., Brown, E., and Nichols, K. (1996). Terahertz measurements of resonant planar antennas coupled to low-temperature-grown GaAs photomixers. *Applied Physics Letters* **69**: 3632–3634.

66 Rumsey, V. (1957). Frequency independent antennas. *IRE International Convention Record* **5**: 114–118.

67 Rumsey, V. (1966). *Frequency Independent Antennas*. New York, NY: Academic Press.

68 Kraus, J.D. (1950). *Antennas*. New York, NY: McGraw-Hil.

69 Dyson, J. (1959). The equiangular spiral antenna. *Antennas Propagation, IRE Transactions* **7** (2): 181–187.

70 DuHamel, R. and Isbell, D. (1957). Broadband logarithmically periodic antenna structures. *IRE International Convention Record* **5**: 119–128.

71 Gregory, I.S., Baker, C., Tribe, W.R. et al. (2005). Optimization of photomixers and antennas for continuous-wave terahertz emission. *IEEE Journal of Quantum Electronics* **41** (5): 717–728.

72 Ravaro, M., Manquest, C., Sirtori, C. et al. (2011). Phase-locking of a 25 THz quantum cascade laser to a frequency comb using a GaAs photomixer. *Optics Letters* **36** (20): 3969.

73 Peytavit, E., Akalin, T., Lampin, J.-F. et al. (2009). THz photomixing: comparison between horn and spiral antennas. 2009 34th International Conference on Infrared, Millimeter, and Terahertz Waves, 1–3.

74 Ito, H. and Ishibashi, T. (2017). Photonic terahertz-wave generation using slot-antenna-integrated uni-traveling-carrier photodiodes. *IEEE Journal of Selected Topics in Quantum Electronics* **23** (4): 1–7.

75 Rutledge, D.B., Neikirk, D.P., and Kasilingam, D.P. (1983). Integrated circuits antennas. In: *Infrared and Millimeter Waves*, vol. **10** (ed. K.J. Button), 1–90. New York, NY: Academic Press.

76 Neikirk, D.P. (1982). Far-infrared imaging antenna arrays. *Applied Physics Letters* **40** (3): 203.

77 Fattinger, C. and Grischkowsky, D. (1989). Terahertz beams. *Applied Physics Letters* **54** (6): 490.

78 Jepsen, P.U. and Keiding, S.R. (1995). Radiation patterns from lens-coupled terahertz antennas. *Optics Letters* **20** (8): 807.

79 Liu, L., Hesler, J.L., Xu, H. et al. (2010). A broadband quasi-optical terahertz detector utilizing a zero bias schottky diode. *IEEE Microwave and Wireless Components Letters* **20** (9): 504–506.

80 Mayorga, I.C. (2008). *Photomixer as Tunable Terahertz Local Oscillators*. Bonn.

81 Sydlo, C., Sigmund, J., Hartnagel, H.L. et al. (2002). Efficient THz-emitters for low-temperature-grown GaAs photomixers. Proceedings, IEEE Tenth International Conference on Terahertz Electronics, 60–62.

82 Beck, A., Akalin, T., Ducournau, G. et al. (2010). Terahertz photomixers based on ultra-wideband horn antennas. *Comptes Rendus Physique* **11** (7–8): 472–479.

83 Prissette, L., Ducournau, G., Akalin, T. et al. (2011). Radiation pattern measurements of an integrated transverse electromagnetic horn antenna using a terahertz photomixing setup. *IEEE Microwave and Wireless Components Letters* **21** (1): 49–51.

84 Reck, T., Chen, L., and Zhang, C. (2011). Micromachined probes for submillimeter-wave on-wafer measurements—Part I: mechanical design and characterization. *IEEE Transactions on Terahertz Science and Technology* **1** (2): 349–356.

85 Reck, T.J., Chen, L., Zhang, C. et al. (2010). Micromachined on-wafer probes. 2010 IEEE MTT-S International Microwave Symposium, 65–68.

86 Popovic, Z. and Grossman, E.N. (2011). THz metrology and instrumentation. *IEEE Transactions on Terahertz Science and Technology* **1** (1): 133–144.

5

Plasmonics-enhanced Photoconductive Terahertz Devices

Mona Jarrahi

Department of Electrical and Computer Engineering, University of California, Los Angeles, CA, USA

5.1 Introduction

Photoconductive antennas (PCAs) are extensively used for generation and detection of terahertz (THz) radiation [1–20]. The concept of photoconductive THz devices was first introduced in 1984 by David H. Auston at Bell Labs for pulsed THz generation [21, 22]. Therefore, PCAs are also called "Auston Switch." A PCA is comprised of an antenna fabricated on a photo-absorbing semiconductor substrate. The antenna is biased to induce an electric field inside the semiconductor substrate. Once illuminated by light, the photo-generated electron–hole pairs in the semiconductor accelerate in the presence of the bias electric field to form an electric current that is coupled to the antenna. In order to radiate THz waves, the generated photocurrent, and therefore the intensity of the incident light, should have THz frequency components.

The design considerations of PCAs are discussed in Section 5.2 of this chapter. The discussion starts with photoconductor physics followed by the impact of the antenna and PCA operation as an emitter and detector of pulsed and continuous-wave (CW) THz radiation. Section 5.3 is dedicated to the introduction of plasmonics-enhanced photoconductive THz devices from fundamentals to practical devices, followed by an outlook of the state-of-the-art plasmonic PCAs.

5.2 Photoconductive Antennas

5.2.1 Photoconductors for THz Operation

A photoconductor is a material (usually a semiconductor) that exhibits an increase in electric conductivity under light illumination due to an increased concentration of free electrons and holes generated from light absorption. In a semiconductor, band-to-band

Fundamentals of Terahertz Devices and Applications, First Edition. Edited by Dimitris Pavlidis.

Companion website: www.wiley.com/go/Pavlidis/FundamentalsofTHz

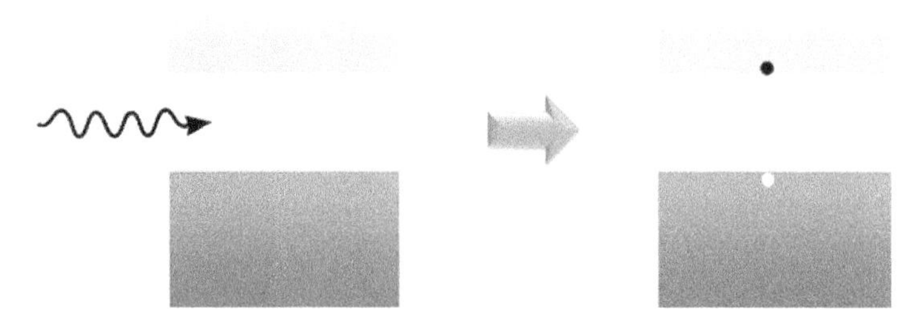

Figure 5.1 Band to band absorption of a photon in a semiconductor, creating a free electron–hole pair that can form a photocurrent when drifted by an electric field.

absorption of a photon can occur if the photon energy is equal to or larger than the bandgap energy, as shown in Figure 5.1. As a result, a free electron and a free hole are generated in the conduction band and valence band, respectively. This electron–hole pair can thus contribute to an electric current upon their movement under an electric field. The conductivity of the material is given by the concentration and mobility of the carriers.

$$\sigma = q\left(\mu_n n + \mu_p p\right) = \sigma_{dark} + q\left(\mu_n \Delta n + \mu_p \Delta p\right) \tag{5.1}$$

where q is the elementary charge, σ_{dark}, also known as the dark conductivity, is the material conductivity under no light illumination, n/p is the concentration of electrons/holes, $\Delta n/\Delta p$ is the concentration of the optically excited excess electrons/holes, and μ_n/μ_p is the electron/hole mobility. The rightmost term in Eq. (5.1) is called the photoconductivity.

The reverse process of carrier generation is recombination, where a free electron and a free hole recombine. The average time for carriers to recombine is called the recombination lifetime. In general, electron and hole recombination lifetimes, τ_n and τ_p, can be very different. The most dominant recombination mechanism is the Shockley–Read–Hall process, where one electron and one hole are captured by the localized recombination centers in the semiconductor, like defects and impurities. These recombination centers usually create energy states deep inside the bandgap, making the process much more energetically favorable compared to direct band-to-band recombination.

In a photoconductor, an external bias voltage is required for the photo-generated electrons and holes to drift and form a photocurrent. The measure of a photoconductor's capability to convert photons to electrons is called external quantum efficiency, η_{EQE}, which is the percentage of the photons incident on the photoconductor that are converted to electrons and collected by the photoconductor electrodes,

$$\eta_{EQE} = \eta_c(1 - R)\left(1 - e^{-\alpha d}\right) \tag{5.2}$$

where η_c is the collection efficiency of the photocarriers, R is the surface reflectivity of the photoconductor, α and d are the absorption coefficient and thickness of the photoconductor. Increasing the electric field inside the photoconductor would help increase the collection efficiency, but photocurrent saturation by space charge screening could also limit the collection efficiency. Another important parameter that quantifies the performance of a photoconductor is the spectral responsivity, R_ν, defined as the ratio of the generated photocurrent and the incident optical power,

$$R_\nu = G\eta_{EQE}\frac{q}{h\nu} \tag{5.3}$$

where $h\nu$ is the photon energy and G is the photoconductive gain, defined as the ratio of the carrier recombination lifetime over transit time. As discussed in the following sections, G of the ultrafast photoconductors used in most photoconductive THz devices is much smaller than 1. Increasing photoconductor responsivity is one of the most important design considerations that directly impacts the efficiency of photoconductive THz devices.

Assuming that the incident photons are uniformly distributed across the photoconductor active area A, and that they are uniformly absorbed within the absorption depth defined as $1/\alpha$, the concentration of the photo-generated excess electrons and holes are governed by the following equations [23, 24],

$$\frac{d\Delta n(t)}{dt} = \frac{\eta_{EQE}\alpha}{h\nu \mathrm{A}}P_{opt}(t) - \frac{\Delta n(t)}{\tau_n} \tag{5.4}$$

$$\frac{d\Delta p(t)}{dt} = \frac{\eta_{EQE}\alpha}{h\nu \mathrm{A}}P_{opt}(t) - \frac{\Delta p(t)}{\tau_p} \tag{5.5}$$

where $P_{opt}(t)$ denotes the time-dependent incident optical power envelope, which is used to carry the desired frequency components in the THz range. In general, $P_{opt}(t)$ can be expressed in terms of a series of sinusoidal components

$$P_{opt}(t) = P_0 + \sum_i P_{opt}(\omega_i)\, sin\,(\omega_i t + \phi_i) \tag{5.6}$$

where ω_i, $P_{opt}(\omega_i)$, ϕ_i are the angular frequency, envelope amplitude, and phase of the ith component, respectively. P_0 denotes the direct current (DC) optical power. In general, THz devices can operate in two regimes: pulsed operation and CW operation. For pulsed THz generation, a femtosecond mode-locked laser is typically used as the optical pump, so $P_{opt}(t)$ would have a broad range of frequency components; while for CW THz generation, $P_{opt}(t)$ would only have a DC component and a single frequency component ω_{THz}, generated from two CW lasers detuned by ω_{THz}. Solving Eqs. (5.4) and (5.5), we have

$$\Delta n(t) = \frac{\eta_{EQE}\alpha\tau_n}{h\nu A}\left[P_0 + \frac{P_{opt}(\omega_1)\, sin\,(\omega_1 t + \varphi_{n,\omega_1})}{\sqrt{1+\omega_1^2{\tau_n}^2}} + \frac{P_{opt}(\omega_2)\, sin\,(\omega_2 t + \varphi_{n,\omega_2})}{\sqrt{1+\omega_2^2{\tau_n}^2}} + \dots\right] \tag{5.7}$$

$$\Delta p(t) = \frac{\eta_{EQE}\alpha\tau_p}{h\nu A}\left[P_0 + \frac{P_{opt}(\omega_1)\, sin\left(\omega_1 t + \varphi_{p,\omega_1}\right)}{\sqrt{1+\omega_1^2{\tau_p}^2}} + \frac{P_{opt}(\omega_2)\, sin\left(\omega_2 t + \varphi_{p,\omega_2}\right)}{\sqrt{1+\omega_2^2{\tau_p}^2}} + \dots\right] \tag{5.8}$$

where $\varphi_n(\omega_i) = \tan^{-1}(-\omega_i\tau_n)$, $\varphi_p(\omega_i) = \tan^{-1}(-\omega_i\tau_p)$. We can clearly see that the concentration of the excess carriers has the same frequency components as $P_{opt}(t)$. In addition, we can see that the carrier concentration at THz frequencies decreases as a function of frequency, which leads to a roll-off in the generated and detected THz radiation at high frequencies. Using the excess carrier concentrations, we can obtain the time-varying conductivity

$$\sigma(t) = \sigma_{dark} + q\left[\mu_n\Delta n(t) + \mu_p\Delta p(t)\right] \tag{5.9}$$

In the following sections, we will show how optically enabled, modulated electric conductivity can induce an ultrafast current in a photoconductor. We will also show how this process can be used in THz emitters and detectors and how the antenna design and material properties would influence the bandwidth and efficiency of these devices.

5.2.2 Photoconductive THz Emitters

When a PCA is pumped by an optical source with THz frequency components, an ultrafast photocurrent is induced and coupled to the antenna to generate THz radiation. Most of the generated THz radiation is emitted through the photoconductive substrate because of its much higher refractive index compared to air. To avoid free carrier absorption of THz waves, it is crucial to use a semiconductor with a very low carrier concentration as the substrate. PCAs are usually mounted on a high-resistivity silicon lens to efficiently couple the generated THz radiation out of the photoconductive substrate and collimate the output radiation for better power collection.

Figure 5.2a shows the schematic diagram of a PCA with a photoconductive active region length and width of l and w ($A = wl$), respectively [25]. The time-domain conductance of the photoconductor $G_p(t)$ can be calculated from the photo-generated conductivity obtained from Eq. (5.9)

(a)

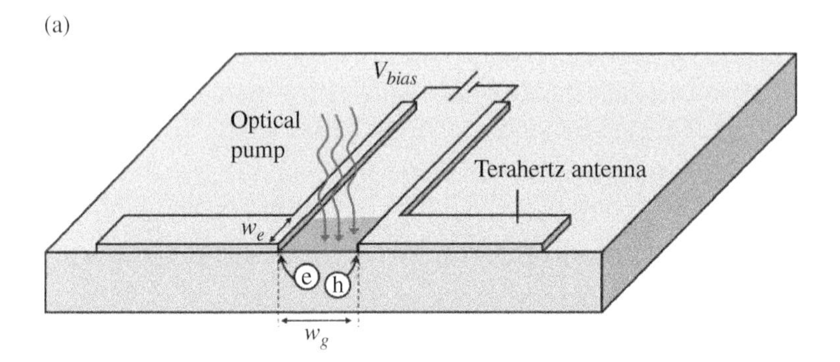

(b)

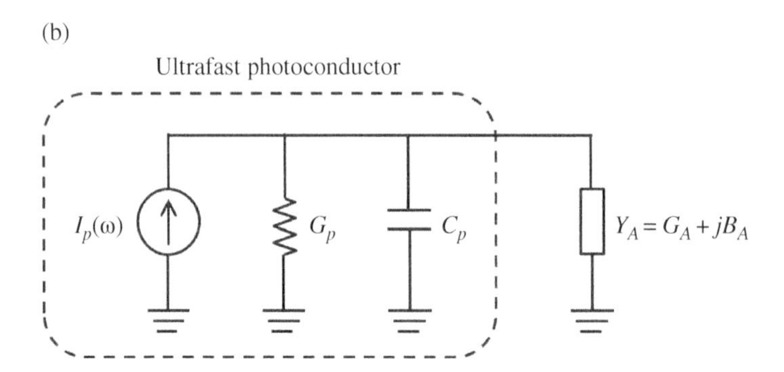

Figure 5.2 (a) Schematic of a PCA in operation. (b) Equivalent circuit model of a PCA-based THz emitter. *Source*: Reprinted by permission from [25]. Copyright (2012) by Springer Nature.

$$G_p(t) = \frac{q\eta_{EQE}\tau\left(\mu_n + \mu_p\right)}{h\nu l^2} \left[P_0 + \frac{P_{opt}(\omega_1)\sin(\omega_1 t + \varphi(\omega_1))}{\sqrt{1+\omega_1^2\tau^2}} + \frac{P_{opt}(\omega_2)\sin(\omega_2 t + \varphi(\omega_2))}{\sqrt{1+\omega_2^2\tau^2}} + \dots\right] + G_{dark} \tag{5.10}$$

where $G_{dark} = w\sigma_{dark}/l\alpha$ is the conductance under no optical illumination. Here, we assume electrons and holes have the same carrier lifetime τ, as shown in the above equation. Many studies have been done to model PCAs [26–29]. Here we use a simplified circuit model shown in Figure 5.2b to describe PCA operation [25]. The photoconductor is represented by a current source $I_p(\omega)$ in parallel with a conductance G_p and a capacitance C_p, while the antenna is represented by a complex admittance $G_A + jB_A$. The current source $I_p(\omega)$ generally can include various frequency components determined by the frequency components of the incident optical pump envelope. The current component at an angular frequency ω is linearly proportional to the optical power

$$I_p(\omega) = R(\omega)P_{opt}(\omega) \tag{5.11}$$

where $R(\omega)$ denotes the photoconductor responsivity at ω given as

$$R(\omega) = \frac{q\eta_{EQE}\tau\left(V_n + V_p\right)}{h\nu l}\frac{1}{\sqrt{1+\omega^2\tau^2}} \tag{5.12}$$

where V_n and V_p are the drift velocity of electrons and holes in the photoconductor.

Assuming the antenna efficiency is ξ, the radiation power from the antenna at the angular frequency ω is given as

$$P_{THz}(\omega) = \frac{1}{2}\left|R(\omega)P_{opt}(\omega)\right|^2 \xi \frac{G_A}{\left(G_A + \overline{G_p}\right)^2 + \left(\omega C_p + B_A\right)^2} \tag{5.13}$$

where $\overline{G_p}$ is the average conductance of the photoconductor given as

$$\overline{G_p} = \frac{q\eta_{EQE}\tau\left(\mu_n + \mu_p\right)P_0}{h\nu l^2} + G_{dark} \tag{5.14}$$

Clearly, the radiated power is maximized when the reactive component of the antenna impedance is minimized, i.e. $\left(G_A + \overline{G_p}\right) \gg \left(\omega C_p + B_A\right)$.

As it can be seen from Eq. (5.12), the recombination lifetime of the photocarriers has a significant impact on the photoconductor responsivity and the THz radiation power at high THz frequencies. As illustrated in Figure 5.3, compared to a long-carrier-lifetime photoconductor, a short-carrier-lifetime photoconductor can achieve larger THz radiation bandwidth and higher THz radiation power at high frequencies.

5.2.2.1 Pulsed THz Emitters

To generate pulsed THz radiation, a femtosecond laser is used to pump the PCA. Typical femtosecond lasers have optical pulses with full width at half maximum (FWHM) of 100 fs or less, which can generate sub-picosecond THz pulses. When a PCA is pumped by a femtosecond laser, the induced transient photocurrent would have a sharp rising edge

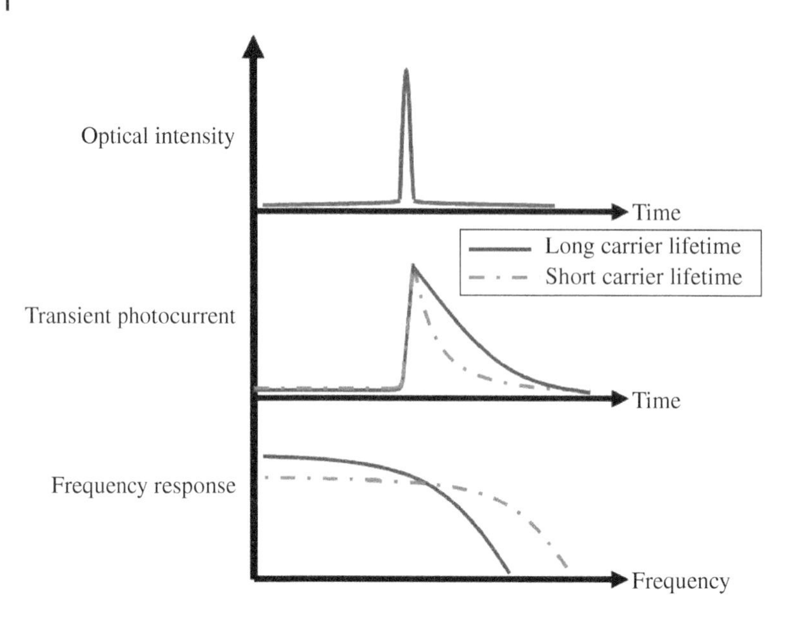

Figure 5.3 Comparison between a short-carrier-lifetime and a long-carrier-lifetime photoconductor. Larger operation bandwidth can be achieved with shorter carrier lifetime.

followed by an exponential decay determined by the carrier recombination lifetime. When this photocurrent pulse feeds the antenna, broadband pulsed radiation is generated with frequency components extended to THz frequencies.

Since the radiated power from an antenna is quadratically proportional to the injected current to the antenna, and the injected current is proportional to the incident optical pump power, the generated THz power is quadratically proportional to the optical pump power. This relation between the generated THz power and the optical pump power sometimes deviates from a quadratic relation at high optical pump powers because of the space-charge screening effect [30]. The space-charge screening happens as the optical pump power increases, resulting in an increase in the concentration of the photo-generated electrons and holes. When these electrons and holes are separated under the applied bias electric field, they induce an electric field in the opposite direction to the bias electric field. In addition, substantial heating of the photoconductor at high optical pump powers could lead to a thermal breakdown. The THz radiation from a PCA is also limited by the carrier velocity saturation in the photoconductor, which occurs due to various carrier scattering mechanisms [31]. Ultimately, the space-charge screening effect and the carrier velocity saturation limit the maximum THz power that can be obtained by a PCA and result in a saturation in the generated THz power when increasing the optical pump power and bias voltage.

5.2.2.2 Continuous-wave THz Emitters

To generate CW THz radiation, two CW lasers with their frequency difference in the THz range are used to pump the PCA. It is crucial to have the polarization of the lasers aligned in the same direction so that the superposition of the two laser beams creates a THz envelope. This process of generating the desired oscillation at the frequency difference of the two inputs lasers is called heterodyning. The PCAs operating in the CW mode under a heterodyning optical pump beam are usually called photomixers [32].

To analyze CW THz generation through photomixing, we assume that the phasor electric fields of two lasers at $\omega_1$and ω_2 angular frequencies are given by

$$E_1 = |E_1|e^{j(\omega_1 t + \phi_1)} \tag{5.15}$$

$$E_2 = |E_2|e^{j(\omega_2 t + \phi_2)} \tag{5.16}$$

where ϕ_1 and ϕ_2 are the phases of the laser electric fields. When the heterodyning optical pump beam is incident on the photomixer, the induced photocurrent is proportional to the squared magnitude of the total phasor field given by

$$\begin{aligned} |E_{total}|^2 &= |E_1 + E_2|^2 = |E_1|^2 + |E_2|^2 + E_1E_2^* + E_2E_1^* \\ &= |E_1|^2 + |E_2|^2 + 2|E_1||E_2| \cos\left[(\omega_2 - \omega_1)t - (\varphi_2 - \varphi_1)\right] \end{aligned} \tag{5.17}$$

Therefore, as it can be seen from Eq. (5.17), the induced photocurrent consists of a DC component and the desired THz oscillation at $\omega_{THz} = |\omega_2 - \omega_1|$. An important observation from the above equation is that at a fixed total optical power, the maximum photomixing efficiency is obtained when the two lasers have the same electric field amplitudes, i.e. $|E_1| = |E_2|$. Therefore, it is essential to keep the powers of the two lasers identical to achieve the largest THz photocurrent and highest radiation power.

A major advantage of CW THz generation through photomixing is the ease of frequency tuning [20, 33]. Thanks to the advancements in laser technology, frequency tunable lasers are readily available at various wavelengths, enabled by either temperature tuning or mechanical tuning. Only a modest frequency tuning range is required for THz applications. For example, at around 1550 nm optical wavelength, a wavelength detuning of only 8 nm for the heterodyning optical pump would offer a 1 THz frequency tunability.

5.2.3 Photoconductive THz Detectors

To detect pulsed and CW THz radiation, PCAs are pumped by pulsed and heterodyning optical pump beams, respectively. The major distinction from the PCA THz emitters is that no bias voltage is applied to the PCA THz detectors. When the incoming THz radiation is received by the antenna, a THz electric field is induced inside the photoconductive active region, which would accelerate the photocarriers to form an output photocurrent proportional to the received THz field.

For pulsed THz detection through PCAs, it is crucial for the incoming THz pulse and the optical pump pulse to have a variable temporal overlap at which short bursts of photocarriers are generated and accelerated by the received THz filed. To achieve this goal, the optical path length needs to be well adjusted and tuned such that the time-domain THz waveform can be accurately acquired and transformed into a frequency spectrum for spectroscopic applications.

For CW THz detection through PCAs, a pair of CW lasers with a fre,quency difference very close to the frequency of the received THz radiation are heterodyned to create a modulated carrier concentration, which is accelerated by the received THz field. Consequently, the received THz signal is down-converted to an intermediate-frequency (IF) signal, which can be easily analyzed by radio-frequency electronics. With the availability of wavelength tunable lasers, these THz detectors can scan through a broad THz frequency band and measure the amplitude and phase spectrum of the received THz radiation.

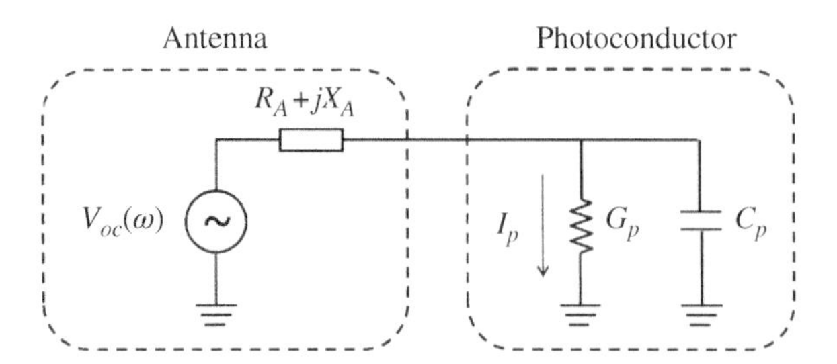

Figure 5.4 Equivalent circuit model of a PCA-based THz detector. *Source*: Reprinted by permission from [34]. Copyright (2013) by Springer Nature.

Figure 5.4 shows the equivalent circuit model of a PCA-based THz detector [34]. The incoming THz radiation induces an open-circuit voltage V_{oc} at the input port of the antenna. The received THz radiation and, thus, the induced open-circuit voltage can have many frequency components. Using this circuit model, the output photocurrent at an angular frequency ω can be obtained as

$$I_p(\omega) = \frac{G_p(\omega)V_{oc}(\omega)}{1 + (R_A + jX_A)(\overline{G_p} + j\omega C_p)} \tag{5.18}$$

where the antenna is represented by a complex impedance $R_A + jX_A$.

To achieve a high signal-to-noise ratio (SNR) THz detection using PCAs, it is crucial to have a high photoconductor responsivity as well as low noise operation. The output noise current of the PCA detectors can be calculated from their spectral noise current density, which is mainly dominated by the Johnson–Nyquist noise and shot noise [34]. Johnson–Nyquist noise comes from the random thermal movements of carriers inside the device; while shot noise originates from the discrete fluctuation of carriers, such as generation and recombination events. Their respective spectral noise currents are given as

$$i_{n-Nyquist} = \sqrt{4k_B T \overline{G_p} B} \tag{5.19}$$

$$i_{n-Shot} = \sqrt{2qI_{DC}B} \tag{5.20}$$

where I_{DC} is the DC current of the PCA, T is the PCA temperature, and B is the PCA detection bandwidth. The total spectral noise current is given by

$$i_n = \sqrt{i_{n-Nyquist}^2 + i_{n-Shot}^2} \tag{5.21}$$

Note that although the detector is unbiased, there could be a DC current as the result of photocarrier diffusion if the optical pump is asymmetrically distributed inside the PCA active area [34].

5.2.4 Common Photoconductors and Antennas for Photoconductive THz Devices

5.2.4.1 Choice of Photoconductor

Except for pulsed THz generation, a short-carrier-lifetime photoconductor is required to achieve efficient CW THz generation as well as pulsed and CW THz detection. The rising edge of the induced photocurrent upon illumination by a femtosecond laser is steep enough to emit strong THz pulses when using a long-carrier-lifetime photoconductor, regardless of the slow falling edge of the induced photocurrent. In addition, the carrier mobility in long-carrier-lifetime photoconductors is often much higher than short-carrier-lifetime photoconductors, enabling faster transport and collection of the photo-generated electrons and holes. When using a short-carrier-lifetime photoconductor, only the carriers that drift to the photoconductor contact electrodes in a time span shorter than the carrier lifetime would contribute to THz generation and detection. Photocarriers generated far away from the photoconductor contacts would recombine.

A commonly used short-carrier-lifetime photoconductor is low-temperature grown GaAs (LT-GaAs). Due to its direct bandgap, GaAs is a very appropriate photo-absorbing semiconductor for optical wavelengths below 870 nm. For most optoelectronics devices, the growth temperature of GaAs is typically around 600 °C to achieve low-defect epitaxy. On the contrary, for THz devices, LT-GaAs is grown at temperatures ranging from 200 °C to 400 °C to intentionally introduce high-density deep-level defects and obtain a sub-picosecond recombination lifetime [35, 36].

Since many high-performance, compact, and cost-effective lasers operate at telecommunication wavelengths, around 1550 nm, there has been a great demand for short-carrier-lifetime photoconductors operating at these wavelengths. Low-bandgap semiconductors like InGaAs or InAs are excellent absorbers of photons at telecommunication wavelengths and have very high carrier mobilities. However, these low-bandgap semiconductors also have very low dark resistivities, which result in an excessive power loss and early thermal breakdown when used as photoconductive semiconductors. ErAs:InGaAs and InAlAs/InGaAs multilayer heterostructures are promising alternatives [7, 10, 37–41]. Both approaches incorporate intentional trapping centers inside InGaAs to drastically decrease the carrier recombination lifetime while increasing the material dark resistivity.

It was recently shown that the use of plasmonics-enhanced PCAs can eliminate the need for a short carrier lifetime substrate [42, 43]. As described in the following sections, the transport time of the photocarriers in plasmonics-enhanced antennas can be significantly reduced to provide the ultrafast device speed required for THz generation and detection. This design flexibility allows utilizing various semiconductors and optical wavelengths in photoconductive THz devices, without being limited by the availability of short-carrier-lifetime photoconductors.

The choices of photoconductive materials for THz PCAs are not limited to bulk semiconductors. 2D materials such as graphene and black phosphorus offer very high carrier mobility and ultra-broadband optical absorption [44–48], which are desirable specifications for developing ultrafast PCAs operating over a broad range of optical pump wavelengths. The use of 2D materials as the photoconductive active region of PCAs has been explored for pulsed THz generation and detection [49–51]. Although the quantum efficiency of these PCAs was limited by the extremely small thickness of the utilized 2D materials, it is anticipated that the use of plasmonic contact electrodes and nanocavities would significantly enhance the efficiency of the THz PCAs based on 2D materials, to be further explained in the following sections.

5.2.4.2 Choice of Antenna

The THz antenna fabricated on the photoconductive substrate transforms the induced photocurrent into THz radiation when used in a PCA emitter and transforms the received THz radiation into photocurrent when used in a PCA detector. Therefore, the antenna specifications should be carefully chosen to offer the required gain, efficiency, impedance, and bandwidth for the desired application.

For generation and detection of pulsed THz radiation, broadband antennas should be used, such as bow-tie, log-periodic, or log spiral antennas. For generation and detection of CW THz radiation, both resonant and broadband antennas can be used. Resonant antennas, such as dipole antennas, usually offer larger radiation powers around their resonance frequency compared to broadband antennas. However, their narrow bandwidth limits their frequency tuning range for CW THz generation and detection.

5.3 Plasmonics-enhanced Photoconductive Antennas

5.3.1 Fundamentals of Plasmonics

Plasmonics is the field that deals with coherent electron oscillations, a.k.a. plasmons, and their interactions with electromagnetic waves. Of particular interest is the plasmons that reside at the interface between metallic and dielectric materials, called surface plasmon polaritons (SPPs). Constrained by the metal-dielectric boundary conditions, only transverse magnetic (TM) SPPs can exist. The dispersion relation of SPPs is shown in Figure 5.5, which shows that SPPs at all energies possess larger wave vectors than free space photons [52]. Due to their large wave vectors, these SPPs can propagate along the metal-dielectric interface with very strong field confinement [53–57]. However, direct excitation of SPPs with free space electromagnetic waves is not allowed because of the mismatch between their wave vectors. Periodic structuring of the surface is a simple way to enable this excitation by effectively providing an additional wave vector. Typically, periodic metallic nanostructures are

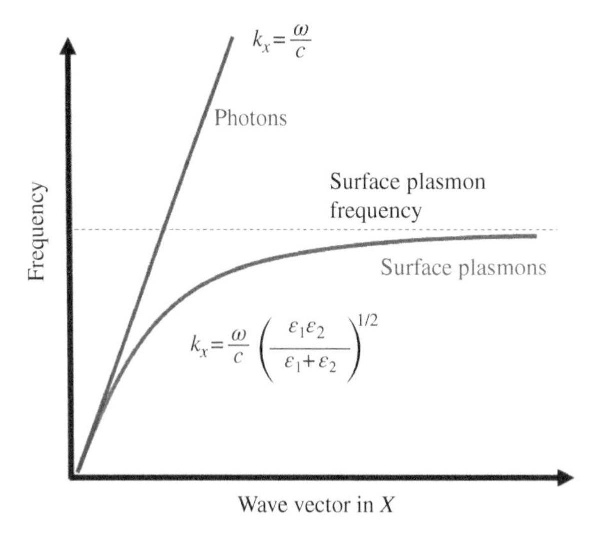

Figure 5.5 Surface plasmon dispersion relation, showing that surface plasmons always have larger wave vector than free-space photons.

fabricated on the surface of a dielectric material to facilitate SPP excitation at optical frequencies.

The unique properties of SPPs have led to breakthroughs in many areas that benefit from the strong electromagnetic field confinement and concentration beyond the diffraction limit. Examples include high-efficiency nonlinear processes [56], high-resolution imaging and spectroscopy [58, 59], deep subwavelength electromagnetic wave focusing and beam shaping [60, 61], high-efficiency photovoltaics [62, 63], photodetectors [64, 65], and modulators [66–71].

5.3.2 Plasmonics for Enhancing Performance of Photoconductive THz Devices

5.3.2.1 Principles of Plasmonic Enhancement

As described in the previous sections, the operation of PCAs relies on the ultrafast drift of photocarriers to the photoconductor contact electrodes. Plasmonic nanostructures are very well suited for shortening the transport path and thus transit time of a large fraction of the photocarriers to the photoconductor contact electrodes by enhancing the optical pump field and photocarrier concentration in close proximity to the contact electrodes. Therefore, the use of plasmonic nanostructures enables achieving high quantum efficiency and ultrafast operation at the same time.

To better understand the impact of plasmonic nanostructures, we describe a specific PCA with plasmonic contact electrode gratings and compare its operation and performance with a comparable conventional PCA without plasmonic contact electrodes [72]. As shown in Figure 5.6, both PCAs are fabricated on an LT-GaAs substrate and use a bow-tie antenna for THz radiation generation and detection when pumped by femtosecond optical pulses at an 800 nm center wavelength. Since the photo-generated electrons have a much higher mobility than the photo-generated holes and since the bias electric field is significantly enhanced near the device contact electrodes, the optical pump beam is focused near the anode contact electrodes to achieve high quantum efficiencies [25].

When a bias voltage is applied to the conventional PCA shown in Figure 5.6a, only the photo-generated electrons that drift to the anode contact electrode within a fraction of the THz oscillation cycle would constructively drive the bow-tie antenna with an ultrafast photocurrent to generate THz pulses. The bandwidth of the generated THz pulses is determined by the antenna bandwidth, pulse-width of the optical pump, and photoconductor carrier lifetime. Since the drift velocity of electrons in the substrate is limited by scattering in the semiconductor lattice, only a small fraction of the photo-generated electrons drift to the anode contact within a fraction of the THz oscillation cycle, even if the optical pump beam is focused down to a diffraction-limited spot size. The remaining photocarriers recombine in the substrate without significant contribution to THz generation, leading to low optical-to-THz conversion efficiencies.

On the other hand, when plasmonic contact electrode gratings are incorporated in the photoconductor contact electrodes as shown in Figure 5.6b, the optical pump intensity is significantly enhanced at the interface between the plasmonic contact electrodes and the LT-GaAs substrate. This field enhancement significantly reduces the average transport path length of the photo-generated electrons to the anode contact electrode, which are in the form of plasmonic gratings. Therefore, the number of the photo-generated electrons that

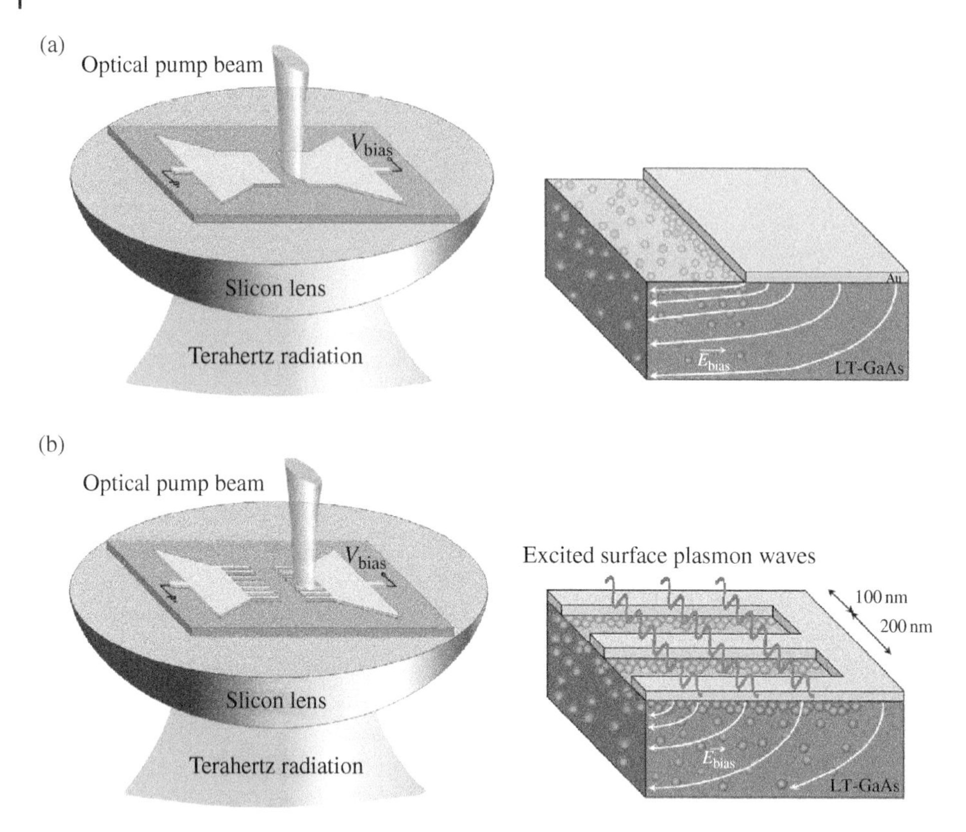

Figure 5.6 Illustrations for (a) conventional and (b) plasmonic PCAs based on a bow-tie antenna. Electrons are shown in blue and holes are shown in red.

drift to the anode contact electrodes within a fraction of the THz oscillation cycle to drive the THz antenna and generate THz pulses is significantly enhanced and much higher optical-to-THz conversion efficiencies are achieved. Similar to the conventional PCAs, the bandwidth of the generated THz pulses is determined by the antenna bandwidth, pulse-width of the optical pump, and photoconductor carrier lifetime.

The geometry of the plasmonic contact electrodes is a crucial factor that directly impacts the optical-to-THz conversation efficiency enhancement. It should be chosen to maximize optical field intensity in close proximity to the plasmonic electrodes inside the substrate. As an example, for the plasmonic PCA shown in Figure 5.6b, a 5/45 nm-thick Ti/Au grating with a 200 nm pitch and a 100 nm spacing covered with a 150 nm-thick SiO_2 anti-reflection coating is used to enhance the intensity of a TM-polarized optical pump beam at an 800 nm center wavelength by exciting SPP waves along the gratings.

Using a finite-element method simulator (COMSOL), the optical field and external bias field distribution is calculated numerically for both the conventional and plasmonic contact electrodes. As shown in Figure 5.7a, when the SPP waves are excited along the gratings, highly confined optical absorption occurs directly under the grating lines. This confirms the essential advantage of the plasmonic grating structure to greatly reduce carrier transport length to the photoconductor contact electrodes. In the conventional design, however, a

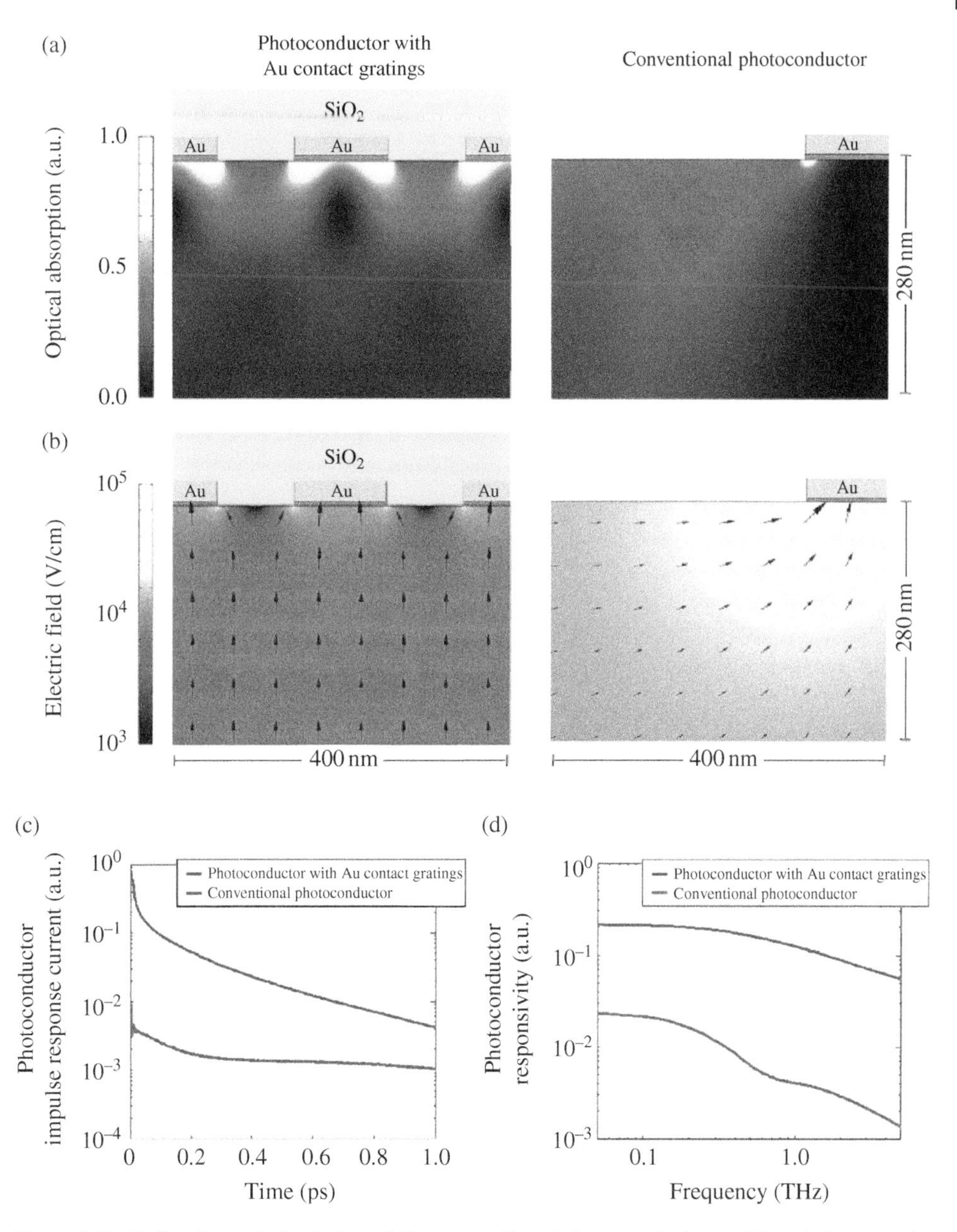

Figure 5.7 Finite element simulation of the conventional photoconductor and the photoconductor with the plasmonic contact electrode gratings pumped at an 800 nm wavelength. The profile of optical absorption and bias electric field is shown in (a) and (b), respectively. The cross section is orthogonal to the gratings for the plasmonics-enhanced photoconductor; and the cross section is orthogonal to the anode edge for the conventional photoconductor. Impulse response and responsivity of the photoconductor are shown in (c) and (d), respectively.

fraction of the optical pump beam is blocked by the contact electrodes and the optical absorption is distributed between the anode and the cathode contacts with a weaker enhancement at the edge of the contact electrode.

The external bias field that drifts the photocarriers is shown in Figure 5.7b, where the bias voltage is set to be high enough to reach the electron saturation velocity over the entire grating area (i.e. electric field higher than 1 kV/cm), but low enough to avoid breakdown (i.e. electric field lower than 100 kV/cm). In the plasmonic design, although the electric field strength is generally lower than the conventional design, it is still well above 1 kV/cm for the majority of the photo-generated electrons to travel at the saturation velocity.

The impulse response photocurrents of the conventional and plasmonic PCAs calculated using the classical drift-diffusion model are shown in Figure 5.7c. For this calculation, the three-dimensional distribution of the photocarrier density is derived from the optical absorption profile and combined with the bias electric field in the semiconductor substrate shown in Figure 5.7a,b, respectively. As expected, the plasmonic PCA offers a significantly stronger impulse response compared with the conventional PCA. The responsivity of the plasmonic and conventional PCAs are calculated by convolving their impulse response with the power envelope of a pair of heterodyned optical beams with a beating frequency ranging from 50 GHz to 5 THz. The obtained responsivity spectra are shown in Figure 5.7d, indicating more than one order-of-magnitude enhancement in responsivity by using the plasmonic contact electrodes. Due to the quadratic dependence of the radiated THz power on the photoconductor responsivity, more than two orders-of-magnitude enhancement in the radiated THz power is expected when using the plasmonic contact electrodes.

Figure 5.8a shows the scanning electron microscopy images of a fabricated conventional PCA and a plasmonic PCA with a magnified view of the utilized plasmonic grating. Both devices are pumped by a Ti:sapphire femtosecond laser with a 200 fs pulse-width, a 76 MHz repetition rate, and an 800 nm center wavelength. Figure 5.8b shows the measured photocurrents of both devices under a 40 V bias voltage when varying the average optical pump power from 0 to 25 mW. About seven times increase in photocurrent is observed for the plasmonic PCA compared to the conventional PCA, due to higher carrier concentration near the plasmonic electrode and shorter carrier transport path lengths. As expected, the plasmonic PCA shows a significantly higher THz power, as shown in Figure 5.8c. Measured by a pyroelectric detector, the THz power is characterized at the same bias voltage and the same range of optical pump powers. The measurement results show that the plasmonic PCA radiates up to 50 times higher THz power levels compared with the conventional PCA when pumped at the same optical power, which is a clear indication of the quadratic relation between the radiated THz power and the ultrafast photocurrent. An additional advantage of the plasmonic PCA is its high tolerance to optical misalignment. In the conventional PCA, the emitted THz power drops dramatically when the optical pump beam is slightly moved from its optimum position. However, in the plasmonic PCA, the optical pump position can move to any spot on the plasmonic grating without a considerable change in the emitted THz power.

The unique properties of plasmonic contact electrodes in enhancing the optical pump intensity near the photoconductor contact electrodes and reducing the average transport path length of the photocarriers to the contact electrodes can be used to enhance THz detection sensitivity of PCAs as well. Referring back to Figure 5.6, both the conventional and

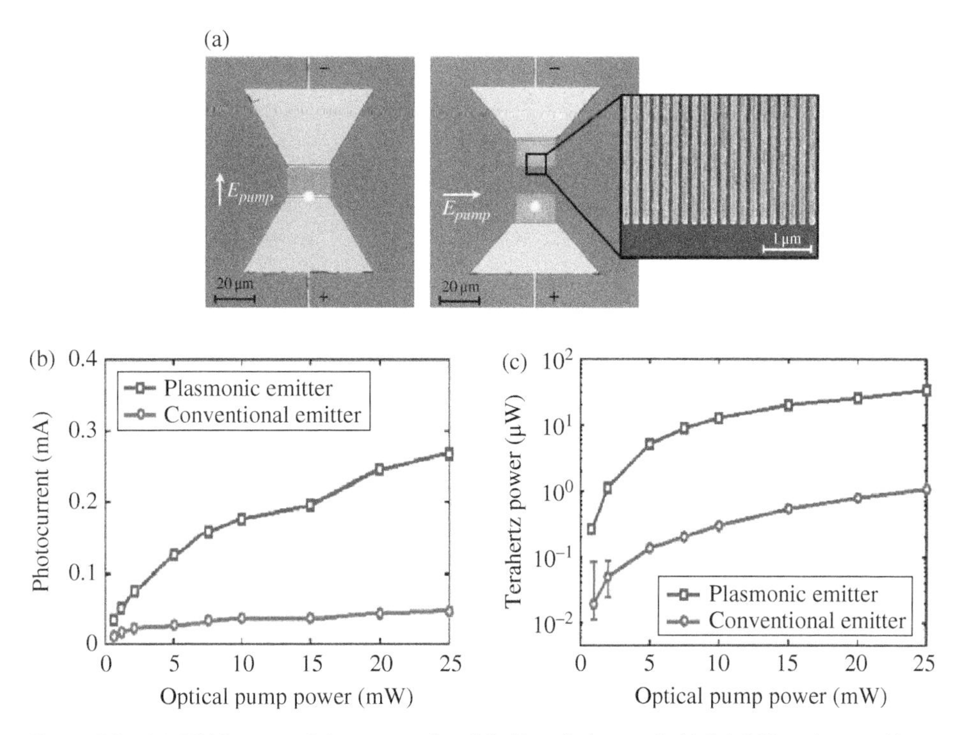

Figure 5.8 (a) SEM images of the conventional (left) and plasmonic (right) PCA emitters with a magnified view of the utilized plasmonic grating. (b) Measured photocurrent for the conventional and plasmonic PCAs at 40 V bias voltage when the average optical pump power varies from 0 to 25 mW. (c) Measured THz radiation power for the conventional and plasmonic PCAs at 40 V bias voltage when the average optical pump power varies from 0 to 25 mW.

plasmonic PCAs can be used for THz detection when a zero bias voltage is applied to them. In the absence of a bias electric field, the received THz signal induces an electric field across the antenna arms, which drifts the photocarriers created by the optical pump beam. As a result, an output photocurrent is generated, which is proportional to the incident THz field. Clearly, the number of the photocarriers that drift to the contact electrodes within the oscillation cycle of the incident THz signal determines the THz detection responsivity, defined as the output photocurrent per unit THz power. As in the case of the conventional PCA emitter, the relatively long transport path length of the photocarriers to the photoconductor contact electrodes reduces the quantum efficiency and responsivity of the conventional PCA detector. However, by using the plasmonic contact electrodes, the average transport path length of the photocarriers to the photoconductor contact electrodes is significantly reduced, leading to significantly higher output photocurrent and responsivity levels for the plasmonic PCA detector.

THz detection performance of both conventional and plasmonic PCAs are assessed by measuring their time-domain response to the radiation from a commercially available PCA emitter inside a time-domain THz spectroscopy setup. The same Ti:sapphire femtosecond laser mentioned above is used to pump the PCA emitter and the THz PCA detectors under test. The optical pump beam is tightly focused near one of the photoconductor

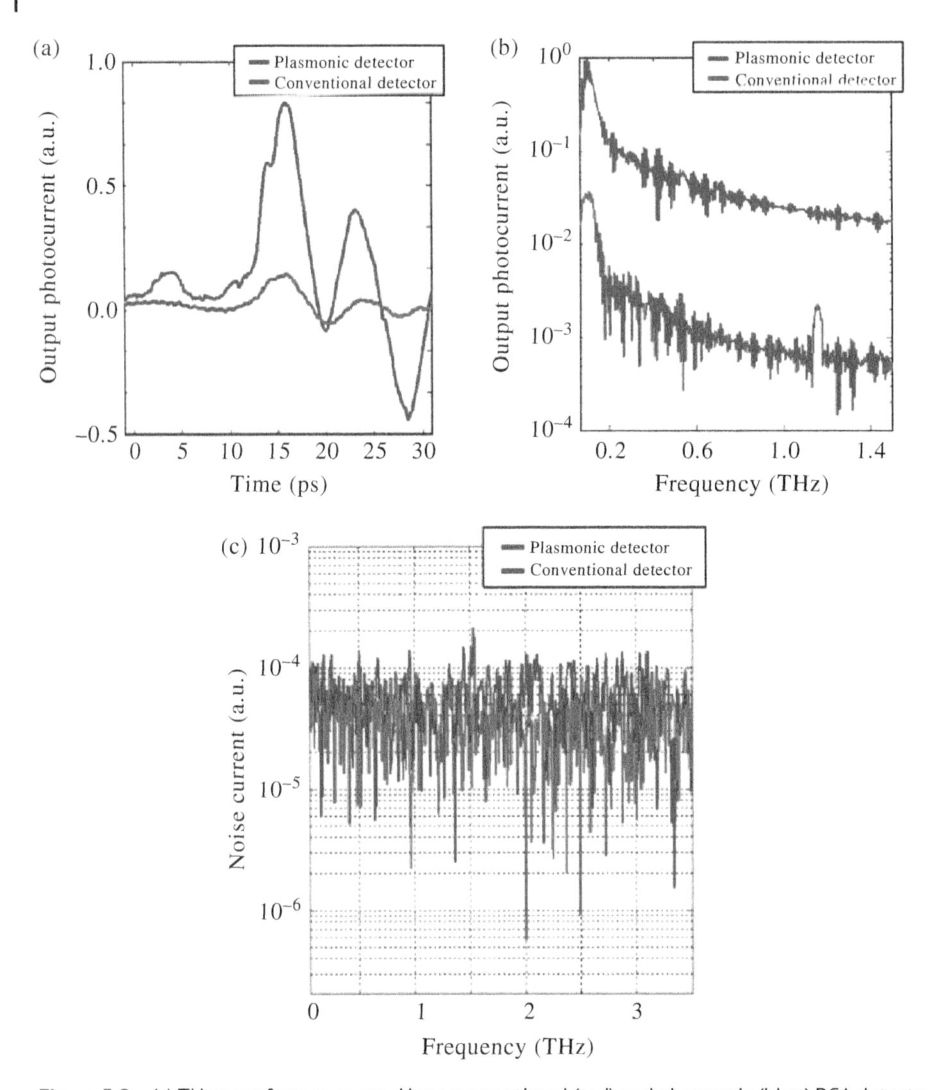

Figure 5.9 (a) THz waveform measured by a conventional (red) and plasmonic (blue) PCA detector in a time-domain THz spectroscopy setup, under the same condition. (b) Output photocurrent of both devices in the frequency domain, obtained by taking the Fourier transform of the temporal signal. (c) Output noise current spectra of both PCA detectors.

contact electrodes, and the exact beam position is independently optimized for the highest output photocurrent for the two PCA detectors. As shown in Figure 5.9a, under the same optical pump power and incident THz electric field, the plasmonic PCA offers 30 times higher output photocurrent. Furthermore, when both time-domain output photocurrents are converted to the frequency domain, the 30-fold responsivity enhancement is maintained over a broad frequency range exceeding 1.5 THz (Figure 5.9b).

To assess the THz detection sensitivity enhancement by the plasmonic PCA detector, the noise currents of both PCA detectors are obtained by measuring their output photocurrent in the absence of any incident THz signal. As shown in Figure 5.9c, the conventional PCA detector and the plasmonic PCA detector have very similar output noise current spectra. This can be explained by the fact that both detector noises are dominated by the Johnson–Nyquist noise and that both detectors have the same conductance values. Consequently, the plasmonic PCA detector offers a 30-fold enhancement of SNR under the same incident THz signal and optical pump power.

THz radiation power and detection sensitivity enhancement through the use of plasmonic nanostructures are not limited to the particular PCA and plasmonic nanostructure discussed in this section. Design considerations for various plasmonic nanostructures and PCA architectures that offer enhanced THz radiation power and detection sensitivity are discussed in the next sections.

5.3.2.2 Design Considerations for Plasmonic Nanostructures

Choosing a proper metal for plasmonics-enhanced PCAs involves a number of important considerations. First, the utilized metal should offer strong optical field enhancement near the plasmonic nanostructures at the optical pump wavelength. The utilized metal should not support any interband transition to prevent the direct absorption of the optical pump in the metal rather than the photoconductive substrate. For these reasons, gold and silver are extensively used in plasmonic devices pumped at visible and near-infrared optical wavelengths. Second, the utilized metal should be chemically stable and adhere well to the photoconductive substrate. Compared to silver, gold is much more resistant to oxidation and its high electrical and thermal conductivities are highly beneficial for operation at high optical pump powers. Therefore, most plasmonics-enhanced PCAs use gold as the plasmonic metal. However, gold requires an adhesion layer to stick to the photoconductive substrate, and this layer plays a significant role in the electrical and optical characteristics of the device. Titanium has been the most commonly used gold adhesion layer in plasmonic PCAs operating at visible and near-infrared optical pump wavelengths. It was recently found that chromium is the most suitable gold adhesion layer with superior electrical and optical properties compared to titanium because electrons encounter a lower energy barrier at the chromium–gold junction and the pump photons experience a lower plasmonic loss in the chromium adhesion layer [73].

5.3.3 State-of-the-art Plasmonics-enhanced Photoconductive THz Devices

5.3.3.1 Photoconductive THz Devices with Plasmonic Light Concentrators

Utilizing plasmonic light concentrators is a promising approach to enhance optical pump intensity and photocarrier density in the PCA active area to boost the generated THz radiation power and detection sensitivity. For instance, a PCA with plasmonic nanorods was fabricated on SI-GaAs and pumped by 800 nm Ti:Sapphire laser pulses polarized orthogonal to the nanorod direction (Figure 5.10a) [74]. This PCA offered about 100% power enhancement over a 0.1–1.1 THz frequency range compared to a similar nonplasmonic design. In another work, a PCA with thermally dewetted plasmonic nanoislands with circular shapes was demonstrated. This PCA delivered comparable THz power enhancement factors with the plasmonic nanorod design, but with a constant radiated power for different pump polarizations [77]. The geometric shape of the plasmonic nanostructures is also proven to significantly impact the plasmonic enhancement factors. As illustrated in

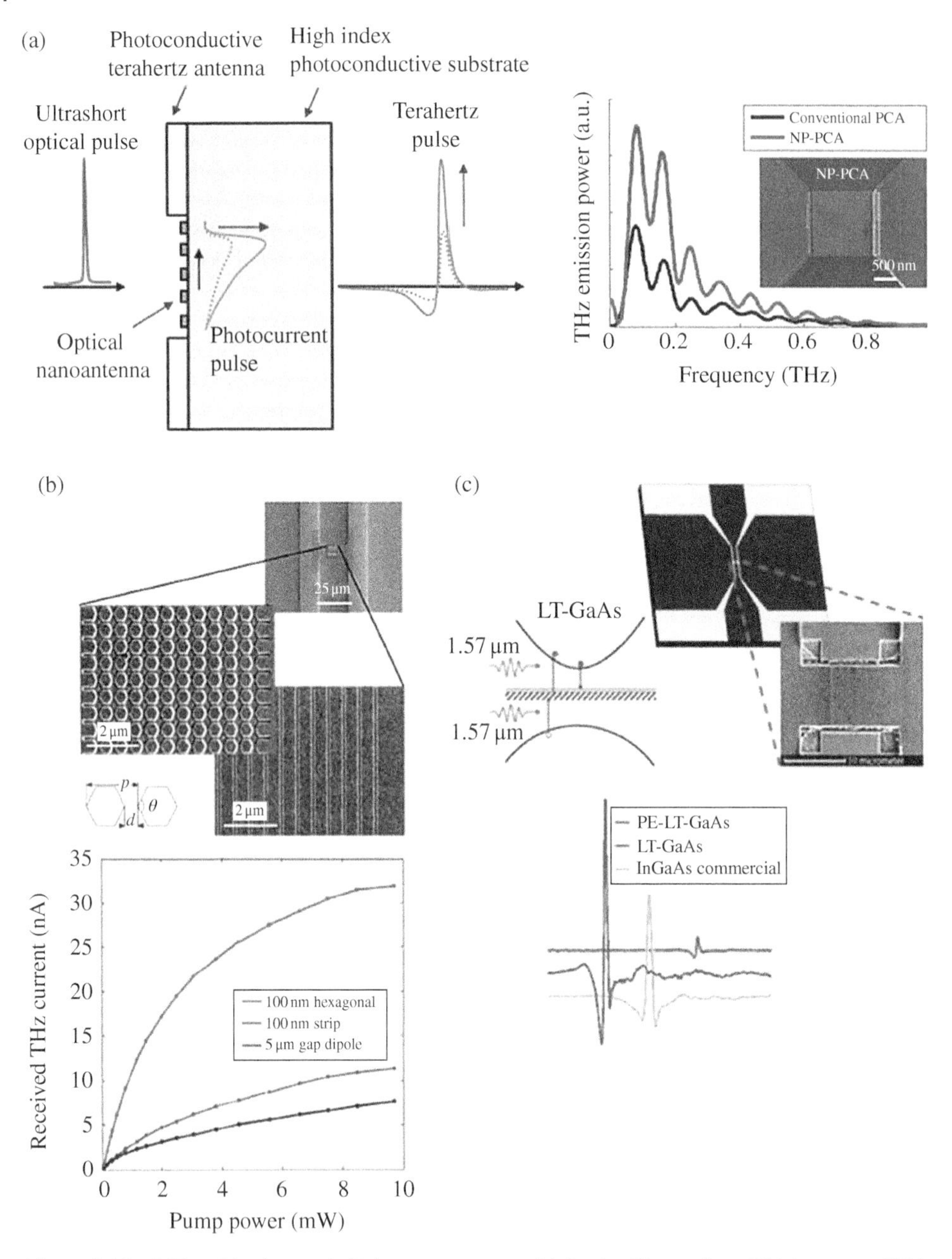

Figure 5.10 PCAs with plasmonic light concentrators. (a) Device illustration, SEM image, and THz emission spectrum of a PCA enhanced by plasmonic nanorods in comparison with a nonplasmonic design. *Source*: Reprinted by permission from [74]. Copyright (2012) by American Chemical Society. (b) SEM images and THz peak current measurement results for PCAs based on plasmonic hexagonal nanostructures and nanostrip arrays. The peak current is also compared to a PCA without plasmonic enhancement. *Source*: Reprinted by permission from [75]. Copyright (2014) by The Optical Society. (c) Device schematic and SEM image of a PCA based on a plasmonic nanoslit array, operating at 1550 nm using LT-GaAs as the photo-absorbing material. Also illustrated are the two-stage optical absorption process and the measured THz electric field in the time domain, compared with an LT-GaAs PCA without a plasmonic nanoslit array and a commercial InGaAs PCA. *Source*: Reprinted by permission from [76]. Copyright (2015) by American Chemical Society.

Figure 5.10b, an SI-GaAs based PCA with arrayed hexagonal nanostructures has offered stronger photocarrier density localization. The use of this design enabled more than three times enhancement of the generated THz field compared to a design based on plasmonic nanostrip arrays, and more than five times enhancement compared to a comparable non-plasmonic design [75]. In another recent work, a PCA fabricated on an LT-GaAs substrate incorporated plasmonic nanoslit arrays to significantly enhance optical absorption at 1550 nm wavelength through LT-GaAs intermediate states [78–80]. Using this design, 10 times higher THz electric fields were achieved when pumped at a 1550 nm optical wavelength, as illustrated in Figure 5.10c [76, 81].

5.3.3.2 Photoconductive THz Devices with Plasmonic Contact Electrodes

The plasmonic light concentrators discussed in the previous section enhance optical pump concentration and photocarrier density inside the PCA active area. However, these nanostructures do not assist with the efficient and ultrafast transport of the photocarriers to the THz antenna. In this section, we describe alternative plasmonic PCAs that utilize metallic plasmonic nanostructures as the PCA contact electrodes to enable ultrafast transport of a large fraction of the photocarriers concentrated near the plasmonic nanostructures to the THz antenna.

The great potentials of plasmonic contact electrodes for enhancing the efficiency of PCAs were first introduced in 2010 [82], and the use of plasmonic contact electrode gratings was explored [83, 84]. The use of plasmonic contact electrode gratings was experimentally demonstrated using a PCA with a dipole antenna array fabricated on a thin InGaAs layer grown on an InP substrate [85]. In a follow-up study, the THz generation and detection sensitivity enhancement of an LT-GaAs-based PCA with a bow-tie antenna was assessed by comparing a PCA with plasmonic contact electrode gratings and a comparable PCA without any plasmonic contact electrodes. As explained in detail in Section 5.3.2, 50-times higher THz radiation power and 30-times higher THz detection sensitivity was offered by the plasmonic PCA, under the same operation conditions [72]. To achieve higher THz radiation powers, an array of 3×3 PCAs with logarithmic spiral antennas and plasmonic contact electrode gratings were developed. A microlens array was used to simultaneously pump the nine PCAs and a 1.9 mW pulsed THz radiation with a 2 THz bandwidth was achieved [86]. Another work demonstrated LT-GaAs-based pulsed THz detectors with interlaced plasmonic contact electrodes, which provided a 40-times higher output photocurrent response compared with a similar nonplasmonic PCA detector (Figure 5.11a) [87].

The design of plasmonic contact electrode gratings is not limited to planar structures only. As shown in Figure 5.11b, a three-dimensional plasmonic grating was designed and embedded in an LT-GaAs substrate to offer a more efficient optical absorption near the plasmonic gratings when pumped at an 800 nm wavelength [88, 89]. The use of this plasmonic contact electrode grating in a PCA with a logarithmic spiral antenna resulted in a record-high optical-to-THz conversion efficiency of 7.5%.

Despite the impressive efficiency enhancement of PCAs with plasmonic contact electrode gratings, these designs have the common drawback of polarization-dependent operation. To achieve a polarization-independent operation, plasmonic nanostructures with a two-dimensional symmetry can be utilized. For example, a recent study demonstrated a THz PCA emitter, which utilized cross-shaped two-dimensional hole arrays as the plasmonic

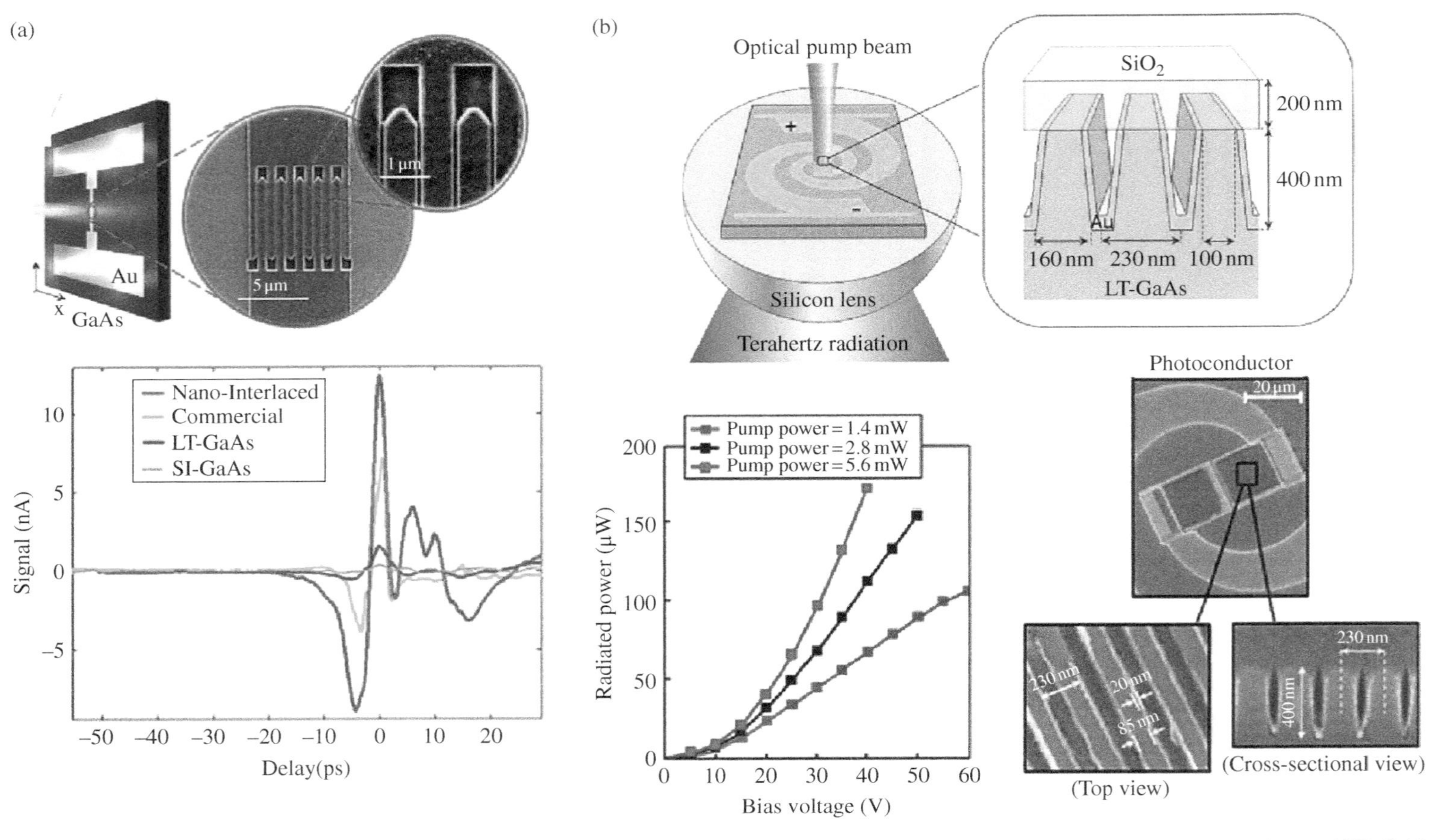

Figure 5.11 Pulsed THz generation and detection using PCAs with plasmonic contact electrodes. (a) Device schematic, SEM images, and the detected THz field by a PCA with interlaced plasmonic contact electrodes. The detected signal is also compared with an LT-GaAs and SI-GaAs PCA without plasmonic contact electrodes. *Source*: Reprinted by permission from [87]. Copyright (2012) by American Chemical Society. (b) Device schematic, SEM images, and THz radiation power of a PCA based on three-dimensional plasmonic contact electrodes gratings.

contact electrodes [90]. It was demonstrated that polarization-independent operation can be achieved with these plasmonic nanostructures while offering high optical-to-THz conversion efficiencies.

The concept of plasmonics-enhanced PCAs is not limited to pulsed THz emitters and detectors and can significantly enhance the efficiency of CW THz emitters and detectors as well. For example, as shown in Figure 5.12a, plasmonic contact electrode gratings were incorporated in an LT-GaAs-based plasmonic photomixer with a logarithmic spiral antenna to enhance the CW THz radiation power at 1 THz to 17 μW at 1 THz [91]. Compared to a similar PCA without plasmonic contact electrodes, this device offered a 10-fold enhancement in the radiation power (Figure 5.12b) [92]. A similar design optimized for operation at a 1550 nm pump wavelength was realized using an ErAs:InGaAs substrate with a carrier lifetime of ~0.85 ps [93]. To prevent thermal breakdown at high optical pump powers, a 2% pump duty cycle was used, and a quasi-CW THz power of 0.8 mW was obtained at 1 THz under an average optical pump power of 150 mW. This plasmonic PCA was also combined with a bimodal laser diode to realize a compact CW THz emitter module [94]. Analyzing the spectral characteristics of the CW THz emission from this plasmonic PCA showed a direct dependence of the THz radiation linewidth and stability on the linewidth and stability of the heterodyning optical pump source [95]. This plasmonic PCA was also combined with a globally stable on-chip microresonator frequency comb laser to generate a 0.6 mW CW radiation around 1 THz with a 1.1% optical-to-THz conversion efficiency [96].

In a recent work, a plasmonic PCA with a logarithmic spiral antenna fabricated on an LT-GaAs substrate was used as a heterodyne THz detector [97]. A heterodyning optical pump beam with a THz beat frequency served as the local oscillator of this heterodyne detector, downconverting a received THz signal to an IF equal to the difference frequency of the received THz signal and the optical pump beam. This THz detector offered quantum-level THz detection sensitivities over a 0.1–5 THz frequency range while operating at room temperatures, enabling spectrometry and precision measurements with unprecedented functionalities [97–101].

5.3.3.3 Large Area Plasmonic Photoconductive Nanoantenna Arrays

Apart from the PCAs with single antennas, large-area plasmonic photoconductive nanoantenna arrays have also shown a great promise for high-power pulsed THz generation and high-sensitivity pulsed THz detection [102, 103]. As shown in Figure 5.13a, these PCAs typically use plasmonic gratings as the THz nanoantennas to confine a large fraction of the photo-generated carriers in close proximity to the nanoantennas. They also use grating lengths much shorter than the THz wavelength to enable broadband THz generation and detection. Due to their large area, the plasmonic photoconductive nanoantenna arrays can be pumped by a high-power optical beam with a large beam spot size on the device, mitigating the carrier screening effect, thermal heating, and alignment difficulty. As an example, a $1 \times 1\ \mathrm{mm}^2$ plasmonic photoconductive nanoantenna array fabricated on an SI-GaAs substrate was developed to deliver a pulsed THz power of 3.8 mW over a 0.1–5 THz frequency range under a 240 mW optical pump power [105].

The impacts of the carrier lifetime, drift velocity, and resistivity of the photoconductive substrate as well as the optical and electrical characteristics of the contact metal on the performance of large-area plasmonic nanoantenna arrays were investigated in detail [73, 106].

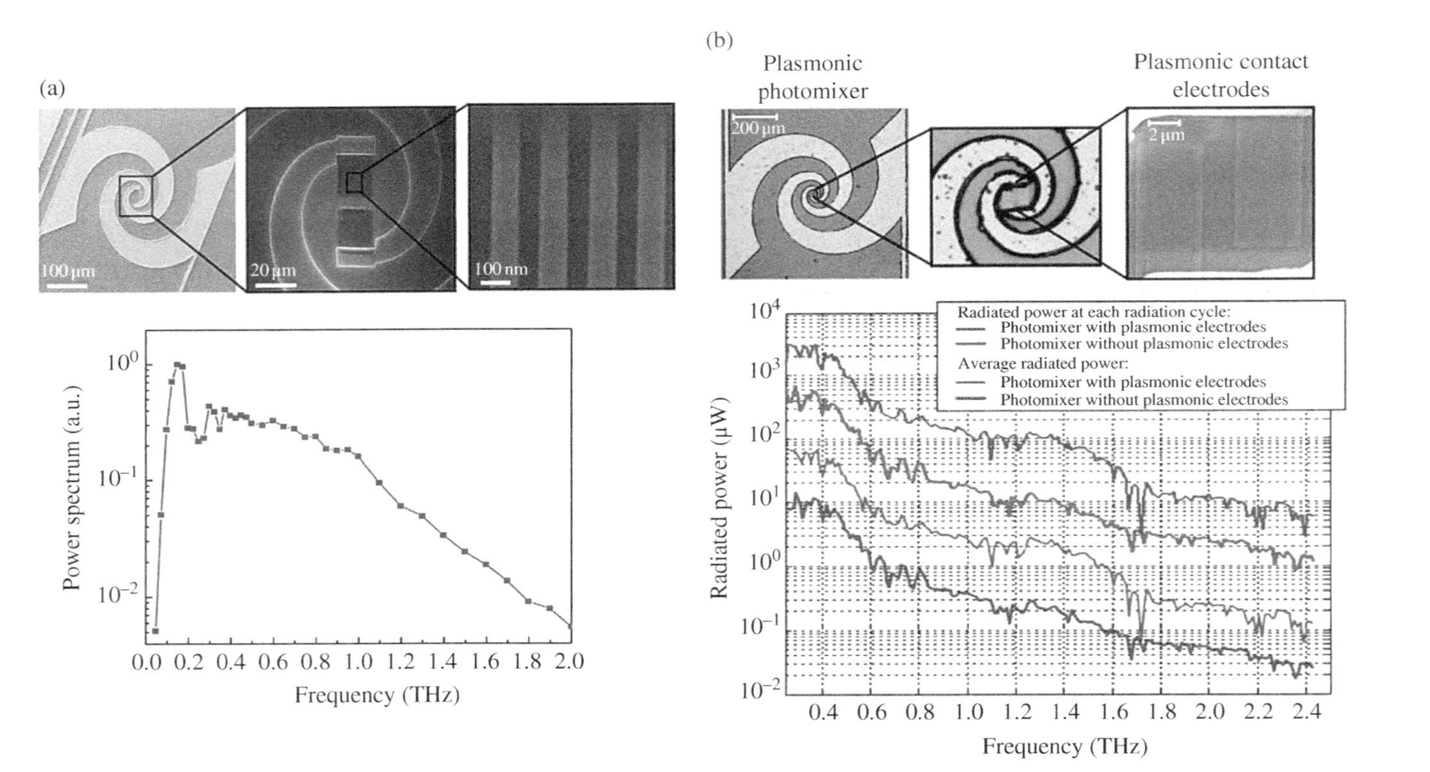

Figure 5.12 CW THz generation using PCAs with plasmonic contact electrodes. (a) SEM images of an LT-GaAs PCA with plasmonic contact electrode gratings, and its measured THz emission spectrum. *Source*: Reprinted by permission from [91]. Copyright (2015) by AIP Publishing. (b) Optical microscopy and SEM images of a PCA with plasmonic contact electrode gratings realized on an ErAs:InGaAs substrate and its THz emission spectrum compared to a similar design without plasmonic contact electrodes. *Source*: Reprinted by permission from [92]. Copyright (2014) by The Optical Society.

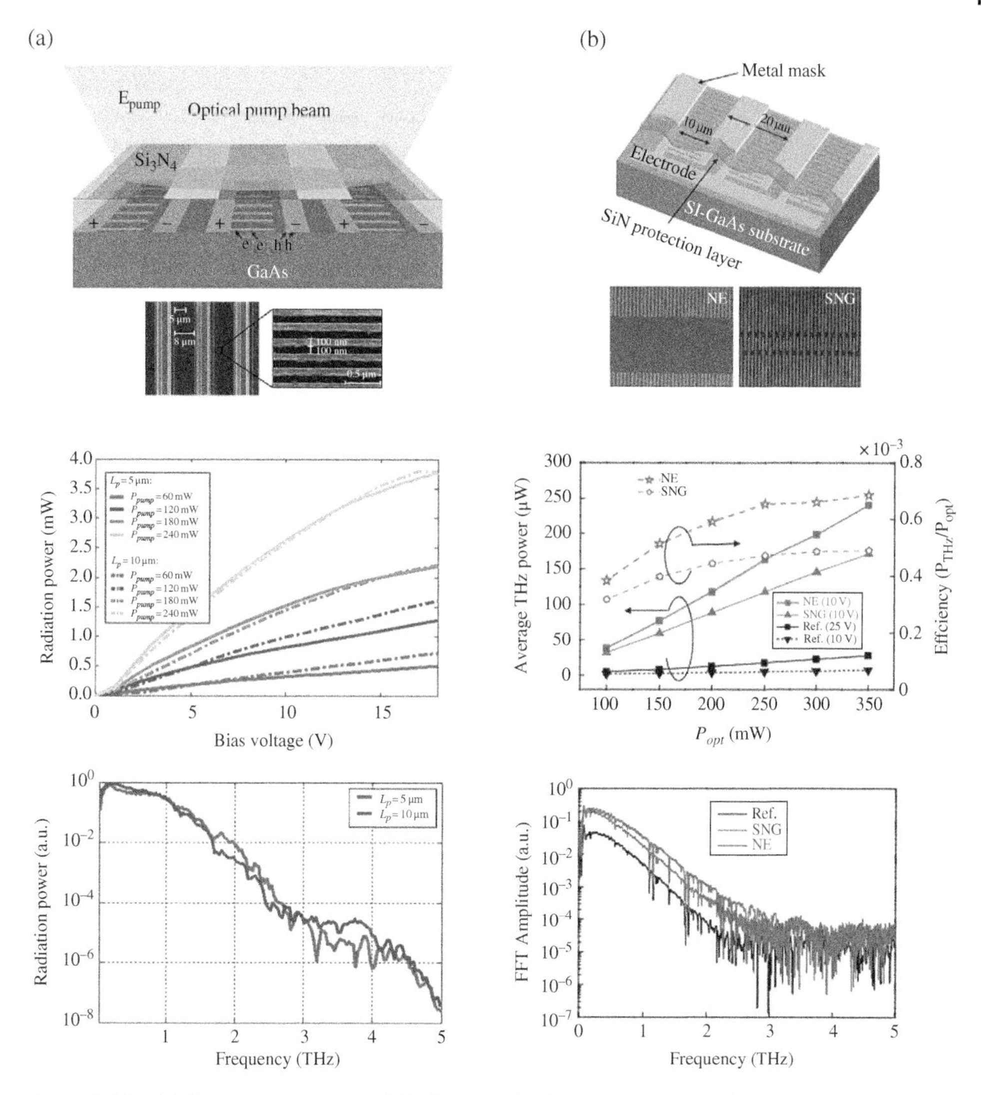

Figure 5.13 (a) Device structure and SEM images of a large-area plasmonic photoconductive nanoantenna array, together with its THz radiation power as a function of the applied bias voltage and the THz radiation spectrum for two grating lengths of 5 μm and 10 μm. (b) Device structure and SEM images of large-area plasmonic photoconductive emitter designs based on nanoelectrodes and shifted nanogaps. Their average radiated THz power, optical-to-THz conversion efficiency, and frequency spectrum are compared to a conventional design. *Source*: Reprinted by permission from [104]. Copyright (2015) by Springer Nature.

Pulsed THz radiation powers exceeding 6.7 mW were demonstrated using an SI-GaAs-based large-area plasmonic nanoantenna array with a chromium adhesion layer for the gold plasmonic contact electrodes under an 800 nm pump wavelength [73]. Similar designs were realized on an ErAs:InGaAs substrate to enable operation at telecommunication-compatible optical pump wavelengths. Pulsed THz radiation powers exceeding 0.3 mW over a 0.1–5 THz bandwidth were obtained when pumping these plasmonic photoconductive nanoantenna arrays with a 400 mW pulsed laser at a center wavelength of 1550 nm [107].

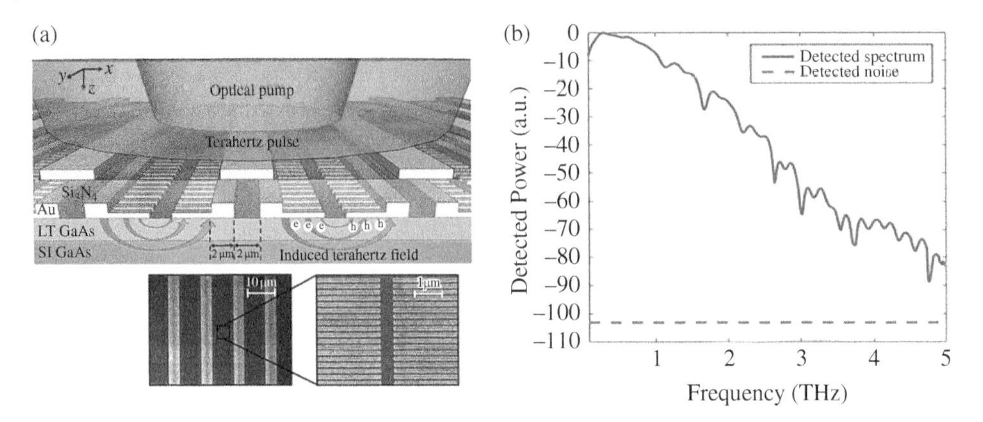

Figure 5.14 Device structure and SEM images of a large-area plasmonic photoconductive nanoantenna array used for pulsed THz detection and its detected THz power spectrum. *Source*: Reprinted by permission from [108]. Copyright (2017) by Springer Nature.

The impact of the geometric configuration of the plasmonic nanoantennas was also investigated (Figure 5.13b). It was shown that among various geometric configurations, the nanoelectrode design offers higher THz radiation efficiency compared with the nanogap and shifted nanogap designs due to its stronger bias field enhancement [104].

Similar to the PCAs with single antennas, the large-area plasmonic photoconductive nanoantenna arrays can be used for THz detection in the absence of an external bias voltage. By engineering the length of the nanoantennas and the gap size between the nanoantenna tips, as shown in Figure 5.14, the interaction of the optical pump beam and the incoming THz wave can be maximized such that the tightly confined photocarriers near the plasmonic nanoantennas can be swept by the received THz electric field in an ultrashort time span, while minimizing the Johnson–Nyquist noise. By using large-area plasmonic photoconductive nanoantenna arrays for pulsed THz generation and detection, a time-domain spectroscopy system with a record high SNR of 107 dB over a 0.1–4 THz frequency range was demonstrated [108, 109].

5.3.3.4 Plasmonic Photoconductive THz Devices with Optical Nanocavities

To further increase the quantum efficiency of plasmonic PCAs, many studies have utilized optical nanocavities to confine a larger fraction of the photo-generated carriers near the device contact electrodes. Most of these optical nanocavities were realized by adding a distributed Bragg reflector (DBR) beneath the photoconductive active layer of plasmonic PCAs. For example, Figure 5.15a shows a plasmonic THz emitter utilizing an optical nanocavity. An array of plasmonic nanoantennas were fabricated on a 190 nm-thick GaAs layer epitaxially grown on top of 25 pairs of AlAs and $Al_{0.33}Ga_{0.67}As$ alternating layers, which form a DBR with a stop band around 770 nm [110]. The use of the optical nanocavity enabled an 80% optical power absorption efficiency within the 190 nm-thick GaAs layer. Pulsed THz radiation powers exceeding 4 mW were obtained over a 0.1–5 THz frequency range, with more than 20-times higher optical-to-THz conversion efficiencies compared to a similar nanoantenna array fabricated on an LT-GaAs substrate without an optical

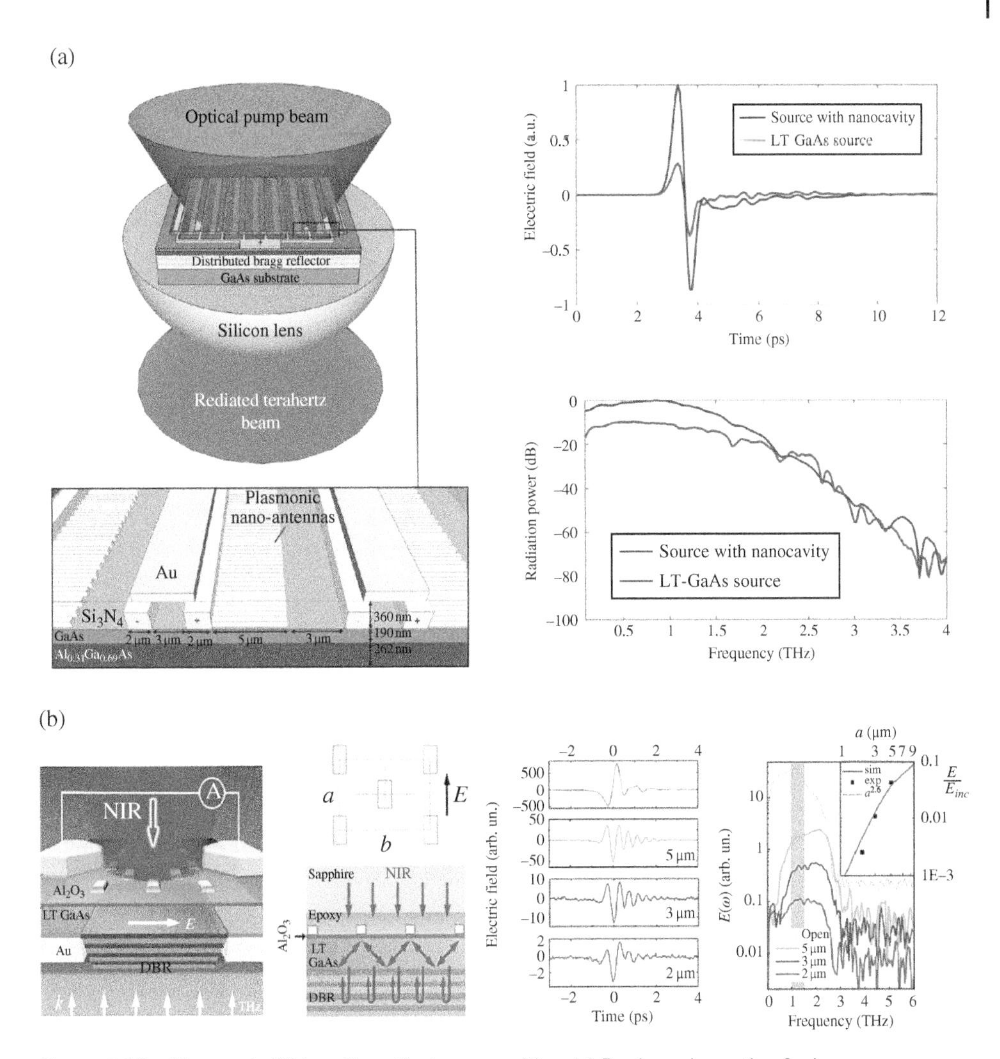

Figure 5.15 Plasmonic PCAs with optical nanocavities. (a) Device schematic of a large-area photoconductive THz emitter with an optical nanocavity, together with its time-domain THz field and power spectrum compared to a similar LT-GaAs emitter without an optical nanocavity. *Source*: Reprinted by permission from [110]. Copyright (2017) by Springer Nature. (b) Device schematic of a PCA detector based on an optical nanocavity, and its detected THz field. *Source*: Reprinted by permission from [111]. Copyright (2015) by American Chemical Society.

nanocavity. Following this work, a plasmonic THz detector with the same optical nanocavity was demonstrated, which detected THz pulses over a 0.1–4.5 THz frequency band with more than a 100 dB dynamic range under a 5 mW optical pump power [112]. Another plasmonic THz detector with an optical nanocavity was demonstrated, which used a gold nanoantenna array and an $AlAs/Al_{0.2}Ga_{0.8}As$ DBR to form the optical nanocavity. A 280 nm-thick LT-GaAs photoconductive layer was embedded inside the nanocavity, as shown in Figure 5.15b [111]. The nanoantenna array was designed to maximize optical transmission over the stop band of the DBR, and it was electrically isolated from the LT-GaAs through a thin layer of Al_2O_3. An 80–85% optical absorption efficiency was achieved over an 800–830 nm pump wavelength range, more than two times higher than the optical absorption without the nanocavity.

5.4 Conclusion and Outlook

The basic operation principles of PCAs and the impact of plasmonic nanostructures on enhancing the THz radiation power and detection sensitivity of PCAs were discussed in this chapter. It was explained that the use of plasmonic nanostructures enhances the photocarrier concentration inside the photoconductive active area and shortens the transport path length of the majority of the photocarriers to the device contact electrodes, leading to several orders of magnitude enhancement in the THz radiation power and detection sensitivity of PCAs [113, 114].

Despite the great promises of plasmonics-enhanced PCAs, realization of these devices usually involves lower throughput, lower yield, and more expensive fabrication processes compared to conventional PCAs. For example, electron beam lithography has been the most commonly used technique for patterning the plasmonic nanostructures, which is a very low throughput and costly process. Various emerging nano-manufacturing techniques could be used to develop future plasmonic PCAs with much lower costs and/or higher throughputs. One example is immersion ultraviolet (UV) lithography, which has been the main enabler for advanced complementary metal–oxide–semiconductor (CMOS) industry. By placing a high refractive index liquid between the UV focusing lens and the substrate, the exposure wavelength could be effectively shortened to achieve resolutions of few tens of nanometers [115]. To realize even smaller feature sizes of sub-10 nm, extreme UV (EUV) lithography can be used with an exposure wavelength of 13.5 nm [116]. Both of these photolithography techniques could achieve high precision and high throughput fabrication. Another candidate is nanoimprint lithography, which is a low-cost and high throughput technology. It uses a stamp to press on a resist to create the desired pattern with a resolution that is determined by the stamp fabrication technique [117]. Soft lithography is also a potential candidate, which uses an elastomeric material as the stamp to realize nanostructures with a high throughput [118, 119]. Similar to nanoimprint lithography, the lithography resolution is determined by the minimum obtainable feature size of the stamp. It is highly expected that these alternative lithography techniques would be significantly more mature within a few years, which would make the mass production of plasmonics-enhanced THz devices economical.

Exercises

E5.1 Calculate the dark conductivity of a p-type GaAs with a hole concentration of $10^{17}/cm^3$, assuming an electron mobility of 8500 cm^2/Vs, a hole mobility of 400 cm^2/Vs, and an intrinsic carrier concentration of 10^6 cm^{-3}.

E5.2 Calculate the quantum efficiency of a 100 nm-thick $In_{0.53}Ga_{0.47}As$ layer grown on an InP substrate when illuminated by a normally incident 1550 nm laser beam, assuming the collection efficiency is 80%, the absorption coefficient is 10^4/cm, and the refractive index is 3.5. Calculate the maximum achievable responsivity if the photoconductive gain is unity.

E5.3 With the same conditions in Exercise 2, assuming the 1550 nm laser beam is collimated to a diameter of 3 mm and has a 1 W power, calculate the steady-state excess electron concentration and photoconductivity of the illuminated $In_{0.53}Ga_{0.47}As$ layer given that the electron and hole recombination lifetimes are both 1 ns and the electron (hole) mobility is 10 000 cm^2/Vs (400 cm^2/Vs).

E5.4 Explain why short-carrier-lifetime semiconductors are typically used for CW photoconductive THz emitters and detectors. Provide examples of typical short-carrier-lifetime semiconductors that absorb at 800 nm and 1550 nm wavelengths.

E5.5 Assume that a GaAs PCA is biased such that the electric field in the active area is 40 kV/cm and the carrier velocity is ~10^7 cm/s. Calculate the average transit time for a photo-generated electron in the active area to travel over a 1 μm distance.

E5.6 Calculate the conductance of the PCA illustrated in Figure 5.2a, assuming an absorption coefficient of $10^4/cm^1$, $w = l = 5$ μm, and an average total conductivity of 100 S/m. Calculate the noise current spectral density in the absence of a bias voltage and at a bias voltage of 10 V at room temperature.

E5.7 Explain why a normally incident optical beam on a metal-dielectric interface cannot excite SPPs. Explain how the setups shown in Figure E1 could allow SPP excitation at the metal-dielectric interface. Find the relation between the values of n_1 and n_2 to enable excitation of SPPs.

E5.8 Using an electromagnetic simulator software, calculate the optical absorption spectrum from 600 to 1000 nm in the top 100 nm layer of an infinitely thick GaAs substrate with a periodic plasmonic gold grating deposited on the substrate. The grating has a 200 nm pitch, a 100 nm width, and a 50 nm height. Does adding an anti-reflection coating affect the optical absorption? Determine the optimum thickness of a Si_3N_4 anti-reflection coating for maximal optical absorption at an 800 nm wavelength.

References

1 Verghese, S., McIntosh, K.A., Calawa, S. et al. (1998). Generation and detection of coherent terahertz waves using two photomixers. *Applied Physics Letters* **73** (26): 3824–3826.

2 Jepsen, P.U., Jacobsen, R.H., and Keiding, S.R. (1996). Generation and detection of terahertz pulses from biased semiconductor antennas. *JOSA B* **13** (11): 2424–2436.

3 Zhao, G., Schouten, R.N., Van Der Valk, N. et al. (2002). Design and performance of a THz emission and detection setup based on a semi-insulating GaAs emitter. *Review of Scientific Instruments* **73** (4): 1715–1719.
4 Matsuura, S., Tani, M., and Sakai, K. (1997). Generation of coherent terahertz radiation by photomixing in dipole photoconductive antennas. *Applied Physics Letters* **70** (5): 559–561.
5 Jarrahi, M. and Lee, T.H. (2008). High-power tunable terahertz generation based on photoconductive antenna arrays. *2008 IEEE MTT-S International Microwave Symposium Digest*. IEEE.
6 Peytavit, E., Lepilliet, S., Hindle, F. et al. (2011). Milliwatt-level output power in the sub-terahertz range generated by photomixing in a GaAs photoconductor. *Applied Physics Letters* **99** (22): 223508.
7 Roehle, H., Dietz, R.J.B., Hensel, H. et al. (2010). Next generation 1.5 μm terahertz antennas: mesa-structuring of InGaAs/InAlAs photoconductive layers. *Optics Express* **18** (3): 2296–2301.
8 Taylor, Z.D., Brown, E.R., Bjarnason, J.E. et al. (2006). Resonant-optical-cavity photoconductive switch with 0.5% conversion efficiency and 1.0 W peak power. *Optics Letters* **31** (11): 1729–1731.
9 Jarrahi, M. (2009). Terahertz radiation-band engineering through spatial beam-shaping. *IEEE Photonics Technology Letters* **21** (13): 830–832.
10 Preu, S., Mittendorff, M., Lu, H. et al. (2012). 1550 nm ErAs: In (Al) GaAs large area photoconductive emitters. *Applied Physics Letters* **101** (10): 101105.
11 Cai, Y., Brener, I., Lopata, J. et al. (1998). Coherent terahertz radiation detection: direct comparison between free-space electro-optic sampling and antenna detection. *Applied Physics Letters* **73** (4): 444–446.
12 Sun, F.G., Wagoner, G.A., and Zhang, X.-C. (1995). Measurement of free-space terahertz pulses via long-lifetime photoconductors. *Applied Physics Letters* **67** (12): 1656–1658.
13 O'Hara, J.F., Zide, J.M.O., Gossard, A.C. et al. (2006). Enhanced terahertz detection via ErAs: GaAs nanoisland superlattices. *Applied Physics Letters* **88** (25): 251119.
14 Liu, T.-A., Tani, M., Nakajima, M. et al. (2003). Ultrabroadband terahertz field detection by photoconductive antennas based on multi-energy arsenic-ion-implanted GaAs and semi-insulating GaAs. *Applied Physics Letters* **83** (7): 1322–1324.
15 Suzuki, M. and Tonouchi, M. (2005). Fe-implanted InGaAs photoconductive terahertz detectors triggered by 1.56 μm femtosecond optical pulses. *Applied Physics Letters* **86** (16): 163504.
16 Liu, T.-A., Tani, M., Nakajima, M. et al. (2004). Ultrabroadband terahertz field detection by proton-bombarded InP photoconductive antennas. *Optics Express* **12** (13): 2954–2959.
17 Castro-Camus, E., Lloyd-Hughes, J., Johnston, M.B. et al. (2005). Polarization-sensitive terahertz detection by multicontact photoconductive receivers. *Applied Physics Letters* **86** (25): 254102.
18 Peter, F., Winnerl, S., Nitsche, S. et al. (2007). Coherent terahertz detection with a large-area photoconductive antenna. *Applied Physics Letters* **91** (8): 081109.
19 Liu, S., Shou, X., and Nahata, A. (2011). Coherent detection of multiband terahertz radiation using a surface plasmon-polariton based photoconductive antenna. *IEEE Transactions on Terahertz Science and Technology* **1** (2): 412–415.
20 Preu, S., Döhler, G.H., Malzer, S. et al. (2011). Tunable, continuous-wave terahertz photomixer sources and applications. *Journal of Applied Physics* **109** (6): 4.

21 Smith, P.R., Auston, D.H., and Nuss, M.C. (1988). Subpicosecond photoconducting dipole antennas. *IEEE Journal of Quantum Electronics* **24** (2): 255–260.

22 Auston, D.H., Cheung, K.P., and Smith, P.R. (1984). Picosecond photoconducting Hertzian dipoles. *Applied Physics Letters* **45** (3): 284–286.

23 Piao, Z., Tani, M., and Sakai, K. (2000). Carrier dynamics and terahertz radiation in photoconductive antennas. *Japanese Journal of Applied Physics* **39** (1R): 96.

24 Ezdi, K., Heinen, B., Jördens, C. et al. (2009). A hybrid time-domain model for pulsed terahertz dipole antennas. *Journal of the European Optical Society-Rapid Publications* **4**.

25 Berry, C.W. and Jarrahi, M. (2012). Principles of impedance matching in photoconductive antennas. *Journal of Infrared, Millimeter, and Terahertz Waves* **33** (12): 1182–1189.

26 Duffy, S.M., Verghese, S., McIntosh, A. et al. (2001). Accurate modeling of dual dipole and slot elements used with photomixers for coherent terahertz output power. *IEEE Transactions on Microwave Theory and Techniques* **49** (6): 1032–1038.

27 Brown, E.R. (2003). THz generation by photomixing in ultrafast photoconductors. In: *Terahertz Sensing Technology: Volume 1: Electronic Devices and Advanced Systems Technology*, 147–195.

28 Brown, E.R., Smith, F.W., and McIntosh, K.A. (1993). Coherent millimeter-wave generation by heterodyne conversion in low-temperature-grown GaAs photoconductors. *Journal of Applied Physics* **73** (3): 1480–1484.

29 Gregory, I.S., Baker, C., Tribe, W.R. et al. (2005). Optimization of photomixers and antennas for continuous-wave terahertz emission. *IEEE Journal of Quantum Electronics* **41** (5): 717–728.

30 Loata, G.C., Thomson, M.D., Löffler, T. et al. (2007). Radiation field screening in photoconductive antennae studied via pulsed terahertz emission spectroscopy. *Applied Physics Letters* **91** (23): 232506.

31 Sze, S.M. (1990). *High-Speed Semiconductor Devices*, 653. New York: Wiley-Interscience p.

32 Yang, S.-H. and Jarrahi, M. (2020). Navigating terahertz spectrum via photomixing. *Optics and Photonics News* **31** (7): 36–43.

33 Bjarnason, J.E., Chan, T.L.J., Lee, A.W.M. et al. (2004). ErAs:GaAs photomixer with two-decade tunability and 12 μW peak output power. *Applied Physics Letters* **85** (18): 3983–3985.

34 Wang, N. and Jarrahi, M. (2013). Noise analysis of photoconductive terahertz detectors. *Journal of Infrared, Millimeter, and Terahertz Waves* **34** (9): 519–528.

35 Warren, A.C., Katzenellenbogen, N., Grischkowsky, D. et al. (1991). Subpicosecond, freely propagating electromagnetic pulse generation and detection using GaAs:As epilayers. *Applied Physics Letters* **58** (14): 1512–1514.

36 Shen, Y.C., Upadhya, P.C., Beere, H.E. et al. (2004). Generation and detection of ultrabroadband terahertz radiation using photoconductive emitters and receivers. *Applied Physics Letters* **85** (2): 164–166.

37 Driscoll, D.C., Hanson, M.P., Gossard, A.C. et al. (2005). Ultrafast photoresponse at 1.55 μm in InGaAs with embedded semimetallic ErAs nanoparticles. *Applied Physics Letters* **86** (5): 051908.

38 Ospald, F., Maryenko, D., von Klitzing, K. et al. (2008). 1.55 μm ultrafast photoconductive switches based on ErAs:InGaAs. *Applied Physics Letters* **92** (13): 131117.

39 Sukhotin, M., Brown, E.R., Gossard, A.C. et al. (2003). Photomixing and photoconductor measurements on ErAs/InGaAs at 1.55 μm. *Applied Physics Letters* **82** (18): 3116–3118.

40 Dietz, R.J.B., Gerhard, M., Stanze, D. et al. (2011). THz generation at 1.55 μm excitation: six-fold increase in THz conversion efficiency by separated photoconductive and trapping regions. *Optics Express* **19** (27): 25911–25917.

41 Dietz, R.J.B., Globisch, B., Gerhard, M. et al. (2013). 64 μW pulsed terahertz emission from growth optimized InGaAs/InAlAs heterostructures with separated photoconductive and trapping regions. *Applied Physics Letters* **103** (6): 061103.

42 Lu, P.-K., Turan, D., and Jarrahi, M. (2020). High-sensitivity telecommunication-compatible photoconductive terahertz detection through carrier transit time reduction. *Optics Express* **28**: 26324–26335.

43 Turan, D., Yardimci, N.T., and Jarrahi, M. (2020). Plasmonics-enhanced photoconductive terahertz detector pumped by ytterbium-doped fiber laser. *Optics Express* **28** (3): 3835–3845.

44 Banszerus, L., Schmitz, M., Engels, S. et al. (2015). Ultrahigh-mobility graphene devices from chemical vapor deposition on reusable copper. *Science Advances* **1** (6): e1500222.

45 Bolotin, K.I., Sikes, K.J., Jiang, Z. et al. (2008). Ultrahigh electron mobility in suspended graphene. *Solid State Communications* **146** (9–10): 351–355.

46 Balci, O., Polat, E.O., Kakenov, N. et al. (2015). Graphene-enabled electrically switchable radar-absorbing surfaces. *Nature Communications* **6**: 6628.

47 Liu, C.-H., Chang, Y.C., Norris, T.B. et al. (2014). Graphene photodetectors with ultra-broadband and high responsivity at room temperature. *Nature Nanotechnology* **9** (4): 273.

48 Chen, X., Lu, X., Deng, B. et al. (2017). Widely tunable black phosphorus mid-infrared photodetector. *Nature Communications* **8** (1): 1672.

49 Prechtel, L., Song, L., Schuh, D. et al. (2012). Time-resolved ultrafast photocurrents and terahertz generation in freely suspended graphene. *Nature Communications* **3**: 646.

50 Hunter, N., Mayorov, A.S., Wood, C.D. et al. (2015). On-chip picosecond pulse detection and generation using graphene photoconductive switches. *Nano Letters* **15** (3): 1591–1596.

51 Mittendorff, M., Suess, R.J., Leong, E. et al. (2017). Optical gating of black phosphorus for terahertz detection. *Nano Letters* **17** (9): 5811–5816.

52 Dionne, J.A., Sweatlock, L.A., Atwater, H.A. et al. (2005). Planar metal plasmon waveguides: frequency-dependent dispersion, propagation, localization, and loss beyond the free electron model. *Physical Review B* **72** (7): 075405.

53 Pendry, J.B., Martin-Moreno, L., and Garcia-Vidal, F.J. (2004). Mimicking surface plasmons with structured surfaces. *Science* **305** (5685): 847–848.

54 Web, I.S.I., Press, H., and York, N. (2012). Plasmonics: merging photonics. *Science* **189**: 189–194.

55 Ebbesen, T.W., Lezec, H.J., Ghaemi, H.F. et al. (1998). Extraordinary optical transmission through sub-wavelength hole arrays. *Nature* **391** (6668): 667.

56 Schuller, J.A., Barnard, E.S., Cai, W. et al. (2010). Plasmonics for extreme light concentration and manipulation. *Nature Materials* **9** (3): 193.

57 Genet, C. and Ebbesen, T.W. (2010). Light in tiny holes. *Nanoscience and Technology: A Collection of Reviews from Nature Journals*: 205–212.

58 Hartschuh, A., Sánchez, E.J., Xie, X.S. et al. (2003). High-resolution near-field Raman microscopy of single-walled carbon nanotubes. *Physical Review Letters* **90** (9): 095503.

59 Frey, H.G., Witt, S., Felderer, K. et al. (2004). High-resolution imaging of single fluorescent molecules with the optical near-field of a metal tip. *Physical Review Letters* **93** (20): 200801.

60 Cubukcu, E., Kort, E.A., Crozier, K.B. et al. (2006). Plasmonic laser antenna. *Applied Physics Letters* **89** (9): 093120.

61 Yu, N., Fan, J., Wang, Q.J. et al. (2008). Small-divergence semiconductor lasers by plasmonic collimation. *Nature Photonics* **2** (9): 564.

62 Atwater, H.A. and Polman, A. (2010). Plasmonics for improved photovoltaic devices. *Nature Materials* **9** (3): 205.

63 Pala, R.A., White, J., Barnard, E. et al. (2009). Design of plasmonic thin-film solar cells with broadband absorption enhancements. *Advanced Materials* **21** (34): 3504–3509.

64 Ishi, T., Fujikata, J., Makita, K. et al. (2005). Si nano-photodiode with a surface plasmon antenna. *Japanese Journal of Applied Physics* **44** (3L): L364.

65 Tang, L., Kocabas, S.E., Latif, S. et al. (2008). Nanometre-scale germanium photodetector enhanced by a near-infrared dipole antenna. *Nature Photonics* **2** (4): 226.

66 Dintinger, J., Robel, I., Kamat, P.V. et al. (2006). Terahertz all-optical molecule-plasmon modulation. *Advanced Materials* **18** (13): 1645–1648.

67 Berry, C.W., Moore, J., and Jarrahi, M. (2011). Design of reconfigurable metallic slits for terahertz beam modulation. *Optics Express* **19** (2): 1236–1245.

68 Unlu, M., Hashemi, M.R., Berry, C.W. et al. (2014). Switchable scattering meta-surfaces for broadband terahertz modulation. *Scientific Reports* **4**: 5708.

69 Unlu, M. and Jarrahi, M. (2014). Miniature multi-contact MEMS switch for broadband terahertz modulation. *Optics Express* **22** (26): 32245–32260.

70 Hashemi, M.R.M., Yang, S.H., Wang, T. et al. (2016). Electronically-controlled beam-steering through vanadium dioxide metasurfaces. *Scientific Reports* **6**: 35439.

71 Hashemi, M.R., Cakmakyapan, S., and Jarrahi, M. (2017). Reconfigurable metamaterials for terahertz wave manipulation. *Reports on Progress in Physics* **80** (9): 094501.

72 Berry, C.W., Wang, N., Hashemi, M.R. et al. (2013). Significant performance enhancement in photoconductive terahertz optoelectronics by incorporating plasmonic contact electrodes. *Nature Communications* **4**: 1622.

73 Turan, D., Corzo-Garcia, S.C., Yardimci, N.T. et al. (2017). Impact of the metal adhesion layer on the radiation power of plasmonic photoconductive terahertz sources. *Journal of Infrared, Millimeter, and Terahertz Waves* **38** (12): 1448–1456.

74 Park, S.-G., Jin, K.H., Yi, M. et al. (2012). Enhancement of terahertz pulse emission by optical nanoantenna. *ACS Nano* **6** (3): 2026–2031.

75 Jooshesh, A., Smith, L., Masnadi-Shirazi, M. et al. (2014). Nanoplasmonics enhanced terahertz sources. *Optics Express* **22** (23): 27992–28001.

76 Jooshesh, A., Bahrami-Yekta, V., Zhang, J. et al. (2015). Plasmon-enhanced below bandgap photoconductive terahertz generation and detection. *Nano Letters* **15** (12): 8306–8310.

77 Park, S.-G., Choi, Y., Oh, Y.J. et al. (2012). Terahertz photoconductive antenna with metal nanoislands. *Optics Express* **20** (23): 25530–25535.

78 Middendorf, J.R. and Brown, E.R. (2012). THz generation using extrinsic photoconductivity at 1550 nm. *Optics Express* **20** (15): 16504–16509.

79 Rämer, J.-M., Ospald, F., von Freymann, G. et al. (2013). Generation and detection of terahertz radiation up to 4.5 THz by low-temperature grown GaAs photoconductive antennas excited at 1560 nm. *Applied Physics Letters* **103** (2): 021119.

80 Tani, M., Lee, K.-S., and Zhang, X.-C. (2000). Detection of terahertz radiation with low-temperature-grown GaAs-based photoconductive antenna using 1.55 μm probe. *Applied Physics Letters* **77** (9): 1396–1398.

81 Fesharaki, F., Jooshesh, A., Bahrami-Yekta, V. et al. (2017). Plasmonic antireflection coating for photoconductive terahertz generation. *ACS Photonics* **4** (6): 1350–1354.

82 Berry, C.W. and Jarrahi, M. (2010). Plasmonically-enhanced localization of light into photoconductive antennas. *Proceedings of the Conference on Lasers and Electro-Optics*, CFI2, 1.

83 Hsieh, B.-Y. and Jarrahi, M. (2011). Analysis of periodic metallic nano-slits for efficient interaction of terahertz and optical waves at nano-scale dimensions. *Journal of Applied Physics* **109** (8): 084326.

84 Berry, C.W. and Jarrahi, M. (2011). Ultrafast photoconductors based on plasmonic gratings. *2011 International Conference on Infrared, Millimeter, and Terahertz Waves*. IEEE.

85 Berry, C.W. and Jarrahi, M. (2012). Terahertz generation using plasmonic photoconductive gratings. *New Journal of Physics* **14** (10): 105029.

86 Berry, C.W., Hashemi, M.R., and Jarrahi, M. (2014). Generation of high power pulsed terahertz radiation using a plasmonic photoconductive emitter array with logarithmic spiral antennas. *Applied Physics Letters* **104** (8): 081122.

87 Heshmat, B., Pahlevaninezhad, H., Pang, Y. et al. (2012). Nanoplasmonic terahertz photoconductive switch on GaAs. *Nano Letters* **12** (12): 6255–6259.

88 Yang, S.-H., Hashemi, M.R., Berry, C.W. et al. (2014). 7.5% optical-to-terahertz conversion efficiency offered by photoconductive emitters with three-dimensional plasmonic contact electrodes. *IEEE Transactions on Terahertz Science and Technology* **4** (5): 575–581.

89 Yang, S.-H. and Jarrahi, M. (2013). Enhanced light–matter interaction at nanoscale by utilizing high-aspect-ratio metallic gratings. *Optics Letters* **38** (18): 3677–3679.

90 Li, X., Yardimci, N.T., and Jarrahi, M. (2017). A polarization-insensitive plasmonic photoconductive terahertz emitter. *AIP Advances* **7** (11): 115113.

91 Yang, S.-H. and Jarrahi, M. (2015). Frequency-tunable continuous-wave terahertz sources based on GaAs plasmonic photomixers. *Applied Physics Letters* **107** (13): 131111.

92 Berry, C.W., Hashemi, M.R., Preu, S. et al. (2014). Plasmonics enhanced photomixing for generating quasi-continuous-wave frequency-tunable terahertz radiation. *Optics Letters* **39** (15): 4522–4524.

93 Berry, C.W., Hashemi, M.R., Preu, S. et al. (2014). High power terahertz generation using 1550 nm plasmonic photomixers. *Applied Physics Letters* **105** (1): 011121.

94 Yang, S.-H., Watts, R., Li, X. et al. (2015). Tunable terahertz wave generation through a bimodal laser diode and plasmonic photomixer. *Optics Express* **23** (24): 31206–31215.

95 Yang, S.-H. and Jarrahi, M. (2015). Spectral characteristics of terahertz radiation from plasmonic photomixers. *Optics Express* **23** (22): 28522–28530.

96 Huang, S.-W., Yang, J., Yang, S.H. et al. (2017). Globally stable microresonator turing pattern formation for coherent high-power THz radiation on-chip. *Physical Review X* **7** (4): 041002.

97 Wang, N., Cakmakyapan, S., Lin, Y.J. et al. (2019). Room-temperature heterodyne terahertz detection with quantum-level sensitivity. *Nature Astronomy* **3** (11): 977–982.

98 Kuwashima, F., Jarrahi, M., Cakmakyapan, S. et al. (2020). Evaluation of high-stability optical beats in laser chaos by plasmonic photomixing. *Optics Express* **28** (17): 24833–24844.

99 Wang, N. and Jarrahi, M. (2020). High-precision millimeter-wave frequency determination through plasmonic photomixing. *Optics Express* **28** (17): 24900–24907.

100 Lin, Y.-J. and Jarrahi, M. (2020). Heterodyne terahertz detection through electronic and optoelectronic mixers. *Reports on Progress in Physics* **83** (6): 066101.

101 Lin, Y.-J., Cakmakyapan, S., Wang, N. et al. (2019). Plasmonic heterodyne spectrometry for resolving the spectral signatures of ammonia over a 1-4. 5 THz frequency range. *Optics Express* **27** (25): 36838–36845.

102 Dreyhaupt, A., Winnerl, S., Dekorsy, T. et al. (2005). High-intensity terahertz radiation from a microstructured large-area photoconductor. *Applied Physics Letters* **86** (12): 121114.

103 Beck, M., Schäfer, H., Klatt, G. et al. (2010). Impulsive terahertz radiation with high electric fields from an amplifier-driven large-area photoconductive antenna. *Optics Express* **18** (9): 9251–9257.

104 Moon, K., Lee, I.M., Shin, J.H. et al. (2015). Bias field tailored plasmonic nano-electrode for high-power terahertz photonic devices. *Scientific Reports* **5**: 13817.

105 Yardimci, N.T., Yang, S.H., Berry, C.W. et al. (2015). High-power terahertz generation using large-area plasmonic photoconductive emitters. *IEEE Transactions on Terahertz Science and Technology* **5** (2): 223–229.

106 Yardimci, N.T., Salas, R., Krivoy, E.M. et al. (2015). Impact of substrate characteristics on performance of large area plasmonic photoconductive emitters. *Optics Express* **23** (25): 32035–32043.

107 Yardimci, N.T., Lu, H., and Jarrahi, M. (2016). High power telecommunication-compatible photoconductive terahertz emitters based on plasmonic nano-antenna arrays. *Applied Physics Letters* **109** (19): 191103.

108 Yardimci, N.T. and Jarrahi, M. (2017). High sensitivity terahertz detection through large-area plasmonic nano-antenna arrays. *Scientific Reports* **7**: 42667.

109 Snively, E.C., Yardimci, N.T., Jacobson, B.T. et al. (2018). Non-invasive low charge electron beam time-of-arrival diagnostic using a plasmonics-enhanced photoconductive antenna. *Applied Physics Letters* **113** (19): 193501.

110 Yardimci, N.T., Cakmakyapan, S., Hemmati, S. et al. (2017). A high-power broadband terahertz source enabled by three-dimensional light confinement in a plasmonic nanocavity. *Scientific Reports* **7** (1): 4166.

111 Mitrofanov, O., Brener, I., Luk, T.S. et al. (2015). Photoconductive terahertz near-field detector with a hybrid nanoantenna array cavity. *ACS Photonics* **2** (12): 1763–1768.

112 Yardimci, N.T., Turan, D., Cakmakyapan, S. et al. (2018). A high-responsivity and broadband photoconductive terahertz detector based on a plasmonic nanocavity. *Applied Physics Letters* **113** (25): 251102.

113 Yardimci, N.T. and Jarrahi, M. (2018). Nanostructure-enhanced photoconductive terahertz emission and detection. *Small* **14** (44): 1802437.

114 Jarrahi, M. (2015). Advanced photoconductive terahertz optoelectronics based on nano-antennas and nano-plasmonic light concentrators. *IEEE Transactions on Terahertz Science and Technology* **5** (3): 391–397.

115 Sanders, D.P. (2010). Advances in patterning materials for 193 nm immersion lithography. *Chemical Reviews* **110** (1): 321–360.

116 Wagner, C. and Harned, N. (2010). EUV lithography: lithography gets extreme. *Nature Photonics* **4** (1): 24.

117 Guo, L.J. (2007). Nanoimprint lithography: methods and material requirements. *Advanced Materials* **19** (4): 495–513.

118 Xia, Y. and Whitesides, G.M. (1998). Soft lithography. *Annual Review of Materials Science* **28** (1): 153–184.

119 Qin, D., Xia, Y., and Whitesides, G.M. (2010). Soft lithography for micro-and nanoscale patterning. *Nature Protocols* **5** (3): 491–502.

6

Terahertz Quantum Cascade Lasers

Roberto Paiella

Department of Electrical and Computer Engineering and Photonics Center, Boston University, Boston, MA, USA

6.1 Introduction

Semiconductor lasers represent a key device technology for the generation of coherent light across a large portion of the optical spectrum. Compared to other types of lasers, they provide several important and desirable features, including small size, low power consumption, compatibility with direct current injection, high speed, and the ability to be combined with several other optical components to form highly integrated photonic circuits. Traditional semiconductor diode lasers rely on interband electronic transitions in bulk films or quantum-confined structures, whereby photons are created through electron–hole recombination between a conduction and a valence band or subband. As a result, the emission wavelength of such devices is essentially fixed by the bandgap of the underlying materials, with only limited tunability provided by quantum confinement if quantum structures are employed in the active layer. For the best-established semiconductor optoelectronic materials, i.e. III–V compounds based on As, P, and/or N, the bandgap energy ranges from several hundred meV to a few eV, which has allowed for the development of high-performance diode lasers operating across the near-infrared and visible spectral regions.

For laser emission at longer wavelengths (into the mid- and far-infrared range including terahertz frequencies), the lack of suitable materials with sufficiently small bandgap energies has stimulated extensive research efforts focused on intersubband (ISB) transitions in quantum wells (QWs), beginning with an early theoretical proposal by Kazarinov and Suris in 1971 [1]. ISB transitions involve quantized electronic states derived from the same energy band (typically the conduction band), and as a result, the emission wavelength can be tailored through the QW design over an extremely broad spectral range, in principle only limited on the short side by the QW band offset. Semiconductor lasers based on these transitions—generally referred to as quantum cascade (QC) lasers—were originally demonstrated in 1994 [2] with a GaInAs/AlInAs multiple-QW active layer emitting at a

Fundamentals of Terahertz Devices and Applications, First Edition. Edited by Dimitris Pavlidis.

Companion website: www.wiley.com/go/Pavlidis/FundamentalsofTHz

mid-infrared wavelength of about 4.3 μm. Since then, QC lasers have experienced explosive growth in performance and capabilities, including the ability to produce room-temperature continuous-wave (cw) emission with output powers of several hundred milliwatts at mid-infrared wavelengths [3, 4], as well as THz operation [5], albeit only at cryogenic temperatures and over an incomplete portion of the THz spectrum. In any case, despite these limitations, THz QC lasers already provide the leading device technology for the generation of high-power narrowband radiation at design-tunable frequencies of a few to several THz.

While several different types of QC gain media have been developed over the years, generally they all consist of a periodic repetition—or cascade—of alternating multiple-QW active stages and injector regions, where electronic transport occurs via tunneling, phonon scattering, and ISB photon emission. For a given materials platform, typically GaInAs/AlInAs or GaAs/AlGaAs QWs, the layer thicknesses and compositions are carefully designed to produce a desired potential-energy profile and sequence of bound states, as needed to enable efficient carrier transport and light emission. With this approach, generally known as bandgap engineering, complex multiple-QW structures can be devised whose electronic and optical properties can be tailored in a highly controlled fashion through the heterostructure design. Experimental samples with the desired characteristics can then be developed with the use of sophisticated epitaxial growth techniques. In particular, the entire field of ISB optoelectronics in general (and QC lasers in particular) has been largely enabled by the development of molecular beam epitaxy (MBE) technology [6]. With this technique, the constituent species are transported to the substrate by atomic or molecular beams under ultrahigh vacuum conditions, which allows monitoring the sample growth in situ using powerful surface analysis techniques such as reflection high-energy electron diffraction (RHEED). These capabilities have made it possible to grow heterostructures with unprecedented interface abruptness and exquisite control of layer thicknesses and compositions, as needed for the initial development and exploration of complex QC active layers. More recently, metal-organic vapor phase epitaxy (MOVPE), which is generally considered more suitable for commercial production, has also been successfully applied to the growth of high-performance QC lasers, leading to significant advances in the commercialization of these devices.

This chapter is focused on the fundamentals of THz QC lasers, with emphasis on their design principles, state-of-the-art performance and limitations, and potential for improvement based on novel materials systems. In Section 6.2, we briefly review the basic physics of ISB transitions in semiconductor QWs, including optical absorption and emission processes and phonon-assisted nonradiative transitions. Sections 6.3 and 6.4 are devoted to the design of the QC gain medium and optical cavity, respectively, including the use of plasmonic waveguides to achieve strong optical confinement at THz wavelengths in epitaxial layers of realistic thicknesses. Next, in Section 6.5, we review the present level of performance of state-of-the-art THz QC lasers at the time of this writing (2018), in terms of spectral coverage, output power, and temperature characteristics. While the best devices to date are based on GaAs/AlGaAs QWs, due to their unparalleled crystalline quality, the presence of THz-range optical phonons in these materials fundamentally limits their ability to cover the entire THz spectrum and to provide laser emission without cryogenic cooling. As a result, alternative materials systems are currently being investigated to overcome these limitations, as described in the last section of the chapter. These novel materials systems include III-nitride QWs (GaN/AlGaN), where the optical phonon frequencies are well above the

THz range, and SiGe which is a nonpolar material and therefore features significantly weaker electron–phonon and photon–phonon interactions compared to III–V compound semiconductors.

6.2 Fundamentals of Intersubband Transitions

Most ISB devices developed so far are based on electronic transitions between quantum-confined states derived from the conduction band in semiconductor QWs or superlattices (although ISB transitions between valence subbands have also been investigated, as reviewed, e.g., in [7]). To illustrate the basic properties of these transitions, Figure 6.1a shows the conduction-band lineup of a single QW, together with the squared envelope functions of its bound subbands. The parameters E_1, E_2, and E_3 introduced in this figure are the energy minima of the ground-state and first two excited conduction subbands (denoted $|1\rangle$, $|2\rangle$, and $|3\rangle$, respectively), i.e. the energies of the lowest three QW bound states with zero in-plane wavevector $\mathbf{k}_{||}$. The subbands dispersion relations (energy versus $\mathbf{k}_{||}$) are plotted in Figure 6.1b and are well approximated by parabolas of roughly the same curvature, i.e. $E_l(\mathbf{k}_|) = E_l + \frac{\hbar^2 k_|^2}{2m_c}$ for $l = 1, 2, 3$, where m_c is the effective mass of the conduction band in the well material. The vertical arrow in this figure illustrates a possible radiative ISB transition, where an electron decays from subband $|3\rangle$ to $|2\rangle$, and correspondingly a photon of energy $\hbar\omega = E_3 - E_2$ is emitted so as to ensure energy conservation. Vice versa, light can also be absorbed through the reverse process, where the energy of each absorbed photon is used to excite an electron into a higher energy subband. In addition, nonradiative ISB transitions can also take place through the absorption and emission of longitudinal optical (LO) phonons (and to a lesser extent through interactions with other phonon modes and through other scattering mechanisms), as illustrated by the diagonal arrow between subbands $|2\rangle$ and $|1\rangle$ in Figure 6.1b. These processes are key to the operation of QC lasers, and therefore they are discussed in some detail below.

All radiative and nonradiative ISB transition rates depend on the QW design through the subband energies $E_l(\mathbf{k}_{||})$ and wave functions $\psi_{l,\mathbf{k}_|}(\mathbf{r})$, which can be widely tailored in complex multiple-QW systems by properly adjusting the various structural parameters such as layer thicknesses, compositions, and doping profile. These energy eigenvalues and eigenfunctions can be readily computed using the effective-mass approximation [8], which allows deriving a simplified Schrödinger equation in which the effect of the rapidly varying crystal-periodic potential (due to the ionized atoms of the crystal lattice) is simply included through the effective-mass parameter. After solving this Schrödinger equation, often self-consistently with a Poisson equation accounting for space charge effects in doped structures, the ISB absorption (or gain) spectrum can be computed as follows:

$$\alpha(\omega) = \frac{\pi q^2}{\varepsilon_0 c n \omega m_0^2} \sum_{l,l'} |\hat{\mathbf{e}}\cdot\hat{\mathbf{z}} p_{l'l}|^2 (N_l - N_{l'}) \frac{\gamma/2\pi}{(E_{l'} - E_l - \hbar\omega)^2 + (\gamma/2)^2} \tag{6.1}$$

In this equation, n is the refractive index of the QW materials, ω and $\hat{\mathbf{e}}$ are, respectively, the angular frequency and polarization unit vector of the light being absorbed (or amplified), $\hat{\mathbf{z}}$ is the unit vector along the QW growth direction, N_l is the electron density of

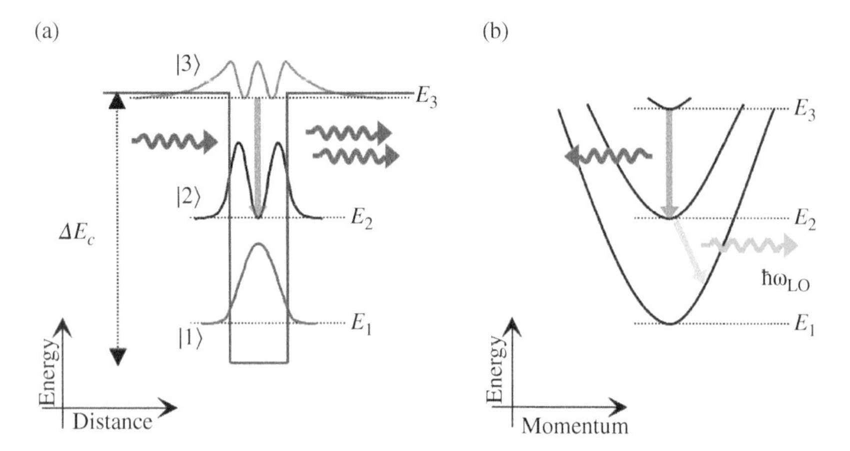

Figure 6.1 (a) Conduction-band lineup of a generic semiconductor QW and squared envelope functions of its electronic bound states, referenced to their respective energy levels. (b) Energy dispersion curves of the subbands of (a). The vertical and diagonal arrows describe photon- and phonon-assisted electronic ISB transitions, respectively.

subband number l, and γ is a phenomenological linewidth that accounts for all electronic scattering processes (e.g. involving phonons, ionized impurities, other carriers, and interface roughness). Finally, $p_{l'l} \equiv \left\langle \varphi_{l',\mathbf{k}_\parallel} \middle| p_z \middle| \varphi_{l,\mathbf{k}_\parallel} \right\rangle$ is the matrix element of the z component of linear momentum between the z-dependent envelope functions of subbands l' and l (the functions $\varphi_{l',\mathbf{k}_\parallel}$ and $\varphi_{l,\mathbf{k}_\parallel}$ whose magnitudes squared are plotted in Figure 6.1a). Under conditions of population inversion between two subbands (e.g. for $N_3 > N_2$), α is negative at the frequency difference between the same two subbands (e.g. at $\omega_{32} = (E_3 - E_2)/\hbar$) and the system provides optical amplification with gain $g = -\alpha$.

Equation (6.1) contains several important pieces of information regarding the nature of ISB optical transitions in QWs. First, these transitions are generally characterized by a narrow Lorentzian lineshape indicative of predominantly homogeneous broadening caused by the aforementioned scattering mechanisms. In fact, inhomogeneous broadening due to QW thickness fluctuations and/or conduction-band nonparabolicities can also contribute to the ISB absorption (gain) linewidth, but its role is typically small in the high-quality QW samples used in QC lasers. Second, because of the $\hat{\mathbf{e}}\cdot\hat{\mathbf{z}}$ term, only TM polarized photons—i.e. linearly polarized light with polarization unit vector along the growth axis—can be absorbed or emitted. This polarization selection rule has important implications for the design of all ISB devices—e.g. in the context of QC lasers, it precludes the use of standard vertical-cavity geometries and requires instead the use of in-plane waveguides supporting strongly confined TM modes. Finally, through the momentum matrix element $p_{l'l}$, the ISB absorption (or gain) strength depends on the mutual symmetry and spatial distribution of the envelope functions of the subbands involved in the optical transitions. In particular, if the QW conduction-band lineup has inversion symmetry along the z-axis, the matrix element of Eq. (6.1) is nonzero only if l and l' have opposite parity. Furthermore, its magnitude can generally be maximized by increasing the spatial overlap between the envelope functions $\varphi_{l',\mathbf{k}_\parallel}$ and $\varphi_{l,\mathbf{k}_\parallel}$.

Regarding nonradiative ISB scattering, in the presence of the relatively low carrier densities typically employed in QC active layers, it tends to be dominated by electron–phonon interactions. In particular, in compound semiconductor materials with partly ionic chemical bonds, electrons are most strongly coupled to LO phonons, since these are the modes of lattice vibrations involving the largest instantaneous displacement of nearest-neighboring atoms. Thus, nonradiative ISB relaxation in typical QC active layers predominantly occurs via LO-phonon emission. The associated ISB scattering rates are among the key design parameters of QC lasers, as they directly affect the operational characteristics (both steady-state and dynamic) of these devices. Similar to the ISB gain coefficient, these scattering rates are related to the active-layer subband structure and thus can be tailored by design using the rules of bandgap engineering, e.g. see [9, 10]. In particular, the probability per unit time $1/\tau_{if}$ that an electron relaxes from subband $|i\rangle$ to subband $|f\rangle$ via LO-phonon emission exhibits a strong dependence on the ISB energy separation $E_i - E_f$ and the LO-phonon energy $\hbar\omega_{LO}$ (about 30 meV or 8 THz in GaInAs/AlInAs and GaAs/AlGaAs QWs). If $E_i - E_f$ is significantly larger than $\hbar\omega_{LO}$ (by at least several ten meV), the nonradiative relaxation lifetime τ_{if} is on the order of a few picoseconds. While this time scale is already ultrafast, even shorter lifetimes of a few hundred femtoseconds are obtained for $E_i - E_f \approx \hbar\omega_{LO}$. Finally, for $E_i - E_f < \hbar\omega_{LO}$ LO-phonon-assisted relaxation from the bottom of subband $|i\rangle$ into $|f\rangle$ is forbidden by the energy-conservation requirement. Under these conditions the ISB relaxation dynamics are dominated by other scattering mechanisms such as the emission of acoustic phonons, with much longer lifetimes on the order of a few hundred picoseconds.

These basic properties of LO-phonon-assisted ISB transitions are exploited in most QC active-layer designs to obtain population inversion. Specifically, light emission in these devices often involves electronic transitions between the second and first excited subbands $|3\rangle$ and $|2\rangle$ of multiple-QW active stages arranged in a periodic fashion. Population inversion (i.e. $N_3 > N_2$) is then ensured through the presence of a ground-state subband $|1\rangle$ in each stage with energy $E_1 \approx E_2 - \hbar\omega_{LO}$, so that the lower laser subbands $|2\rangle$ are depopulated via LO-phonon scattering on a particularly fast (sub-picosecond) time scale. In THz QC lasers, the emission photon energy, and therefore the energy separation $E_3 - E_2$ of the laser subbands, are smaller than the LO-phonon energy. This general property has several important implications on the design and high-temperature performance of these devices, as explained below.

6.3 Active Material Design

In general, laser action involves the combination of two key ingredients—an optical gain medium where light is emitted and amplified, and a cavity providing optical feedback. In QC lasers, the gain medium consists of a cascade of several identical repeat units (typically 20–30 but in some cases as many as 100), each comprising an active stage and an injector region. The active stages are suitably designed coupled-QW systems where population inversion is established across two consecutive subbands leading to amplified light emission. The injector regions consist of additional QWs (often comprising a short-period superlattice), where electrons are collected from the ground state of the active stage upstream and reinjected into the upper laser state of the active stage downstream. This periodic gain-medium

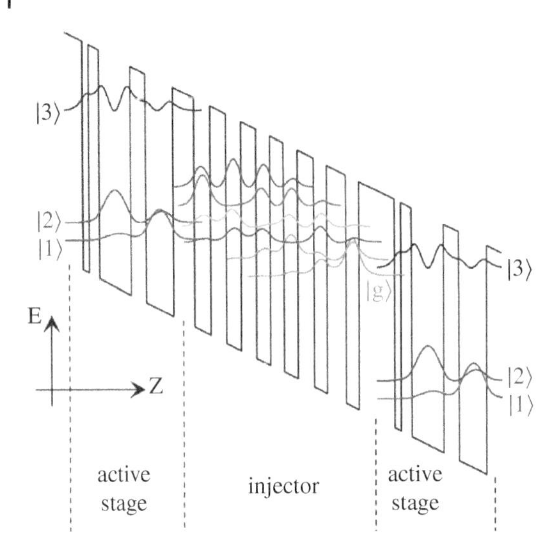

Figure 6.2 Conduction-band diagram of a representative bound-to-bound QC gain medium based on the $Ga_{0.47}In_{0.53}As$/$Al_{0.48}In_{0.52}As$:InP QW materials system. The layer thicknesses in Å of one repeat unit, from left to right and starting from the injection barrier (i.e. the barrier between each active stage and the injector region upstream), are 50/9/15/47/22/40/30/23/23/22/22/20/20/20/23/19/28/19. The externally applied field is 76 kV/cm. The variables E and z indicate, respectively, electron energy and position along the growth axis.

arrangement leads to the basic cascading (or electron recycling) property of QC lasers, whereby each electron injected into the active material can potentially generate N_p photons (N_p being the total number of repeat units). Thus, unlike the case of conventional diode lasers, the quantum efficiency of QC lasers can in principle be larger than one photon per electron, which is of course favorable for high-power laser emission.

Since the original demonstration of QC lasers [2], many different types of ISB active layers have been developed, as reviewed, e.g., in [11]. Their basic building blocks consist of coupled QWs, i.e. adjacent QWs separated by sufficiently thin barriers so that their bound states of similar energy can mix with one another via tunneling, to produce "bonding" and "antibonding" eigenstates whose properties are widely tunable by design. To illustrate the basic features of typical QC gain media, here we begin by describing a relatively early design that has produced several high-performance devices emitting throughout the mid-infrared spectral region [12, 13]. This design is illustrated in Figure 6.2, where we plot the conduction-band lineup and the relevant squared envelope functions (referenced to their respective energy levels) of a portion of the QC gain medium of [12]. All well and barrier layers consist of $Ga_{0.47}In_{0.53}As$ and $Al_{0.48}In_{0.52}As$, respectively, both of which are lattice matched to the InP growth substrate. While this design is optimized for mid-infrared light emission (at a wavelength of 5 μm in Figure 6.2), its description will also help introduce the key design challenges specific to THz QC lasers, as discussed below.

Each active stage of this QC structure consists of three strongly coupled QWs, where stimulated emission takes place between subbands $|3\rangle$ and $|2\rangle$. The laser output photon energy is, therefore, determined by the energy separation $E_3 - E_2$, which can be varied over a broad range with the same materials system (and the same design strategy) by properly varying the well and barrier thicknesses: in fact, this is one of the most distinctive and unique properties of QC lasers. For these and similar three-level gain media, it can be shown using a simple rate-equation analysis that the fundamental condition for population inversion is approximately $\tau_2 < \tau_{32}$, where τ_2 is the lifetime of the lower laser subband $|2\rangle$ and $1/\tau_{32}$ is the nonradiative decay rate from $|3\rangle$ to $|2\rangle$—see Problem 1 below. In the design of Figure 6.2, the main nonradiative decay channel for the electrons in the upper laser subbands |3⟩ is provided by LO-phonon-assisted relaxation into the lower-lying subbands |2⟩ and |1⟩, with corresponding decay lifetimes on the order of a few picoseconds. Similarly, the lower laser subbands |2⟩ are mainly depopulated via LO-phonon-assisted transitions into the ground states |1⟩ and via tunneling into the injector downstream. To maximize the speed of the former process, by design the energy separation between subbands |2⟩ and |1⟩ in each active stage is approximately equal to the LO-phonon energy (~ 32 meV in GaInAs). As a result, the overall lifetime τ_2 of subband |2⟩ is sub-picosecond, which ensures the feasibility of population inversion under proper injection.

The injector region in Figure 6.2 can be described as a superlattice designed to facilitate the transport of electrons from the lower states |2⟩ and |1⟩ of the active stage upstream to the upper laser state |3⟩ of the stage downstream. To that purpose, its lowest "miniband" (strictly speaking a manifold of six delocalized states derived from the ground states of its six wells) is by design lined up with the aforementioned states |1⟩ and |2⟩ upstream and |3⟩ downstream. Transport in and out of the miniband occurs by tunneling [14], with characteristic time scales on the order of a few picoseconds. At the same time, the injector lowest "minigap" lies in front of the upper laser state |3⟩ upstream, to minimize electron escape from this state into the injector, and thus help maintain a relatively long lifetime τ_3. In addition, unlike the active-stage QWs, some of the injector QWs (typically the middle ones) are doped n-type. As a result, as the injected electrons traverse the gain medium their negative charge is compensated by that of the positively ionized donor impurities, leading to an overall state of charge neutrality. This avoids the formation of space-charge domains in the active layer that would otherwise cause an inhomogeneous field distribution along the growth direction, and therefore spoil the electrical transport characteristics and the optical gain.

Alternative QC designs that are particularly suitable for high-power operation employ superlattice-based active stages, each consisting of several [e.g. 5 or 6] strongly coupled QWs [15–17]. As an example, the structure of Tredicucci et al. [17] is plotted in Figure 6.3. In these devices, light emission involves electronic transitions from the bottom of the upper miniband to the top of the lower miniband of each active-stage superlattice.

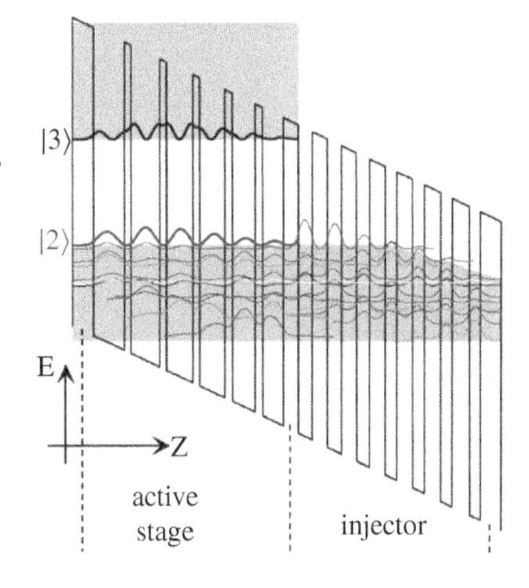

Figure 6.3 Conduction-band diagram of a representative superlattice QC gain medium based on the $Ga_{0.47}In_{0.53}As/Al_{0.48}In_{0.52}As$:InP QW materials system. The layer thicknesses in Å of one repeat unit, from left to right and starting from the injection barrier, are 35/51/11/48/11/44/11/41/12/38/13/35/25/ 23/25/ 23/25/22/26/20/26/19/27/19/29/18. The externally applied field is 46 kV/cm. The shaded areas denote the superlattice minibands. The emission wavelength is 7.2 μm.

Specific advantages of this approach compared to the "bound-to-bound" gain media illustrated in Figure 6.2 include (i) larger oscillator strength of the laser transitions, due to the larger spatial extent of the superlattice envelope functions; (ii) faster depopulation of the lower laser states by LO-phonon-assisted intraminiband relaxation; and (iii) more efficient electron extraction out of each active stage into the injector downstream. At the same time, however, selective electron injection into the upper laser states is not as effective as in the devices of Figure 6.2, due to the large density of states in the upper miniband into which electrons can be injected. A variation of the superlattice scheme designed to avoid this drawback is the "bound-to-continuum" QC structure, where a discrete state is introduced by design within the first minigap of the active-stage superlattice and used as the upper laser state [18].

The ISB gain media just described were originally designed for the development of mid-infrared semiconductor lasers. As already pointed out, their emission wavelengths can be tuned by design over a broad spectral range, only limited on the short side by the finite conduction-band offset of the underlying QW materials system. However, the extension of these devices to THz frequencies (below the LO phonon frequency) is far from straightforward. In particular, the ultrafast nature of resonant electron/LO-phonon ISB scattering can no longer be directly exploited in simple coupled-QW structures to ensure population inversion. For example, in a structure based on the design approach of Figure 6.2 with a small, THz-range transition energy $E_3 - E_2$, if the $|2\rangle$–$|1\rangle$ transition is near resonant to the LO-phonon energy, so is the $|3\rangle$–$|1\rangle$ transition. As a result, both laser states $|3\rangle$ and $|2\rangle$ would be very efficiently depopulated by LO-phonon scattering, leading to negligibly small optical gain. The selective injection and extraction of electrons in discrete subbands separated from one another by THz-range energies is also quite challenging, and often limited to a small range of values of the applied bias.

The first QC laser operating at THz wavelengths was demonstrated in 2002 with a GaAs/AlGaAs superlattice active-layer design [5], conceptually similar to the structure of Figure 6.3. Its emission frequency was 4.4 THz, well below the LO-phonon frequency of GaAs (8.7 THz). Unlike the structure of Figure 6.3, in this THz design, the width of the lower miniband is smaller than the LO-phonon energy, so as to avoid any LO-phonon-assisted relaxation for the electrons in the upper laser states $|3\rangle$. Depopulation of the lower laser states $|2\rangle$ then relies on electron/electron scattering, although LO-phonon emission still plays an important role. Specifically, electron/electron scattering creates a "hot" carrier distribution in subband $|2\rangle$, whereby many electrons acquire sufficient excess kinetic energy so that LO-phonon-assisted relaxation into lower-lying states of the same miniband becomes allowed [19]. It should be noted that this process of thermally activated LO-phonon emission is also responsible for enhanced nonradiative relaxation between the laser subbands $|3\rangle$ and $|2\rangle$, which ultimately limits the maximum operating temperature of THz QC lasers to values well below room temperature. This key limitation is discussed in more detail in Section 6.5.

In addition to this superlattice approach, two other design schemes have mainly been utilized in the development of THz QC lasers. One is the bound-to-continuum structure [20], which generally provides improved carrier injection into the upper laser states as described above. The other is commonly referred to as the resonant-phonon design [21, 22] and is illustrated schematically in Figure 6.4, where we plot the gain medium of a 2-THz GaAs/AlGaAs device based on this approach. Each repeat unit of this structure consists of only three

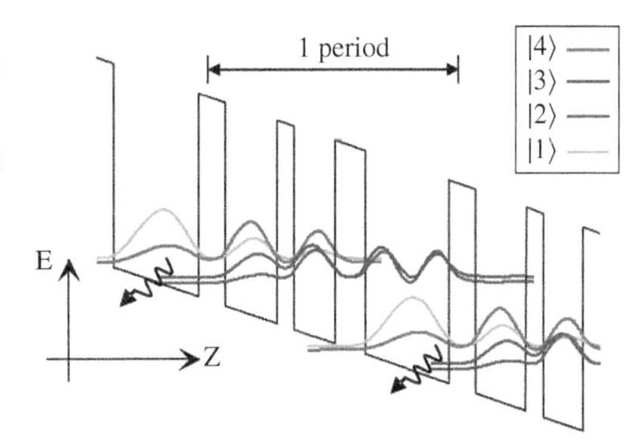

Figure 6.4 Conduction-band diagram of a representative resonant-phonon THz QC gain medium based on the GaAs/$Al_{0.15}Ga_{0.85}As$: GaAs QW materials system. The layer thicknesses in Å of one repeat unit, from left to right and starting from the injection barrier, are 49/94/33/74/56/156. The externally applied field is 12 kV/cm.

coupled QWs. The laser transitions occur between the strongly coupled states labeled $|4\rangle$ and $|3\rangle$ in the two leftmost QWs of each period. Efficient depopulation of the lower laser states then occurs via resonant tunneling into the first excited state $|2\rangle$ of the adjacent "injector" well, followed by LO-phonon-assisted relaxation into the ground state $|1\rangle$ of the same well. Finally, electrons are injected into the upper laser state of the following repeat unit, again by resonant tunneling. As in the bound-to-bound QC structure of Figure 6.2, the energy separation between the two states involved in the LO-phonon-assisted transitions ($|2\rangle$ and $|1\rangle$) is approximately equal to the LO-phonon energy, which maximizes the corresponding transition rate. However, in the present design, the processes of light emission and LO-phonon assisted relaxation are spatially separated from each other, so that depopulation of the upper laser states $|4\rangle$ via LO-phonon emission is limited. In other words, even though the $|4\rangle$–$|1\rangle$ transition is nearly resonant to the LO-phonon energy, its probability rate is relatively small due to its reduced envelope-function overlap. This general design strategy has led to the highest operating temperature of about 200 K reported to date for THz QC lasers [23].

6.4 Optical Waveguides and Cavities

In many semiconductor lasers, the cavity simply consists of an optical waveguide with cleaved facets acting as the mirrors of a Fabry–Perot resonator. The waveguide is typically based on a dielectric slab, where light is confined vertically in a lower-bandgap (i.e. higher-refractive-index) core layer containing the gain medium and sandwiched between higher-bandgap cladding regions. Lateral confinement of the laser mode is then provided by a ridge structure, often coated with an overgrown high-bandgap epitaxial layer to improve heat dissipation from the active region. This basic device geometry is also commonly used with standard mid-infrared QC lasers. For example, in the important case of GaInAs/AlInAs devices (currently the best performing QC lasers at mid-infrared wavelengths), the core layer typically consists of the gain medium embedded between two relatively thick (a few hundred nm) GaInAs guiding layers. These films are used to increase the average refractive index difference between the core and cladding regions, and hence to maximize the spatial overlap between the laser mode and the gain medium (i.e. the optical confinement factor Γ). Finally, the InP growth substrate, epitaxial AlInAs layers, and an overgrown InP film are used as the cladding materials.

In the case of QC lasers emitting at THz wavelengths, the dielectric waveguide geometry just described is largely unpractical for two main reasons. First, the epitaxial layer thicknesses needed to produce sufficiently strong optical confinement at such long wavelengths are prohibitively large, especially for MBE. Second, the high doping densities required in the cladding layers to ensure good electrical characteristics would lead to exceedingly large free-carrier absorption losses at THz frequencies (since the free-carrier absorption coefficient increases as the square of the optical wavelength). To overcome these challenges, more innovative waveguides have been developed based on surface plasmon polaritons (SPPs), i.e. guided electromagnetic waves supported by the interface between an insulator and a metal (or a highly doped semiconductor with negative dielectric constant), associated with collective electronic oscillations in the latter material. Unlike the guided modes of dielectric waveguides, SPPs can be confined in sub-wavelength dimensions and therefore do not require the growth of exceedingly thick guiding layers, even in THz devices. Furthermore, their penetration depth into the lossy conductive layer can be very small at THz wavelengths. Finally, SPPs are intrinsically TM-polarized and thus are naturally compatible with ISB transitions.

The SPP waveguide used in the initial demonstration of THz QC lasers [5] is illustrated in Figure 6.5a. In this device, the ISB gain medium (consisting of low-doped GaAs/AlGaAs QWs) is sandwiched between the top metal contact and a highly doped GaAs guiding layer grown on a semi-insulating GaAs substrate. By design, the doping level of the guiding layer is large enough that its dielectric constant has negative real part at the laser emission frequency. As a result, its top and bottom surfaces can both support SPPs at this frequency. The fundamental laser mode along the growth direction is then derived from these guided waves combined with the SPPs supported by the gain-medium/top-metal-contact interface. The resulting modal intensity profile, which can be computed from the Helmholtz equation, is shown schematically in Figure 6.5a. Finally, waveguiding in the lateral direction in these devices can be obtained with a simple ridge structure, etched through the gain medium, and into the highly doped guiding layer.

An alternative SPP waveguide design for THz QC lasers is the double-metal geometry of Figure 6.5b [24, 25], which is conceptually analogous to a millimeter-wave microstrip transmission line. Here the gain medium is sandwiched directly between the two metal contacts (without any additional core or cladding layers), and the guided mode is derived from the SPPs at both semiconductor/metal interfaces. This design suffers from somewhat larger free-carrier absorption losses compared to the "central-layer-guided" structure of Figure 6.5a. At the same time, it features several distinctive advantages, including (i) a maximally large confinement factor Γ close to unity, as opposed to a few ten percent in the design of Figure 6.5a; (ii) stronger sub-wavelength mode confinement, which allows for smaller device geometries with lower drive currents and improved heat dissipation; (iii) larger "impedance mismatch" between its guided modes and free-space radiation, which is advantageous in terms of low-threshold and high-temperature operation (see Problem 2). An important drawback of the latter property, however, is that it limits the maximum amount of optical power that can be extracted from the output facet. Furthermore, the strong intracavity mode confinement also leads to highly diverging and asymmetric output beams, which complicates the external collection and collimation of the laser light.

In addition to the simple Fabry–Perot devices described so far, THz QC lasers based on distributed feedback (DFB) have also been investigated extensively, as a way to enforce

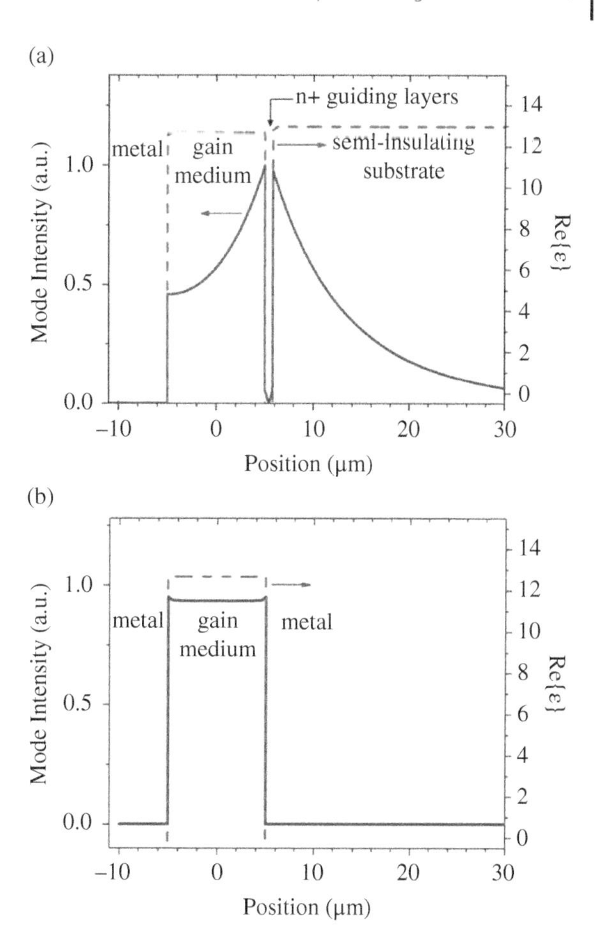

Figure 6.5 Real part of the dielectric constant ε and mode intensity profile along the growth direction of (a) a "central-layer-guided" and (b) a double-metal THz QC-laser waveguide.

single-longitudinal-mode (i.e. single-frequency) emission. In general, a DFB cavity consists of a waveguide where the effective refractive index of the laser mode n_{eff} is modulated periodically along the propagation axis. Correspondingly, strong optical feedback can be obtained if the reflections from the different repeat units of the periodic index perturbation add up to one another in phase. Since this condition is only satisfied at widely-spaced discrete wavelengths $\lambda_p = 2n_{eff}\Lambda/p$ (where Λ is the period and p a positive integer), single-longitudinal-mode operation with large side-mode suppression ratios can be obtained. In THz QC lasers, such distributed Bragg reflectors can be introduced by patterning the top metal contact with a periodic series of slits [26], or by periodically varying the ridge width [27]. The latter geometry is shown in Figure 6.6. A variation of this geometry has also been used to demonstrate a widely tunable THz wire laser, where a movable object (e.g. a silicon micromachined plunger) is used to vary the resonance frequency of a leaky guided mode of the adjacent DFB cavity [28].

Several other types of optical resonators have also been used in THz QC lasers, particularly as a way to address the low outcoupling efficiency and highly diverging far-field profile of edge-emitting devices based on the double-metal waveguide of Figure 6.5b. For example,

Figure 6.6 Metal–metal corrugated ridge DFB THz QC laser. *Source*: Reprinted with permission from [27]. Copyright (2005), OSA Publishing.

second-order gratings (where the emission wavelength is given by the aforementioned resonance λ_p with $p = 2$) have been used to obtain vertical surface emission [29, 30], which in QC lasers is otherwise forbidden by the TM-polarized nature of ISB transitions. Photonic-crystal THz QC lasers have also been developed for the same purpose, with a hexagonal array of air holes etched through the top electrode of a double-metal waveguide [31]. More recently, surface emission has also been obtained with a vertical-external-cavity THz QC laser based on an amplifying metasurface reflector, which allows for high beam quality as well as electrically switchable polarization of the output light [32]. At the same time, considerable research efforts have also been devoted to the development of QC lasers based on microcavities, where light is confined within microscale dimensions in all directions, including microrings [33] and more complex geometries conceptually similar to an electronic LC resonant circuit [34].

6.5 State-of-the-Art Performance and Limitations

Most THz QC lasers demonstrated to date are based on GaAs/AlGaAs QWs grown by MBE on semi-insulating GaAs substrates, although other materials systems have also been employed including GaInAs/AlInAs, GaInAs/GaAsSb, and InAs/AlAsSb [35–37]. The key advantage of the GaAs/AlGaAs system is its ability to provide unparalleled interface quality, material purity, and doping control. Furthermore, AlGaAs is nearly lattice matched to GaAs over its entire alloy composition, and therefore the conduction-band offset of GaAs/AlGaAs QWs can be tuned by design over a broad range (~0–300 meV for direct-bandgap alloys) without introducing undesirable strain in the active material. This property is important for THz devices where the laser subbands tend to lie near the bottom of the QWs, so that relatively low energy barriers are needed to obtain efficient interwell coupling and tunneling transport. At the same time, the (Al)(Ga)InAs materials system features lighter electronic effective masses compared to (Al)GaAs (e.g. $0.043 \times m_0$ in $Ga_{0.47}In_{0.53}As$ versus $0.067 \times m_0$ in GaAs), which is desirable in QC lasers for the purpose of minimizing the threshold current density.

After the initial demonstration of THz QC lasers [5], extensive research efforts have focused on extending their maximum operating temperature, output power, and spectral coverage. Regarding the latter parameter, so far THz QC lasers have been able to cover the frequency range from 1.2 [38] to 5.2 THz [39] (i.e. the wavelength range from 250 to 58 μm). Representative emission spectra from [38] are shown in Figure 6.7. Operation at higher frequencies is complicated by electron/LO-phonon and ultimately photon/LO-phonon interactions, as described in detail below. Operation at lower frequencies becomes increasingly challenging due to larger free-carrier absorption losses and more stringent requirements on selective carrier injection and is ultimately limited by the ISB transition broadening. A possible approach to extend the covered spectral range involves the use of a large magnetic field (a few T) parallel to the QW growth axis. The presence of such field forces the in-plane electronic motion into Landau orbits and therefore breaks up the active subbands into discrete energy levels. Correspondingly, LO-phonon-assisted nonradiative relaxation can be strongly suppressed depending on the magnetic field intensity, due to the reduced density of available final states. With this approach, THz QC lasers with lower threshold current densities have been demonstrated [40, 41], as well as emission at frequencies as small as 740 GHz (over 400-μm wavelength) [42]. On the other hand, the use of magnetic fields also clearly limits the use of these devices for practical applications.

For high-temperature operation, the best results so far have been obtained with devices based on the resonant-phonon gain-medium design and double-metal waveguide geometry, with a maximum reported heat-sink temperature of nearly 200 K in pulsed mode at a frequency of about 3 THz [23]. These results are illustrated by the light–current–voltage (LIV) characteristics and emission spectra shown in Figure 6.8. At lower temperatures of a few 10 K, high-performance THz QC lasers can typically deliver pulsed output powers on the order of several ten mW. Recently, total (two-facet) output powers approaching and surpassing one Watt at ~5 K have also been reported [43, 44] with further optimization of the device design, growth, and fabrication, including improved heat dissipation and the use of thicker active regions. It should be noted that these record-high-power devices employ "central-layer-guided" plasmonic waveguides, rather than double-metal structures. These results, therefore, confirm the distinct tradeoff between high-temperature and high-power

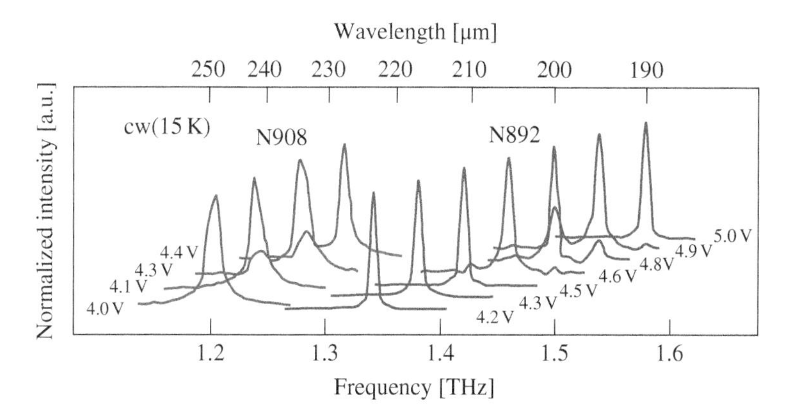

Figure 6.7 Voltage-tunable cw emission spectra measured at 15 K with the lowest-emission-frequency THz QC lasers reported to date. *Source*: Reprinted with permission from Walther et al. [38]. © 2007, AIP Publishing LLCC.

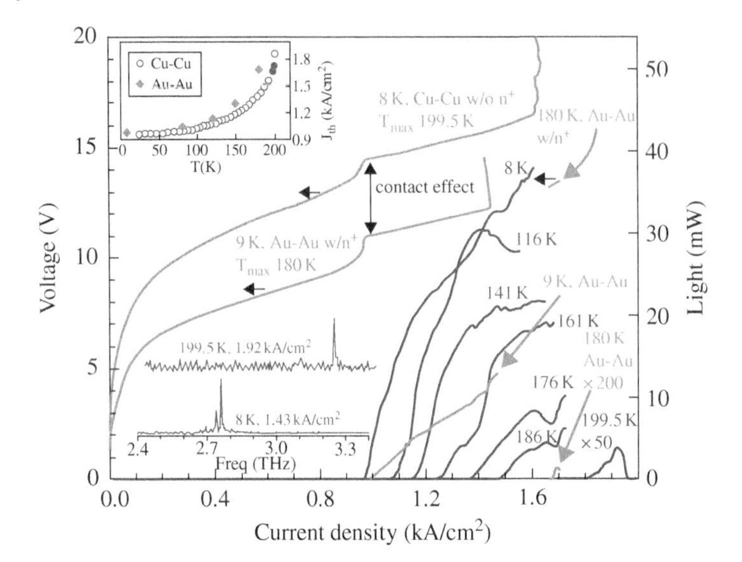

Figure 6.8 Temperature-dependent LIV characteristics measured with the highest-operating-temperature THz QC lasers reported to date. Upper inset: threshold current density versus temperature. Lower inset: temperature-dependent emission spectra. *Source*: Reprinted with permission from Fathololoumi et al. [23]© 2012, The Optical Society.

operation in terms of waveguide design, as discussed above and elaborated in more detail in Problem 2.

Other important recent developments include the demonstration of widely tunable single-frequency emission (e.g. with a tuning range of ~140 GHz at 3.8 THz using the aforementioned wire-laser approach [28]), active mode-locking with pulse widths ~10 ps [45], and frequency combs [46]. Furthermore, coherent THz emission at room temperature has been obtained by intracavity difference-frequency generation in a dual-wavelength mid-infrared QC laser [47–49]. This approach nicely leverages the impressive level of performance of mid-infrared QC laser technology and the design flexibility of bandgap engineering. However, the achievable output power levels are fundamentally limited by the nonlinear conversion efficiency, although significant improvements have been reported recently, including nearly 2-mW emission at ~3.5 THz in pulsed mode at room temperature [49].

The state-of-the-art THz QC lasers described in this section have already reached a sufficient level of performance and robustness that they are beginning to be employed for a variety of imaging and spectroscopic applications, e.g., see [50–52]. At the same time, however, the need for cryogenic cooling (and the associated penalties in terms of compactness, portability, and cost) significantly limits their widespread use in many instances. In addition, their inability to cover frequencies beyond ~5 THz (where electronics-based solutions are impractical) represents another significant technological drawback. Importantly, both of these limitations are fundamental, as they are related to an intrinsic material property of As-based semiconductors—namely, the presence of THz-range optical phonons. For example, in GaAs the transverse-optical (TO) and LO phonon energies are 8.0 and 8.7 THz, respectively. Due to the polar nature of III-V compound semiconductors, these phonons can interact strongly with both electronic ISB transitions and photons of comparable THz-range energies, and therefore have a substantial impact on the operation of THz QC lasers.

In particular, the maximum operating temperature of these devices is fundamentally limited to cryogenic values by the process of thermally activated LO-phonon emission, illustrated schematically in Figure 6.9a. In this plot, the traces labeled $|u\rangle$ and $|l\rangle$ denote the upper and lower laser subbands of a THz QC structure emitting at a frequency ν below the LO-phonon frequency ν_{LO}. At cryogenic temperatures, a large population inversion can be established since most electrons in $|u\rangle$ occupy states near the bottom of the subband, from which they cannot decay nonradiatively into $|l\rangle$ via LO-phonon emission because of the requirement of energy conservation. As the temperature is increased, however, more and more electrons in $|u\rangle$ gain enough thermal energy that scattering into $|l\rangle$ via LO-phonon emission becomes allowed. Since these scattering processes are ultrafast, the end result is a dramatic reduction in population inversion and optical gain. A quantitative analysis of this effect has been carried out via Monte Carlo simulations and is summarized in Figure 6.9b [54]. The dotted line in this figure is the calculated population inversion Δn versus temperature T for the resonant-phonon GaAs/AlGaAs THz QC structure of Figure 6.4 (the solid line corresponds to an equivalent GaN/AlGaN device described below). A rapid decrease in Δn with increasing T is observed across the entire temperature range. Furthermore, Δn drops below the estimated value needed to reach threshold in this device (about 16% of the total carrier density in the active layer [53]) for T larger than about 200 K—in agreement with the maximum operating temperature reported to date [23].

The optical-phonon energies also play a key role in determining the accessible spectral coverage of THz QC lasers. In particular, in any polar crystalline material, the coupling between optical phonons and photons of comparable energy produces a forbidden frequency gap between $h\nu_{TO}$ and $h\nu_{LO}$ where light cannot propagate (the Reststrahlen

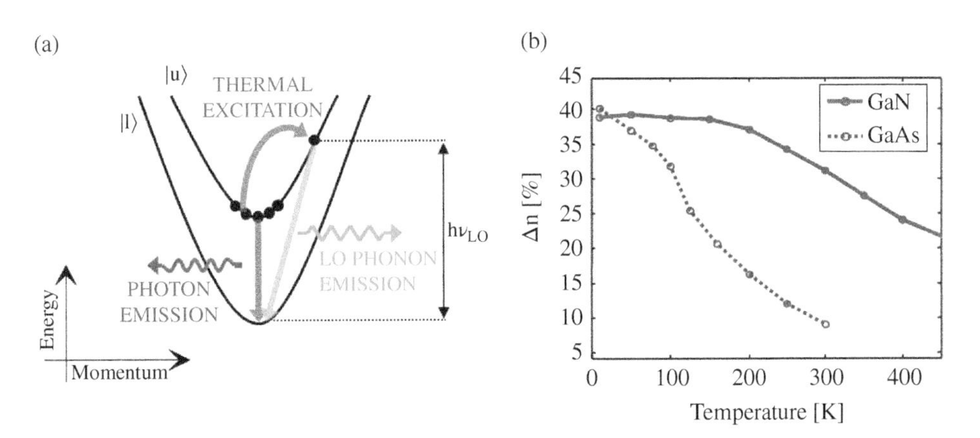

Figure 6.9 (a) Schematic illustration of ISB relaxation between two QW subbands $|2\rangle$ and $|1\rangle$ with energy separation $E_2 - E_1 < h\nu_{LO}$ via the process of thermally activated LO-phonon emission. (b) Calculated fractional population inversion versus temperature for the resonant-phonon GaAs/AlGaAs THz QC structure of Figure 6.4 (dotted line) and for a GaN/AlGaN gain medium based on the same design, shown in Figure 6.11 below (solid line). *Source*: Reprinted with permission from Bellotti et al. [53]. © 2008, AIP Publishing LLC.

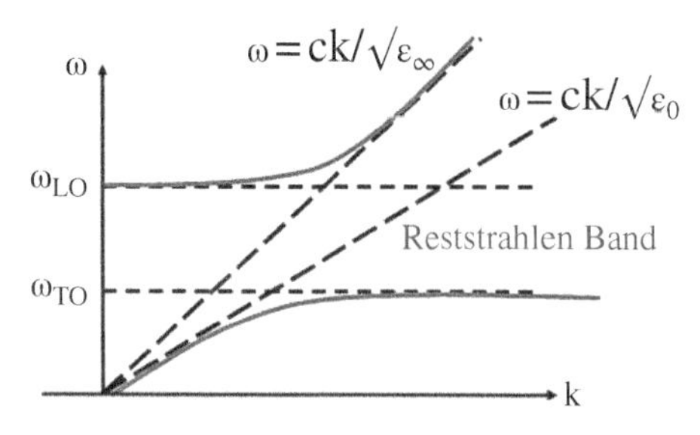

Figure 6.10 Dispersion of light in a polar crystal in the spectral vicinity of the optical-phonon frequencies, showing the formation of the Reststrahlen gap between $h\nu_{TO}$ and $h\nu_{LO}$.

band—see Figure 6.10). As a result, QC lasers are fundamentally precluded for operation in this spectral range. Furthermore, laser action at nearby photon energies is also very challenging, due to the particularly fast (thermally activated) LO-phonon-assisted relaxation across the lasing subbands. These considerations explain the inability of existing GaAs THz QC lasers (where the Reststrahlen band extends roughly between 8 and 8.7 THz) to cover frequencies above ~6 THz.

6.6 Novel Materials Systems

The arguments just presented suggest that any substantial improvement in THz QC laser performance (at least in terms of spectral coverage and maximum operating temperature) may require the use of novel semiconductor materials, either featuring much higher optical-phonon frequencies or with nonpolar character. In the following, we discuss two such systems that have been investigated widely for this purpose in recent years, namely III-nitride and SiGe QWs.

6.6.1 III-Nitride Quantum Wells

III-nitride semiconductors, including InGaN, GaN, and AlN, have emerged in the past few decades as the leading materials system for (interband) optoelectronic device applications at visible and UV wavelengths—for a recent review, see [55]. The investigation of ISB transitions in GaN/Al(Ga)N QWs was initially motivated by their large conduction-band offsets (up to 1.75 eV in GaN/AlN) [56], which allows for record short ISB transition wavelengths throughout the mid-infrared and well into the near-infrared spectral region. In particular, this capability was explored for the demonstration of ISB devices operating at fiber-optics communication wavelengths near 1.55 μm, including photodetectors [57, 58] and ultrafast all-optical switches [59, 60]. More recently, it has been argued that the GaN/AlGaN QW system is also ideally suited to the development of THz QC lasers by virtue of its large optical-phonon frequencies [53, 61, 62], e.g., in GaN $\nu_{TO} = 16$ THz and $\nu_{LO} = 22$ THz.

In particular, the potential of III-nitride semiconductors in this context is illustrated in Figure 6.9b above. As already discussed, the dotted line in this plot shows the population inversion of the GaAs/AlGaAs THz QC structure of Figure 6.4, computed as a function of temperature using Monte Carlo methods. The solid line corresponds to the QC structure shown in Figure 6.11, which consists of GaN/AlGaN QWs but is otherwise based on the

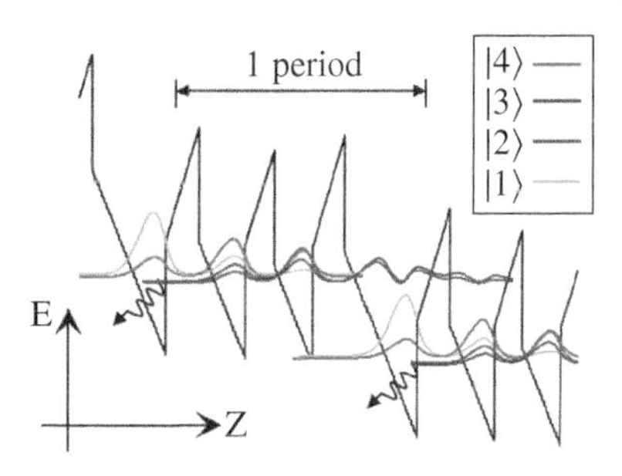

Figure 6.11 Conduction-band diagram of a resonant-phonon THz QC gain medium based on the GaN/$Al_{0.15}Ga_{0.85}N$ QW materials system. The layer thicknesses in Å of one repeat unit, from left to right and starting from the injection barrier, are 26/37/22/31/26/59. The externally applied field is 73 kV/cm. The laser transitions occur between subbands |4⟩ and |3⟩, whereas the energy separation between subbands |2⟩ and |1⟩ is nearly resonant with the LO-phonon energy of GaN.

same design strategy and has the same emission frequency (2 THz) as the device of Figure 6.4. The comparison between these two traces shows that the GaN device exhibits a much weaker temperature sensitivity and therefore can provide superior high-temperature performance. In fact, its maximum operating temperature obtained from an estimate of the threshold population inversion (about 22% of the total carrier density) is even above room temperature [53]. These results are a direct consequence of the larger LO-phonon energy of GaN compared to GaAs, whereby a larger amount of thermal energy (i.e. a higher temperature) is required to activate LO-phonon-assisted relaxation across the lasing subbands. Similar improvements in high-temperature THz-QC-laser performance over GaAs were also predicted by Monte Carlo simulations for the ZnO/MgZnO QW system [54], where $\nu_{LO} \approx 17$ THz.

These basic arguments have motivated extensive research efforts in recent years toward the development of III-nitride THz QC lasers. The feasibility of ISB light emission with the GaN/Al(Ga)N materials system has already been demonstrated by optical pumping in multiple-QW samples [63, 64] and, more recently, via electrical injection in a QC structure [65], all at short-wave mid-infrared wavelengths. However, further progress toward a working QC laser, at any wavelength, has so far been hindered by two key challenges specific to III-nitride QWs. First, while QC lasers generally require active layers of the highest crystalline quality, III-nitride heterostructures notoriously feature high densities of threading defects that can significantly degrade their electrical and optical properties. This issue still remains a formidable challenge, although recent progress in the fabrication of high-quality free-standing GaN substrates is quite promising for future progress. Second, III-nitride QWs oriented along their most common growth direction (the polar crystallographic *c*-axis of their wurtzite crystal structure) contain strong intrinsic electric fields that can significantly distort their band lineup. Specifically, as a result of these internal fields, the conduction-band (and valence-band) edges of the GaN wells and Al(Ga)N barriers are strongly tilted in the opposite directions, as illustrated in the QC structure of Figure 6.11. In turn, this arrangement significantly complicates the design of THz ISB devices (as the internal fields tend to blue-shift the ISB transition energies) and makes robust interwell tunneling transport rather problematic.

Initial work aimed at the measurement of THz ISB transitions in GaN has focused on this design challenge through the development of step-QW structures [66], where two or more layers of different (Al)GaN compositions are used in each well to create a more rectangular potential energy profile. In [67], an ISB photodetector operating near 13 THz was demonstrated based on a variation of this approach, shown in the inset of Figure 6.12a. In this

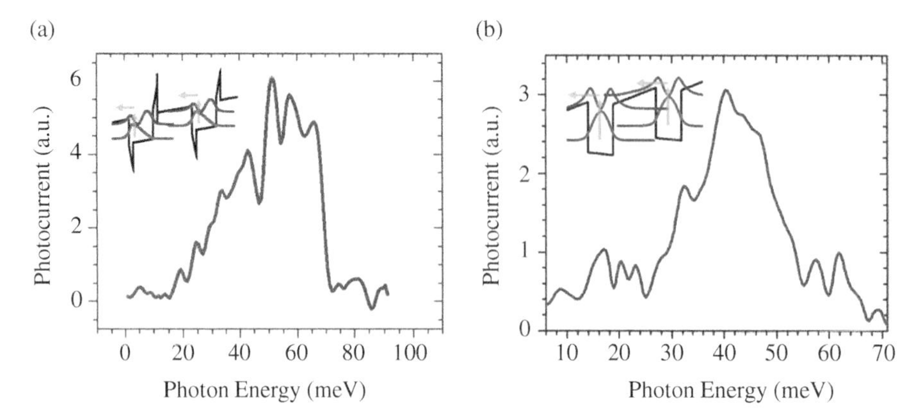

Figure 6.12 (a) Photocurrent spectrum of a double-step III-nitride ISB photodetector measured at 20 K under an applied voltage bias of +0.8 V. The sharp cutoff near 70 meV is due to the Reststrahlen band of GaN. *Source*: Reprinted with permission from Sudradjat et al. [67]. © 2012, AIP Publishing LLC. (b) Photocurrent spectrum of a semi-polar III-nitride device measured at 10 K under an applied voltage of 1.2 V. *Source*: Reprinted with permission from Durmaz et al. [68]. © 2016, AIP Publishing LLC. The conduction-band diagrams of both devices are shown in the insets.

structure, two different AlGaN layers are also used in each barrier, which allows for the design of QWs with the first-excited subband nearly resonant to the top of the barriers, as needed for efficient extraction of the photocarriers. The resulting device features a peak responsivity of 7 mA/W, which is reasonable for ISB photodetectors, with a pronounced but relatively broad low-temperature photocurrent spectrum (Figure 6.12a). An alternative approach to address the same design challenge is to use QWs grown along semi-polar (or nonpolar) crystallographic directions, where the undesirable internal electric fields are significantly reduced (or completely eliminated). This approach has already been used for the measurement of THz ISB absorption [68, 69] and photodetection. The latter work is summarized in Figure 6.12b, where a significantly narrower photocurrent spectrum is observed compared to the double-step device of Figure 6.12a. This difference is likely due to reduced scattering from interface roughness in QWs consisting of only two epilayers per period. Importantly, it should also be noted that the spectra of both devices of Figure 6.12 fully overlap the Reststrahlen band of GaAs (~33–36 meV), clearly illustrating the ability of GaN devices to cover this frequency range that has so far been inaccessible to ISB optoelectronics.

Finally, the other key ingredient for the development of III-nitride THz QC lasers is the ability to obtain efficient tunneling transport. This transport mode has already been extensively studied in the context of GaN/Al(Ga)N double-barrier structures [70–72], where stable and repeatable negative differential resistance has eventually been measured using samples with minimal defect densities. Sequential tunneling transport in simple GaN/AlGaN QC structures has also been investigated through the measurement of highly nonlinear electrical characteristics, with the experimental turn-on voltages well reproduced by simulation results [73]. In the same report, a weak electroluminescence peak near 10 THz was also presented, consistent with a picture of photon-assisted tunneling (i.e. interwell ISB light emission) in the active QWs. However, the measured signal level was too low to allow for an unambiguous identification. Significantly stronger emission at 1.4 THz with a linewidth as small as 0.13 THz was measured in a different QC structure in [74] and explicitly attributed to ISB transitions (although not the intended transitions in the active layer design). In any case, we believe that the overall progress just reviewed, while still rather early-stage, is quite promising toward the future realization of III-nitride THz QC lasers.

6.6.2 SiGe Quantum Wells

The group-IV semiconductors Si, Ge, and their alloy SiGe are nonpolar materials, and as a result, their available spectral range for device applications is not limited by Reststrahlen absorption. Furthermore, for the same reason, nonradiative ISB relaxation in SiGe QWs is dominated by the generally weaker deformation-potential (rather than polar) phonon scattering mechanism, leading to slower decay lifetimes which are favorable for the development of THz QC lasers with improved high-temperature performance. SiGe QWs are also compatible with the CMOS microelectronics platform, and therefore in general are promising for the large-scale monolithic integration of photonic and electronic devices. To date, ISB electroluminescence (but no laser emission) has been demonstrated with SiGe/Si QWs, both at mid-infrared wavelengths in full-fledged QC structures [75–77] and at THz frequencies in simple multiple-QW samples [78]. All these experiments as well as preceding theoretical proposals [79] employ hole ISB transitions, motivated by the exceedingly small conduction-band offsets of standard SiGe/Si QWs grown on Si (100) substrates. The valence-band diagrams of the type of structures used in [76, 78] are plotted in Figure 13a and b, respectively, where the vertical axes indicate hole (rather than electron) energy E_h. In these structures, the SiGe layers provide the wells, the Si layers act as the barriers, and bound states derived from both heavy-hole (HH) and light-hole (LH) valence bands (indicated by the blue and red traces, respectively) play a role in the device operation.

Similar to the case of III-nitride ISB devices, the demonstration of QC laser emission in SiGe is complicated by materials quality issues, specifically related to the large (4.2%) lattice mismatch between Si and Ge. In devices requiring a large number of QWs, such as QC lasers, the resulting accumulated strain generally relaxes through the formation of undesirable misfit dislocations and other structural defects. To address this issue, more recent experimental realizations of SiGe/Si QC light emitters [77] have been grown on composition-graded SiGe buffer layers plastically relaxed on Si. This approach allows growing thicker multiple-QW active layers, with small (ideally zero) net strain per period if the final composition of the graded buffer matches the average composition of the QWs. The structural quality of the resulting strain compensated samples, however, was still inadequate for the demonstration of QC-laser action, which can be ascribed to the nonnegligible density of misfit dislocations that are necessarily created in the SiGe buffer during its growth and strain relaxation [80]. Such dislocations tend to propagate into the overlaying epitaxial layers, and also produce lateral strain inhomogeneities as well as a rough surface morphology.

A more innovative and potentially promising strain-management approach has been reported more recently, based on nanomembrane technology [81]. With this approach, illustrated schematically in Figure 6.14a, "lattice-matched" substrates are fabricated by first growing a few periods of the desired SiGe/Si QW structure (with overall thickness well below the threshold for plastic relaxation) on a Si-on-insulator (SOI) wafer. The resulting epitaxial stack is then released from the handle wafer by etching away the underlying

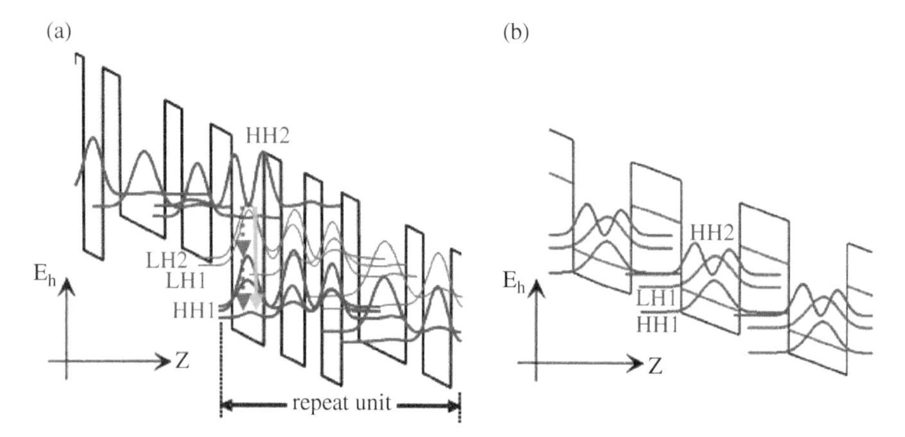

Figure 6.13 (a) Valence-band lineup of the SiGe/Si QC active material of Demichel et al. [29]. The thick blue and thin red lines are the squared envelope functions of the relevant HH and LH bound states, respectively. The solid and dotted arrows indicate possible radiative and phonon-assisted nonradiative transitions between the lasing subbands HH2 and HH1. *Source*: Modified from Dehlinger et al. [76]. (b) Valence-band lineup of a SiGe/Si multiple-QW structure similar to the device of Lynch et al. [78]. *Source*: Modified from Lynch et al. [78]. In both plots, the variable E_h indicates hole energy.

SiO_2 layer, to form a free-standing nanomembrane. In the process, the compressive strain in the SiGe layers (caused by their lattice mismatch with the Si growth template) is partially relaxed through the introduction of tensile strain in the Si layers, such that the global average strain in the stack becomes zero. Next, the nanomembrane is transferred and bonded onto a new Si host wafer and finally used as the substrate for the subsequent growth of an arbitrary number of additional identical QW periods. Under these conditions, the sample growth is then fully strain compensated, so that no net strain accumulates in the entire epitaxial film. The initial SiGe/Si multiple-QW samples fabricated with this approach exhibited strong far-infrared ISB absorption near 15 THz with linewidths considerably narrower than those of similar QWs grown on a standard Si substrate (Figure 6.14b), indicative of improved crystalline structure [81]. More recently, a mid-infrared ISB photodetector was also reported based on the same process, showing improved performance compared to otherwise identical devices grown simultaneously on the supporting Si substrate [82].

Finally, we note that the development of SiGe QC lasers based on hole ISB transitions is also complicated by the intricacy of the valence-band structure. In particular, it is difficult to achieve population inversion in these devices as the presence of HH and LH subbands at comparable energies results in strongly enhanced nonradiative decay from the upper laser states. This issue is clearly illustrated in the valence-band diagram of the QC structure of Figure 6.13a. While the radiative transitions in this device involve consecutive HH subbands (labeled HH2 and HH1 in the figure), LH states exist at intermediate energies that provide an additional decay channel for the upper subband HH2 (indicated by the dotted vertical arrows).

To address this design issue, more recently n-type QC lasers based on electronic ISB transitions in the L valleys of Ge/SiGe QWs have also been proposed and theoretically investigated [83]. A representative QC structure based on this approach (designed for emission at

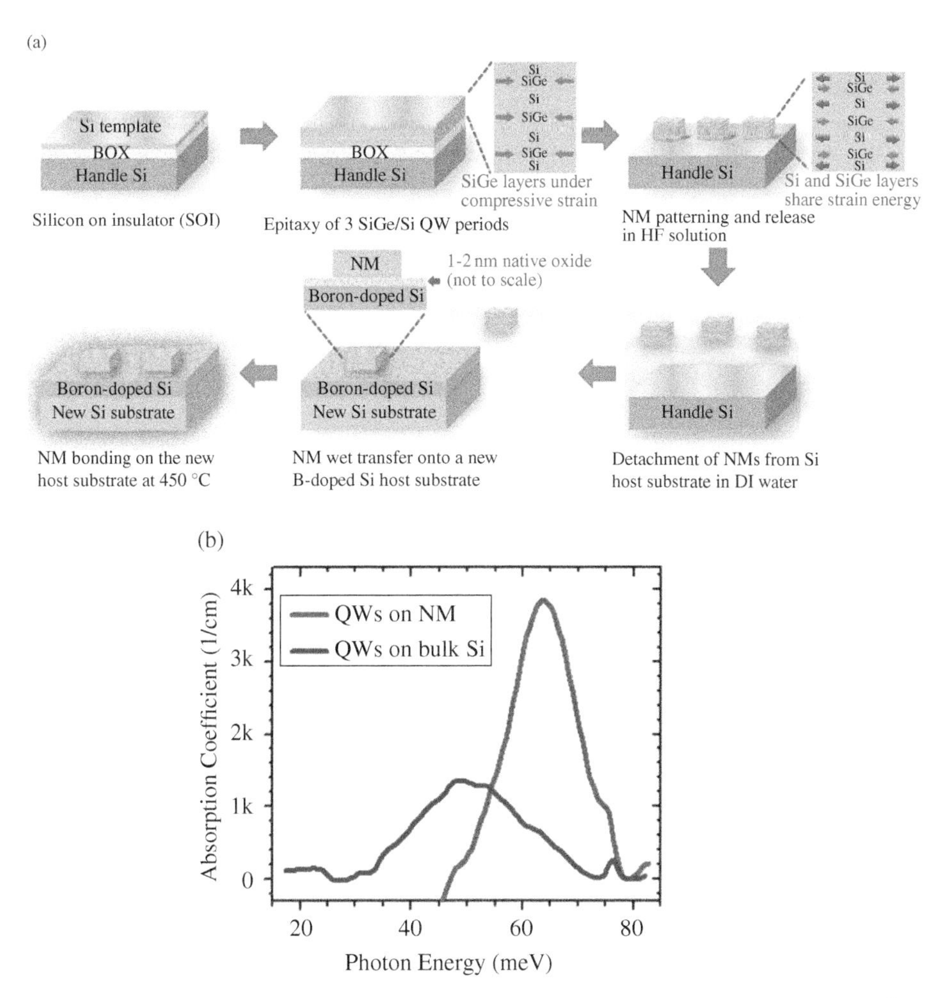

Figure 6.14 (a) Fabrication process developed in [81, 82] for the formation of high-quality "lattice-matched" SiGe/Si nanomembranes substrates. *Source*: Reprinted with permission from Durmaz et al. [82]. © 2016, American Chemical Society. (b) Low-temperature (20 K) ISB absorption spectrum of a SiGe/Si multiple-QW sample developed with the process of (a) (solid line) and of a similar QW structure grown on a bulk Si substrate (dashed line). *Source*: Reprinted with permission from Sookchoo et al. [81]. © 2013, American Chemical Society.

6 THz) is plotted in Figure 6.15, where the solid lines show the conduction-band lineup and bound-state envelope functions of the L valleys, while the dotted lines correspond to the $\Delta 2$ valleys (i.e. the two Δ valleys along the (001) growth direction). As shown in this figure, strong quantum confinement in the Ge wells can be achieved for the L-valley electrons, provided that the SiGe barriers are sufficiently thin so that their $\Delta 2$ bound states lie at sufficiently high energies. Otherwise, L-to-$\Delta 2$ intervalley scattering would lead to extremely fast nonradiative decay out of the L-valley states involved in the device operation. Importantly, the L valleys also have relatively small electronic effective mass along the (001) growth direction, as well as nonzero off-diagonal terms of the inverse effective-mass tensor.

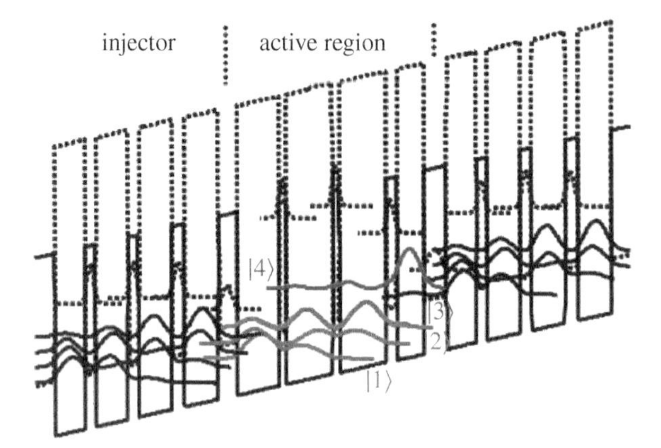

Figure 6.15 Conduction-band lineup and squared envelope functions of the relevant bound states of an n-type Ge/$Si_{0.22}Ge_{0.78}$ QC structure [on a $Si_{0.12}Ge_{0.88}$ (001) growth template] designed for emission near 6 THz. The solid and dotted lines correspond to the L and $\Delta 2$ valleys, respectively. The layer thicknesses in Å of one repeat unit, from right to left and starting from the injection barrier, are 36/43/17/76/11/74/11/69/30/54/20/54/18/53/17/49. The externally applied field is −11 kV/cm. Laser emission in this device involves ISB transitions between the L-valley states labeled |4⟩ and |3⟩ in each active stage, followed by deformation-potential phonon scattering into states |2⟩ and |1⟩ and tunneling into the injector region downstream. *Source*: Reprinted with permission from Driscoll and Paiella [83]. © 2006, AIP Publishing LLC.

The former feature is advantageous for the purpose of maximizing the ISB optical gain. The latter feature is also desirable as it enables surface-normal ISB light emission, which may allow overcoming the low outcoupling efficiency and large beam divergence of THz-QC lasers based on plasmonic waveguides. Numerical simulations based on a simple rate-equation model suggest that this design approach can provide high performance across a wide portion of the far-infrared spectrum [84]. Experimental progress so far with n-doped Ge/SiGe QWs include the observation of THz ISB transitions by transmission spectroscopy [85] and the report of optically pumped THz emission in a pump-probe configuration [86]. Altogether, this approach appears to be particularly promising for the realization of SiGe THz QC lasers, again as long as suitable high-quality growth substrates can be developed.

6.7 Conclusion

This chapter has reviewed the operation principles, design, and state-of-the-art performance of THz QC lasers. In many respects, these devices have already reached a sufficiently high level of technological maturity that they can begin to be deployed in real-world applications. At the same time, they still suffer from major performance limitations, particularly related to the need for cryogenic cooling. It now appears unlikely that room-temperature operation may be reached with incremental improvements of existing devices, and radically new solutions will likely be required. The use of III-nitride or SiGe QWs is quite compelling based on the general theoretical arguments just presented, but it will also require significant advances in terms of materials quality and design strategy. In any case, it is clear that QC lasers have already found a major and lasting role in THz science and technology, as the leading solid-state sources of coherent radiation at frequencies beyond 1 THz.

Acknowledgments

It is a pleasure to acknowledge extensive collaborations with T. D. Moustakas (Boston University) and M. G. Lagally (University of Wisconsin—Madison) focused on ISB transitions in III-nitride and SiGe QWs, respectively, which have produced some of the results reviewed in this chapter. These activities have been funded by NSF under Grants ECS-0824116 and DMR-0907296, and by AFOSR under Grant FA9550-14-1-0361. The contribution from several students involved in this research at Boston University, including Yan Li, Kristina Driscoll, Faisal Sudradjat, Habibe Durmaz, and Yuyu Li, is also gratefully acknowledged.

Exercises

E6.1 QC-laser threshold and transparency currents.

Consider a QC laser based on a three-level active-stage design (e.g. as in Figure 6.2). The operation of this device can be described with the following set of rate equations:

$$\begin{cases} \dfrac{dN_3}{dt} = \dfrac{\eta_i I}{qV_{st}} - \dfrac{N_3}{\tau_3} - v_g g_N (N_3 - N_2) N_p & (1) \\ \dfrac{dN_2}{dt} = \dfrac{N_3}{\tau_{32}} + v_g g_N (N_3 - N_2) N_p - \dfrac{N_2 - N_2^{(0)}}{\tau_2} \\ \dfrac{dN_p}{dt} = \Gamma v_g g_N (N_3 - N_2) N_p - \dfrac{N_p}{\tau_p} & (3) \end{cases} \tag{6.2}$$

where we have neglected spontaneous emission. In these equations, N_3 and N_2 are the carrier densities in the upper and lower laser subbands $|3\rangle$ and $|2\rangle$, respectively, N_p is the photon density in the laser cavity, η_i the injection efficiency, I the drive current, q the electron charge, V_{st} the volume of each active stage, τ_3 and τ_2 the lifetimes of the upper and lower laser subbands, respectively, v_g the group velocity of the laser mode, g_N the ISB differential gain, $1/\tau_{32}$ the decay rate from $|3\rangle$ to $|2\rangle$, $N_2^{(0)}$ the equilibrium value of N_2, Γ the optical confinement factor, and τ_p the photon lifetime. When the laser is biased at or above threshold, the modal gain is clamped to the total cavity losses, i.e.

$$g_{mod} = \Gamma g_N (N_3 - N_2) = \alpha_{tot} \tag{6.3}$$

a) Use these equations to derive the following expression for the steady-state threshold current

$$I_{th} = \frac{V_{st} q}{\eta_i} \frac{1}{\tau_3} \frac{\alpha_{tot}/(\Gamma g_N) + N_2^{(0)}}{1 - \tau_2/\tau_{32}} \tag{6.4}$$

You can assume that at threshold the photon density inside the cavity is small enough that it can be neglected.

b) Use the results of (a) to derive the key condition for QC laser action, i.e. $\tau_2 < \tau_{32}$.
c) Use the same equations above, together with the definition of transparency (i.e. $g_{mod} = 0$), to calculate a similar expression for the transparency current I_{tr} of these devices.

E6.2 Performance tradeoff on waveguide design in THz QC lasers.
The slope efficiency above threshold of a QC laser can be written as follows:

$$\frac{dL}{dI} = \frac{\hbar\omega}{q} N_{st}\eta_i \frac{\alpha_{mirr}}{\alpha_{tot}}, \tag{6.5}$$

where L is the laser output power, ω the emission angular frequency, N_{st} the number of repeat units in the active material, α_{mirr} the contribution to α_{tot} due to transmission through the output facet, and all other parameters are defined in Problem 1. For a Fabry–Perot cavity of length d, the output-mirror loss coefficient is

$$\alpha_{mirr} = \frac{1}{2d} \log \frac{1}{R}, \tag{6.6}$$

where the reflection coefficient of the output facet R depends on the effective index of the laser mode inside the cavity n_{eff} as

$$R = \left| \frac{n_{eff} - 1}{n_{eff} + 1} \right|^2. \tag{6.7}$$

Use these formulas, together with the expressions for the threshold current and threshold gain from Problem 1, to discuss the relative merits and drawbacks of the central-layer-guided and double-metal waveguide designs of Figure 5a and b.

References

1 Kazarinov, R.F. and Suris, R. (1971). Possibility of amplification of electromagnetic waves in a semiconductor with a superlattice. *Soviet Physics – Semiconductors* **5**: 707–709.
2 Faist, J., Capasso, F., Sivco, D.L. et al. (1994). Quantum cascade laser. *Science* **264**: 553–556.
3 Beck, M., Hofstetter, D., Aellen, T. et al. (2002). Continuous wave operation of a mid-infrared semiconductor laser at room temperature. *Science* **295**: 301–305.
4 Lyakh, A., Pflügl, C., Diehl, L. et al. (2008). 1.6 W high wall plug efficiency, continuous-wave room temperature quantum cascade laser emitting at 4.6 μm. *Applied Physics Letters* **92**: 111110.

5 Köhler, R., Tredicucci, A., Beltram, F. et al. (2002). Terahertz semiconductor-heterostructure laser. *Nature* **417**: 156–159.
6 Cho, A.Y. (ed.) (1994). *Molecular Beam Epitaxy*. Woodbury, NY: AIP Press.
7 Helm, M. (2000). The basic physics of intersubband transitions. In: *Intersubband Transitions in Quantum Wells: Physics and Device Application I* (eds. H.C. Liu and F. Capasso), 1–100. London: Academic Press.
8 Chuang, S.L. (2009). *Physics of Optoelectronic Devices*, 2e. New York, NY: Wiley.
9 Harrison, P. (2002). *Quantum Wells, Wires, and Dots*. Chichester: Wiley.
10 Ridley, B.K. (1997). *Electrons and Phonons in Semiconductor Multilayers*. Cambridge University Press.
11 Paiella, R. (2011). Quantum cascade lasers. In: *Comprehensive Semiconductor Science & Technology*, vol. **5** (eds. P. Bhattacharya, R. Fornari and H. Kamimura), 683–723. Amsterdam: Elsevier.
12 Faist, J., Capasso, F., Sirtori, C. et al. (1996). High power mid-infrared ($\lambda \geq 5$ μm) quantum cascade lasers operating above room temperature. *Applied Physics Letters* **68**: 3680–3682.
13 Gmachl, C., Tredicucci, A., Capasso, F. et al. (1998). High-power $\lambda \approx 8$ μm quantum cascade lasers with near optimum performance. *Applied Physics Letters* **72**: 3130–3132.
14 Sirtori, C., Capasso, F., Faist, J. et al. (1998). Resonant tunneling in quantum cascade lasers. *IEEE Journal of Quantum Electronics* **34**: 1722–1729.
15 Capasso, F., Tredicucci, A., Gmachl, C. et al. (1999). High-performance superlattice quantum cascade lasers. *IEEE Journal of Selected Topics in Quantum Electronics* **5**: 792–807.
16 Scamarcio, G., Capasso, F., Sirtori, C. et al. (1997). High-power infrared (8-micrometer wavelength) superlattice laser. *Science* **276**: 773–776.
17 Tredicucci, A., Capasso, F., Gmachl, C. et al. (1998). High performance interminiband quantum cascade lasers with graded superlattices. *Applied Physics Letters* **73**: 2101–2103.
18 Faist, J., Beck, M., Aellen, T., and Gini, E. (2001). Quantum-cascade lasers based on bound-to-continuum transition. *Applied Physics Letters* **78**: 147–149.
19 Köhler, R., Iotti, R.C., Tredicucci, A., and Rossi, F. (2001). Design and simulation of terahertz quantum cascade lasers. *Applied Physics Letters* **79**: 3920–3922.
20 Scalari, G., Ajili, L., Faist, J. et al. (2003). Far-infrared ($\lambda \approx 87$ μm) bound-to-continuum quantum-cascade lasers operating up to 90 K. *Applied Physics Letters* **82**: 3165–3167.
21 Luo, H., Laframboise, S.R., Wasilewski, Z.R. et al. (2007). Terahertz quantum-cascade lasers based on a three-well active module. *Applied Physics Letters* **90** 041112.
22 Williams, B.S., Callebaut, H., Kumar, S. et al. (2003). 3.4-THz quantum cascade laser based on LO-phonon scattering for depopulation. *Applied Physics Letters* **82**: 1015–1017.
23 Fathololoumi, S., Dupont, E., Chan, C.W.I. et al. (2012). Terahertz quantum cascade lasers operating up to $\sim$ 200 K with optimized oscillator strength and improved injection tunneling. *Optics Express* **20**: 3866–3876.
24 Unterrainer, K., Colombelli, R., Gmachl, C. et al. (2002). Quantum cascade lasers with double metal-semiconductor waveguide resonators. *Applied Physics Letters* **80**: 3060–3062.
25 Williams, B.S., Kumar, S., Callebaut, H. et al. (2003). Terahertz quantum-cascade laser at $\lambda \sim$ 100 μm using metal waveguide for mode confinement. *Applied Physics Letters* **82**: 2124–2126.
26 Mahler, L., Tredicucci, A., Köhler, R. et al. (2005). High-performance operation of single-mode terahertz quantum cascade lasers with metallic gratings. *Applied Physics Letters* **87**: 181101.

27 Williams, B.S., Kumar, S., Hu, Q., and Reno, J.L. (2005). Distributed-feedback terahertz quantum-cascade lasers with laterally corrugated metal waveguides. *Optics Letters* **30**: 2909–2911.

28 Qin, Q., Williams, B.S., Kumar, S. et al. (2009). Tuning a terahertz wire laser. *Nature Photonics* **3**: 732–737.

29 Demichel, O., Mahler, L., Losco, T. et al. (2006). Surface plasmon photonic structures in terahertz quantum cascade lasers. *Optical Express* **14**: 5335–5345.

30 Fan, J.A., Belkin, M.A., Capasso, F. et al. (2006). Surface emitting terahertz quantum cascade laser with a double-metal waveguide. *Optical Express* **14**: 11672–11680.

31 Chassagneux, Y., Colombelli, R., Maineult, W. et al. (2009). Electrically pumped photonic-crystal terahertz lasers controlled by boundary conditions. *Nature* **457**: 174–178.

32 Xu, L., Chen, D., Curwen, C.A. et al. (2017). Metasurface quantum-cascade laser with electrically switchable polarization. *Optica* **4**: 468–475.

33 [33]Fasching, G., Benz, A., Unterrainer, K. et al. (2005). Terahertz microcavity quantum-cascade lasers. *Applied Physics Letters* **87**: 211112.

34 Walther, C., Scalari, G., Amanti, M.I. et al. (2010). Microcavity laser oscillating in a circuit-based resonator. *Science* **327**: 1495–1497.

35 Ajili, L., Scalari, G., Hoyler, N. et al. (2005). InGaAs–AlInAs⁄InP terahertz quantum cascade laser. *Applied Physics Letters* **87**: 141107.

36 Brandstetter, M., Kainz, M.A., Zederbauer, T. et al. (2016). InAs based terahertz quantum cascade lasers. *Applied Physics Letters* **108**: 011109.

37 Deutsch, C., Benz, A., Detz, H. et al. (2011). Terahertz quantum cascade lasers based on type II InGaAs/GaAsSb/InP. *Applied Physics Letters* **97**: 261110.

38 Walther, C., Fischer, M., Scalari, G. et al. (2007). Quantum cascade lasers operating from 1.2 to 1.6 THz. *Applied Physics Letters* **91**: 131122.

39 Chan, C.W.I., Hu, Q., and Reno, J.L. (2012). Ground state terahertz quantum cascade lasers. *Applied Physics Letters* **101**: 151108.

40 Alton, J., Barbieri, S., Fowler, J. et al. (2003). Magnetic field in-plane quantization of population inversion in a THz superlattice quantum cascade laser. *Physical Review B* **68**: 081303.

41 Scalari, G., Blaser, S., Faist, J. et al. (2004). Terahertz emission from quantum cascade lasers in the quantum Hall regime: evidence for many-body resonances and localization effects. *Physical Review Letters* **93** (237403).

42 Scalari, G., Turčinková, D., Lloyd-Hughes, J. et al. (2010). Magnetically assisted quantum cascade laser emitting from 740 GHz to 1.4 THz. *Applied Physics Letters* **97** (081110).

43 Brandstetter, M., Deutsch, C., Krall, M. et al. (2013). High power terahertz quantum cascade lasers with symmetric wafer bonded active regions. *Applied Physics Letters* **103**: 171113.

44 Li, L.H., Zhu, J.X., Chen, L. et al. (2015). The MBE growth and optimization of high performance terahertz frequency quantum cascade lasers. *Optics Express* **23**: 2720.

45 Barbieri, S., Ravaro, M., Gellie, P. et al. (2011). Coherent sampling of active mode-locked terahertz quantum cascade lasers and frequency synthesis. *Nature Photonics* **5**: 306–313.

46 Burghoff, D., Kao, T.-Y., Han, N. et al. (2014). Terahertz laser frequency combs. *Nature Photonics* **8**: 462–467.

47 Belkin, M.A., Capasso, F., Belyanin, A. et al. (2007). Terahertz quantum-cascade-laser source based on intracavity difference-frequency generation. *Nature Photonics* **1**: 288–292.

48 Jung, S., Kim, J.H., Jiang, Y. et al. (2017). Terahertz difference-frequency quantum cascade laser sources on silicon. *Optica* **4**: 38–43.

49 Razeghi, M., Lu, Q.Y., Bandyopadhyay, N. et al. (2015). Quantum cascade lasers: from tool to product. *Optics Express* **23**: 8462–8475.

50 Williams, B.S. (2007). Terahertz quantum cascade lasers. *Nature Photonics* **1**: 517–525.

51 Dean, P., Valavanis, A., Keeley, J. et al. (2014). Terahertz imaging using quantum cascade lasers—a review of systems and *applications*. *Journal of Physics D: Applied Physics* **47**: 374008.

52 Yang, Y., Burghoff, D., Hayton, D.J. et al. (2016). Terahertz multiheterodyne spectroscopy using laser frequency combs. *Optica* **3**: 499–502.

53 Bellotti, E., Driscoll, K., Moustakas, T.D., and Paiella, R. (2008). Monte Carlo study of GaN versus GaAs terahertz quantum cascade structures. *Applied Physics Letters* **92**: 101112.

54 Bellotti, E., Driscoll, K., Moustakas, T.D., and Paiella, R. (2009). Monte Carlo simulation of terahertz quantum cascade laser structures based on wide-bandgap semiconductors. *Journal of Applied Physics* **105**: 113103.

55 Moustakas, T.D. and Paiella, R. (2017). Optoelectronic device physics and technology of nitride semiconductors from the UV to the terahertz. *Reports on Progress in Physics* **80**: 106501.

56 Gmachl, C., Ng, H.M., Chu, S.-N.G., and Cho, A.Y. (2000). Intersubband absorption at 1.55∼λ μm in well- and modulation-doped GaN/AlGaN multiple quantum wells with superlattice barriers. *Applied Physics Letters* **77**: 3722.

57 Hofstetter, D., Schad, S.S., Wu, H. et al. (2003). GaN/AlN-based quantum-well infrared photodetector for 1.55 μm. *Applied Physics Letters* **83**: 572–574.

58 Vardi, A., Bahir, G., Guillot, F. et al. (2008). Near infrared quantum cascade detector in GaN/AlGaN/ AlN heterostructures. *Applied Physics Letters* **92**: 011112.

59 Iizuka, N., Kaneko, K., and Suzuki, N. (2006). All-optical switch utilizing intersubband transition in GaN quantum wells. *IEEE Journal of Quantum Electronics* **42**: 765–771.

60 Li, Y., Bhattacharyya, A., Thomidis, C. et al. (2007). Ultrafast all-optical switching with low saturation energy via intersubband transitions in GaN/AlN quantum-well waveguides. *Optics Express* **15**: 17922–17927.

61 Jovanović, V.D., Indjin, D., Ikonić, Z., and Harrison, P. (2004). Simulation and design of GaN/ AlGaN far-infrared 34∼(λ)μm quantum-cascade laser. *Applied Physics Letters* **84**: 2995.

62 Sun, G., Soref, R.A., and Khurgin, J.B. (2005). Active region design of a terahertz GaN/ $Al_{0.15}Ga_{0.85}N$ quantum cascade laser. *Superlattices and Microstructures* **37**: 107.

63 Driscoll, K., Liao, Y., Bhattacharyya, A. et al. (2009). Optically pumped intersubband emission of short-wave infrared radiation with GaN/AlN quantum wells. *Applied Physics Letters* **94**: 081120.

64 Nevou, L., Tchernycheva, M., Julien, F.H. et al. (2007). Short wavelength m)μ=2.13λ (intersubband luminescence from GaN/AlN quantum wells at room temperature. *Applied Physics Letters* **90**: 121106.

65 Song, A.Y., Bhat, R., Allerman, A.A. et al. (2015). Quantum cascade emission in the III-nitride material system designed with effective interface grading *Appl. Physics Letters* **107**: 132104.

66 Machhadani, H., Kotsar, Y., Sakr, S. et al. (2010). Terahertz intersubband absorption in GaN/ AlGaN step quantum wells. *Applied Physics Letters* **97**: 191101.

67 Sudradjat, F.F., Zhang, W., Woodward, J. et al. Far-infrared intersubband photodetectors based on double-step III-nitride quantum wells 2012. *Applied Physics Letters* **100** (241113).

68 Durmaz, H., Nothern, D., Brummer, G.C. et al. (2016). Terahertz intersubband photodetectors based on semi-polar GaN/AlGaN heterostructures. *Applied Physics Letters* **108**: 201102.

69 Edmunds, C., Shao, J., Shirazi-HD, M. et al. (2014). Terahertz intersubband absorption in non-polar m-plane AlGaN/GaN quantum wells. *Applied Physics Letters* **105**: 021109.

70 Bayram, C., Vashaei, Z., and Razeghi, M. (2010). Reliability in room-temperature negative differential resistance characteristics of low-aluminum content AlGaN/GaN double-barrier resonant tunneling diodes. *Applied Physics Letters* **97**: 181109.

71 Kikuchi, A., Bannai, R., Kishino, K. et al. (2002). AlN/GaN double-barrier resonant tunneling diodes grown by rf-plasma-assisted molecular-beam epitaxy. *Applied Physics Letters* **81**: 1729–1731.

72 Li, D., Tang, L., Edmunds, C. et al. (2012). Repeatable low-temperature negative-differential resistance from $Al_{0.18}Ga_{0.82}N$/GaN resonant tunneling diodes grown by molecular-beam epitaxy on free-standing GaN substrates. *Applied Physics Letters* **100**: 252105.

73 Sudradjat, F., Zhang, W., Driscoll, K. et al. (2010). Sequential tunneling transport characteristics of GaN/AlGaN coupled-quantum-well structures. *Journal of Applied Physics* **108**: 103704.

74 Terashima, W. and Hirayama, H. (2011). Spontaneous emission from GaN/AlGaN terahertz quantum cascade laser grown on GaN substrate. *Physica Status Solidi C: Current Topics in Solid State Physics* **8**: 2302.

75 Bormann, I., Brunner, K., Hackenbuchner, S. et al. (2002). Midinfrared intersubband electroluminescence of Si/SiGe quantum cascade structures. *Applied Physics Letters* **80**: 2260–2262.

76 Dehlinger, G., Diehl, L., Gennser, U. et al. (2000). Intersubband electroluminescence from silicon-based quantum cascade structures. *Science* **290**: 2277–2280.

77 Diehl, L., Menteşe, S., Müller, E. et al. (2002). Electroluminescence from strain-compensated $Si_{0.2}Ge_{0.8}$/Si quantum-cascade structures based on bound-to-continuum transitions. *Applied Physics Letters* **81**: 4700.

78 Lynch, S.A., Bates, R., Paul, D.J. et al. (2002). Intersubband electroluminescence from Si/SiGe cascade emitters at terahertz frequencies. *Applied Physics Letters* **81**: 1543–1545.

79 Sun, G., Friedman, L., and Soref, R.A. (1995). Intersubband lasing lifetimes of SiGe/Si and GaAs/AlGaAs multiple quantum well structures. *Applied Physics Letters* **66**: 3425–3427.

80 Gallas, B., Hartmann, J.M., Berbezier, I. et al. (1999). Influence of misfit and threading dislocations on the surface morphology of SiGe graded-layers. *Journal of Crystal Growth* **201**: 547.

81 Sookchoo, P., Sudradjat, F.F., Kiefer, A.M. et al. (2013). Strain-engineered SiGe multiple-quantum-well nanomembranes for far-infrared intersubband device applications. *ACS Nano* **7**: 2326–2334.

82 Durmaz, H., Sookchoo, P., Cui, X. et al. (2016). SiGe nanomembrane quantum well infrared photodetectors. *ACS Photonics* **3**: 1978–1985.

83 Driscoll, K. and Paiella, R. (2006). Silicon-based injection lasers using electronic intersubband transitions in the L valleys. *Applied Physics Letters* **89**: 191110.

84 Driscoll, K. and Paiella, R. (2007). Design of n-type silicon-based quantum cascade lasers for terahertz light emission. *Journal of Applied Physics* **102**: 093103.
85 Ortolani, M., Stehr, D., Wagner, M. et al. (2011). Long intersubband relaxation times in *n*-type germanium quantum wells. *Applied Physics Letters* **99**: 201101.
86 Sabbagh, D., Schmidt, J., Winnerl, S. et al. (2016). Electron dynamics in silicon–germanium terahertz quantum fountain structures. *ACS Photonics* **3**: 403–414.

7

Advanced Devices Using Two-Dimensional Layer Technology

Berardi Sensale-Rodriguez

Department of Electrical and Computer Engineering, The University of Utah, Salt Lake City, UT, USA

7.1 Graphene-Based THz Devices

Investigation of the terahertz properties of two-dimensional (2D) materials started in the early 2010s with studies in the graphene materials system [1–3]. Graphene is an intrinsically 2D material, which is composed of sp2-bonded carbon atoms arranged in a hexagonal lattice. In contrast to traditional semiconductors, with parabolic bands, graphene holds a linear energy-momentum dispersion. Large mobility in graphene has been reported [4], which makes this material attractive for high-frequency and terahertz applications. Furthermore, graphene can be easily synthesized, e.g., by chemical vapor deposition (CVD) techniques [5, 6], and transferred to arbitrary substrates employing simple techniques. This section discusses graphene, its electromagnetic properties relevant to terahertz technologies, and several device applications proposed and demonstrated in this material.

7.1.1 THz Properties of Graphene

In materials, the imaginary part of permittivity is associated with electromagnetic absorption and losses. Fundamentally, the imaginary part of permittivity is proportional to the real part of the material's optical conductivity. From this perspective, to understand the optical properties of graphene, it is necessary to go through the physical mechanisms determining its optical conductivity.

In general, for any materials system, including graphene, optical transitions can be classified in intra- and inter-band transitions. Depicted in Figure 7.1a is a detail of the band structure of graphene and a sketch showing representative intra- (long arrows) and inter (short arrows)-band transitions [7]. The real part of the optical conductivity of graphene can be derived based on the linearization of its tight binding Hamiltonian

Fundamentals of Terahertz Devices and Applications, First Edition. Edited by Dimitris Pavlidis.

Companion website: www.wiley.com/go/Pavlidis/FundamentalsofTHz

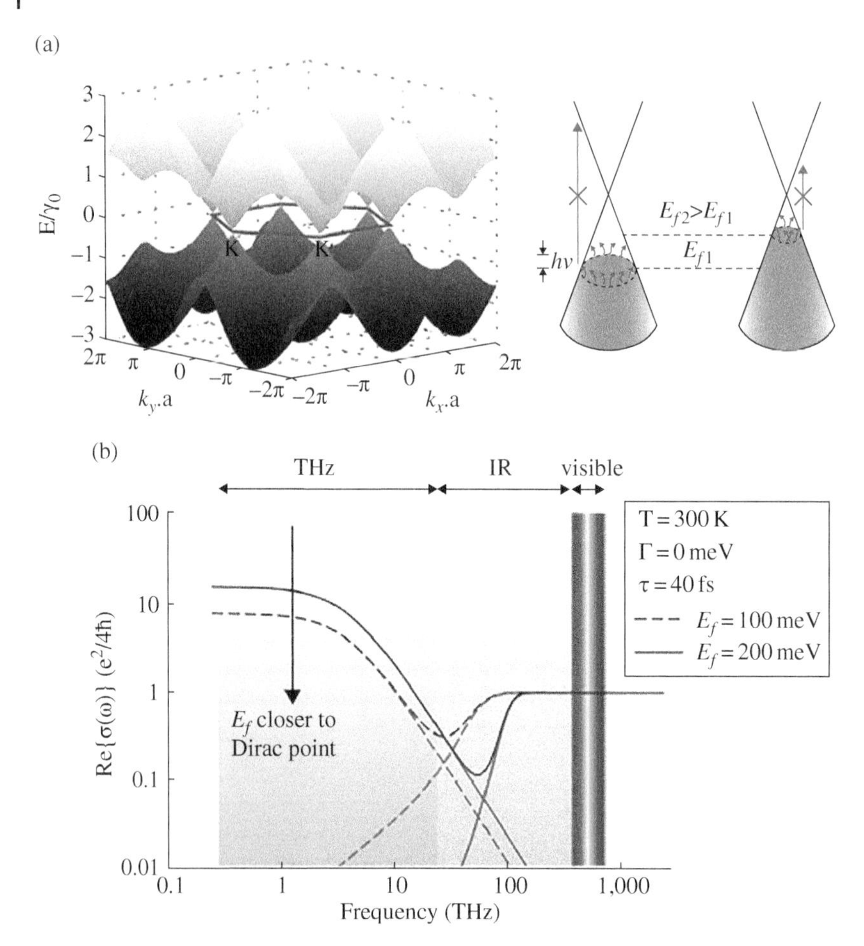

Figure 7.1 (a) Band structure of graphene and sketch of the possible optical transitions: inter-band (long arrows) and intra-band (short arrows); in the terahertz range intra-band transitions are dominant and optical conductivity depends on the density of states available for intra-band transitions, thus Fermi level. (b) Real part of the optical conductivity in graphene versus frequency (according to Eq. (7.1)). The contribution from intra-band transitions is plotted in blue, while the contribution from inter-band transitions is plotted in red. The dependence of optical conductivity with Fermi level, thus tunability, is high in the THz/IR region and reduces to zero for the visible range or higher frequencies. *Source*: Figure extracted from Ref. [7].

near the Dirac points as the sum of contributions arising from these two optical transitions [8]:

$$\sigma(\omega) = \frac{e^2\tau E_f}{\pi\hbar^2\left((\omega t)^2 + 1\right)} + \frac{\pi e^2}{2h}\left[tanh\left(\frac{\hbar\omega + 2E_f}{4k_bT}\right) + tanh\left(\frac{\hbar\omega - 2E_f}{4k_bT}\right)\right], \quad (7.1)$$

where e is electron charge, $\hbar$ is reduced Planck constant, $h = 2\pi\hbar$ is Planck constant, ω is angular frequency, E_f is Fermi level (relative to the Dirac point), τ is momentum relaxation time, T is temperature, and k_B is Boltzmann's constant. Figure 7.1b shows the calculated real part of optical conductivity versus frequency for two Fermi levels corresponding to 100 and 200 meV in energy from the Dirac point. Whereas at high frequencies, including the visible range, the contribution of inter-band transitions is dominant, at low frequencies, including the terahertz range, the optical conductivity of graphene is mainly determined by intra-band transitions. Furthermore, as can be noticed in Eq. (7.1), the optical conductivity of graphene becomes a strong function of its Fermi level and therefore can be controlled through altering the charge density in graphene. This can be done in practice in many ways, electrically, optically, or chemically. In particular, via neglecting the contribution of inter-band transitions in Eq. (7.1), the real part of graphene optical conductivity at terahertz wavelengths can be modeled through a Drude-dispersion as:

$$\sigma(\omega) = \frac{\sigma_0}{1 + (\omega\tau)^2}, \tag{7.2}$$

where σ_0 is its zero-frequency optical conductivity. In this regard, a difference between semiconductors (with classical parabolic energy dispersion) and Dirac semimetals, such as graphene, having a linear energy-dispersion, is on the dependence of its zero-frequency optical conductivity on charge density. Whereas in the first case optical conductivity is linearly proportional to charge density, in graphene this dependence is nonlinear (square root dependence) [9]. The computation of the optical conductivity of graphene and effect of Fermi level and electron momentum relaxation time on it are further discussed in Section 7.4.1.

7.1.2 How to Simulate and Model Graphene?

Graphene can be electromagnetically modeled on the basis of its constitutive parameters, with $\sigma(\omega)$ as described by Eq. (7.1) or Eq. (7.2). It is worth mentioning that conductivity, as defined in these equations, is a “sheet-conductivity” rather than a “bulk-conductivity” thus having units of [S] rather than [S/m]. Commercially-available full-wave electromagnetic solvers, such as ANSYS HFSS [10], allow for the definition of materials as layered impedances, treating graphene as zero-thickness and modeling its conductivity directly through Eqs. (7.1) and (7.2) is possible (see e.g. [11]). Alternatively, graphene can be geometrically modeled as a thin layer with a small finite thickness, e.g. 1 nm, with bulk conductivity (in [S/m]) given by the ratio between the quantity in Eqs. (7.1) or (7.2) and the assumed thickness of the graphene layer [12].

In many problems, from an analytical perspective, the effect of graphene on the terahertz beam propagation can be evaluated using the scattering matrix formalism. In this regard, an air/graphene/dielectric interface (see Figure 7.2) can be modeled with Fresnel coefficients (S) by treating graphene as a zero-thickness conductive medium [13]:

$$S = \begin{bmatrix} \dfrac{2n_{air}}{n_{air} + n_d + Z_0\sigma} & \dfrac{n_d - n_{air} - Z_0\sigma}{n_{air} + n_d + Z_0\sigma} \\ \dfrac{n_{air} - n_d - Z_0\sigma}{n_{air} + n_d + Z_0\sigma} & \dfrac{2n_d}{n_{air} + n_d + Z_0\sigma} \end{bmatrix}, \tag{7.3}$$

where $Z_0 = 377\,\Omega$ is the characteristic impedance of free-space, σ is the graphene optical conductivity, and n_{air} and n_d are the refractive indexes of air ($n_{air} = 1$) and the dielectric,

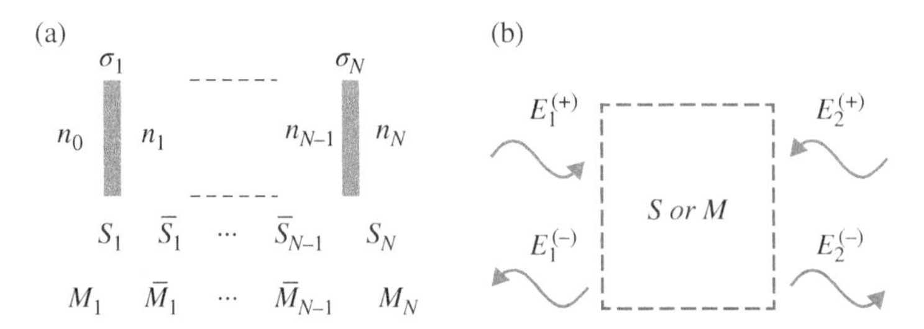

Figure 7.2 (a) Definition of parameters in a generic structure with N conductive sheets separated by dielectric interlayers. (b) Definition of the electric field vectors of incident and reflected waves at two ports of a generic optical system. *Source:* Figure extracted from Ref. [13].

respectively. In general, in a problem having multiple interfaces, the following formulas for Fresnel coefficients can be employed so to model the interface between medium i and medium j [14]:

$$S_{ij} = \begin{bmatrix} \dfrac{2n_i}{n_i + n_j} & \dfrac{n_j - n_i}{n_i + n_j} \\ \dfrac{n_i - n_j}{n_i + n_j} & \dfrac{2n_j}{n_i + n_j} \end{bmatrix}, \tag{7.4}$$

where n_i and n_j represent the refractive indexes of medium i and j, respectively. Furthermore, the scattering matrix representing propagation through a distance d in a uniform dielectric medium (i) is represented by [14]:

$$\widetilde{S}_i = \begin{bmatrix} e^{-in_ik_0d} & 0 \\ 0 & e^{-in_ik_0d} \end{bmatrix}, \tag{7.5}$$

where $k_0 = 2\pi/\lambda_0$ with $\lambda_0 = c/f$ and c is the speed of light in vacuum. By converting these S matrices into M matrices, as discussed, e.g. in Ref. [13], such a multilayer terahertz transmission problem can be represented and analyzed analytically. Examples in the Section 7.4 go through the calculation of terahertz transmission through graphene-based structures and the use of the transfer matrix approach.

7.1.3 Terahertz Device Applications of Graphene

Based on the properties mentioned above, a series of reconfigurable graphene-based devices have been proposed and demonstrated [7, 15–23]. This section focuses on a subset of these devices, namely amplitude and phase modulators and active filters.

7.1.3.1 Modulators

7.1.3.1.1 Broadband Structures

Intrinsically broadband modulator structures can be realized by directly employing large-area graphene layers. Such devices consist of one (or multiple) capacitively coupled graphene-semiconductor or graphene–graphene pairs, as depicted in Figure 7.3, in a way such that charge carriers of different species are alternately induced in each of the layers when

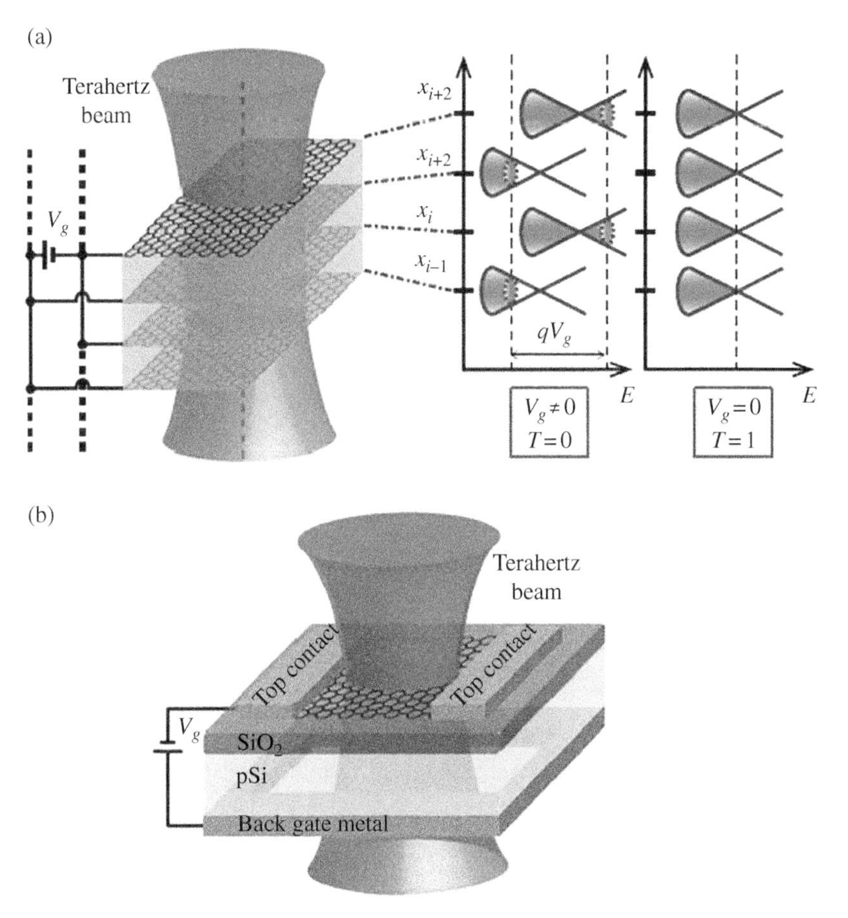

Figure 7.3 (a) Schematic of the reconfigurable region in a terahertz modulator structure, which is comprised of stacked graphene pairs with capacitively coupled 2DHG in the valence band and 2DEG in the conduction band. (b) Schematic of a proof-of-concept intrinsically-broadband graphene terahertz modulator. The structure is comprised of single layer graphene, which is transferred on top of a SiO_2/Si substrate. *Source*: Figure extracted from Ref. [24].

applying a DC voltage [24]. Shown in Figure 7.3a is a schematic of the reconfigurable region in such a terahertz modulator structure. The structure is comprised of stacked graphene pairs with capacitively coupled two-dimensional hole gas (2DHG) in the valence band and 2DEG in the conduction band (to overcome potentially limited modulation by single-layer graphene). Also shown are the energy-band diagrams of two graphene pairs in the middle of the stack (x_j stands for the vertical position of the jth graphene layer). To provide minimum signal attenuation, or insertion loss (IL), the Fermi level in all the graphene layers should be at Dirac point at zero-bias. At zero bias ($Vg = 0$ V), the Fermi level is at the Dirac point for all graphene layers, resulting in near-unity terahertz transmission (T). With an applied bias, the Fermi level moves into the valence and conduction band in the graphene pair, and charge carriers of opposite species accumulate in each of the layers forming 2D electron and hole gases, leading to near-zero terahertz transmission. Such principle of

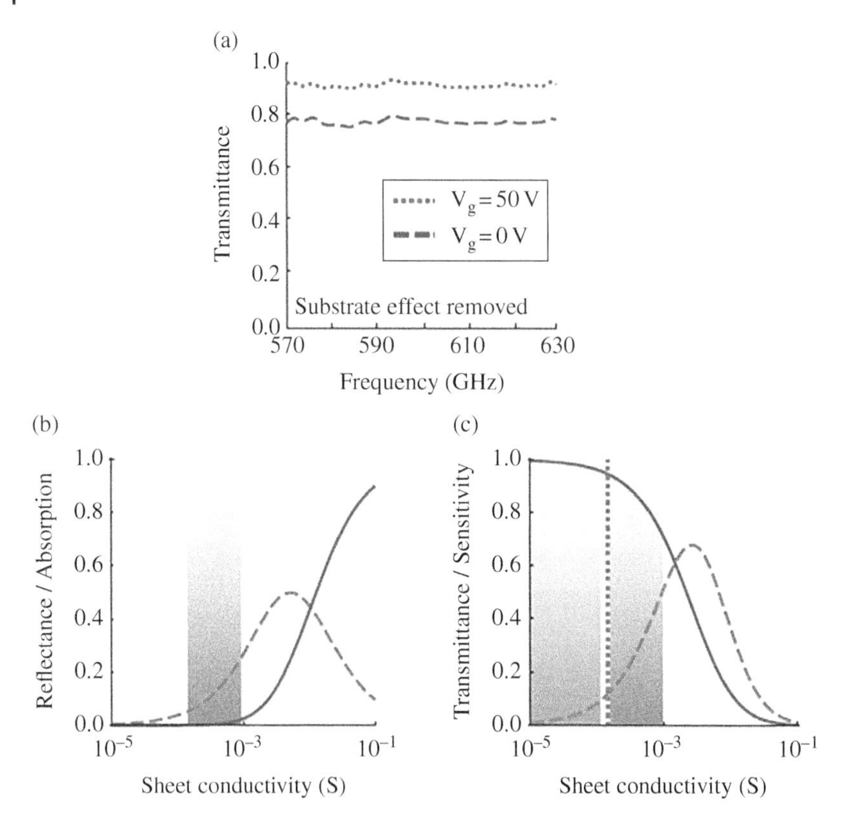

Figure 7.4 Experimental results in broadband modulator structures. (a) Transmittance after subtracting both the cavity effect and free carrier absorption due to the lightly doped bulk *p*-Si substrate for the structure depicted in Figure 7.3(b). (b) Modelled absorption (dashed line) and reflectance (solid line) as a function of graphene optical conductivity. The pink-shaded area indicates the range of graphene DC conductivity observed in this study. (c) Modelled intensity transmittance as a function of graphene optical conductivity (solid line) and the associated transmission sensitivity (dashed line), defined as the derivative of transmittance with respect to the logarithm of conductivity. Also marked on the graph are a typical minimum DC conductivity of graphene ~0.15 mS (orange line), and the range of Si 2D hole gas DC conductivity (shaded area). *Source*: Figure extracted from Ref. [24].

operation was demonstrated by Sensale-Rodriguez et al. [24] in a structure consisting of a graphene layer on top of a SiO_2/*p*-Si substrate as depicted in Figure 7.3b. Top contacts were used to monitor the graphene conductivity, and a ring-shaped bottom gate was used to tune the graphene Fermi level. Terahertz transmission spectroscopy and mapping were performed in the 570–630 GHz frequency band.

Experimentally these structures were fabricated from large-area single-layer CVD graphene, which was transferred using polymethyl methacrylate (PMMA) and wet-etch methods [25]. Depicted in Figure 7.4 are the reported experimental results in these broadband modulator structures. Figure 7.4a shows measured transmittance after subtracting both the cavity effect (Fabry-Perot resonances) and free carrier absorption due to the lightly doped bulk *p*-Si substrate. After removing these effects, a flat intensity modulation depth of ~16% was observed when applying a gate-voltage swing between 0 and 50 V (see Figure 7.4a) [24]. The electromagnetic response of the analyzed structure was modeled through the scattering matrix formalism (as discussed in Section 7.1.2) (see Supplementary Information of Ref. [24] and Section 7.4.3); the calculated absorption, reflection, and transmission are shown in Figure 7.4b, c as a function of graphene conductivity (with the substrate effects removed). It is observed that in transmission mode, the maximum absorption possible by a graphene layer is 50% of the incoming terahertz power. Such extraordinary large absorption results from a large number of allowed intra-band transitions, which are due to the large number of available carriers and density of states (DOS) in graphene at high conductivities [24].

In addition to a large modulation-depth, an important practical feature of these broadband modulator structures is that the minimum conductivity of graphene does not introduce appreciable IL. As observed in Figure 7.4a, the experimentally measured IL in these devices was ~5% [24]. In general, in transmission mode, the IL can be <0.2 dB per graphene layer. However, when accounting for substrate losses and reflection, larger losses are observed.

7.1.3.1.2 Electromagnetic-cavity Integrated Structures

Attaining larger modulation depths, as well as total absorption of terahertz light is possible in an electromagnetic-cavity integrated structure. As reported by Sensale-Rodriguez et al. [26], an example of such structure consists of a single layer of graphene on top of a SiO_2/Si substrate with a continuous metal back gate. Illustrated in Figure 7.5a is a schematic of the terahertz modulator. Top and bottom metal contacts were employed to tune the graphene conductivity. The back-metal contact electromagnetically acts as a reflector, as depicted in Figure 7.5a. The operation mechanism in this structure is similar to the one previously discussed in broadband transmittance modulators. Since the back-metal contact acts both as an electrode and as a reflector, the terahertz wave intensity is zero at this point. At the position where the graphene layer is located, the wave strength thus depends on two parameters: the substrate optical thickness and the terahertz wavelength. Typical band diagrams for various voltages (V) are depicted on the right panel of Figure 7.5a. When the Fermi level in graphene is tuned to Dirac point ($V = V_{CNP}$), the reflectance is at its maxima. Shown in Figure 7.5b is the calculated power reflectance as a function of the substrate optical thickness normalized to terahertz wavelength for different graphene conductivities. As discussed in [26] and depicted in Figure 7.5b, it can be easily demonstrated that when the substrate optical thickness is an odd-multiple of the terahertz wavelength, the field intensity in graphene is maxima. As a result, a substantial modulation of terahertz absorption takes place when the graphene conductivity is tuned through applying a bias. However, in the other hand, if the substrate optical thickness is an even-multiple of the terahertz wavelength, the field-intensity becomes zero at the position where the graphene layer is located; therefore, graphene does not absorb terahertz light regardless of its conductivity, and no modulation is possible. From this perspective, the physical mechanism leading to an "extraordinary" modulation is also responsible for making this modulation to be intrinsically narrow-band. Another interesting aspect of this structure is that total terahertz absorption is possible through proper impedance matching. This is the case when the conductivity of graphene approaches $1/Z_0 = 1/377\,\Omega$ ~2.7 mS; under this condition, as discussed in Ref. [27] operation with a 100% modulation depth is possible.

Shown in Figure 7.6 is a summary of the experimental results obtained in fabricated samples following the methods discussed in [26]. Measurements were performed in the 0.57 to

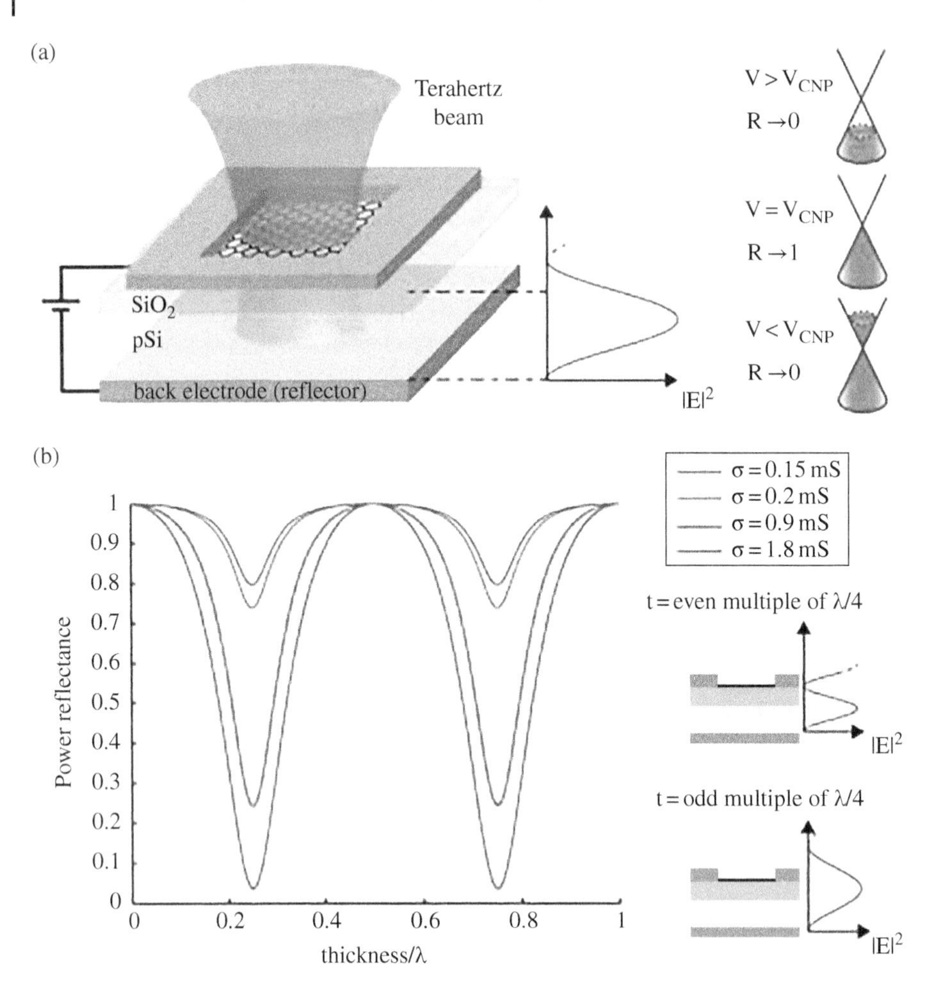

Figure 7.5 Structure of an electromagnetic-cavity integrated graphene electro-absorption modulator and operating principle: (a) Schematic of the terahertz modulator structure, which operates in reflection mode. Typical band diagrams for various voltages (V) are depicted on the right. When the Fermi level in graphene is tuned to Dirac point ($V = V_{CNP}$), reflectance is at its maxima. (b) Calculated power reflectance as a function of the substrate optical thickness normalized to terahertz wavelength for different graphene conductivities. When the substrate optical thickness is an even-multiple of a quarter wavelength, the electric field intensity vanishes in the graphene layer, therefore leading to zero absorption. On the other hand, if the substrate optical thickness is an odd-multiple of a quarter-wavelength, absorption can be extraordinarily tuned. *Source*: Figure extracted from Ref. [26].

0.63 THz frequency band and depicted in Figure 7.6a is the measured power reflectance as a function of frequency at different voltages. It is observed that reflectance closely follows the discussed theoretical trends (see Figure 7.5b). At a frequency of ~590 GHz, high transmission and no modulation is observed. This frequency corresponds to a situation where the substrate optical thickness is an even-multiple of a quarter wavelength. In the other hand, at ~620 GHz a very strong modulation is observed. The IL at 620 GHz is slightly less than 2 dB. Figure 7.6b shows the normalized reflectance ($R/R[V_{CNP}]$) at 620 GHz. The Dirac point

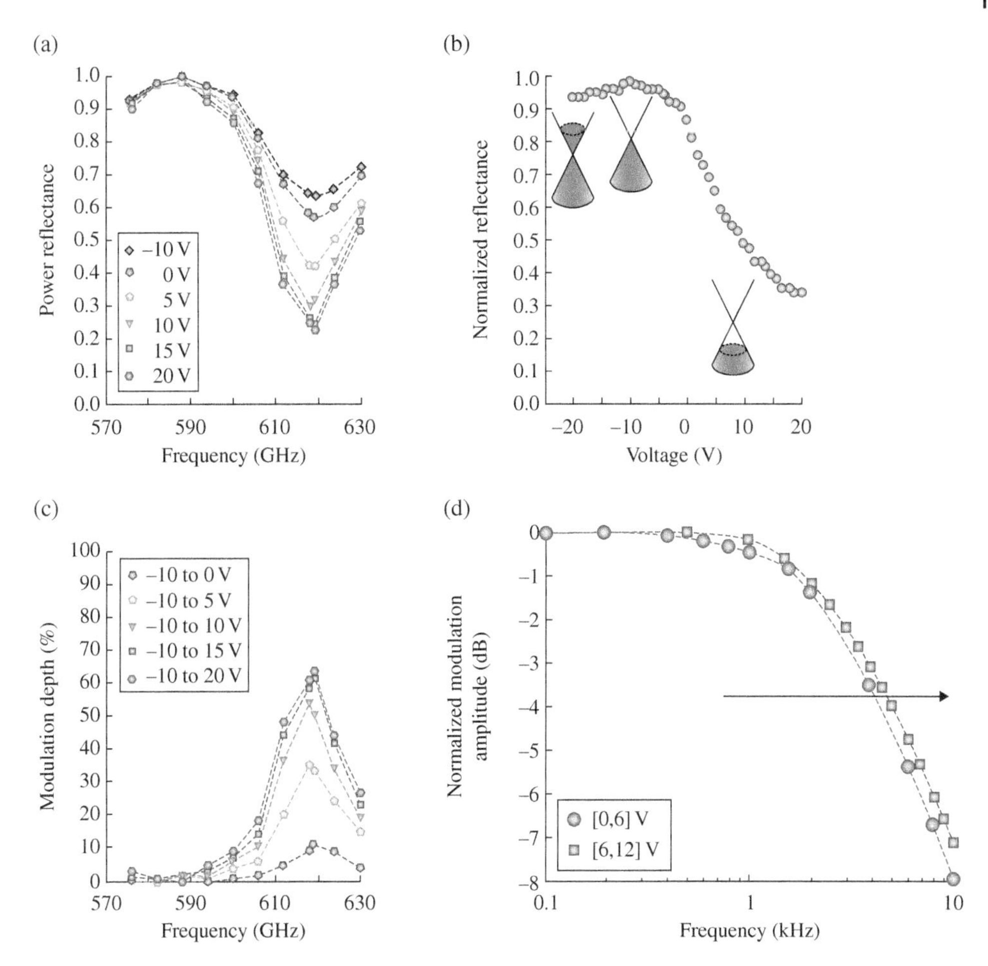

Figure 7.6 Experimental results on electromagnetic-cavity integrated modulators. (a) Power reflectance as a function of frequency at different voltages. (b) Normalized reflectance ($R/R(V_{CNP})$) at 620 GHz. Band diagrams of graphene are sketched for three bias regimes. (c) Modulation depth as a function of frequency for different voltage swings, the maximum measured modulation depth was 64%. (d) Modulator dynamic characteristics, i.e. normalized modulation amplitude versus switching frequency for two different voltage swings, for a carrier frequency ~620 GHz. *Source*: Figure extracted from Ref. [26].

voltage V_{CNP} was found to be −10 V. Band diagrams of graphene are sketched for three bias regimes so to illustrate the dependence of Fermi level with bias. In this study, the voltages applied were limited to the −20 to 20 V range due to oxide leakage. The computed modulation depth, defined as $100 \times |R(V_{CNP}) - R(V)|/R(V_{CNP})$, is plotted as a function of frequency in Figure 7.6c. The maximum measured modulation depth was 64%. This modulation is around 4× larger than what was observed for a similar conductivity swing in the transmission mode broadband structure discussed in the previous section.

At this end, it is worth mentioning that in all the device structures discussed thus far, the speed of the device is limited by RC time constants. A comprehensive analysis of these effects is provided in Ref. [26], wherein the modulation amplitude was measured as a function of frequency for various voltage swings. These results are illustrated in Figure 7.6d where normalized modulation amplitude is plotted versus switching frequency for two different voltage swings (6–12 V and 0–6 V). The voltage swing of 6–12 V (a higher average graphene conductivity) is further away from V_{CNP} than that of 0–6 V, which leads to a decrease in the device RC time constant thus a higher 3 dB cutoff frequency. In these cases, the 3 dB cutoff frequencies were estimated to be on the kHz range and observed to increase with increasing graphene conductivity thus lower equivalent R. In general, the speed performance of these devices can be improved by reducing the device area or by employing graphene with higher mobilities. As to be discussed in forthcoming sections in general trade-offs exist between speed, modulation depth, losses, and terahertz spectral bandwidth of operation.

7.1.3.1.3 Graphene/Metal-Hybrid Metamaterials

An alternative way in which to attain considerable modulation depth is via employing graphene-based metamaterials, e.g. [12, 19, 28, 29]. These structures typically consist of a passive metallic frequency selective surface (FSS) and one (or multiple) layers of graphene. Whereas the graphene layer has an active role and provides reconfiguration to the whole structure, the FSS is passive and provides for field enhancement in the plane of graphene. Therefore, leading to a very strong light-matter interaction in graphene. The strength of this field enhancement can be controlled via optimizing the geometry of the FSS or through controlling the spacing between the graphene layers and the FSS (see Ref. [11]). In general, for a given graphene conductivity swing, the larger the field enhancement, the higher the modulation depth. However, because of a stronger light-matter interaction, this is accompanied by higher losses. From this perspective, there exists a trade-off between modulation depth and IL, which needs to be carefully engineered [11, 30].

To provide for a comprehensive experimental study of these effects, a series of samples were fabricated employing CVD graphene. These layers were transferred and separated by a polyimide spacer layer from a passive metallic cross-slot FSS [11]. The separation (d) between the graphene layer and the FSS was varied using changing the polyimide thickness. Furthermore, the conductivity of graphene was varied through chemical doping using HNO_3 [31]. Depicted in Figure 7.7a, b are schematics of the analyzed device structures consisting of a metallic FSS embedded in a polyimide (PI) film with nonpatterned and patterned graphene layers on top. The right half of the sample (control region), which contains unpatterned graphene on top of PI, is used to monitor the conductivity of graphene. An optical image of a fabricated terahertz modulator on a flexible PI substrate rolled on a glass pipet is shown in Figure 7.7c. An optical image showing a detail of the FSS structure and a sketch showing the dimensions of the FSS unit cell are depicted in Figure 7.7d, e. Besides providing for a comprehensive analysis of the trade-offs between modulation depth and IL, an important observation in this study is that due to the ability of the FSS to confine electromagnetic waves into subwavelength volumes, the graphene area required to obtain a given modulation performance can be significantly reduced. As a result, this metamaterial approach can lead to an improvement of the device operation speed with respect to the two structures discussed in previous sections as a result of a reduced RC constant [11].

Depicted in Figure 7.8 is a summary of the results on graphene/metal-hybrid metamaterials. Shown in Figure 7.8a, b is the measured terahertz transmission spectra for a series of

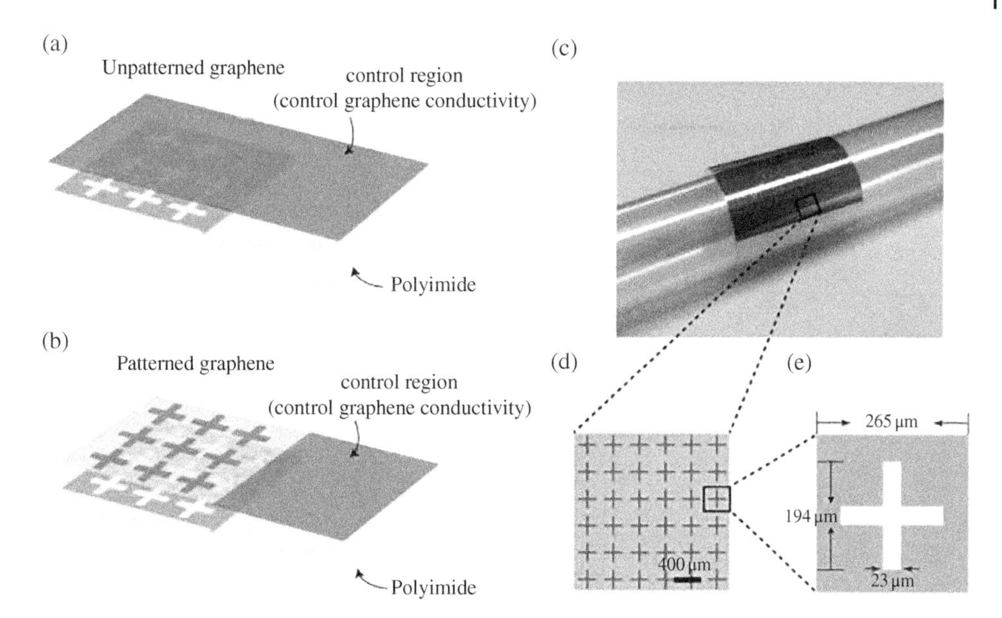

Figure 7.7 Example work on graphene/metal-hybrid metamaterial structures so to analyze the effect of the graphene placement and field enhancement arising from the metallic FSS on the electromagnetic response of the structure. (a, b) Schematic of the analyzed device structures consisting of a metallic FSS embedded in a polyimide (PI) film with nonpatterned and patterned graphene layers on top. The right half of the sample (control region), which contains un-patterned graphene on top of PI, is used to monitor the conductivity of graphene. (c) Optical image of a fabricated THz modulator on a flexible PI substrate rolled on a glass pipet. (d) Optical image showing detail of the FSS structure. (e) Sketch showing the dimensions of the FSS unit cell. *Source:* Figure extracted from Ref. [11].

samples when systematically altering the graphene conductivity and polyimide spacer thickness. The curves in Figure 7.8a represent transmission for samples containing one, two, and three layers of graphene and d = 16.5 μm. Depicted in Figure 7.8b is the measured intensity transmission spectra for devices with one layer of graphene and various polyimide spacer thicknesses, d = 16.5, 9.4, and 4.5 μm, respectively. In both cases, transmission drops as either the spacer thickness is reduced or the graphene conductivity is increased. Both trends can be understood on the basis of an increase in an overall "effective" equivalent-circuit conductivity. Depicted in Figure 7.8c is the electric field distribution at the graphene plane for different separations between the graphene layer and the FSS. As the graphene layer is placed closer to the FSS, it lays in planes with stronger fields, and thus the light-matter interaction can be enhanced.

The electromagnetic response of the analyzed structure can be modeled through the equivalent circuit represented in Figure 7.9a (see Section 7.4.7), where the graphene layer is represented by a parallel impedance of value $Z = 1/\sigma$ and the FSS is depicted by a lossy LC network—with resistance R in series with the inductor L representing the loss in the metal. To model the effect of the electric field enhancement and dielectric environment on the response of the structure, two parameters are introduced: $\alpha 1$ and $\alpha 2$ (see Figure 7.9a). Due to the near-field enhancement, the effective sheet conductivity of graphene is enhanced to $\sigma_g{}' = \alpha 1 \sigma_g$ where $\alpha 1$ (>1) represents a nonlinear enhancement factor induced by the FSS.

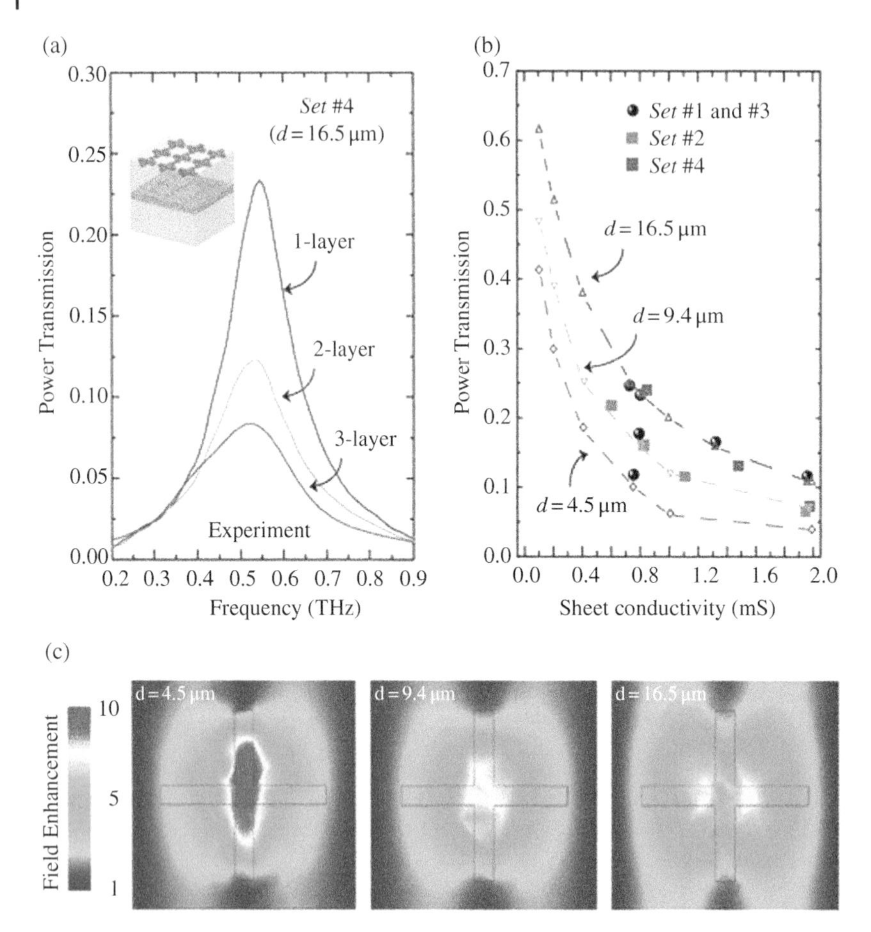

Figure 7.8 Summary of results on graphene/meta-hybrid metamaterials when changing the graphene conductivity and polyimide spacer thickness. (a) Experimentally measured intensity transmission spectra for devices containing one, two, and three layers of graphene and d = 16.5 μm. (b) Experimentally measured intensity transmission spectra for devices with one layer of graphene and various polyimide spacer thicknesses, d = 16.5, 9.4, and 4.5 μm, respectively. (c) Simulated electric filed enhancement distribution inside the plane of the graphene for different values of *d*. It could be clearly seen that enhancement gets smaller when graphene gets farther away from the FSS plane. *Source:* Figure extracted from Ref. [11].

This parameter $\alpha 1$, therefore, arises from the electric-field enhancement at the graphene plane, which translates into an enhancement of the graphene conductivity in the equivalent circuit model. In the other hand, the parameter $\alpha 2$ results from the change in capacitance due to the introduction of the graphene layer and change in the dielectric spacer thickness. Depicted in Figure 7.9b are the extracted values of $\alpha 1$ and $\alpha 2$ versus graphene conductivity for a structure with fixed polyimide spacer thickness; $\alpha 1$ and $\alpha 2$ are observed to be independent of the graphene conductivity in the simulations as well as in the experiments. Figure 7.9c, in turn, shows the extracted values of $\alpha 1$ and $\alpha 2$ when varying the spacer thickness for a structure with fixed graphene conductivity; as d is increased $\alpha 1$ exponentially decays directly following the field enhancement in the graphene plane, whereas $\alpha 2$ follows a nonlinear trend that physically arises from a larger effective permittivity in the structure. From this perspective, the effective

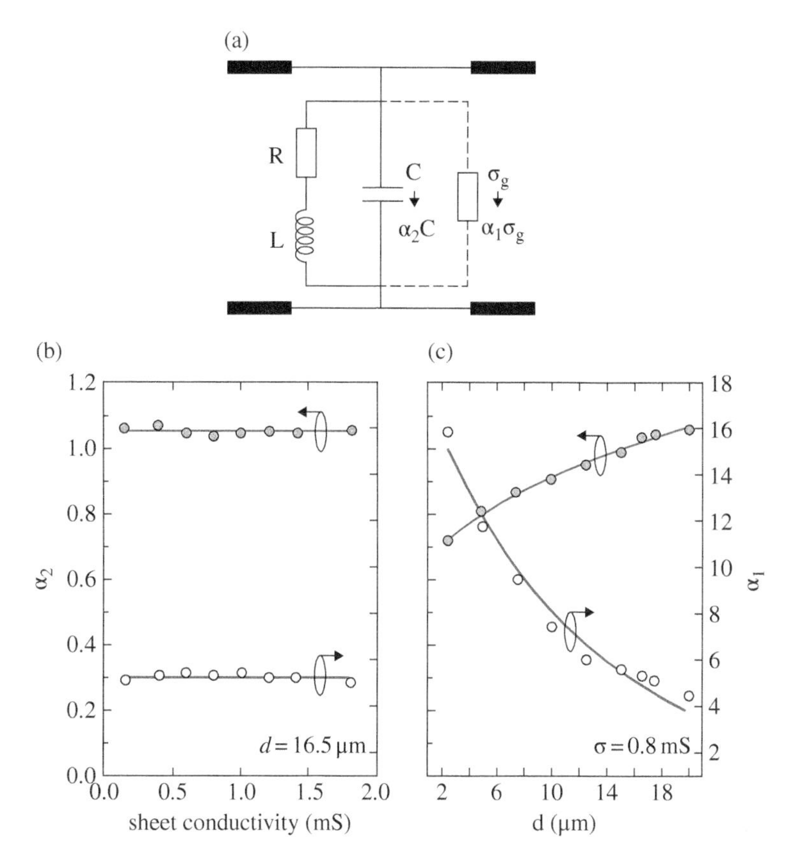

Figure 7.9 Graphene/metal-hybrid metamaterials: transmission line equivalent circuit model and fitting of experimental results to the model. (a) Schematic of the equivalent circuit model. The graphene layer is represented via inclusion of a parallel impedance element with sheet conductivity of value σ_g to the original equivalent circuit RLC model of the FSS. (b) Extracted $\alpha1$ and $\alpha2$ (symbols) versus sheet conductivity (for a structure with d = 16.5 μm); $\alpha1$ and $\alpha2$ are observed to be independent of the graphene conductivity in our simulations as well as in our experiments. (c) Extracted $\alpha1$ and $\alpha2$ (symbols) versus spacer thickness *d* (for a structure with σ_0 = 0.8 mS) and their fittings (solid lines); as *d* is increased, $\alpha1$ exponentially decays, whereas $\alpha2$ follows a nonlinear trend that physically arises from a larger effective permittivity in the structure. *Source:* Figure extracted from Ref. [11].

equivalent circuit conductivity of graphene and therefore the electromagnetic response of the structure can be arbitrarily tailored through optimizing the polyimide spacer thickness as well as through engineering the FSS geometry so to attain the desired field enhancement in graphene. In general, there are trade-offs between loss, modulation depth, and bandwidth when altering these parameters (see discussion in Ref. [11]).

7.1.3.1.4 Graphene/Dielectric-Hybrid Metamaterials

We have discussed the effect of field enhancement in graphene/metal-hybrid structures as a mechanism leading to effective conductivity enhancements in graphene. Enhanced light-matter interaction leading to large modulation depths and large terahertz absorption by

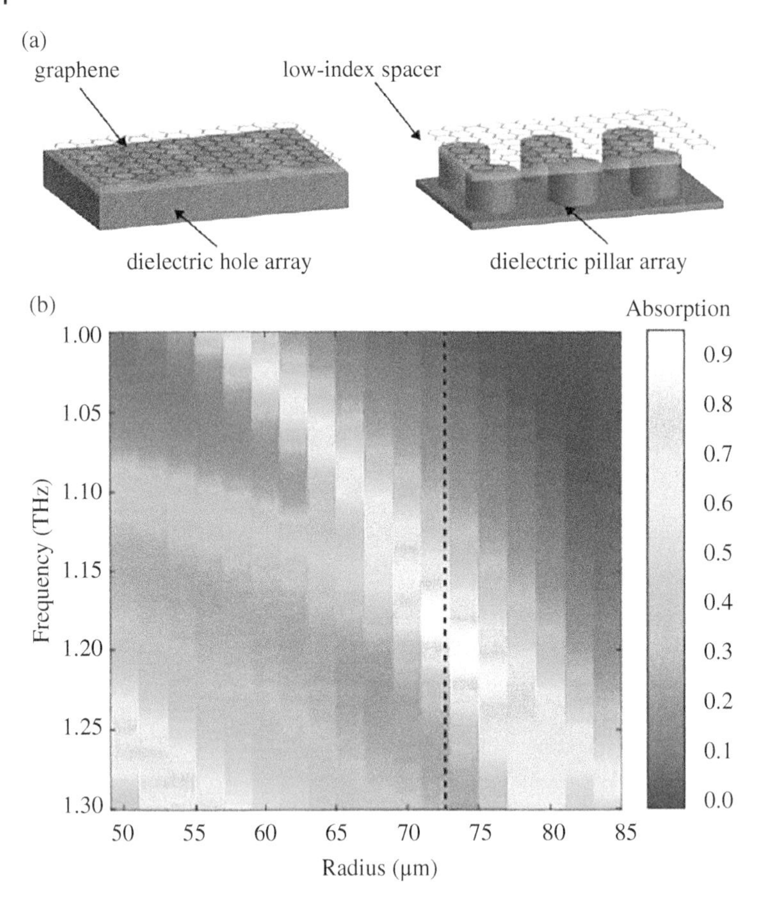

Figure 7.10 (a) Sketch of the analyzed graphene-dielectric integrated metasurfaces, which are comprised of passive dielectric hole or pillar arrays and a reconfigurable graphene sheet. (b) Design space exploration. Absorption versus hole radius and frequency for a fixed graphene sheet conductivity of 1 mS. *Source:* Figure extracted from Ref. [32].

graphene is also possible in graphene/dielectric-hybrid metamaterials, as proposed by Arezoomandan et al. [32]. Such structures can be realized employing Si pillar- or (circular)hole- arrays as example passive lossless dielectric patterns as depicted in Figure 7.10a. For the purpose of structural support, a dielectric spacer (e.g. SiN) is located between the dielectric pattern and the graphene sheet. As in all the previously discussed devices, dynamically varying the conductivity of the graphene layer is the active mechanism leading to modulation of the terahertz absorption.

Via optimizing the geometric dimensions of the dielectric patterns, it is possible to attain almost complete terahertz absorption at an arbitrary frequency of interest. Depicted in Figure 7.10b is a design space exploration for a structure comprising Si pillar-arrays. Colormaps depict absorption versus hole radius and frequency for a fixed graphene sheet conductivity of 1 mS. Maximum absorption approaches 100% at ~1.2 THz for a radius of 72 μm as shown in Figure 7.10b [32]. This "extraordinary" absorption can be understood on the basis of an electric-field enhancement at the interface where graphene is located. In general, the periodic dielectric structure can support multiple resonant modes [33]. Such a mechanism has been exploited in lossy dielectrics as a way of attaining perfect absorption [34]. However, whereas in previous works the absorption takes place in the (lossy) dielectric itself, in the structures analyzed in Ref. [32] the employed dielectrics are lossless, and the graphene layer placed on-top of the dielectric is responsible for absorption. In this case, resonant modes having maximum field intensity at the graphene plane are to be excited in the dielectric.

Two proposed ways in which to alter the sensitivity of terahertz absorption to variations in graphene sheet conductivity in graphene/dielectric-hybrid metamaterials are depicted in Figure 7.11. The sensitivity of absorption to variations in graphene conductivity can be fine-tuned through the following two ways: (i) via controlling the thickness, d, of a low-index dielectric spacer located between the dielectric pattern and the graphene sheet (see Figure 7.11a), or (ii) via choosing a material of appropriate index of refraction for the dielectric pattern (see Figure 7.11b). Whereas the first effect has a similar origin and physical explanation to that previously discussed in graphene/metal-hybrid structures, the second approach can be understood on the basis of stronger mode confinement when employing high-index dielectrics which leads to larger fields at the interface.

7.1.3.2 Active Filters

In addition to structures capable of controlling the transmission, reflection, and absorption of terahertz radiation using graphene, when actively tuning the conductivity of graphene it is also possible to actively alter the frequency response of a metamaterial structure. These structures can operate as "active filters" and therefore might constitute an essential building block for future terahertz systems [35]. The detailed sketch of a proposed graphene-based active filter is depicted in Figure 7.12. The device consists of an array of split-ring resonators loaded with graphene. Each ring has four gaps where active graphene stripes are placed. In general, the ring geometry should be engineered in order so to tailor the frequency response of the device.

In early work by Yang et al. [36], such devices were fabricated on glass substrates employing CVD grown graphene and standard transfer and lithographic processes. Depicted in Figure 7.12b is a sketch of a unit cell of the array; in experiments, the inner radius ($R1$) is 200 μm, and the outer radius (R_2) is 220 μm. The gap width is 1 μm. Graphene is placed in between the small gaps. The physical operation mechanism of this device can be understood as follows. A normally incident terahertz beam induces an inductive current flow through the rings. When the frequency of the incoming terahertz wave approaches the characteristic resonance frequency of the structure, absorption becomes maximum, and the terahertz transmittance exhibits a minimum. The resonance frequency of the structure is controlled through varying the conductivity of the graphene layers inside the ring gaps and thus can be actively tuned. Shown in Figure 7.12c is an optical micrograph of the gap region of a fabricated device.

Shown in Figure 7.13a is the simulated and measured terahertz transmittance versus frequency (as a function of graphene conductivity) for a structure designed to exhibit resonances in the 0.1–0.3 THz frequency range. A shift between two-extreme resonant-states (corresponding to ring-closed to ring-open) is observed when the conductivity of graphene is

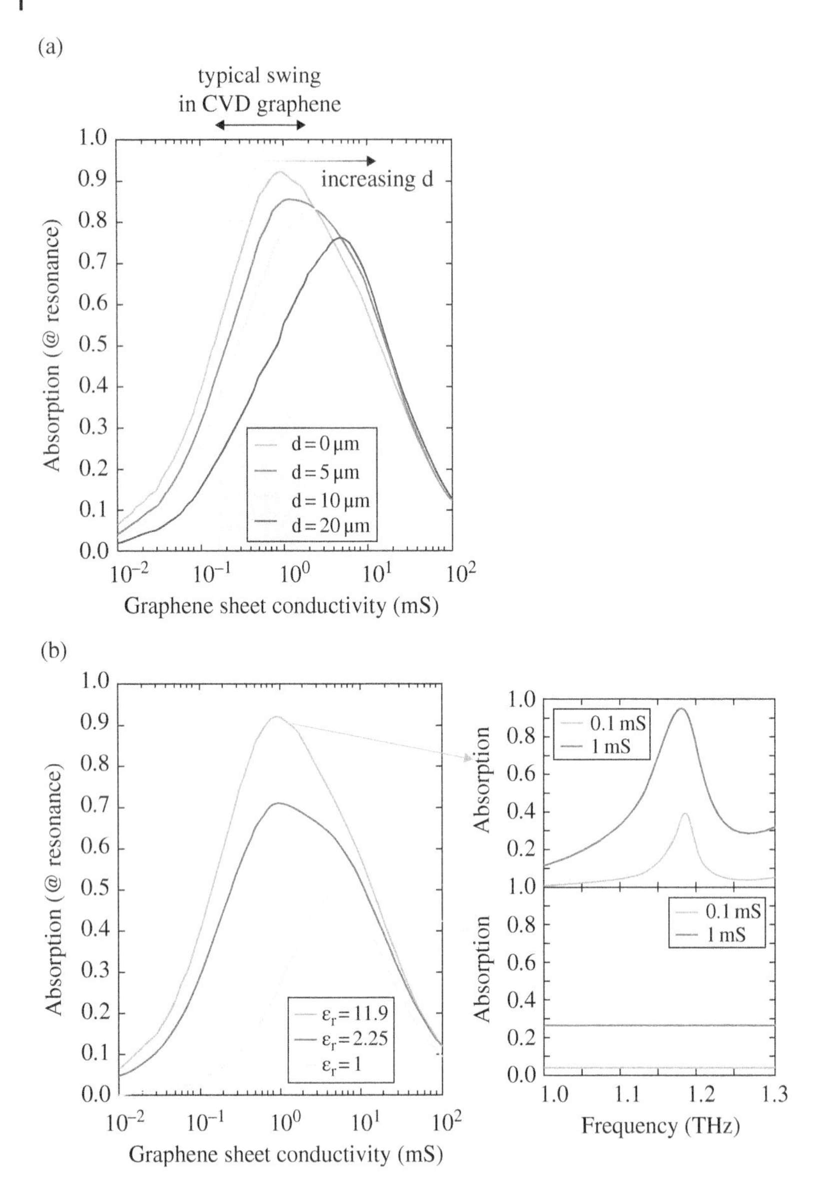

Figure 7.11 Two proposed ways in which to alter the sensitivity of terahertz absorption to variations in graphene sheet conductivity in graphene/dielectric-hybrid metamaterials. (a) Via controlling the thickness of a low-index dielectric spacer located between the dielectric pattern and the graphene sheet, and (b) Engineering the relative permittivity of the dielectric pattern itself. (a) and the left panel in (b) depict absorption versus sheet conductivity in graphene (at resonance frequency); the right panels in (b) depict absorption versus frequency for structures comprising dielectric patterns with ε_r = 11.9 and 1, respectively. *Source*: Figure extracted from Ref. [32].

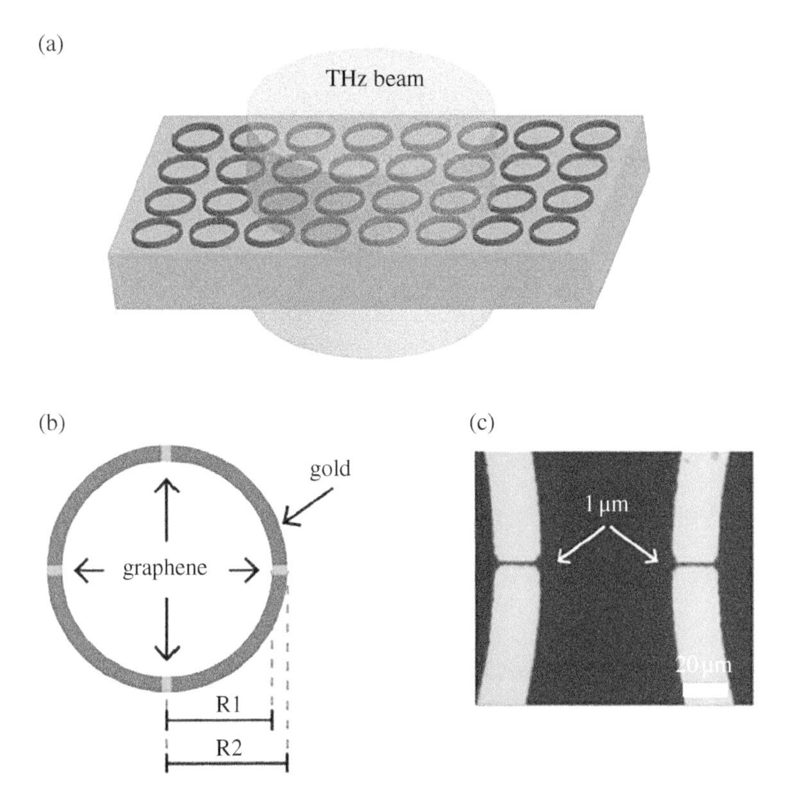

Figure 7.12 Graphene-based active terahertz filters. (a) Sketch of the device structure which consists of an array of terahertz split-ring resonators with four gaps. (b) Sketch of a unit cell of the array; in experiments, the inner radius (*R*1) is 200 μm and the outer radius (*R*2) is 220 μm. The gap width is 1 μm. Graphene is placed in between the small gaps. Its conductivity is experimentally tuned by stacking multiple graphene layers. (c) Optical micrograph of the gap region of a fabricated device. *Source*: Figure extracted from Ref. [36].

varied. Although in the proof-of-principle experimental work reported in [36] conductivity was varied through varying number of graphene layers, electrical or optical actuation is also possible. It is worth highlighting that a critical design parameter in these devices is the gap dimension, which controls the sensitivity of the electromagnetic response of the structure to variations in the graphene conductivity. This parameter needs to be carefully optimized to achieve optimal device switching performance between the two resonant states as illustrated in Figure 7.13c, d. Figure 7.13c depicts simulated transmittance versus frequency for different gap sizes (0.1, 1, 5, and 20 μm) when varying the graphene conductivity from 0.15 to 1.2 mS. From this data, resonance shift versus gap size is extracted when the conductivity is varied between 0.15 and 1.20 mS as illustrated in Figure 7.13d. In general, the larger the gap, the harder it is for the filter to shift away from its high-frequency resonant state; the smaller the gap, the harder it is for the filter to shift away from the low-frequency resonance. It is observed that a 1 μm gap gives the best performance.

An additional important aspect of this device is its potential for high-speed operation. In the reported structures, the ratio between active area and unit cell area is <0.1% [36], which

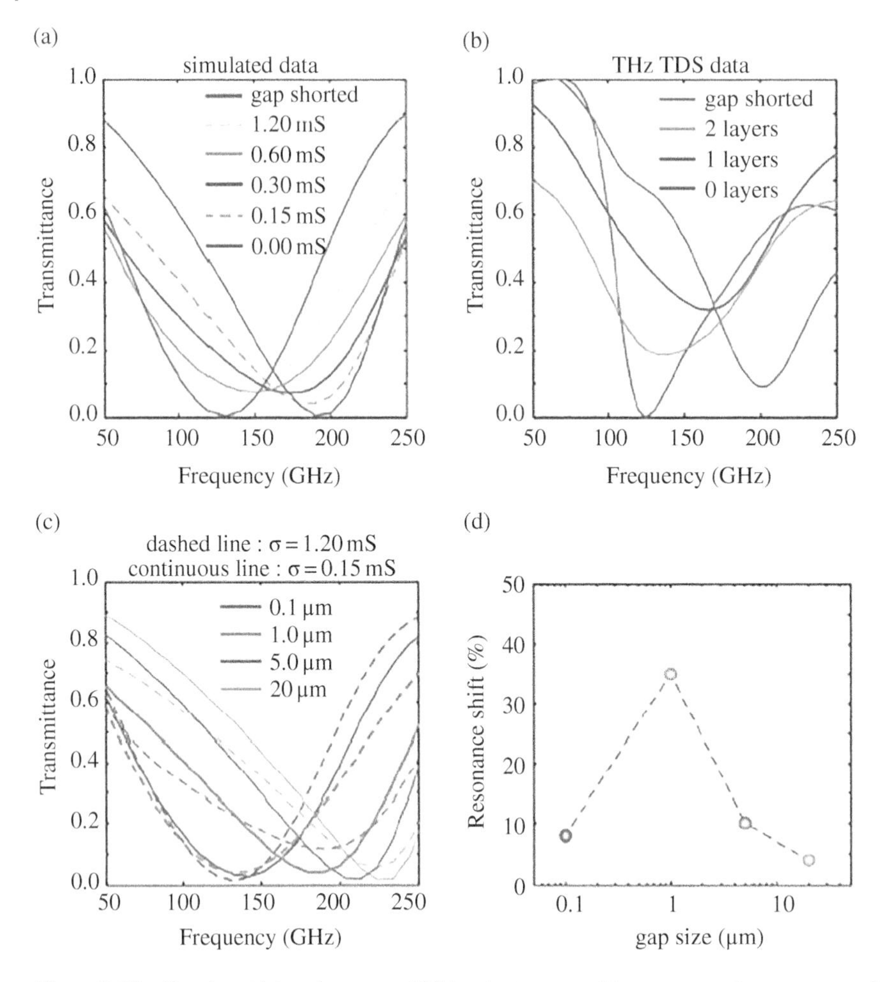

Figure 7.13 Simulated (a) and measured (b) terahertz transmittance versus frequency as a function of graphene conductivity (number of graphene layers). Terahertz time domain spectroscopy was employed in order to characterize the experimental samples. The extracted optical conductivity was 0.3 mS per graphene layer. (c) Simulated transmittance versus frequency for different gap sizes (0.1, 1, 5, and 20 μm) when varying the graphene conductivity from 0.15 to 1.2 mS. (d) Resonance shift versus gap size when the conductivity is varied between 0.15 and 1.20 mS, a 1 μm gap gives the best performance. *Source:* Figure extracted from Ref. [36].

is small enough to enable GHz switching speeds. As discussed in previous sections, the device-speed is limited by RC time constants and reducing graphene area is key so to enhance speed.

7.1.3.3 Phase Modulation in Graphene-Based Metamaterials

In previous sections, we have focused our discussion on the amplitude of a transmitted (or reflected) terahertz beam. In addition to amplitude control, phase modulation is essential for the development of future terahertz technologies. For instance, beam shaping applications such as beam steering and beam focusing rely on the construction of arbitrary phase gradients. Such phase gradients can be constructed through reconfiguring the individual response of each unit-cell in a metamaterial phase modulator. In this context, unit-cells on an ideal metamaterial geometry should provide: (i) large phase modulation, (ii) large transmittance levels, and (iii) small unit cell to wavelength ratio. From the perspective of (iii), efficient deeply-scaled metamaterials are desired for these applications. Depicted in Figure 7.14a is the structure of a graphene-based deep-subwavelength phase modulator proposed by Arezoomandan et al. [37]. The device consists of an array of spiral resonators with an active graphene layer that is placed in the center of each unit cell. Shown in Figure 7.14b is a detail of one element of the array (meta-cell) and the gating structure (electrode configuration) for electrically controlling the graphene conductivity. As discussed in previous sections, the graphene conductivity and therefore the effective properties of the device can be controlled via controlling the Fermi level in graphene.

Through an optical element, e.g. the unit-cells comprising a metasurface, the terahertz transmission amplitude, and phase are not independent of each other but related by Kramers–Kronig (KK) relations [38]. Near frequencies where the transmission amplitude has no strong dependence on the graphene conductivity, the phase experiences a maximum shift and vice-versa. Depicted in Figure 7.14c, d is the simulated electromagnetic response of the phase modulator proposed in Ref. [37]. Amplitude (Figure 7.14c) and phase (Figure 7.14d) transmission are computed versus frequency for different graphene conductivities. The graphene conductivity is varied from 0.1 to 4 mS. A strong phase modulation of 42° is possible at a frequency of 0.32 THz. In these structures the unit-cell dimension is $54 \times 54\ \mu m^2$ thus unit cell to wavelength ratio is just 0.06. Such highly-scaled unit-cell dimensions are among the smallest reported in the literature and thus result promising for beam shaping applications [37].

Later follow-up work by Arezoomandan et al. [39] analyzed the general geometric trade-offs when designing these deep-subwavelength structures. For this purpose, two types of deeply-scaled meta-cell geometries were analyzed and compared, namely: multi-split-ring resonators (MSRRs), and multispiral resonators (MSRs). Two figures of merit, related to (a) the loss (FoM_1) and (b) the degree of reconfigurability achievable by such metamaterials (FoM_2) were introduced with:

$$FoM_1 = PM \times T/([360°] \times [100\%]). \tag{7.6}$$

$$FoM_2 = L/\lambda_P \tag{7.7}$$

where PM is phase-modulation, T is the transmittance at the frequency where maximum phase-modulation takes place, L is the edge-length of the meta-cell and λ_p is the wavelength associated with the frequency at which maximum phase modulation takes place. For an ideal meta-cell FoM_2 should approach zero; the smaller the FoM_2 the most suitable a metamaterial geometry is for beam steering since a continuous phase profile can be better discretized and thus superior efficiency is possible. Furthermore, for an ideal meta-cell FoM_1 should approach unity (since PM and T are bounded by 360° and 100%, respectively); the larger the FoM_1 the most suitable a metamaterial geometry is for beam steering, i.e. the less loss the device will provide.

Systematic simulations of these two types of deep-subwavelength geometries by varying the number of turns in the spirals in MSRs or adding more rings in MSRRs show that there

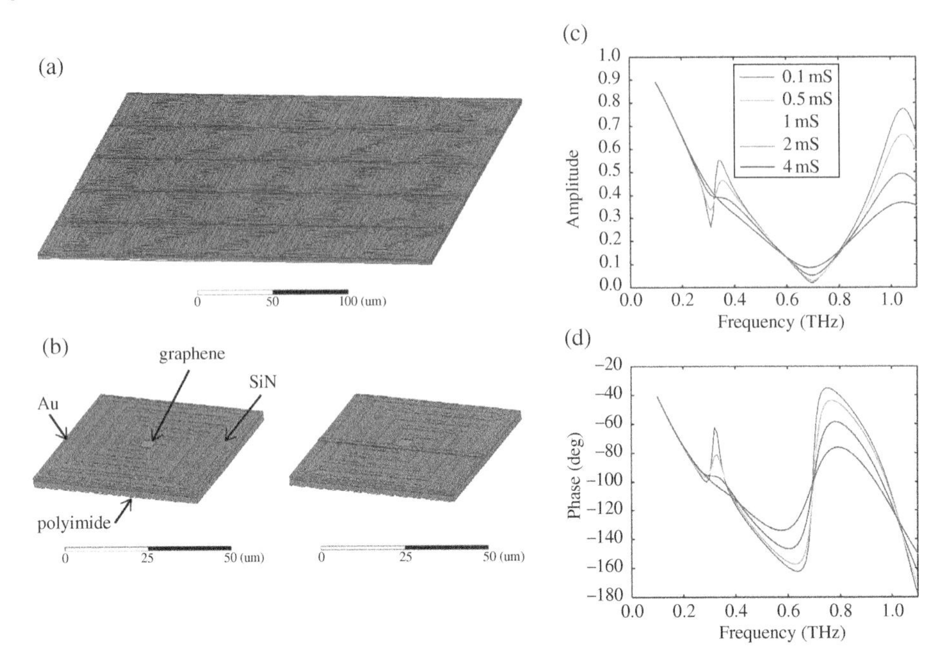

Figure 7.14 Deep-subwavelength metamaterial phase modulators. (a) Sketch of the device structure. (b) Detail of one element of the array (meta-cell) and gating structure (electrode configuration) for electrically controlling the graphene conductivity. (c) Amplitude and (d) phase transmission vs. frequency, for different graphene conductivities. Graphene conductivity is varied from 0.1 mS to 4 mS. *Source:* Figure extracted from Ref. [37].

is an optimal metal coverage-fraction that provides for the best trade-off in terms of loss versus degree of reconfigurability, as depicted in Figure 7.15. Multi-spiral resonators with metal coverage-fraction 64%, 42%, 37%, and 30%, and MSRRs with metal coverage-fraction 62%, 48%, 39%, and 24% were analyzed. The coverage-fraction is defined as the ratio between the area covered by metal and the total area of a metacell. Figures of merit (FoM_1 and FoM_2) vs. metal coverage-fraction were computed for both analyzed metamaterial geometries. In both geometries, the best trade-off between FoM_1 and FoM_2 occurs when the area covered by the metal is around 40% of the total area (shaded region in Figure 7.15). In these cases, a unit-cell to wavelength ratio on the order of 1/20 leads to a good trade-off between loss, phase modulation, and projected far-field performance [39]. This observation sets a constraint on how much unit-cells can be practically scaled.

7.2 TMD Based THz Devices

Reconfigurable terahertz devices, e.g. metamaterials, can also be constructed employing 2D materials beyond graphene. In this context, several studies have reported on the use of transition-metal dichalcogenides (TMDs) as the reconfigurable element, e.g. [40–42]. Although the carrier mobility in these materials is quite small, as compared to that in graphene (see, e.g. [43]), the finite bandgap and parabolic band structure in these materials, allows for a large induced charge carrier density through both electrical and optical injection. Furthermore, the use of undoped materials can lead to low-loss with respect to what is possible in graphene. From these perspectives, TMDs have emerged as promising 2D materials for terahertz applications.

Depicted in Figure 7.16a is the schematic of a proposed MoS_2/metal-hybrid metamaterial terahertz modulator [44]. The structure consists of a metallic FSS separated from the reconfigurable MoS_2 film through a polyimide spacer. Following the discussion in Ref. [11], see Section 7.1.3, the light-matter interaction in the MoS_2 film can be controlled by varying the

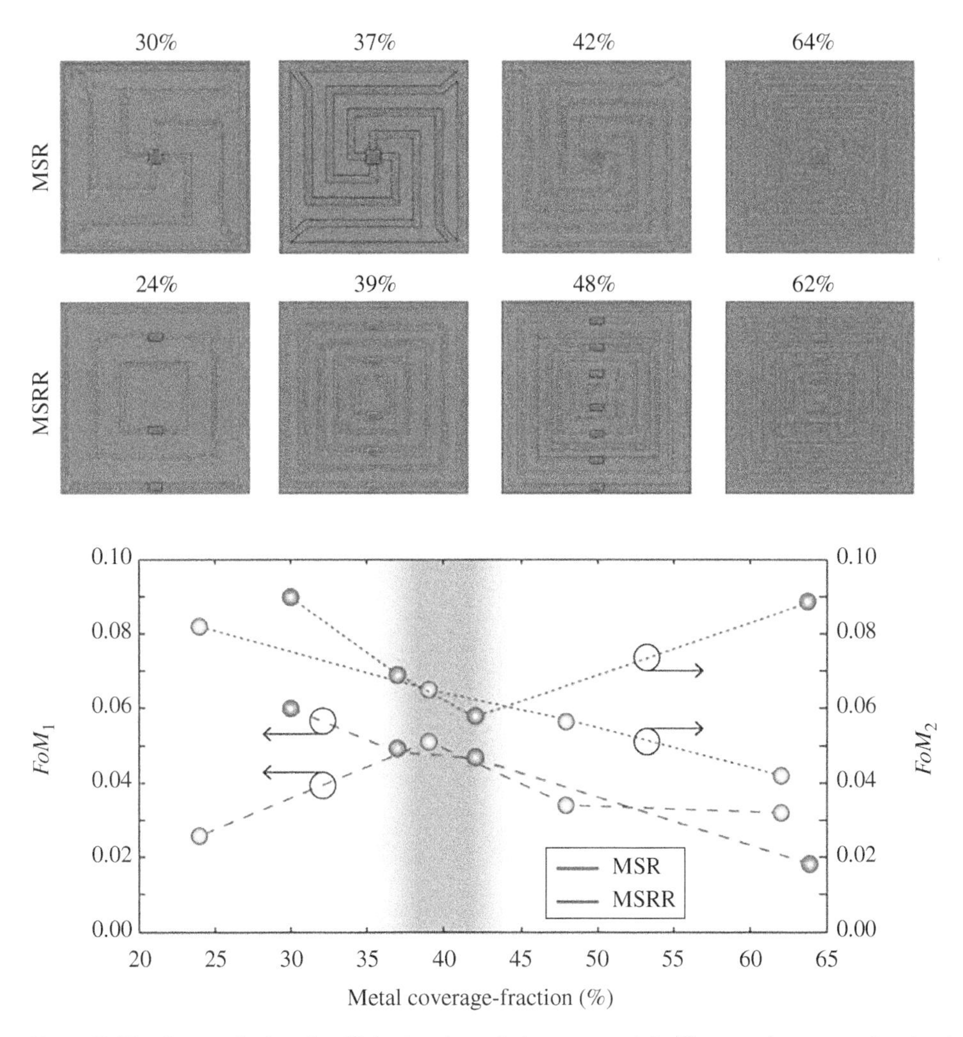

Figure 7.15 Geometrical trade-offs in deeply-scaled metamaterials. The metal coverage-fraction is defined as the ratio between the area covered by metal and the total area of a metacell. Figures of merit (FoM_1 and FoM_2) are computed for various metal coverage-fractions in MSR and MSRR's. In both cases, the best trade-off occurs when the metal coverage-fraction is around 40% (shaded region). *Source:* Figure extracted from Ref. [39].

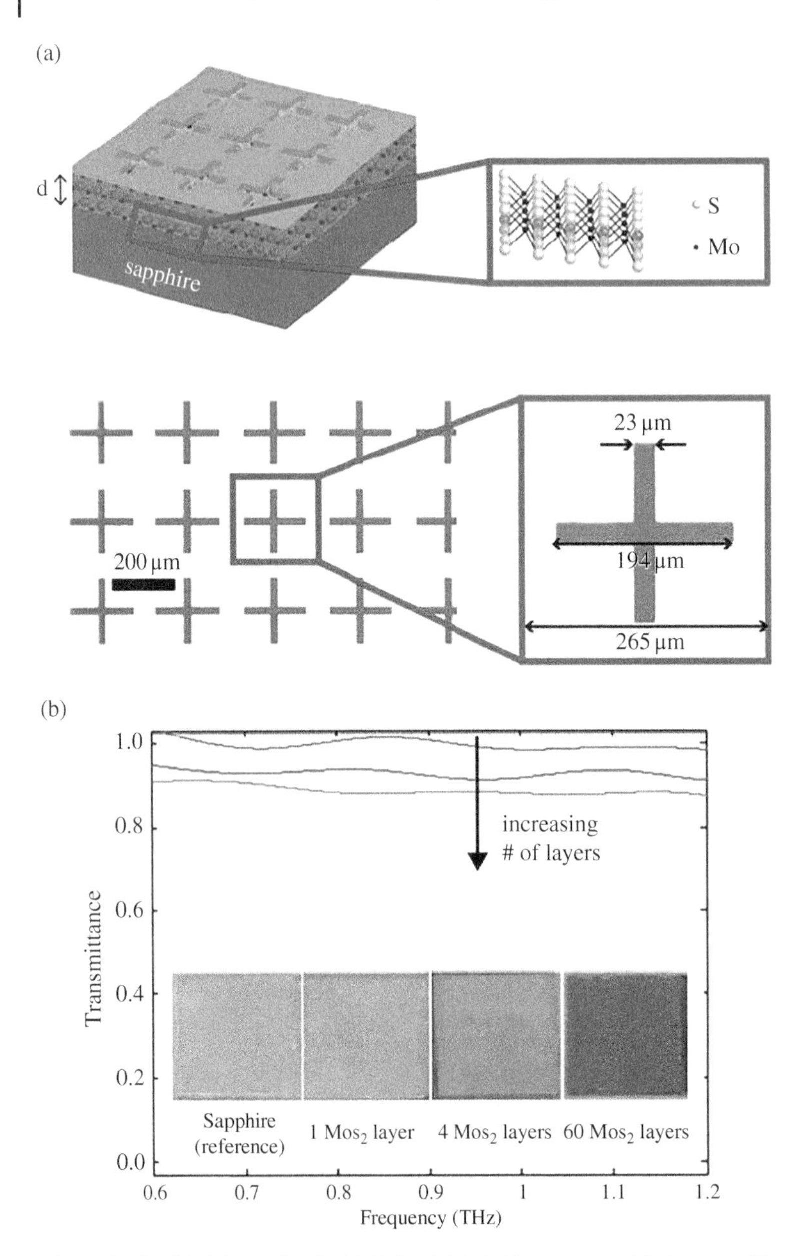

Figure 7.16 (a) Schematic of a MoS_2/metal-hybrid metamaterial structure. The polyimide spacer thickness, *d*, is varied in our experiments and simulations. An optical image of a section of the metamaterial and detail of a unit cell is depicted in the bottom. (b) Terahertz transmittance versus frequency for samples with varied number of MoS_2 layers under optical excitation. The measured transmittance under no optical excitation (not shown) is unity for all the analyzed samples, which is an evidence of the as-grown material being intrinsically undoped. Depicted in the inset is an optical image of various MoS_2 films grown in sapphire substrates. *Source:* Figure extracted from Ref. [44].

spacer thickness. Samples were fabricated employing pulsed laser deposition (PLD) grown MoS_2 on top of which the polyimide dielectric spacer was spin-coated; then, the metallic FSS was deposited and patterned through standard optical lithography processes [44]. The polyimide spacer thickness, d, was varied in our experiments and simulations. An optical image of a section of the metamaterial and detail of a unit cell is depicted in the bottom panel of Figure 7.16a. In general, as a result of a large optical absorption for light with energy above its optical bandgap, the terahertz conductivity in TMDs can be varied through photoexcitation of carriers with an optical pump. To optically excite the analyzed MoS_2 films, a broadband quartz tungsten halogen lamp was employed. Figure 7.16b shows the measured terahertz transmittance versus frequency for samples consisting of the as-grown films with varied number of MoS_2 layers under optical excitation. The measured transmittance under no optical excitation (not shown) is unity for all the analyzed samples, which is an evidence of the as-grown material being intrinsically undoped. Depicted in the inset is an optical image of various MoS_2 films grown on sapphire substrates.

Terahertz transmittance was measured with and without illumination in metamaterial structures with different spacer thickness in samples with varied number of MoS_2 layers. Depicted in Figure 7.17a is the measured terahertz transmittance (with and without illumination) for metamaterial structures with different spacer thickness ($d = 0$, 2 and 6 μm) for a sample with 60 atomic layers of MoS_2. In accordance with the discussion in Section 7.1.3, the closer the MoS_2 films are to the FSS, the largest the observed modulation depth. This observation is a result of the strong light-matter interaction in the near field [11].

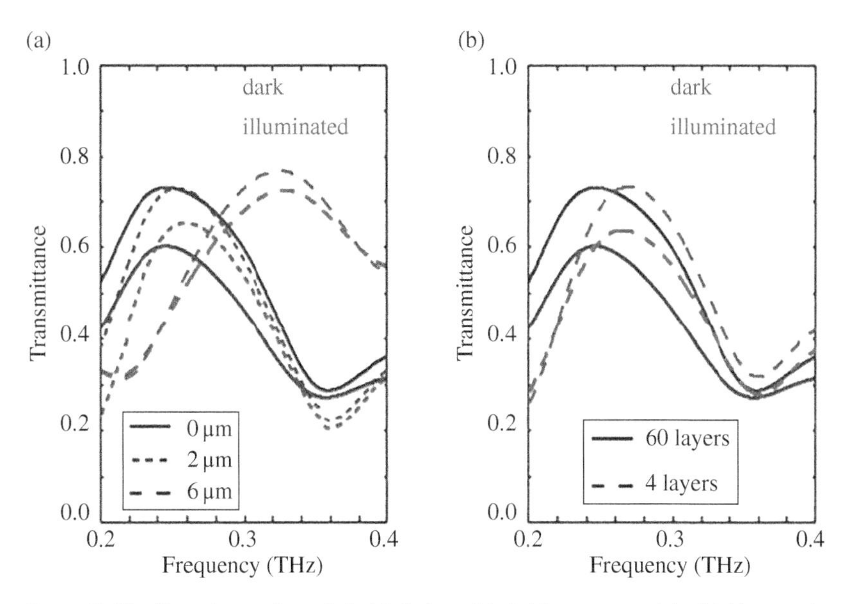

Figure 7.17 Experimental result in MoS_2/metal hybrid metamaterials. (a) Measured terahertz transmittance (with and without illumination) for metamaterial structures with different spacer thickness ($d = 0$, 2, and 6 μm) for a sample with 60 atomic layers of MoS_2. (b) Measured terahertz transmittance (with and without illumination) for samples with no PI spacer and 4 and 60 MoS_2 layers. In all cases the IL <3 dB. *Source*: Figure extracted from Ref. [44].

Furthermore, when analyzing transmittance through samples with different number of MoS_2 layers, it is observed that samples having more MoS_2 layers can provide larger modulation depths. Figure 7.17b shows the measured terahertz transmittance (with and without illumination) for samples with no PI spacer and 4 and 60 MoS_2 layers. This can be under stood as a result of a larger total conductivity in thicker samples. From these perspectives, using multilayer TMD films and placing these films in the proximity of the FSS can yield the largest modulation depths. Contrary to what is the case in graphene-based metamaterials, in TMDs there is no trade-off between losses and modulation depth since TMD films can provide no absorption in off-state.

In general, when employing an optical excitation as the mechanism providing for reconfiguration, materials with fast carrier relaxation times are highly desirable so to attain high-speed operation. In this regard, TMD monolayers exhibit picosecond carrier lifetimes and therefore could be harnessed for high-speed applications. In these materials, the carrier lifetimes are strongly influenced by the number of layers. While mono- to few-layers exhibit lifetimes that are on the order of a few picoseconds [45], bulk crystals show carrier lifetimes exceeding the hundreds of *ps* [46, 47]. However, although monolayers can offer faster switching speeds, which is positive from an applications perspective, they absorb only a small portion of the incident optical excitation and consequently have smaller conductivity modulations. In general, there is a trade-off between carrier lifetime (thus device-speed) and modulation-depth.

In this context, as a result of a smaller carrier effective mass than MoS_2, and a very high absorption coefficient of ~10^5 cm^{-1} at 800 nm [48, 49], that is at least an order of magnitude larger than other semiconductors such as silicon (Si), gallium arsenide (GaAs) and MoS_2 [50], WSe_2 is an exciting platform for terahertz optically reconfigurable terahertz devices. From this perspective, recent work by Gopalan et al. [51] focused on studying the ultrafast terahertz dynamics of a series of custom-growth WSe_2 thin films. The main objective of this study was to elucidate on the trade-offs between film thickness and grain domains with carrier-lifetime and modulation depths. Therefore, bridge the materials aspects with the device practical performance metrics. Depicted in Figure 7.18a are optical-pump/THz-probe (OPTP) measurements for samples with different grain sizes [51]. Results on these samples indicate significant differences in the recombination lifetimes when altering the grain dimensions. All samples exhibit two decays: a fast one and a slow one. In general, the smaller the grains, the faster the fast decay components. However, the smaller the grains, the smaller the modulation on the transmitted terahertz field. These observations indicate that the size of the grains relative to the probe length plays a significant role in the induced terahertz photoconductivity and its dynamics. Smaller grains have a faster response but a smaller value of induced photoconductivity (at the same optical fluence). This effect is schematically illustrated in Figure 7.18b. When analyzing the observed trade-off between photoconductivity and carrier lifetime with grain-size, an optimum window exists between the two, which can enable high-speed terahertz modulators with simultaneous large modulation depth.

7.3 Applications

As a result of its transparency in nonmetallic materials, such as plastics, ceramics, polymers, and so on, where other types of radiation might not propagate, terahertz waves are

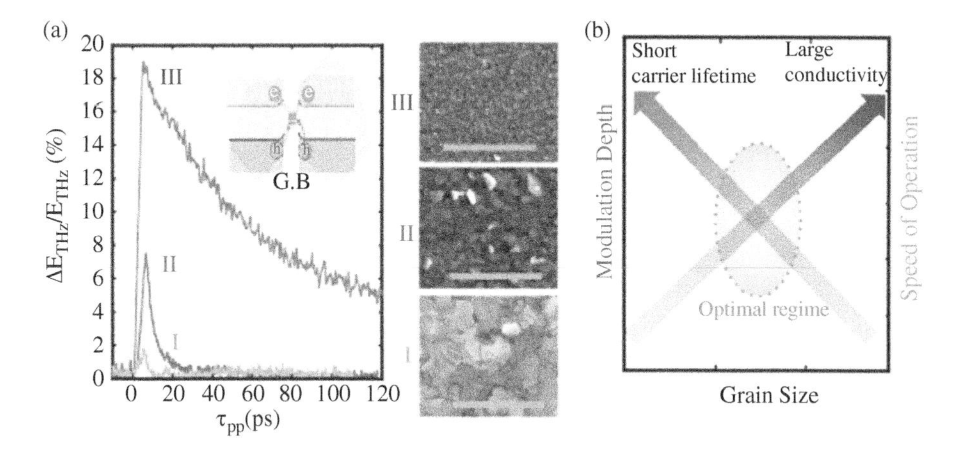

Figure 7.18 Ultrafast dynamics in WSe_2 thin films. (a) OPTP measurements of different grain sized samples show significant differences in recombination lifetimes. This indicates that the size of the grains (relative to the probe length) plays a significant role in the induced THz photoconductivity. Smaller grains have a faster response but a smaller value of induced photoconductivity at the same optical fluence. The schematic depicted in the inset illustrates the role of grain boundary as recombination centers. The scale bar in all the SEM images is 2 μm. (b) Schematic illustration showing the observed trade-off between photoconductivity and carrier lifetime with grain-size. An optimum window exists between the two, which can enable high-speed terahertz modulators with simultaneous large modulation depth. *Source:* Figure extracted from Ref. [51].

appealing for imaging in several applications including security, integrated circuit counterfeit detection, dentistry, etc. From this perspective, it is not surprising that among the most demanded terahertz systems are compact, high-performance, low-cost imagers operating at room temperature.

Several systems and methods have been proposed for terahertz imaging (e.g. [52–54]). However, the discussion in this section will be focused on imaging systems where a terahertz source is used to illuminate the object under detection and the role of 2D materials-based devices as the active element in such imagers. Although scanning systems, which employ a single source and a single detector and scan either the object or the detector to obtain an image, are cost-effective and straightforward to implement, the image acquisition rates in such systems are very low because of the reliance on mechanical stages and sequential acquisition of pixels. On the other hand, systems employing arrays of sources or detectors can achieve real-time operation (e.g. the focal-plane detector arrays used in RF and IR/Vis systems where high-performance and cost-effective components are available). However, in the terahertz range, these systems are highly complex and expensive.

Furthermore, the performance of detectors used in focal plane arrays is generally compromised in comparison to what is possible on a single optimized detector. Single-detector imaging systems are particularly attractive since they can benefit from the superior detection sensitivity of a single detector as well as the spatial coherence of a single-point source

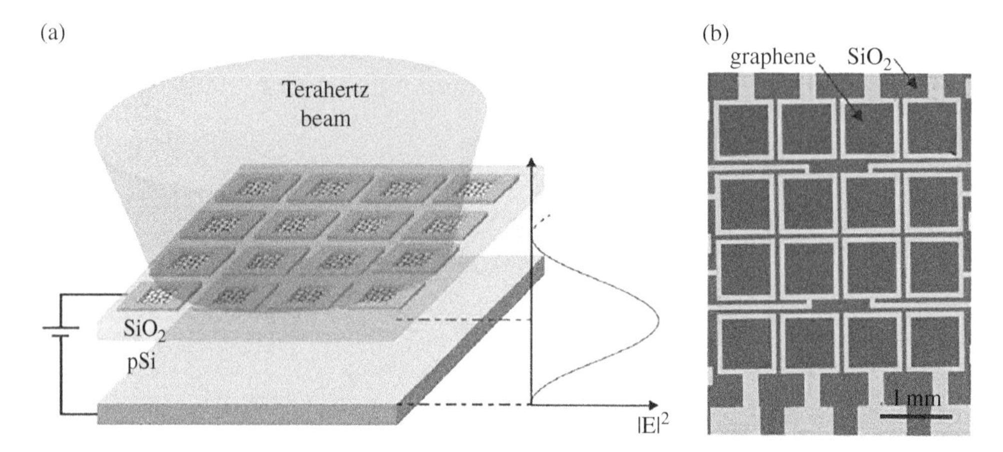

Figure 7.19 (a) Schematic and (b) optical image of a graphene reflection-mode modulator array used as active element in a THz imager. The pixel size, i.e. size of each array-element, is $0.7 \times 0.7\ mm^2$. As shown in (a) the electric field intensity at the active graphene is enhanced by four times when the substrate optical thickness is an odd multiple of the quarter wavelength of the incoming THz radiation, thus leading to augmented modulation in reflectance in comparison to the transmission mode. *Source:* Figure extracted from Ref. [56].

[54, 55]. Because of the ease of fabrication, high switching speed, and relatively low manufacturing cost, arrays of 2D material-based terahertz modulators constitute an ideal platform for real-time terahertz single-detector systems. This section discusses the application of graphene-based modulator arrays in terahertz imaging [56].

Following the fabrication process and device structure discussed in Section 7.1.3. (*electromagnetic-cavity integrated structures*) a prototype device was constructed on a SiO_2/p-Si substrate [56]. The transferred large-area single-layer CVD graphene was patterned into an array of 16 graphene squares with an area of $0.8 \times 0.8\ mm^2$ per square and a 100 μm separation between squares. Etch of graphene was performed through reactive ion etching (RIE) employing O_2 as the etchant gas. A continuous metal layer (Ti/Au) was deposited into the back of the substrate to act both as a reflector and electrode following the previous discussions. Each pixel (graphene square) can be independently biased via top ring metal contacts (Ti/Au). A schematic of the device structure and an optical image of the fabricated device are presented in Figure 7.19.

Imaging experiments were carried out employing the setup depicted in Figure 7.20 [56]. The terahertz frequency at which imaging experiments were performed was chosen so that the substrate thickness is an odd multiple of a quarter wavelength; thus, modulation is maxima [56]. The device was placed a small distance away from the focal point, so that contribution to modulation from all the pixels was observed. Since the focal length (in the order of ~20 cm) is much larger than both: the dimensions of the device (~ 3.5 mm × 3.5 mm) and the minimum beam waist (< 1 mm), the beam divergence produced by defocusing is not significant [56]. In this case, as pictured in Figure 7.20, each pixel contributes to a different part of the reflected collimated terahertz beam; therefore, modulation

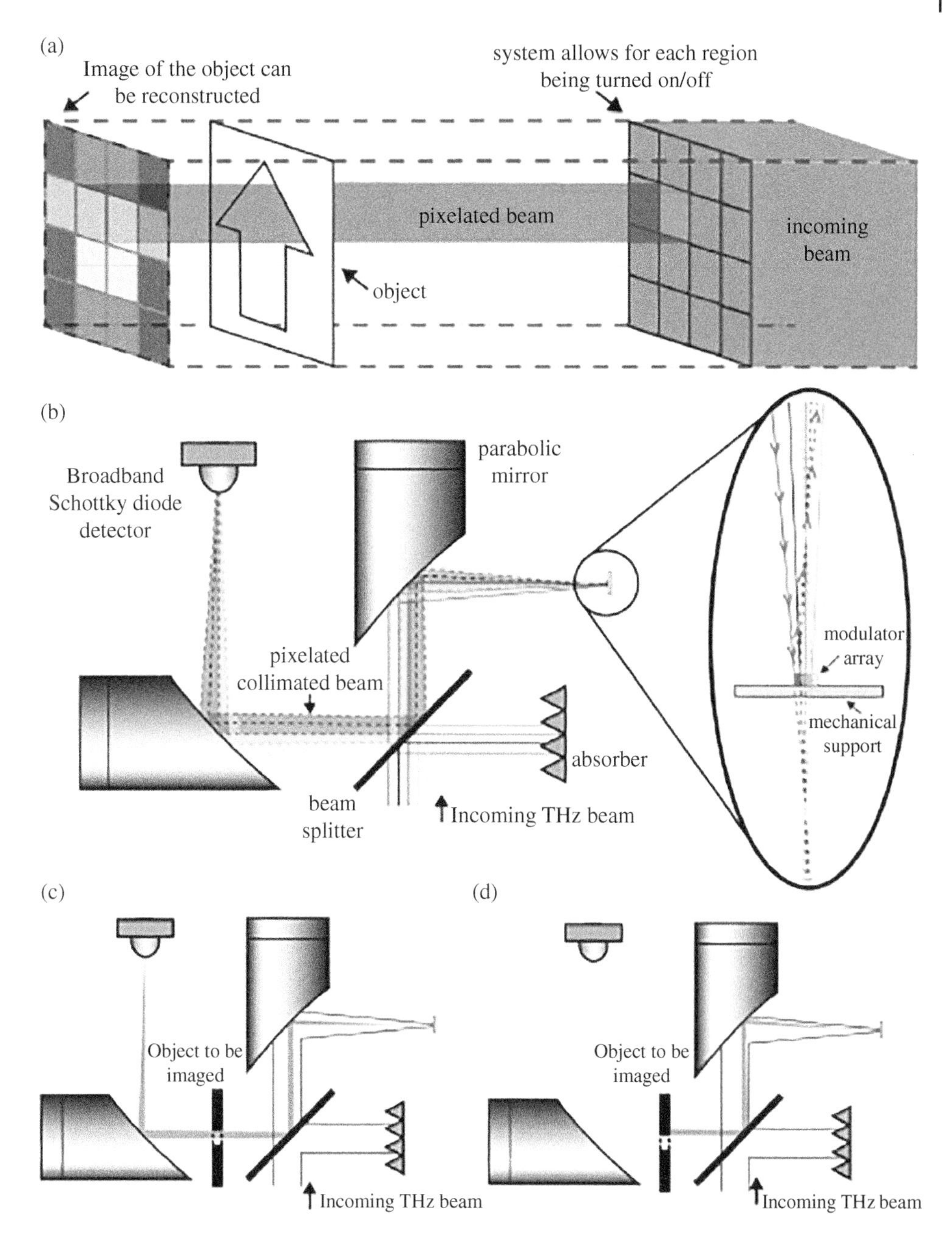

Figure 7.20 Principle of the imaging experiment using graphene-modulator arrays. (a) Sketch of an object in a pixelated and collimated illumination that is transformed by the modulator array. Each pixel of the illumination, denoted by different colors, can be turned on and off by the modulator array. (b) Sketch of the beam path showing the expanded beam is about the same size with the modulator array placed away from the focal plane so that the beam toward the detector is pixelated. No object is placed in the path of the beam. (c, d) Two pixels of illumination when an object is placed in the path of the beam. *Source*: Figure extracted from Ref. [56].

of the spatial intensity distribution of this collimated beam can take place when controlling the reflectance of each pixel by applying a voltage into each pixel in the array.

Objects of different shapes, made of terahertz absorber, were placed in the region where the beam is collimated as indicated in Figure 7.20. The operation mechanism of the imaging setup can be understood as follows [56]:

i) With *no-object* placed in the path of the beam, each element of the array is set to *low reflectance* while leaving the other ones in the low reflective state; the readings of the detector are recorded and stored in a matrix (A_1).
ii) With *no-object* placed in the path of the beam, each element of the array is set *to high reflectance* while leaving the other ones in a low reflective state; the readings of the detector are recorded and stored in a matrix (A_2).
iii) The matrices A_2 and A_1 are subtracted, and the result is stored in a new matrix $A = A_2 - A_1$.
iv) With *an object placed* in the path of the beam, each element of the array is set to *low reflectance* while leaving the other ones in a low reflective state; the readings of the detector are recorded and stored in a matrix (B_1).
v) With *an object placed* in the path of the beam each element of the array is set to *high reflectance* while leaving the other ones in a low reflective state; the readings of the detector are recorded and stored in a matrix (B_2).
vi) The matrices B_2 and B_1 are subtracted, and the result is stored in a new matrix $B = B_2 - B_1$.
vii) A new matrix (C), is computed where $C\ (i, j) = B\ (i, j)/A\ (i, j)$.

The larger $C\ (i, j)$ the larger the terahertz transmittance through the spatial region of the collimated beam controlled by the (i, j) element of the array. Therefore, matrix C constitutes "a map of the spatial terahertz transmittance" through an object.

Several examples of reconstructed images (spatial maps of $C\ (i, j)$) are depicted in Figure 7.21 for objects of different shapes; sketches of the objects are shown below each map. The reconstructed maps closely resemble the shape of the objects. The system spatial resolution, determined by the "pixel" dimension of the pixelated collimated beam, was reported to be ~1.1 cm. The image acquisition rate was limited by the switching speed of each element of the array, which was found to be ~6 kHz [56]. This speed is enough for video-rate imaging for systems with up to 16×16 pixels (i.e. a 16×16 pixel array might image at ~24 frames per second). It is worth mentioning that the detector readings were taken employing lock-in measurements, which allows for large dynamic ranges even though the modulation achievable by each pixel alone was found to be on the order of ~3–10% only [56]. Furthermore, the dynamic range of the system was found to be ~26 dB [56]. This imaging technique exhibits a trade-off between dynamic range and spatial resolution. The larger the number of pixels, the smaller the area of the collimated beam controlled by an individual element of the array, therefore, the smaller the system dynamic range. However, this experimental demonstration of an imaging system based on graphene modulators arrays shows the promise of 2D material-based devices in practical terahertz applications.

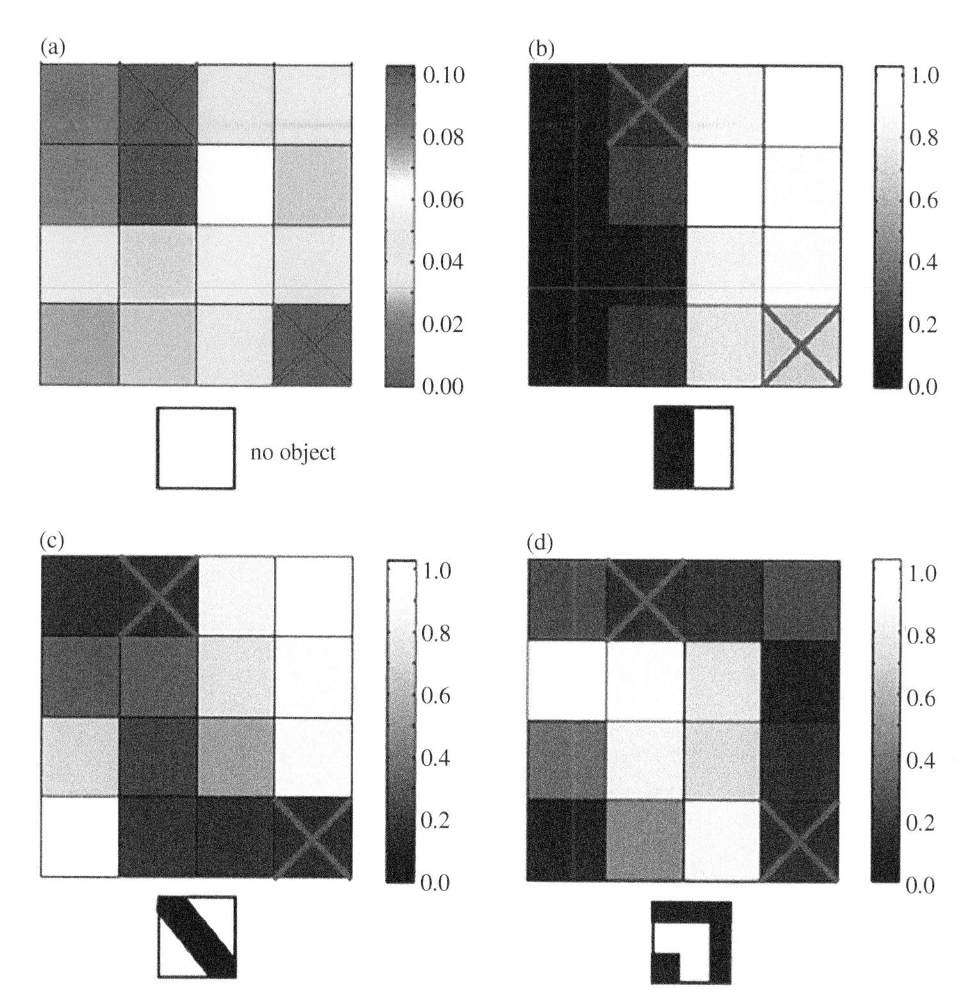

Figure 7.21 (a) Map of "pixelated illumination" without an object: $\Delta R_{i0}/R_{0,low}$, where $R_{0,low}$ is the measured total reflectance when all the pixels are off, i.e. the minimum reflectance of the modulator array. (b, d) Map of $\Delta R_i/\Delta R_{i0}$ for three different objects placed in the path of the reflected beam. ΔR_i is the detected difference when switching the ith pixel on and off while keeping all other pixels off. The two red crossed pixels crossed did not show modulation due to fabrication issues. The sketches of the objects made from the absorber material are shown below each map. The close resemblance between the map and the object indicates the graphene modulator array can be used for imaging. *Source*: Figure extracted from Ref. [56].

Exercises

E7.1 Computation of the Optical Conductivity of Graphene

Plot the real part of the optical conductivity of graphene, from the optical to the terahertz range assuming: $\tau = 0.01$, 0.1, and 1 ps, $E_f = 100$, 200, and 500 meV, and $T = 300$ K. Discuss the effect of altering τ and E_f in this response.

E7.2 Terahertz Transmission Through a 2D Material Layer Placed at an Optical Interface

Assume that a 2D material layer, with zero-frequency conductivity σ_{2D} (and no frequency dispersion), is (a) suspended and (b) placed at the interface between air and a semi-infinite lossless silicon substrate. Assuming the relative permittivity of silicon to be 11.9, plot the transmittance through the structure as a function of σ_0 in the 10 μS to 10 mS range.

E7.3 Transfer Matrix Approach for Multi-layer Transmission Problems

Employing the transfer-matrix formalism derive an expression for the transmission and reflection through a structure consisting of a layer of graphene placed on top of a lossless substrate of finite thickness. (a) Assuming the substrate to be lossless silicon (thickness $t = 300\ \mu m$) and the graphene layer to have $\sigma_0 = 1\ mS$ and $\tau = 0$ ps, plot transmission, reflection, and absorption as a function of frequency. (b) if a metal reflector is placed at the back of the substrate, calculate and plot reflection and absorption as a function of frequency.

E7.4 A Condition for Perfect Absorption

Demonstrate that perfect absorption is possible in a reflection mode geometry consisting of graphene on top of a quarter-wavelength thick substrate that is coated with metal on the back (as depicted in Figure 7.5a).

E7.5 Terahertz Plasmon Resonances in Periodically Patterned Graphene Disk Arrays

A layer of graphene placed at the interface between air and a lossless silicon substrate is patterned into disks (with radius d). Assuming this pattern to be periodic, employing Maxwell-Garnett effective medium theory, determine an analytical expression for the transmittance through the structure. Assuming the following Drude parameters for graphene: $\sigma_0 = 2\ mS$ and $\tau = 0.5$ ps, a filling factor of 0.5, and a disk radius of 5 μm, plot transmittance versus frequency. Analyze how does (a) changing the disk radius, and (b) changing τ (keeping the charge density in graphene constant), affect the transmittance spectra.

E7.6 Electron Plasma Waves in Gated Graphene

In this example, we will analyze the properties of electron plasma waves in gated-graphene. The geometry to be analyzed in this problem is depicted in Figure 7.26 and represents a graphene field effect transistor. The graphene layer is spatially located above a metal electrode (gate). The gate and the graphene layer are separated by a dielectric (thickness t). The metallic gate and the graphene layer are capacitively coupled so that the gate electrode can control the charge density in the graphene layer through field effect. We will also assume that the device is biased with a very small drain to source voltage so that the charge density could be considered as uniform in the graphene layer.

a) *Assuming the gate dielectric not being leaky, find distributed parameters, i.e., R [Ω/m], L [H/m], and C [F/m], so to model the device as a lossy transmission line. Write equations for these as a function of the charge density in graphene, Fermi level, and Fermi velocity as well as on the materials parameters and geometric dimensions depicted in the figure {e.g. ε, t, W}.*

b) *Derive the condition that should be satisfied so that the electron plasma wave propagation could be considered as low-loss. Notice that this condition sets a lower bound for the frequencies at which electron plasma waves can be excited, i.e. this is a very high-frequency phenomenon that is dictated by τ (typically fs to ps scale).*

c) *Find the phase velocity for these waves.*

d) *Find an equation for the attenuation constant for these waves.*

e) *Find an equation for the quality factor (Q) of the plasmonic resonances.*

E7.7 Equivalent Circuit Modeling of 2D Material-Loaded Frequency Selective Surfaces

The FSS is depicted in Figures 7.7a and 7.16a can be modeled through the R, L, C equivalent circuit shown in Figure 7.9a, where the values of the parameters are given. (a) Plot transmission versus frequency for this structure. (b) Assuming that a layer of graphene is placed in the proximity of the FSS and that near field enhancement leads to an extracted effective conductivity enhancement factor $\alpha 1 = 4$, plot transmission versus frequency in the 0.1–2 THz range for the following graphene conductivity levels: $\sigma = 0.3$, 1, and 2 mS. Assume no frequency dispersion for the optical conductivity of graphene.

E7.8 Maximum Terahertz Absorption in 2D Material-Loaded Frequency Selective Surfaces

Demonstrate that the maximum possible absorption in a 2D material-loaded FSS is fundamentally limited to < 50% of the incoming terahertz intensity. Assume the metal constituting the FSS to be a perfect electric conductor, i.e. $R = 0$.

References

1 Tomaino, J.L., Jameson, A.D., Kevek, J.W. et al. (2011). Terahertz imaging and spectroscopy of large-area single-layer graphene. *Optics Express* **19** (1): 141–146.

2 Horng, J., Chen, C.F., Geng, B. et al. (2011). Drude conductivity of dirac fermions in graphene. *Physical Review B* **83** (16): 165113.

3 Docherty, C.J. and Johnston, M.B. (2012). Terahertz properties of graphene. *Journal of Infrared, Millimeter, and Terahertz Waves* **33** (8): 797–815.

4 Bolotin, K.I., Sikes, K.J., Jiang, Z. et al. (2008). Ultrahigh electron mobility in suspended graphene. *Solid State Communications* **146** (9–10): 351–355.

5 Li, X., Cai, W., An, J. et al. (2009). Large-area synthesis of high-quality and uniform graphene films on copper foils. *Science* **324** (5932): 1312–1314.

6 Muñoz, R. and Gómez-Aleixandre, C. (2013). Review of CVD synthesis of graphene. *Chemical Vapor Deposition* **19**: 297–322.

7 Sensale-Rodriguez, B., Yan, R., Liu, L. et al. (2013). Graphene for reconfigurable terahertz optoelectronics. *Proceedings of the IEEE* **101** (7): 1705–1716.

8 Falkovsky, L.A. (2008). Optical properties of graphene. *Journal of Physics: Conference Series* **129** (1): 012004.

9 Falkovsky, L.A. and Pershoguba, S.S. (2007). Optical far-infrared properties of a graphene monolayer and multilayer. *Physical Review B* **76** (15): 153410.

10 https://www.ansys.com/products/electronics/ansys-hfss

11 Yan, R., Arezoomandan, S., Sensale-Rodriguez, B., and Xing, H.G. (2016). Exceptional terahertz wave modulation in graphene enhanced by frequency selective surfaces. *ACS Photonics* **3** (3): 315–323.

12 Yan, R., Sensale-Rodriguez, B., Liu, L. et al. (2012). A new class of electrically tunable metamaterial terahertz modulators. *Optics Express* **20** (27): 28664–28671.

13 Sensale-Rodriguez, B., Fang, T., Yan, R. et al. (2011). Unique prospects for graphene-based terahertz modulators. *Applied Physics Letters* **99** (11): 113104.

14 Saleh, B.E., Teich, M.C., and Saleh, B.E. (1991). *Fundamentals of Photonics*, vol. **22**. New York: Wiley.

15 Rana, F. (2008). Graphene terahertz plasmon oscillators. *IEEE Transactions on Nanotechnology* **7** (1): 91–99.

16 Ju, L., Geng, B., Horng, J. et al. (2011). Graphene plasmonics for tunable terahertz metamaterials. *Nature Nanotechnology* **6** (10): 630.

17 Vicarelli, L., Vitiello, M.S., Coquillat, D. et al. (2012). Graphene field-effect transistors as room-temperature terahertz detectors. *Nature Materials* **11** (10): 865.

18 Otsuji, T., Tombet, S.B., Satou, A. et al. (2012). Graphene-based devices in terahertz science and technology. *Journal of Physics D: Applied Physics* **45** (30): 303001.

19 Lee, S.H., Choi, M., Kim, T.T. et al. (2012). Switching terahertz waves with gate-controlled active graphene metamaterials. *Nature Materials* **11** (11): 936.

20 Tamagnone, M., Gomez-Diaz, J.S., Mosig, J.R., and Perruisseau-Carrier, J. (2012). Reconfigurable terahertz plasmonic antenna concept using a graphene stack. *Applied Physics Letters* **101** (21): 214102.

21 Yang, K., Arezoomandan, S., and Sensale-Rodriguez, B. (2013). The linear and nonlinear THz properties of graphene. *International Journal of Terahertz Science and Technology* **6** (4): 223–233.

22 Tassin, P., Koschny, T., and Soukoulis, C.M. (2013). Graphene for terahertz applications. *Science* **341** (6146): 620–621.

23 Tredicucci, A. and Vitiello, M.S. (2014). Device concepts for graphene-based terahertz photonics. *IEEE Journal of Selected Topics in Quantum Electronics* **20** (1): 130–138.

24 Sensale-Rodriguez, B., Yan, R., Kelly, M.M. et al. (2012). Broadband graphene terahertz modulators enabled by intraband transitions. *Nature Communications* **3**: 780.

25 Li, X., Zhu, Y., Cai, W. et al. (2009). Transfer of large-area graphene films for high-performance transparent conductive electrodes. *Nano Letters* **9** (12): 4359–4363.

26 Sensale-Rodriguez, B., Yan, R., Rafique, S. et al. (2012). Extraordinary control of terahertz beam reflectance in graphene electro-absorption modulators. *Nano Letters* **12** (9): 4518–4522.

27 Sensale-Rodriguez, B., Yan, R., Rafique, S. et al. (2012). Exceptional tunability of THz reflectance in graphene structures. In: *2012 37th International Conference on Infrared, Millimeter, and Terahertz Waves*, 1–3. IEEE.

28 Gao, W., Shu, J., Reichel, K. et al. (2014). High-contrast terahertz wave modulation by gated graphene enhanced by extraordinary transmission through ring apertures. *Nano Letters* **14** (3): 1242–1248.
29 Degl'Innocenti, R., Jessop, D.S., Shah, Y.D. et al. (2014). Low-bias terahertz amplitude modulator based on split-ring resonators and graphene. *ACS Nano* **8** (3): 2548–2554.
30 Tamagnone, M., Fallahi, A., Mosig, J.R., and Perruisseau-Carrier, J. (2014). Fundamental limits and near-optimal design of graphene modulators and non-reciprocal devices. *Nature Photonics* **8** (7): 556.
31 Shin, D.W., Lee, J.H., Kim, Y.H. et al. (2009). A role of HNO3 on transparent conducting film with single-walled carbon nanotubes. *Nanotechnology* **20** (47): 475703.
32 Arezoomandan, S., Quispe, H.C., Chanana, A. et al. (2018). Graphene–dielectric integrated terahertz metasurfaces. *Semiconductor Science and Technology* **33** (10): 104007.
33 Liu, X., Fan, K., Shadrivov, I.V., and Padilla, W.J. (2017). Experimental realization of a terahertz all-dielectric metasurface absorber. *Optics Express* **25** (1): 191–201.
34 Cardin, A., Fan, K., and Padilla, W. (2018). Role of loss in all-dielectric metasurfaces. *Optics Express* **26** (13): 17669–17679.
35 Arezoomandan, S., Quispe, H.O.C., Ramey, N. et al. (2017). Graphene-based reconfigurable terahertz plasmonics and metamaterials. *Carbon* **112**: 177–184.
36 Yang, K., Liu, S., Arezoomandan, S. et al. (2014). Graphene-based tunable metamaterial terahertz filters. *Applied Physics Letters* **105** (9): 093105.
37 Arezoomandan, S., Yang, K., and Sensale-Rodriguez, B. (2014). Graphene-based electrically reconfigurable deep-subwavelength metamaterials for active control of THz light propagation. *Applied Physics A* **117** (2): 423–426.
38 Chen, H.T., Padilla, W.J., Cich, M.J. et al. (2009). A metamaterial solid-state terahertz phase modulator. *Nature Photonics* **3** (3): 148.
39 Arezoomandan, S. and Sensale-Rodriguez, B. (2015). Geometrical tradeoffs in graphene-based deeply-scaled electrically reconfigurable metasurfaces. *Scientific Reports* **5**: 8834.
40 Zheng, W., Fan, F., Chen, M. et al. (2016). Optically pumped terahertz wave modulation in MoS_2-Si heterostructure metasurface. *AIP Advances* **6** (7): 075105.
41 Cao, Y., Gan, S., Geng, Z. et al. (2016). Optically tuned terahertz modulator based on annealed multilayer MoS_2. *Scientific Reports* **6**: 22899.
42 Fan, Z., Geng, Z., Lv, X. et al. (2017). Optical controlled terahertz modulator based on tungsten disulfide nanosheet. *Scientific Reports* **7** (1): 14828.
43 Podzorov, V., Gershenson, M.E., Kloc, C. et al. (2004). High-mobility field-effect transistors based on transition metal dichalcogenides. *Applied Physics Letters* **84** (17): 3301–3303.
44 Arezoomandan, S., Gopalan, P., Tian, K. et al. (2017). Tunable terahertz metamaterials employing layered 2-D materials beyond graphene. *IEEE Journal of Selected Topics in Quantum Electronics* **23** (1): 188–194.
45 Docherty, C.J., Parkinson, P., Joyce, H.J. et al. (2014). Ultrafast transient terahertz conductivity of monolayer MoS_2 and WSe_2 grown by chemical vapor deposition. *ACS Nano* **8** (11): 11147–11153.
46 Strait, J.H., Nene, P., and Rana, F. (2014). High intrinsic mobility and ultrafast carrier dynamics in multilayer metal-dichalcogenide MoS_2. *Physical Review B* **90** (24): 245402.

47 He, C., Zhu, L., Zhao, Q. et al. (2018). Competition between free carriers and excitons mediated by defects observed in layered WSe_2 crystal with time-resolved terahertz spectroscopy. *Advanced Optical Materials* **6** (19): 1800290.

48 Frindt, R.F. (1963). The optical properties of single crystals of WSe_2 and $MoTe_2$. *Journal of Physics and Chemistry of Solids* **24** (9): 1107–1108.

49 Beal, A.R., Liang, W.Y., and Hughes, H.P. (1976). Kramers-Kronig analysis of the reflectivity spectra of 3R-WS_2 and 2H-WSe_2. *Journal of Physics C: Solid State Physics* **9** (12): 2449.

50 Roxlo, C.B., Chianelli, R.R., Deckman, H.W. et al. (1987). Bulk and surface optical absorption in molybdenum disulfide. *Journal of Vacuum Science & Technology A: Vacuum, Surfaces, and Films* **5** (4): 555–557.

51 Gopalan, P., Chanana, A., Krishnamoorthy, S. et al. (2019). Ultrafast THz modulators with WSe_2 thin films. *Optical Materials Express* **9** (2): 826–836.

52 Hu, B.B. and Nuss, M.C. (1995). Imaging with terahertz waves. *Optics Letters* **20** (16): 1716–1718.

53 Zimdars, D. (2005). High speed terahertz reflection imaging. In: *Advanced Biomedical and Clinical Diagnostic Systems III*, vol. **5692**, 255–260. International Society for Optics and Photonics.

54 Xu, J. and Zhang, X.C. (2006). Terahertz wave reciprocal imaging. *Applied Physics Letters* **88** (15): 151107.

55 Chan, W.L., Moravec, M.L., Baraniuk, R.G., and Mittleman, D.M. (2008). Terahertz imaging with compressed sensing and phase retrieval. *Optics Letters* **33** (9): 974–976.

56 Sensale-Rodriguez, B., Rafique, S., Yan, R. et al. (2013). Terahertz imaging employing graphene modulator arrays. *Optics Express* **21** (2): 2324–2330.

57 Allen, S.J. Jr., Störmer, H.L., and Hwang, J.C.M. (1983). Dimensional resonance of the two-dimensional electron gas in selectively doped GaAs/AlGaAs heterostructures. *Physical Review B* **28** (8): 4875.

8

THz Plasma Field Effect Transistor Detectors

Naznin Akter[1], Nezih Pala[1], Wojciech Knap[2] and Michael Shur[3]

[1] *Department of Electrical and Computer Engineering, Florida International University, Miami, FL, USA*
[2] *CENTERA Laboratories, IHPP-Polish Academy of Sciences, Warsaw, Poland and Laboratoire Charles Coulomb, CNRS & Université de Montpellier, Montpellier, France*
[3] *Electrical, Computer, and Systems Engineering Department and Physics, Applied Physics and Astronomy, Rensselaer Polytechnic Institute, New York, USA*

8.1 Introduction

The terahertz (10^{12} Hz) region of electromagnetic spectrum covers the frequency range from roughly 100 GHz to 10 THz. Potential applications for terahertz technology in communication, imaging, and sensing are wide ranging. THz wavelengths have several properties that could promote their use as sensing and imaging tool. There is no ionization hazard for biological tissue, and Rayleigh scattering of electromagnetic radiation is many orders of magnitude less for THz wavelengths than for the neighboring infrared and optical regions of the spectrum [1]. THz radiation can also penetrate nonmetallic materials, such as fabric, leather, and plastic, which makes it useful in security screening for concealed weapons and hardware cybersecurity applications. The THz frequencies correspond to energy levels of molecular rotations and vibrations of DNA [2] and proteins [3], as well as other organic molecules and explosives [4], and these may provide characteristic fingerprints to differentiate biological tissues in a region of the spectrum not previously explored for medical use or detect and identify trace amounts of explosives. THz wavelengths are particularly sensitive to water and exhibit absorption peaks, which makes the THz technique very sensitive to hydration state [5] and can indicate tissue condition. THz radiation has also been used in the characterization of semiconductor materials [5–7], and in testing and failure analysis of VLSI circuits [8, 9]. The emerging 5G and beyond (6G) WiFi technology with its promise of increasing the network speed by up to 100 and reducing latency by a factor of 20 will enable a plethora of applications including but not limited to virtual reality sufficient for remote surgeries as well as intense high resolution video communications and connecting a much larger number of devices, such as sensors for Internet of Things than conventional 4G WiFi technology [10]. It has been argued that 0.1–10 THz band is the most advantageous

Fundamentals of Terahertz Devices and Applications, First Edition. Edited by Dimitris Pavlidis.

Companion website: www.wiley.com/go/Pavlidis/FundamentalsofTHz

one to achieve Terabit per second (Tbps) data rates compared to the neighboring frequency bands particularly for compact transceivers [11]. Millimeter wave and photonics communities have already started to explore the next frontier moving from 28 GHz 5G communications to 240–320 GHz band with a window for THz absorption in atmosphere with the first applications for the transmission from a fiber to a curb [12]. The 500–600 GHz atmospheric window is also of interest. However, it is believed that, with the increasing demand, upper region of the THz band (1–10 THz) will soon be allocated especially for indoor wireless communication links [13]. Indeed, the need for a higher communication frequency and a larger bandwidth addressed by moving to beyond 5G is clear from the expected data travel over the next few years projected by CISCO systems to go from 22 Exabyte in 2019 to nearly 50 in 2021 [14, 15]. The available bandwidth at THz frequencies, which is orders of magnitude larger compared to conventional microwave frequencies, makes them uniquely suited for Wireless Local Area Networks (WLAN) and Wireless Personal Area Networks (WPAN) applications with both Line-of-Sight (LoS) and Non-LoS settings. Moreover, spread spectrum techniques enabling low power directional networking making THz frequencies to be attractive for next-generation high throughput covert networks with resistance to eavesdropping and anti-jamming. THz spectrum has the potential of making Tbps and communication at distances ranging from thousands of miles in space to submicron on chips. Along with enormous opportunities, there exist many challenges both in the device and the communication sides that require innovative solutions [11, 16]. Particularly, compact and high output power sources and high-responsivity/low-noise detectors that can operate at THz band are needed to overcome the high path-loss at these frequencies. Such devices not only revolutionize our lives through beyond-5G wireless communication with Tbps data rates but also help bridging infamous "THz-gap" effectively and make other THz technologies such as THz security, medical imaging as well as high-resolution automotive radar for autonomous vehicles possible. A post-COVID19 pandemic world will need THz communication technology because of enormous increase of online applications.

8.2 Field Effect Transistors (FETs) and THz Plasma Oscillations

The idea of using field effect transistors (FETs) for THz applications was first proposed by Dyakonov and Shur in the early 1990s [17]. Their theoretical work predicted that a steady current flow in an asymmetric FET channel can lead to instabilities against spontaneous generation of plasma waves, which are the oscillations of electron density in space and time. These instabilities which are now called Dyakonov-Shur (DS) instabilities can, in turn, produce the emission of electromagnetic radiation at the plasma wave frequency. Later, they showed that the nonlinear properties of the 2D plasma in FET channels can be used for detection and mixing of THz radiation [18]. FETs operating in the THz range (called TeraFETs) are already competitive for THz detection or mixing applications and there is an intense interest and research to apply the TeraFET arrays for compact and efficient THz electronic sources.

8.2.1 Dispersion of Plasma Waves in FETs

Plasma waves—oscillations of the electron density in time and space—have properties that depend on the electron density and on the dimensions and geomctry of the electronic system. For a three-dimensional (3D) case, the plasma oscillation frequency is nearly independent of the wavelength, whereas in a gated two-dimensional electron gas (2DEG), the plasma waves have a linear dispersion law similar to that light in vacuum or sound waves or shallow water waves. In this case, the plasma wave velocity, s, is proportional to the square root of the electron sheet density, which can be easily tuned by the gate bias.

We will start from a general discussion of plasma waves in low dimensional structures including an FET channel, and of the boundary conditions that determine the frequencies of the plasma modes. This discussion will be followed by the analysis of the THz detection in different modalities.

The dispersion relations for plasma waves in the systems of different dimensions can be derived in a simple manner by neglecting collisions and considering only the average drift velocity, $\boldsymbol{v}$. In this case, neglecting the electron scattering, the small-signal equation of motion and the continuity equation are

$$\frac{\partial \boldsymbol{j}}{\partial t} = \boldsymbol{E}\frac{e^2 n}{m} \tag{8.1}$$

$$\frac{\partial \rho}{\partial t} + \nabla \cdot \boldsymbol{j} = 0 \tag{8.2}$$

where $\boldsymbol{j} = en\boldsymbol{v}$ is the current density, e is the electronic charge, n is the electron density, m is the electronic effective mass, and ρ is a small-signal charge density related to a deviation of n from its equilibrium value, $\boldsymbol{E}$ is the small-signal electric field. Differentiating Eq. (8.2) with respect to time and using Eq. (8.1)

$$\frac{\partial^2 \rho}{\partial t^2} + \frac{e^2 n}{m}\nabla \cdot \boldsymbol{E} = 0 \tag{8.3}$$

These equations are valid for any dimensionality of the problem. For the 3D case, the $\boldsymbol{j}$, n, and ρ are current per unit area, electron concentration, and electric charge per unit volume, respectively satisfying the relation

$$\nabla \cdot \mathbf{E} = \rho/\varepsilon \tag{8.4}$$

where ε is the dielectric constant of the material. Substituting Eq. (8.4) into Eq. (8.3) results a harmonic oscillator equation for the charge density with the bulk plasma frequency

$$\omega_p = \sqrt{\frac{e^2 n}{\varepsilon m}} \tag{8.5}$$

For the 2D case which represents the channel of an FET, the electric field vector, $\boldsymbol{E}$, in Eqs. (8.1) and (8.3) should be understood as the in-plane electric field having only two components, E_x and E_y, since the third component of the electric field (which is perpendicular to the plane, xy, of the 2DEG) does not contribute to the in-plane current and, hence, does not enter into Eq. (8.1). Accordingly, in the 2D case, the $\boldsymbol{j}$, n, and ρ are current per unit length, electron concentration per unit area, and the electric charge per unit area, respectively.

For a gated 2DEG (i.e. an FET), the relation between the electron concentration and electric potential is given by

$$\rho = en = CU \tag{8.6}$$

where $U = U_g - U_c - U_T$, U_T is the threshold voltage, $U_g - U_c$ is the potential difference between the gate and the channel, $C = \varepsilon/d$ is the gate-to-channel capacitance per unit area, and d is the gate-to-channel separation. This equation is valid when the gradual channel approximation is valid that is U changes along the channel on the scale large compared to d. Hence, the in-plane electric field is given by

$$\mathbf{E} = -\nabla U = -\frac{\nabla \rho}{C} \tag{8.7}$$

which can be substituted into Eq. (8.3) to obtain

$$\frac{\partial^2 \rho}{\partial t^2} + s^2 \nabla^2 \rho = 0 \tag{8.8}$$

where

$$s = \sqrt{\frac{e^2 nd}{m^* \varepsilon}} \tag{8.9}$$

is the velocity of the surface plasma waves. The solution of Eq. (8.8) corresponds to waves with a linear dispersion law:

$$\omega = sk \tag{8.10}$$

where ω and k are the frequency and the wave vector of the plasma waves, respectively. It was shown in Ref. [17] that the nonlinear hydrodynamic equations describing the electrons in the channel of a FET are exactly the same as the shallow water equations in conventional hydrodynamics. The term "shallow water" refers to a situation when the wavelength, or more generally, the spatial scale of variation of the water level is much greater than the depth h. This is analogous to the assumption that the wavelength is much greater than the gate-to-channel separation d, or $kd \ll 1$. In the opposite case of short wavelengths ($kd \gg 1$) the existence of the gate is of no importance which is equivalent to the "ungated" scheme, and plasma waves have the dispersion relation similar to the "deep water" case:

$$\omega = \sqrt{\frac{e^2 n}{2m^* \varepsilon} k} \tag{8.11}$$

Using the same approach, one can obtain the dispersion law for the plasma waves propagating along a one-dimensional wire:

$$\omega_p = sk \sqrt{ln \frac{1}{kr}} \text{ with } s = \sqrt{\frac{e^2 n}{m^* \varepsilon}} \tag{8.12}$$

where r is the radius of the wire, s is the velocity of the one-dimensional plasma waves.

Plasma wave frequencies and dispersion relations for systems of different dimensions and geometry are schematically shown in Figure 8.1. We note the similarity between the

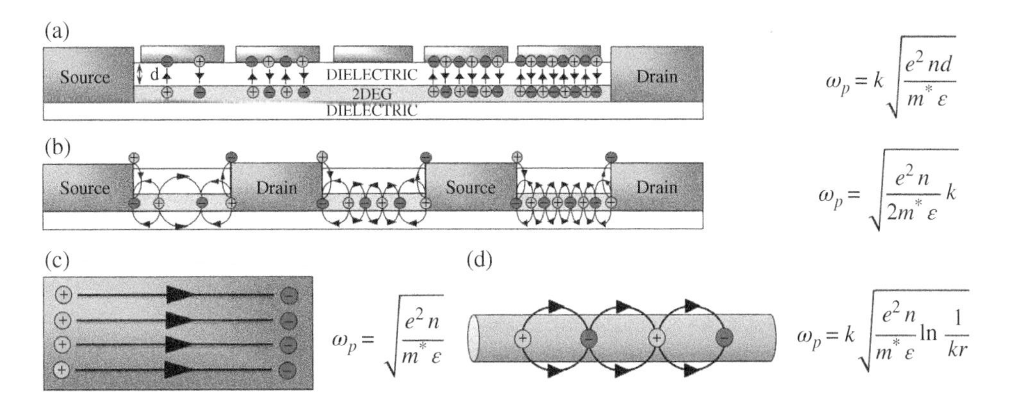

Figure 8.1 Plasma wave frequencies for different sample geometries. (a) Gated 2DEG that corresponds to an FET; (b) ungated 2DEG; (c) 3D bulk; (d) 1D nanowire.

dispersion relations for the gated 2DEG case and ungated 1DEG case. This similarity means that the results for the gated 2DEG should equally apply to ungated quantum wires.

8.2.2 THz Detection by an FET

The idea of using a FET for detection of THz radiation was proposed in Ref. [18]. Such a detection is made possible by the nonlinear properties of the transistor, which rectifies an alternating current (AC) induced by the incoming THz radiation. As a result, a photoresponse appears in the form of a direct current (DC) voltage between source and drain, which is proportional to the radiation power (photovoltaic effect). This mechanism requires an asymmetry between the source and drain to induce a DC voltage. Such an asymmetry can be achieved in different ways. Asymmetric design of the source and drain contact pads or some external (parasitic) capacitances may result in the difference in the source and drain boundary conditions. A special antenna design—that is connected to the one side of the transistor channel can result even a better asymmetry. The asymmetry can be naturally obtained if a DC is passed between source and drain, creating a depletion of the electron density on the drain side of the channel.

Let us consider an FET shown in Figure 8.2 where the incoming radiation creates an AC voltage with amplitude U_{AC} only between the source and the gate and there is no DC

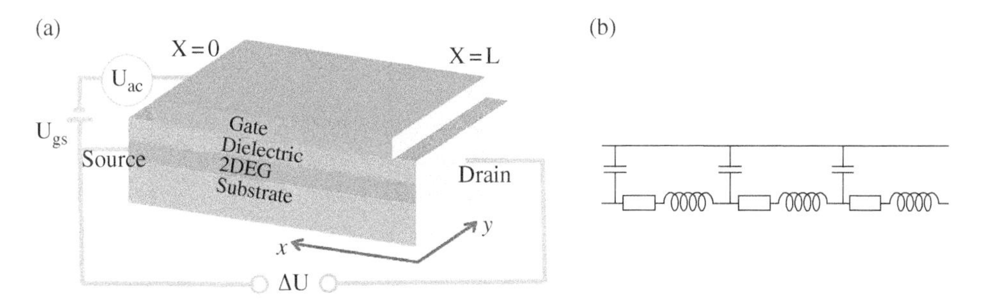

Figure 8.2 Schematics of a FET as a THz detector (a); and the equivalent circuit for the channel (b).

between the source and drain. In such an FET, the channel can be modeled by the distributed gate-to-channel capacitance, distributed inductance, and the channel resistance, which depends on the gate and drain voltage through the electron concentration in the channel. When the scale of the spatial variation of $U(x)$ is larger than the gate-to-channel separation d (the gradual channel approximation) the electron concentration in the channel is given by Eq. (8.6). The distributed inductance represents the so-called kinetic inductance, which is due to the electron inertia and is proportional to m, the electron effective mass. The electron inertia is accounted by the Drude model

$$Z = R(1 + j\omega\tau) = R + j\omega L \tag{8.13}$$

here Z is impedance, R is the resistance, ω is frequency, τ is the momentum relaxation time and $L = R\,\tau$ is the kinetic inductance.

Under static conditions and in the absence of the drain current, $U = U_0 = U_g - U_{th}$, where U_0 is the static voltage swing. Depending on the frequency ω, one can distinguish two regimes of operation, and each of them can be further divided into two sub-regimes depending on the gate length L (see Figure 8.3).

1) ***High-frequency regime*** **($\omega\tau > 1$):** In this case, the electron momentum relaxation time τ determines the conductivity in the channel $\sigma = ne^2\tau/m$, and the kinetic inductances in Figure 8.2 are important. Plasma waves that are analogous to the waves in an RLC

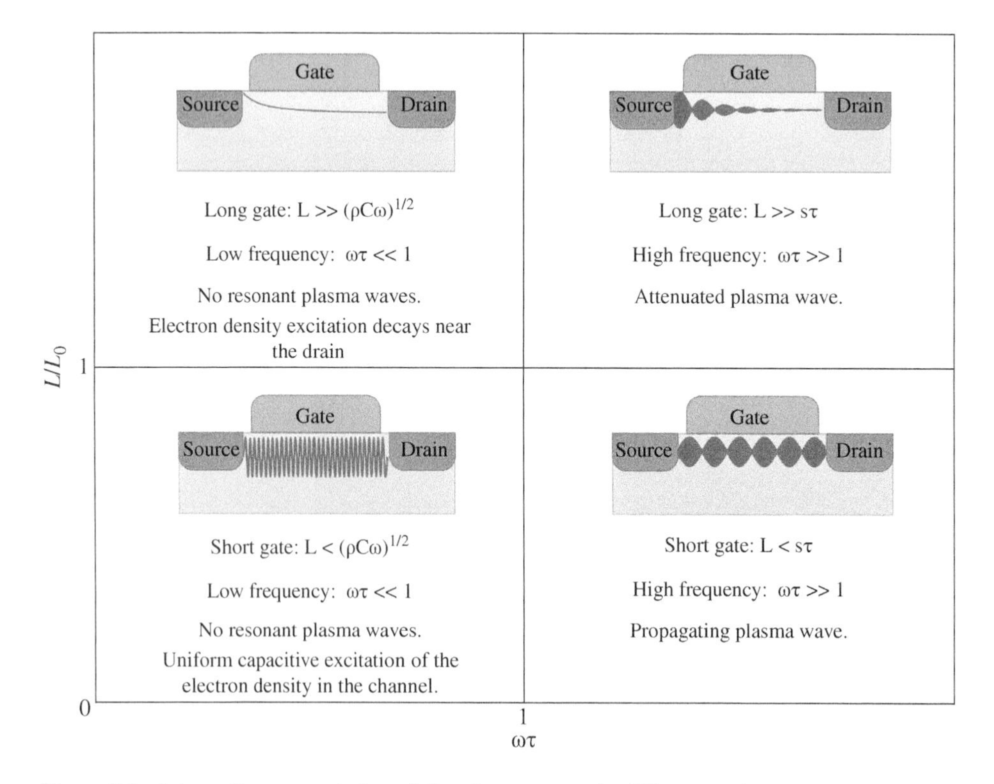

Figure 8.3 Schematic representation of the plasma waves in different regimes.

transmission line are excited. The plasma waves have a velocity $s = (eU_0/m)^{1/2}$ and a damping time τ. Thus, their propagation distance is $s\tau$.

1.*a*. *Short gate case* ($L < s\tau$): The plasma wave reaches the drain side of the channel, gets reflected, and forms a standing wave with enhanced amplitude, so that the channel serves as a high-quality resonator for plasma oscillations. The fundamental mode has the frequency ~ s/L, with a numerical coefficient depending on the boundary conditions.

1.*b*. *Long gate case* ($L \gg s\tau$): The plasma waves excited at the source will decay before reaching the drain, so that the AC will exist only in a small part of the channel adjacent to the source.

2) ***Low-frequency regime* ($\omega\tau \ll 1$):** In this case, the plasma waves cannot exist because of overdamping. At these low frequencies, the inductance in Figure 8.2 becomes simply short-circuits which leads to an RC line. Its properties further depend on the gate length, the relevant parameter being $\omega\tau_{RC}$, where τ_{RC} is the RC time constant of the whole transistor. Since the total channel resistance is $L\rho/W$, and the total capacitance is CWL (where W is the gate width and $\rho = 1/\sigma$ is the channel resistivity), one finds $\tau_{RC} = L^2\rho C$.

2.*a*. *Short gate case* ($L < [\rho C\omega]^{1/2}$): This means that $\omega\tau_{RC} < 1$, so that the AC goes through the gate-to-channel capacitance practically uniformly on the whole length of the gate. This is the so-called "resistive mixer" regime. For the THz frequencies, this regime can apply only for transistors with extremely short gates.

2.*b*. *Long gate case* ($L >> [\rho C\omega]^{1/2}$): Now $\omega\tau_{RC} \gg 1$, and the induced AC will leak to the gate at a small distance/from the source, such that the resistance $R(l)$ and the capacitance $C(l)$ of this piece of the transistor channel satisfy the condition $\omega\tau_{RC}(l) = 1$, where $\tau_{RC}(l) = R(l)C(l) = l^2\rho C$. This condition gives the value of the "leakage length" l on the order of $(\rho C\omega)^{1/2}$ (which can also be rewritten as $s[\tau/\omega]^{1/2}$). If $l \ll L$, then neither AC voltage, nor AC will exist in the channel at distances beyond a few distances l from the source.

The characteristic length where the AC exists is $s\tau$ for high-frequency regime ($\omega\tau > 1$), and $s(\tau/\omega)^{1/2}$ for low-frequency regime ($\omega\tau < 1$). Let us now make some estimations of the characteristic length for the different cases presented above. For $\tau = 30$ fs ($\mu = 300$ cm^2/[V·s] in Si MOSFET) and $s = 10^8$ cm/s the regime 1 will be realized for the radiation frequencies f greater than 5 THz; the regime 1.*a* corresponds to $L < 30$ nm. For $f = 0.5$ THz (regime 2), one finds the characteristic gate length distinguishing regimes 2.*a* and 2.*b* to be around 100 nm. If the conditions of the case 1.*a* are satisfied, the photoresponse will be resonant, corresponding to the excitation of discrete plasma oscillation modes in the channel. Otherwise, the FET will operate as a broad-band detector. For a long gate, there is no qualitative difference between the low-frequency regime ($\omega\tau \ll 1$), when plasma waves do not exist (the case 2.*b*) and the high-frequency regime ($\omega\tau \gg 1$), where plasma oscillations are excited (the case 1.*b*).

Response of plasmonic THz detectors can be analyzed by the hydrodynamic approach [18] since the continuous size reduction in the semiconductor devices and the current technology improvements allow a very high values of electron densities in FET channels, which can be described as a two-dimensional (2D) electron fluid. As described earlier, within the gradual channel approximation conditions, the surface charge concentration, n, in such FET channel is related to the local gate-to-channel voltage swing, $U = U_g - U_c - U_T$ through

$$n_s = \frac{CU}{e} \tag{8.14}$$

By neglecting electron–electron collisions and considering only the average electron drift velocity, FET channel can be modeled by the Euler equation and the continuity equation:

$$\frac{\partial v}{\partial t} + v\frac{\partial v}{\partial x} + \frac{e}{m}\frac{\partial U}{\partial x} + \frac{v}{\tau} = 0 \tag{8.15}$$

$$\frac{\partial U}{\partial t} + \frac{\partial (Uv)}{\partial x} = 0 \tag{8.16}$$

where $\partial U/\partial x$ is the longitudinal electric field in the channel, $v(x, t)$ is the local electron velocity, and m is the electron effective mass. The last term in the Euler equation accounts for electronic collisions with phonons and/or impurities. The incident THz radiation induces gate voltage oscillations of amplitude U_a at frequency ω which leads to excitation of electron density oscillations (plasma waves) in the device channel with a fixed amplitude at the source side with the same value of U_a. Hence, these two equations can be solved together assuming open-drain boundary conditions:

$$U(0,t) = U_0 + U_a \cos \omega t \quad \text{at x} = 0 \tag{8.17}$$

$$j(L,t) = 0 \quad \text{at x} = \text{L} \tag{8.18}$$

where U_0 is the DC gate-to-source voltage swing, $U_{AC} = U_a$, $\cos \omega t$ is the external AC voltage induced between the gate and source by the incident electromagnetic wave, and $j = CUv$ is the electron current per unit width. The solutions can be found by assuming series form:

$$v = \overline{v} + v_1 + v_2 + \cdots \tag{8.19}$$

$$U = \overline{U} + U_1 + U_2 + \cdots \tag{8.20}$$

where the first terms represent the time-averaged electron velocity and channel potential, respectively, and the following terms vary with time with the frequency $n\omega$, where ω is the frequency of the incident wave.

Assuming that the input signal, U_a is relatively small, v_1 and U_1 are proportional to U_a and hence, for the first order of U_a Eqs. (8.15) and (8.16) becomes:

$$\frac{\partial v_1}{\partial t} + \frac{\partial u_1}{\partial x} + \frac{v_1}{\tau} = 0 \tag{8.21}$$

$$\frac{\partial u_1}{\partial t} + s^2\frac{\partial v_1}{\partial x} = 0 \tag{8.22}$$

where $u_1 = eU_1/m$ and $s = (eU_0/m)^{1/2}$ is the plasma wave velocity. The boundary conditions given in Eqs. (8.17) and (8.18) can be written for the above equations as:

$$u_1(0) = \frac{e}{m}U_a \cos \omega t \tag{8.23}$$

$$\overline{u}(0) = \frac{e}{m}U_0 \tag{8.24}$$

$$\overline{v}(L) = v_1(L) = 0 \tag{8.25}$$

which result in

$$\Delta u = \overline{u}(L) - \overline{u}(0) - \frac{1}{2}\left\langle v_1^2(0)\right\rangle - \frac{1}{\tau}\int_0^L \overline{v}dx \tag{8.26}$$

$$\overline{v} = -\frac{\langle u_1 v_1\rangle}{s^2} \tag{8.27}$$

where the angular brackets denote the time averaging over the period $2\pi/\omega$. The DC response of the plasmonic detector to an incident wave proportional to $exp\ (ikx{-}i\omega t)$ is given by Δu, which can be found by solving the Eqs. (8.21) and (8.22) for u_1, v_1 with the boundary conditions given by Eqs. (8.23)–(8.25) and substituting these solutions into (8.26) and (8.27). Hence, remembering the relation between u_1 and U_1 given above, the drain-to-source DC voltage induced by the incident radiation can be found as the detector response $\Delta U = m\Delta u/e$

$$\Delta U = \frac{U_a^2}{4U_0} f(\omega) \tag{8.28}$$

where

$$f(\omega) = 1 + \beta - \frac{1 + \beta\cos\left(2k_0'L\right)}{\sinh^2\left(k_0''L\right) + \cos^2\left(k_0'L\right)} \tag{8.29}$$

with

$$\beta = \frac{2\omega\tau}{\sqrt{1 + (\omega\tau)^2}} \tag{8.30}$$

and are the real and imaginary parts of the wave vector k are

$$k_0' = \frac{\omega}{s}\left(\frac{\sqrt{(1 + \omega^{-2}\tau^{-2})} + 1}{2}\right)^{1/2} \tag{8.31}$$

$$k_0'' = \frac{\omega}{s}\left(\frac{\sqrt{(1 + \omega^{-2}\tau^{-2})} - 1}{2}\right)^{1/2} \tag{8.32}$$

Equation (8.28) describes the detector response for all frequencies and device lengths. The result depends on the values of two dimensionless parameters: $\omega\tau$ and $s\tau/L$. Let us look at the response at different regimes.

8.2.2.1 Resonant Detection

For $\omega\tau \gg 1$, the damping of the plasma oscillations excited by the incident radiation is negligible which results in

$$\beta = 2, k_0' = \omega/s, k_0'' = 1/(2s\tau) \tag{8.33}$$

and

$$f(\omega) = \frac{3\sinh^2\left(\frac{L}{2s\tau}\right) + \sin^2\left(\frac{\omega L}{s}\right)}{\sinh^2\left(\frac{L}{2s\tau}\right) + \cos^2\left(\frac{\omega L}{s}\right)} \tag{8.34}$$

For a short sample, such that $s\tau/L \gg 1, f(\omega)$ exhibit sharp resonances at the fundamental plasma frequency

$$\omega_0 = \frac{\pi s}{2L} \tag{8.35}$$

and its odd harmonics. Near the resonant frequencies, the response can be found as

$$\Delta U = \frac{U_a^2}{U_0}\left(\frac{s\tau}{L}\right)^2 \frac{1}{4(\omega - n\omega_0)^2\tau^2 + 1} \tag{8.36}$$

where the factor $s\tau/L$ determines the quality of the plasma resonator. This equation describes probably the most important feature and advantage of the model that an FET can be used for tunable resonant detection of electromagnetic radiation.

8.2.2.2 Broadband Detection

For $\omega\tau \ll 1$ the plasma oscillations are overdamped, and the device operates as broadband detector. In this case:

$$\beta \ll 1, k_0' = k_0'' = \frac{1}{s}\sqrt{\frac{\omega}{2\tau}} \tag{8.37}$$

and

$$f(\omega) = \frac{\sinh^2(\kappa) - \sin^2(\kappa)}{\sinh^2(\kappa) + \cos^2(\kappa)} \tag{8.38}$$

where $\kappa = (L/s)(\omega/2\tau)^{1/2}$. For a long enough device, $\kappa \gg 1$ which results in $f = 1$. In the opposite case, $\kappa \ll 1$ which leads to $f = 2\kappa^4/3 = L^4\omega^4/(6s^4\tau^2) \ll 1$.

It should be noted that Eq. (8.38) describes the response for gate biases only above the threshold voltage. This approach was later extended to the subthreshold regime by calculating plasma wave velocity directly from the channel carrier concentration and by taking into account the gate leakage current [19]. In this case, the response of an FET is given by

$$\Delta U = \frac{eU_a^2}{4ms^2}\left\{\frac{1}{1 + Q\exp\left(-\frac{eU_0}{\eta k_B T}\right)} - \frac{1}{\left[1 + Q\exp\left(-\frac{eU_0}{\eta k_B T}\right)\right]^2\left[\sinh^2\kappa + \cos^2\kappa\right]}\right\} \tag{8.39}$$

where

$$Q = \frac{j_0 L^2 me}{2C\tau\eta^2 k_B^2 T^2} \tag{8.40}$$

with the j_0 is the gate leakage current density. This model, however, failed to explain the response at gate bias values much below the subthreshold voltage that is when the electron carrier density becomes extremely small in the device channel. Later, Stillman et al. [20] introduced the concept of external impedance loading effect using a simple voltage divider model, which leads to a better fitting in the subthreshold regime. Based on this loading effect, the measured change in drain-to-source voltage can be expressed as:

$$V_M = \frac{V_U}{1 + (R_{Ch}/R_L)} \tag{8.41}$$

where V_U is the unloaded device response and V_M is the measured response, where R_L is the external load resistor, and R_{Ch} is the device channel resistance calculated from the measured I–V characteristics.

8.2.2.3 Enhancement by DC Drain Current

Effect of source-to-drain DC on the detection of the terahertz radiation both in resonant and nonresonant regime was also studied in detail theoretically and it was demonstrated experimentally that source-to-drain current I_d leads to a very sharp increase of the detection efficiency by a factor of two orders compared to the zero current case [21]. The analysis starts with the hydrodynamic equations in (8.15) and (8.16) above but with the boundary conditions

$$U(0,t) = U_0 + U_a \cos \omega t \text{ at x} = 0 \tag{8.42}$$

$$U(L)v(L) = \frac{j_d}{C} \text{ at x} = 0 \tag{8.43}$$

where $j_d = |I_d|/W$ is the absolute value of the current density (since the electron moves from source to drain, the current is negative), and W is the channel width. Following a similar approach that is outlined above, two different results are obtained for long and short channel devices in the nonresonant regime.

For a short channel ($L < s\tau$) device in the nonresonant ($\omega\tau \ll 1$) regime the response is given by

$$\delta U_0(L) = \frac{U_a^2}{4U_g(1-\lambda)^{3/2}}\left(-\lambda + \frac{4\epsilon^2}{15}\frac{5 + 4\sqrt{1-\lambda} + 1 - \lambda}{\left(1+\sqrt{1-\lambda}\right)^4}\right) \tag{8.44}$$

where $\lambda = j_d/j_{sat}$ and $\epsilon = (L^2\omega)/(s^2\tau)$. It should be noted that the first term in the equation does not depend on radiation frequency. Considering the limiting case $\omega \to 0$ ($\epsilon \to 0$) the instant value of the voltage at the drain can be found using stationary current–voltage characteristic of the channel.

For a long channel ($L > s\tau$) device the response becomes

$$\delta U_0(L) = \frac{U_a^2}{4U_g}\frac{1}{\sqrt{1-\lambda}} \tag{8.45}$$

We see that in both short channel and long channel cases, the response sharply increases when $\lambda = j_d/j_{sat} \to 1$. This increase is caused by the increase in the nonuniformity of the potential and field distribution in the channel that increases the nonlinear properties of the FET.

On the other hand, for the resonant ($\omega\tau \gg 1$) regime the response for an FET with DC drain current is given by

$$\delta U_0(L) = \frac{U_a^2}{4U_g}\frac{\omega_0^2}{(\omega-\omega_0)^2 + \left(\frac{1}{2\tau} + \frac{v_d}{L}\right)^2} \tag{8.46}$$

It shows that response is a resonant function of the ω centered at $\omega = \omega_0$. The width of the resonant peak is given by

$$\frac{1}{2\tau_{eff}} = \frac{1}{2\tau} - \frac{v_d}{L} \tag{8.47}$$

Since $v_d \sim j_d$ this width decreases with the current and results an infinitely sharp resonant peak at $v_d/L = 1/2\tau$.

Short channel FETs supporting decaying or resonant plasma waves have the promise of expanding high-frequency electronics into THz spectrum by providing efficient THz and sub-THz detectors and sources. Such transistors are sometimes called "TeraFETs" [10]. TeraFET detectors using silicon complementary metal-oxide-semiconductor (CMOS) FETs [22], Si FinFET devices [23], Silicon-on-Insulator MOSFETs [24], InGaAs High Electron Mobility Transistors (HEMTs) [25], AlGaN/GaN HEMTs [26], and graphene [27] operated in the 0.1–22 THz range [28] with Noise Equivalent Power (NEP) as low as 0.5 pW/Hz$^{1/2}$ [29]. Recently, it was shown that p-diamond might have advantages for implementing TeraFETs operating in 240–320 GHz range for beyond 5G WiFi operation [30]. In the following sections a brief survey of different THz detector technologies and applications will be presented were demonstrated.

8.3 THz Detectors Based on Silicon FETs

The first experimental evidence of the plasma wave detection by silicon FETs was reported by Knap et al. in [22]. The n-type (MOSFETs) used in the experiments had a 1.2 nm gate oxide, with a 1200 Å N$^+$ polygate. The test structures consisted of transistor arrays with common source and gate contacts and variable gate lengths (ranging from 800 to 30 nm). The gate width, W, was 10 μm. All the transistors had approximately the same threshold voltage of $V_t \sim 0.3$ V. Using the relaxation time determined from the mobility it was estimated that for frequencies below 1 THz for all the devices the condition $\omega\tau \ll 1$ is fulfilled, meaning that plasma oscillations were overdamped, and the photoresponse of FET was expected to be a smooth function of ω as well as of the gate voltage (nonresonant broadband detection). The open-circuit source-drain voltage was measured as photoresponse using a 120 GHz back-wave oscillator (BWO) radiation system with the maximum output power of 45 mW. The measured photoresponse is shown in Figure 8.4 for the devices different gate lengths along with the predictions (solid curves) of the plasma wave theory. As seen, the theory reproduces relatively well the overall shape of the experimental data, including the position and the width of the observed maximum. Note that the frequency used in these experiments is much higher than the "classical" cutoff frequency for devices with $L > 300$ nm. This result clearly demonstrates the possibility of plasma wave THz operation of these nanometer-scale devices and sub-50 nm silicon transistors can be used as selective and voltage tunable detectors of THz radiation operating at room temperature (RT). The authors later reported on responsivity and NEP of the same Si MOSFETs as $R_V = 33$ V/W and NEP $= 10^{-10}$ W/Hz$^{1/2}$, respectively [31].

Shortly after, nonresonant detection of terahertz radiation by silicon-on-insulator MOSFETs was reported for the devices with the gate width, W of 10 μm and the gate length L of ~130 nm at the temperature range of 8–350 K [24]. It was observed that the position of the peak response increased with decreasing temperature following the threshold voltage of the device, which was a clear evidence of the overdamped plasma wave THz detection. Moreover, the amplitude of the response increased with increasing drain current which was also consistent with the mechanism linking the photoresponse to the excitation of the overdamped plasma waves in the transistor channel.

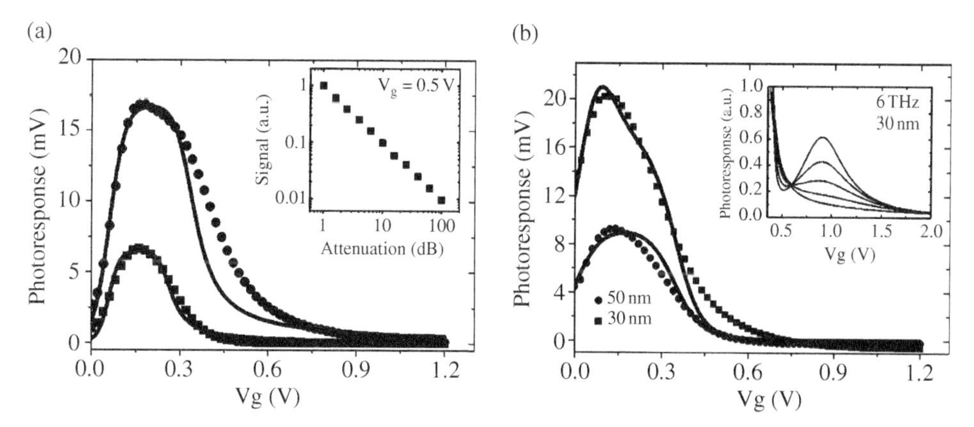

Figure 8.4 (a) Detected drain-source signal as a function of the gate voltage for two transistors with 600 nm (squares) and 800 nm (circles) gates. Radiation frequency f = 120 GHz, T = 300 K. Inset shows the signal versus attenuation for the gate voltage 0.5 V. (b) Detected drain-source signal as a function of the gate voltage for two transistors [30 nm (squares) and 50 nm (circles)]. Radiation frequency f = 119 GHz. T = 300 K. The inset shows the calculation of the photoresponse for higher mobilities—with increasing mobility (100 cm^2/V s, 200 cm^2/V s, 300 cm^2/V s, 400 cm^2/V s, 500 cm^2/V s) signal rises and changes from a nonresonant response to a resonant response. *Source*: © IEEE. Reprinted, with permission, from [22].

The first extremital demonstration of terahertz and sub-terahertz detection by nanometer-scale CMOS devices was reported in 2007 [32]. The devices used in the experiments were supplied by IBM Microelectronics, from the technology described by Steegen et al. [33] Regular V_t (R V_t), low V_t (L V_t), and high V_t (H V_t) (where V_t is the threshold voltage) devices with gate widths ranging from 1 to 5 μm, physical oxide thickness t_{ox} ~ 2.0 nm, and drawn gate lengths ranging from 50 to 180 nm were measured. Figure 8.5a, b shows the responses of RV_t NFETs and RV_t PFETs, respectively. The PFET response is significantly lower than that of the NFETs, due to the larger effective mass of the holes compared with the electrons which reduces the (unloaded) maximum response, and to the larger channel resistance of the PFETs, which brings about earlier roll-off in the sub-threshold region. Averaging response over several devices in order to ameliorate observed differences in coupling efficiency, NFET responsivity was estimated to be in the range of 1 V/W. The PFET responsivity is seen to be 20% that of the NFET, which was in agreement with the model predictions. Measurements with a drain current resulted gate voltage-dependent enhancement ranging from 10 to 25 times the open-drain response at 0.2 THz and 0.6 THz in both NFETs and PFETs (see Figure 8.5c). The devices were expected to have a modulation frequency 40GHz at the device threshold, and 30GHz at the response maximum.

Since coupling typically involves the gate bonding pad, the device responsivity is decreased by distributive effects of the induced THz current along the gate. Due to this effect, only a section of the device close to the bonding pad participates in the THz detection, with the remaining device width effectively shunting the load and decreasing the response. The width of the active section at high frequencies is inversely proportional to the square root of frequency and at high frequencies (above one THz) might be as small as a fraction of a micron. At such small widths, however, narrow channel effects might become detrimental

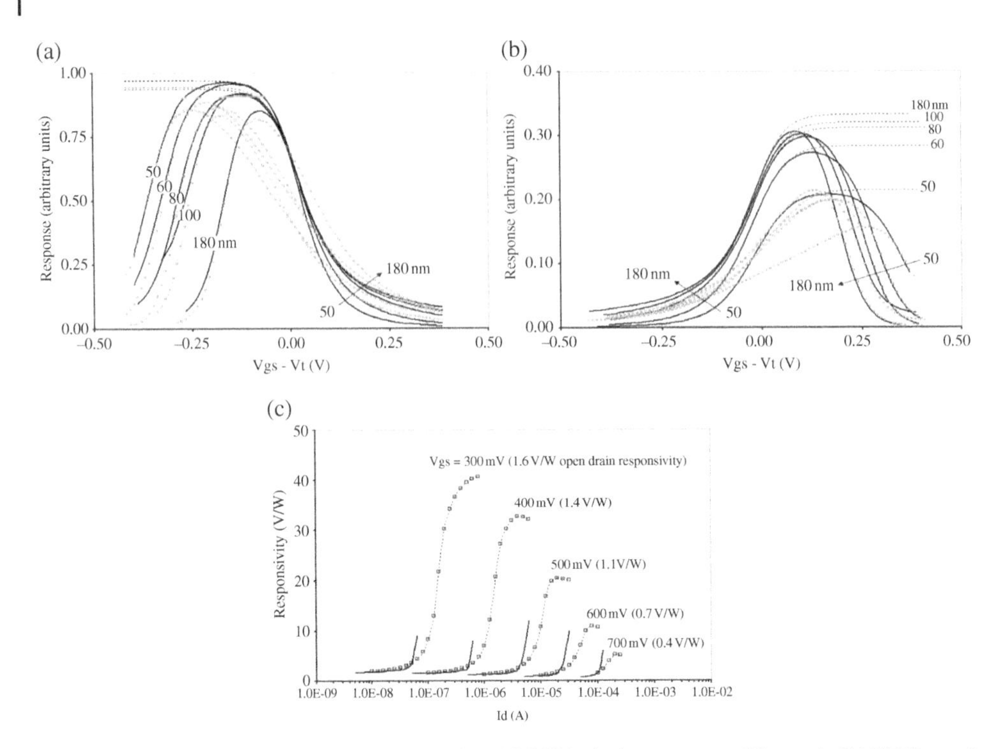

Figure 8.5 (a) Measured (squares) and calculated 0.2 THz drain response of 2 μm × L RV_t NFETs. and (b) RV_t PFETs. Unloaded response (dashed lines) and response with an 8.2 MΩ preamplifier input impedance load (heavy solid lines) are calculated using the theory. (c) 0.2 THz responsivity of 2 × 60 nm HV_t NFET vs. drain current for selected gate bias values. The values in parentheses are the open-drain responsivity measured for this device. Heavy lines illustrate the onset of response enhancement predicted by the theory. *Source*: © IEEE. Reprinted, with permission, from [32].

to the performance of conventional devices. The FinFET geometry avoids these effects facilitating significant improvements in both responsivity and NEP [23]. Stillman et al. [23] measured FinFET devices, which has shown schematically in Figure 8.6a, with 2, 20, or 200 fins of 40 nm height and designed widths from 40 to 100 nm. The designed gate length range was from 40 to 100 nm. Fin widths are typically reduced in processing on the order of 10–20 nm; gate lengths are expected to be approximately 5 nm shorter than as designed. Typical response dependence upon gate bias and frequency are shown in Figure 8.6b, c, respectively along with the theoretical predictions. While there are anomalies, theoretical curves fit reasonable, especially in the response attenuation at and above 0.6 THz. In addition, FinFET devices in many cases exhibited considerably greater responsivity than standard CMOS FETs.

With these successful demonstrations, interest in Si MOSFET based THz detectors increased. CMOS technology has seen widespread use over the last decade for the fabrication of broadband terahertz detectors [34–38]. In [39, 40], the first antenna-coupled MOSFET detectors and few-pixel focal plane arrays (FPA) were demonstrated for 645 GHz using a 250 nm CMOS foundry process without any additional fabrication steps. The fully integrated 3 × 5 pixel array with differential on-chip patch antennas, NMOS direct detectors,

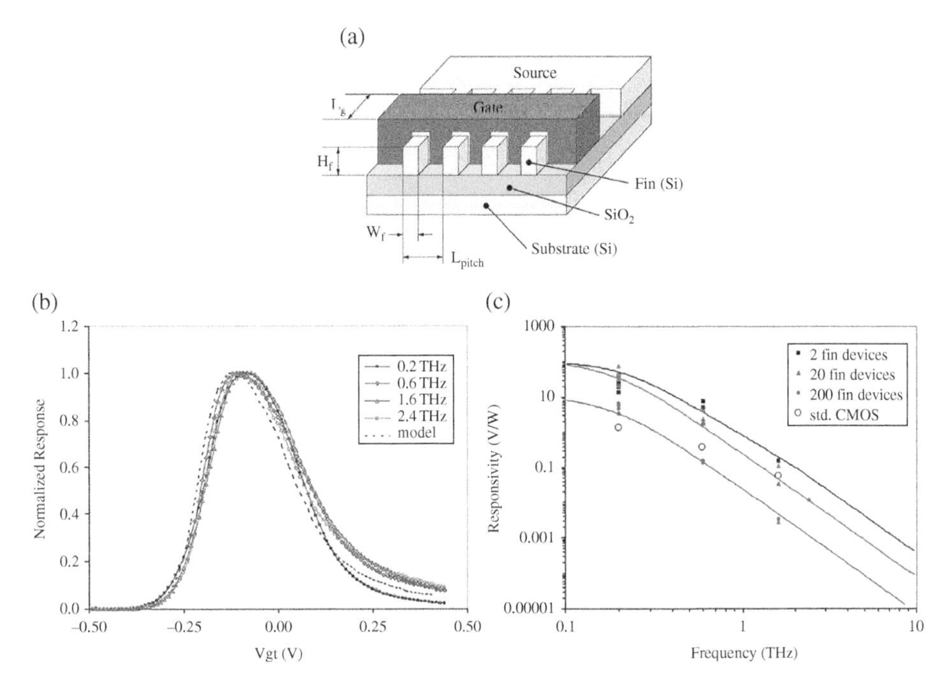

Figure 8.6 (a) Schematic illustration of FinFET device structure (the drain region is omitted from the foreground for clarity). The vertical channel fins are surrounded by the gate dielectric and gate conductor forming an effective dual-gate MOSFET. Note that the dielectric along the top surfaces of the fin may be increased in order to reduce parasitic capacitance. (b) 20 fin NFET open-drain response, normalized to response maxima, across several incident frequencies. $W_f = 40$ nm; $L_g = 100$ nm. Dashed line represents theoretical prediction with $f = 0.2$ THz. (c) Open-drain responsivity of several FinFET devices at various incident frequencies. Filled symbols are measured data; lines are theoretical predictions. Open symbols are measured responsivity for standard CMOS FETs reported in the literature. *Source*: © IEEE. Reprinted, with permission, from [23].

and integrated 43-dB voltage amplifiers achieved a responsivity of 80 kV/W and a NEP of 300 pW/Hz$^{1/2}$. Detectors with cost-efficient 130 nm [36] and 150 nm [41] CMOS technology, as well as more advanced 65 nm [42] SOI process technology have also been demonstrated.

In an impressive demonstration, 32 × 32 pixel array consisting of 1024 differential on-chip ring antennas coupled to NMOS direct detectors fully integrated into a 65-nm CMOS bulk process technology was shown in Figure 8.7 [37]. NMOS FETs operated well beyond their cutoff frequency based on the principle of distributed resistive self-mixing. The chip also included row and column select and integrate-and-dump circuitry capable of capturing terahertz videos up to 500 fps and was packaged together with a 41.7-dBi silicon lens. It achieved a responsivity of about 115 kV/W (incl. a 5-dB VGA gain) and a total NEP of about 12 nW integrated over its 500-kHz video bandwidth at 856 GHz in video mode without lock-in. At a 5-kHz chopping frequency, a single pixel provided a maximum responsivity of 140 kV/W (incl. a 5-dB VGA gain) and a minimum NEP of 100 pW/Hz$^{1/2}$ at 856 GHz.

For the realization of research projects based on the CMOS process, there are several production tools such as MPW service (multi-project wafer—a service integrating a number of different circuit designs from various teams within one substrate and offering a certain

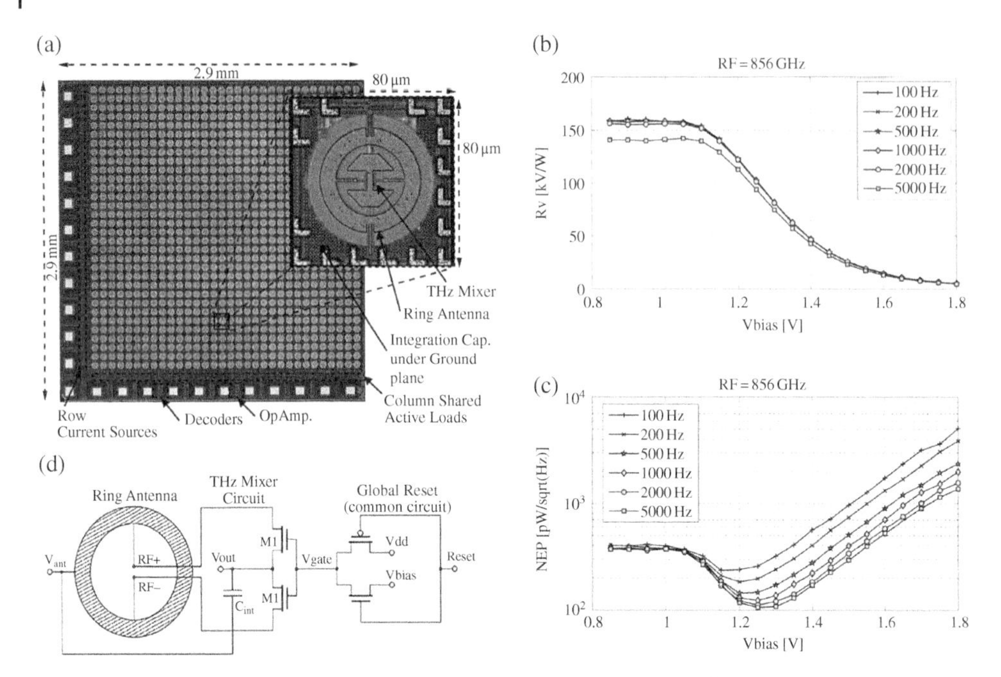

Figure 8.7 (a) 32 × 32 FPA chip complete die micrograph (2.9 × 2.9 mm^2) and a single detector topography (80 × 80 μm^2). (b) Measured camera responsivity (R_V) at 856 GHz versus gate bias and at 1-m distance from the source for different chopping frequencies. The data includes a VGA gain of 5-dB. (c) Measured at 856 GHz versus gate bias at 1-m distance from the source for different chopping frequencies. (d) Single terahertz mixer schematic with its ring antenna and reset circuitry. *Source:* © IEEE. Reprinted, with permission, from [37].

amount of chips for a reasonable price. It should be pointed out that CMOS technology allows for an easy integration of the THz detectors with the readout circuitry, image processing units, and other electronic components. However, there are a number of challenges associated with the CMOS integration schemes, which are the subject of current active research [43].

For instance, the large area required by the monolithically integrated antennas of each single detector remains as a major challenge. This area dominates and cannot be reduced because it is related to the THz radiation wavelength (sub-mm range). That makes R&D studies by means of the MPW services very expensive, especially for the detector FPAs. Moreover, the proximity of the substrate with its high dielectric constant (and low resistivity used in most of MPW offers) leads most of incoming THz radiation to penetrate to the substrate and propagate in it resulting in important losses of detector efficiency, as well as detectors' cross-talk effects. The impact of the substrate can be minimized by using a metal layer (ground plane) that shields the antenna from the substrate [44] or by thinning the detector substrate [45].

Layout of transistors also plays an important role in the value of parasitic elements affecting the efficiency of energy transfer from the antenna [46]. It has been shown that for standard operating conditions, the channel conductivity had no significant influence on the input impedance Z_{gs}. The real part of this impedance is dominated by contact resistances and is typically well matched with the values of simple antennas integrated within the chips, the imaginary part being related to the overlap capacitances between gate metallization and source. The measurements done on relatively large transistors showed that smaller transistors had always larger detectivity. The near hyperbolic dependence of responsivity on the width of the transistor indicates the dominant role of the source-gate capacitance shunting the incoming THz signal. For this reason, in the MPW services, which provide submicron devices, the smallest available transistors have usually the highest responsivity [43].

8.4 Terahertz Detection by Graphene Plasmonic FETs

Graphene, a semi-metallic, 2D material with a single atomic thickness has been the most widely researched material owing to its superior electrical, mechanical and optical properties [47]. Gapless nature of graphene enables charge carrier generation over a wide energy spectrum from ultraviolet (UV) to terahertz (THz) making it a unique material for photonics and optoelectronics during the recent past. The collective oscillations of the 2D correlated quasiparticles in graphene make it a particularly attractive medium for plasmonics [48], with applications ranging from tight-field-enhanced modulators, detectors, lasers, and polarizers to biochemical sensors [49–52]. Unlike conventional noble metal plasmons, graphene plasmons are dominant in the terahertz and far-infrared frequencies [53]. Conventional plasmonic materials, mostly noble metals, suffer from the difficulty in controlling and varying their permittivity functions and the existence of material losses—especially at visible wavelengths—which degrade the quality of the plasmon resonance and limit the relative propagation lengths of SPP waves along the interface between such metals and dielectric materials. On the other hand, graphene's complex conductivity depends on the radian frequency ω, charged particle scattering rate Γ representing the loss mechanism, temperature T, and chemical potential μ_c. The chemical potential depends on the carrier density and can be controlled by gate voltage, electric field, magnetic field, and/or chemical doping [54, 55]. The imaginary part of graphene conductivity can attain negative and positive values in different ranges of frequencies depending on the level of chemical potential [56]. Such tunability of its optical characteristics combined with its high carrier mobility ($\approx$200 000 cm^2/V/s) graphene has emerged as a prominent material suitable for THz plasmonic applications [57–59]. As large-area graphene samples with good electrical characteristics have become widely available, graphene-based THz devices have attracted great interest of the researchers.

The electron energy dispersion in undoped graphene is linear

$$E = \nabla V_{F.}k \tag{8.48}$$

where k is the electron momentum and V_F is the 2D Fermi velocity, which is a constant for graphene ($V_F = 10^6$ m/s). The linear electron energy dispersion results in zero effective electron mass in graphene. The electron "inertia" in graphene is described by a fictitious "relativistic" effective mass $m_F = E_F/V_F{}^2$, where E_F is the Fermi energy. The relation between the Fermi energy and sheet carrier density in a graphene FET can be given as

$$E_F = \hbar V_F \sqrt{2\pi N_S} \tag{8.49}$$

where $N_S = \varepsilon(V_g - V_D)/ed$ is the sheet electron density in a graphene layer with ε is the permittivity, d is the dielectric thickness, e is the electron charge, V_g is the gate voltage and V_D is the voltage corresponding to Dirac point at which the graphene sheet has charge neutrality. Hence the electron effective mass becomes

$$m_F = \frac{\hbar}{V_F}\sqrt{\frac{2\pi\varepsilon}{ed}V_0} \tag{8.50}$$

where $V_0 = V_g - V_D$. For a typical value of $V_0 = 1$ Eq. (8.23) yields $m_F = 0.03\, m_0$ where m_0 is the free electron mass. It should be noted that m_F decreases to zero with the gate voltage V_g approaches to the Dirac voltage. The electron relaxation rate in graphene with an electron mobility μ can be estimated as $1/\tau = e/\mu m_F$. Therefore, $\omega_0\tau \gg 1$ is satisfied for frequencies above 2 THz even for low mobility graphene with $\mu = 1600\ \mathrm{cm^2/Vs}$. Plasmon dispersion in a gated graphene can be formally written in the same form as in a conventional semiconductor structure with substituting the effective electron mass by the "relativistic" effective mass m_F:

$$\omega = q\sqrt{\frac{eV_0}{m_F}} \tag{8.51}$$

where q is the plasmon wavevector, which is determined by the grating gate period L, $q = 2\pi n/L$ ($n = 1,2,3$ ...). Substituting Eq. (8.50) into Eq. (8.51) results in

$$\omega = q\sqrt[4]{N_S\frac{V_F^2 e^4 d^2}{2\pi\hbar^2\varepsilon^2}} \tag{8.52}$$

which gives the plasmon resonant modes in graphene [60].

The first study of tunable plasmon excitations and light–plasmon coupling at terahertz frequencies in graphene micro-ribbon arrays were reported in [57]. The investigated structures were based on graphene nanoribbons which were fabricated by plasma etching. It was shown that graphene plasmon resonances could be tuned over a broad terahertz frequency range of 1–10 THz by changing micro-ribbon width and in situ electrostatic doping. Carrier concentration in graphene ribbon arrays was controlled by using an ion-gel top gate, which allowed a large doping range through the electrostatic gating. Figure 8.8a, b show the schematic view of a typical gated device. The ribbon width and carrier doping dependences of graphene plasmon frequency demonstrated power-law behavior characteristic of the 2D massless Dirac electrons. Despite the relatively low-quality factors, which were attributed to the high carrier scattering rate and limited mobility due to the plasma damages induced during the etching process, the structures presented prominent absorption peaks at room temperature.

Later, a large number of studies on graphene THz plasmonics have been reported. In one of the earlier ones, THz detection at room temperature by antenna-coupled single layer graphene field effect transistors (SLG-FET) and bilayer graphene field effect transistors (BLG-FET) was demonstrated [58]. The devices exploit the nonlinear response to the oscillating radiation field at the gate electrode, with contributions of thermoelectric and photoconductive origin. Devices attached to log-periodic planar metal antenna achieved the responsivity of $R_V \sim 100$ mW/W and NEP $\sim 30\ \mathrm{nWHz^{-1/2}}$ at 0.3 THz and at room

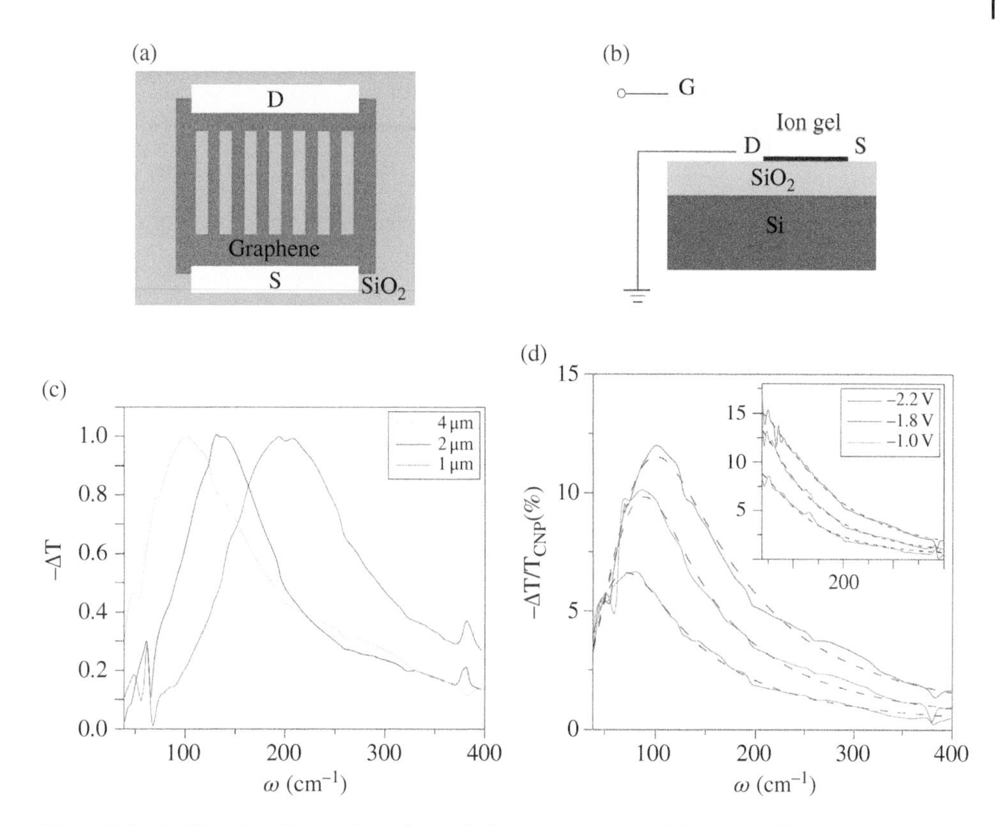

Figure 8.8 (a) Top-view illustration of a typical graphene micro-ribbon array. The array was fabricated on transferred large-area CVD graphene using optical lithography and plasma etching. (b) Side view of a typical device incorporating the graphene micro-ribbon array on a Si/SiO_2 substrate. The carrier concentration in graphene is controlled using the ion-gel top gate. (c) Change of transmission spectra (tuning) with different graphene micro-ribbon widths for the doping concentration of 1.5×10^{13} cm^{-2} (d) Control of terahertz resonance of plasmon excitations through electrical gating. Terahertz radiation was polarized perpendicular to the graphene ribbons. The plasmon resonance shifts to higher energy and gains oscillator strength with increased carrier concentration. For comparison, the inset shows corresponding spectra due to free carrier absorption for terahertz radiation polarized parallel to the ribbons. *Source:* Reprinted by permission from Nature/Springer [57] © 2011.

temperature (Figure 8.9). Using these devices for scanning coffee capsules resulted in a reasonably good spatial resolution of the THz images. This was achieved using 200×550 scanned points collected by raster-scanning the objects in the beam focus with the integration time of 20 ms per point. The results proved that the graphene THz devices could be used in realistic settings, enabling large-area, fast imaging of macroscopic samples.

One of the challenges in graphene plasmonic devices is the mobility reduction when processed and integrated with substrates. Remembering the resonance conditions presented above, resonant excitation of plasmons becomes exceedingly difficult for the THz frequencies and can only be achieved if the momentum relaxation rate is below the plasmon frequency, which, in turn, requires ultra-high electron mobility. That restriction resulted that the plasma waves in early graphene-based THz detectors [58, 61–66] were overdamped, and

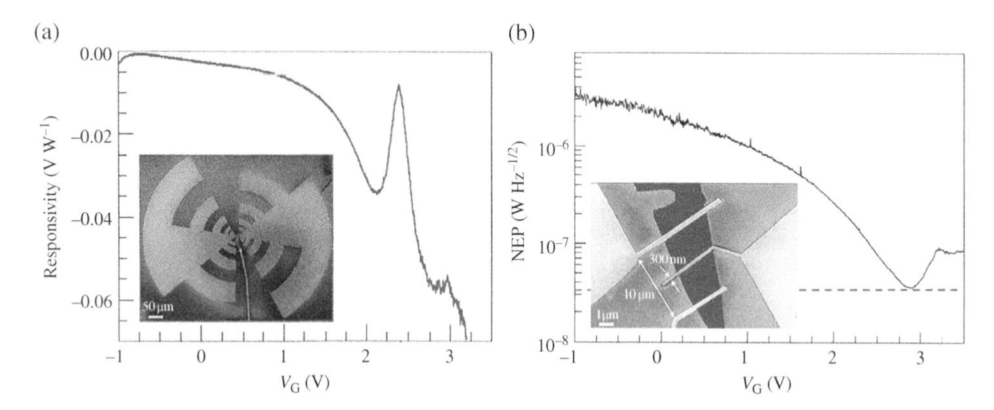

Figure 8.9 (a) Room temperature responsivity as a function of the gate bias for detectors based on bi-layer graphene FET (BLG-FET) Different background colors identify regions below and above the charge neutrality point. The inset shows the detector consisting of a log-periodic circular-toothed antenna patterned between the source and gate of a GFET. (b) NEP as a function of VG for detectors based on BLG-FET assuming a noise level dominated by the thermal Johnson–Nyquist contribution. A dashed horizontal line is drawn at the value of minimum NEP. The inset shows the details of the detector with a drain which is a metal line running to the bonding pad and a 10 μm–long channel and a L_G ~ 300 nm-wide gate. *Source:* Reprinted by permission from Nature/Springer, [58] © 2012.

the devices exhibited only a broadband (nonresonant) photoresponse [27]. Only relatively recently, Bandurin et al. [27] demonstrated resonant regime in graphene FETs based on high-quality van der Waals heterostructures with graphene encapsulated between hexagonal boron nitride (hBN) crystals which have been shown to provide the cleanest environment for long-lived graphene plasmons. The devices were consisted of high-mobility bilayer graphene (BLG) channels sandwiched between two relatively thin (d ≈ 80 nm) slabs of hBN with 3–6 μm in length L and from 6 to 10 μm in width W and integrated with broadband bow-tie antenna and a hemispherical silicon lens (Figure 8.10). The unique structure allowed the mobility of the devices to exceed 100 000 cm^2/Vs and remain above 20 000 cm^2/Vs at temperatures $T = 10$ K and 300 K, respectively at the characteristic carrier density $n = 10^{12}\ cm^{-2}$.

The devices were tested for both broadband and resonant detection characteristics. For the broadband (nonresonant) detection, low end of the sub-THz domain ($f = 0.13$ THz), where the plasma oscillations are overdamped was used for irradiation. An example of the responsivity $R_a = \Delta U/P$, where ΔU is the emerging source-to-drain photovoltage and P is the incident radiation power, as a function of the top gate voltage, V_g is shown in Figure 8.10b. It is interesting to observe that R_a increases in magnitude upon approaching the charge neutrality point (NP) where it flips sign because of the change in charge carrier type. Measurements at different temperatures showed that R_a grows with decreasing T (bottom inset of Figure 8.11b and reaches its maximum $R_a \approx 240$ V/W at $T = 10$ K due to a steeper $F(V_g)$ at this T. This performance was further improved by increasing the nonlinearity by taking advantage of the gate-tunable band structure of BLG which resulted a bandgap opening and a steep $F(V_g)$ dependence that, in turn, causes a drastic enhancement of R_a exceeding 3 kV/W with the NEP of about 0.2 pW/$Hz^{1/2}$.

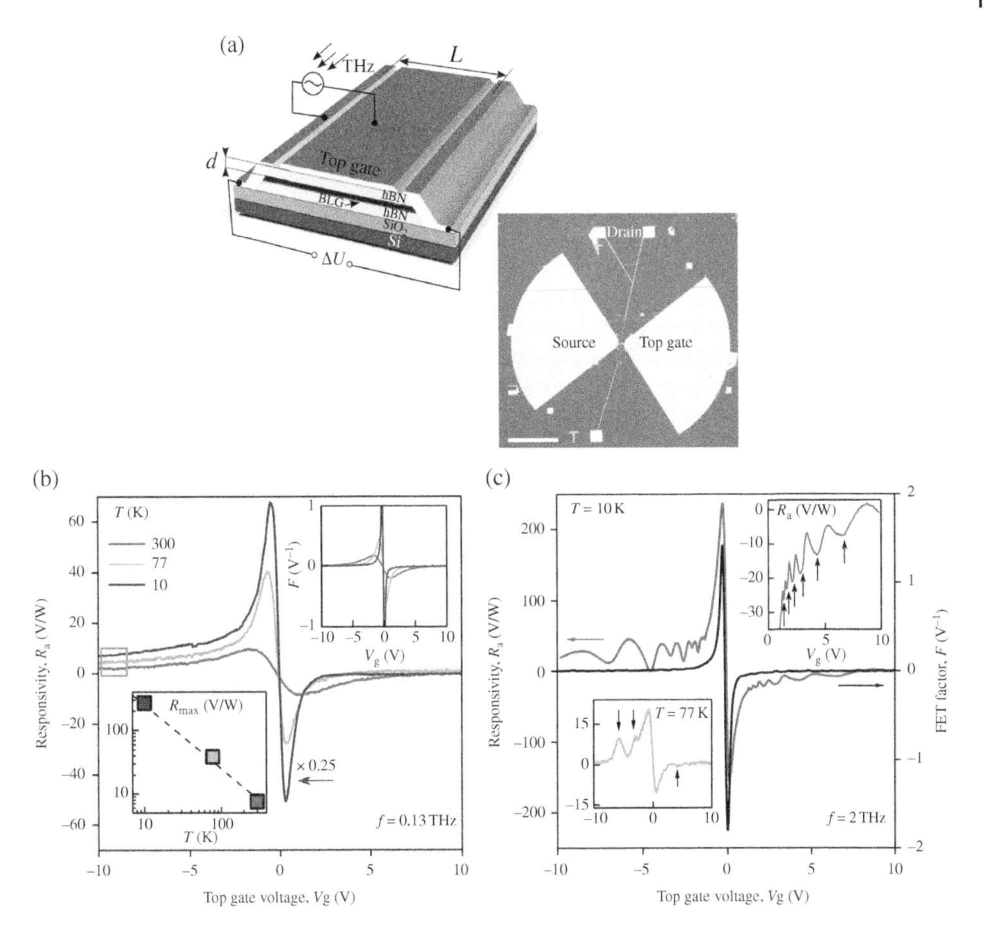

Figure 8.10 (a) Schematics of the encapsulated BLG FET (top) and optical photograph of one of the detectors (bottom) used in the work (Scale bar is 200 μm). (b) Broadband (nonresonant) responsivity measured at $f = 130$ GHz and three representative temperatures. The hollow rectangle frame at low voltages highlights an offset stemming from the rectification of incident radiation at the p-n junction between the p-doped graphene channel and the n-doped area near the contact. Upper inset: FET-factor F as a function of V_g at the same T. Lower inset: maximum R_a as a function of T. (c) Gate dependence of the resonant responsivity recorded under 2 THz radiation. The upper inset shows a zoomed-in region of the photovoltage for electron doping. Resonances are indicated by the arrows. Lower inset: resonant responsivity at liquid-nitrogen temperature. *Source*: Reprinted by permission from Nature/Springer [27] © 2018.

Resonant detection characteristics of the devices were studied under 2 THz irradiation. The gate voltage dependence of R_a shown in Figure 8.10c exhibits prominent oscillations, despite the fact that F as a function of V_g is featureless (dark curve). The oscillations are clearly visible for both electron and hole doping and well discerned at 10 K, although they persist up to liquid nitrogen T, especially for $V_g < 0$. Measurements at intermediate frequencies showed that the resonant operation of the reported devices onsets in the middle of the sub-THz domain and the resonances are already well developed at $f = 460$ GHz. The results clearly show that graphene FETs can be used resonant detection of THz provided that their

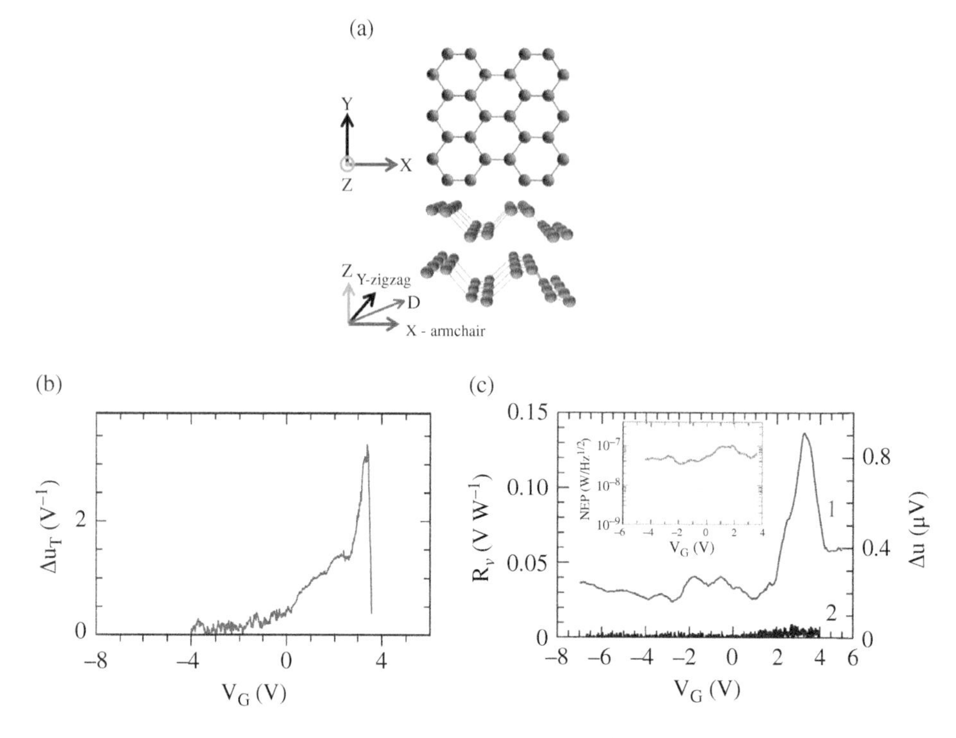

Figure 8.11 (a) BP atoms are arranged in puckered honeycomb layers bounded together by Van der Waals forces; the armchair (*x*) and zigzag (*y*) crystal axis are shown on the graph. *D*-axis is between the *x*- and *y*-axis with 45° angle from each. (b) Predicted photoresponse as a function of V_G, under the overdamped plasma-wave regime. (c) Gate bias dependence of the experimental RT responsivity (R_ν)/ photovoltage (Δu). Curve 1 was measured by impinging the THz beam on the detector surface; the curve 2 was measured while blanking the beam with an absorber. Inset: NEP as a function of V_G, extracted from the relation N_{th}/R_ν. *Source*: Reprinted by permission from Nature/Springer [67] © 2015.

mobility is high enough. Although it was not enough to achieve resonances at room temperature, BLG/hBN superlattices helped to attain very large mobility and my offer a promising approach to demonstrate room temperature resonance THz detection.

8.5 Terahertz Detection in Black-Phosphorus Nano-Transistors

Impressive results of graphene research, particularly in photonics applications, stimulated the investigation of other 2D materials such as single-unit cell thick layers of transition-metal dichalcogenides (TMDCs: MoS_2, $MoSe_2$, WS, WSe_2, etc.). Similar to graphene, 2D TMDCs can be obtained from bulk crystals by employing the micromechanical exfoliation method, but they show a direct band gap (0.4–2.3 eV), which enables applications that well complement graphene capabilities. However, their relatively low mobility ($\leq$200 cm^2 V^{-1} s^{-1}) poses a major challenge for high-frequency applications. A good trade-off between graphene and TMDCs is represented by a novel class of atomically thin

2D elemental materials: silicene, [68] germanene, [69] and phosphorene [70]. Among them, phosphorene has single- or few-layer form called black phosphorus (BP). Unlike silicon and germanium, BP, the most stable allotrope of the phosphorus element in standard conditions, shows a layered graphite-like structure, where atomic planes are held together by weak Van der Waals forces of attraction, thus allowing the application of standard micro-mechanical exfoliation techniques. As in graphene, each BP atom is connected to three neighbors, forming a stable layered honeycomb structure with an interlayer spacing of ≈5.3 Å. In contrast with graphene, the hexagonally distributed phosphorus atoms are arranged in a puckered structure rather than in a planar one (Figure 8.11a). This property generates an intrinsic in-plane anisotropy that results in angle-dependent conductivity [71] and an intrinsic optical anisotropy at visible and near-IR frequencies [72]. Bulk BP has a small direct band gap of ≈0.3 eV. However, the reduction of the flake thickness leads to quantum confinement and enhances the band gap up to $E_g \approx 1.0$ eV in the limit case of phosphorene (a single layer of BP) which provides huge carrier density tunability. These electronic properties along with the large hole mobilities exceeding 650 $cm^2/V/s$ at RT and well above 1,000 $cm^2/V/s$ at 120 K make BP very attractive for THz detection applications [71].

One of the first THz detectors based on BP was reported in [67]. The top-gated FET detectors were fabricated using mechanically exfoliated 10 nm-thick (corresponding to ≈16 layers) BP flakes and integrated with THz asymmetric antennas designed to enhance the sensitivity. A lower bound for the field-effect mobility was estimated from the transconductance measurements as $\mu_{FE} = 470\ cm^2/V/s$. Photoresponse of the devices was measured under the illumination of 0.26–0.38 THz and analyzed considering two detection. In the first, detection was assumed to be the rectification by the self-mixing of overdamped charge density waves. In this case, expected generated photovoltage can be estimated from the transfer characteristics [6, 21–23]

$$\Delta u_T \propto -\frac{1}{\sigma}\frac{d\sigma}{dV_G}\left[\frac{R_L}{\frac{1}{\sigma} + R_L}\right] \tag{8.53}$$

where the minus sign accounts for the hole majority carriers, and R_L represents the loading effects arising from the finite impedance of the measurement setup. It should be noted that under the given experimental conditions, the THz-induced plasma waves propagating in the transistor channel are expected to be strongly damped, being $2\pi\nu\tau = 1$, where, $\tau = m_x^{*}\mu/e$ is the hole relaxation time (≤200 fs) and m_x^{*} is the hole effective mass along the armchair direction.

The second mechanism was considered to be thermoelectric effect that, when illuminated by THz radiation, would result a photothermoelectric voltage is proportional to the Seebeck coefficient (S):

$$S = -\frac{\pi^2 k_B^2 T}{3e}\left(\frac{1}{\sigma}\frac{d\sigma}{dV_G}\right)\frac{dV_G}{dE_f} \tag{8.54}$$

where k_B is the Boltzmann constant, e is the electron charge, and E_f is the Fermi energy.

Theoretically calculated photovoltage Δu_T as a function of V_G is shown in Figure 8.11b. A comparison with the experimental responsivity/photovoltage curve (Figure 8.11c) reveals a good qualitative agreement, indicating that the rectification of THz-induced overdamped plasma waves, in the BP channel, might be the dominant detection process while there is contribution from thermoelectric effect since R_ν does not decrease to zero. A maximum responsivity $R_\nu = 0.15$ V/W was achieved, which is slightly higher than that was obtained in BLG-FETs fabricated under identical geometries. The minimum NEP was estimated as 40 nWHz$^{1/2}$ using the ratio N_{th}/R_ν and by assuming that the main contribution to the noise figure is the thermal Johnson–Nyquist noise. This is comparable with the NEP of the best BLG-based THz photodetectors, operating at similar frequencies.

In a later publication, by the same group reported on antenna-coupled nanodetectors based on thin flakes of exfoliated BP, to selectively activate plasma-wave, thermoelectric, and bolometric THz detection modes by taking advantage of the inherent electrical and thermal in-plane anisotropy of the flakes [73]. The flakes had the thickness h in the range of 8–14 nm and prepared following the same method described in [67]. Two types of devices were fabricated. The device-A had $h \sim 14$ nm (~ 23 layers) and source (S)—drain (D) channel defined along the D-axis (see Figure 8.11a), i.e. at a 45° angle in-between the armchair (x) and zigzag (y) directions. The device-B had $h \sim 9$ nm (corresponding to ~15 layers) and S-D channel defined along the y-axis. Moreover, in device-A, the S and gate (G) electrodes were patterned in the shape of a half 110° bow-tie antenna, introducing a strong asymmetry whereas, in device-B, a 110° bow-tie antenna was symmetrically placed at the S and D electrodes (Figure 8.12a). Carrier concentrations extracted from the transconductance measurements for device-A and device-B were $n \sim 8.0 \times 10^{18}$ cm^{-3} and $n \sim 2.2 \times 10^{19}$ cm^{-3} and $\mu \approx 330$ cm^2/V/s whereas the mobilities were $\mu \approx 330$ cm^2/V/s and $\mu \approx 380$ cm^2/V/s, respectively. Photoresponse of both devices was measured as a function of gate voltage V_G under the illumination of 0.29 THz beam with $V_{SD} = 0$ V.

Measured and predicted responses of the devices are presented in Figure 8.12c–g. For the device-A, (i) the thermoelectric effects heating of the metallic contacts due to the THz-driven local currents and (ii) the excitation of plasma oscillations along the channel were expected to result in the photoresponse. In the first mechanism, the THz-induced carrier distribution gradient would generate a diffusive flux of holes from the hot-side (S) to the cold-side (D) of the channel and produce (positive) thermoelectric value Δu_{pe}. For $V_{\text{SD}} \neq 0$, the voltage generated drain current $I_{\text{SD,off}}$ would flow in the opposite direction with respect to the light-induced generated photothermoelectric current (I_{pe}). This results in a negative $\Delta u_{\text{pe}}* = (I_{\text{pe}} - I_{\text{SD,off}})/\sigma$ if $I_{\text{SD,off}} > I_{\text{pe}}$. However, this was not observed when the incident radiation $f = 0.29$ THz. For the second mechanism, the excited carrier density would be pushed toward one channel side or the other depending on the structure asymmetry. The THz-induced current will then sum up with the pre-existent DC, leading to $\Delta u_{pw}{}^* > 0$, which was in agreement with the experimental data proving that device-A behaved like a plasma-wave THz detector. Considering that $\omega \mu m_d{}^*/e \ll 1$, where $m_d{}^*$ is the hole effective mass along the D-axis, e is the electron charge and $\omega/2\pi = 0.29$ THz, which in the given geometry corresponded to $f \ll 0.85$ THz, it was concluded operating in the nonresonant overdamped regime. Interestingly, when the frequency was shifted to 0.32 THz the impedance matching between the device and the antenna was lost leading to cease excitation of plasma waves (Figure 8.12g) and thermoelectric effect became dominant mechanism of detection.

For device-B, only bolometric detection could be expected since both plasma-wave and thermoelectric effects were prevented by the inherent symmetry. The estimated bolometric photovoltage $V_B = \sigma^{-1} d\sigma/dT$ (Figure 8.12f) was in good agreement with the measured

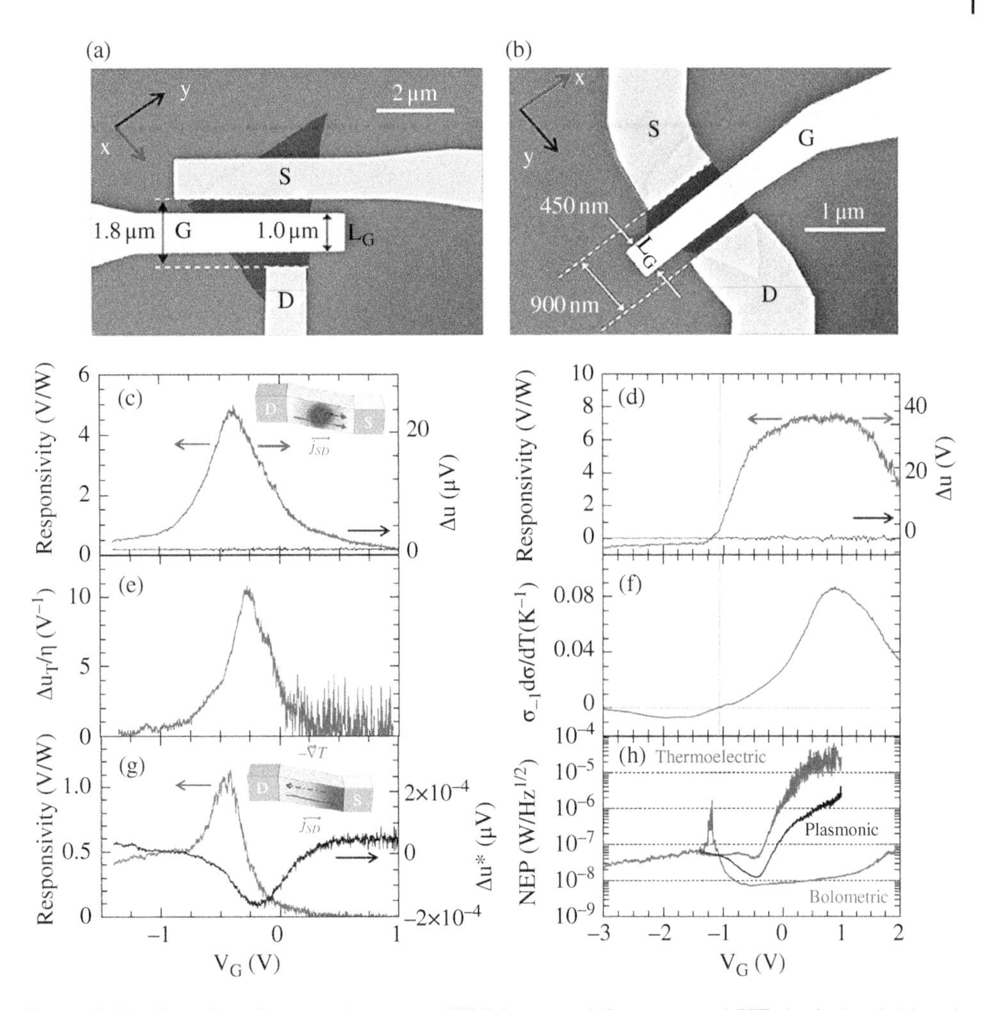

Figure 8.12 Scanning electron microscope (SEM) images of the top-gated FETs in device-A (a) and device-B (b), respectively. Gate-bias dependence of the experimental RT responsivity (R_υ)/ photovoltage (Δu) for device-A (c) and B (d). The lines for higher responsivity measured by impinging the 0.29 THz beam on the detector surface; the lines for close-to-zero responsivity were collected while blanking the beam with an absorber. Inset: schematics of the overdamped plasma-wave dynamics. (e) Predicted photoresponse of device-A as a function of V_G, under the overdamped plasma-wave regime. (f) Predicted bolometric trend, estimated from the gate voltage dependence of the ratio $\sigma^{-1}d\sigma/dV_G$ for device-B. (g) Left vertical axis: Gate bias dependence of the experimental RT responsivity of device-A collected while funneling the antenna resonant-frequency detuned 0.32 THz beam on the FET channel. Right vertical axis: DC photovoltage trend vs V_G, measured from the difference between the $I_{SD,on}$ and $I_{SD,off}$, multiplied by the channel resistance. Inset: schematics of the thermoelectric dynamics (h) Noise equivalent power (NEP) as a function of V_G, for the plasma-wave (device-A, 0.29 THz), thermoelectric (device-A, 0.32 THz), and bolometric (device-B) detectors extracted from the ratio between the thermal noise spectral density $N_{th} = (4k_BT\sigma - 1)^{½}$ and the device responsivity. *Source*: Reprinted by permission from Nature/Springer [73] © 2016.

response (Figure 8.12d) proving the bolometric detection. The choice of the *y*-orientation was ideal for the activation of an efficient, single, and selective bolometric detection process.

The maximum R_v of 5.0 V/W and 7.8 V/W have been reached for device-A and device-B, respectively, significantly larger than those reported in exfoliated graphene FETs [58, 62]. The responsivity reduced to 1.1 V/W when device-A operated in the thermoelectric mode. Figure 8.12h shows the extrapolated NEP curves, whose minimum reaches 7 nW/Hz$^{1/2}$, 10 nW/ Hz$^{1/2}$, and 45 nW/Hz$^{1/2}$ for the BP-bolometer, plasma-wave, and thermoelectric detector, respectively (Figure 8.12h). The versatility provided by the BP anisotropy and its tunable band gap unveils its potential fully switchable THz response, promising exceptional impact for devising active and passive THz devices and components [73].

8.6 Diamond Plasmonic THz Detectors

Diamond has unique physical properties such as a wide bandgap of 5.46 eV, a high breakdown field of 10 MV/cm, and a high thermal conductivity of 23 W/cmK which makes it a very attractive material for high voltage, [74] high-power, [75, 76] and high-temperature applications [77]. Diamond transistors using delta [78] and transfer doping [79, 80] with SiN [81] and Al_2O_3 [82] dielectric layers have been reported and operated at 200 °C, 250 °C, [83] and 350 °C [84]. At high energies, diamond is an order of magnitude more radiation hard than silicon [85]. The highest reported values of the diamond electron and hole mobilities determined by time-resolved cyclotron resonance are 7300 cm^2/V/s and 5300 cm^2/V/s, respectively [86]. The time-of-flight measurements yielded somewhat smaller values of $\mu_e = 4500$ cm^2/V/s and $\mu_h = 3800$ cm^2/V/s, respectively) [87]. Typical mobilities in the Two-Dimensional Hole Gas (2DHG) range from 30 to 150 cm^2/V/s. Very large surface heavy hole effective mass values, in the range of 0.74–2.12, have been reported [88–90]. The optical phonon energy in diamond is also very large. This results in suppression of optical phonon scattering and, consequently, in a very large momentum relaxation time τ limited by optical phonon scattering. This makes p-diamond very promising for sub-THz and THz plasmonic devices. The reported values of the low field mobility and hole effective mass vary so much. For the highest reported values of these parameters, the room temperature momentum relaxation time could be higher than 5 ps, in which case the plasmonic resonance quality factor $\omega_p\tau$ could exceed unity at frequencies as low as 60 GHz.

Recently, p-diamond has been proposed for plasmonic THz applications and its ultimate potential for THz and IR plasmonic detection [30]. As mentioned above, the key parameter determining the plasma response in diamond—the hole momentum relaxation time, τ—could vary from 0.013 to 6.4 ps at room temperature. Therefore, the expected response was evaluated for the values of τ ranging from 0.01 to 6 ps for the gate lengths of 20 nm, 60 nm, and 100 nm. The calculation is done based on theory developed in [18] which was outlined in Section 8.1 above.

The calculated plasma velocity for 2DHG in diamond as a function of the gate voltage swing and the hole effective mass is shown in Figure 8.13a. As seen, the plasma velocity is relatively small because the hole effective mass is large. This makes it possible to achieve "plasmonic boom" instability in 2DHG [91, 92].

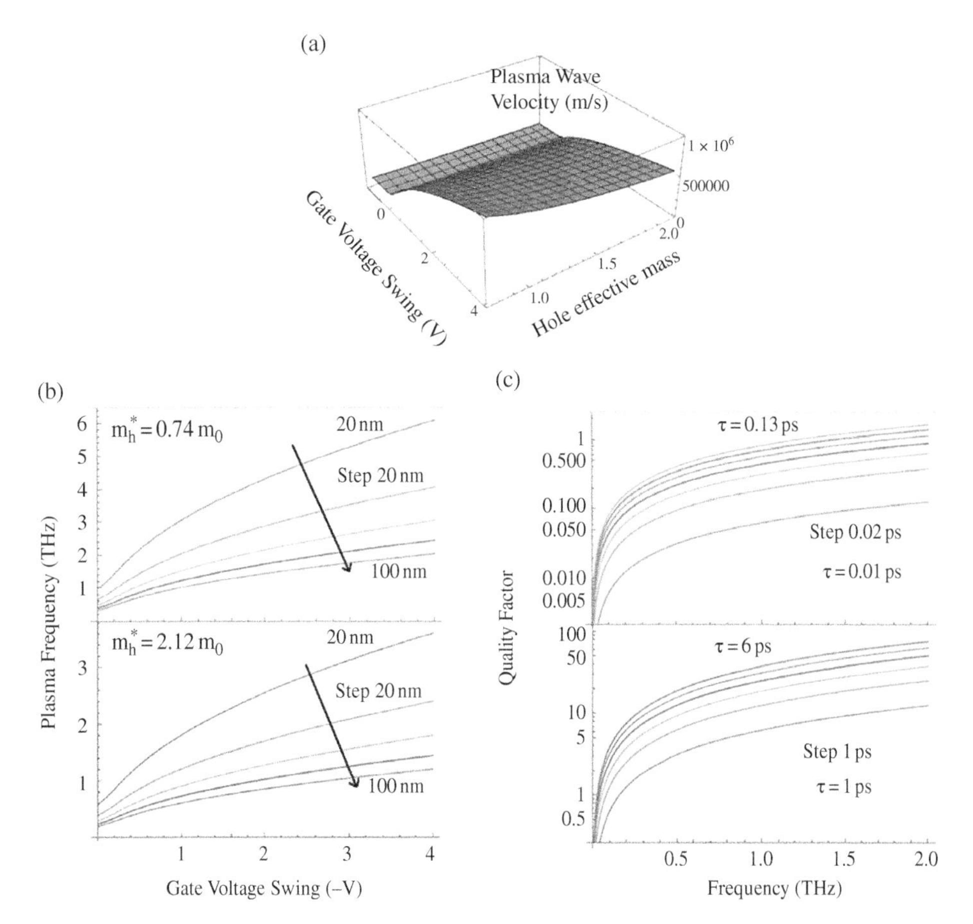

Figure 8.13 (a) Plasma velocity in diamond 2DHG as a function of hole effective mass and gate voltage swing; (b) Fundamental plasma frequency versus gate voltage swing for hole effective mass of $0.74m_0$ and $2.12m_0$; (c) Quality factor vs fundamental plasma frequency.

Figure 8.13b shows the fundamental plasma frequency versus gate voltage swing for the hole effective mass of 0.74 and 2.12, respectively, for the gate length ranging from 20 to 100 nm. The calculation was done for the boundary conditions of the open drain and the THz signal applied between the source and gate $f_p = s/4\,L$ where L is the gate length, s is the plasma velocity which was calculated using [19]

$$s = \sqrt{\frac{\eta V_{th}}{m_{eff} m_0} \left(1 + e^{-\frac{U_0}{\eta V_{th}}}\right) \, ln\left(1 + e^{-\frac{U_0}{\eta V_{th}}}\right)} \tag{8.55}$$

where U_0 is the gate voltage swing, V_{th} is the thermal voltage, η is the subthreshold ideality factor, m_{eff} is the hole effective mass, and m_0 is the free electron mass.

The calculated quality factor at the fundamental plasma frequency versus gate length is shown in Figure 8.13c. The calculations are done for the two ranges of the momentum relaxation time typical for low mobility and high mobility diamond FETs, respectively: from 0.01 to 0.13 ps and from 1 to 6 ps, respectively. As seen from the comparison of Figures 8.13b, c, a

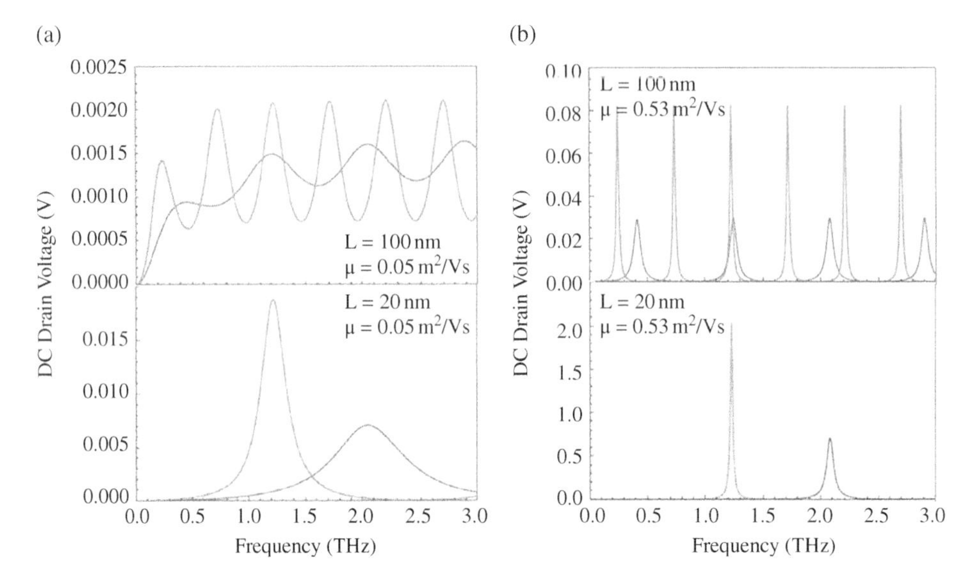

Figure 8.14 DC voltage response to the THz voltage of the 0.01 V amplitude as a function of frequency. Lines with larger and smaller DC response are calculated for the hole effective mass of 2.12 and 0.74, respectively. The gate voltage swing is 0.0254 V.

sharp plasma resonance is expected in the high mobility samples and the high mobility FETs could even operate in the resonance mode in the 200–300 GHz atmospheric window. This makes the diamond plasmonic transistors to be very appealing candidates for the THz communication applications for WiFi beyond 5G and for the wireless links to optical fibers. This conclusion is further confirmed by the calculations of the diamond transistors response to the sub-THz and THz radiation presented in Figure 8.14, which shows the calculated DC voltage response to the THz voltage of the 0.01 V amplitude as a function of frequency assuming the open-drain boundary conditions.

The ultimate predicted performance of diamond plasmonic detectors shows that these devices are a viable contender, especially for high temperature, high voltage, and radiation hard applications, including the 200–300 GHz atmospheric window that is of the special interest for the beyond 5G THz communications. New design and approaches based on the diamond transistor physics are required to achieve the transistor performance with high hole field-effect mobility and for minimizing the effect of parasitics, such as the perforated channel [93] plasmonic crystal [94], and plasmonic ratchet [95] designs. When such p-channel diamond transistors are achieved, diamond will become the material of choice for plasmonic THz applications in beyond 5G and IoT communications, defense, and bio-medical and industrial sensing.

8.7 Conclusion

The THz technology for sensing, imaging, and communications requires detectors with high responsivity, selectivity, and large bandwidth and compact and efficient THz sources.

Table 8.1 A summary of recent plasmonic THz detectors.

Material and device type	Responsivity (V/W)	NEP (pW/Hz$^{1/2}$)	Operation temp.	Frequency (THz)	Mode	Reference
GaN/AlGaN HEMTs	15 000	25–31	RT	0.49–0.65	Broadband	[96]
		0.48	RT	0.14	Broadband	[97]
GaAs HEMTs	70–80	15–1000	RT	0.24	Broadband	[98]
InP HEMTs	22 700	0.48	RT	0.2	Resonant	[29]
65 nm Si CMOS	140	100	RT	0.7–1.1	Broadband	[37]
65 nm Si CMOS	2200	14	RT	0.724	Resonant	[99]
150 nm Si CMOS	350, 50, 5	42, 487	295 K	0.595, 2.9, 4.1	Resonant	[44]
90 nm Si CMOS	75	20, 59, 63, 85, 110,404	RT	0.59, 0.216, 2.52, 3.11, 4.254.75	Resonant	[100] [101]
90 nm Si CMOS	220	48–70	RT	0.6–1.5	Broadband	[102]
Si CMOS, single pixel	19 000	535	RT	0.2	Broadband	[103]
10 × 10 Si CMOS FPA	16 400	216×10^3	RT	0.2	Broadband	[103]
Graphene FETs	14 0.1	515 3×10^4, 2×10^5	300 K	0.6 0.3	Broadband	[58][104]
Black phosphorous FETs	0.15	40,000	RT	0.298		[67]
MoS_2	2[a]		RT	2.52		[105]

[a] For comparison—original data given as current responsivity (mA/W) is rescaled to V/W using the reported 200 Ω device resistance.

The plasma wave electronics detectors based on different material systems have demonstrated excellent performance reaching responsivities up to tens of kV/W and NEP down to the sub-pW/Hz$^{1/2}$ range. A summary of the performance of plasmonic detectors reported recently in the literature is provided in Table 8.1.

A big advantage of plasma wave electronics detectors is that they are tunable by applied bias voltage and can be modulated at very high frequencies (up to 200 GHz and above) [106]. This makes this technology very attractive for potential applications in ultrahigh-speed wireless communications for ultrahigh-resolution video with the advantages of big reductions in cost system size, and power consumption [107, 108]. If this potential is achieved the plasmonic electronics could become a dominant THz electronics technology for THz applications.

Future development of the plasmonic THz technology might focus on the plasmonic crystal structures with each unit cell having deep submicron feature sizes but the entire structure having dimensions comparable to the wavelength of the THz radiation. Variable width plasmonic structures should allow for synchronizing the response of the individual cells [109–111]. With the minimum feature sizes of Si VLSI reaching 3 nm [112] there are new opportunities for realizing in Si the THz plasmonic crystal components as well as the THz vector plasmonic detector implemented in Si that detect both the signal intensity and phase [112]. Multilayered graphene heterostructures [113] might compete with more conventional materials systems for superior plasmonic THz detectors and sources.

Given recent achievements in the plasmonic THz technology and emerging new avenues for further dramatic improvement in performance we expect that by the time of the future deployment of beyond 5G sub-THz WiFi THz plasmonic technology will be ready for this application.

Exercises

E8.1 According to the Drude model, the dielectric permittivity of a 3D conducting layer is

$$\varepsilon = \varepsilon_o(\varepsilon_1 + j\varepsilon_2), \text{where } \varepsilon_1 = \varepsilon_r\left(1 - \frac{\omega_p^2}{\omega^2 + \gamma^2}\right) \text{ and } \varepsilon_2 = \varepsilon_r \frac{\gamma\omega_p^2}{\omega(\omega^2 + \gamma^2)} \tag{1}$$

Here ε_o is the permittivity of free space, ε_o is the dielectric constant, ω_p is the plasma frequency, $\gamma = \frac{1}{\tau} = \frac{q}{m*\mu}$, where $m*$ is the effective mass, and q is the electronic charge. For the three dimensional layer

$$\omega_p^2 = \frac{q^2 n}{\varepsilon_0 \varepsilon_r m} \tag{2}$$

The absorption in the layer is determined by imaginary part of the wave vector

$$\alpha = \frac{\varepsilon_2 \omega}{\left(\varepsilon_1/2 + (\varepsilon_1^2 + \varepsilon_2^2)^{0.5}/2\right)^{0.5} c} \tag{3}$$

The normalized transmitted intensity, considering only absorption, is given as

$$T = e^{-\alpha t} \tag{4}$$

where t is the layer thickness. Using the parameters for Si: $m* = 0.19$, $\varepsilon_r = 11.7$, $n = 5 \times 10^{23}\ \text{m}^{-3}$, $\mu = 0.06\ \text{m}^2/\text{Vs}$ and assuming $t = 0.01$ m, calculate T in the frequency range of 10 GHz to 100 THz.

E8.2 Using the parameter table given below calculate and compare fundamental plasma frequencies of Si, InGaAs/GaAs and AlGaN/GaN TeraFETs for 20 nm, 60 nm, and 130 nm gate lengths. Use the barrier thicknesses of 2 nm for Si and 10 nm for other materials systems for the boundary condition of the grounded source and open drain (so that the fundamental plasma frequency fp = s/(4L), where s is the plasma velocity and L is the channel length.

Material	n_s (m^{-2})	m^*	ε_r
Si	1×10^{16}	0.19	11.7
InGaAs/GaAs	4×10^{16}	0.041	12.1
AlGaN/GaN	1×10^{17}	0.23	8.9

E8.3 By equating the transit time for AlGaN/GaN HEMT $\tau_t = L^2/(\mu V_{ds})$ for the drain-to-source voltage of 100 mV to the momentum relaxation time $\tau = (m^*\mu)/q$ determine the critical mobility μ_{cr} of switching form the collision dominated to the ballistic regime for the gate length of $L = 60$ nm and plot it as a function of the gate length.

E8.4 Calculate plasmonic quality factors $\omega_p\tau$ for Si, GaN, and InGaAs FETs for 20 nm, 60 nm, and 130 nm gate lengths.

E8.5 To account for the terahertz field spatial distribution in the device channel, the Si MOS model has to split the channel into the segments with a minimum number of the segments being $3L/L_0$, where L is the channel length, and $L_0 = L_o = \sqrt{\mu V_{gte}/(2\pi f)}$ is the characteristic plasmonic decay distance (see W. Knap, D. B. But, N. Dyakonova, D. Coquillat, A. Gutin, O. Klimenko, S. Blin, F. Teppe, M. S. Shur, T. Nagatsuma, and S. D. Ganichev, "Recent Results on Broadband Nano-transistor Based THz Detectors," in Proc. THz Secur. Appl., pp. 189–209, Mar. 2014.) Here f is the THz radiation frequency and V_{gte} is the effective gate voltage swing. Calculate the minimum number of such segment for Si for 20 nm and 130 nm gate lengths as a function of frequency (between 0.1 THz and 6 THz) and gate voltages between 0.1 and 0.3 V.

E8.6 Calculate the effective mobility accounting for the ballistic scattering for Si, InGaAs, and GaN HEMTs with the gate length of $L_G = 20$ nm at 77 K and 300 K. ($1/\mu_{eff} = 1/\mu_{balistic} + 1/\mu_{scattering}$)

E8.7 Calculate mean free path for Si, GaN, mad GaAs FETs as a function of temperature between 30 K and 300 K assuming the degenerate 2DEG.

E8.8 Calculate the plasma velocity in gated graphene FET where the gate is separated from the graphene channel with a 100 nm SiO_2 layer, for the gate bias range of 0 – 3 V.

E8.9 Electron-electron collision time could be estimated as $\tau_e = (\hbar E_F)/(3.4(kT)^2)$ where E_F is the Fermi energy, k is the Boltzmann constant and T is the temperature. Compare this time as a function of temperature 77 K< T < 300 K with the momentum relaxation time. Assume $E_F = 0.1$ eV, mobility of ~ 0.1 (T_o/T) m^2/Vs, T_o= 300 K, and effective mass of 0.23.

E8.10 What is a plasma wave and what factors affect its characteristics?

E8.11 What is the basic principle of plasmonic THz detection by an FET?

E8.12 What are the advantages of graphene for plasmonic THz detectors?

E8.13 What properties of p-diamond make it promising material for plasmonic THz detectors?

References

1 Arnone, D., Ciesla, C., and Pepper, M. Terahertz imaging comes into view. *Physics World* **13** (4): 35–40.

2 Markelz, A.G., Roitberg, A., and Heilweil, E.J. Pulsed terahertz spectroscopy of DNA, bovine serum albumin and collagen between 0.1 and 2.0 THz. *Chemical Physics Letters* **320** (1–2): 42–48. https://doi.org/10.1016/S0009-2614(00)00227-X.

3 Walther, M., Fischer, B., Schall, M. et al. Far-infrared vibrational spectra of all-trans, 9-cis and 13-cis retinal measured by THz time-domain spectroscopy. *Chemical Physics Letters* **332** (3–4): 389–395. https://doi.org/10.1016/S0009-2614(00)01271-9.

4 Liu, H.-B., Chen, Y., Bastiaans, G.J., and Zhang, X.-C. (2006). Detection and identification of explosive RDX by THz diffuse reflection spectroscopy. *Optics Express* **14** (1): 415–423.

5 Mittleman, D.M., Cunningham, J., Nuss, M.C., and Geva, M. Noncontact semiconductor wafer characterization with the terahertz Hall effect. *Applied Physics Letters* **71** (1): 16–18. https://doi.org/10.1063/1.119456.

6 Ganichev, S.D., Ziemann, E., Yassievich, I.N. et al. (2001). Characterization of deep impurities in semiconductors by terahertz tunneling ionization. *Materials Science in Semiconductor Processing* **4** (1–3): 281–284. https://doi.org/10.1016/S1369-8001(00)00120-7.

7 Zhang, W., Azad, A.K., and Grischkowsky, D. (2003). Terahertz studies of carrier dynamics and dielectric response of n-type, freestanding epitaxial GaN. *Appl Phys Lett* **82** (17): 2841–2843. https://doi.org/10.1063/1.1569988.

8 Yamashita, M., Kawase, K., Otani, C. et al. (2005). Imaging of large-scale integrated circuits using laser terahertz emission microscopy. *Optics Express* **13** (1): 115–120. https://doi.org/10.1364/Opex.13.000115.

9 Kiwa, T., Tonouchi, M., Yamashita, M., and Kawase, K. (2003). Laser terahertz-emission microscope for inspecting electrical faults in integrated circuits. *Optics Letters* **28** (21): 2058–2060. https://doi.org/10.1364/Ol.28.002058.

10 Shur, M., Rudin, S., Rupper, G. et al. (2019). Tera FETs for terahertz communications. *Photonics Newsletter* **33** (3): 4–7.

11 Akyildiz, I.F., Jornet, J.M., and Han, C. (2014). Terahertz band: next frontier for wireless communications. *Physical Communication* **12**: 16–32.

12 Dat, P.T., Kanno, A., Yamamoto, N., and Kawanishi, T. (2018). Seamless convergence of fiber and wireless systems for 5G and beyond networks. *Journal of Lightwave Technology* **37** (2): 592–605.

13 Yazgan, A., Jofre, L., and Romeu, J. (2017). The state of art of terahertz sources: a communication perspective at a glance. In: *40th International Conference on Telecommunications and Signal Processing (TSP)*, 810–816. IEEE.

14 Jastrow, C., Mu, K., Piesiewicz, R. et al. (2008). 300 GHz transmission system. *Electronics Letters* **44** (3): 213–214.

15 Song, H.-J., Ajito, K., Hirata, A. et al. (2009). 8 Gbit/s wireless data transmission at 250 GHz. *Electronics Letters* **45** (22): 1121–1122.

16 Pirinen, P. (2014). A brief overview of 5G research activities. In: *1st International Conference on 5G for Ubiquitous Connectivity*, 17–22. IEEE.

17 Dyakonov, M. and Shur, M. (1993). Shallow water analogy for a ballistic field effect transistor: new mechanism of plasma wave generation by dc current. *Physical Review Letters* **71** (15): 2465.

18 Dyakonov, M. and Shur, M. (1996). Detection, mixing, and frequency multiplication of terahertz radiation by two-dimensional electronic fluid. *Ieee T Electron Dev* **43** (3): 380–387.

19 Knap, W. et al. (2002). Nonresonant detection of terahertz radiation in field effect transistors. *Journal of Applied Physics* **91** (11): 9346–9353.

20 Stillman, W., Shur, M., Veksler, D. et al. (2007). Device loading effects on nonresonant detection of terahertz radiation by silicon MOSFETs. *Electronics Letters* **43** (7): 422–423.

21 Veksler, D., Teppe, F., Dmitriev, A. et al. (2006). Detection of terahertz radiation in gated two-dimensional structures governed by dc current. *Physical Review B* **73** (12): 125328.

22 Knap, W. et al. (2004). Plasma wave detection of sub-terahertz and terahertz radiation by silicon field-effect transistors. *Applied Physics Letters* **85** (4): 675–677.

23 Stillman, W. et al. (2011). Silicon Fin FETs as detectors of terahertz and sub-terahertz radiation. *International Journal of High Speed Electronics and Systems* **20** (01): 27–42.

24 Pala, N., Teppe, E., Veksler, D. et al. (2005). Nonresonant detection of terahertz radiation by silicon-on-insulator MOSFETs, " (in English). *Electronics Letters* **41** (7): 447–449. https://doi.org/10.1049/el:20058182.

25 Otsuji, T., Hanabe, M., and Ogawara, O. (2004). Terahertz plasma wave resonance of two-dimensional electrons in In Ga P/In Ga As/Ga as high-electron-mobility transistors. *Applied Physics Letters* **85** (11): 2119–2121.

26 El Fatimy, A. et al. (2006). Terahertz detection by GaN/AlGaN transistors, " (in English). *Electronics Letters* **42** (23): 1342–1344. https://doi.org/10.1049/cl:20062452.

27 Bandurin, D.A. et al. (2018). Resonant terahertz detection using graphene plasmons. *Nature Communications* **9** (1): 1–8.

28 Regensburger, S., Mittendorff, M., Winnerl, S. et al. (2015). Broadband THz detection from 0.1 to 22 THz with **large area** field-effect transistors. *Optics Express* **23** (16): 20732–20742.

29 Kurita, Y. et al. (2014). Ultrahigh sensitive sub-terahertz detection by InP-based asymmetric dual-grating-gate high-electron-mobility transistors and their broadband characteristics. *Applied Physics Letters* **104** (25): 251114.

30 Shur, M., Rudin, S., Rupper, G., and Ivanov, T. (2018). p-Diamond as candidate for plasmonic terahertz and far infrared applications. *Applied Physics Letters* **113** (25): 253502.

31 Tauk, R. et al. (2006). Plasma wave detection of terahertz radiation by silicon field effects transistors: responsivity and noise equivalent power. *Applied Physics Letters* **89** (25): 253511.

32 Stillman, W. et al. (2007). Nanometer scale complementary silicon MOSFETs as detectors of terahertz and sub-terahertz radiation, " (in English). *Ieee Sensor*: 934–937. https://doi.org/10.1109/Icsens.2007.4388556.

33 Steegen, A. et al. (2005). 65nm CMOS technology for low power applications. In: *IEEE International Electron Devices Meeting, 2005. IEDM Technical Digest.*, 64–67. IEEE.

34 Pfeiffer, U.R. and Ojefors, E. (2008). A 600-GHz CMOS focal-plane array for terahertz imaging applications. In: *ESSCIRC 2008-34th European Solid-State Circuits Conference*, 110–113. IEEE.

35 Schuster, F. et al. (2011). A broadband THz imager in a low-cost CMOS technology. In: *2011 IEEE International Solid-State Circuits Conference*, 42–43. IEEE.

36 Schuster, F. et al. (2011). Broadband terahertz imaging with highly sensitive silicon CMOS detectors. *Optics Express* **19** (8): 7827–7832.

37 Al Hadi, R. et al. (2012). A 1 k-pixel video camera for 0.7–1.1 terahertz imaging applications in 65-nm CMOS. *IEEE Journal of Solid-State Circuits* **47** (12): 2999–3012.

38 Lisauskas, A. et al. (2012). Detectors for terahertz multi-pixel coherent imaging and demonstration of real-time imaging with a 12x12-pixel CMOS array. In: *Terahertz Emitters, Receivers, and Applications III*, vol. **8496**, 84960J. International Society for Optics and Photonics.

39 Lisauskas, A., Pfeiffer, U., Öjefors, E. et al. (2009). Rational design of high-responsivity detectors of terahertz radiation based on distributed self-mixing in silicon field-effect transistors. *Journal of Applied Physics* **105** (11): 114511.

40 Ojefors, E., Pfeiffer, U.R., Lisauskas, A., and Roskos, H.G. (2009). A 0.65 THz focal-plane array in a quarter-micron CMOS process technology. *IEEE Journal of Solid-State Circuits* **44** (7): 1968–1976.

41 Boppel, S., Lisauskas, A., Krozer, V., and Roskos, H. (2011). Performance and performance variations of sub-1 THz detectors fabricated with 0.15 μm CMOS foundry process. *Electronics Letters* **47** (11): 661–662.

42 Ojefors, E., Baktash, N., Zhao, Y. et al. (2010). Terahertz imaging detectors in a 65-nm CMOS SOI technology. In: *2010 Proceedings of ESSCIRC*, 486–489. IEEE.

43 Marczewski, J. et al. (2018). THz detectors based on Si-CMOS technology field effect transistors–advantages, limitations and perspectives for THz imaging and spectroscopy. *Opto-Electronics Review* **26** (4): 261–269.

44 Boppel, S. et al. (2012). CMOS integrated antenna-coupled field-effect transistors for the detection of radiation from 0.2 to 4.3 THz. *IEEE Transactions on Microwave Theory and Techniques* **60** (12): 3834–3843.

45 Coquillat, D., Marczewski, J., Kopyt, P. et al. (2016). Improvement of terahertz field effect transistor detectors by substrate thinning and radiation losses reduction. *Optics Express* **24** (1): 272–281.

46 Kopyt, P., Salski, B., Marczewski, J. et al. (2015). Parasitic effects affecting responsivity of sub-THz radiation detector built of a MOSFET. *Journal of Infrared, Millimeter, and Terahertz Waves* **36** (11): 1059–1075.

47 Novoselov, K.S. et al. (2004). Electric field effect in atomically thin carbon films. *Science* **306** (5696): 666–669.

48 Grigorenko, A., Polini, M., and Novoselov, K. (2012). Graphene plasmonics. *Nature Photonics* **6** (11): 749–758.

49 Ansell, D., Radko, I., Han, Z. et al. (2015). Hybrid graphene plasmonic waveguide modulators. *Nature Communications* **6** (1): 1–6.

50 Freitag, M., Low, T., Zhu, W. et al. (2013). Photocurrent in graphene harnessed by tunable intrinsic plasmons. *Nature Communications* **4** (1): 1–8.
51 Chakraborty, S., Marshall, O., Folland, T. et al. (2016). Gain modulation by graphene plasmons in aperiodic lattice lasers. *Science* **351** (6270): 246–248.
52 Rodrigo, D. et al. (2015). Mid-infrared plasmonic biosensing with graphene. *Science* **349** (6244): 165–168.
53 Yao, B. et al. (2018). Broadband gate-tunable terahertz plasmons in graphene heterostructures. *Nature Photonics* **12** (1): 22–28.
54 Gusynin, V., Sharapov, S., and Carbotte, J. (2006). Magneto-optical conductivity in graphene. *Journal of Physics: Condensed Matter* **19** (2): 026222.
55 Li, Z. et al. (2008). Dirac charge dynamics in graphene by infrared spectroscopy. *Nature Physics* **4** (7): 532–535.
56 Vakil, A. and Engheta, N. (2011). Transformation optics using graphene. *Science* **332** (6035): 1291–1294.
57 Ju, L. et al. (2011). Graphene plasmonics for tunable terahertz metamaterials. *Nature Nanotechnology* **6** (10): 630.
58 Vicarelli, L. et al. (2012). Graphene field-effect transistors as room-temperature terahertz detectors. *Nature Materials* **11** (10): 865–871.
59 Xia, F., Mueller, T., Lin, Y.-M. et al. (2009). *Ultrafast graphene photodetector. Nature Nanotechnology* **4** (12): 839–843.
60 Karabiyik, M., Al-Amin, C., and Pala, N. (2013). Graphene-based periodic gate field effect transistor structures for terahertz applications. *Nanoscience and Nanotechnology Letters* **5** (7): 754–757. https://doi.org/10.1166/nnl.2013.1622.
61 Generalov, A.A., Andersson, M.A., Yang, X. et al. (2017). A 400-GHz graphene FET detector. *Ieee T Thz Sci Techn* **7** (5): 614–616.
62 Spirito, D. et al. (2014). High performance bilayer-graphene terahertz detectors. *Applied Physics Letters* **104** (6): 061111.
63 Tong, J., Muthee, M., Chen, S.-Y. et al. (2015). Antenna enhanced graphene THz emitter and detector. *Nano Letters* **15** (8): 5295–5301.
64 Qin, H. et al. (2017). Room-temperature, low-impedance and high-sensitivity terahertz direct detector based on bilayer graphene field-effect transistor. *Carbon* **116**: 760–765.
65 Cai, X. et al. (2014). Sensitive room-temperature terahertz detection via the photothermoelectric effect in graphene. *Nature Nanotechnology* **9** (10): 814.
66 Auton, G. et al. (2017). Terahertz detection and imaging using graphene ballistic rectifiers. *Nano Letters* **17** (11): 7015–7020.
67 Viti, L. et al. (2015). Black phosphorus terahertz photodetectors. *Advanced Materials* **27** (37): 5567–5572.
68 Tao, L. et al. (2015). Silicene field-effect transistors operating at room temperature. *Nature Nanotechnology* **10** (3): 227–231.
69 Dávila, M., Xian, L., Cahangirov, S. et al. (2014). Germanene: a novel two-dimensional germanium allotrope akin to graphene and silicene. *New Journal of Physics* **16** (9): 095002.
70 Liu, H. et al. (2014). Phosphorene: an unexplored 2D semiconductor with a high hole mobility. *ACS Nano* **8** (4): 4033–4041.

71 Xia, F., Wang, H., and Jia, Y. (2014). Rediscovering black phosphorus as an anisotropic layered material for optoelectronics and electronics. *Nature Communications* **5** (1): 1–6.
72 Yuan, H. et al. (2015). Polarization-sensitive broadband photodetector using a black phosphorus vertical p–n junction. *Nature Nanotechnology* **10** (8): 707.
73 Viti, L., Hu, J., Coquillat, D. et al. (2016). Efficient terahertz detection in black-phosphorus nano-transistors with selective and controllable plasma-wave, bolometric and thermoelectric response. *Sci Rep-Uk* **6**: 20474.
74 Twitchen, D. et al. (2004). High-voltage single-crystal diamond diodes. *IEEE Transactions on Electron Devices* **51** (5): 826–828.
75 Chao, P.-C. et al. (2015). Low-temperature bonded GaN-on-diamond HEMTs with 11 W/mm output power at 10 GHz. *Ieee T Electron Dev* **62** (11): 3658–3664.
76 Kawarada, H. et al. (2017). Durability-enhanced two-dimensional hole gas of CH diamond surface for complementary power inverter applications. *Scientific Reports* **7**: 42368.
77 Vescan, A., Daumiller, I., Gluche, P. et al. (1997). Very high temperature operation of diamond Schottky diode. *IEEE Electron Device Letters* **18** (11): 556–558.
78 Balmer, R.S. et al. (2013). Transport behavior of holes in boron delta-doped diamond structures. *Journal of Applied Physics* **113** (3): 033702.
79 Nebel, C.E. (2007). Surface-conducting diamond. *Science* **318** (5855): 1391–1392.
80 Maier, F., Riedel, M., Mantel, B. et al. (2000). Origin of surface conductivity in diamond. *Physical Review Letters* **85** (16): 3472.
81 Wang, W. et al. (2015). Diamond based field-effect transistors of Zr gate with dielectric layers. *Journal of Nanomaterials* **2015**: 124640.
82 Hirama, K., Sato, H., Harada, Y. et al. (2012). Thermally stable operation of H-terminated diamond FETs by NO_2 Adsorption and Al_2O_3 Passivation. *Ieee Electr Device L* **33** (8): 1111–1113.
83 Aleksov, A. et al. (2003). Properties of (111) diamond homoepitaxial layer and its application to field-effect transistor. *Diamond and Related Materials* **12**: 391–392.
84 Iwasaki, T. et al. (2013). High-temperature operation of diamond junction field-effect transistors with lateral pn junctions. *Ieee Electr Device L* **34** (9): 1175–1177.
85 Sato, Y. et al. (2015). Radiation hardness of a single crystal CVD diamond detector for MeV energy protons. *Nuclear Instruments and Methods in Physics Research Section A: Accelerators, Spectrometers, Detectors and Associated Equipment* **784**: 147–150.
86 Akimoto, I., Naka, N., and Tokuda, N. (2016). Time-resolved cyclotron resonance on dislocation-free HPHT diamond. *Diamond and Related Materials* **63**: 38–42.
87 Isberg, J. et al. (2002). High carrier mobility in single-crystal plasma-deposited diamond. *Science* **297** (5587): 1670–1672.
88 Gheeraert, E., Koizumi, S., Teraji, T. et al. (1999). Electronic states of boron and phosphorus in diamond. *Physica Status Solidi* **174** (1): 39–51.
89 Löfås, H., Grigoriev, A., Isberg, J., and Ahuja, R. (2011). Effective masses and electronic structure of diamond including electron correlation effects in first principles calculations using the GW-approximation. *AIP Advances* **1** (3): 032139.

90 Naka, N., Fukai, K., Handa, Y., and Akimoto, I. (2013). Direct measurement via cyclotron resonance of the carrier effective masses in pristine diamond. *Physical Review B* **88** (3): 035205.

91 Kachorovskii, V.Y. and Shur, M. (2012). Current-induced terahertz oscillations in plasmonic crystal. *Applied Physics Letters* **100** (23): 232108.

92 Aizin, G., Mikalopas, J., and Shur, M. (2016). Current-driven plasmonic boom instability in three-dimensional gated periodic ballistic nanostructures. *Physical Review B* **93** (19): 195315.

93 Simin, G.S., Islam, M., Gaevski, M. et al. (2014). Low RC-constant perforated-channel HFET. *Ieee Electr Device L* **35** (4): 449–451.

94 Petrov, A.S., Svintsov, D., Ryzhii, V., and Shur, M.S. (2017). Amplified-reflection plasmon instabilities in grating-gate plasmonic crystals. *Physical Review B* **95** (4): 045405.

95 Rozhansky, I., Kachorovskii, V.Y., and Shur, M. (2015). Helicity-driven ratchet effect enhanced by plasmons. *Physical Review Letters* **114** (24): 246601.

96 Bauer, M. et al. (2019). A high-sensitivity AlGaN/GaN HEMT terahertz detector with integrated broadband bow-tie antenna. *IEEE Transactions on Terahertz Science and Technology* **9** (4): 430–444.

97 Hou, H., Liu, Z., Teng, J. et al. (2016). A sub-terahertz broadband detector based on a GaN high-electron-mobility transistor with nanoantennas. *Applied Physics Express* **10** (1): 014101.

98 Watanabe, T. et al. (2012). InP-and GaAs-based plasmonic high-electron-mobility transistors for room-temperature ultrahigh-sensitive terahertz sensing and imaging. *IEEE Sensors Journal* **13** (1): 89–99.

99 Pfeiffer, U., Grzyb, J., Sherry, H. et al. (2013). Toward low-NEP room-temperature THz MOSFET direct detectors in CMOS technology. In: *2013 38th International Conference on Infrared, Millimeter, and Terahertz Waves (IRMMW-THz)*, 1–2. IEEE.

100 Bauer, M. et al. (2014). Antenna-coupled field-effect transistors for multi-spectral terahertz imaging up to 4.25 THz. *Optics Express* **22** (16): 19235–19241.

101 Zdanevičius, J. et al. (2018). Field-effect transistor based detectors for power monitoring of THz quantum cascade lasers. *Ieee T Thz Sci Techn* **8** (6): 613–621.

102 Ikamas, K., Čibiraitė, D., Lisauskas, A. et al. (2018). Broadband terahertz power detectors based on 90-nm silicon CMOS transistors with flat responsivity up to 2.2 THz. *Ieee Electr Device L* **39** (9): 1413–1416.

103 Zak, A. et al. (2014). Antenna-integrated 0.6 THz FET direct detectors based on CVD graphene. *Nano Letters* **14** (10): 5834–5838.

104 Xie, Y. et al. (2020). Defect engineering of MoS_2 for Room-temperature terahertz photodetection. *ACS Applied Materials and Interfaces* **12** (6): 7351–7357.

105 Wang, Q. et al. (2017). Ultrafast broadband photodetectors based on three-dimensional Dirac semimetal Cd_3As_2. *Nano Letters* **17** (2): 834–841.

106 Kachorovskii, V.Y. and Shur, M. (2008). Field effect transistor as ultrafast detector of modulated terahertz radiation. *Solid State Electronics* **52** (2): 182–185.

107 Kachorovskii, V.Y., Rumyantsev, S., Knap, W., and Shur, M. (2013). Performance limits for field effect transistors as terahertz detectors. *Applied Physics Letters* **102** (22): 223505.

108 Otsuji, T. and Shur, M. (2014). Terahertz plasmonics: good results and great expectations. *IEEE Microwave Magazine* **15** (7): 43–50.

109 Aizin, G., Mikalopas, J., and Shur, M. (2018). Plasmons in ballistic nanostructures with stubs: transmission line approach. *IEEE Transactions on Electron Devices* **66** (1): 126–131.

110 Aizin, G., Mikalopas, J., and Shur, M. (2018). Current-driven Dyakonov-Shur instability in ballistic nanostructures with a stub. *Physical Review Applied* **10** (6): 064018.

111 Shur, M., Mikalopas, J., and Aizin, G. (2020). Compact design models of cryo and room temperature Si MOS, GaN, InGaAs, and p-diamond HEMT teraFETs. pp. 209–212.

112 Mujtaba, H. (2020). Intel lays down 2019-2029 manufacturing roadmap – 1.4nm in 10 years, 2 year cadence with back porting on advanced ++ nodes. https://wccftech.com/intel-2021-2029-process-roadmap-10nm-7nm-5nm-3nm-2nm-1nm-back-porting/

113 Ryzhii, V., Otsuji, T., and Shur, M. (2020). Graphene based plasma-wave devices for terahertz applications. *Applied Physics Letters* **116** (14): 140501.

9

Signal Generation by Diode Frequency Multiplication

Alain Maestrini and Jose V. Siles

Jet Propulsion Laboratory, California Institute of Technology, Pasadena, CA, USA

9.1 Introduction

Frequency multiplication is a widely used technique to generate a coherent signal at high frequency from a lower frequency fundamental source. It is used in electronic systems whenever it brings an advantage in terms of performance or cost. At centimeter or millimeter wavelengths, frequency multipliers can be either passive or active. The first type multipliers use diodes, the last ones require transistors. At submillimeter wavelengths, though, diodes are currently the only devices available for building high-efficiency frequency multipliers. In particular, GaAs Schottky diodes, which can exhibit dynamic cutoff frequencies of 4 THz or more, have been instrumental during the past two decades in pushing the frontiers of room-temperature electronic sources well above 1 THz.

High-performance terahertz multiplier sources are frequency agile, tunable over more than 10% of relative bandwidth, while broader bandwidth sources can be designed to cover up to 40% of relative bandwidth. They are compact, radiation hardened, and robust enough to withstand the toughest environment like the vicinity of Jupiter and its icy moons. Terahertz frequency multiplied sources are indeed used in many practical systems ranging from terahertz active imagers and radars to spaceborne heterodyne instruments for studying the interstellar medium or the atmosphere of planets.

The progress of terahertz multiplied sources is closely tied to the progress of power amplifiers (PAs) at millimeter wavelength, in particular at W-band (75–110 GHz), since PAs set the input power of the terahertz multiplier chain that follows them. Starting from about 150 mW from a single GaAs W-band PA in the early 2000s to more than 1 W from a GaN counterpart in the late 2010s, the input power of each stage of the terahertz multiplier chains has been dramatically increased, leading to a tenfold increase in output power levels in the 1–3

The research was carried out at the Jet Propulsion Laboratory, California Institute of Technology, under a contract with the National Aeronautics and Space Administration (80NM0018D0004).

Fundamentals of Terahertz Devices and Applications, First Edition. Edited by Dimitris Pavlidis.

Companion website: www.wiley.com/go/Pavlidis/FundamentalsofTHz

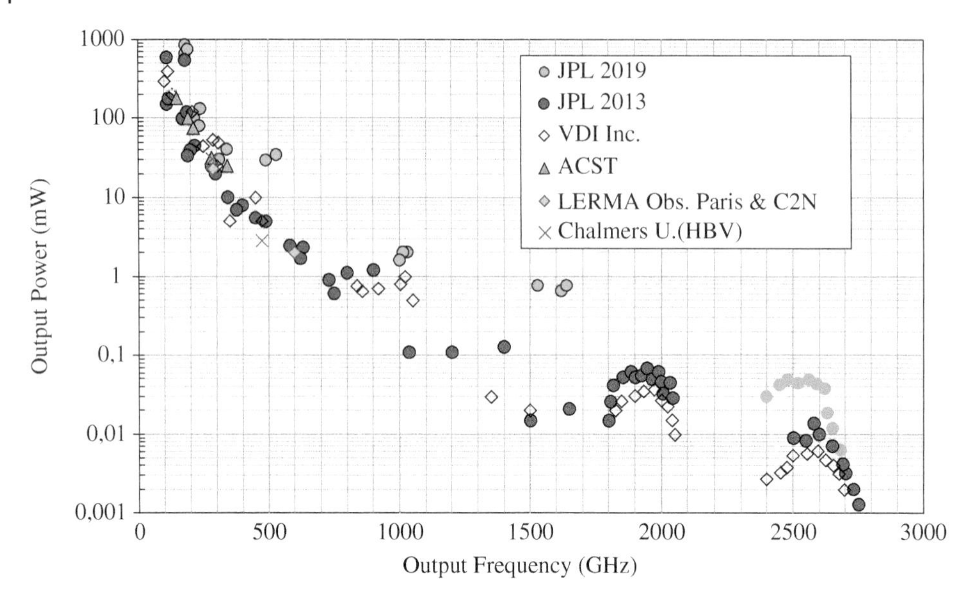

Figure 9.1 State-of-the-art of frequency multiplier sources at room temperature (2019).

THz range [1, 2]. The size and mass of terahertz multiplier sources have also dramatically decreased since a single 1 W $20 \times 20 \times 20\,mm^3$ W-band GaN PA can replace a four-way power-combined GaAs PA that requires more than 10 times the same mass and volume.

This chapter gives a brief state-of-the-art of terahertz multipliers in Section 9.2 and presents the most important results in terms of output power versus output frequency in Figure 9.1. Section 9.3 introduces a practical approach to the design of frequency multipliers. Section 9.4 presents the evolution of THz frequency multiplier technology, with an emphasis on the building of the local oscillators of the Heterodyne Instrument for the Far-Infrared (HIFI) [3] on board of the Herschel Space Observatory, and the latest results toward building milliwatt-level tunable electronic sources working above 1 THz. Section 9.5 of this chapter presents the most recent power-combining techniques for millimeter and sub-millimeter wave frequency multipliers (Figure 9.2).

9.2 Bridging the Microwave to Photonics Gap with Terahertz Frequency Multipliers

Lack of tunable, broadband, robust, and reliable power sources in the submillimeter-wave frequency range has been a major limiting factor in developing applications in this part of the spectrum. The range from 0.3 THz, where transistors show only limited gain, to about 10 THz, where room temperature solid-state lasers become available, is of increasing scientific interest where sources are much needed [4–9].

Photonic solutions to coherent generation at terahertz frequencies have dominated the field for decades starting with far-infrared lasers able to produce tens of milliwatts of

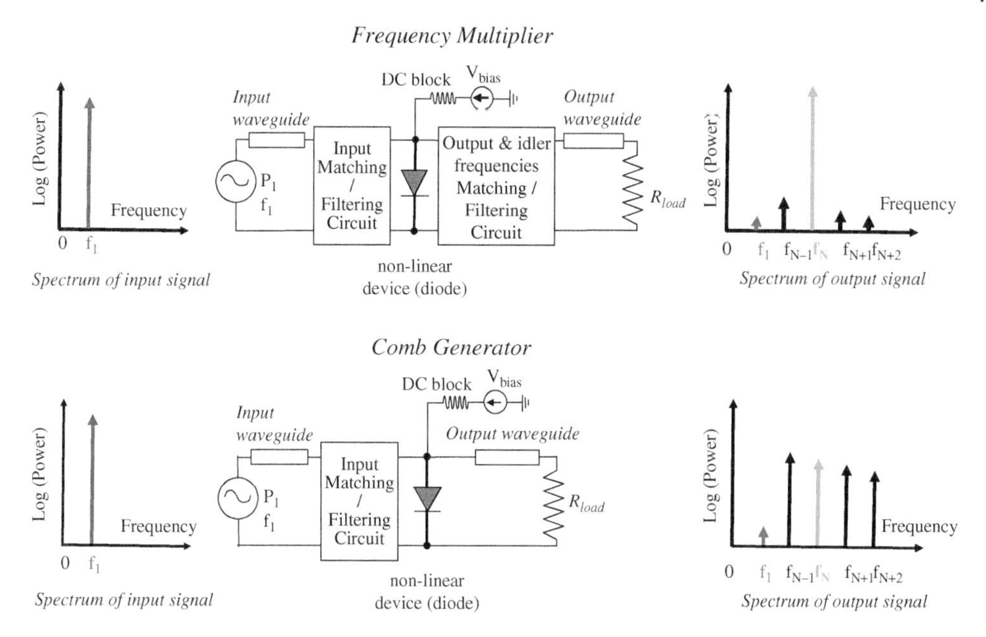

Figure 9.2 Frequency multiplier vs comb-generator.

coherent power, to femtosecond infrared lasers and photoconductors that enable broadband terahertz sources suited for numerous spectro-imagery applications [10]. Photomixers are also an attractive solution for generating coherent terahertz CW waves thanks to their wide frequency tunability [11, 12]. Recently, quantum cascade lasers have made incursions into the sub-terahertz domain and are routinely delivering milliwatts or tens of milliwatts in the 1 to 5 THz range [9, 13, 14], albeit at cryogenic temperatures and with limited bandwidth. These lasers have been successfully phase-locked and used to build the local oscillator of heterodyne receivers based on a super-conducting hot-electron-bolometer (HEB) mixers working at 2.8 THz [15] and at 4.75 THz [16].

In contrast, electronic sources in the terahertz region are scarce; if we put aside non-solid-state sources like the power-hungry and heavy backward-wave oscillators (BWOs) that can work to about 1.2 THz, there is indeed only one proven solution: frequency multiplier chains from the microwave region to the terahertz.

Until 2010, the power of terahertz frequency multipliers was measured in microwatts rather than milliwatts. At that date, the state-of-the-art at room temperature was 100 μW at 1.2 THz [17–19], 15–20 μW at 1.5–1.6 THz [1, 20], and 3 μW at 1.9 THz [1]. As predicted in [21], these powers improve dramatically upon cooling: the same sources produce respectively, 200, 100, and 30 μW at 120 K.

Despite relatively low output power levels, frequency-multiplier sources have some decisive advantages that make them the technology of choice for building the local oscillators of heterodyne receivers: firstly, they are inherently phase lockable and frequency agile, secondly, they work at room temperature, or at cryogenic temperatures that can be reached by passive cooling in space for enhance performance; thirdly, multiplier sources are robust enough, compact enough, and use a sufficiently low level of dc power to claim several years of heritage in the selective world of space technologies. From AURA [22] to the Herschel Space Observatory [3], frequency multipliers have demonstrated their real-word operability and are proposed for even more challenging missions to the outer planets [23].

The prospect of having a milliwatt-level broadband terahertz frequency-multiplied source seemed farfetched. In 2010, a fully solid-state source based on a cascade of two frequency triplers delivered more than 1 mW at 900 GHz at room temperature [24]. This level of power has enabled the demonstration of an 840–900 GHz fundamental balanced Schottky receiver with state-of-the-art noise and conversion loss [25]. It can also enhance terahertz imaging applications by driving frequency multipliers to even higher frequencies [26–28], such as the 2.5–2.7 THz band.

Considering these results, as well as advances in thermal management of frequency multipliers [29], the continuous progress of PAs around or above 0.1 THz [30], and the prospect of high-breakdown-voltage GaN Schottky diodes for submillimeter-wave multipliers [31, 32], it was clear that electronic coherent sources had the potential to deliver milliwatts of tunable single-mode power well into the terahertz range.

It took indeed near a decade to reach that point. Figure 9.1 shows the state-of-the-art of frequency multiplier sources at room temperature. It includes data from Jet Propulsion Laboratory (JPL) and Virginia Diodes Inc. in the USA and from several groups in Europe. The following sections will detail the steps that made possible the design and the building of a new generation of room temperature THz sources with near 1 mW of power at 1.6 THz and 50 μW at 2.5 THz [2, 33].

9.3 A Practical Approach to the Design of Frequency Multipliers

9.3.1 Frequency Multiplier Versus Comb Generator

A frequency multiplier of order N is a nonlinear electronic device that generates a specific harmonic of an input signal. If f_1 and P_1 are respectively the frequency and the power of the input signal, the output signal frequency will be $f_N = N \times f_1$ and the output power P_N. Practical frequency multipliers generate undesired harmonics at frequencies $f_k = k \times f_1$ with $k \neq N$ and power P_k with $\sum_{k\neq 1, k\neq N}^{+\infty} P_k \ll P_N$.

A frequency multiplier is, therefore, different from a comb generator that generates a series of harmonics which power decreases (usually) with the increasing frequency: $P_k \geq P_{k+1}$.

9.3.2 Frequency Multiplier Ideal Matching Network and Ideal Device Performance

To maximize the transfer of energy from the fundamental signal to its Nth harmonic, optimum matching networks at the input, output, and idler frequencies must be provided at the nonlinear device terminals. For example, a frequency tripler must be properly matched to the second harmonic to effectively produce third harmonic power. Determining the maximum output power that a nonlinear device can produce at a given harmonic is a task that can be solved analytically only in a very limited number of cases, and with simplified device

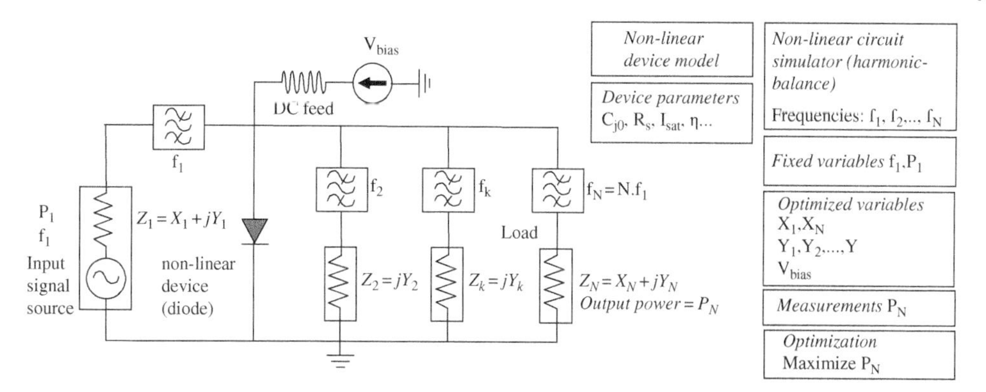

Figure 9.3 Frequency multiplier ideal matching network and optimization.

models [34, 35]. With realistic device models, only numerical simulations can be used. This paragraph will present a method to extract the ideal imbedding impedances of a diode frequency multiplier and calculate its maximum conversion efficiency, using a simplified diode model detailed in Section 9.3.7.

The diode terminals are connected to a sinusoidal source and a set of complex impedances that are isolated from each over by ideal filters (see Figure 9.3). If f_1 is the frequency of the source, $f_k = k \times f_1$, $k = 2 \dots N-1$, the idler frequencies and $f_N = N \times f_1$, the output frequency, these ideal filters present an impedance equal to zero over the bandwidth $\Delta f_k = k \times \Delta f_1$ around the center frequency f_k, with $\Delta f_1 < f_1/N$, and infinite impedance elsewhere. The source impedance Z_1 and the load impedance Z_N are complex with a strictly positive real part, while the impedances at the idler frequencies are purely complex, since for best conversion efficiency no power should be dissipated at the idler frequencies. A DC block is used to bias the diode with the voltage V_{bias} that can be either positive or negative, depending on the type of diode and the input power P_1.

The analysis is made for a fixed frequency f_1 and a fixed input power P_1. To optimize the physical structure of the diode with the ideal imbedding impedances of the frequency multiplier, it is necessary to derive the electrical parameters, like the zero-bias junction capacitance C_{j_0}, the series resistance R_S, the saturation current I_{sat} and the ideality factor η, from the physical parameters of the diode, such as the anode and the ohmic contact geometry, the epilayer structure, and doping levels, with a simple set of analytical equations. If the physical structure of the diode is fixed, this setup gives only the ideal embedding impedance and the maximum conversion efficiency of the frequency multiplier.

Figure 9.3 schematizes a generic bench that can be easily implemented in commercial circuit simulators. It relies on an harmonic balance code and an optimizer, that will maximize the output power P_N by adjusting the value of the impedances $Z_1, \dots, Z_k, \dots Z_N$ and the bias voltage V_{bias}.

Synthetizing an optimum matching network for each frequency involved in the frequency multiplier circuit is a delicate task that can be greatly simplified by the use of symmetrical structures at the device level or at the circuit level.

9.3.3 Symmetry at Device Level Versus Symmetry at Circuit Level

Symmetries introduced at the device level like in heterostructure barrier varactor (HBV) frequency multipliers greatly simplify the circuit topology since only odd harmonics are produced and no bias voltage is applied to the device. This eliminates the need for a matching circuit at even idler frequencies. High order frequency multipliers can, therefore, be more easily designed. A practical advantage of zero-bias operation of HBV frequency multipliers is the absence of DC connection, which removes the risk of device destruction by electrostatic discharge (ESD). Although, HBVs are effective devices at millimeter wavelengths [36–41], (Stake et al., 2017), they never dethroned Schottky diodes nor were demonstrated at terahertz frequencies. HBV diodes require indeed a very precise engineering of the epilayer and a precise material etch to define the active area, which is harder to achieve for very small dimensions than to create a metal-semiconductor contact. In addition, once integrated into a microwave monolithic integrated circuits (MMICs)-like circuit, HBVs diodes are designed to work for a specific input power, with little flexibility, due to the impossibility of biasing them and consequently adjusting their frequency-dependent complex impedances.

Another example of symmetries introduced at the device level can be found in [42]: two anti-serial Schottky diodes are used as nonlinear varactor. The single GaAlAs/GaAs devices with *nonuniform doping profile* optimized for the desired frequency are quasi-monolithically integrated into a microstrip circuit fabricated on a quartz substrate. These devices were used for a 228 GHz frequency tripler that exhibits a state-of-the-art conversion efficiency of 22%. Like for HBV didoes, this tripler is biasless.

Symmetries introduced at the circuit level in Schottky-diode frequency multiplier can, however, result in the elimination of even or odd harmonics, hence enabling broadband operations and high conversion efficiencies with the possibility of biasing the devices for increased bandwidth and operational flexibility.

9.3.4 Classic Balanced Frequency Doublers

At millimeter wavelengths, an efficient balanced doubler topology was introduced by Neal Erickson in [43–46]. This design became a standard and was scaled later up to terahertz frequencies [47–49]. Figure 9.4 shows a picture of a 400 GHz balanced doubler used for the HIFI of the Herschel Space Observatory, launched in 2009 by the European Space Agency (ESA) [3].

9.3.4.1 General Circuit Description

This circuit features four diodes of about 11 fF each fabricated on a 20 μm-thin GaAs membrane and implanted on a split-waveguide block. The diodes are located in the input waveguide cavity and are symmetrical with respect to the central conductor line. The fundamental signal propagates on a TE_{10} mode while the second harmonic is generated on a quasi-transverse electromagnetic (TEM) mode that cannot be coupled to the input waveguide due to its reduced dimensions (details are provided in Section 9.3.4.2). The output signal is then channeled toward the output waveguide through a short section of suspended micro-strip line – under-cutoff for the input signal – and a transition (E-probe.) A bias voltage can be applied evenly to the diodes through the central conductor line that features an

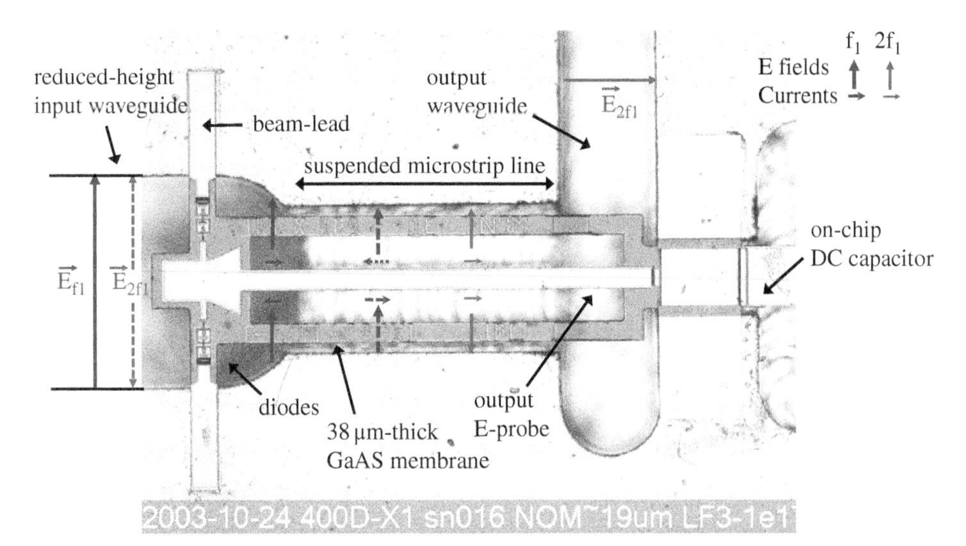

Figure 9.4 JPL 400 GHz four-anode doubler with substrate-less technology (design by E. Schlecht). In this design, the diodes are cathode grounded, and a negative voltage is applied to the bias port to properly operate the doubler in a varactor mode. This doubler was part of the LO chain of the 1.2 THz channel of HIFI. The input matching circuit features additional sections of waveguide of high and low impedance (not shown in the picture.) $\vec{E}_{f_1}$ and $\vec{E}_{2f_1}$ stand respectively for the electric field at the fundamental frequency f_1 and at the output frequency $2f_1$. The electric fields and the current lines are represented for f_1 (thick blue lines), and $2f_1$ (light red lines). Plain lines are for guided modes while dashed lines are for modes under cutoff.

integrated capacitor at its end for DC/RF decoupling. The input and output matching circuits are completed by additional low and high impedance waveguide sections.

Other designs rely on a suspended microstrip or coaxial low-pass filter instead of on-chip capacitors for biasing. On-chip capacitors provide an advantage in terms of chip length at the expense of the complexity of device manufacturing and device quality control. On the other hand, standard suspended microstrip low-pass filters made with low/high impedance sections of propagating lines, or compact hammer low-pass filters (see Figure 9.5) can be modeled with accuracy at terahertz frequencies and are both effective and reliable solutions for DC/RF decoupling.

9.3.4.2 Necessary Condition to Balance the Circuit

Figure 9.6 shows a schematic of the diode area of a four-anode balanced doubler. The two branches of the device feature two anodes each and are perfectly symmetrical with respect to the central suspended microstrip or coaxial line. This condition can always be met for doublers, regardless of the physical dimensions of the diodes.

The fundamental signal at the pulsation $\omega_1 = 2\pi \times f_1$, where f_1 is the frequency, is the input pump signal and set the frequency of each component of the electrical field and of the currents flowing in the multiplier. It creates the electrical field $\vec{E}_1(\omega_1 t)$ in the vicinity of the diodes and propagates in the rectangular input waveguide in TE_{10} mode. The

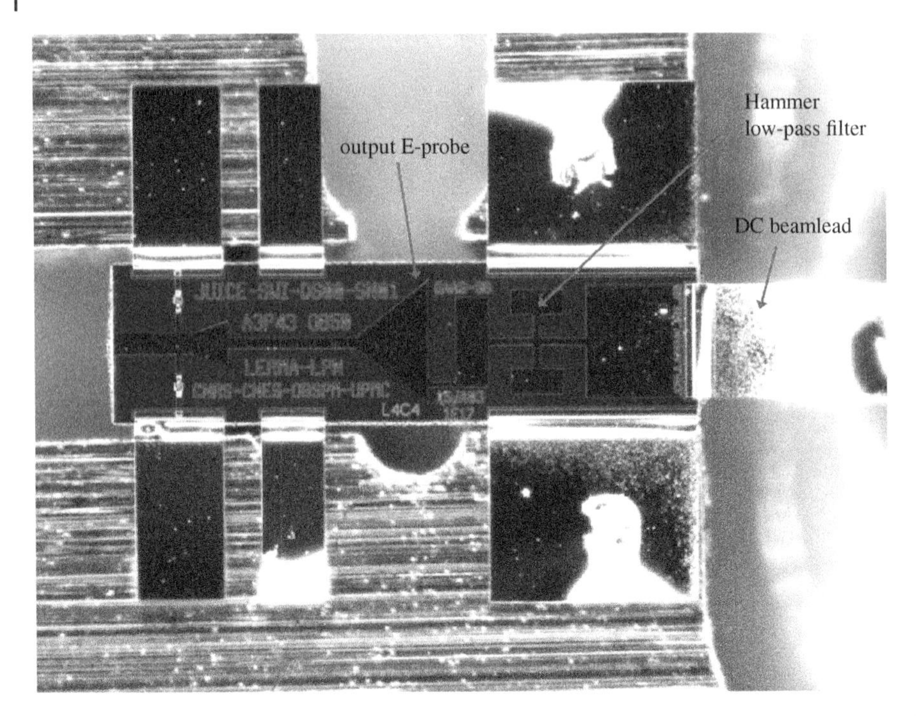

Figure 9.5 LERMA-C2N demonstration model of the 600 GHz two-anode balanced doubler of SWI (design by Jiang-Qiao Ding and A. Maestrini at LERMA-Observatoire de Paris.) The DC/RF decoupling is made with a hammer filter implanted on a 5 μm-thin GaAs membrane. The thin membrane thickness allows for broadband filtering and low-loss with a single filtering cell. No on-chip capacitor is used.

symmetry of the device causes a phase difference of 180° between $i_a(t)$ and $i_b(t)$, respectively the currents flowing in the upper and lower branches of the doubler:

$$i_b(t) = i_a(t + T_1/2)\forall t \tag{9.1}$$

where $T_1 = 1/f_1$ is the period of the pump signal. According to [35, 50] and Eq. (9.1) $i_a(t)$ and $i_b(t)$ can be written as:

$$i_a(t) = \sum_{n=-\infty}^{+\infty} c_n \cdot e^{i \cdot n \cdot \omega_1 \cdot t} \ \forall t \tag{9.2}$$

$$i_b(t) = \sum_{n=-\infty}^{+\infty} c_n \cdot e^{i \cdot n \cdot \omega_1 \cdot (t + T_1/2)} \ \forall t \tag{9.3}$$

where $(C_n)_{n \in \mathbb{Z}}$ are the complex Fourier coefficients with $C_{-n} = C_n^*$ for $n > 0$. In practice, at submillimeter wavelengths, only a limited number of terms are relevant ($n \leq 4$ suffices for doublers). The currents $i_a(t)$ and $i_b(t)$ can be broken down into a series of even and odd harmonics:

$$i_a(t) = \sum_{n=-\infty}^{+\infty} c_{2n} \cdot e^{i \cdot 2n \cdot \omega_1 \cdot t} + \sum_{n=-\infty}^{+\infty} c_{2n+1} \cdot e^{i \cdot (2n+1) \cdot \omega_1 \cdot t} \ \forall t \tag{9.4}$$

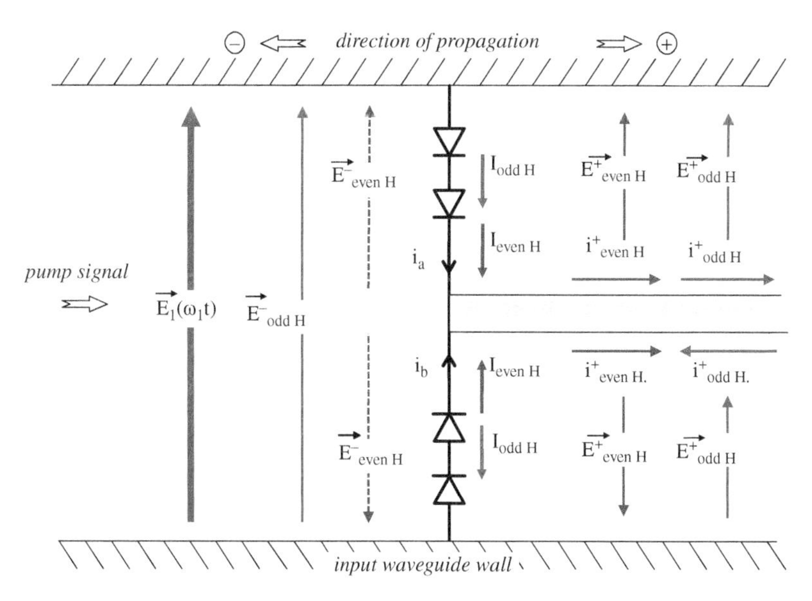

Figure 9.6 Currents and electrical fields in the vicinity of the diodes of a four-anode balanced doubler. In this schematic, the diodes are anode-grounded, therefore, a positive bias is applied to the bias port to operate the doubler in a varactor mode. The polarity of the diodes does not change anything in the doubler except the polarity of the bias voltage that is applied. The designations "even H" and "odd H" refer respectively to even harmonics and odd harmonics. The dashed electrical field lines at the even harmonic indicate that the corresponding mode is under cutoff and cannot propagate (this condition must be met at the second harmonic, but is not met usually at the fourth harmonic).

$$i_b(t) = \sum_{n=-\infty}^{+\infty} c_{2n} \cdot e^{i\cdot 2n\cdot\omega_1\cdot t} \cdot e^{i\cdot 2n\cdot\omega_1\cdot T_1/2} + \sum_{n=-\infty}^{+\infty} c_{2n+1} \cdot e^{i\cdot(2n+1)\cdot\omega_1\cdot t} \cdot e^{i\cdot(2n+1)\cdot\omega_1\cdot T_1/2} \ \forall t \tag{9.5}$$

After simplification, Eq. (9.5) becomes:

$$i_b(t) = \sum_{n=-\infty}^{+\infty} c_{2n} \cdot e^{i\cdot 2n\cdot\omega_1\cdot t} - \sum_{n=-\infty}^{+\infty} c_{2n+1} \cdot e^{i\cdot(2n+1)\cdot\omega_1\cdot t} \ \forall t \tag{9.6}$$

With the following notations:

$$i_{even\ H} = \sum_{n=-\infty}^{+\infty} c_{2n} \cdot e^{i\cdot 2n\cdot\omega_1\cdot t} \ \forall t \tag{9.7}$$

$$i_{odd\ H}(t) = \sum_{n=-\infty}^{n=-\infty} c_{2n+1} \cdot e^{i\cdot(2n+1)\cdot\omega_1\cdot t} \ \forall t \tag{9.8}$$

Eqs. (9.4) and (9.6) become:

$$i_a(t) = i_{even\ H}(t) + i_{odd\ H}(t) \ \forall t \tag{9.9}$$

$$i_b(t) = i_{even\ H}(t) - i_{odd\ H}(t) \ \forall t \tag{9.10}$$

The currents $i_a(t)$ and $i_b(t)$ flowing in the branches of the frequency doubler create electromagnetic fields which try to propagate in two opposite directions, toward the entry of the circuit (marked with a sign "−") and to the output waveguide (marked with a "+" sign). The currents $i_{even\ H}$ create the fields $\vec{E}^{-}_{even\ H}$ with a symmetry compatible with the TM_{11} mode, toward the input waveguide and the fields $\vec{E}^{+}_{even\ H}$ with a symmetry compatible with the TEM or quasi-TEM mode toward the output waveguide. The currents $i_{odd\ H}$ create the fields $\vec{E}^{-}_{odd\ H}$ compatible with the TE_{10} mode toward the input waveguide and the fields $\vec{E}^{+}_{odd\ H}$ compatible with a TE mode toward the output waveguide.

The TM_{11} mode of the input rectangular waveguide must, therefore, be under cutoff at the output frequency $f_2 = 2.f_1$, so that the output signal can only propagate toward the output waveguide. Cutting of TM_{11} mode can only be done by reducing the height of the input waveguide since its width defines the cutoff frequency of TE_{10} mode. According to Collin [51], p.189, the cutoff frequencies of TE_{10} and TM_{11} modes are given by:

$$\lambda_{cutoff}[TE_{10}] = 2a \tag{9.11}$$

$$\lambda_{cutoff}[TM_{11}] = 2ab/\sqrt{a^2 + b^2} \tag{9.12}$$

where a and b are respectively the width and the height of the input rectangular waveguide.

The third harmonic of the input signal at $f_3 = 3.f_1$ is generated in the input waveguide in TE_{10} mode and can, therefore, leak to the circuit input. However, it cannot leak to the output waveguide, due to the suspended-microstrip or coaxial line section which separates the input and output waveguides, and which is under cutoff for TE mode. This insures a very good spectral purity of the output signal. In addition, power generation at the third harmonic has no significant impact on the conversion efficiency.

9.3.5 Balanced Frequency Triplers with an Anti-Parallel Pair of Diodes

Balanced triplers featuring an anti-parallel pair of diodes were demonstrated at submillimeter wavelengths by Erickson [43]. The pair of diodes traps the RF currents at even harmonics in a closed loop so the multiplier can generate only odd harmonics. Opening the loop at DC is, however, necessary for biasing the diodes and optimizing the power handling capabilities as well as the conversion efficiency of the frequency tripler. This can be achieved by implanting a DC capacitor on top of one of the mesas to insulate the diodes from each other (see Figure 9.7). This frequency tripler uses a split-block waveguide design where the Schottky planar varactor diodes are monolithically fabricated on a GaAs-based substrate. The microelectronic device is inserted between the input and the output waveguides in a channel. An E-plane probe located in the input waveguide couples the signal at the fundamental frequency to a microstrip line having two radial stubs to prevent the third harmonic from leaking into the input waveguide. The third harmonic produced by the diodes is coupled to the output waveguide by the output E-plane probe which is a section of the on-chip capacitor used for RF/DC decoupling. This design was used for the local oscillators (LOs) of the sub-THz channels of HIFI.

9.3.6 Multi-Anode Frequency Triplers in a Virtual Loop Configuration

Another topology of balanced tripler was adopted for building several THz multipliers used in the local oscillators of HIFI [1, 17, 18, 52, 53] and all the multiplier stages of the first 2.7

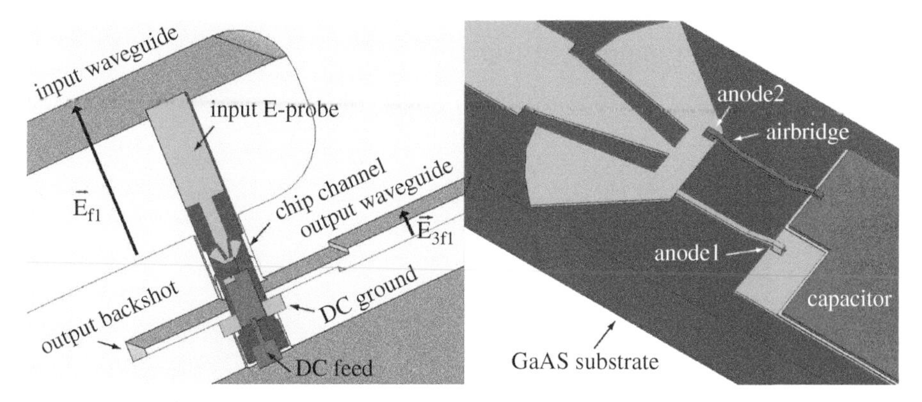

Figure 9.7 Schematics of a sub-millimeter wave frequency tripler using an anti-parallel pair of diodes (design by N. Erickson). (a) Bottom block with the integrated device. (b) Detail of the anode area. $\vec{E}_{f_1}$ and $\vec{E}_{3f_1}$ stand respectively for the electric field at the fundamental frequency f_1 and at the output frequency $3f_1$. Multi-anode frequency triplers in a virtual loop configuration.

THz solid-state source able to deliver tens of microwatts at room temperature [24, 54], (Siles et al., 2016).

9.3.6.1 General Circuit Description

This type of frequency triplers uses an even number of diodes, that are in series at DC but that form a virtual loop at RF (see Figure 9.8). As for the anti-parallel diode configuration, the virtual loop traps the second harmonic of the fundamental, so the multiplier can efficiently produce the third harmonic.

This configuration has the advantage of enabling the building of balanced frequency triplers with four diodes or more, for increased power handling capabilities while keeping a good phase balance between the diodes.

The frequency tripler uses a split-block waveguide design where the Schottky planar varactor diodes are monolithically fabricated on a GaAs-based substrate. The microelectronic device is inserted between the input and the output waveguides in a channel. An E-plane probe located in the input waveguide couples the signal at the fundamental frequency to a suspended microstrip line. This line has several sections of low and high impedance used to match the diodes at the input and output frequency and to prevent the third harmonic from leaking into the input waveguide. The third harmonic produced by the diodes is coupled to the output waveguide by a second E-plane probe. The circuit features additional matching elements in the input and output waveguides, made with a succession of waveguide sections of different heights and lengths.

To increase the spectral purity of the frequency tripler, the dimensions of the output waveguide are chosen to cutoff any second harmonic leakage that could result of a circuit unbalance. In addition, the balanced geometry of the circuits ensures that power at the fourth harmonic of the input is strongly suppressed. The closest harmonic that can leak is the fifth, but, given the fact that the capacitance of the diodes is optimized for the third harmonic, no significant power is expected at the fifth harmonic.

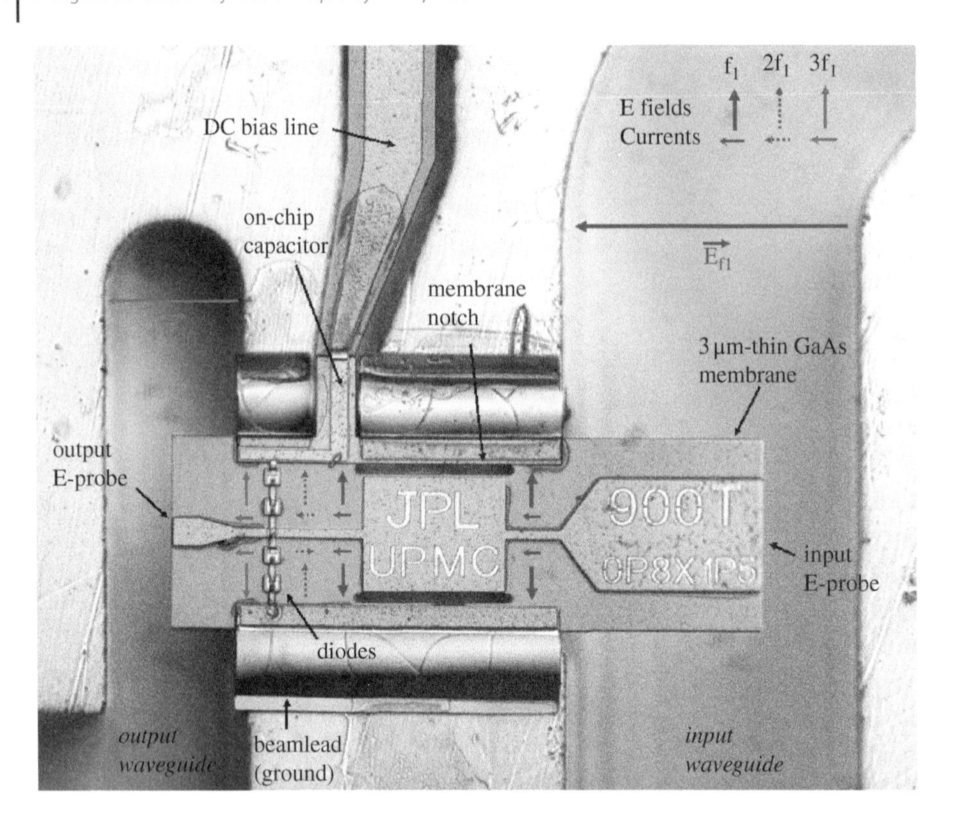

Figure 9.8 Picture of a four-anode 900 GHz balanced tripler (designed by A. Maestrini at Université Pierre et Marie Curie and Observatoire de Paris, fabrication by JPL). This tripler uses a split-block waveguide design and features additional waveguide sections of different impedances and lengths (not shown) which are used for the input and output matching. $\vec{E}_{f_1}$ and $\vec{E}_{3f_1}$ stand respectively for the electric field at the fundamental frequency f_1 and at the output frequency $3f_1$. The electric fields and the current lines are represented for the fundamental frequency f_1 (thick plain blue lines), the idler frequency $2f_1$ (dashed thin purple line), and the output frequency $3f_1$ (light plain red lines).

As detailed in Section 9.3.6.2, this circuit configuration requires that the dimensions of the channel that runs between the input and the output waveguides are set to cutoff the TE mode of the suspended microstrip line [52] at the second harmonic, i.e. only a quasi-TEM mode can propagate inside this channel at the idler frequency. This condition is generally extended in frequency up to the third harmonic to avoid any risk of coupling part of the output signal to an undesirable mode which could be excited by asymmetries present in the real circuit.

In this design, the on-chip capacitor used to decouple the DC voltage from the RF signals is not implanted on the mesa as in the classic balanced tripler in Figure 9.7. It is located on the side of the microelectronic device. The DC circuit requires, therefore, a dedicated channel, orthogonal to the chip-channel, and located between the input and output waveguides. This channel is, therefore, quite delicate to implement for triplers working above 1.2 THz. Despite this drawback, this lateral bias circuit has the advantage of being completely independent of the input or output E-probes, giving more degrees of freedom to the design.

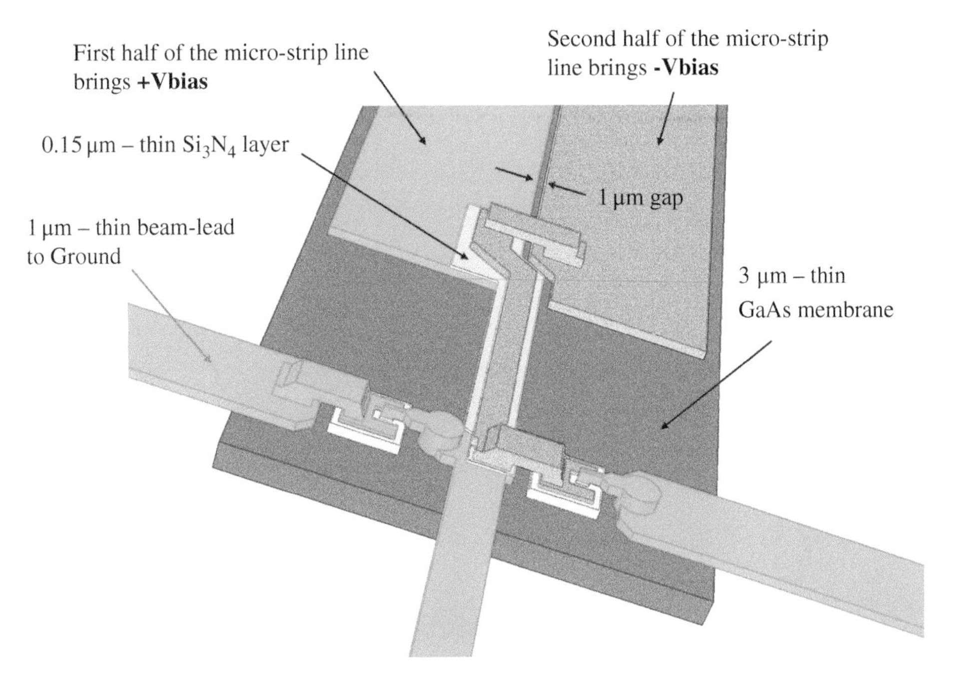

Figure 9.9 Detail of the diode area of a the balanced 1.5 THz tripler with a dual bias line at the center of the chip (design by A. Maestrini at JPL, 2002).

An alternate approach for biasing frequency triplers in an open-loop configuration is to bias the diodes through the center quasi-TEM line, which needs to be split (see Figure 9.9) Each diode is biased by an independent bias line with opposite voltages. The polarization lines must pass through the input waveguide, so an on-chip low-pass filter is necessary to decouple the DC signals from the input RF signal.

9.3.6.2 Necessary Condition to Balance the Circuit

Figure 9.10 shows a schematic of the diode area of a four-anode balanced tripler using the virtual-loop configuration. The same notations as in Section 9.3.4.2 and Figure 9.6 apply but the polarity of the diodes is different. In this design, the input signal at the fundamental frequency propagates along the suspended microstrip line on a TEM or quasi-TEM mode and creates, in the vicinity of the diodes, an excitation field $\vec{E}_1(\omega_1 t)$ that generates the currents $i_a(t)$ and $i_b(t)$) in the diodes. If the diodes can be assimilated as point sources, there is a phase difference of 180° between the currents $i_a(t)$ and $i_b(t)$:

$$i_b(t) = i_a\left(t + \frac{T_1}{2}\right) \ \forall t \tag{9.13}$$

According to Manley and Rowe [50] and Eq. (9.11) i_a (t) and i_b (t) can be written as:

$$i_a(t) = \sum_{n=-\infty}^{+\infty} c_n \cdot e^{i \cdot n \cdot \omega_1 \cdot t} \ \forall t \tag{9.14}$$

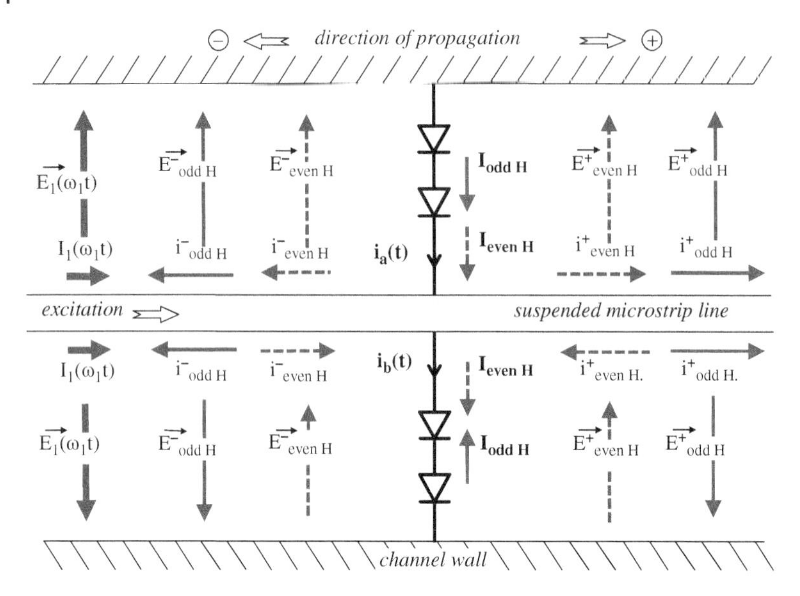

Figure 9.10 Currents and electrical fields in the vicinity of the diodes of a four-anode balanced tripler. For simplicity, the DC bias circuit is not represented. The designations "even H" and "odd H" refer respectively to even harmonics and odd harmonics. The dashed electrical field lines at the even harmonic indicate that the corresponding mode is under cutoff and cannot propagate (this condition must be met at the second harmonic, but is not met usually at the fourth harmonic).

$$i_b(t) = \sum_{n=-\infty}^{+\infty} c_n \cdot e^{i \cdot n \cdot \omega_1 \cdot (t + T_1/2)} \ \forall t \tag{9.15}$$

where $(C_n)_{n \in \mathbb{Z}}$ are the complex Fourier coefficients with $C_{-n} = C_n^*$ for $n > 0$. In practice, at submillimeter wavelengths, only a limited number of terms are relevant ($n \leq 5$ suffices for triplers). These coefficients depend on the input frequency f_1 and on the input power. With the following notations and the same simplification as in Section 9.3.4.2:

$$i_{even\ H}(t) = \sum_{n=-\infty}^{n=+\infty} c_{2n} \cdot e^{i \cdot 2n \cdot \omega_1 \cdot t} \ \forall t \tag{9.16}$$

$$i_{odd\ H}(t) = \sum_{n=-\infty}^{n=-\infty} c_{2n+1} \cdot e^{i \cdot (2n+1) \cdot \omega_1 \cdot t} \ \forall t \tag{9.17}$$

Eqs. (9.12) and (9.13) become:

$$i_a(t) = i_{even\ H}(t) + i_{odd\ H}(t) \ \forall t \tag{9.18}$$

$$i_b(t) = i_{even\ H}(t) - i_{odd\ H}(t) \ \forall t \tag{9.19}$$

Equations (9.18) and (9.19) are respectively identical to Eqs. (9.9) and (9.10) of balanced doublers, but the polarity of the diodes of the lower branch of the frequency multiplier is opposite, leading to the excitation of different modes of propagation. The currents $i_a(t)$ and $i_b(t)$ flowing in the branches of the frequency doubler create electromagnetic fields which try to propagate in two opposite directions, toward the entry of the circuit (marked with a sign "−") and to the output waveguide (marked with a "+" sign). The currents $i_{even\ H}$ create the fields $\vec{E}^{-}_{even\ H}$ and $\vec{E}^{+}_{even\ H}$ with a symmetry compatible with the TE mode of the suspended microstrip line. The currents $i_{odd\ H}$ create the fields $\vec{E}^{-}_{odd\ H}$ and $\vec{E}^{+}_{odd\ H}$ with a symmetry compatible with the quasi-TEM mode of the same suspended-microstrip line.

The TE mode of the suspended-microstrip line must, therefore, be under cutoff at the idler frequency $f_2 = 2.f_1$, so that it can be confined in the virtual-loop formed by the diodes. This is a necessary condition for optimizing the conversion efficiency of the frequency tripler. As explained in Section 9.3.7, this is not a sufficient condition. The third harmonic of the input signal at $f_3 = 3.f_1$ propagates in a quasi-TEM mode like the input signal. Therefore, only a low-pass filter can prevent it to leak toward the circuit input.

Due to the physical dimensions of the diodes, the balance of this design cannot be perfect, in particular at THz frequencies. Therefore, the phase difference between the currents in two branches of the devices departs from the optimal 180° assumed in Eqs. (9.11). To mitigate the circuit imbalance, the size of the mesas should be made as small as possible and their location with respect to the central microstrip line should be precisely optimized.

9.3.7 Multiplier Design Optimization

This section will present a practical methodology that was used to design several wide-band, fix-tuned, state-of-the-art frequency triplers in the 0.3 to 2.7 THz range [1, 24, 52–55]. While a number of concepts defined in Tuovinen et al. [46] for balanced doublers are utilized, a number of significant points must be addressed for the balanced triplers presented in this section.

9.3.7.1 General Design Methodology

The first step in the design of a frequency multiplier is to determine the physical characteristics of the diodes that best suit the application. It consists of optimizing the active layer thickness and doping level, the anode dimensions, and the bias voltage for a given input power and operative temperature. To perform this task, physics-based numerical models based on drift-diffusion or Monte Carlo techniques are recommended [56]. These models, however, are usually in-house tools and difficult to have access to, so simplified analytical electric models can be used for the Schottky diodes. The major limitation of these analytical models is that they do not include important physical effects such as the electron velocity saturation at high frequencies. Furthermore, these models have to be empirically adjusted to be used for THz circuit simulation. As an example, the measured DC series resistance of the diodes cannot be directly included in simple analytical models of the Schottky diodes, which do not account for the physics of the device (e.g. the Schottky diode model that is included in commercial circuit simulators like the advanced design system (ADS) suite from Keysight[1]). An approximate rule commonly used in practice consists of fixing the product $R_s \times C_{j_0}$, which is derived empirically (or through physics-based numerical models) and depends on the temperature, the doping level, etc. Another problem of these models is the impossibility of predicting the actual temperature of the junctions. Therefore, there will be always an uncertainty in the results when simplified models are used.

The great advantage of physics-based numerical CAD tools over-analytical circuit models lies in the possibility of performing a concurrent optimization of both the internal structure of the Schottky diode and the external circuit. All the design parameters play an important role during the optimization process and cannot be assumed independent one from each other. Furthermore, there might not be a unique solution to the optimization process as different combinations of the design parameters might lead to similar performances. In these cases, the final choice would depend on technological aspects or other design constraints. A brief description of the design parameters to be optimized for the design of millimeter and submillimeter-wave Schottky multipliers is provided below.

9.3.7.1.1 Bias Voltage

In an ideal case, the optimum bias point for Schottky diodes is halfway between avalanche breakdown and forward conduction:

$$V_{bias} \approx V_{bd} + \frac{\varphi_b}{2} \tag{9.20}$$

where V_{bd} is the DC breakdown voltage and $\emptyset_b$ is the Schottky barrier height. Thus, the maximum capacitance modulation is achieved (the ratio C_{max}/C_{min} is maximized). Nevertheless, when the available input power is not enough to guarantee a voltage swing from V_{bd} to forward conduction, better efficiencies are obtained for bias voltages around 0 V due to the stronger nonlinearity of the diode capacitance around this operation point (see Figure 9.11a).

9.3.7.1.2 Input Power

When enough input power is available, it is possible to maximize the output power at the expense of the conversion efficiency. However, maximization of the output power when input power is limited involves maximization of the efficiency. A high input power implies a high voltage amplitude in the device. Therefore, greater capacitance modulation and efficiency are achieved as long as breakdown and/or conduction are not reached. However, there is a limit for the input power in order not to push the diode into breakdown or forward conditions. On the one hand, entering the breakdown regime limits the lifetime of the multiplier device. A good margin of security is to guarantee that the power delivered to the Schottky diode does not produce a voltage peak beyond the DC breakdown voltage. It is known that the RF breakdown voltage is higher (in modulus) than the DC breakdown voltage [58]. The reason is that the impact ionization that causes the avalanche breakdown does not occur instantaneously. Hence, the voltage sweep has to enter the breakdown regime during a sufficient period of time, which implies a voltage amplitude greater than the DC breakdown voltage. On the other hand, when forward conduction is reached ($V \approx \emptyset_b$) for a percentage of time during one period of the voltage excitation, the multiplier will present a certain varistor mode of operation due to the nonlinearity of the resistance, as shown in Figure 9.11b. Under these operating conditions, the conversion efficiency of the multiplier drops as a consequence of the major contribution of the series resistance (flat region in Figure 9.11b). Nevertheless, the device gets chiefly resistive and better bandwidths may be achieved [59, 60].

9.3.7.1.3 Epitaxial Layer Length and Doping

On the one hand, the lowest series resistance is obtained when the epitaxial layer is fully depleted. For doublers, the best conversion efficiencies are obtained when this occurs for

(a)

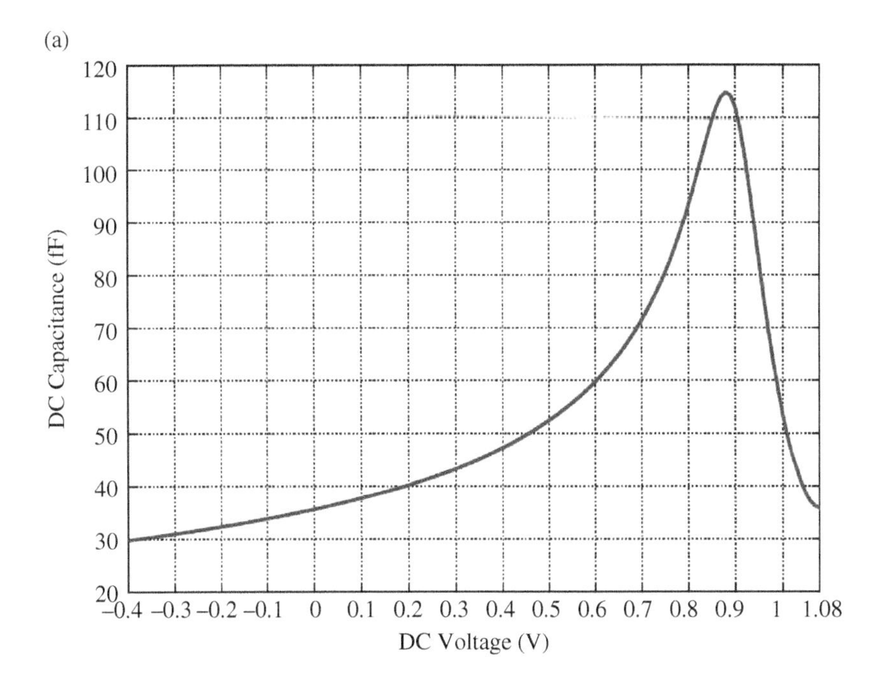

(b)

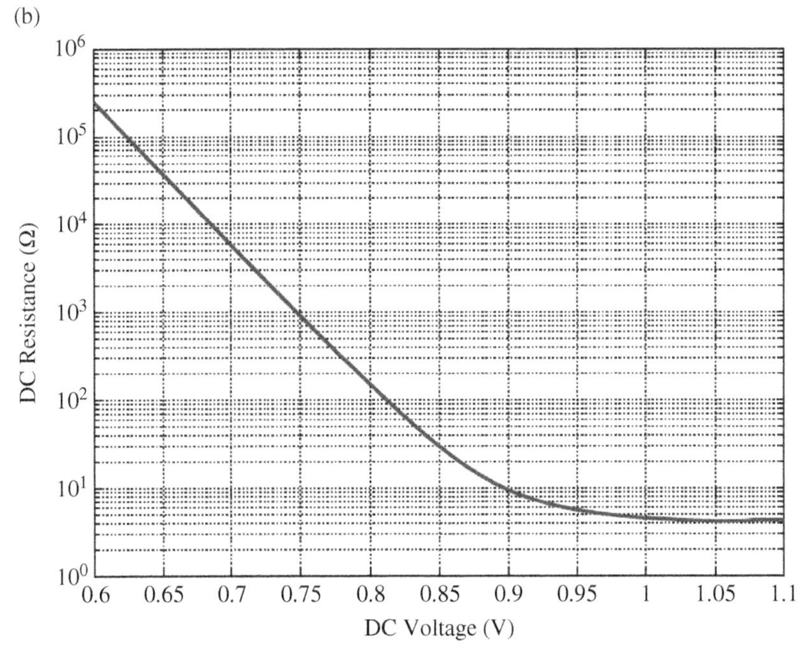

Figure 9.11 DC capacitance Vs voltage (a) and DC resistance Vs. Voltage (b) obtained using physics-based drift-diffusion methods for a typical Schottky diode (A = 36 μm^2, Nd = 1.10^{17} cm^{-3}). *Source*: Reprinted from [57].

the maximum voltage excursion in the diode. For triplers, the epilayer should be selected so that it is fully depleted at the bias voltage. On the other hand, a higher doping implies a lower series resistance and, at high frequencies, a reduction in the influence of the carrier velocity saturation. Only the nondepleted region of the epilayer has a contribution to the series resistance. The resistivity (ρ) in this zone can be calculated as,

$$\rho = 1/(q \cdot N_D \cdot \mu(E, N_D) \tag{9.21}$$

where, N_D is the doping level of the epilayer, q is the charge of the electron, and $\mu(E, N_D)$ is the electron mobility, which is a function of the electric field and the doping level. Apparently, the series resistance decreases with the doping as they are inversely proportional. However, the mobility also decreases with the doping contributing to increase the resistance Figure 9.12a). Then, it is necessary to evaluate which effect dominates in order to know the real dependence of the series resistance on the doping. The results are presented in Figure 9.12b for different doping levels. It can be seen that the global impact of an increase in the doping level is a reduction in the series resistance.

At high frequencies, high doping levels are used in order to mitigate the effect of carrier velocity saturation. However, the higher the doping, the more input power will be required to fully deplete the epilayer as a consequence of the reduction of the space charge region W:

$$W(V) = \sqrt{\frac{2\varepsilon_s}{qN_D} \cdot (V_{bi} - V)} \tag{9.22}$$

$$V_{bi} = \varphi_{b0} + (k_B T/q) \cdot ln\ (N_D/N_C) \tag{9.23}$$

V being the external applied voltage, V_{bi} the built-in potential, ε_s the semiconductor electric permittivity, and N_D the doping level of the epilayer. The built-in potential V_{bi}, assuming Maxwell–Boltzmann statistics, can be obtained with Eq. (9.23), where φ_{b0} is the ideal barrier height, k_B is the Boltzmann's constant. T is the temperature, and N_C is the effective density of states of the conduction band.

Since the available LO power is limited at high frequencies, it is not possible to increase the thickness of the depletion region by increasing the input power. Hence, the epitaxial layer thickness has to be reduced in order to reduce the series resistance. Another limiting factor that must be taken into account is the reduction of the breakdown voltage when the thickness of the epilayer is reduced and/or the doping level is increased (Figures 9.13a and b). Lower breakdown voltages imply a lower bias voltage and a lower input power in the design. In Figure 9.13a, the theoretical breakdown voltage has been obtained assuming and infinite thickness of the epilayer, according to the analytical approach provided in [61]. In Figure 9.13b, a full depletion of the epilayer is assumed. The constant region of the curves corresponds to the case when the space charge region does not exceed the epitaxial layer: $\leq L_{epi}$. Otherwise, i.e. $W > L_{epi}$, the highly doped substrate prevents the space charge region from extending beyond the epilayer, causing considerable reduction of the breakdown voltage.

9.3.7.1.4 Anode Area

This parameter is connected to the power handling capabilities of the Schottky diode. The input power required to achieve maximum efficiency for a certain Schottky multiplier is

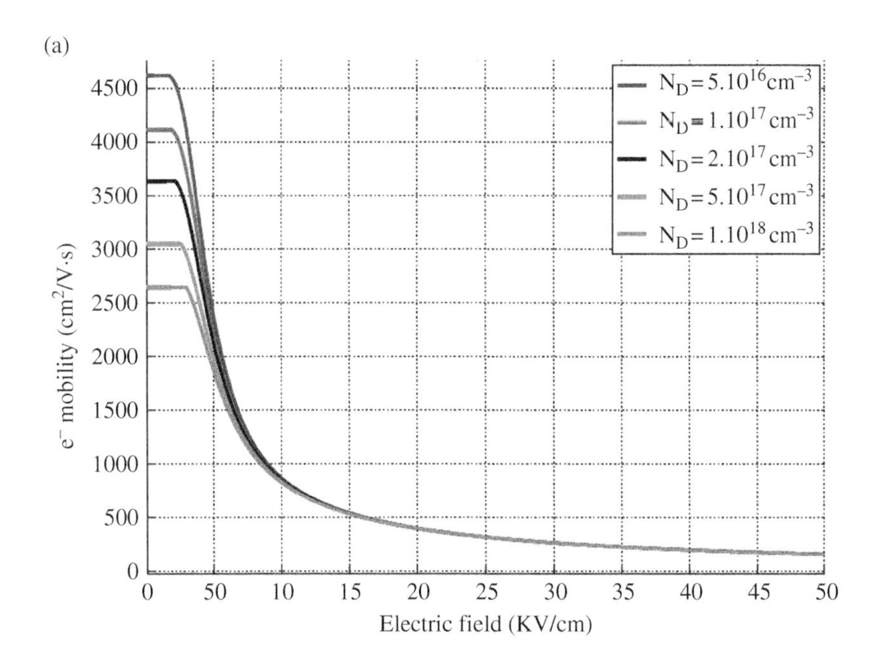

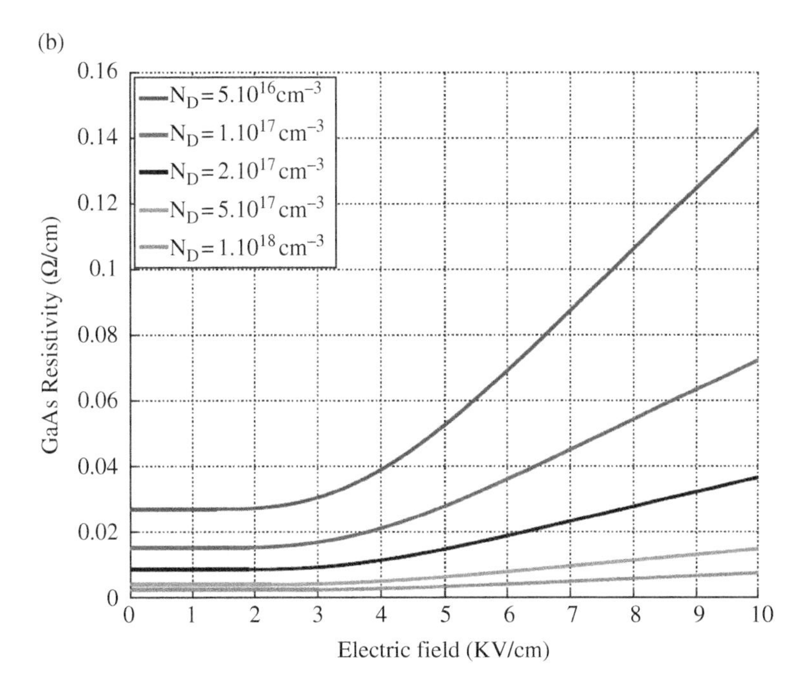

Figure 9.12 GaAs electron mobility and resistivity as a function of the electric field for different doping levels. *Source*: Reprinted from [57].

(a)

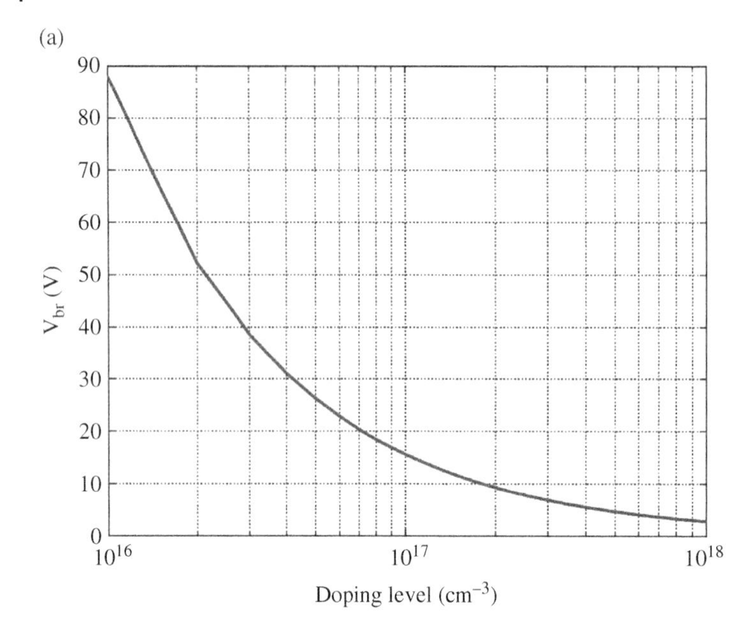

(b)

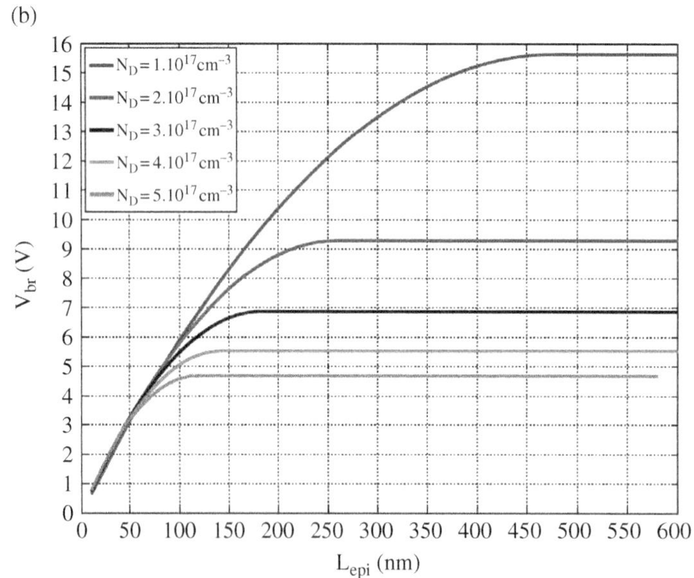

Figure 9.13 Theoretical DC breakdown voltage as a function of the epilayer doping assuming an infinite epilayer thickness (a), and as a function of the epilayer thickness (b).

proportional to the anode area of the diodes. Hence, it is possible to situate the peak of efficiency at the available input power just by modifying the anode area. The optimum anode area necessary to maximize the multiplier efficiency for a given frequency and input power depends on design parameters such as the epitaxial layer thickness and doping, the bias voltage, etc; and high-frequency effects like the electron velocity saturation. Physics-based

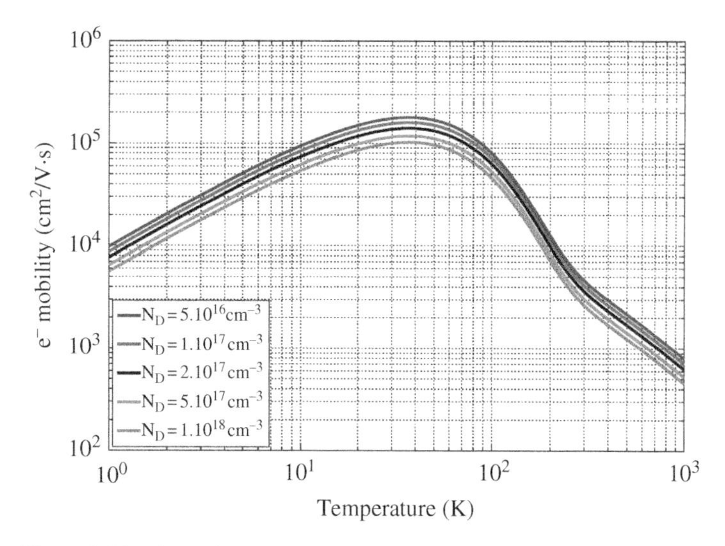

Figure 9.14 GaAs electron mobility as a function of the ambient temperature for different doping levels.

models based on drift-diffusion or Monte Carlo models are best suited to calculate this parameter [56]. Simplified analytical models like those included in commercial software might not provide the best solution due to the difficulty to account for the specific diode structure and high-frequency effects.

9.3.7.1.5 Temperature

The lower the temperature, the higher the carrier mobility. Hence, the series resistance decreases resulting in better efficiencies. The theoretical low-field mobility as a function of the temperature for different doping levels in the epilayer is provided in Figure 9.14 [62].

9.3.7.1.6 Carrier Velocity Saturation

At high frequencies in the submillimeter-wave band, the performance of Schottky multipliers is limited by the carrier velocity saturation [63], i.e. the edge of the depletion region (W) cannot move faster than allowed by the maximum electron velocity (v_{max}), which is determined by the material. The negative effect of this physical phenomenon on multiplier can be mitigated by choosing a high doping level for the epilayer. This way, it can be guaranteed that the maximum current density supported by the material (J_{max}) is larger than the displacement current (J_{disp}):

$$J_{max} = q.N_D.v_{max} \geq J_{disp} = \frac{\partial Q}{\partial t} = qN_D\frac{\partial W}{\partial t} \tag{9.24}$$

$$N_D \geq \frac{J_{disp}}{q.v_{max}} \tag{9.25}$$

For a fixed doping, Eq. (9.24) gives also the maximum varying velocity of the edge of the space charge region to avoid the carrier velocity saturation. The carrier velocity saturation causes a lower capacitance modulation, and thereby, a loss of conversion efficiency. This is illustrated in Figure 9.15 for different frequencies of the excitation waveform [58].

To summarize, the design parameters cannot be individually optimized since a variation in one parameter has an impact on the optimum values of the rest of parameters. An example of this is represented in Figure 9.16, where the space charge region of a GaAs Schottky

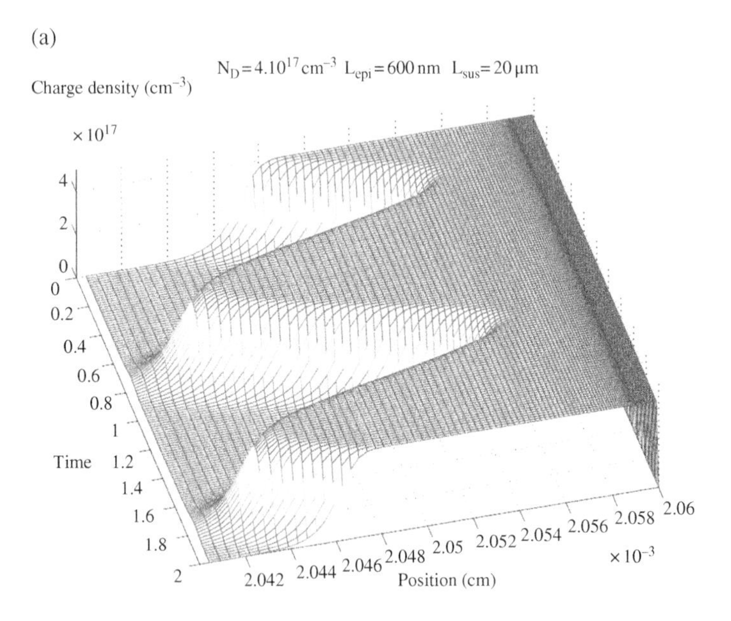

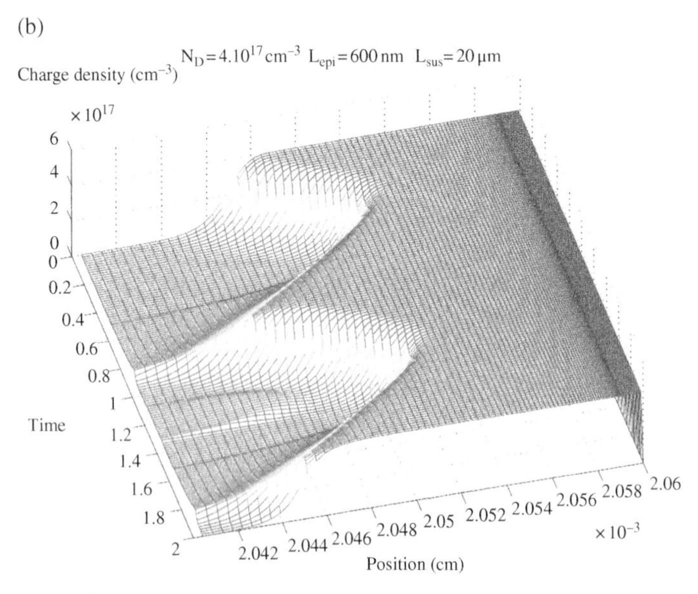

Figure 9.15 Variation of the space charge region for different frequencies of the applied voltage: (a) 100 GHz, (b) 500 GHz, (c) 800 GHz. The characteristics of the diodes are: Area = 36 μm^2, L_{epi} = 600 nm, $N_D = 4 \times 10^{17}$ cm^{-3}. *Source*: Reprinted from [57].

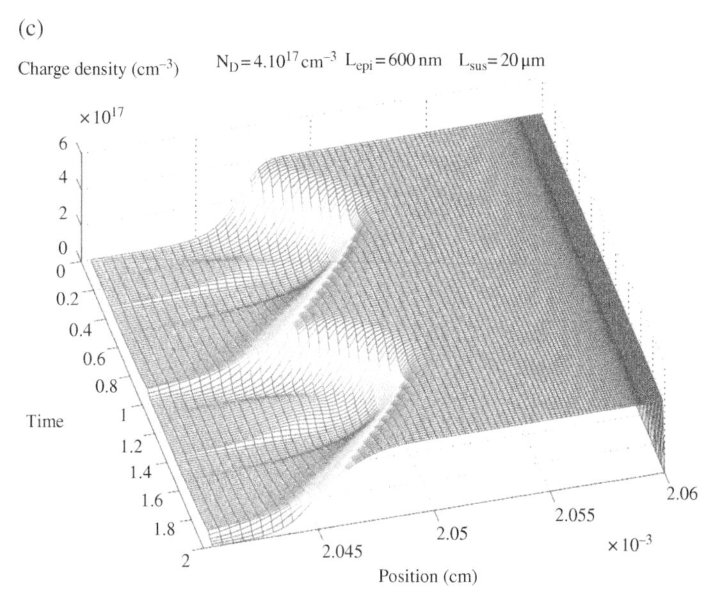

Figure 9.15 (Continued)

diode has been depicted for different sinusoidal excitations at 100 GHz using a physics-based drift-diffusion simulator [57]. The first case (Figure 9.16a) is nearly optimum as just a small portion of the epilayer remains undepleted for the maximum amplitude of the applied voltage. In the second case (Figure 9.16b), the excessive amplitude of the applied voltage pushes the diode into forward conduction, which can be noticed in the space charge region by the presence of a empty charge region at the Schottky junction. The third case (Figure 9.16c) corresponds to the situation where the input power is not high enough to fully deplete the epilayer. In this situation, the big contribution to the series resistance of the nondepleted region will result in a high reduction of the conversion efficiency. One solution, assuming that there is a limited available input power, might consist of reducing the thickness of the epilayer. But this leads to lower breakdown voltages so a modification of the bias point might be also required, which also has an impact on the thickness of the space charge region (W). Another possibility would be to decrease the doping in order to increase the space charge region. This results in a higher resistivity and, at high frequencies, an increment in the effect of carrier velocity saturation. The last example (Figure 9.16d) shows the consequences of an excessively high (in modulus) reverse bias voltage together with a high input power. On the one hand, the high voltage peak of the applied excitation leads to the situation: $W > L_{epi}$. Hence, the charge movement is blocked by the substrate resulting in the saturation that can be observed at the edge of the depletion region in Figure 9.16d. Also, the breakdown is reached and can be seen as an increase in the charge density at the Schottky junction as a consequence of the current flowing from the metal to the semiconductor.

Once these parameters are fixed, and the optimum embedding impedances of the diodes determined using the method presented earlier, a linear circuit can be synthesized [60]. The optimization of the linear embedding circuit can be done with S-parameter simulations at the input and output frequencies using the ideal impedances of the diodes. This approach

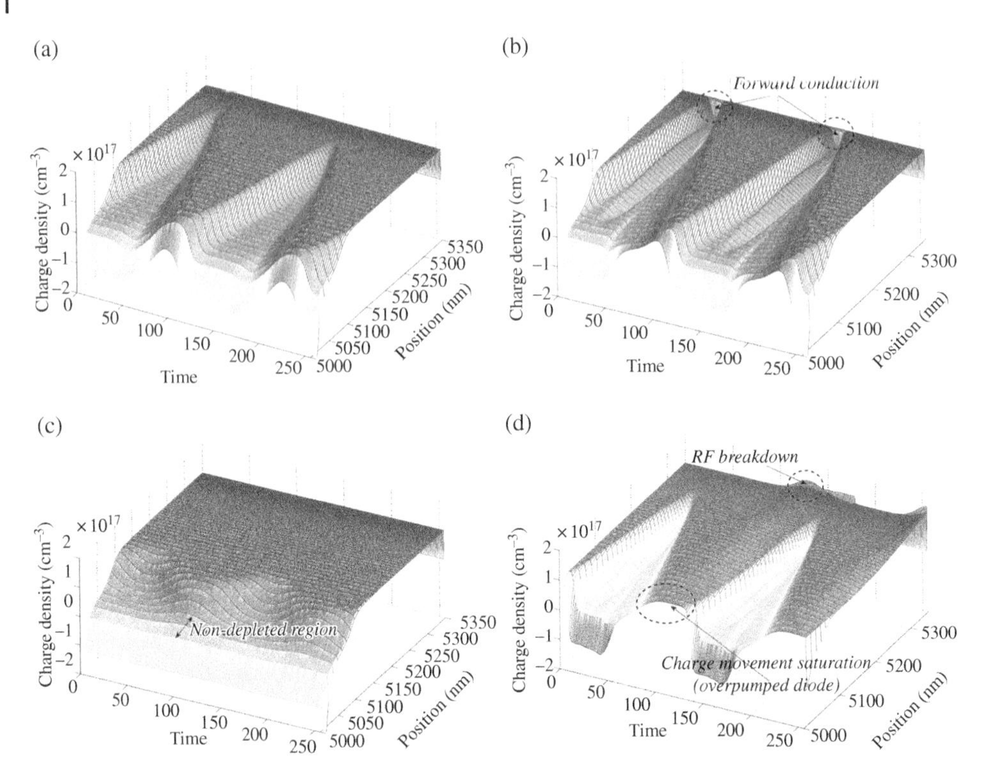

Figure 9.16 Charge density within the epilayer for different operation conditions: (a) $v(t) = -4{+}4.5{\cdot}\sin(2\pi.100\ \text{GHz}\ t)$, (b) $v(t) = -4{+}5.5{\cdot}\sin(2\pi.100\ \text{GHz}{\cdot}t)$, (c) $v(t) = -4{+}1.0{\cdot}\sin(2\pi{\cdot}100\ \text{GHz}{\cdot}t)$, (d) $v(t) = -10{+}10.5{\cdot}\sin(2\pi{\cdot}100\ \text{GHz}{\cdot}t)$. The characteristics of the Schottky diode are: Area = 36 μm^2, L_{epi} = 350 nm, $N_D = 1 \times 10^{17}$ cm^{-3}, L_{subs} = 5 μm, $N_{subs} = 2 \times 10^{18}$ cm^{-3}, T = 300 K. *Source*: Reprinted from [57].

was particularly effective for optimizing balanced doublers of the local oscillators of HIFI [20, 47, 49, 64].

However, for the design of the frequency triplers mentioned above, a different approach was employed. The topology of the diodes-cell, which is the heart of the frequency tripler has been optimized up to that an appropriate compromise is reached between conversion efficiency and bandwidth using exclusively nonlinear circuit simulations. Moreover, the main electrical characteristic of the diode, i.e. its zero-bias junction capacitance has been optimized together with the diodes cell, starting from the initial value obtained by the method presented in Section 9.3.2. More details on this method are given latter in this section.

The main reason for this change in methodology was the difficulty of accurately predicting the impact of changes in the 3D structure of the diodes cell on the actual performance of the tripler, based on S-parameters simulations at the input, idler, and output frequencies. Co-optimization of the zero-bias junction capacitance of the diodes together with the topology of the diodes cell has proven more practical in designing these particular frequency triplers for wide bandwidth (up to 25% of relative bandwidth). However, for maximum conversion efficiency, it is preferable to design the frequency triplers using the diode parameters determined by the method presented in Section 9.3.2, and to fine-tune the zero-bias junction capacitance of the diodes at the very end of the design. The following paragraphs will detail this method.

9.3.7.2 Nonlinear Modeling of the Schottky Diode Barrier

Optimizing the multi-anode diodes cell of a frequency multiplier requires a large number of nonlinear simulations. A reliable device model implemented in a stable and fast-converging circuit simulator is, therefore, essential. For designing a frequency multiplier, that operates mainly in varactor mode, the most important part of the device model is the description of the junction capacitance as a function of the voltage.

For all the designs of the 0.3 to 2.7 THz frequency triplers mentioned previously, a simplified electrical model of the Schottky diode was used, consisting of a nonlinear junction capacitance C_j in parallel with a nonlinear conductance G_j and in series with a resistance R_S.

i) For varactor Schottky diodes, the junction capacitance was classically modeled as follows [65]:

$$\text{For } V \leq \frac{V_{bi}}{2}, C_j(V) = A.\frac{\varepsilon_S}{W(V)} \tag{9.26}$$

V is the bias voltage, V_{bi} is the built-in potential, ε_s is the semiconductor electric permittivity, A is the junction area, $W(V)$ is the thickness of the depletion layer given in Eq. (9.22). For GaAs Schottky diodes, the built-in potential varies between 0.7 and 1 V, depending on the device fabrication process. For Schottky diodes fabricated at JPL, V_{bi} is approximately 0.75 V at room temperature.

For $V \geq \frac{V_{bi}}{2}$, $C_j(V)$ is defined by a linear extrapolation of Eq. (9.26) from $V = \frac{V_{bi}}{2}$, to avoid the singularity at $V = V_{bi}$. This simplification has no practical impact since the nonlinear conductance of the diode short-circuits the nonlinear capacitance in this bias voltage range.

For anodes of a few square microns or smaller, a correction term should be added to $C_j(V)$ to take into account edge effects. For circular anodes, the correction terms given in [66] are a function of the anode radius. For rectangular anodes, as those used in the multipliers fabricated at JPL, a different formulation like the one proposed in [67] is necessary.

ii) The nonlinear conductance G_j is derived from the classic equations of thermionic emission in Schottky contacts [65]. Saturation effects [63, 66, 68] and breakdown effects are not directly included in the simulation. However, time-domain simulations are performed to check that the voltage across the diodes never enters breakdown to minimize the risk of damaging the diodes [69].

iii) The series resistance R_S of the Schottky diode is a major limiting factor in the conversion efficiency of frequency multipliers. Underestimating its value affects the design, which would be based on diodes with unrealistic dynamic cutoff frequencies, i.e. on diodes with junction capacitances too high for the operating frequency.

DC measurements of R_S give a clear indication of the quality of the diodes, but the measured values are usually too low to be used in RF simulations based on this simplified diode model. On the other hand, calculations of the series resistance have to take into account the particular topology of the planar diode, where skin effects play a major role. In addition, as mentioned earlier, any physical model of the Schottky barrier has to be properly implemented in a stable and fast-converging circuit simulator able to simulate multi-anode circuits, unless only linear impedances of the diodes are used.

A practical and effective approach to design sub-millimeter wavelength frequency multipliers is based on the empirical rule introduced in [70] which consists in fixing the product $R_S \times C_j(0)$. For submillimeter-wave multipliers working at room temperature and for a doping of $1 \times 10^{17}\ \text{cm}^{-3}$, $R_S \times C_j(0)$ is set to:

$$R_S \times C_j(0) = 120\,\Omega \times fF \tag{9.27}$$

In the case of the four-anode 600 GHz frequency tripler described in [52], this rule gives $R_S \approx 20\,\Omega$ per diode compared to the DC measured value of $R_S = 8\,\Omega$ per diode. However, for frequency multipliers featuring diodes fabricated on substrates doped at higher concentration, a lower value of the product $R_S \times C_j(0)$ is used. The experimental results obtained with the 2.7 THz tripler described in [54] show that this product is indeed four times lower than that defined in Eq. (9.27).

This empirical approach relies by essence on measurements obtained from previously built frequency multipliers and is consequently only effective in a specific context. It has proven to be one of the keys to the successful design of HIFI terahertz frequency multipliers and state-of-the-art 2.7 THz multiplier sources. However, physics-based diode models are much needed for designing frequency multipliers in uncharted territories. Abundant literature is available about the modeling of Schottky diodes at millimeter and submillimeter wavelengths working at room temperature [56, 57, 63, 64, 66, 68, 71, 72] and at cryogenic temperatures [21]. A practical and effective approach consists in relying on a physics-based diode model to derive a simplified model, like the one presented in this section, which will in turn be used for the intensive nonlinear simulations necessary for the optimization of the frequency multiplier. This translates in relying on physics-based diode models to determine the product $R_S \times C_j(0)$ that best reproduces the behavior of the real diode, when using the simplified model proposed above.

9.3.7.3 3D Modeling of the Extrinsic Structure of the Diodes

While the Schottky barrier is modeled in a nonlinear circuit simulator, the extrinsic structure of the diode, i.e. the surround of the Schottky barrier, is modeled in a 3D linear electromagnetic simulator. For the frequency multipliers mentioned in this chapter, the ANSYS HFSS[2] finite element method is used owing to its ability to simulate circuits with very small details compared to their overall dimensions. The diode physical structure is simplified by merging all the semiconductor layers and the ohmic contact layer in a single metallic layer with the same finite conductivity employed for the anode and the airbridge.

The electromagnetic field around the Schottky barrier is measured with a virtual coaxial probe, placed at the exact location of the Schottky contact. This coaxial probe is defined as an internal wave-port in HFSS (see Figure 9.17) defined by a rectangular surface that surrounds the anode and that lies on the top face of the mesa. The mesa terminates the probe on one end since it is modeled as a conductor. The anode defines the inner conductor of the coaxial probe while the outer conductor is defined by the edges of the port area. The width of

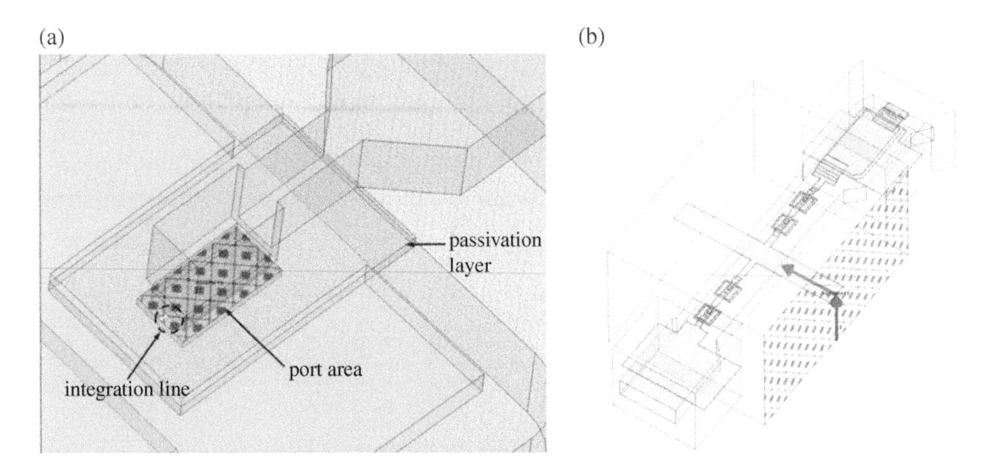

Figure 9.17 (a) Definition of the HFSS wave-port at the exact location of the Schottky contact. The integration line defines the polarity of the diode. All the diodes in the circuits need to be should have the same polarity (b) HFSS model of the diode cell and definition of one of the suspended micro-strip ports. *Source*: Figure republished from [52].

the gap between the anode and the edges of the port area should not exceed the thickness of the active layer, to avoid underestimating the extrinsic parasitic capacitance of the diode. With this port, the intrinsic junction capacitance of the diode is not modeled, as desired. From that port, outgoing electromagnetic waves are generated and incoming waves are received so S-parameters can be calculated. These parameters are then incorporated into the circuit simulator and connected to the nonlinear model of the Schottky barrier to reproduce the behavior of the full structure. The dimensions of the coaxial probe set the impedance of the HFSS wave-port, which is usually of a few Ohms only. However, this has no consequence since the S-parameters calculated by the 3D-EM simulator are re-normalized to 50 Ω before being used by the circuit simulator.

The probe has no length and therefore no de-embedding is necessary to set the reference plane at the Schottky contact location. The wave-port uses an integration line that runs from the outer-conductor to the anode to properly set the polarity of the diode. The definition of the port and the meshing around the diode are critical to get accurate results. The 3D geometrical structure of the diode must also be drawn accurately. Details such as the passivation layers greatly contribute to the parasitic capacitances and must be included in any accurate 3D representation of the diode.

9.3.7.4 Modeling and Optimization of the Diode Cell

The diode cell is the heart of the frequency multiplier and is optimized first before more elements are added to the circuit. An initial diode cell is drawn based on the chip topology. For balanced triplers in a virtual loop configuration (see Figure 9.10), the diode cell consists of a symmetric section of the circuit around the diodes which features the suspended microstrip line (see Figure 9.17b). Wave-ports are defined as in Section 9.3.7.3 for each diode of the circuit, and two additional wave-ports are added for each end of the suspended microstrip line. For these two ports, the de-embedding plane is located at the center of the diode cell.

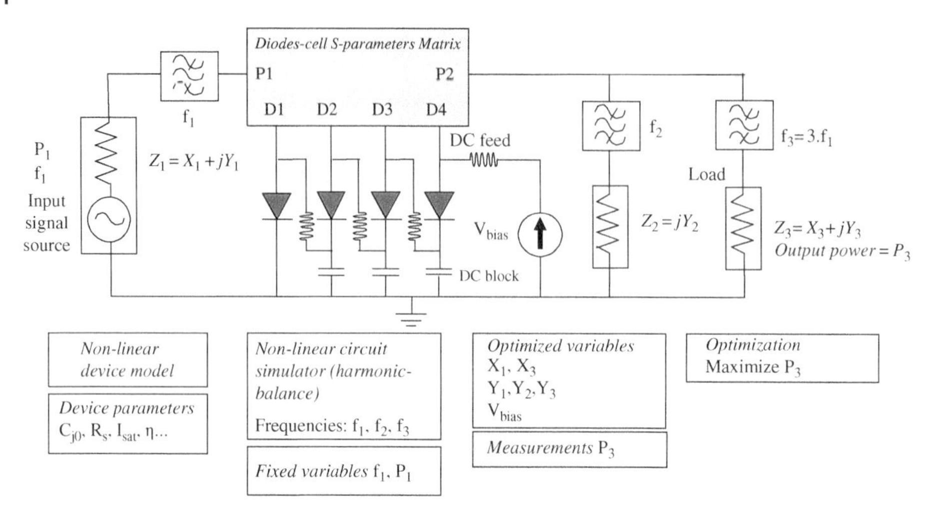

Figure 9.18 Optimization of the diodes-cell of a four-anode frequency tripler. The S-parameters matrix should provide an infinite impedance at 0 Hz or DC blocks should be added at each of the diode ports D1 to D4. P1 and P2 are the suspended-microstrip ports.

The S-parameters of this initial structure are calculated with the 3D EM simulator and then used in harmonic-balance simulations. As mentioned earlier, the initial electrical parameters of the diodes are set using the method described in Section 9.3.2.

The diode cell alone cannot be an efficient tripler; thus, harmonic-dependent complex impedances are connected to the ports of the nonlinear simulation bench that correspond to the HFSS wave-ports of the suspended microstrip line (see Figure 9.18). The complex impedances are optimized for the center of the band. To have a first indication of the instantaneous bandwidth, the frequency is swept across the band. The optimization is an iterative process consisting in direct modifications of the 3D structure of the diode cell in HFSS and nonlinear simulations to verify their impact on performances. The phase and power balance between the diodes are closely monitored both with linear simulations in HFSS and with subsequent nonlinear circuit simulations; the location of each diode on the circuit must be finely tuned. This task is made easier when the dimensions of the mesas are small with respect to the other dimensions of the circuit and when there are only a limited number of diodes on the chip. Actually, as mentioned in Section 9.3.6.2, a perfect balance between the diodes can only be achieved if the diodes can be assimilated as point sources. Moreover, the addition of a bias circuit can further degrade the balance as it is located on one side of the circuit only. Therefore, the balance between the diodes tends to degrade with the increasing frequency.

As the diode cell is defined by a large set of parameters, this task can be extremely time consuming and unproductive if not guided by a set of effective design rules. To properly match the diodes at the three relevant frequencies, the capacitance of each diode needs to be compensated by adjusting the length, the width and the thickness of the fingers, the size of the mesas, and the dimensions of the cross-section of the chip-channel. Diodes with small junction capacitance require longer fingers (implying wider channels) or higher channels than diodes with large junction capacitances. For this frequency tripler topology, both the width and the height of the channel are equally critical parameters. During this process, it is necessary to ensure that the dimensions of the channel and the dimensions of other circuit elements guarantee that only the quasi-TEM mode can propagate to the second harmonic, and preferably up to the third harmonic as well. In addition, losses of the propagating lines are to be kept as low as possible, therefore, one of the main drivers for the optimization of the diode cell is the minimization of the RF losses inside the chip channel.

9.3.7.5 Input and Output Matching Circuits

Once the diode cell and the size of the anodes are fixed, the different sections of the suspended-microstrip line and the input and output E-probes are optimized to maximize the conversion efficiency and the input coupling. The design is driven by the necessity to minimize the number of on-chip matching elements in order to reduce both the chip dimensions and the RF losses. At this stage of the design, most of the multiplier is already in place and a fine-tuning of the diode cell and anode size can be performed. Upon completion of this step, the chip topology is fixed.

To further extend the bandwidth, a succession of sections of high and low impedance is added to the input waveguide. As they have no impact on the output match, it is possible to use only linear simulations by linearizing the diodes. To broaden the output match, the same method is applied to the output waveguide. Broadband input and output frequency matching networks are indeed better synthesized using sections of waveguides rather than on-chip sections of TEM lines due to much lower RF losses per wavelength.

9.4 Technology of THz Diode Frequency Multipliers

This section gives a brief history of the tremendous technological improvements that made possible the building of the local oscillators of the HIFI [3] onboard of Herschel Space Observatory that was launched by the ESA in 2009 and the building of the first solid-state electronic source at 2.7 THz. Therefore, it focuses on GaAs membrane frequency multipliers developed at the JPL.

9.4.1 From Whisker-Contacted Diodes to Planar Discrete Diodes

Frequency multipliers using whisker-contacted Schottky diodes played an important role in the development of heterodyne receivers for radio astronomy and planetary sciences. As predicted by Räisänen [73], they appeared to be the only available solution for the LOs of space-borne submillimeter-wave heterodyne instruments in the years 1995–2000. Actually, ODIN, launched in February 2001, was the first satellite to embark heterodyne receivers in the 486–580 GHz band using whisker-contacted Schottky multipliers as final stages of the LOs [74]. But in 1992, this already mature technology was still unable to pass the 1 THz milestone [73]. Progress toward the terahertz region was reported by Crowe and Zimmerman [75] before Peter Zimmerman reached 1.135 THz in 1998 with an all-solid-state source that produced 40 μW of output power [76]. The Schottky diode was usually mounted in a crossed-waveguide structure featuring several mechanical tuners. The input and output signals were decoupled through a low-pass filter, which was either coaxial (Räisänen/Erickson's design) or a stripline structure (Takada/Archer's design) – see Ref. [73]. At submillimeter wavelengths and until the year 2000, whisker-contacted diodes outperformed Schottky planar diodes introduced in the mid-eighties by Cronin and Law [77] at the University of Bath, UK, and shortly later by Bishop et al. [78] at the University of Virginia (UVa), USA, due to their lower parasitic capacitances and lower series resistances. However, at millimeter-wavelengths Schottky planar discrete diodes started to give better performance due to the use of multiple-anodes in balanced configurations. Erickson's balanced doubler, proposed and demonstrated in [43–45], has become a standard topology for frequency multiplication due to its excellent performance [47–49, 79].

In the nineties, GaAs-based HBVs were introduced by Kollberg and Rydberg at the University of Chalmers [38], Sweden, as alternate diodes. They were initially made to be whisker-contacted before being made planar. HBV diodes produce only odd harmonics of an incident signal due to their internal symmetry. Thus, they are attractive devices to design high order odd harmonic multipliers such as triplers [37, 39] or quintuplers that can reach conversion efficiencies up to 5% at 210 GHz [41] or 11% at 100 GHz [40]. HBV technology took a significant turn in the late nineties when Lippens and Mélique at IEMN, France, introduced InP-based multiple-barrier devices [37]. The results obtained on a 250 GHz waveguide tripler (11% efficiency and 9.5 mW of output power) demonstrated that HBV technology was a serious challenger to the classic and simpler Schottky technology. However, despite further efforts by IEMN, Chalmers, and UVa, HBV multipliers did not reach the level of performance of Schottky multipliers. A comprehensive and recent study on the physics, the technology, and the applications of HBV multipliers can be found in Stake et al. (2017). Another technique to build devices that exhibit internal symmetries was explored by Krach et al. [42] (See Section 9.3.3).

9.4.2 Semi-Monolithic Frequency Multipliers at THz Frequencies

In the mid-nineties, the release of powerful commercial three-dimensional (3D) field-solvers (Ansoft-now-Ansys HFSS) and nonlinear circuit simulators (HP-now-Keysight MDS-now-ADS) transformed the way frequency multipliers were designed and built. These codes greatly increased the accuracy and the speed of the calculations necessary to optimize frequency multipliers. Erickson and Tiovunen pioneered the way by designing a four-anode balanced doubler at 170 GHz entirely with HFSS and MDS [46]. In this article, they gave rationale to justify the use of 3D field-solvers instead of traditional RF measurements performed on scaled-models: "*Conventional scale model measurements, because of the wide range of sizes (>1000 : 1), are difficult when one considers the smallest important features on the diode relative to the size of a waveguide mount. Another major problem is providing the small coaxial probes to the diode locations.... The advantages of numerical analysis are that one may easily study dielectric thickness effects, optimum inductances in the diode package, power balance between the diodes, and the origin of the parasitic effects.*" Erickson's and Tiovunen's design methodology was rapidly adopted by other researchers and opened the way, several years later, to the design of highly integrated fixed-tuned waveguide multipliers working well above 1 THz.

Within the years 1995–2000, it became clear that discrete planar diodes were limited in frequency due to their size and the difficulty to connect them to the circuit with sufficient precision. Their integration on a circuit featuring several matching elements and providing

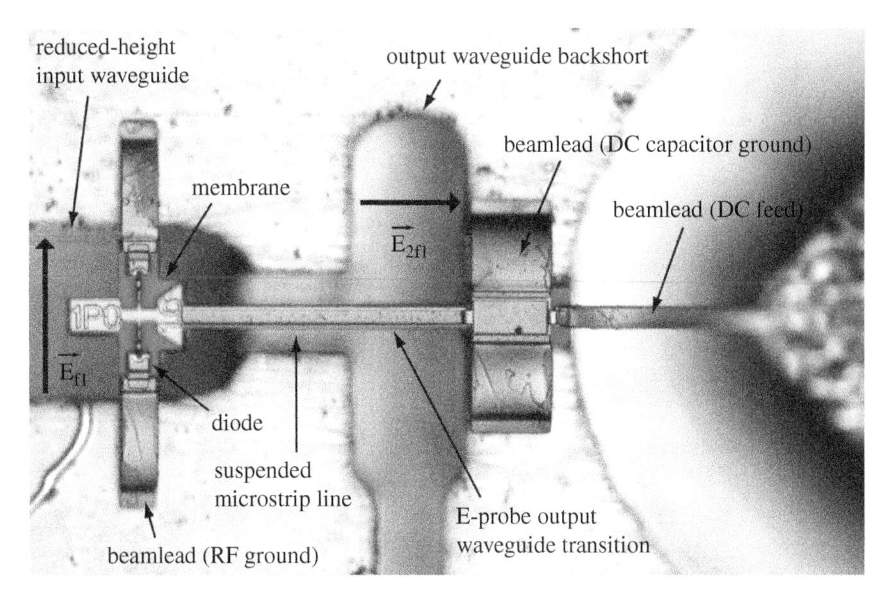

Figure 9.19 JPL 1500 GHz doubler with JPL frame-less membrane technology (design by Goutam Chattopadhyay). This Doubler was used to build the 1.5 THz channel of the heterodyne instrument of the Herschel Space Observatory.

precise connections to the waveguide block was necessary. However, MMIC-like submillimeter-wave circuits on GaAs substrate presented the inconvenience of being lossy and dispersive due to the high dielectric constant of GaAs (or InP for IEMN HBV diodes). To solve this difficulty several device fabrication technologies were proposed. One consists of transferring the epilayer on quartz (or some other application-optimized substrate) to decrease the losses and dispersion [36, 42], or on high thermal conductivity substrates to address heat dissipation issues [29, 48, 79].

An alternative approach introduced by Mehdi and Smith at the JPL, USA, is to decrease dielectric loading by removing most of the substrate from the chip [47, 49, 52, 80], or by using GaAs membrane technology [1, 17–20, 80–83]. The first solution, called substrate-less technology is used at JPL for sub-THz circuits with substrate thickness ranging from 12 to 50 μm depending on the frequency (see Figure 9.4). For THz circuits, only the membrane process combined with e-beam lithography is used. JPL membranes are 3–5 μm thick and can be made with no supporting frame (see Figures 9.8 and 9.19). The introduction of beam leads (metal membranes) to facilitate chip handling and placement and provide more precise RF and DC grounding brought significant further improvement to this technology.

It is important to mention that the precision of the machining of the waveguide blocks plays a fundamental role in the operation of THz frequency multipliers. For instance, JPL THz frequency multipliers fabricated for HIFI required a precision of 2-to-3 μm for the alignment of the chip in the channel and the alignment of the two halves of the block. Terahertz frequency multipliers should be designed on the basis of a clear understanding of the limitations of the fabrication of waveguide blocks and the assembly procedure. Therefore, parametric analysis of the impact of key design features on RF performance should be

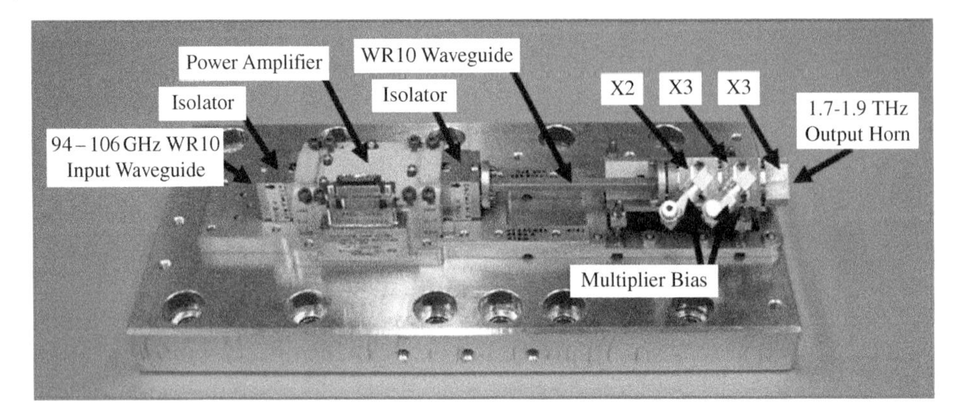

Figure 9.20 JPL flight model of the 1.7–1.9 THz local oscillator chain for HIFI-Herschel. *Source*: Reproduced from [83].

incorporated at the start of design optimization to help define the structure least sensitive to mechanical and assembly deviations. This is indeed the most important and time-consuming task in the design phase. In addition, careful planning of RF test assembly and test workflows is fundamental to a successful program in this frequency range, as testing multiple multipliers is often necessary to achieve the required performance.

9.4.3 THz Local Oscillators for the Heterodyne Instrument of Herschel Space Observatory

One of the greatest challenges for HIFI was achieving 10% electronically tunable sources in the 1.6–1.9 THz range with sufficient power to each pump a pair of HEB mixers, which translates to being able to produce more than about 2 μW of output power. Due to the great improvement of the performances of frequency multipliers upon cooling, a passive cooling scheme was implanted in the LO unit to bring the multipliers at an ambient temperature of 120–150 K. The W-band PAs used in the chain [84] were left at about 290 K and were separated from the multipliers by 5 cm stainless-steel piece of rectangular waveguide (see Figure 9.20).

A ×2×3×3 multiplication scheme was envisioned to cover the desired band in two sub-bands, 1.6–1.7 and 1.7–1.9 THz. The last stage frequency triplers are waveguide biasless balanced frequency triplers in an open-loop configuration featuring two Schottky diodes on a few micrometer thick GaAs membrane (see Figure 9.21a). These 1.6–1.7 and 1.7–1.9 THz frequency triplers were designed and built following the successful design of HIFI band 5 1.2 THz frequency tripler, which was the first membrane frequency multiplier without a supporting frame, and also the first semi-monolithic frequency multiplier working above 400 GHz [17, 18]. The 1.6–1.7 and 1.7–1.9 THz frequency triplers use the same device, except for the size of the anodes (around 0.3 μm^2 of surface area and 1 fF of intrinsic zero bias capacitance per anode). However, the waveguide input and output matching networks are optimized for each sub-band, therefore, two different waveguide blocks are necessary.

Figure 9.21b shows the output power produced by the two chains at room temperature and at 120 K. Cooling the chain produces an appreciable increase of bandwidth and output

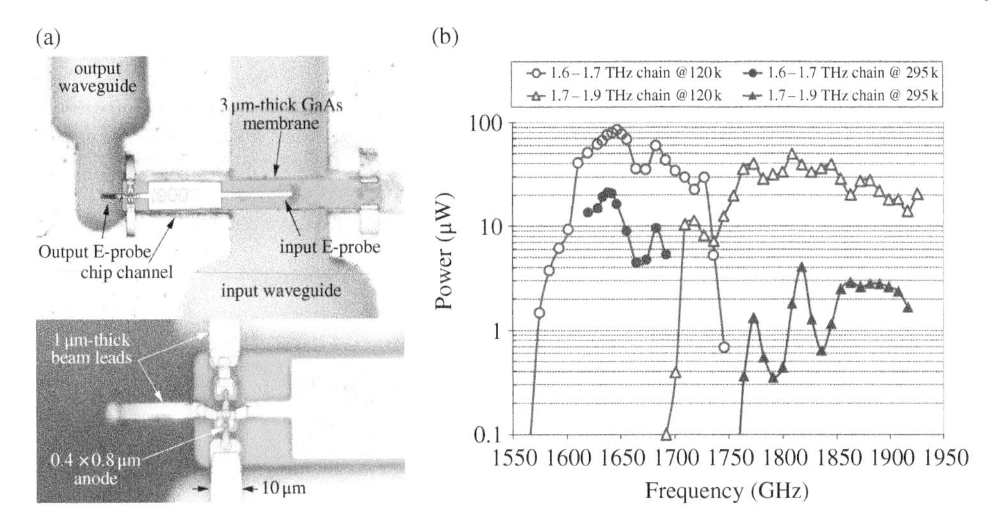

Figure 9.21 (a) Last stage frequency tripler used for the 1.6–1.7 and 1.7–1.9 THz local oscillator chains of the heterodyne instrument of the Herschel Space Observatory (designed by Alain Maestrini at JPL). The top picture shows the chip placed inside the waveguide block. The bottom pictures show a close-up of the chip. (b) Performance of 1.6–1.7 THz (circles) and 1.7–1.9 THz (triangles) local oscillator chains developed for the heterodyne instrument of the Herschel Space Observatory at room temperature (filled markers) and at cryogenic temperature (open markers).

power. There is a three-fold reason for this drastic improvement. Firstly, as the device is cooled the GaAs mobility improves thus improving the intrinsic performance of each diode. Secondly, ohmic losses associated with the waveguides and the on-chip matching circuits decrease due to the decrease in phonon scattering. Thirdly, as the drive power increases, the efficiency of the last stage increases significantly since, at room temperature, the last stage is under-pumped. The 1.6–1.7 THz chain produced a record room-temperature output power of 21 μW and an estimated conversion efficiency of 1.5% at 1647 GHz. A record output power of 86 μW with an estimated efficiency of 3% was measured at 1647 GHz at 120 K [53]. The 1.7–1.9 THz chain produced an output power of 4 μW at room temperature with an estimated efficiency of 0.4% at around 1818 GHz. At 120 K, the measured peak power was 50 μW around 1809 GHz and the estimated efficiency was 1% [1, 83].

It must be pointed out that these results far exceeded expectations. Until December 2002, when the first tests of the 1.9 THz tripler were performed, it was unclear if the 1.9 THz channel would ever be able to provide the required 2 μW of power to pump the HEB mixers. The paradoxical consequence for the instrument was the late addition of a fixed optical attenuator at the output of the local oscillator chain! It was actually not possible to sufficiently decrease the local oscillator power by adjusting – as initially planned – the bias voltages of the lower stage frequency multipliers or the drain voltage of the PAs. Trying to do so would have put the frequency multipliers in a forbidden bias regime or the amplifiers would have become unstable. HIFI, however, was a tremendous scientific success with a near continuous coverage from 480 to 1912 GHz.

As shown in Section 9.2 of this chapter, in the mid-2010s, power levels in the range of 1.6 to 1.9 THz were increased 10-fold or more at room temperature thanks to several factors: the

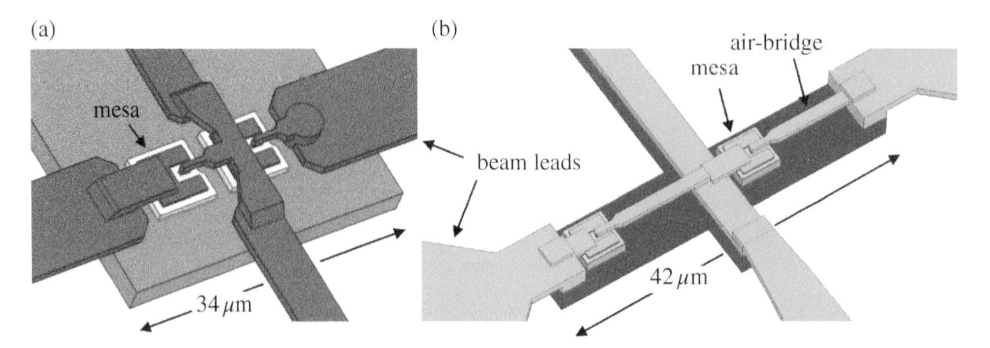

Figure 9.22 (a) Detail of the diode area of the Herschel 1.9 THz frequency tripler (designed by A. Maestrini at JPL); (b) the post-Herschel re-designed version (designed by A. Maestrini at Observatoire de Paris.) In the later, the mesas area has been reduced by 40%, the beam leads land now much closer to the membrane edges allowing longer air-bridges to be implemented in the circuit. The thickness of the GaAs membrane has been kept unchanged but the dielectric has been removed around the diodes to decrease dielectric loading.

increase of the available power at W-band, due to the availability of watt-level GaN amplifiers, the improvement of the thermal management and the design of lower stages of the frequency multiplication chain, and, decisively, a much better understanding of the limits of the technology of THz frequency multipliers, that allowed for an important reduction of the size of the mesas and of the anodes (see Figure 9.22) The second generation of frequency triplers JPL 1.9 THz and its 2.06 THz variant could, therefore, be designed with an improved phase and power balance between the diodes and junction capacities better suited to the operating frequency, leading to conversion efficiency significantly improved and with a wider bandwidth (see Figure 9.23). These technological improvements were in fact first implemented on the design of a 2.7 THz frequency tripler to build the first room temperature solid state source working in the 2.5–2.7 THz band with more than ten microwatts of power.

9.4.4 First 2.7 THz Multiplier Chain with More Than 10 μW of Power at Room Temperature

The design of JPL post-HIFI 2.7 THz multiplier chain was primarily intended for the construction of the local oscillator of airborne or spaceborne heterodyne receivers dedicated to the observation at 2.675 THz of the important $J = 1–0$ line of HD. In astrophysics, the HD abundance is a tracer of the cosmologically significant D/H ratio. This molecule is also important to study star formation, the evolution of giant planets, and the atmospheric evolution of rocky planets and moons. Actually, HIFI initially included a channel at 2.675 THz in its baseline but it was descoped in 2000 due to lack of technological maturity.

This chain is based on a cascade of three frequency triplers pumped by a four-way power-combined W-Band source [54, 86, 87]. Although, never demonstrated at terahertz frequencies before, the $3 \times 3 \times 3$ multiplication scheme was chosen based on the successful design of the $2 \times 3 \times 3$ frequency multiplier chain at 1.9 THz for HIFI. The first two triplers introduced in-phase power combining for sub-millimeter wave frequency multipliers. The designs of these triplers will be presented in a dedicated section of this chapter (see Sections 9.5.1.1 and 9.5.1.2).

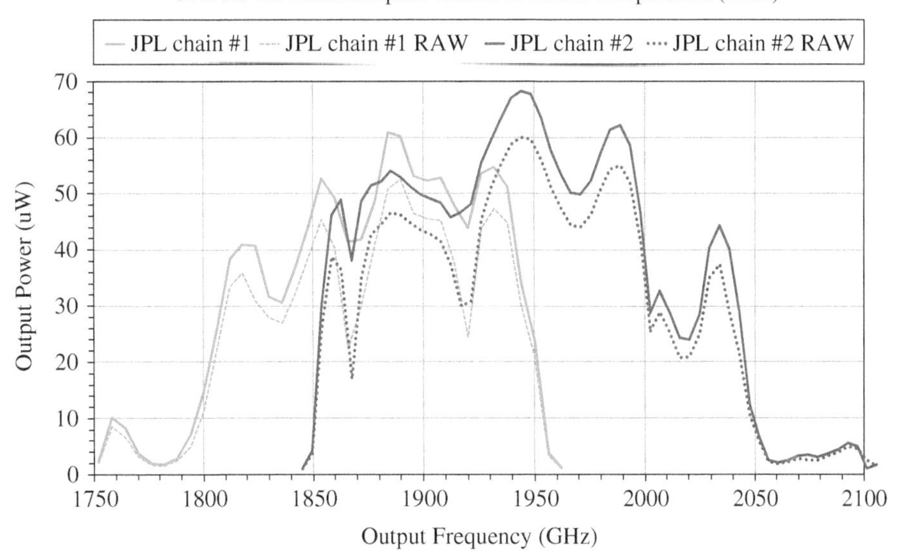

Figure 9.23 JPL 1.8–2.1 THz multiplier sources at room temperature measured in air (2013.) The multiplier source #1 uses WR10 GaAs PAs, a first-stage frequency 210 GHz doubler form HERSCHEL-HIFI, a new generation 650 GHz tripler, and a new generation 2.06 THz tripler. The multiplier source #2 uses WR8 GaAs PAs, a new generation 225 GHz doubler, and the same new generation 650 GHz tripler and new generation 2.06 THz trip as in the multiplier chain #1. In green: multiplier chain #1; in red: multiplier chain #2. Thick plain lines: data corrected for water absorption in a 50 mm air path at 40% relative humidity at 22 °C, and waveguide losses (0.5 dB). Dashed lines: RAW data. Design of 1.8–2.1 THz triplers by A. Maestrini and J. Siles. Design of new generation driver chains by J. Siles.

The 2.7 THz frequency tripler (see Figure 9.24) uses the same configuration as the 1.9 THz frequency tripler of HIFI and was designed for 1 mW of input power at 900 GHz. It is biasless, i.e. the diodes are grounded. For pump power levels around 1 mW, operating the frequency tripler at zero bias gives near optimum conversion efficiency and greatly simplify the circuit layout. On the one hand, the absence of a DC port prevents the destruction of the device by ESD; on the other hand, the absence of such a port makes it impossible to monitor the currents flowing through the diodes during operations and preclude any inspection of the Schottky diodes by current measurements as a function of the voltage. The latter must be taken into account when defining a quality control process for any system using such devices.

The 2.7 THz frequency tripler features two Schottky planar varactor diodes with nominal anode area of around 0.15 μm^2 deposited on an epilayer of GaAs doped at high concentration ($>2 \times 10^{17}$ cm^{-3}) to mitigate the effect of carrier velocity saturation at high frequencies. The nonlinear response of the diodes is simulated using the model proposed in Section 9.3.7.2. An approximate value for the series resistance is calculated assuming that the epilayer of the diodes is fully depleted at the optimum operating condition and that the actual path of the current flow is equivalent to a vertical path through the thin n^+-layer (the ohmic contact resistance was considered small enough to not have any significant impact). This yields a series resistance of around 50 Ω using the mobility-field characteristics of

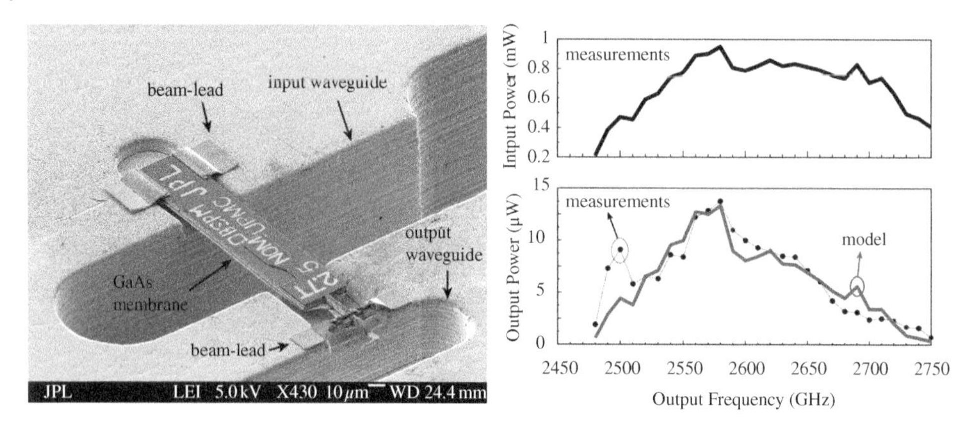

Figure 9.24 SEM image of the 2.7 THz balanced frequency tripler chip mounted on the bottom half of the waveguide block (left). Right: Measured performance of JPL 2010 2.7 THz source (bottom) compared to simulations of the final frequency tripler (bottom) accounting for the measured available input power in the 823–917 GHz band (top). For these measurements, the power at W-band was limited to a flat 350 mW for most of the band and between 150 and 300 mW for the low-end of the band. *Source*: Figure reproduced from [54].

n-doped GaAs [61] and the analytical equations in [88]. This value gives a good estimate of the achievable peak efficiency. Realistic metal losses have been accounted for in the simulations by including high-frequency gold conductivities as indicated in [89] ($\sigma \sim 1 \times 10^7$ S/m for evaporated gold and $\sigma \sim 2 \times 10^7$ S/m for electroplated gold).

Measured input and output power of the 2.7 THz multiplier chain at room temperature and in dry air are shown in Figure 9.24. With an input power at W-band of 150–300 mW from 91.9 to 93 GHz and 350 mW from 93.3 to 102.2 GHz, this all-solid-state, frequency-agile source produces over 1 μW across the entire 2.48–2.75 THz band, with a peak power of 14 μW at 2.58 THz, and up to 18 μW when the W-band power is increased to 450 mW at that frequency. Simulated performance of the 2.7 THz tripler for the measured input power is also plotted in Figure 9.24. The agreement between simulations and measurements is excellent except around 2.5 THz, where a resonance is observed, possibly the result of an interaction between the driver stage and the final tripler. The power levels obtained in 2010 at JPL in the 2.5–2.7 THz range are still two times higher than the power levels obtained by any other group in 2020, see [90]. However, JPL 2.7 THz chain was recently upgraded by replacing the four-way power-combined GaAs W-band amplifier by a single GaN W-band amplifier and by replacing the two first-stage frequency triplers by newly designed higher power ones [2]. With more available input power, the same 2.7 THz tripler produces up to 50 μW at room temperature (see Figure 9.25) This new LO chain will be used to pump an array of HEB mixers at 2.5–2.7 THz for the next NASA astrophysics balloon mission ASTHROS[3].

9.4.5 High Power 1.6 THz Frequency Multiplied Source for Future 4.75 THz Local Oscillator

A high power ×2×3×3 multiplier chain to 1.6 THz, which produces unprecedented output power levels approaching a milliwatt, was designed and fabricated at JPL, as driver chain

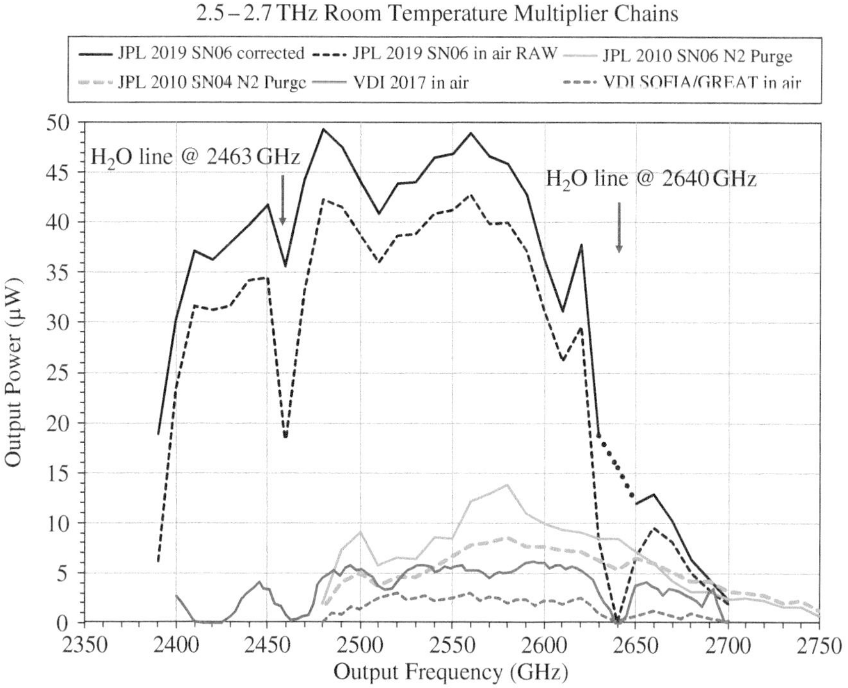

Figure 9.25 State-of-the-art of 2.5–2.7 THz solid-state sources at room temperature (2019). Thick plain black line: corrected measured powers of 2019 JPL 2.5–2.7 THz source SN06 with GaN W-band amplifiers, redesigned 300 and 900 GHz tripler stages, the correction is made for 50 mm air path, 40% relative humidity at 22 °C and 0.5 dB of waveguide losses. Dashed thin black lines: RAW data of JPL 2019 SN06 in air. Thick plain green line: measured power in pure nitrogen of the original 2010 JPL 2.7 THz source SN06. Thick dashed green line: measured power in pure nitrogen of the original 2010 JPL 2.7 THz source SN04. Red thick plain line: VDI Inc. 2.5–2.7 THz source 2017 measured in air. Red thin dashed line: VDI Inc. 2.5–2.7 THz source for GREAT/SOFIA measured in air.

of future 4.75 THz local oscillators for high sensitivity spaceborne heterodyne receivers dedicated to atmospheric science or astrophysics. Actually, global atmospheric measurements of wind vectors and temperature profiles over the whole Earth's atmosphere is critically needed for understanding the dynamics of the exchanges between the different layers of the atmosphere [91, 92]. The atomic oxygen (OI) emissions at 4.7448 THz (63.184 μm) is the brightest emission line (in terms of photons/cm^2/s) in the terrestrial thermosphere, (the other is the OI line at 2.06 THz or 145.525 μm), which makes it an ideal tracer of wind velocity, temperature, and atomic oxygen density[4]. In astrophysics, the OI line at 4.75 THz is one of the major cooling lines with [CII] at 1.908 THz, and [NII] at 1.46 THz.

While quantum cascade lasers can produce milliwatts of power at this frequency [16], which constitutes a prime advantage other competing technologies, an all-electronic 4.75 THz local oscillator with an output power of about 2 μW, associated with a superconducting high sensitivity mixer, could lead to smaller size and lower power instruments. Such an electronic chain is to be demonstrated, but a decisive step in that direction has been made.

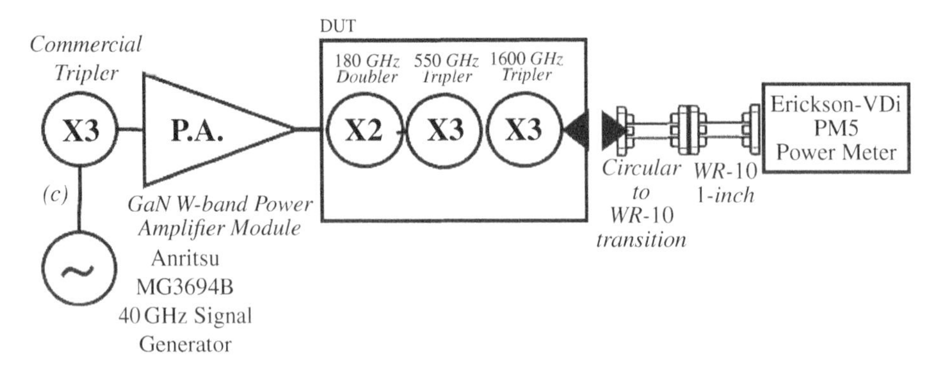

Figure 9.26 Schematics of JPL 1.6 THz high power multiplier chain and test setup. An Erickson-VDI PM5 power meter is used. *Source*: Figure reproduced from [2].

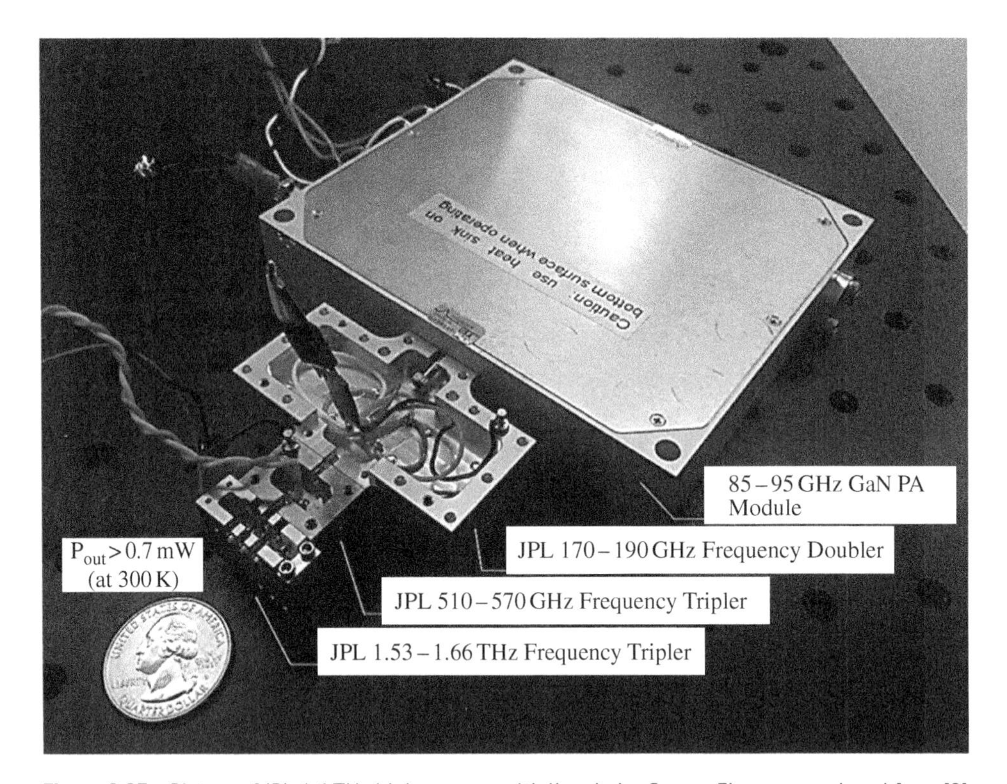

Figure 9.27 Picture of JPL 1.6 THz high power multiplier chain. *Source*: Figure reproduced from [2].

JPL 1.6 THz high power source schematics are given in Figures 9.26 and 9.27. It is based on a high power GaN PA external module that can deliver 1–2 W in the 84–96 GHz band, a high power 180 GHz frequency doubler that can deliver up to 400 mW (see description of the multiplier chip at Section 9.5.3.3), a high power 550 GHz frequency tripler that can deliver 30–40 mW (see Section 9.5.3.1), and finally a 1.6 THz four-anode bias-able balanced frequency tripler in a virtual loop configuration (Figure 9.28).

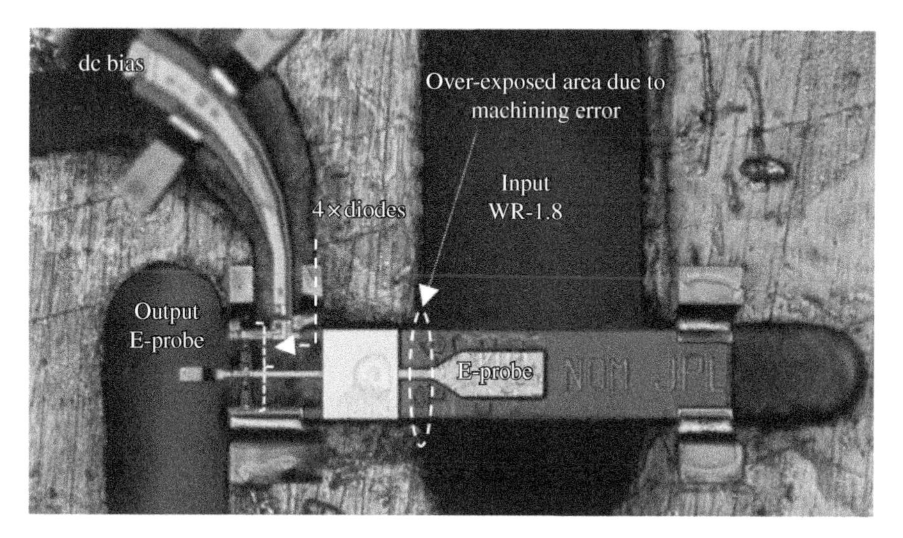

Figure 9.28 Picture of JPL 1.6 THz frequency tripler. This Frequency tripler is the first four-anode bias-able frequency multiplier above 1 THz. *Source*: Figure reproduced from [2].

The 1.6THz frequency triler chips are based on the monolithic membrane diode (MOMED) process introduced in [82]. They feature four anodes of ~1 fF zero-voltage junction capacitance each, for high input power handling. For lowering the thermal resistance from the diodes to the waveguide block, the diodes are integrated on a 5-μm GaAs membrane instead of a thinner membrane that is traditionally employed in this frequency range (see Figure 9.27) The implementation of four anodes and a bias line in a 1.6 THz frequency tripler constitutes a major step forward with respect to HIFI terahertz frequency multiplier generation, since it requires extremely precise and very small features on the chip that are at the limits of the fabrication process. Unfortunately, the waveguide blocks were not fabricated to specifications, and the input matching of the frequency tripler was degraded (see Figure 9.27) However, the recorded output power approaches 1 mW (see Figure 9.29) and are more than 20 times higher than any other sources in this frequency range. The performance of this source is expected to improve significantly with waveguide blocks at the specifications and with the improved 180 GHz frequency doubler module shown in Figure 9.43, which features integrated GaN PAs instead of an external PA module.

9.5 Power-Combining at Sub-Millimeter Wavelength

The practical limit of the output power of a frequency multiplier is typically either the power beyond which conversion efficiency drops off due to saturation effects or the device lifetime becoming unacceptably short due to thermal or reverse-breakdown effects [69]. To increase power handling, the device doping can be optimized and the number of anodes per chip can be increased. Additionally, the epilayer can be transferred to a high thermal conductivity substrate [29, 48, 79, 93, 94]. Multi-anode frequency doublers were first introduced [44–47, 49, 79], before frequency triplers with more than two anodes were demonstrated.

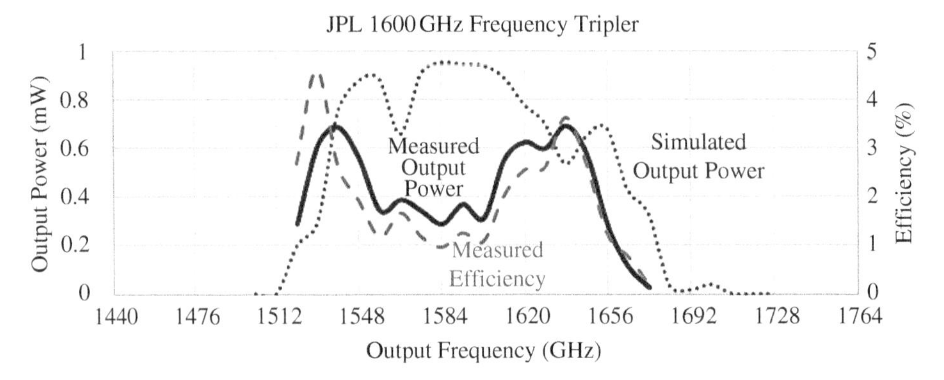

Figure 9.29 Performance of JPL 1.6 THz frequency tripler. RF performance in the mid-band is degraded due to a machining error during the fabrication of the waveguide blocks. *Source*: Figure reproduced from [2].

For example, wideband and high-efficiency balanced triplers at 300 and 600 GHz featuring respectively six and four Schottky anodes per chip have been presented by JPL in [52, 95], and unbalanced high-efficiency multi-anode frequency triplers on high-thermal conductivity substrates at 200 and 400 GHz have been presented by Virginia Diodes Inc. in [96]. However, there is a practical limit to the number of anodes based on the chip size, the device impedance, and coupling efficiency. As the number of anodes is increased, compromises must be made between an optimum and even input coupling to the anodes, an optimum matching of each anode at the idler frequencies and optimum matching at the output frequency.

9.5.1 In-Phase Power Combining

A complementary approach to increase the power handling of a given source is to power-combine two or more parallel stages. However, for efficient power combining this approach requires increasing care at short wavelengths to keep the parallel paths well-matched despite fabrication and assembly tolerances and to minimize losses in the additional circuits required for dividing and recombining the signal.

The design of power-combined frequency multipliers starts with the design of a single-chip version for half the input power. Once this design is completed and the dimensions of the microelectronic circuit fixed, the design of the waveguide circuit of the power-combined version can start.

9.5.1.1 First In-Phase Power-Combined Submillimeter-Wave Frequency Multiplier

The first demonstrated in-phase power-combined submillimeter-wave frequency multiplier is a 300 GHz frequency tripler in a virtual loop configuration [55]. This frequency multiplier is based on two mirror-image microelectronic circuits, totalizing 12 Schottky anodes monolithically integrated on a 5 μm-thin GaAs membrane, and inserted in a single split-waveguide block. Each anode has an intrinsic zero-bias capacitance of about 20 fF. A compact Y-junction splitter, and a compact Y-junction combiner, are incorporated to

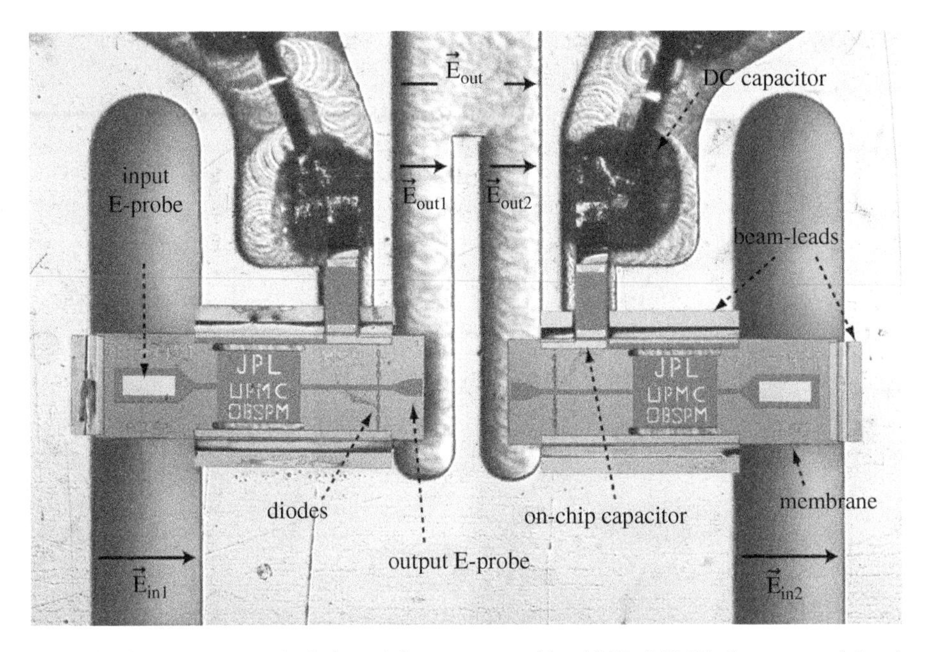

Figure 9.30 Close-up vertical view of the power-combined 260–340 GHz frequency tripler showing the two mirror-image GaAs integrated circuits. The E-field vectors in the input and output waveguides are indicated by plain arrows. The E-fields generated by the two sub-circuits are combined in-phase in the output waveguide. *Source*: Figure reproduced from [55].

respectively the input and the output matching networks of the frequency multiplier, yielding a power combining efficiency of 100%.

Figure 9.30 shows a photograph of the area where the microelectronic circuits are mounted. This tripler uses two independent DC bias lines, so the balance between the two microelectronic circuits can be monitored. By applying asymmetrical bias voltages, the effects of possible asymmetries between them can be potentially compensated for improved RF performance. This frequency tripler covers the band 260–340 GHz and produces up to 26 mW at 318 GHz with 250 mW of input power at 106 GHz. The design and fabrication of this dual-chip tripler was the first step of the construction of JPL state-of-the-art ×3×3×3 2.7 THz multiplier chain.

This in-phase power-combined dual-chip frequency tripler was actually further power-combined using input and output 3 dB/90° hybrid couplers (design by John S. Ward, JPL). Power-combining with hybrid couplers provides a natural isolation between the two stages, limiting undesired interactions between the stages. In addition, for a multiplier chain, it provides inter-stage isolation which reduces power-fluctuations versus frequency due to improved matching. However, hybrid couplers tend to be delicate to machine at sub-millimeter wavelengths, introduce more RF losses than compact Y-junctions, and increase the overall size of the waveguide block. Nevertheless, for power-combining four frequency triplers, the configuration is shown in Figure 9.31 has proven to be an excellent choice. When pumped with 330–500 mW, this quad-chip frequency tripler delivers 29–48 mW in the 276–321 GHz band [55].

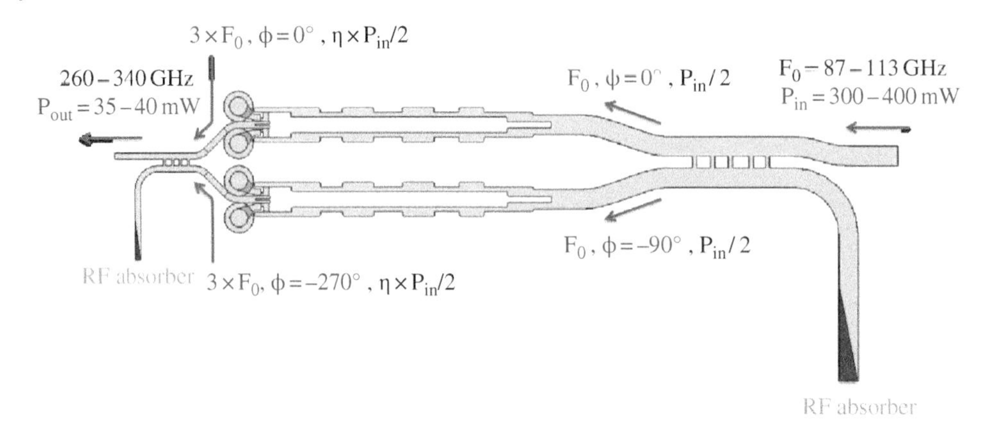

Figure 9.31 Quad-chip 260–340 GHz tripler designed by John Ward at Jet Propulsion Laboratory based on [55].

9.5.1.2 In-Phase Power Combining at 900 GHz

Based on the previous in-phase power combing scheme at 300 GHz, an in-phase power-combined frequency tripler working at 900 GHz was designed to be used as a second stage driver of JPL 2.7 THz frequency multiplier chain. Due to the difference of frequency, anode size, and input power, the diodes require a different matching circuit than a simple scaled model of the 300 GHz dual-chip matching circuit. In addition, the bias circuit requires extra space and forces the use of special bended waveguides that add some constraints in the design.

Figure 9.32 shows an overall photograph of the tripler including the input matching circuit and two different close-ups of the device area. The tripler uses a symmetrical split-block waveguide design with one device mounted in each half block. Each device features four Schottky varactor diodes monolithically integrated on a 3-μm thin GaAs membrane in a balanced configuration and biased in series. Each anode has an intrinsic zero bias capacitance of about 4 fF.

The driver chain of the 900 GHz in-phase power-combined frequency tripler is constituted by a W-band synthesizer followed by a four-way power-combined W-band GaAs amplifier, and the quad-chip power-combined 300 GHz frequency tripler presented in Figure 9.31. The 900 GHz power-combined frequency tripler produces over 1 mW from 840 to 905 GHz at room temperature with a pick power of 1.3 mW at 860 GHz. The conversion efficiency is in the range of 2.1% to 2.5% in the same frequency range.

9.5.1.3 In-Phase Power-Combined Balanced Doublers

Contrary to balanced frequency triplers, it is not possible to use input and output Y-junctions to power-combine frequency doublers in a perfectly symmetrical design. It would results in a perfect null at the output. Instead, a 3 dB/90° hybrid coupler at the input and a Y junction at the output of the frequency doubler are required for best RF performance across the full intrinsic bandwidth of the individual doublers (see Figure 9.33).

An ultra-compact design of such a doubler was made for the 300 GHz stage of the local oscillator chain of submillimetre wave instrument (SWI) 1.2 THz channel [97, 98] based on

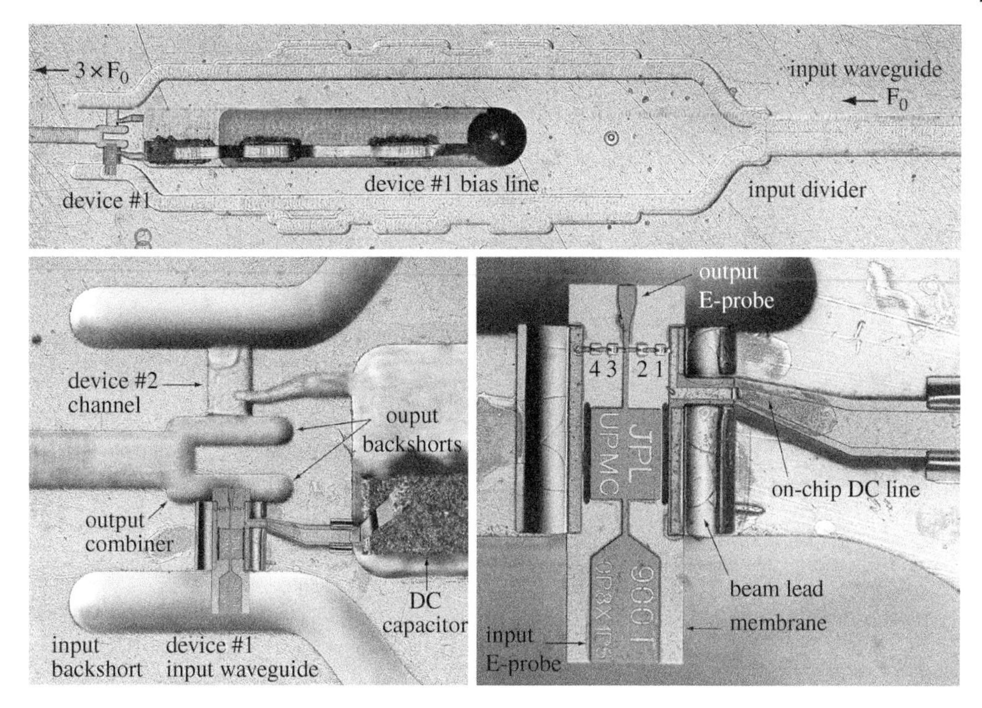

Figure 9.32 Picture of the bottom half of the power-combined 900 GHz frequency tripler showing device #1 (top view). Close-up view of the output combiner, device #1 and the DC capacitor (bottom left view.) Close-up view of device #1 with labels for diodes #1–4 (bottom right view.). The tripler chip is approximately 300 μm long and 100 μm wide. *Source*: Reproduced from [24].

the design presented in [99]. The input 3 dB/90° hybrid coupler set the separation of the two balanced doubler devices and the dimensions of the output compact Y-junction, which is integrated in the output matching network (see Figure 9.34).

Each individual device features four anodes with ~20 fF of zero-bias junction capacitance monolithically integrated on a 5 μm-thick GaAs membrane. It covers the 270–320 GHz band with a conversion efficiency of ~20% at room temperature when pumped with 60 mW, identical to the conversion efficiency of the single-chip version of this doubler (designed for the 600 GHz channel of SWI), when pumped with 30 mW. The devices were fabricated at Centre for Nanoscience and Nanotechnology-Marcoussis, France [100].

9.5.2 In-Channel Power Combining

Power combining at the "chip level" is already in place with multi-anode design. One complementary solution that was never exploited in this frequency range until recently [101] consists in designing double-sided circuits, perfectly symmetrical. Technologically it is a very challenging proposition, though not impossible: right-handed and left-handed chips can be assembled using a very thin layer of a nonconductive low-RF-loss glue. A similar proposition is to leave a well-controlled air-gap in between the two circuits, removing

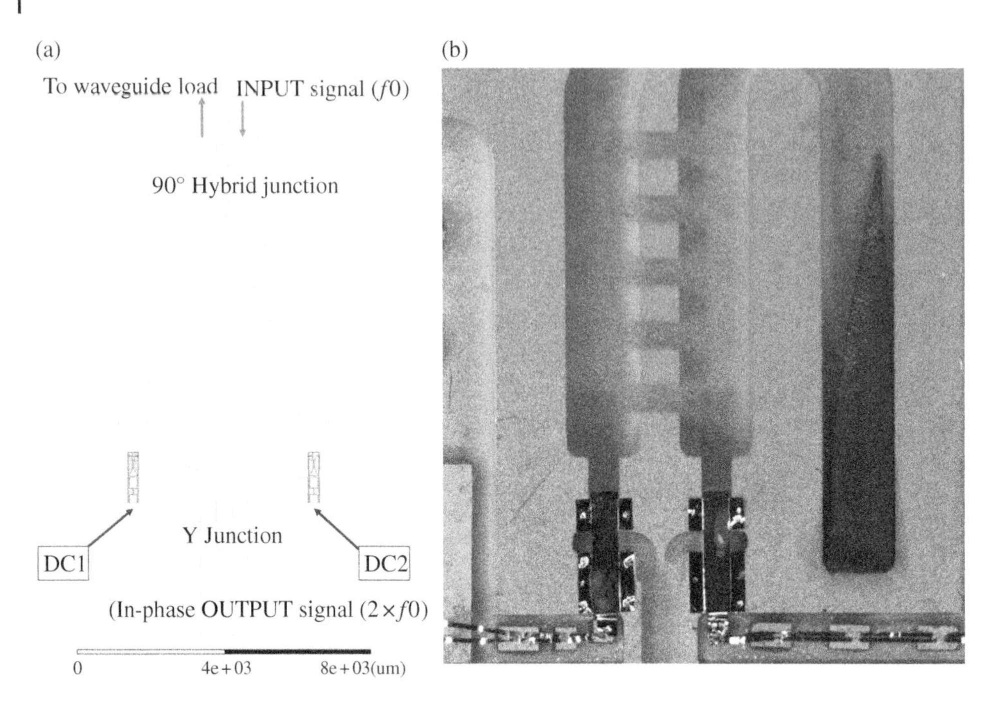

Figure 9.33 Schematics of an in-phase power-combined balanced frequency doubler (a) and its optimized implementation for the demonstration model of the 300 GHz doubler of SWI 1.2 THz channel local oscillator (b). While each device has an individual bias line, they were connected to the same DC port to simplify the implementation and the operations of the instrument. Designed by A. Maestrini, J. Treuttel and F. Yang at LERMA – Observatoire de Paris, France. The fabrication was done by LERMA in partnership with C2N-Marcoussis, France.

the need of a complex assembly procedure and allowing each chip being mounted in one-half of the waveguide blocks, independently of the other.

This power combining scheme can be applied to different types of frequency multipliers and is perfectly complementary to the other combining schemes.

In terms of design complexity, it is, however, much more challenging than classic in-phase power-combining, in part due to the addition of new propagating modes and to the doubling of the number of devices in the nonlinear simulations at the first stage of the circuit optimization.

This idea was indeed first demonstrated at the Université Pierre et Marie Curie (Now Sorbonne Université, Paris) & Observatoire de Paris with a dual-symmetry 190 GHz balanced frequency doubler using the Schottky process of the Franco-German commercial foundry United Monolithic Semiconductors [102] (see Figures 9.35–9.37). This frequency doubler differs from the classic balanced doubler introduced by Erickson [43]: the diodes are in a similar configuration as multi-anode balanced triplers presented in Section 9.3.6, but they are located in the output waveguide, not in a channel that can propagate only a quasi-TEM mode; the second harmonic of the input signal can, therefore, be generated. The third harmonic is also suppressed. This 190 GHz doubler is fabricated with a Schottky process suited for mixers, not high power frequency multipliers, and it is operated at zero bias (the diodes

Figure 9.34 Detail of the demonstration model of the 300 GHz frequency doubler for SWI 1.2 THz channel local oscillator. Both stages of the power-combined doubler use identical chips and not mirror-imaged chips as for the 300 GHz tripler presented in Figure 9.30. As the output E-probe design is asymmetrical, one of the chip needs to be mounted upside-down on the bottom block of the doubler. In this design, the DC/RF decoupling is made with a low-pass filter featuring sections of suspended-microstrip lines of low and high impedance.

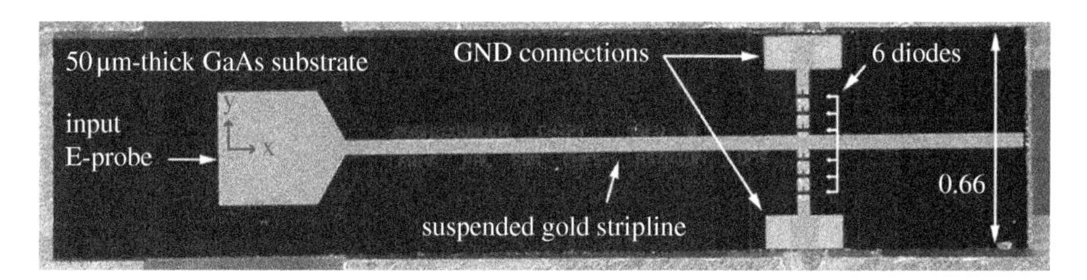

Figure 9.35 Photograph of a dual-symmetry 190 GHz balanced frequency doubler chip using UMS BES Schottky diodes process. (design José Vicente Siles & Alain Maestrini, at Université Pierre et Marie Curie – Observatoire de Paris). *Source*: Photo reproduced from [101].

are grounded, to ease the assembly), however the conversion efficiency reaches 10% at 100 mW of input power.

9.5.3 Advanced on-Chip Power Combining

In phase-power combining was further developed at sub-millimeter wavelength at JPL by the introduction of a novel chip topology. It consists in combining several individual

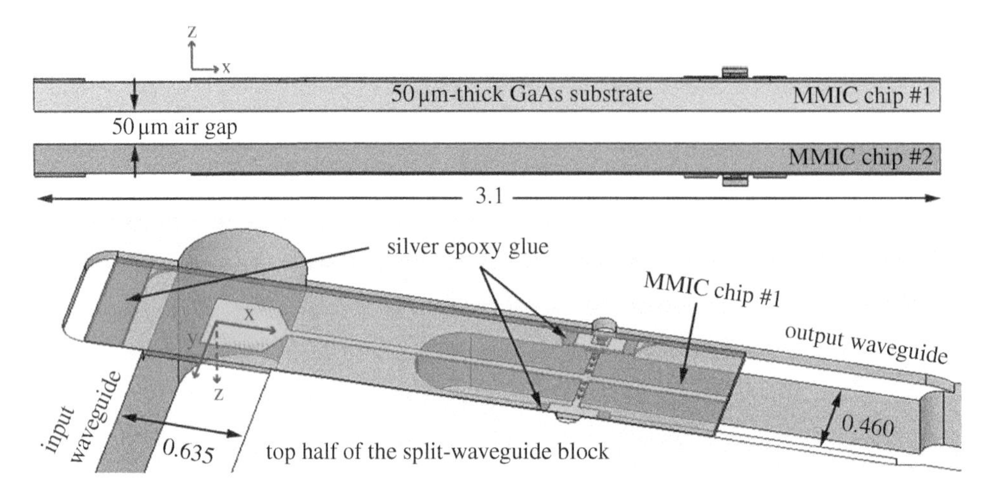

Figure 9.36 Dual symmetry 190 GHz balanced frequency doubler using UMS Schottky diodes. Top: lateral view of the two MMIC chips. Bottom: view shows one MMIC chip assembled in the top half of the split-waveguide block. For this version of the dual-symmetry 190 GHz balanced doubler, the diodes were grounded at DC. *Source*: Figure reproduced from [101].

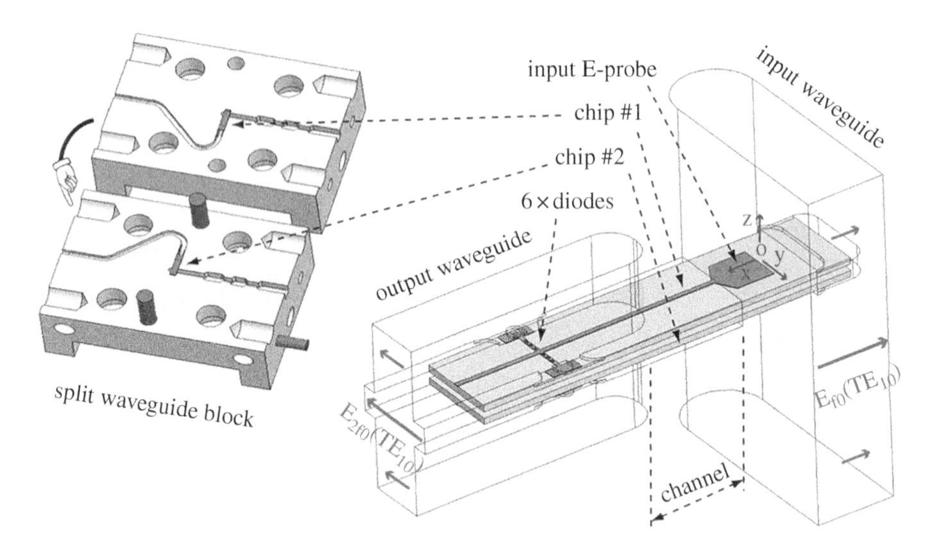

Figure 9.37 Dual symmetry 190 GHz balanced frequency doubler using UMS Schottky diodes. The left view shows the waveguide structure of the multiplier. The right view shows the two symmetrical chips in the same channel. *Source*: Figure reproduced from [101].

multiplier devices in a single one to insure both a better Schottky diode uniformity across the overall multiplier, owing to the proximity of the diodes on the wafer, and an ultra-precise alignment of each channel, owing to the precision of the lithography. This idea is actually *patented* under US Patent US9143084B2 [103].

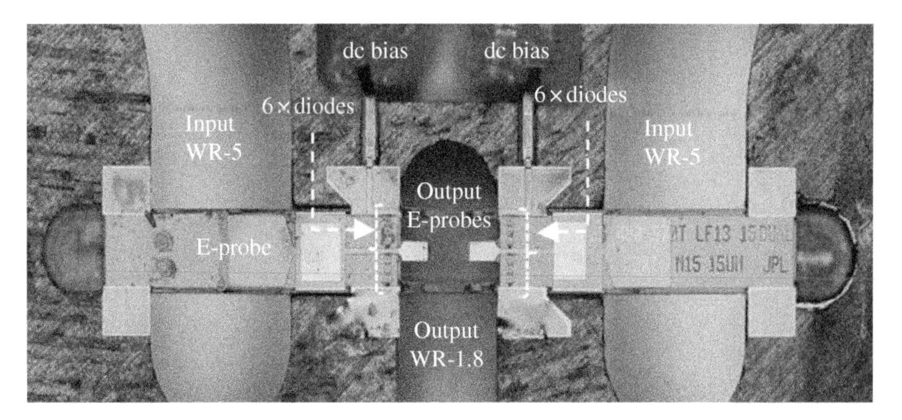

Figure 9.38 Detail of JPL on-chip in-phase power combined 490–560 GHz frequency tripler. *Source*: Picture reproduced from [2].

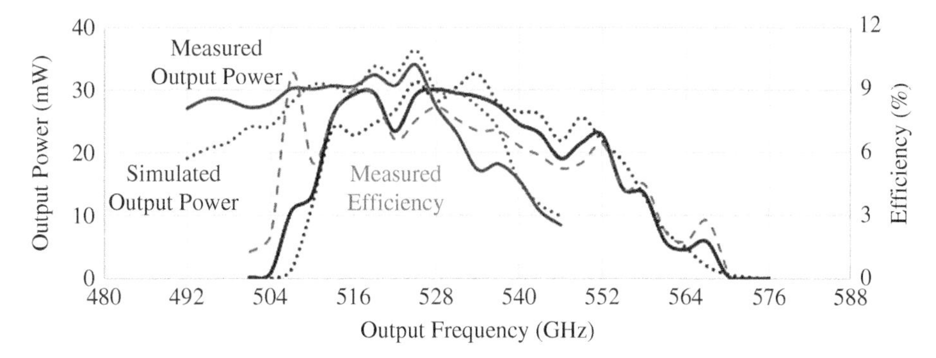

Figure 9.39 Performance of JPL 490–560 GHz frequency tripler. Two PAs modules are used to cover the band. *Source*: Figure reproduced from [2].

9.5.3.1 High Power 490–560 GHz Frequency Tripler

Using this concept, a high power in-phase power-combined frequency tripler at 490–560 GHz was designed (see Figure 9.38). Each side of the symmetrical device features six anodes in a virtual loop configuration and monolithically integrated on a 15 μm-thick GaAs membrane. The thickness of the membrane insures a good thermal management and is suited for high power handling capabilities. Two symmetrical E-probes couple the third harmonic of the input signal to the output waveguide. With this design, no compact Y-junction is needed at the output waveguide, greatly simplifying the mechanical design of the waveguide block. This tripler gave a record output power of 35 mW at 526 GHz with a conversion efficiency of 7% see Figure 9.39 [2].

9.5.3.2 Dual-Output 550 GHz Frequency Tripler

The same approach as in Section 9.5.3.1 was even pushed further, with the design at JPL of a quad-channel single input/dual output 550 GHz balanced frequency tripler. Four E-probes are located at the center of a single input rectangular waveguide in order to drive four

Figure 9.40 Detail of JPL on-chip in-phase power combined 550 GHz single input/dual output frequency tripler in a silicon micro-machined waveguide block. *Source*: Picture reproduced from [2].

multiplying structures with two diodes each in a virtual loop configuration (see Figure 9.40). The output power produced by individual triplers is then recombined at the output using four E-probes in two independent rectangular waveguides. The four multiplication structures are physically connected on a single chip so that the alignment and symmetry of the circuits can be very well preserved as this is controlled by the high precision of MMIC lithography. This design cannot use a split-waveguide block configuration since it would break the balance between the two bottom and the two top frequency triplers by introducing a frequency-dependent phase-shift between them, both at the input and at the output frequency. The input and output waveguides propagation axis are, therefore, perpendicular to the device. Hence, the multiplier block can be fabricated using either electroforming or silicon micromachining technology [104, 105]. The two independent outputs can further be in-phase power combined using a compact Y-junction as proposed in Figure 9.41.

9.5.3.3 High-Power Quad-channel 165–195 GHz Frequency Doubler

On-chip power combining at millimeter-wavelengths requires large wafer areas since both the size of each individual multiplying sub-circuit and the supporting structure are several millimeters in length. The Schottky diode process developed at the JPL is suited for 4-inch GaAs wafers, hence, tens of large devices can be produced the same time. This is an essential technological aspect to take into account for space application since the qualification requires tens of devices *of a same lot* for the many different tests. For commercial applications, the individual cost of the devices fabricated with a similar process would be still

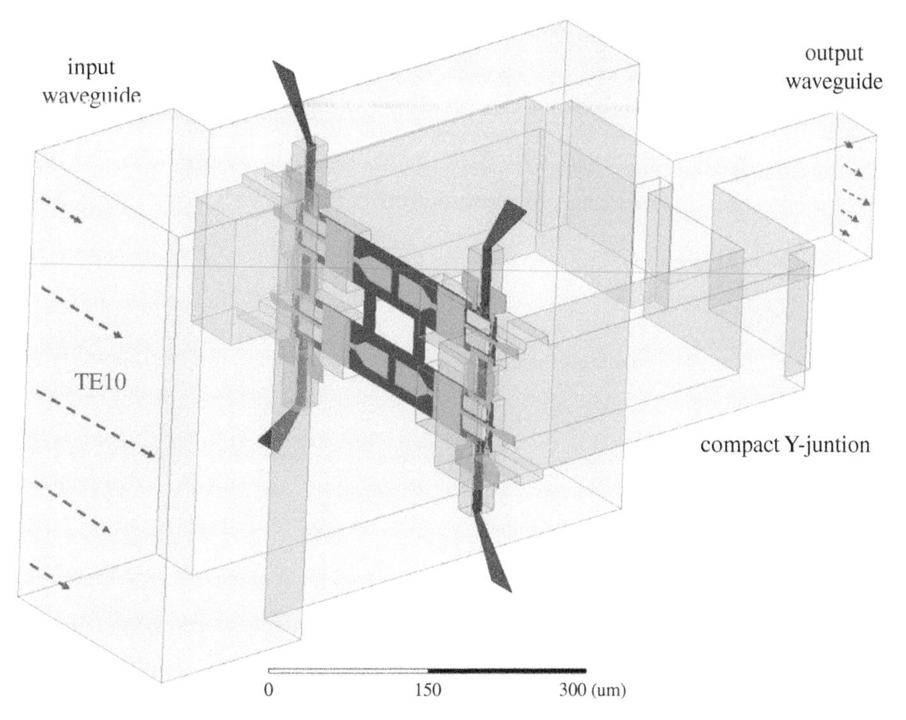

Figure 9.41 Schematics of JPL on-chip in-phase power combined 550 GHz quad-channel frequency tripler. A compact Y-junction at the output waveguide recombines in-phase the two output signals. This Y-junction is entirely part of the output matching network of the individual channels. Micro-machined waveguide blocks made out of silicon can be made to implement this topology. Electroforming is an alternate possibility with redesigned waveguide sections to avoid sharp angles. *Source*: Figure reproduced from [2].

reasonable. However, on-chip power combining at millimeter wavelengths is not suited, or is much too expensive, for processes that can only deal with small size wafers.

Making full use of JPL Micro Device Laboratory capabilities, a high-power 165–195-GHz frequency doubler featuring 24 anodes on a single 50-μm thick GaAs chip was designed as the first multiplier stage of a high power 1.6 THz chain [2], see Section 9.4.5. It comprises four multiplying structures featuring six anodes each, with a zero-voltage junction capacitance of ~60 fF per anode on a Si-doped epitaxial layer. The doubler is based on the original doubler developed for the Band 5 of the HIFI instrument onboard the Herschel Space Observatory [3] based on the architecture proposed in [43] and [44], and modified to increase its power-handling capabilities [2]. This is achieved by optimizing the anode size for higher input power and using a "full-substrate" approach instead of the "substrate-less" topology [49, 80], in order to reduce the thermal resistance from the diodes to the waveguide blocks (see Figure 9.42).

This quad-doubler topology, Figure 9.43 (top), cannot be considered an on-chip power-combined topology, strictly speaking, since the input signal is split into four waveguide inputs using hybrid couplers in order to drive separately each of the four multiplying structures. The four outputs are recombined in-phase using traditional Y-junction combiners

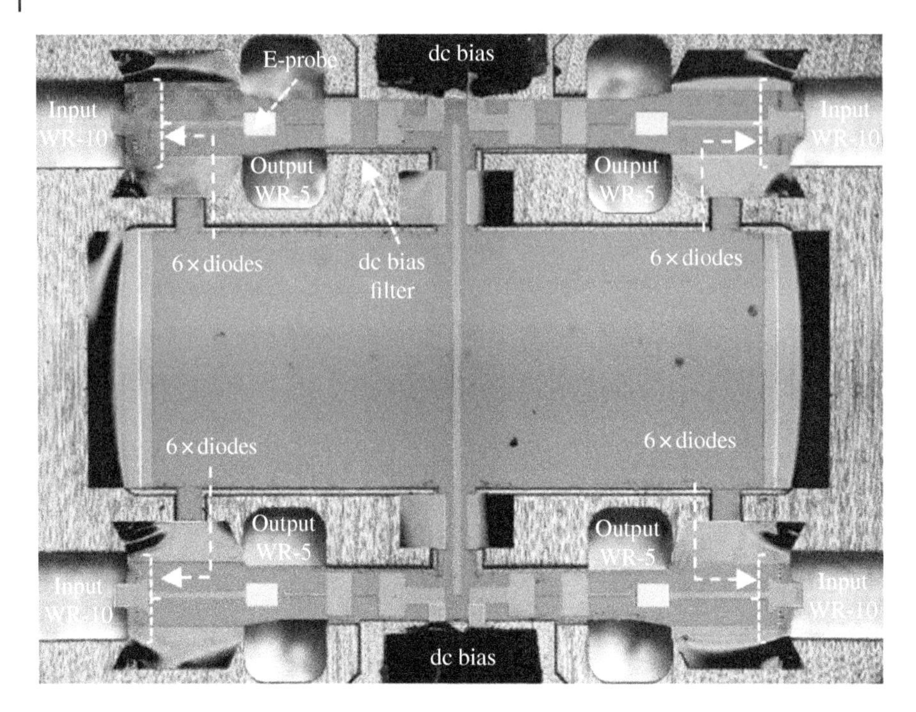

Figure 9.42 Photograph of JPL high power 165–195 GHz frequency doubler featuring four multiplying structures of six anodes each. *Source*: Picture reproduced from [2].

(see Figure 9.43). The quad-device is able to handle up to ~3 W input power, providing a maximum output power of 700–800 mW when PAs modules are directly integrated on the frequency doubler block, greatly reducing waveguide losses (see Figure 9.43).

9.6 Conclusions and Perspectives

In this chapter, we have presented several aspects of the design and fabrication of modern terahertz frequency multipliers. It focuses on the work over two decades of a team of researchers associated with JPL, who revolutionized the field by introducing new technologies and concepts that inspired the work of many groups in the world. A new generation of room-temperature terahertz Schottky diode-based frequency multiplier sources is now close to 1 mW of output power at 1.6 THz [2], 10 years after the first milliwatt-level multiplier sources at 900 GHz [55]. This achievement was only possible thanks to the integration of the manufacturing and assembly constraints into practical designs and by pushing technological limits in the right direction.

The measured conversion efficiencies of this new generation of frequency multipliers, especially at frequencies above 1 THz, follow perfectly the theoretical limit predicted by physics-based numerical models [32, 57]. This yields a very significant increase in performance above 1 THz in both conversion efficiency and generated output power. The purity of these sources is extremely good, as discussed in [54], and no oscillations have been noticed.

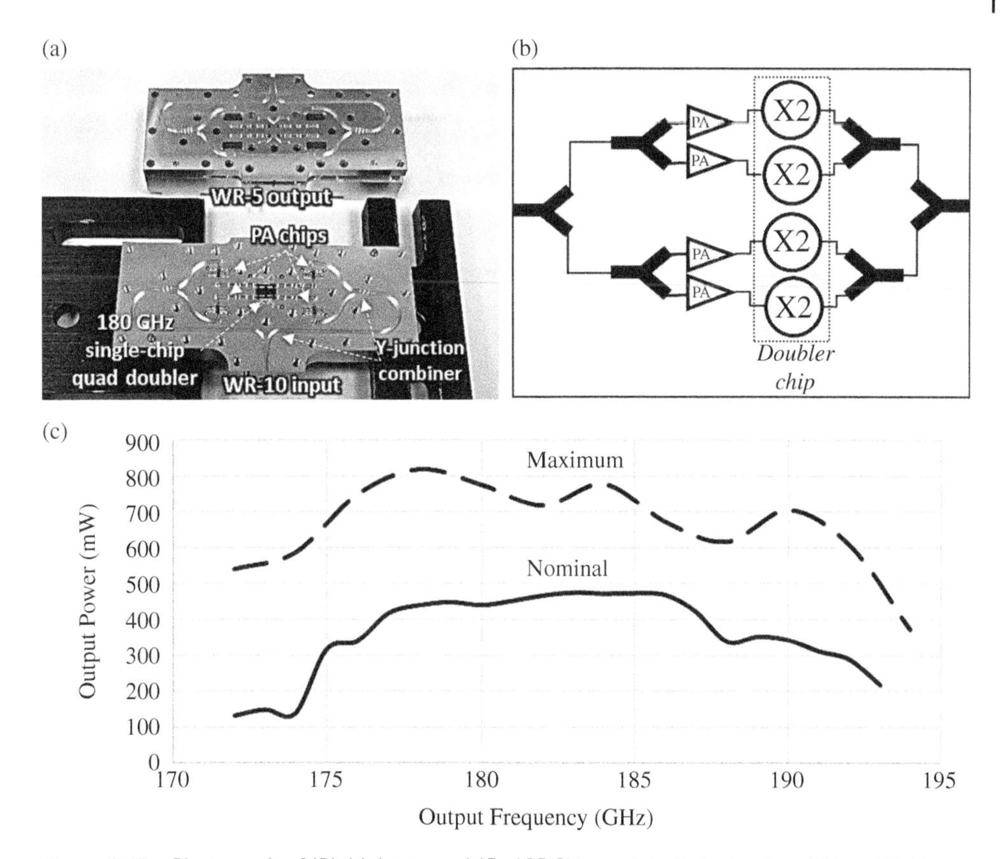

Figure 9.43 Photograph of JPL high power 165–195 GHz module featuring four W-band GaN amplifiers and an on-chip power-combined frequency doubler (a). Functional schematics of JPL 165–195 GHz high-power module (b). Measured RF performance of JPL 165–195 GHz high-power module (c). *Source*: Picture and graphs reproduced from [2].

This new generation of frequency-multiplied sources opens up the possibility of building larger array receivers in the terahertz range and makes it possible to expand the frequency operation of this technology up to 4.75 THz as JPL 1.6-THz source produces enough power to pump a 4.75 THz frequency tripler. Moreover, with >0.5 mW available beyond 1.5 THz, sufficient to pump Schottky diode-based mixers, room- temperature high spectral resolution heterodyne receivers can be now envisioned beyond 2–3 THz for the first time. Terahertz radar imagers and communication systems operating in the 100–300-GHz range can also benefit enormously from sources providing power levels in the 200–500 mW range.

Exercises

E9.1 Balanced Frequency Doublers and Third Harmonic

In section 5-2, 4th paragraph, it is stated that "This frequency doubler, differs from the classic balanced doubler introduced by Neal Erickson [43]: the diodes are in a similar configuration as multi-anode balanced triplers presented in paragraph 3.6, but they are located in the output waveguide, not in a channel that can propagate only a quasi-TEM mode; the second harmonic of the input signal can therefore be generated. The third harmonic is also suppressed" – Demonstrate both propositions in the case that only one device is used (use only Fig. 9.35 & Fig. 9.37).

E9.2 In-Phase Power Combining of Balanced Frequency Doublers

"Contrary to balanced frequency triplers, it is not possible to use input and output Y-junctions to power-combine frequency doublers in a perfectly symmetrical design. It would result in a perfect null at the output. Instead, a 3dB/90° hybrid coupler at the input and a Y junction at the output of the frequency doubler are required for best RF performance across the full intrinsic bandwidth of the individual doublers (see Fig. 5-4)." – Demonstrate both propositions by using only schematics.

E9.3 In-Phase Power Combining of Balanced Frequency Triplers

Using only schematics, demonstrate that the actual design proposed in [55] and shown in Fig. 9.30 can work.

E9.4 Waveguide Schottky Diode Frequency Multipliers Versus MMIC Diode Frequency Multipliers

Waveguide Schottky diode frequency multipliers tend to exhibit much higher conversion efficiency and bandwidth that their MMIC counterparts for a given frequency band. This chapter focuses on waveguide THz frequency multipliers since they define the state-of-the-art. However, MMIC frequency multipliers, in microstrip technology for instance, have some key advantages for applications where compactness and volume of production are key. Give key reasons why waveguide diode-based frequency multipliers are inherently better in terms of RF performance than their MMIC counterparts?

E9.5 To Balance or not to Balance the Frequency Multiplier?

In this chapter, the emphasis was put on balanced frequency multipliers since they define the state-of-the-art. In which cases one should consider designing a single-ended frequency multiplier instead of a balanced one?

Explanatory Notes (see superscripts in text)

1. Advanced design system, Keysight Technologies, https://www.keysight.com.
2. High frequency structure simulator, Ansys Inc., https://www.ansys.com.
3. Astrophysics stratospheric telescope for high-spectral resolution observations at submillimeter-wavelengths.
4. The OI line at 4.7 THz is 12 times more intense at the temperature in the lower thermosphere and E region ionosphere (300–900 K), and could provide better wind velocity resolution than the OI line at 2.06 THz.

References

1 Maestrini, A., Ward, J., Gill, J. et al. (2004). A 1.7 to 1.9THz local oscillator source. *IEEE Microwave and Wireless Components Letters* **14** (6): 253–255.
2 Siles, J.V., Cooper, K.B., Lee, C. et al. (2018). A new generation of room-temperature frequency-multiplied sources with up to 10× higher output power in the 160-GHz–1.6-THz range. *IEEE Transactions on Terahertz Science and Technology* **8** (6): 596–604.
3 de Graauw, T., Helmich, F.P., Phillips, T.G. et al. (2010). The herschel-heterodyne instrument for the far-infrared (HIFI). *Astronomy and Astrophysics* **518**: id.L6.
4 Mehdi, I., Siles, J.V., Lee, C., and Schlecht, E. (2017). THz diode technology: status, prospects, and applications. *Proceedings of the IEEE* **105** (6): 990–1007.
5 Mittleman, D. (2003). Terahertz imaging. In: *Sensing With Terahertz Radiation* (ed. D. Mittleman), 117–153. Berlin, Germany: Springer-Verlag.
6 Siegel, P.H. (2002). Terahertz technology. *IEEE Transactions on Microwave Theory and Techniques* **50** (3): 910–928.
7 Siegel, P.H. (2004). Terahertz technology in biology and medicine. *IEEE Transactions on Microwave Theory and Techniques* **52** (10): 2438–2447.
8 Woolard, D.L., Brown, E., Pepper, M., and Kemp, M. (2005). Terahertz frequency sensing and imaging: a time of reckoning future applications? *Proceedings of the IEEE* **93** (10): 1722–1743.
9 Dhillon, S.S., Vitiello, M.S., Linfield, E.H. et al. (2017). The 2017 terahertz science and technology roadmap. *Journal of Physics D: Applied Physics* **50**: 043001.
10 Tonouchi, M. (2007). Cutting-edge terahertz technology. *Nature Photonics* **1**: 97–105.
11 Ito, H., Nakajima, F., Furuta, T., and Ishibashi, T. (2005). Continuous THz-wave generation using antenna-integrated uni-travelling-carrier photodiodes. *Semiconductor Science and Technology* **20**: S192–S198.
12 Sartorius, B., Schlak, M., Stanze, D. et al. (2009). Continuous wave terahertz systems exploiting 1.5 μm telecom technologies. *Optics Express* **17** (17): 15001–15007.
13 Belkin, M.A., Fan, J.A., Hormoz, S. et al. (2008). Terahertz quantum cascade lasers with copper metal-metal waveguides operating up to 178 K. *Optics Express* **16** (5): 3242–3248.
14 Williams, B.S. (2007). Terahertz quantum-cascade lasers. *Nature Photonics* **1**: 517–525.
15 Gao, J.R., Hovenier, J.N., Yang, Z.Q. et al. (2005). Terahertz heterodyne receiver based on a quantum cascade laser and a superconducting bolometer. *Applied Physics Letters* **86**: 244104.
16 Hagelschuer, T., Richter, H., Wienold, M. et al. (2019).Towards a 4.75-THz local oscillator based on a terahertz quantum-cascade laser with a back-facet mirror. *2019 44th International Conference on Infrared, Millimeter, and Terahertz Waves (IRMMW-THz)*, Paris (1–6 September 2019).
17 Bruston, J., Maestrini, A., Pukala, D. et al. (2001). A 1.2 THz planar tripler using GaAs membrane based chips. *Proceedings of the 12th International Symposium on Space Terahertz Technology*, San Diego, CA, 310–319.

18 Maestrini, A., Bruston, J., Pukala, D. et al. (2001). Performance of a 1.2 THz frequency tripler using a GaAs frameless membrane monolithic circuit. *Proceedings of the IEEE MTT-S International* (20–25 May 2001) vol. 3, Phoenix, Arizona, 1657–1660.

19 Maiwald, F., Schlecht, E., Maestrini, A. et al. (2002). THz frequency multiplier chains based on planar Schottky diodes. *Invited Paper – Proceedings of SPIE: International Conference on Astronomical Telescopes and Instrumentation* (22–28 August 2002), vol. 4855, Waikoloa, Hawaii, 447–458.

20 Chattopadhyay, G., Schlecht, E., Ward, J. et al. (2004). An all solid-state broadband frequency multiplier chain at 1500 GHz. *IEEE Transactions on Microwave Theory and Techniques* **52** (5): 1538–1547.

21 Louhi, J.T., Räisänen, A.V., and Erickson, N.R. (1993). Cooled Schottky varactor frequency multipliers at submillimeter wavelengths. *IEEE Transactions on Microwave Theory and Techniques* **41** (4): 565–571.

22 Barath, F.T., Chavez, M.C., Cofield, R.E. et al. (1993). The upper atmosphere research satellite microwave limb sounder instrument. *Journal of Geophysical Research Atmospheres* **98** (D6): 10751–10762.

23 Hartogh, P., Barabash, S., Bolton, S.J. et al. (2009). Submillimeter wave instrument for EJSM. *Presented at the Europa Jupiter System Mission Instrument Workshop* (15–17 July 2009). Maryland, USA (http://opfm.jpl.nasa.gov).

24 Maestrini, A., Ward, J.S., Gill, J.J. et al. (2010). A frequency-multiplied source with more than 1 mW of power across the 840–900 GHz band. *IEEE-MTT Special Issue on "THz Technology: Bridging the Microwave-to-Photonics Gap"* **58** (7): 1925–1932.

25 Thomas, B., Maestrini, A., Gill, J. et al. (2009). A broadband 835–900 GHz fundamental balanced mixer based on monolithic GaAs membrane Schottky diodes. Submitted to IEEE-MTT Special Issue on "THz Technology: Bridging the Microwave-to-Photonics Gap".

26 Appleby, R. and Wallace, H.B. (2007). Standoff detection of weapons and contraband in the 100 GHz to 1 THz region. *IEEE Transactions on Antennas and Propagation* **55**: 2944–2956.

27 Cooper, K.B., Dengler, R.J., Chattopadhyay, G. et al. (2008). A high-resolution imaging radar at 580 GHz. *IEEE Microwave and Wireless Components Letters* **18** (1): 64–66.

28 Mehdi, I., Ward, J., Maestrini, A. et al. (2009). Broadband sources in the 1–3 THz range. *Proceedings of the 34th International Conference on Infrared, Millimeter, and Terahertz Waves* (September 2009) Busan, Korea.

29 Lee, C., Ward, J., Lin, R. et al. (2009). A wafer-level diamond bonding process to improve power handling capability of submillimeter-wave Schottky diode frequency multipliers. *Microwave Symposium Digest, 2009 IEEE MTT-S International* (7–12 June 2009) Boston, 957–960.

30 Pukala, D., Samoska, L., Gaier, T. et al. (2008). Submillimeter-wave InP MMIC amplifiers from 300–345 GHz. *IEEE Microwave and Wireless Components Letters* **18** (1): 61–63.

31 Schwierz, F. (2005). An electron mobility model for wurtzite GaN. *Solid-State Electronics* **48** (6): 889–895.

32 Siles, J.V. and Grajal, J. (2008). Capabilities of GaN Schottky multipliers for LO power generation at millimeter-wave bands. 19th International Symposium on Space Terahertz Technology (April 2008) 504–507.

33 Siles, J.V., Schlecht, E., Lin, R. et al. (2015). High-efficiency planar Schottky diode based submillimeter-wave frequency multipliers optimized for high-power operation. *Proceedings of the 2015 40th International Conference on Infrared, Millimeter, and Terahertz waves (IRMMW-THz)* (23–28 August 2015) Hong Kong, China, 2835–2843.
34 Diamond, B.L. (1963). Idler circuits in varactor frequency multipliers. *IEEE Transactions on Circuit Theory* **10** (1): 35–44.
35 Penfield, P. and Rafuse, R. (1962). *Varactor Applications*. Cambridge, MA, Chapter 8. Harmonic Multipliers: The M.I.T. Press.
36 David, T., Arscott, S., Munier, J.-M. et al. (2002). Monolithic integrated circuits incorporating InP-based heterostructure barrier varactors. *IEEE Microwave and Wireless Components Letters* **12** (8): 281–283.
37 Mélique, X., Maestrini, A., Mounaix, P. et al. (2000). Fabrication and performance of InP-based heterostructure barrier varactors in a 250 GHz waveguide tripler. *IEEE Transactions on Microwave Theory and Techniques* **48** (6): 1000–1006.
38 Rydberg, A., Grönqvist, H., and Kollberg, E. (1990). Milllimeter – and submillimeter-wave multipliers using quantum barrier-varactor (QBV) diodes. *IEEE Electron Device Letters* **11**: 373–375.
39 Sağlam, M., Schumann, B., Duwe, K. et al. (2003). High-performance 450-GHz GaAs-based heterostructure barrier varactor tripler. *IEEE Electron Device Letters* **24** (3): 138–140.
40 Bryllert, T., Olsen, A., Vukusic, J. et al. (2005). 11% efficiency 100 GHz InP-based heterostructure barrier varactor quintupler. *Electronics Letters* **41** (3): 131–132.
41 Xiao, Q., Duan, Y., Hesler, J.L. et al. (2004). A 5 mW and 5% efficiency 210 GHz InP-based heterostructure barrier varactor quintupler. *IEEE Microwave and Wireless Components Letters* **14** (4): 159–161.
42 Krach, M., Freyer, J., and Claassen, M. (2003). An integrated ASV frequency tripler for millimeter-wave applications. Proceedings of the 33rd European Microwave Conference (7–9 October 2003), vol. 3, 1279–1281.
43 Erickson, N.R. (1990). High efficiency submillimeter frequency multipliers. Proceedings of the IEEE MTT-S International, 1301–1304.
44 Erickson, N.R. and Rizzi, B.J. (1993). A high power doubler for 174 GHz using a planar diode array. Proceedings of the 4th International Symposium on Space Terahertz Technology, 287–295.
45 Rizzi, B.J., Crowe, T., and Erickson, N.R. (1993). A high-power millimeter-wave frequency doubler using a planar diode array. *IEEE Microwave and Guided Wave Letters* **3** (6): 188–190.
46 Tuovinen, J. and Erickson, N.R. (1995). Analysis of a 170 GHz frequency doubler with an array of planar diodes. *IEEE Transactions on Microwave Theory and Techniques* **43** (4): 962–968.
47 Chattopadhyay, G., Schlecht, E., Gill, J. et al. (2002). A broadband 800 GHz Schottky balanced doubler. *IEEE Microwave and Wireless Components Letters* **12** (4): 117–118.
48 Porterfield, D., Hesler, J., Crowe, T. et al. (2003). Integrated terahertz transmit/receive modules. *Proceedings of the 33rd European Microwave Conference* (7–9 October 2003) Munich, Germany, 1319–1322.

49 Schlecht, E., Chattopadhyay, G., Maestrini, A. et al. (2001). 200, 400 and 800 GHz Schottky diode 'Substrateless' multipliers: design and results. *Proceedings of the IEEE MTT-S International* (May 2001), Phoenix, AZ, 1649–1652.

50 Manley, J.M. and Rowe, H.E. (1956). Some general properties of nonlinear elements – Part I. General energy relations. Proceedings of the IRE (July 1956) 904–913.

51 Collin, R.E. (1992). *Foundations for Microwave Engineering*, 2e. McGraw-Hill.

52 Maestrini, A., Ward, J., Gill, J. et al. (2005). A 540–640 GHz high efficiency four anode frequency tripler. *IEEE Transactions on Microwave Theory and Techniques* **53**: 2835–2843.

53 Maestrini, A., Ward, J.S., Javadi, H. et al. (2005). Local oscillator chain for 1.55 to 1.75 THz with 100 μW peak power. *IEEE Microwave and Wireless Components Letters* **15** (12): 871–873.

54 Maestrini, A., Mehdi, I., Siles, J.V. et al. (2012). Design and characterization of a room temperature all-solid-state electronic source tunable from 2.48 to 2.75 THz. *IEEE Transactions on Terahertz Science and Technology* **2** (2): 177–185.

55 Maestrini, A., Ward, J.S., Tripon-Canseliet, C. et al. (2008). In-phase power-combined frequency triplers at 300 GHz. *IEEE Microwave and Wireless Components Letters* **18** (3): 218–220.

56 Siles, J.V. and Grajal, J. (2010). Physics-based design and optimization of Schottky diode frequency multipliers for terahertz applications. *IEEE Transactions on Microwave Theory and Techniques* **58** (7): 1933–1942.

57 Siles, J.V. (2008). Design and optimization of frequency multipliers and mixers at millimeter and submillimeter-wave bands. Thesis doctoral. Universidad Politécnica de Madrid (April 2008).

58 Grajal, J., Krozer, V., Gonzalez, E. et al. (2000). Modeling and design aspects of millimeter-wave and submillimeter-wave Schottky diode varactor frequency multipliers. *IEEE Transactions on Microwave Theory and Techniques* **48** (4): 700–712.

59 Erickson, N.R., Smith, R.P., Martin, S.C. et al. (2000). High efficiency MMIC frequency triplers for millimeter and submillimeter wavelengths. *Proceedings of the IEEE MTT-S International*, vol. 2, Boston, MA, 1003–1006; 11–16.

60 Faber, M., Chramiec, J., and Adamski, M. (1995). Microwave and millimeter-wave diode frequency multipliers. *Artech House*, Norwood, MA.

61 Sze, S.M. and Ng, K.K. (2007). *Physics of Semiconductor Devices*, 3e. Wiley.

62 Krozer, V. (1997). Material parameters of InGaAs. TU Chemnitz, internal report (available at request).

63 Kollberg, E.L., Tolmunen, T.J., Frerking, M.A., and East, J.R. (1992). Current saturation in submillimeter wave varactors. *IEEE Transactions on Microwave Theory and Techniques* **40** (5): 831–838.

64 Schlecht, E., Chattopadhyay, G., Maestrini, A. et al. (2002). Harmonic balance optimization of terahertz Schottky diode multipliers using an advanced device model. *Proceedings of the 13th International Symposium on Space Terahertz Technology* (March 2002) Cambridge, MA, 187–196.

65 Massobrio, G. and Antognetti, P. (1993). *Semiconductor Device Modeling With SPICE*, 2e. New York: McGraw-Hill.

66 Louhi, J.T. and Räisänen, A.V. (1995). On the modeling and optimization of Schottky varactor frequency multipliers at submillimeter wavelengths. *IEEE Transactions on Microwave Theory and Techniques* **43** (4): 922–926.

67 Moro-Melgar, D., Maestrini, A., Treuttel, J. et al. (2016). Monte Carlo study of 2-D capacitance fringing effects in GaAs planar Schottky diodes. *IEEE Transactions on Electron Devices* **63** (10): 3900–3907.

68 Crowe, T.W., Peatman, W.C.B., Zimmermann, R., and Zimmermann, R. (1993). Consideration of velocity saturation in the design of GaAs varactor diodes. *IEEE Microwave Guided Wave Letters* **3** (6): 161–163.

69 Maiwald, F., Schlecht, E., Ward, J. et al. (2003). Design and operational considerations for robust planar GaAs varactors: a reliability study. *Proceedings of the 14th International Symposium on Space Terahertz Technology* (April 2003) Tucson, AZ.

70 Erickson, N.R. (1998). Diode frequency multipliers for terahertz local-oscillator applications. *Proceedings of the SPIE: Advanced Technology MMW, Radio, and Terahertz Telescopes*, vol. 3357, Kona, HI, 75–84.

71 Louhi, J.T. and Räisänen, A.V. (1996). Dynamic shape of the depletion layer of a submillimeter-wave Schottky varactor. *IEEE Transactions on Microwave Theory and Techniques* **44** (12): 2159–2165.

72 Pardo, D., Grajal, J., Pérez-Moreno, C.G., and Pérez, S. (2014). An assessment of available models for the design of Schottky-based multipliers up to THz frequencies. *IEEE Transactions on Terahertz Science and Technology* **4** (2): 277–287.

73 Räisänen, A.V. (1992). Frequency multipliers for millimeter and submillimeter wavelengths. *Proceedings of the IEEE* **8** (11): 1842–1852.

74 von Schéele, F. (1996). The Swedish Odin satellite to eye heaven and earth. Proceedings of the 47th International Astronautical Congress IAF (October 1996).

75 Crowe, T. and Zimmermann, R. (1996). Progress toward solid state local oscillators at 1 THz. *IEEE Microwave and Guided Wave Letters* **6** (5): 207–208.

76 Zimmerman, P. (1998). Multipliers for terahertz local oscillators. *Proceedings fo the SPIE: Advanced Technology MMW, Radio, and Terahertz Telescopes* (March 1998), vol. 3357, Kona, HI, 152–158.

77 Cronin, N.J. and Law, V.J. (1985). Planar millimeter-wave diode mixer. *IEEE Transactions on Microwave Theory and Techniques* **33** (9): 827–830.

78 Bishop, W.L., McKinney, K., Mattauch, R.J. et al. (1987). A novel whiskerless Schottky diode for millimeter and submillimeter wave applications. *IEEE MTT-S International Microwave Symposium Digest* **II**: 607–610.

79 Porterfield, D., Crowe, T., Bishop, W. et al. (2005). A high-pulsed-power frequency doubler to 190 GHz. *Proceedings of the 30th International Conference on Infrared and mm-Waves* (19–23 September 2005), Williamsburg, Virginia, USA, 78–79.

80 Martin, S., Nakamura, B., Fung, A. et al. (2001). Fabrication of 200 GHz to 2700 GHz multiplier devices using GaAs and metal membranes. *Proceedings of the IEEE MTT-S International* (20–25 May 2001) Phoenix, Arizona.

81 Erickson, N.R., Narayanan, G., Grosslein, R. et al. (2002). 1500 GHz tunable source using cascaded planar frequency doublers. *Proceedings of the 13th International Symposium on Space Terahertz Technology*, Cambridge, MA, 177–186.

82 Siegel, P.H., Smith, R.P., Martin, S., and Gaidis, M. (1999). 2.5 THz GaAs mono- lithic membrane-diode mixer. *IEEE Transactions on Microwave Theory and Technology* **47** (5): 596–604.

83 Ward, J.S., Schlecht, E., Chattopadhyay, G. et al. (2005). Local oscillators from 1.4 to 1.9 THz. *Proceedings of the 16th International Symposium on Space Terahertz Technology* (May 2005) Göteborg, Sweden.

84 Samoska, L.A., Gaier, T.C., Peralta, A. et al. (2000). MMIC power amplifiers as local oscillator drivers for FIRST. *Proceedings of the SPIE: UV, Optical, and IR Space Telescopes and Instruments* (August 2000), vol. 4013, San Diego, CA, 275–284.

85 Crowe, T.W., Hesler, J.L., Retzloff, S.A., and Kurtz, D.S. (2016). Higher power multipliers for terahertz sources. 2016 41st International Conference on Infrared, Millimeter, and Terahertz waves (IRMMW-THz).

86 Maestrini, A., Mehdi, I., Ward, J. et al. (2011). A 2.5–2.7 THz room temperature electronic source. *Proceedings of the 22nd International Symposium on Space Terahertz Technology* (26–28 April 2011) Tucson, AZ, 49–52.

87 Pearson, J.C., Drouin, B.J., Maestrini, A. et al. (2011). Demonstration of a room temperature 2.48–2.75 THz coherent spectroscopy source. *Review of Scientific Instruments* **82**: 093105. https://doi.org/10.1063/1.3617420.

88 Crowe, T.W., Mattauch, R.J., Röser, H.P. et al. (1992). GaAs Schottky diodes for THz mixing applications. *Proceedings of the IEEE* **80** (11): 1827–1841.

89 Lamb, J. (1996). Miscellaneous data on materials for millimeter and submillimetre optics. *International Journal of Infrared and Millimeter Waves* **17** (12): 1997–2034.

90 Crowe, T.W., Hesler, J.L., Retzloff, S.A., and Kurtz, D.S. (2017). Higher power terahertz sources based on diode multipliers. 2017 42nd International Conference on Infrared, Millimeter, and Terahertz waves (IRMMW-THz).

91 Ochiai, S., Baron, P., Nishibori, T. et al. (2017). SMILES-2 mission for temperature, wind, and composition in the whole atmosphere. *SOLA* **13A**: 13–18. https://doi.org/10.2151/sola.13A-003.

92 Wu, D.L., Yee, J.-H., Schlecht, E. et al. (2016). THz limb sounder (TLS) for lower thermospheric wind, oxygen density, and temperature. *Journal of Geophysical Research: Space Physics* **121**: 7301–7315. https://doi.org/10.1002/2015JA 022314.

93 Cojocari, O., Moro-Melgar, D., and Oprea, I. (2019). High-power MM-wave frequency multipliers. 2019 44th International Conference on Infrared, Millimeter, and Terahertz Waves (IRMMW-THz), Paris, 1–6.

94 Moro-Melgar, D., Cojocari, O., Oprea, I. et al. (2018). First generation of high power discrete diodes based doublers beyond 300 GHz. Proceedings of the International Symposium on Space THz Technology (March 2018), 1–4.

95 Maestrini, A., Tripon-Canseliet, C., Ward, J.S. et al. (2006). A high efficiency multiple-anode 260–340 GHz frequency tripler. *Proceedings of the 17th International Symposium on Space Terahertz Technology*, Paris, France (10–12 May 2006).

96 Porterfield, D. (2007). High-efficiency terahertz frequency triplers. *Proceedings of the IEEE MTT-S International* (3–8 June 2007) Honolulu, Hawaii, 337–340.

97 Maestrini, A., Gatilova, L., Treuttel, J. et al. (2016). 1200 GHz and 600 GHz Schottky receivers for JUICE-SWI. *Proceedings of the 27th International Symposium on Space Terahertz Technology* (12–15 April 2016), Nanjing, China, 3.

98 Maestrini, A., Gatilova, L., Treuttel, J. et al. (2018). The 1200 GHz receiver frontend of the submillimetre wave instrument of ESA JUpiter ICy moons explorer. *Proceedings of 2018 43rd International Conference on Infrared, Millimeter, and Terahertz Waves (IRMMW-THz)* (9–14 September 2018) Nagoya, Japan.

99 Treuttel, J., Gatilova, L., Yang, F. et al. (2014). A 330 GHz frequency doubler using European MMIC Schottky process based on E-beam photolithography. *General Assembly and Scientic Symposium, 2011 XXXIth URSI* (13–20 August 2014), Beijing, China.

100 Gatilova, L., Maestrini, A., Treuttel, J. et al. (2019). Recent progress in the development of French THz Schottky diodes for astrophysics, planetology and atmospheric study. *Proceedings of the 2019 44th International Conference on Infrared, Millimeter, and Terahertz Waves (IRMMW-THz)*, Paris (1–6 September 2019).

101 Siles, J.V., Maestrini, A., Alderman, B. et al. (2011). A single-waveguide in-phase power-combined frequency doubler at 190 GHz. *IEEE Microwave and Wireless Component Letters* **21** (6): 332–334.

102 Siles, J.V., Maestrini, A., Alderman, B.J. et al. (2009). A novel dual-chip single-waveguide power combining scheme for millimeter-wave frequency multipliers. *Proceedings of the 20th Int. Symp. on Space THz Technology* (20–22 April 2009) Poster P1B, Charlottesville, VA.

103 Siles, J.V., Chattopadhyay, G., Lee, C. et al. (2015). On-chip power-combining for high-power Schottky diode based frequency multipliers. US Patent 914,308,4B2 (September 2015).

104 Jung, C., Lee, C., Thomas, B. et al. (2010). Silicon micromachining technology for THz applications. Proceedings of the 35th International Conference on Infrared, Millimeter, and Terahertz Waves (5–10 September 2010).

105 Jung, C., Thomas, B., Lee, C. et al. (2011). Compact submillimeter-wave receivers made with semiconductor nano-fabrication technologies. *Microwave Symposium Digest*, 2011 IEEE MTT-S International Microwave Symposium (5–10 June 2011).

10

GaN Multipliers

Chong Jin[1] *and Dimitris Pavlidis*[2]

[1] *ams AG, Premstätten, Austria*
[2] *College of Engineering and Computing, Department of Electrical and Computer Engineering, Florida International University, Miami, FL, USA*

10.1 Introduction

10.1.1 Frequency Multipliers

The development of frequency multipliers has a long history. Millimeter-wave multipliers' efforts increased in the 1980s with whisker contacted discrete diodes. In the 1990s, the planar Schottky diodes were developed to replace unreliable whisker contacted diodes. Efforts were made to replace hybrid mounting of diodes on quartz substrate by monolithic microwave integrated circuit (MMIC) like chips in the ~10 years. GaAs has been the material of choice over several decades. In recent years, the development of frequency multipliers has been greatly enhanced by advanced design tools and fabrication technology. A review of frequency multipliers performance demonstrated in recent years can be found in [1], while a review for the frequency multipliers developed before 2002 can be found in [2].

Among the record frequency results with frequency multipliers are 2.7 THz [3] with a tripler from 840 to 900 GHz, input power around 1 mW, and a maximum output power of 18 μW at 2.58 THz. State-of-the-art room temperature results include an output power of 1 mW at 1 THz, above 10 mW at 300–500 GHz and above 100 mW at 100 GHz.

Both the frequency and output power increase demonstrated during the ~10 years are based on technological advances. Integration of diodes and matching circuits on the same substrate allows better controlling of the device parasitics and circuit optimization [1, 4]; Computer numerical control milling machines are employed for waveguide block machining with micrometer precision [5]; Full 3-D electromagnetic software and nonlinear circuits simulators are commercially available, which eases the co-simulation of the nonlinear diodes and linear matching circuits. Besides, one of the driving forces is to enhance the power handling capability of GaAs Schottky diodes and multipliers built with them, which are limited by their relatively low breakdown voltage.

Fundamentals of Terahertz Devices and Applications, First Edition. Edited by Dimitris Pavlidis.

Companion website: www.wiley.com/go/Pavlidis/FundamentalsofTHz

Following the demonstration of single diode multipliers with maximum power, it is natural to seek solutions for higher power using several diodes connected in series. By connecting n identical diodes in series, one obtains a breakdown voltage n times higher than for a single diode. The utilization of 2, 4 even 6 diodes in series has been demonstrated. However, the maximum number of diodes is limited by the waveguide dimensions [6]. Besides, the thermal dissipation problem becomes more serious as the diode number increases. Electrothermal models have to be employed for design optimization [7]. Substrates with high thermal conductivity like diamond also provide a good solution [8]. The recently developed power combining multipliers is another option. The utilization of power combination techniques assures adequate input pump power level for the aforementioned 2.7 THz tripler. Furthermore, all the first stages of the multiplier chain rely on power combination techniques; an output of ~500 mW at W-band was provided by combining four GaAs power amplifier chips, ~40 mW by quad-chip power combined tripler at 300 GHz and ~1 mW by double-chip power combined tripler at 900 GHz [9, 10].

A wide bandgap material like GaN with high breakdown voltage is thus a suitable candidate. Theoretical studies indicate that eight GaAs diodes are required for a 200 GHz doubler with input power of 150 mW, while one GaN diode with similar anode area is capable of handling this input power [11], i.e. the power handling capability of a GaN Schottky diode is almost one order higher than its GaAs counterpart. Diodes of this type are the focus of this chapter.

10.1.2 Properties of Nitride Materials

III-nitrides have the crystal structures of wurtzite, zincblende, and rock salt. The thermodynamically stable structure is wurtzite. The Ga-face is representative of metal organic chemical vapor deposition (MOCVD) grown GaN epitaxial films on c-plane sapphire. The wurtzite structure has a hexagonal unit cell and thus two lattice constants c and a with values of 5.185 and 3.189 Å, respectively. The properties of GaN are listed in Table 10.1 together with those of GaAs for comparison.

GaN is a wide-bandgap material with a large bandgap of 3.4 eV and thus greater than Si (1.12 eV) and GaAs (1.42 eV). Benefiting from the larger bandgap, the breakdown electric field of GaN is larger than 5 MV/cm. The intrinsic carrier concentration at room temperature is $2.8 \times 10^{-10}/\text{cm}^3$ for GaN [13], compared to $1 \times 10^{10}/\text{cm}^3$ for Si and $2.1 \times 10^6/\text{cm}^3$ for GaAs and increases with temperature.

The low intrinsic carrier concentration of GaN allows its use under very high ambient or junction temperature, without being affected by thermally generated carriers. Both the high breakdown field and low intrinsic carrier concentration are crucial attributes for the enhancement of power handling capability of GaN Schottky diodes.

A high electron drift velocity is favorable for frequency multiplier applications. Due to the multi-valley band structure of GaAs, its electron velocity is high at relatively low fields but reduces at values beyond the critical field. This has found to be the reason for lower multiplication efficiency, since current saturation lead to increased equivalent series resistance [14]. However, this is not an issue for GaN. The electron drift velocity is even higher than GaAs if the pump power is high enough to provide an electric field over 12 kV/cm.

Table 10.1 GaN and GaAs material properties.

	GaN	GaAs
Density (g/cm^3)	6.15	5.32
Dielectric constant	8.9	12.9
Bandgap (eV)	3.39	1.424
Effective mass	$0.2\ m_0$	$0.063\ m_0$
Electron mobility ($cm^2/V/s$)	1000	8000
saturation velocity (cm/s)	2.5×10^7	1×10^7
Peak velocity (cm/s)	3.1×10^7	2×10^7
Peak velocity field (kV/cm)	150	3.5
Breakdown field (MV/cm)	>5	0.4
Thermal conductivity (W/cm/K)	>2.3	0.55

Source: Shur et al. [12] and IOFFE [13].

Thermal conductivity is another parameter of importance. A thermal conductivity of 2.3 WK/cm [15] has been reported for GaN, while as this value is 0.55 W/cmK for GaAs. Thanks to the elevated thermal conductivity, the aforementioned electron-thermal model may not be needed for GaN multiplier diodes/chip design, thus the design procedure is simplified. The superior properties of GaN have been utilized successfully to develop high-electron-mobility transistors (HEMTs) for RF power amplifier applications. The current gain cutoff frequency of GaN-based HEMTs achieved so far is 370 GHz [16], and devices of this type have the potential of operating up to 500 GHz. The RF output power of W-band GaN-based HEMTs has been reported to be more than 1 W for a 600 μm gate width [17], resulting in a power density of 1.7 W/mm. As a comparison, the output power density for a GaAs-based p-HEMT with the same gate width is 0.28 W/mm [18]. By power combining, an output power of 3 W [19] and 5 W has been reported [20] for GaN-based HEMTs.

W-band amplifiers are of interest for the work reported in this chapter since they are used as first stage pump source for THz multiplier chains. GaAs or InP-based Gunn diodes have traditionally been the choice for this purpose, but fail to provide very high output power. It has been estimated theoretically that an output power of 1400 kW/cm^2 can be obtained from GaN Gunn diodes as compared to 4.9 kW/cm^2 from GaAs diodes [21]. Despite the numerous papers on theoretical studies of GaN Gunn diodes, experimental studies are still in the beginning and only preliminary results are available [22]. One of the remaining big challenges is the high dislocation density of epitaxial GaN, which may limit the performance of GaN Schottky diodes.

10.1.3 Motivation and Challenges

The superior material properties of Nitrides are a key motivation factor for exploring the properties of multipliers based on them. However, the mobility of nitrides is lower than that of arsenides and one may expect the operation frequency of nitride-based devices to be limited. The work reported here is mainly motivated by the expectation of higher output power by utilizing the previously discussed superior material properties. In this work, the fabrication technology for multiplier diodes is first demonstrated with optical lithography. Dielectric bridges were used to demonstrate first devices. Micron-size diodes were characterized to understand their properties. Replacement of dielectric bridges by air bridges is also considered to allow higher cutoff frequency by decreasing the parasitic capacitances. Modeling of such devices was performed and allowed the multiplier performance to be predicted. E-beam lithography was employed for further studies to ease the realization of smaller anode size. A technology based on E-beam lithography is also described.

Since multiplier diodes are working under large-signal excitation, their performance cannot be known until diodes of this type are used in well-designed waveguide circuits. Studies of this type involve considerable fabrication, mounting, and packaging. On-wafer large-signal characterization is, therefore, proposed in order to provide a quick feedback to the device studies.

10.2 Theoretical Considerations of GaN Schottky Diode Design

This section discusses the design aspects of GaN Schottky diodes. Based on the current–voltage/charge–voltage relationship, the frequency multiplication performance of Schottky diodes can be evaluated. Figures of merit like cutoff frequency (f_c), maximum input power (P_{in_max}), and conversion efficiency (η) are used. Taking the material properties of GaN into account, Schottky diodes can then be studied analytically or using a commercial semiconductor device simulator allowing to do physical device modeling. The advantage of GaN over GaAs will be discussed and the design aspects of GaN-based Schottky multiplier diodes will also be addressed.

10.2.1 Analysis by Analytical Equations

10.2.1.1 Nonlinearity and Harmonic Generation

A nonlinear element converts a sinusoidal input signal to a periodic output signal containing high order harmonics. The I–V and C–V characteristics of Schottky diodes are both nonlinear and can be considered as a nonlinear resistor and nonlinear capacitor, respectively. I–V and C–V characteristics representative of Schottky diodes are shown in Figure 10.1a and b.

Manley–Rowe equations [23] provide the fundamental limit of harmonic generation by nonlinear elements. Consideration of power conservation implies that the sum of power into and out of a nonlinear resistor is larger or equal to 0. The n^2 coefficient in the resulted Eq. (10.1) [24] corresponds to severe loss at higher harmonic. It has been shown that n-order harmonics generated by a nonlinear resistor have at best a power of $1/n^2$ with respect to the pumping power, corresponding, therefore, to 1/4, 1/9, and 1/16 for second order, third order, and fourth order harmonic generation, respectively.

$$\sum_{n=0}^{\infty} n^2 P_n > 0 \tag{10.1}$$

For a lossless nonlinear capacitor, the sum of all harmonic power is 0 simply because of power conservation and the lossless device, as indicated in Eq. (10.2). The absence of n^2

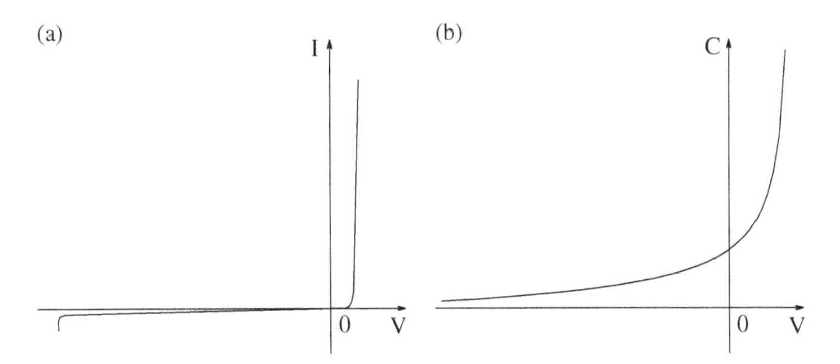

Figure 10.1 I–V and C–V characteristics of an ideal Schottky contact.

coefficient suggests that the generated high order harmonic has in principle the same power as the pump signal. Based on the discussion above, one can conclude that the nonlinear C–V characteristics of Schottky diodes are preferred than I–V characteristics for harmonic generation.

$$\sum_{n=0}^{\infty} P_n = 0 \tag{10.2}$$

The operation principle of harmonic generation from nonlinear capacitance is described in Figure 10.2. The total amount charge Q is modulated by the applied voltage, and it is written in form of power series of voltages as follows:

$$Q(V(t)) = A_0 + A_1V(t) + A_2V(t)^2 + A_3V(t)^3 + \cdots \tag{10.3}$$

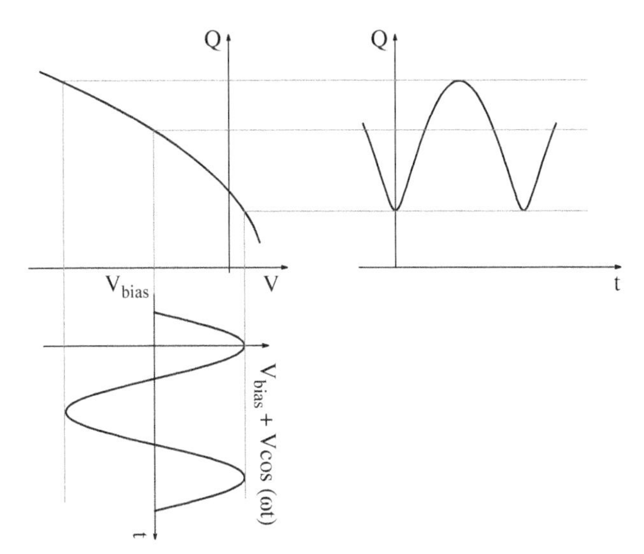

Figure 10.2 Harmonic generation from nonlinear Q–V.

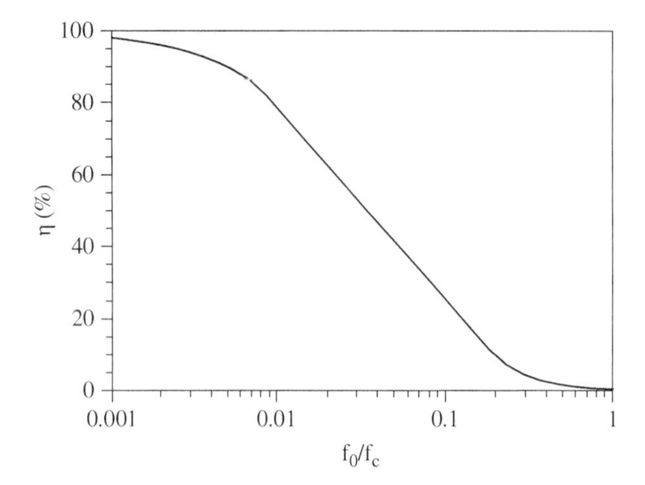

Figure 10.3 The dependence of multiplication efficiency on f_0/f_c.

Since the current is the time derivative of charge, one obtains

$$I(t) = \frac{dQ(t)}{dt} = \left[A_1 + 2A_2V(t) + 3A_3V(t)^2 + \cdots\right]\frac{dV(t)}{dt} \tag{10.4}$$

A n-order frequency multiplier is achieved by embedding the Schottky diode in a properly designed circuit, which should behave as filter or idler at the unwanted frequencies. Frequency multiplication efficiency (η) is defined by the resulting n-order output signal power over the input fundamental signal power. The presence of a resistance is inevitable in real Schottky diodes making the theoretical frequency multiplication efficiency of 100% unachievable. Early research has related the efficiency η to the cutoff frequency (f_c) of a Schottky diode, where f_c is given by

$$f_c = \frac{1}{2\pi R_s C_{j0}} \tag{10.5}$$

In terms of obtaining efficient frequency multiplication, f_c of the Schottky diode has to be much higher than the operation frequency. Figure 10.3 shows the dependence of theoretical efficiency of frequency doublers [25] on the operation frequency over the cutoff frequency. The efficiency is steadily decreasing as the operation frequency increases.

10.2.1.2 Nonlinearity of Ideal Schottky Diode

Harmonic generation relies on the nonlinearity of the device. In the case of Schottky diodes, the nonlinear junction capacitance is the main source of the nonlinearity. Good understanding of the nonlinear behavior of junction capacitance is essential for its application in harmonic generation. Formulas in close form can be derived [26] to show the correlation between the doubler performance (maximum input power, efficiency) and the nonlinear capacitance of varactor physical parameters (elastance $S = 1/C$, barrier height eφ and junction capacitance at 0 V C_{j_0}). These can be obtained under various conditions, i.e. constant doping epitaxial layer, presence of only fundamental and second harmonics, diode being fully driven, and losses arising only from a constant R_s.

For a Schottky diode with constant doping profile, the capacitance–voltage characteristics can be adopted to

$$C = \frac{1}{S} = \frac{C_{j_0}}{\sqrt{1 - \frac{v}{\varphi}}} \tag{10.6}$$

Where C_{j_0} is the junction capacitance at zero voltage and φ is $V_{bi} - V_T$.

The elastance modulation factor is defined as the coefficients in the series representation of all harmonic components in Eq. (10.7) describing the nonlinearity of elastance.

$$s = S_0\left(1 + m_1 e^{j\omega t} + m_1{*}e^{-j\omega t} + m_2 e^{j2\omega t} + m_2{*}e^{-j2\omega t} + \cdots\right), \tag{10.7}$$

By limiting the harmonics up to second order (ideal frequency doubler), the elastance modulation factor for the fundamental and second harmonic can be derived with a self-consistent method to be $|m_1| = 0.502$ and $|m_2| = 0.166$, respectively. The fully driven assumption requires that the fundamental signal is sinusoidal with an amplitude of $\varphi - V_{bias}$. The relation between V_{bias} and S_{bias} is given by

$$V_{bias} = \varphi - \varphi(C_0 S_{bias})2\left(1 + 2|m_1|_2 + 2|m_2|_2\right) \tag{10.8}$$

The power handled by such a nonlinear capacitance satisfying the above assumptions is expressed by

$$P_c = 4\varphi_2 C_{j_0}\left(C_{j_0} S_{bias}\right)3\omega|m_1|_2 \mid m_2 \mid \tag{10.9}$$

where the S_{bias} represents the elastance at the chosen operation point. The bias voltage and breakdown voltage play their role through S_{bias}. The resulted output power has the same expression due to Manley–Rowe energy relations, when no lossy elements exist. For the case of constant R_s, the power losses of fundamental and second harmonic are

$$\begin{cases} P_{loss_1} = 4\varphi^2 C_{j_0}{}^2\left(C_{j_0} S_{bias}\right)^2\omega^2|m_1|^2 R_s \\ P_{loss_2} = 16\varphi^2 C_{j_0}^2\left(C_{j_0} S_{bias}\right)^2\omega^2|m_2|^2 R_s \end{cases} \tag{10.10}$$

Overall, the resulted input power, output power, and conversion efficiency of a Schottky diode under the above assumptions are

$$\begin{cases} P_{in} = P_c + P_{loss_1} + P_{loss_2} \\ P_{out} = P_c \\ \eta = \dfrac{P_{out}}{P_{in}} \end{cases} \tag{10.11}$$

Having the above relationship unfolded, the performance of frequency doubling for a diode with specific capacitance–voltage characteristics can be easily obtained. A diode with barrier height ($e\varphi$) of 0.85 eV, junction capacitance at zero volte $\left(C_{j_0}\right)$ of 50 fF, series resistance (R_s) of 5 Ω was assumed to represent the above relationship by way of an example. The capacitance–voltage characteristics are given in Figure 10.4 assuming a breakdown voltage of 20 V. With this breakdown voltage of 20 V set, the maximum bias voltage allowed is about

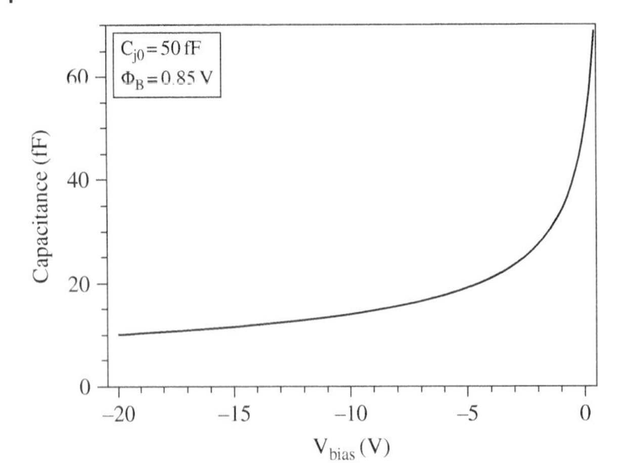

Figure 10.4 C–V characteristics of a diode with $\varphi = 0.85$ V, $C_{j_0} = 50$ fF.

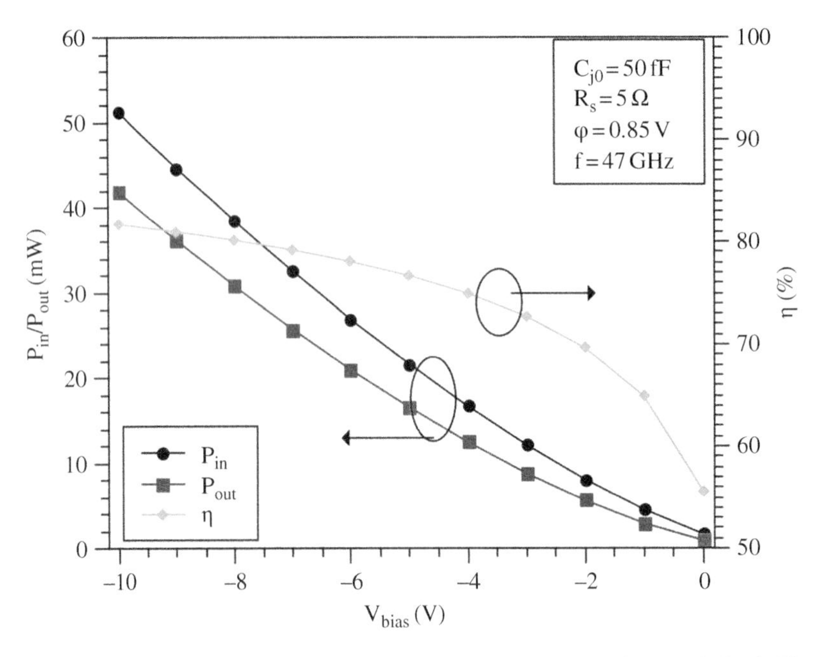

Figure 10.5 Doubler input and output power using the diode characteristics in Figure 10.4.

−10 V. Using this diode as frequency doubler, the resulting input and output power versus bias voltage are given in Figure 10.5.

As shown in the figure, the input power increased from 21 to 51 mW for a bias voltage of −5 and −10 V, respectively. This requires the breakdown voltage to be −10 and −20 V, respectively. The resulting 2.5 times increase of power for a change of maximum bias voltage from −5 to −10 V suggests a dependence $P_c \propto {V_{br}}^{1.3}$. As expected from Eqs. (10.8) and (10.9), the power that can be handled by such a diode is proportional to the breakdown

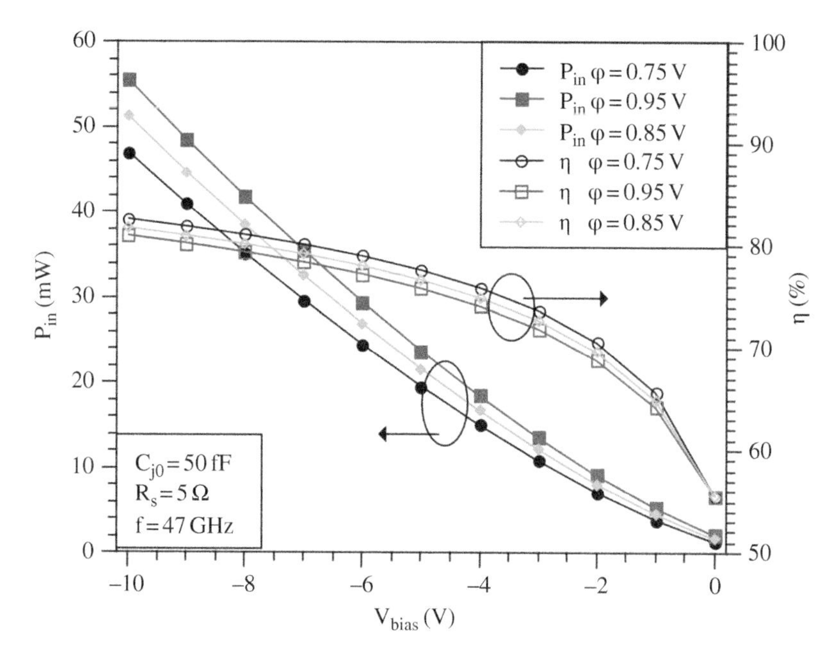

Figure 10.6 The dependence of input power and efficiency on φ.

voltage to the power of 1.5, when (φ) is relatively small compared to V_{bias}. Thus, the power handling capability of a diode increases considerably with breakdown voltage as expected.

The output power P_{out} in Figure 10.5 poses the same trend as P_{in}. This can be explained by the dissipation power $P_{loss} \propto V_{br}$ as shown in Eq. (10.10). However, this is not true in practice since a larger R_s is the by-product of enhancing breakdown voltage, as will be discussed later.

In the same way, the impact of φ and C_{j_0} on the frequency doubler was also examined and is shown in Figures 10.6 and 10.7. The increase of both φ and C_{j_0} results in larger power absorbed by the diode. This is to be expected given the fact that a larger capacitance (also charge) resulted from increased φ or C_{j_0}. Meanwhile, the current becomes higher and results in higher loss. This is consistent with the rule that higher conversion efficiency can be achieved for diodes with larger cutoff frequency.

The analytical approach discussed here depicts nicely the operation of harmonic generation from a nonlinear capacitance, which is correlated directly to the diode design. The R_s, on which the conversion loss relies, was however set without considering its physical significance. A study of R_s for given diode structure is, therefore, presented in the next section.

10.2.1.3 Series Resistance

The role of series resistance R_s and its effect on doubler performance will be discussed in this section to allow for further understanding of its impact on diode performance and permit its minimization. A planar diode with the cross section shown in Figure 10.8 is discussed here. The path of current flow has been divided into several segments and the R_s of the diode is calculated by the sum of all the resistances corresponding to each segment. The analytical equations given below are based on previous reports [27, 28].

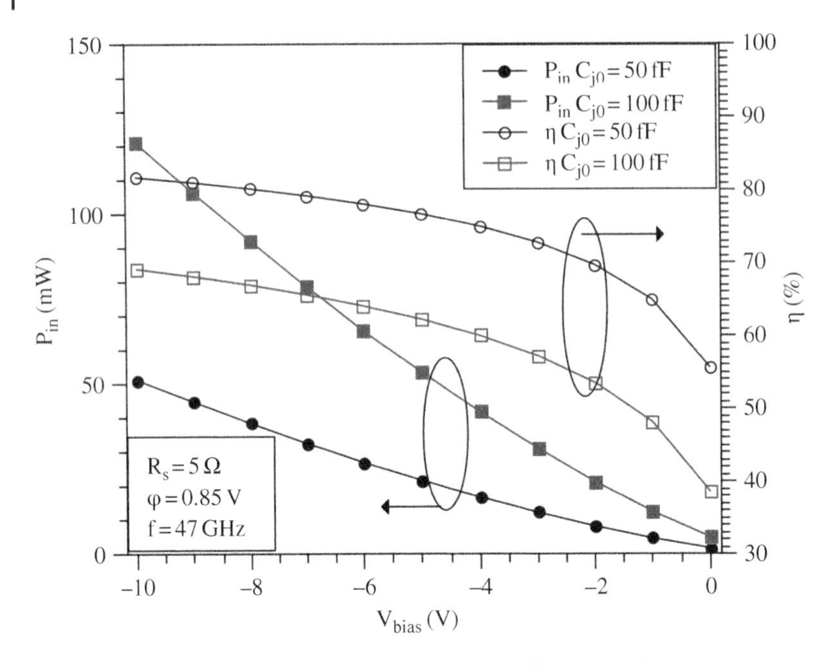

Figure 10.7 The dependence of input power and efficiency on C_{j_0}.

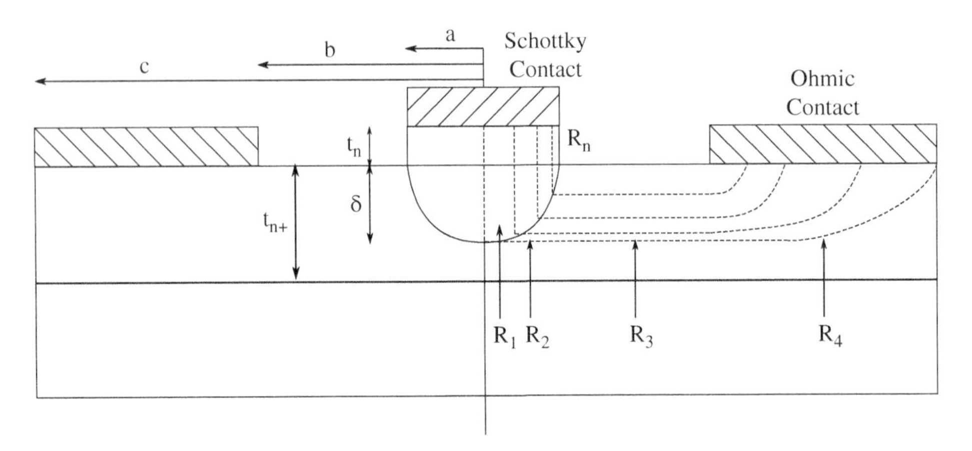

Figure 10.8 Schematic for determining series resistance. *Source*: Redrawn from Chen et al. [27].

The resistance of the N^- layer is considered to correspond to a cylinder where the area perpendicular to current flow is the same as that of the mesa, and its thickness t_{n^-} equals the undepleted N^- layer.

$$R_n = \frac{(t_{n^-} - t_d)\rho_{n^-}}{\pi a^2} \tag{10.12}$$

where t_d is the depletion layer thickness, ρ_{n^-} is the resistivity of the N^- layer which depends on the doping concentration N_D and carrier mobility μ; Other resistance components to consider include the following:

- R_1 represents the spherical cap under the mesa, with base radius a and cap-height δ_{n^+}.
- R_2 is associated with a cylinder under the mesa but without the spherical cap of R_1.
- R_3 represents the access region between mesa and ohmic metallization.
- R_4 is the resistance of ohmic metallization, which is expressed by Bessel functions.

The series resistance was calculated for a GaAs diode structure based on the above equations and the results are shown in Table 10.2 [27]. The resulting total resistance is 0.915 Ω and was found to be slightly different from the measured value of 0.87 Ω in [27]. The resistance of a GaN diode with the same dimensions was evaluated by replacing the GaAs material parameters by those of GaN. The resulting total resistance is about 10 times larger for GaN than GaAs. The individual segments of the resistance were also larger, except R_4, which is less dependent on semiconductor material resistivity.

The dependence of series resistance on Schottky contact dimensions is shown in Figure 10.9, where the anode radius was varied from 1 to 10 μm, while the cathode radius and space between anode and cathode were maintained at 10 and 2 μm, respectively. R_n was found to be the dominant part of the total series resistance when the diode is scaled down. R_2 is almost constant, contributing mainly to the total resistance of large diodes but less for

Table 10.2 Example of R_s estimation for GaN and GaAs.

Diode dimensions (μm)	**$a = 8$ $b = 10$ $c = 20$** **$t_n = 0.65$ $t_{n+} = 2.5$**	
Metal thickness (μm)	$t_m = 1$	
Resistivity (Ω cm)	$\rho_m = 5 \times 10^{-5}$	
Specific resistivity (Ω cm^2)	$\rho_c = 1 \times 10^{-6}$	
Material	GaAs	GaN
Epsilon	13.1	8.9
Nn (cm^{-3})	7×10^{16}	1×10^{17}
μ_n (cm^2/V/s)	4780	400
N_n+ (cm^{-3})	8×10^{18}	3×10^{18}
μ_n+ (cm^2/V/s)	1280	200
Calculated R		
R_n (Ω)	0.476	4.31
R_1 (Ω)	0.0343	0.589
R_2 (Ω)	0.193	3.31
R_3 (Ω)	0.0859	1.48
R_4 (Ω)	0.127	0.651
R_{total} (Ω)	0.915	10.3

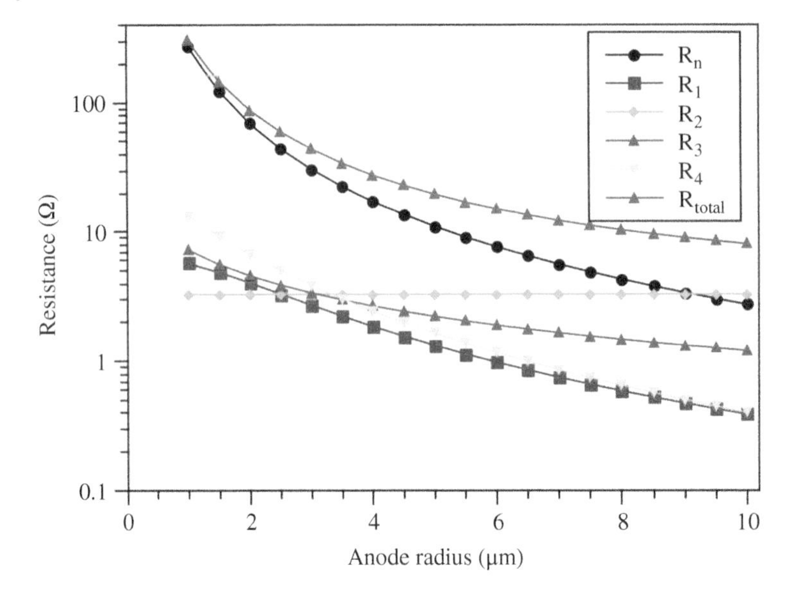

Figure 10.9 Evaluated resistance of GaN diode versus diode anode size.

smaller diodes. The other three diode segments increase steadily with decreasing diode radius. It can be concluded that smaller diodes require better attention of the N^- layer design. The analysis of R_S shown above assumes the Schottky contact being enclosed entirely by ohmic metallization, while for most practical cases the ohmic metal is not a complete ring. Thus the series resistance value is in practice higher than estimated above (Table 10.2).

10.2.2 Analysis by Numeric Simulation

Although useful conclusions can be drawn from the analytical considerations made above, these are limited to specific designs as in the case discussed here, i.e. constant doping profile for the device and only second harmonic generation considered for the multiplier circuit. However, in practice, an arbitrary doping profile may be required and Schottky diodes can in such a case be used as a tripler (for third order harmonic generation). Numerical simulation offers a universal approach in such a situation. Moreover, numerical simulation provides insight into the internal physical mechanisms associated with device operation. Techniques of this type are discussed in this section and used to analyze different device structures.

10.2.2.1 Introduction of Semiconductor Device Numerical Simulation

A flowchart of the numerical simulation procedure used, applicable to semiconductor devices in general is shown in Figure 10.10. In addition to defining the device geometry and employing an appropriate mesh, material parameters and good physical understanding are a key for successful simulation. Following the definition of device geometry, material parameters, and physical models e.g. for mobility and impact ionization, the device can be analyzed by solving a series of partial differential equations such as Poisson, Current

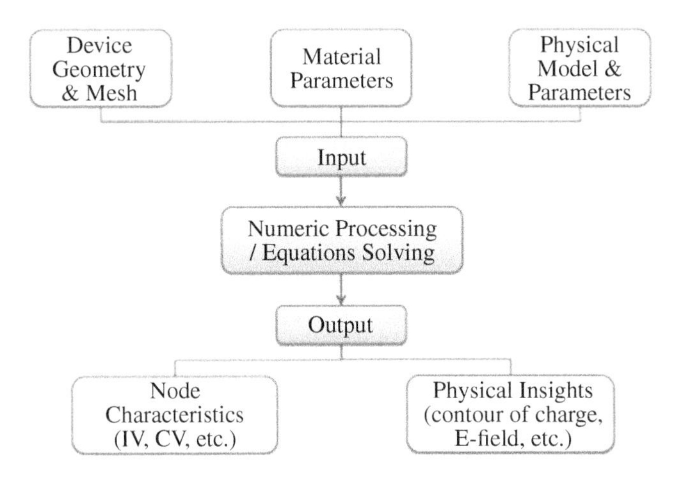

Figure 10.10 Flowchart of semiconductor device numerical simulation.

Table 10.3 Material parameters used in numerical simulation (at 300 K).

Material	GaAs	GaN
Dielectric constant	13.2	8.9
Bandgap (eV)	1.42	3.43
Electron affinity (eV)	4.07	4.31
Conduction band density of states (cm^{-3})	4.35×10^{17}	2.24×10^{18}
Valence band density of states (cm^{-3})	1.29×10^{19}	2.51×10^{19}
Intrinsic carrier density (cm^{-3})	2.67×10^{6}	1.06×10^{-10}
Electron thermal velocity (cm/s)	4.51×10^{7}	2.61×10^{7}
Hole thermal velocity (cm/s)	1.46×10^{7}	1.17×10^{7}

Continuity, and Carrier Transport (also referred as Current Density) Equations. These coupled equations are solved numerically by either finite element or finite difference method (Table 10.3).

The electrical characteristics at each node of the device can be obtained from the numerical simulation, including static current–voltage, capacitance–voltage, transient current–voltage characteristics, and S-parameters under high-frequency operation. Physical insights into device operation are given by numerical simulation results, such as charge distribution, E-field strength, etc. These complement the understanding of device operation obtained through experimentation.

10.2.2.2 Parameters for GaN-Based Device Simulation

Bandgap material parameters are listed below. Parameters for GaAs are also given, so that GaAs Schottky diodes will also be simulated numerically for comparison. Since no ternary compound was used in this study, only the physical model dependence on doping concentration and temperature are discussed. The temperature dependence of bandgap was also taken into consideration.

The bandgap is a function of temperature for both GaN and GaAs and can be expressed by Eq. (10.13)

$$E_G(GaN) = 3.507\text{eV} - \frac{0.909 \times 10^{-3}\text{eV/K} \times T^2}{T + 830\,0\text{K}} \tag{10.13}$$

$$E_G(GaAs) = 1519\text{eV} - \frac{5.405 \times 10^{-4}\text{eV/K} \times T^2}{T + 204.0\text{K}} \tag{10.14}$$

The electron mobility model of GaN is based on Monte Carlo simulation as reported in [29, 30], where both low field and high field mobility were fitted with an empirical equation. The low field mobility was considered as a function of temperature and doping concentration. In case of operation under high E-field, the electron velocity saturation was taken into account and the high field mobility was expressed as a function of low field mobility and E-field. Based on equations of this type, the low field mobility and velocity were calculated at various temperature, doping concentration, and a wide range of E-field values. The results obtained are shown in Figures 10.11 and 10.12, which illustrate the mobility and velocity E-field dependence at various temperatures and doping concentration.

In case of GaAs, the low field mobility used for a comparative study was expressed as a function of doping concentration. The temperature dependence was also taken into consideration. The high field mobility was calculated as a function of saturation velocity. intervalley scattering in GaAs leads in negative differential mobility (NDM). As explained in [30], the NDM model may introduce instability in the numerical process. For the GaAs Schottky diode simulated in this study, the saturation velocity is more important than the NDM. A modified mobility model was, therefore, used for this purpose.

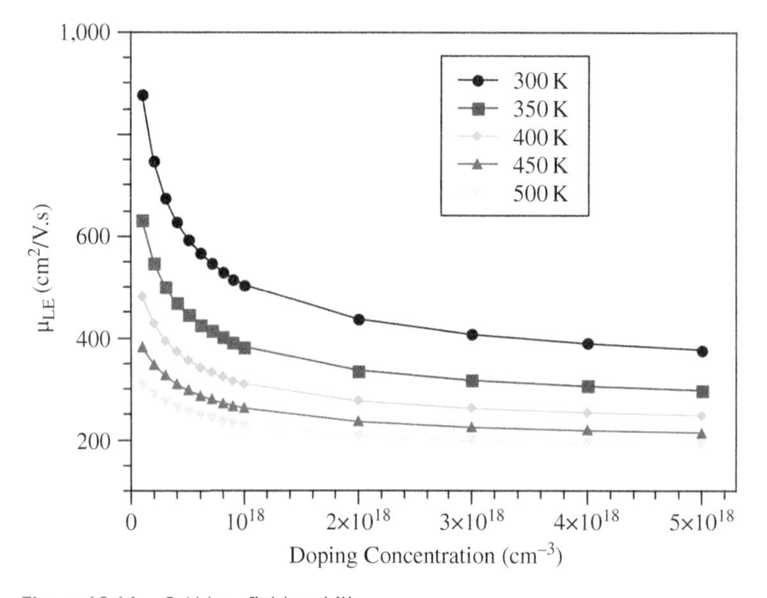

Figure 10.11 GaN low field mobility.

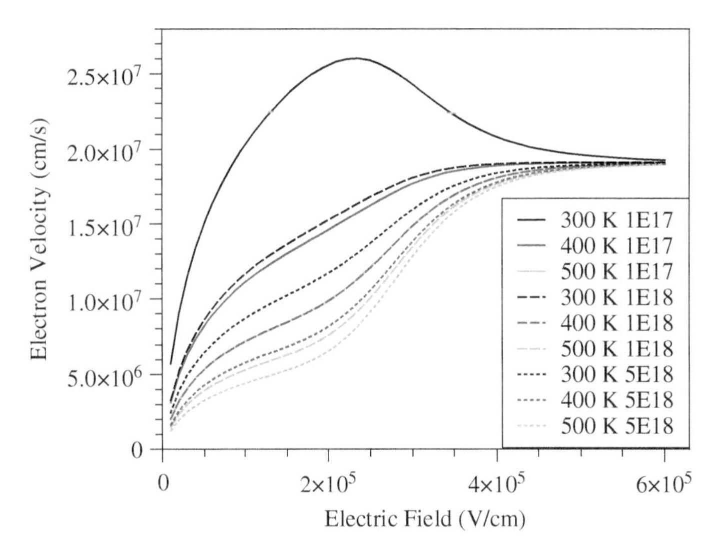

Figure 10.12 GaN electron velocity.

The breakdown voltage is evaluated by calculating the impact ionization integral. The ionization integral is introduced as a criterion for determining when avalanche is triggered. The integration of generation rate along a distance corresponds to the multiplication efficiency of newly generated electron-hole pairs over this distance. Once the integral value exceeds 1, the number of electron-hole pairs increases resulting in high current which then leads to breakdown.

In the simulation, the E-field over the entire device structure is calculated first, followed by the derivation of the E-field dependent impact ionization generation rate, based on which the integral is further evaluated along E-field lines. This method avoids current equation solving, which very often causes instability in the numerical process. The generation rate is described by Selberherr's model for both GaN and GaAs material, but with different coefficients. The temperature dependence of this model is given by temperature-dependent coefficients; all the coefficients listed in Table 10.4 are temperature dependent. However,

Table 10.4 Parameters of Selberherr's model used in simulation (at 300 K).

Material	GaAs	GaN
AN (cm^{-1})	1.889×10^5	2.52×10^8
B_N (V/cm)	5.57×10^5	3.41×10^7
β_N	1.82	1
AP (cm^{-1})	2.215×10^5	5.37×10^6
B_P (V/cm)	6.57×10^5	1.96×10^7
β_P	1.75	1

only the breakdown voltage at room temperature will be discussed, since the temperature dependence of these coefficients was not available in literature.

Both thermionic emission (TE) theory and thermionic field emission (tunneling) were considered in the simulation. TE theory describes the current flow through the metal-semiconductor (MS) contact due to electrons with energy above the barrier height $q\Phi_b$. The current depends, therefore, on barrier height exponentially. The tunneling current can be expressed by considering the tunneling probability and the Maxwell Boltzmann distribution functions in semiconductor and metal. Both the tunneling probability and electron distribution are given as a function of energy. In order to calculate J_T for a device structure represented by discrete grids, one needs to find the local generation rate by applying a gradient on J_T, which provides the generation rate for a specific grid. This generation rate will be further included into the current continuity equations [31]. The tunneling probability as a function of grid location is given by the Wentzel–Kramers–Brillouin (WKB) approximation by assuming a triangular barrier shape. A key parameter of this model in Silvaco [30] is the distance from electrode, within which all the grid points will be calculated in a way shown above. The default value of this distance is 10 nm, which is found sufficient for our study.

10.2.2.3 Simulation Results

10.2.2.3.1 Device Structure

The vertical structure shown in Figure 10.13(a) was used as common reference for comparing GaN over GaAs. This simplified structure contains a N^- layer for Schottky contact and a N^+ layer for Ohmic contact. Various thickness and doping concentration values were considered for the N^- layer, while the N^+ layer was maintained to be 1 μm thick with a doping concentration of $3 \times 10^{18}/\text{cm}^3$ for both GaAs and GaN diodes. A typical structure cross section generated by simulation is shown in Figure 10.13(b), where the dense mesh grid covering the entire N^- layer guaranteed the simulation accuracy.

Though planar diodes are the workhorses for most studies, a simple vertical diode structure will be discussed here as applied to the frequency conversion and signal control studies conducted in this chapter. This simple structure allows to focus on the material properties without being impacted by the more complex design of planar devices due for example to horizontal distribution of electric fields, anisotropic mobility, etc. Numeric efficiency is also enhanced due to the simplified structure.

10.2.2.3.2 Breakdown Voltage

The calculated ionization integral and corresponding maximum electric field are shown in Figure 10.14 for both GaAs and GaN diodes with a N^- layer of 300 nm thickness and $1 \times 10^{17}/\text{cm}^3$ doping. The simulated breakdown voltages were 12.8 and 129.6 V for GaAs and GaN, respectively, corresponding to an electric field of 0.63 and 4.25 MV/cm. The greater breakdown voltage for GaN diodes allows a higher RF input power to be applied to the diodes. N^- layers with different thickness and doping concentration were simulated and the results are listed in Table 10.5. In all cases, the breakdown voltage of GaN diodes is about 10 times larger than GaAs. The difference is even greater for thicker N^- layers.

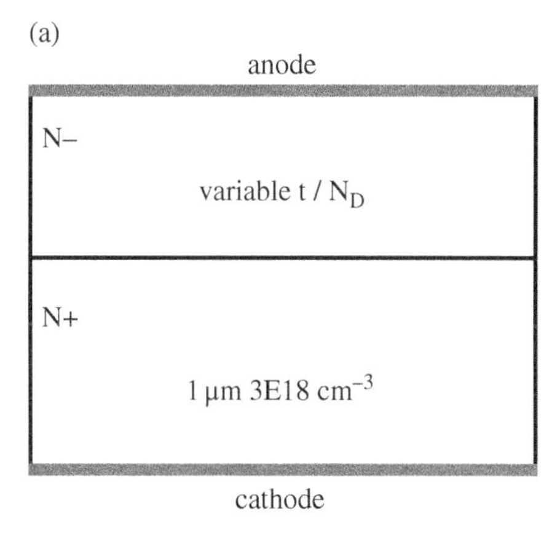

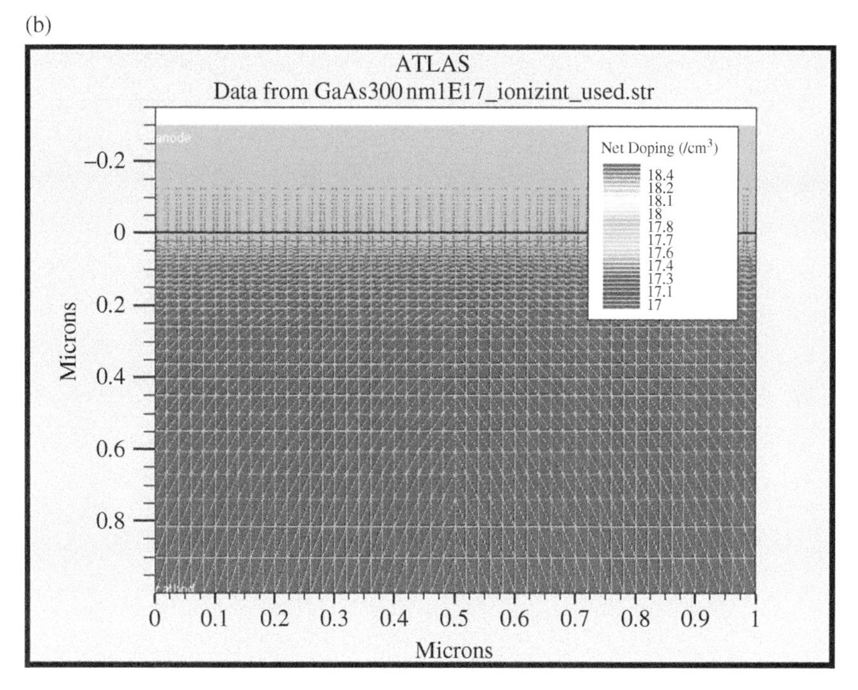

Figure 10.13 (a) Schematic of epi-structure used in simulation (b) an example of formed structure and the mesh grids.

10.2.2.3.3 I–V Characteristics

The simulated static I–V characteristics are shown in Figure 10.15, for variable N^- layer doping, thickness as well as temperature. In general, the current is greater in both the forward and reverse region when the doping or temperature is higher. The scale of the current axis is the same, for easier comparison. Under the same structure conditions, i.e. same barrier height, N^- layer doping, thickness, temperature, GaN diodes show slightly higher forward current than GaAs. This can be explained by the greater Richardson constant of GaN over GaAs. Another feature observed from the forward I–V characteristics R_s of GaN diodes

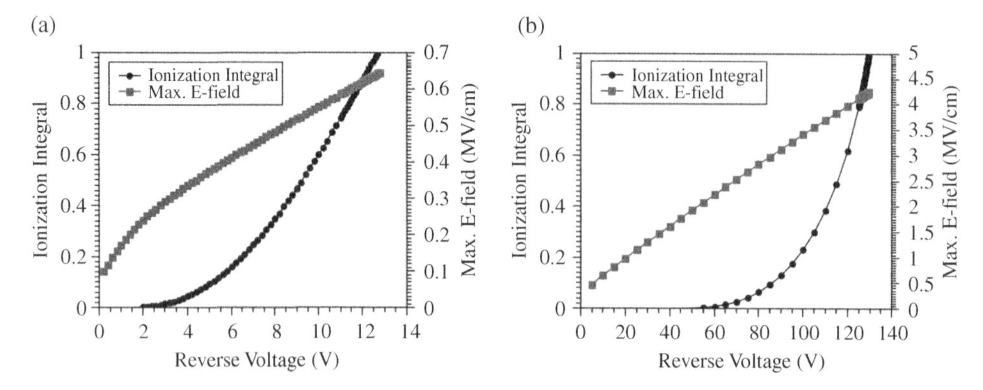

Figure 10.14 Simulated ionization integral and electric field vs. bias voltage.

Table 10.5 Breakdown voltage based on Ionization Integral (at 300 K).

t & N_D	GaAs (V)	GaN (V)
300 nm 5×10^{16} cm^{-3}	13.2	132.8
300 nm 1×10^{17} cm^{-3}	12.8	129.6
300 nm 2×10^{17} cm^{-3}	11.2	122.8
400 nm 1×10^{17} cm^{-3}	15.1	168.2
500 nm 1×10^{17} cm^{-3}	16.5	204.5

is their smaller saturation voltage in the forward region. This results naturally from the lower mobility of GaN. In the reverse region, the voltage applied to GaAs is limited to −10 V while it's −20 V for GaN. Thanks to the wide bandgap, GaN leads in lower reverse leakage current than GaAs. The difference between the two technologies is one order of magnitude when comparing them at −10 V.

10.2.2.3.4 Series Resistance

R_s is one of the key diode figures-of-merit and represents how lossy a diode is. The R_s of the above-simulated diodes was extracted from the forward I–V characteristics and is shown in Figure 10.16. GaN diodes have undoubtedly a larger R_s than GaAs, due to the inferior mobility of GaN.

Considering the diode resistance at 300 K, the values of GaN-based diodes are two to three times higher than for GaAs with the same N^- layer doping. This difference increases to 6–18 times when the temperature is 500 K. The R_s temperature dependence for GaAs is not monotone; the R_s increases when the temperature is elevated from 300 to 400 K, while R_s decreases when the temperature increases further to 500 K. This is due to the two temperature-dependent mechanisms involved, namely mobility and thermally activated carriers, that have reverse temperature dependence and compensate each other at a medium temperature. The R_s increase of GaN at elevated temperature relies on mobility drop, since the

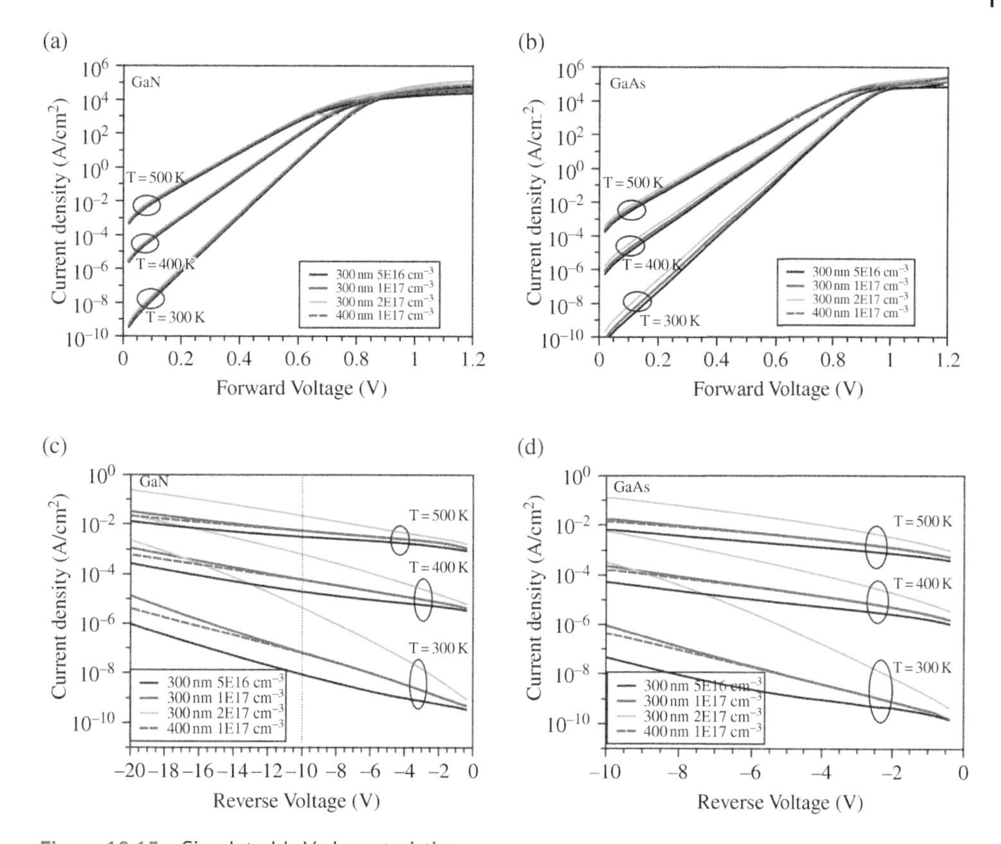

Figure 10.15 Simulated I–V characteristics.

contribution of thermally activated carriers can be neglected due to the properties of wide bandgap materials.

10.2.2.3.5 C–V Characteristics

Quasi-static C–V characteristics were obtained by AC analysis and shown in Figure 10.17. For the same barrier height, N^- layer doping, thickness, and bias voltage, GaAs diodes show a larger capacitance than GaN. This is due to the different depletion region width arising from the dielectric constant difference.

C_0 (with bias voltage of 0 V) is larger for diodes with higher doping concentration. Despite the doping concentration, for given N^- layer thickness C_{min} remains practically constant independent of carrier concentration. This corresponds to the case of completely depleted N^- layer. For the same reason, the 400 nm thick N^- layer exhibits the same C_0 but smaller C_{min} than the 300 nm thick N^- layer with same doping concentration.

10.2.2.3.6 Time-domain Transient Analysis

Transient analysis was performed by applying a sinusoidal voltage stimulus to the diodes. Consistent with previous simulations, the operation of GaAs diodes was limited to a peak-to-peak amplitude of 10 V and offset of −5 V. GaN diodes were simulated with the same

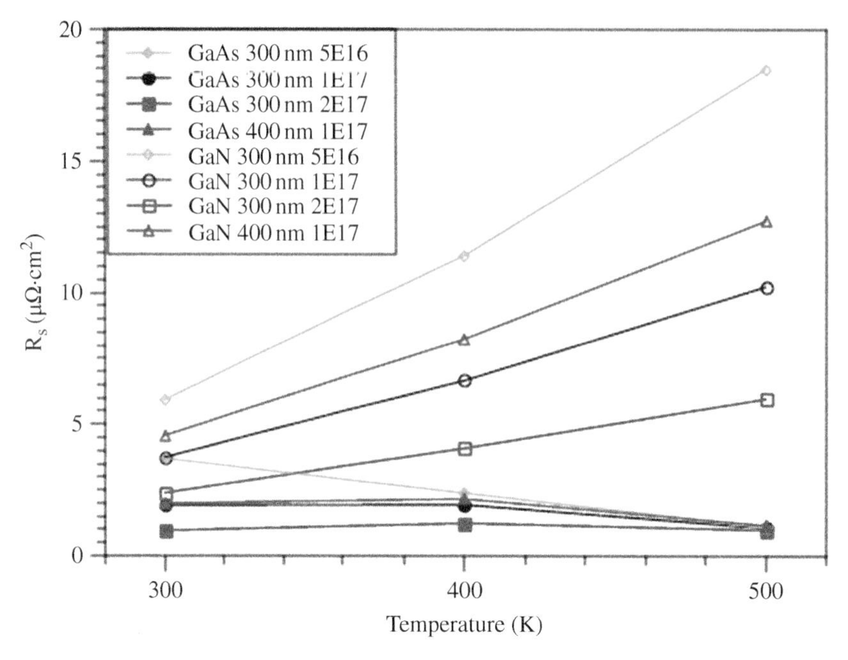

Figure 10.16 Series resistance extracted from forward IV curves in Figure 10.15a and b.

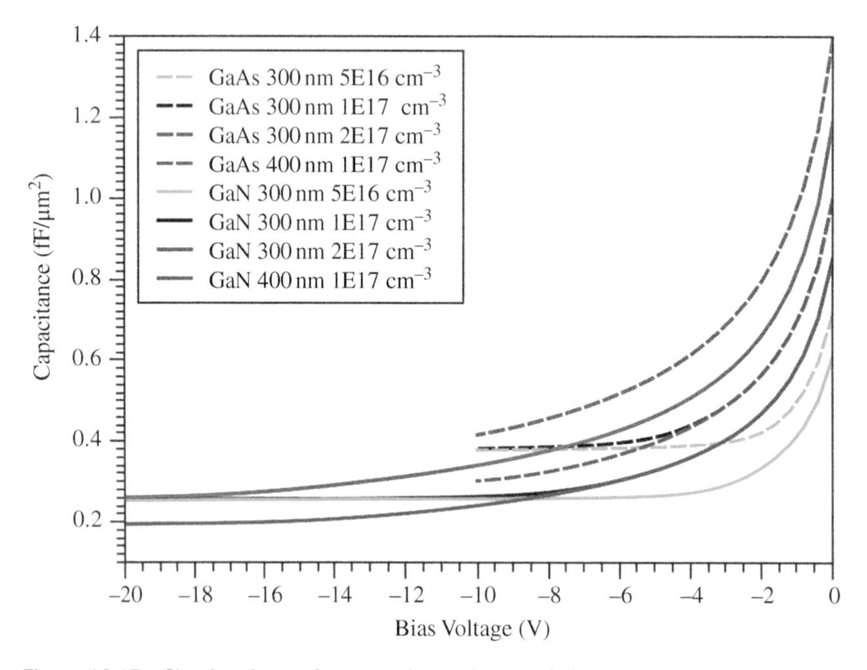

Figure 10.17 Simulated capacitance–voltage characteristics.

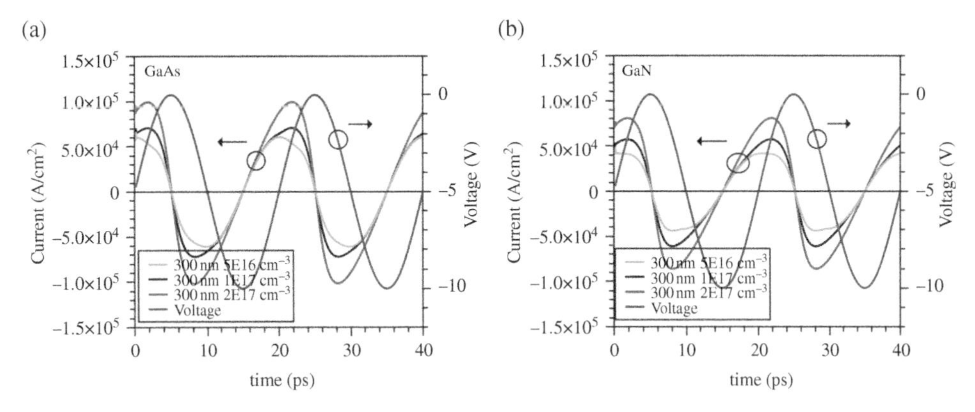

Figure 10.18 Transient $V(t)$,$I(t)$ characteristics.

stimulus condition for comparison. A 20 V amplitude and offset of −10 V were applied to GaN diodes for revealing the advantage of power handling. With a 50 GHz stimulus, the obtained current waveforms are shown in Figure 10.18 for 300 nm thick N^- layer and variable doping concentration. Lower doping corresponds to a lower current since less electrons participate in the process. For the same doping concentration, GaN diodes lead to a smaller transient current amplitude than GaAs diodes. This can be explained by the lower electron mobility of GaN material. As a result, diodes with lower current have smaller power exchange with the signal source in one signal cycle. The instant power is given in Figure 10.19.

The time-domain waveforms contain useful information, such as power handling capability and nonlinearity, which requires further data processing. Applying Fourier transform on the current waveform $I(t)$, the amplitude of current on each harmonic order (I_1, I_2, I_3...) can be obtained. The squared current amplitude corresponds to power, and the diode nonlinearity can be evaluated by $\frac{I_i^2}{\sum_1^n I_i^2}$.

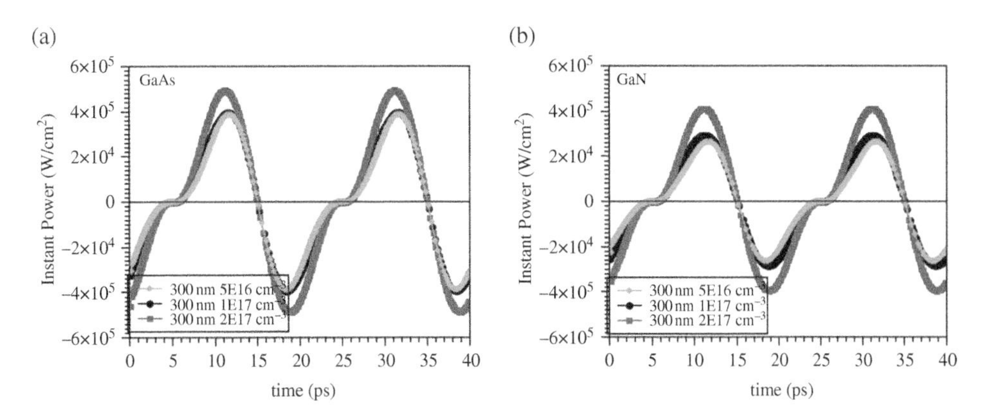

Figure 10.19 Transient $P(t)$ characteristics corresponds to Figure 10.18.

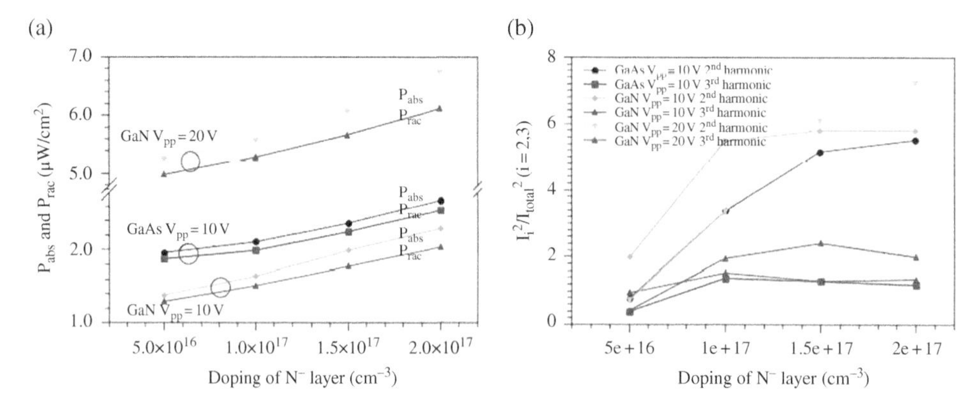

Figure 10.20 Power and nonlinear performance of 300 nm thick N⁻ layer with various doping concentration. (a) absorbed power P_{abs} and reactive power P_{rac} in one sinusoid stimulus cycle (b) power of second and third harmonics, relative to sum of all harmonics power.

The positive instant power corresponds to power absorption of the diode from the source, while the negative instant power corresponds to power provided by the diode. Thus, the integration over one time period reveals the power handling capability of each diode, where the positive half-cycle represents the power handling capability and the negative half-cycle represents a reactive power stored in the diode over the positive half-cycle. The difference between the two corresponds to power loss. It must be pointed out that the results were obtained from one operation cycle and the signal frequency has to be taken into account when comparing with experimental results. Considering a 300 nm thick, 1×10^{17} cm^{-3} doped N$^-$ layer as reference structure, and 10 V V_{pp} (−5 V offset) sinusoidal voltage as reference stimulus, one obtains the results discussed next. A 20 V V_{pp} (−10 V offset) stimulus was applied to GaN diodes in order to evaluate the power enhancement due to the increased breakdown voltage.

As shown in Figure 10.20(a), higher absorbed power, reactive power as well as loss can be observed for heavier doped diodes. The loss, denoted by P_{abs}–P_{rac}, was larger for GaN than GaAs devices. When doping was increased to 2×10^{17} cm^{-3}, the loss remained small at increased absorbed power levels, which indicates that the doping can be further increased.

The nonlinearity comparison shown in Figure 10.20(b) suggests that GaN diodes produce greater harmonics than GaAs. This can be attributed to the larger C_0/C_{min} ratio of GaN diodes. The second harmonic ratio of GaN diodes driven by 10 V V_{pp} stimulus (green diamonds in the figure) showed a fast increase when the doping increased from 5×10^{16} cm^{-3} to 1×10^{17} cm^{-3} but remained nearly unchanged when the doping increased further. Similar "saturation" trends were observed for all third and second harmonics generated by GaAs diodes; the absolute amplitude of second harmonic generated was increased due to the increased absorbed power. The dependence of power performance on N$^-$ layer thickness is shown in Figure 10.21(a), where the doping was fixed as 1×10^{17} cm^{-3}. It was found that P_{abs} and P_{rac} were smaller for thicker N$^-$ layer, while the loss increased with N$^-$ layer thickness.

The dependence of nonlinearity on N$^-$ layer thickness was found to be weak, except for the second harmonics generated by GaN diodes driven by a 20 V V_{pp} stimulus and GaAs

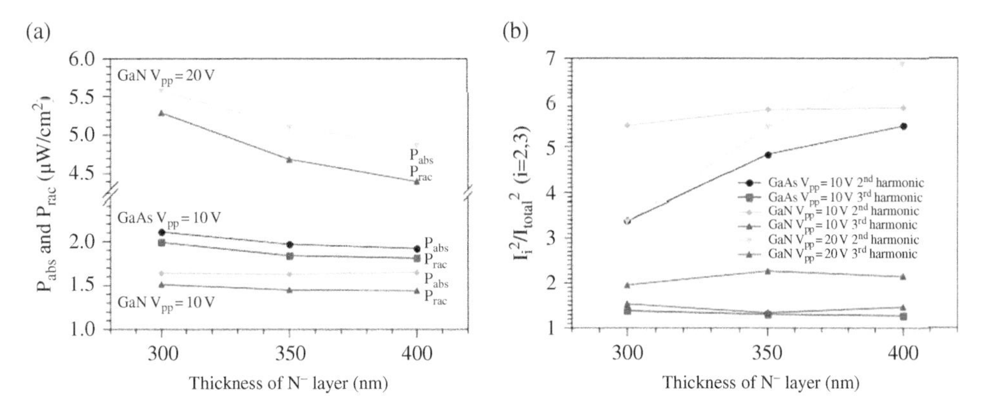

Figure 10.21 Power and nonlinear performance of 1×10^{17} cm^{-3} doped N^- layer with thickness varying from 300 to 400 nm. (a) Absorbed power P_{abs} and reactive power P_{rac} in one sinusoid stimulus cycle (b) power of second and third harmonics, relative to sum of all harmonics power.

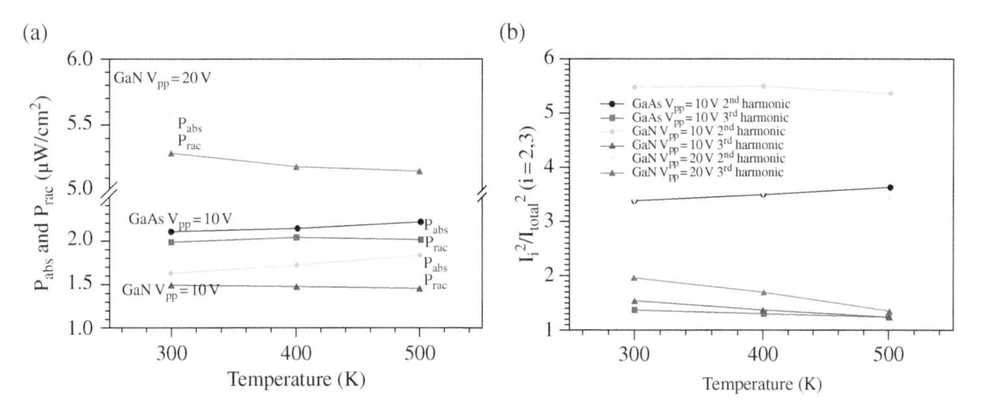

Figure 10.22 Power and nonlinear performance of 1×10^{17} cm^{-3} doped 300 nm thick N^- layer, with temperature from 300 to 500 K. (a) Absorbed power P_{abs} and reactive power P_{rac} in one sinusoid stimulus cycle (b) power of second and third harmonics, relative to sum of all harmonics power.

diodes driven by a 10 V V_{pp} stimulus. This was explained by the smaller C_{min} arising from thicker N⁻ layer, which was present at the highest peak of negative voltage swing of each stimulus cycle.

The temperature dependence of power performance is shown in Figure 10.22(a). Due to mobility drop at higher temperature, the loss becomes higher. The P_{abs} was found to increase while the P_{rac} decreased. This implied that the increased part of the P_{abs} has been consumed by loss, instead of storage in P_{rac}.

It is worth noting that the simulation was performed under isothermal condition. The elevated temperature was set within a range of values permitting to estimate the general trends of device characteristics replicating self-heating but not determined by such an effect. Considering the fact that the thermal conductivity of GaN is at least four times higher than

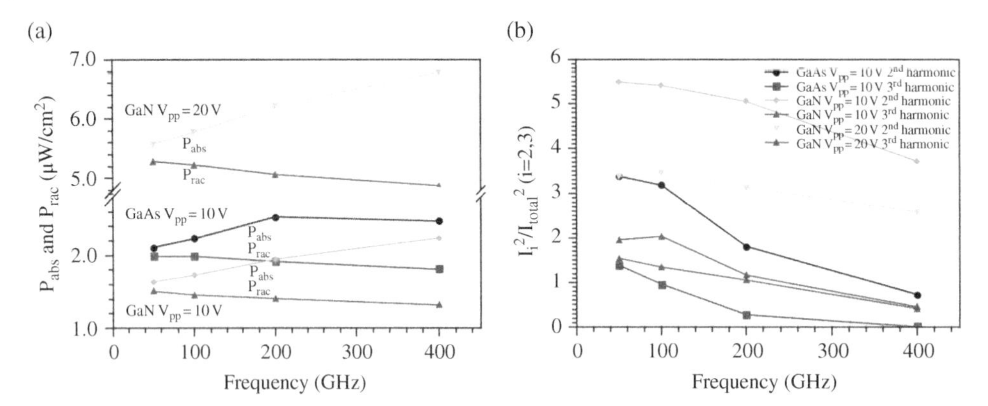

Figure 10.23 Power and nonlinear performance of 1×10^{17} cm^{-3} doped 300 nm thick N^- layer, with operation frequency from 50 to 400 GHz. (a) Absorbed power P_{abs} and reactive power P_{rac} in one sinusoid stimulus cycle (b) power of second and third harmonics, relative to sum of all harmonics power.

GaAs, one expects that for the same dissipated power, GaN didoes will have a lower junction temperature than their GaAs equivalent. A full thermal-electrical simulation would be helpful for revealing the junction temperature of GaN diodes subjected to large-signal operation and thus high power dissipation conditions.

Overall, GaN materials have been demonstrated to be excellent candidates for high power high temperature applications. This study indicates that performance variation has to be taken into consideration in designing such a diode. The temperature had less impact on nonlinearity, as shown in Figure 10.22(b).

Current saturation has been reported to limit GaAs diodes performance under high power and high-frequency operation [14]. As shown in Figure 10.23(a), P_{abs} was found to be increased with higher operation frequency for GaN diodes. But in GaAs case, the P_{abs} was slightly reduced from 200 to 400 GHz. As a result of increased loss at higher frequency, P_{rac} became smaller. It is also worth noting that in Figure 10.23(b), the harmonic power ratio dropped much faster with increased operation frequency for GaAs than GaN. This indicates the potential of using GaN Schottky diodes for higher frequency, despite the lower mobility of this material.

Based on all the results shown above, it was concluded that,

1) The power handling capability of GaN Schottky diodes is two to three times higher than that of GaAs diodes, when the input stimulus voltage is doubled.
2) The increase of loss is compensated by increased absorbed power.
3) GaN diodes are more nonlinear than GaAs diodes, and this nonlinearity provides them an advantage for higher power stimulus.
4) The nonlinearity of GaN diodes drops less than in GaAs diodes under high-frequency operation.
5) Thanks to the higher input power that can be applied, GaN Schottky diodes are expected to be robust and working at higher operation temperature. Designers need also to pay attention to the power variation with operation temperature.

10.2.3 Conclusions on Theoretical Considerations of GaN Schottky Diode Design

The approach used for exploring the power handling capability, losses and nonlinearity of GaN Schottky diodes have been described in detail. An analytical approach based on ideal Schottky diode C–V and I–V characteristics allowed performance correlation such as power handling capability, losses, and nonlinearity to the diode parameters (C_0, V_{bias}, R_s, etc.). Numerical simulation based on technology computer-aided design (TCAD) tools allowed accurate verification of the performance for a specific design and provides physical insights into the diode operation. These two approaches together define an effective way of understanding the operation principles of Schottky diodes and further analyzing their performance. The results of the analysis demonstrate, the superior power handling capability of GaN Schottky diodes. The potential of GaN Schottky diodes in high-frequency has also been shown. The presented results allow better understanding of the characteristics of fabricated devices that will be discussed in the following sections.

10.3 Fabrication Process of GaN Schottky Diodes

The material properties of GaN lead to differences compared with GaAs-based technology. For example, GaAs is easily etched by wet solutions, permitting flexibility in fabrication. Trench isolation can in this way be done after interconnection metallization, allowing beam-lead realization [32]. Devices can also be made using titanium as Schottky contact offering the possibility of combining Schottky and pillar metallization in the same step. The work described here discusses the various steps involved in GaN diode fabrication using E-beam lithography technology.

10.3.1 Fabrication Process

The steps necessary for fabricating GaN Schottky diodes are listed in Table 10.6, which also includes a comparison of various fabrication technologies adopted for contacting the diode anodes. A general description of the technologies used is provided first, followed by a detailed discussion on each step.

The materials used for the study were grown by an in-house MOCVD facility. The GaN epitaxial structure used for Schottky diodes in this study employs GaN layers with two different doping concentrations. A N^+ layer was grown first on sapphire substrate and served as cathode contact of the Schottky diodes. A doping concentration between 3×10^{18} and 5×10^{18} cm^{-3} was used in order to obtain ohmic contacts with minimum resistance. The thickness of the N^+ layer was 2 μm. The N^- layer grown on top of the N^+ layer acts as the active layer. The Schottky metal is deposited on this layer at a later step to form a Schottky contact. The nonlinear properties of the device depend both on the N^- GaN layer and MS interface established on it. The diode analysis presented earlier showed that a doping concentration as low as 10^{16} cm^{-3} is necessary in order to obtain higher barrier height and lower leakage current required for higher power handling capability. A doping concentration of this value is in the limit of what can be achieved by MOCVD growth of GaN, and cannot be easily

Table 10.6 Process flow of the three technologies used in this study.

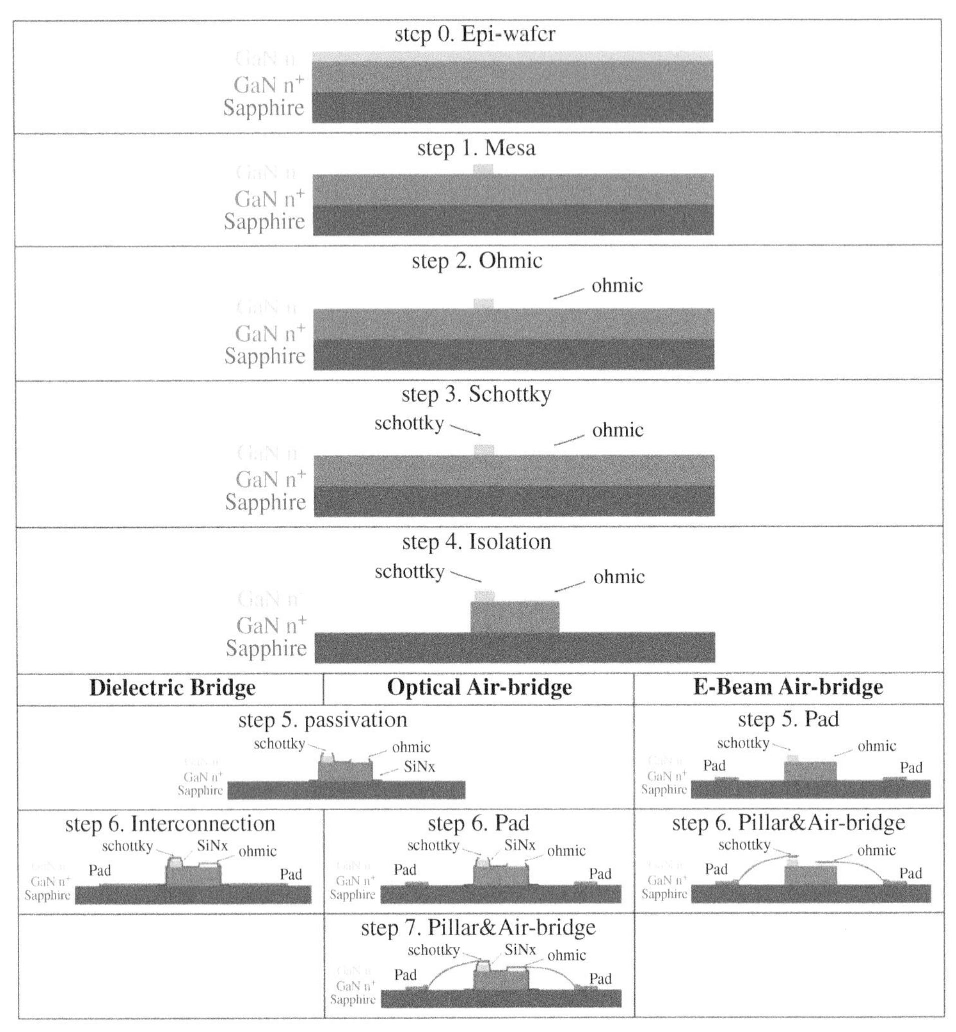

controlled, leading often to resistive material. It was, therefore, decided to have the N^- layer grown for this study at a doping concentration of 1×10^{17} cm^{-3}. Considering the two-layer (N^+ and N^-) epitaxial structure, the cross section of the processed diode can be represented as shown in step 4 of Table 10.6. To obtain a device of this type, the following fabrication steps are necessary.

1) Mesa etching: to remove the N^- layer outside the active area of the diode and allow the formation of Ohmic contacts on the N^+ layer.
2) Ohmic contact metallization to obtain Ohmic contacts on the N^+ layer.
3) Schottky contact metallization: for establishing Schottky contacts on the N^- layer.
4) Isolation: to completely remove the GaN layer down to sapphire substrate and ensure isolation of a diode from those adjacent to it.

10.3.2 Etching

Wet etchants for III-nitride materials are limited. Under normal conditions, only molten salts like KOH or NaOH at elevated temperature are capable to etch GaN at reasonably high rates [33]. Moreover, etching of this type requires an appropriate etching mask, which in most cases is not photoresist. Thus, the application of wet etching in III-nitride device processing is limited and dry etching has been developed.

Dry etching of III-nitrides normally employs a plasma generated in a vacuum chamber coupled to an RF signal. The removal of III-nitrides is accomplished by either physical sputtering, chemical reaction, or combination of both. The typical gas used for physical etching of GaN is argon. The achieved etching rate depends on the plasma density and energy of accelerated Ar^+ ions. In contrast to the above, the mechanism of chemical reaction etching relies on the chemical reaction of the surface material with reactive ions formed in the plasma. Chemical reaction provides normally a good morphology due to the low energy of the reactive ions. Chlorine-based gases, like Cl_2 or BCl_3, may be used for this purpose. Ion bombardment has found to be the dominant mechanism due to the high bond energy of III-nitrides. Thus the energy of ionized Cl^- has to be high enough to act properly. The combination of both physical sputtering and chemical reaction is expected to lead in faster etching rate as well as anisotropic profile. A key part of this technology is the definition of an optimized condition where the physical and chemical processes are balanced. Considering the various approaches necessary for plasma generation, one can define three types of instrumentation. These are reactive ion etching (RIE), inductively coupled plasma (ICP), and electron cyclotron resonance (ECR).

Oxford Plasmalab 80 RIE and Oxford Plasmalab 100 ICP were used for etching. RIE allowed to perform Ar physical sputtering, with an etching rate of 20 nm/min. A nickel layer is often used as hard mask but photoresist is more attractive due to faster processing. AZ4562 resist was, therefore, used and was found to resist in Ar plasma for more than one hour. For ICP etching, a Cl_2/Ar gas mixture was used and resulted in an etching rate of 200 nm/min. Figure 10.24 presents scanning electron microscope (SEM) images of etched mesas

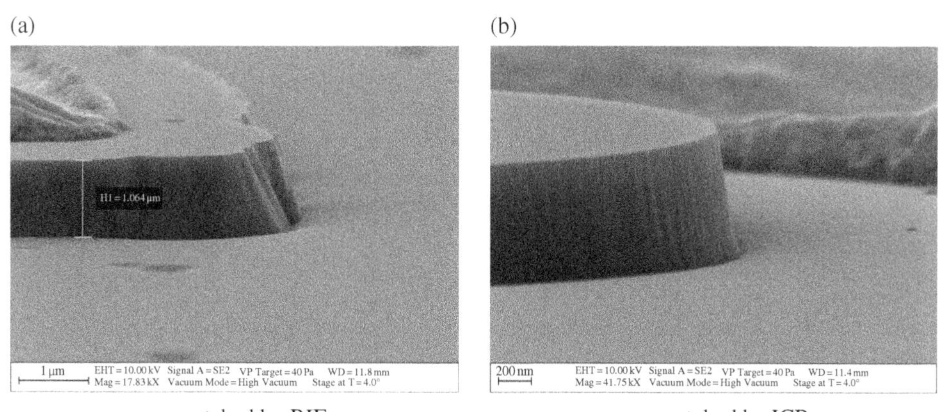

Figure 10.24 SEM images of etched mesa by (a) Ar sputtering only in RIE and (b) Cl_2/Ar mixture in ICP.

by Ar sputtering in RIE and Cl_2/Ar mixture in ICP. Good etched profiles and surface roughness has been achieved by both approaches.

10.3.3 Metallization

10.3.3.1 Ohmic Contacts on GaN

For III-nitrides, a metal scheme with a four-layer Ti/Al/X/Au stacks has often been reported, where X can be Ti, Ni, Pd, and Pt. The first titanium layer is believed to react with nitrogen atoms on the GaN surface. The resulting semiconductor surface tends to be highly doped due to the *N*-vacancies. A thin barrier is thus formed due to the highly doped layer, which increases the probability for electrons to tunnel through. Contact formation benefits also from the titanium layer due to its capability of dissolving the native oxide and improving adhesion. Better Ohmic contact quality can be achieved without native oxide, and better mechanical stability is possible by adhesion improvement.

The importance of aluminum layer has been evidenced by the dependence of its contact resistivity on the ratio of Al over Ti thickness [34, 35]. Al prevents the underlaying Ti layer from oxidizing by forming a Al_3Ti layer at elevated temperature. Transverse electromagnetic (TEM) studies of the annealed microstructures [36] showed that micro-voids tend to be formed at the interface when only the Ti layer is present on GaN. The use of a Ti/Al bilayer leads in Al_3Ti formation as predicted by the Ti–Al bulk phase diagram, thus making the Ti-GaN reaction less aggressive.

Both Ti and Al can be oxidized easily, preventing the successful formation of Ohmic contact. It has been shown that the oxygen content in the annealing ambient has a pronounced impact on the resulting contact resistance [35]. An Au layer has often been used to protect them from oxidization. The conductivity of the metal stacks can in this way be increased. However, a blocking metal layer is required in this case to prevent the diffusion of Au to the Al layer as well as, the out-diffusion of Al. This blocking layer is noted above by X, since several metals like Ti, Ni, Pd, and Pt may be used.

The transfer length method (TLM) was used to evaluate the quality of Ohmic contact. By providing information on contact resistance (R_c, Ω·mm), specific contact resistance (ρ_c, Ω·cm^2) as well as sheet resistance (R_{sh}, Ω)of the semiconductor film. A Ti (25 nm)/Al (150 nm)/Ti (25 nm)/Au (150 nm) stack for optical process and Ti (12 nm)/Al (200 nm)/Ni (40 nm)/Au (100 nm) for E-beam process were employed in this study. The same rapid thermal annealing (RTA) condition was applied to both metal stacks. This was 850 °C annealing in N_2 ambient for 30 seconds. Typical Ohmic contact results achieved in this study correspond to a R_c of 10^{-1} $\Omega\ mm$, and ρ_c about $3 \times 10^{-6}\Omega$ cm^2.

10.3.3.2 Schottky Contacts on GaN

Schottky contacts are the most critical part of the diodes. From the MS contact theory introduced in the last chapter, we know that the metal has to have a high work function to form a Schottky contact on GaN. The conventionally used metals for GaN Schottky contacts include Ni (W_m = 5.15 eV), Pt (W_m = 5.65 eV), Pd (W_m = 5.12 eV), and Au (W_m = 5.1 eV). Pt was chosen in this study due to the possibility of forming a high barrier. A comparative study was also made with Ni Schottky contact.

Oxygen plasma treatment is necessary to remove the residual resist film [37] often formed at the surface and is often used for semiconductor and quartz mask cleaning. However, plasma may cause surface defects and its power needs to be optimized.

Schottky contact quality relies strongly on surface treatment before metal deposition. HF and HCl are the commonly used solutions for removing the surface native oxide [38–40]. The alkaline counterpart for this purpose is NH_4OH [37, 41]. Aqua regia ($HNO_3 : HCl = 1 : 3$) [42] and $(NH_4)_2S_x$ [43] were also suggested in literature. Besides, fluoride-based plasma treatment has also been implemented [44]. The study reported here showed that HF and HCl lead in clean surface.

10.3.3.2.1 Analysis of Schottky Contact Characteristics

The quality of Schottky contacts is normally described by the parameters n (ideality factor) and Φ_B (barrier height). Both parameters can be extracted from I–V characteristics. Ideal MS contacts obey TE theory, while in practice the current transport across Schottky contacts involves also other mechanisms such as tunneling. The deviation from pure TE transport is evaluated by n. The "hypothesis testing" method is applied in practice to analyze the IV curves of Schottky diodes. Assuming TE is the dominant mechanism, a value of n is obtained by fitting the measured IV with this equation. Obviously, $n = 1$ corresponds to pure TE transport, and therefore the obtained n is in practice often larger than 1.

It has been reported that Schottky contacts annealed at an appropriate temperature show a barrier height enhancement. A Ni/GaN Schottky contact was found to have a barrier height of 0.86 eV and ideality factor of 1.19 was obtained after 5 minutes annealing at 600 °C in N_2 ambiance, compared with the as-deposited 0.69 eV barrier height and ideality factor of 1.47 [45]. This has been attributed to the formation of Gallium Nickel (Ga_4Ni_3) [46]. Barrier height enhancement after annealing was also found on Pt/GaN contacts [47], where 30 minutes 500 °C annealing in N_2 ambiance was employed.

Both Ni and Pt Schottky contacts on GaN were tested in this work. The epi-layer was 300 nm thick with a surface doping concentration of 3×10^{17} cm^{-3}. Oxygen plasma together with buffered hydrofluoric acid (BHF) etching was applied to clean the GaN surface before metallization. Each chip was annealed in N_2 ambiance for five minutes but with temperature ranging from 350 to 500 °C.

The results obtained are shown in Figure 10.25. The Ni contacts have in general higher current than Pt contacts in both the forward and reverse region, due to their lower workfunction. For both Ni and Pt contacts, an appropriate annealing temperature resulted in the largest measured barrier height and lowest ideality factor, as well as reverse saturation current, which was 450 °C for Ni and 500 °C for Pt in this study. It's also noted that before annealing at the desired temperature, the barrier height of annealed contacts was lower than that of the as-deposited contacts.

Different ideality factor, barrier height, and series resistance were obtained when extracting in different regions such as in the lower bias range (0.1–0.2 V) and the higher (0.5–0.08 V) bias range. The larger ideality factor resulting from Cheung's method indicated that the deviation from TE mechanism was severe under higher bias. This deviation resulted, therefore, in a lower effective barrier height.

As a result of increased barrier height after post-annealing, the reverse leakage current density was considerably decreased. However, the best barrier height obtained (0.625 eV

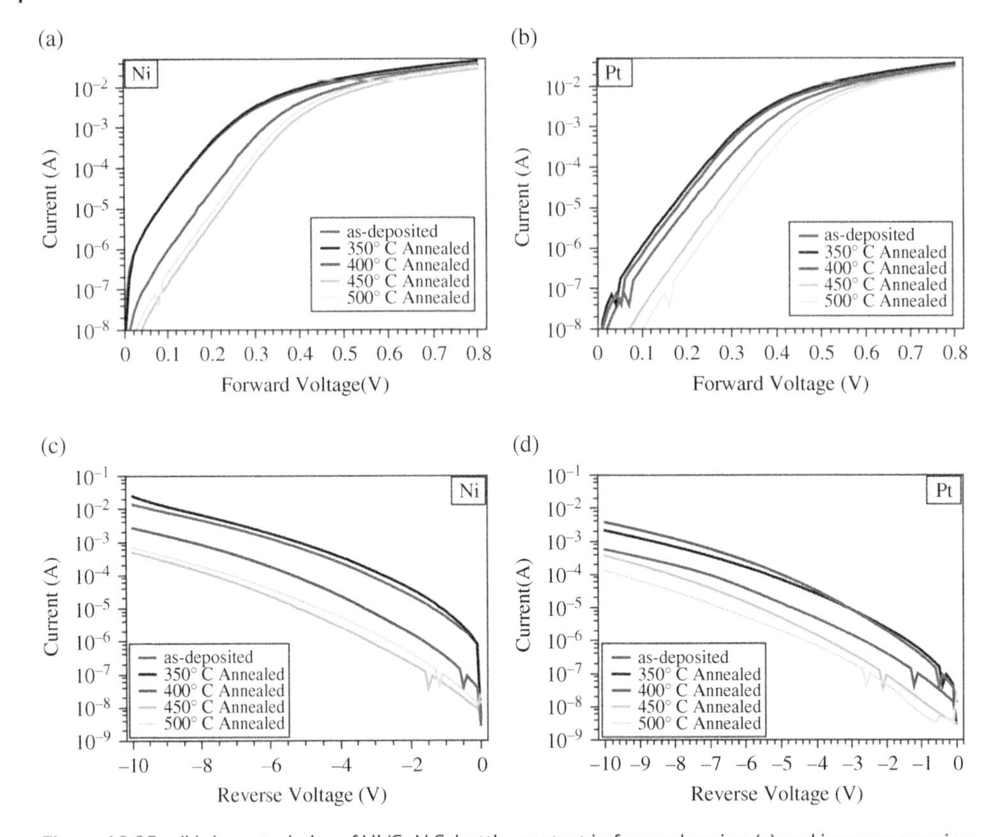

Figure 10.25 IV characteristics of Ni/GaN Schottky contact in forward region (a) and in reverse region (c); Pt/GaN Schottky contacts in forward region (b) and in reverse region (d).

for Ni and 0.68 for Pt) was still far from the values calculated from work-function difference. Consideration of the process employed indicated that surface cleaning before Schottky metallization could potentially cause surface damage.

10.3.3.2.2 Oxygen Plasma Before Schottky Metallization

As described earlier, Oxygen plasma treatment is necessary to ensure fabrication of Pt Schottky contacts. Pt is known for its poor adhesion on semiconductors and the presence of a residual resist layer is expected to lead to even worse characteristics. Appropriate processing using oxygen plasma is necessary for removing the residual resist layer without considerable damage of the semiconductor surface. The epi-layer used had a nominal doping concentration of 1×10^{17} cm^{-3}. The devices subjected to 100 W 60 seconds O_2 plasma treatment showed higher current in both forward and reverse regions, while the 20 W 30 seconds O_2 plasma-treated device showed a current similar as the reference (without plasma treatment) sample. This finding suggested that O_2 plasma excited with a power of 100 W was too energetic and could introduce additional defects on the GaN surface before Schottky metallization. A power of 20 W was verified to be appropriate for our application.

This appears to be "strong" enough for removing the residual resist but also "soft" enough for keeping the GaN surface free of additional damage.

The barrier height was found to be lower and the ideality factor was larger for the device subjected to 100 W O_2 plasma treatment, reflecting, therefore, the presence of nonideal TE transport mechanism due to surface defects. Despite the similar ideality factor and barrier height, the I_s and R_s for the non-plasma-treated reference device were two times higher than that for 20 W plasma-treated device. This may be explained as an effect of random distribution of residual resist on the GaN surface without plasma treatment. Compared with other samples treated by 100 W O_2 plasma in previous tests, such as the post-annealing test samples, one sees an improved barrier height for the current sample. This is believed to be due to material quality. A reverse current density as low as 10^{-5} A/cm^2 under -10 V bias could be achieved. The state of the art low leakage current density also reflected the material quality. The dependence of reverse leakage current on the barrier height was examined by artificially altering the barrier height by ± 0.05 eV around 1 eV. The measured leakage current was found to be two orders higher than the theoretically obtained results, indicating that other mechanisms dominate the re-verse leakage instead of tunneling. A possible explanation is the presence of dislocation related leaky paths [48], which would allow larger leaky current density due to their low barrier height.

10.3.4 Bridge Interconnects

Three different technologies are possible for diode multiplier technology:

10.3.4.1 Dielectric Bridge

After isolation etching, a SiN_x layer is deposited all over the wafer. The resulting dielectric is rounded off compared with the shape of etched mesa step. One step lithography and SF_6 etching in RIE followed this step to open the Schottky and Ohmic contact metal for accessing it. The bridge metal was realized by conventional lift-off at the last step. The resulted device can be seen in Figure 10.26a.

10.3.4.2 Optical Air-bridge

The optical lithography-based air-bridge process starts with a pad metallization followed by pillar definition and bridge realization. The photoresist used in this step must be thick enough to offer a planar surface over the nearly 2 μm thick mesa. The areas where the bridge pillar will rest were then defined by lithography. A reflow of this photoresist layer was performed after to achieve a smooth profile. This was important to assure the continuity of the seed layer for electroplating of the gold bridge. A thin metal layer which acts as plating seed layer was then deposited by either evaporation or sputtering. This metal layer has the additional function of isolating the pillar photoresist from the one used for the bridges.

The photoresist for bridge definition was then deposited on top of the metal layer. During gold plating, the current must be controlled to avoid fast deposition, which may shorten the device anode and cathode. Releasing of the bridges is also critical. The top layer photoresist must be removed by flood exposure and development, rather than remover or solvent. The seed metal layer was removed by a gold etchant. The time of etching must be long enough to

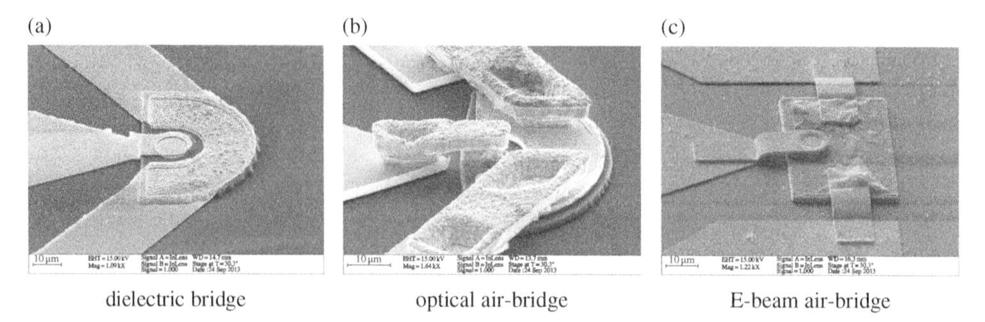

Figure 10.26 Devices realized in this study.

assure no metal left, which allows easier removal of pillar photoresist layer. But the etching time has also to be as short as possible to avoid the gold bridge being etched.

10.3.4.3 E-beam Air-bridge

The process of E-beam air-bridge is less time consuming compared with optical air-bridge process for two reasons: first, the alignment of lithography is guaranteed and there is no need for passivation; second, the bridge is based on conventional lift-off, which eliminates the plating step as well as the complicated bridge releasing step.

10.3.5 Conclusion on Fabrication Process of GaN Schottky Diodes

The technology approaches used for processing GaN-based Schottky diodes were presented. Approaches for interconnection were also discussed, including dielectric bridges for rapid diode evaluation, air-bridge with optical and E-beam lithography. Optical lithography presents low-cost advantages, while E-beam lithography opens the possibility for sub-micron anode realization, meeting the requirements of certain high-frequency diode applications.

10.4 Small-signal High-frequency Characterization of GaN Schottky Diodes

Equivalent circuit (EC) model is necessary for analyzing the high-frequency diode characteristics. The parameter values of elements, as well as their dependence on bias and frequency, are linked to the device physically. They are also correlated to the material parameters, device structure, and fabrication processes. The DC and small-signal characteristics of GaN Schottky diodes are presented in the next sections. The intrinsic elements, i.e. junction capacitance and junction resistance are also discussed.

10.4.1 Current-voltage Characteristics

The fabricated diodes had diameters between 2.5 and 20 μm, while the epi-structures had a N^- layer thickness of 300 and 600 nm. Typical results are shown in Figure 10.27, where diodes with various diameters have been measured.

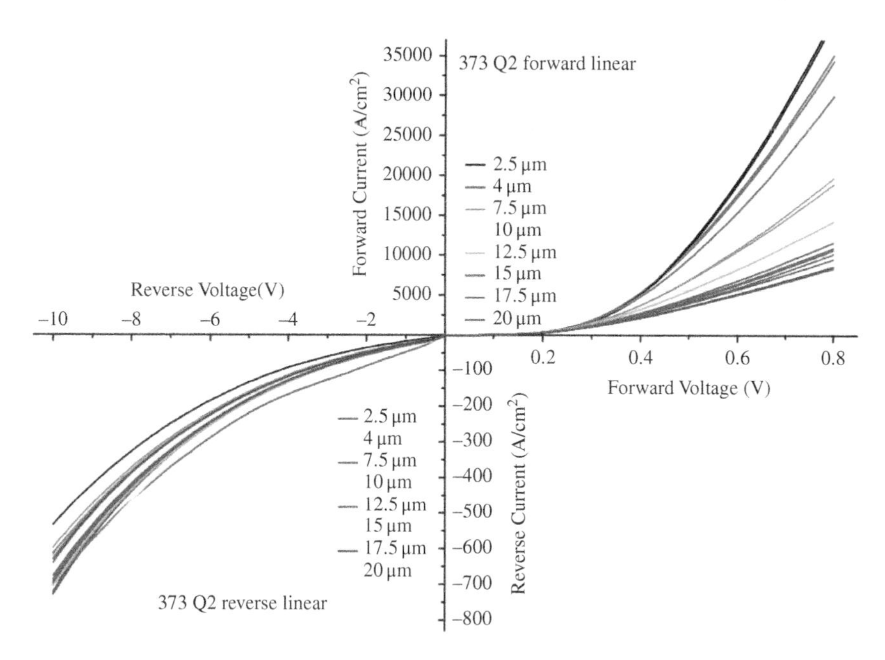

Figure 10.27 I–V characteristics of diodes with various diameter.

All curves have been normalized to the nominal anode area, so that a comparison can be made based on current density. As can be seen, the forward current density depends very much on anode diameter, and diodes with smaller diameter have higher current density. The current density of 2.5 μm devices is about 4.6 times higher than for 20 μm diodes. This indicates that periphery effects have to be taken into account.

In general, electric field lines show crowding at the edge of the Schottky metal which causes the current density at the metal periphery to be higher than in the rest of the contact area. This can be better explained by numerically simulated results as shown in Figure 10.28, where a current density about two times higher was found to exist at the metal edge. The periphery effect is in principle present regardless of the bias. The leakage current of a reverse-biased diode depends much more on the edge current, given the high tunneling rate resulting from the electric field peak at the anode edge. However, the measured reverse I–V shown in Figure 10.27 showed that the dependence on diode diameter was less significant. This indicates that the dominant current transport mechanism under reverse bias is dislocation related other than simply thermionic field emission.

Conventional SPICE diode models have an area factor, which accounts for model scaling. The existence of a periphery effect and its bias dependence makes it difficult to obtain a scalable model. In other words, diodes with different diameters need to be modeled independently. The device selected for discussion in this chapter had only 2.5 to 10 μm diameter.

10.4.2 Small-signal Characterization and Equivalent Circuit Modeling

The high-frequency characteristics of the fabricated diodes were obtained with a vector network analyzer (VNA). A short, open, load, and thru (SOLT) calibration procedure

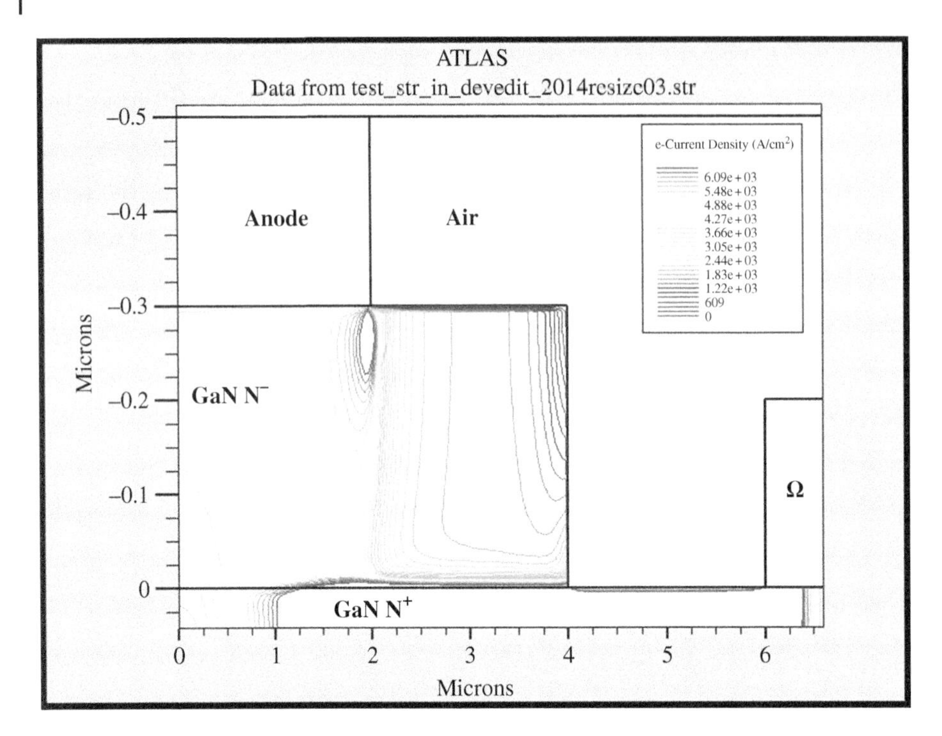

Figure 10.28 Simulated current density contour of a diode.

was applied before each measurement. Given the fact that the diodes are placed in series between the signal ports of the VNA, it is expected that S_{12}/S_{21} provides information about the diode itself, while S_{11} and S_{22} are affected by the parasitics at port 1 and port 2, respectively.

Based on the measured S-parameters, one can obtain the diode EC. The GaN Schottky diodes fabricated as described earlier have the cross section shown in Figure 10.29, which also includes the EC.

L_1/L_2, R_1/R_2, C_1/C_2 correspond to the parasitic inductance, resistance, and capacitance respectively related to the anode and cathode pads and interconnects, while C_P is the parasitic capacitance between the anode and cathode. The intrinsic diode characteristics are represented by three bias-dependent elements, namely the junction resistance (R_j), junction capacitance (C_j), and series resistance (R_s).

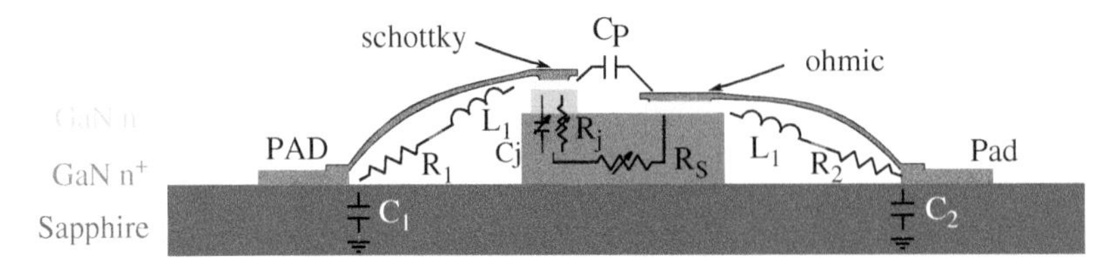

Figure 10.29 Cross section of GaN Schottky diode and the corresponding equivalent circuit model.

The parameter extraction procedure for field effect transistors (FETs) has been well established. All the elements can be expressed to be linear functions of port characteristics (S/Y/Z parameters) and can thus be derived by a set of operations of S/Y/Z parameter matrixes [49]. In contrast to FETs, such a straightforward extraction method is still lacking for diodes. This is due to their elements being entangled to each other and therefore not being possible to be expressed individually as a function of port characteristics. Optimization by tuning the elements toward the measured S-parameter values is the most commonly used method. In a recent paper [50], the authors claimed to provide a straightforward analytical method to extract parameters for high-frequency diodes. However, fitting was used to derive the junction capacitance (C_j) and a linear regression method was used to derive finger inductance.

The modeling strategy in this study is based on a combination of extraction and fitting to derive the initial values for the elements, followed by optimization and fine tuning of the element values for minimum error. This procedure is described below in detail with example data from one of the characterized diodes.

10.4.2.1 Step 1. Parasitic Elements

On-wafer high-frequency measurements generally employ coplanar waveguides (CPWs) to access the active region. The effects of the CPW have to be removed to obtain accurate device characteristics. The "removal" of the access CPW is implemented by de-embedding structures, as shown in Figure 10.30. Parasitic elements of the diode EC correlate directly to the access section of the CPW and can, therefore, be derived from de-embedding structures.

The de-embedding structures are ideally designed with the same dimension as the device, in order to avoid any discrepancy. The open, short, and through lines shown here were all adapted from the layout of 4 μm diameter diode. Open structures are realized by simply

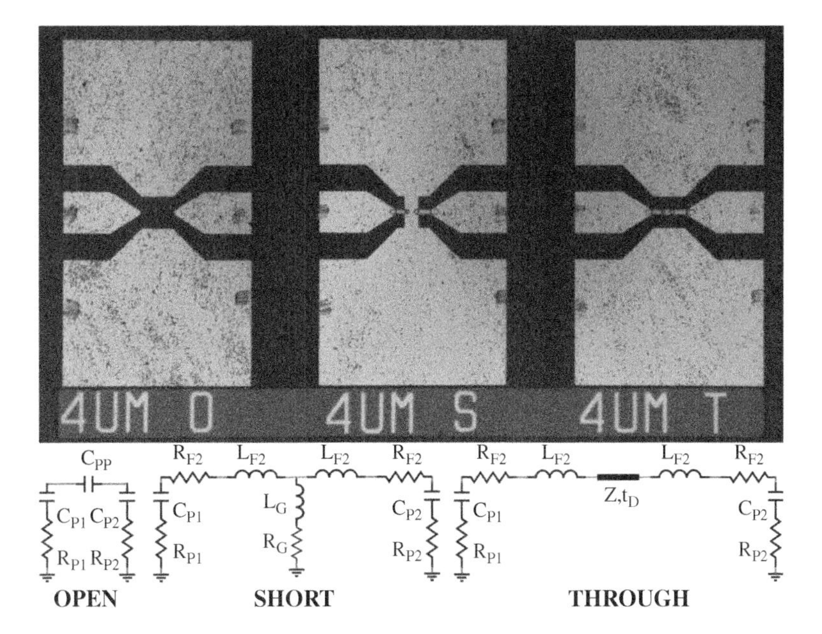

Figure 10.30 De-embedding structures and the corresponding equivalent circuits.

removing from the diode layout all other layers except the pad. Short structures shorted the signal pad to ground through bridge fingers. Through lines are simple lines which are used to verify the de-embedding structure quality.

The validity of the de-embedding process was tested by the through line, the characteristics of which should correspond to a delay line with an impedance Z and delay time t_D after deembedding. A physically short transmission line can be seen after de-embedding the open and short. The resulting line was fitted to an impedance of 40 Ω and a delay time of 73 fs.

The EC elements can be easily obtained by fitting to measured S-parameters. Values of 10.74 and 10.57 fF for C_{P_1} and C_{P_2} respectively were found to fit well the Phase (S_{ii}) of open structures, while 5 Ω for both R_{P_1} and a R_{P_2} were found to fit the Magnitude. A C_{PP} of 0.75 fF was found to satisfy the Phase (S_{ij}) indicating weak coupling between the two signal pads. Both measurement and fitted data are shown in the figure below.

As far as the pad capacitance is concerned, it is convenient to utilize an extraction rather than optimization method. With the assumption of no pad resistance existing (i.e. R_{P_1} and R_{P_2} being omitted), the pad capacitance can be simply expressed as

$$C_{Pi} = \frac{Im(Y_{ii} + Y_{ij})}{\omega} \tag{10.15}$$

where i and j denote pad 1 or 2, and ω is the angular frequency.

Using the EC fitted here, the validity of the extraction method can be checked by artificially tuning R_{Pi}. While keeping C_{P_1} constant at 10.74 fF, the S-parameters of the EC were calculated for R_{P_1} being artificially set to 5, 25, 50 Ω. The pad capacitance was extracted based on Eq. (10.15) and compared with the real value of 10.74 fF. The deviation of the extracted from the real value becomes greater for larger R_{P_1} and higher frequency. However, a maximum error of 0.3 fF resulted for $R_{P_1} = 50$ Ω and a frequency of 50 GHz. This indicates that the directly extracted pad capacitance discussed here is accurate enough.

For short structures, the elements of concern are the finger inductance and resistance, which can be extracted from the Z-parameters after de-embedding from the "open" structure. The corresponding Z-parameters are denoted by the suffix "S-O" in the formulas below:

$$\begin{aligned}
Y_{S-O} &= Y_{Short} - Y_{Open} \\
L_{F1} &= \frac{Im(Z_{S-O,11} - Z_{S-O,12})}{\omega} \\
R_{F1} &= Re(Z_{S-O,11} - Z_{S-O,12}) \\
L_{F2} &= \frac{Im(Z_{S-O,22} - Z_{S-O,21})}{\omega} \\
R_{F2} &= Re(Z_{S-O,22} - Z_{S-O,21}) \\
L_G &= \frac{Im(Z_{S-O,12})}{\omega} = \frac{Im(Z_{S-O,21})}{\omega} \\
R_G &= Re(Z_{S-O,11} - Z_{S-O,12})
\end{aligned} \tag{10.16}$$

The values extracted from the short structure are used to find the equivalent inductance of the fingers is about 45 pH and the resistance about 0.5 Ω. The path from finger to ground was found to be equivalent to an inductance of 3 pH with a resistance below 0.1 Ω.

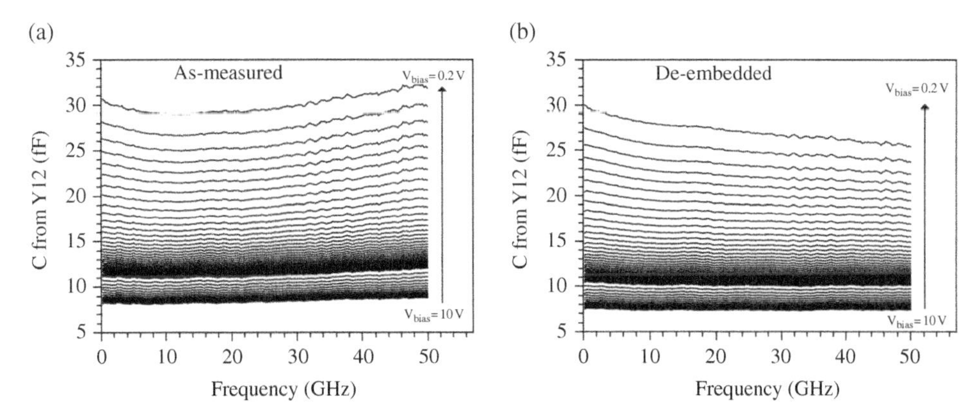

Figure 10.31 Extracted $C_{Y\,12}$ from both as-measured and de-embedded *Y*-parameters.

10.4.2.2 Step 2. Junction Capacitance

The junction capacitance can be extracted from the imaginary part of Y_{ij} in the low-frequency range, as shown by Eq. (10.17) [50], given the fact that junction capacitance dominates the impedance at this frequency range while series resistance and inductance are negligible. This is justified by simply calculating the impedance at lower frequency for a capacitor, resistor, and inductor with values representative of a diode.

$$C_{Yij} = \frac{-Im(Y_{ij})}{\omega} = C_{j0}\left(1 - \frac{V}{V_{bi}}\right)^{-\frac{1}{2}} + C_{pp} \tag{10.17}$$

The capacitance extracted from Y_{12} (denoted as $C_{Y\,12}$) of the 4 μm diameter diode studied here is shown in Figure 10.31, where the results after de-embedding are also given. The extracted $C_{Y\,12}$ is not exactly constant over the entire frequency range measured, due to the greater impact of inductance (as measured) versus frequency or series resistance (de-embedded). In any case, the resulting values of $C_{Y\,12}$ at lower frequency remain without considerable change, which validates this method. The $C_{Y\,12}$ at 250 MHz was used for further C–V fitting.

The resulting $C_{Y\,12}$ vs. V_{bias} is shown in Figure 10.32. It is worth noticing from the diode EC that $C_{Y\,12}$ is the combination of C_j and C_{PP}, and both have to be derived by fitting the curve using Eq. (10.17) [50]. The fitted results were listed in the inset of the figure. The junction capacitance was obtained simply by subtracting the (C_{PP}) from $C_{Y\,12}$. The fitted junction capacitance at 0 V (C_{j_0}) and built-in voltage (V_{bi}), together with the grading factor of 0.5 for constant doping profile, describe the C–V characteristics of the diode understudy in the SPICE model.

10.4.2.3 Step 3. Optimization

The analysis of DC IV characteristics and high-frequency small signal S-parameters allows the complete derivation of the EC elements. The deviation of the fitted S-parameters from the measured ones is described by the error function of Eq. (10.18),

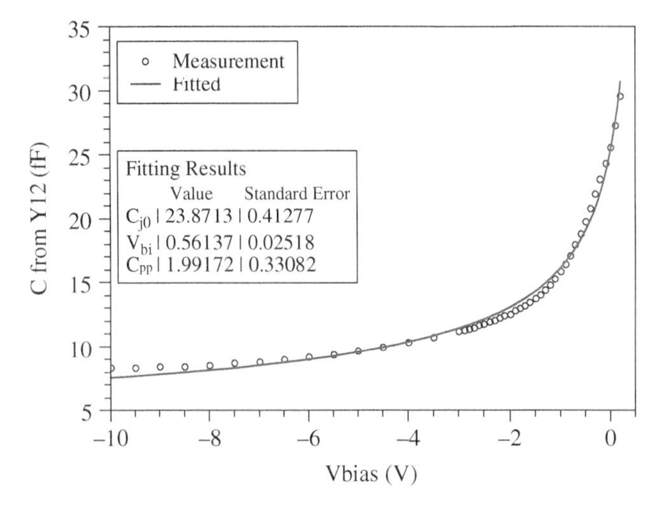

Figure 10.32 Extracted $C_{Y\,12}$ versus bias voltage.

$$Error_{Sij} = MaG\left(S_{ij}^{simu} - S_{ij}^{meas}\right) \tag{10.18}$$

This error function represents physically the distance between two S-parameter points on the polar plane. By employing this error function, one can perform a fine tuning of the element values to further decrease the error. It is worth mentioning that R_s was assumed to be a fixed value in the above analysis, which physically relates to the resistance of the N^+ layer and contact resistance. The resulted errors indicate that the above EC fits the measurements very well. However, the undepleted N^- layer also plays a role in the derivations and results in a bias-dependent resistance [51]. Such a feature can be fitted by tuning R_s at each bias voltage toward smaller errors.

10.4.2.4 Summary

The modeling flow used in this work is summarized in Figure 10.33 and was implemented with the help of in Keysight integrated circuit characterization and analysis program (ICCAP). Once again, it is worth pointing out that the nature of the intrinsic parameters requires an optimization-based modeling method using a circuit simulator, where all the intrinsic elements are varied together to obtain a minimum error from measured data. This impacts the modeling procedure in two ways.

(i) Taking the bias dependence of C_j into consideration, a standard diode SPICE model would be beneficial for performing optimization in ICCAP. Without it, the use of a lump capacitor for representing C_j will require lengthy optimization, depending on the total number of bias points. (ii) Due to the GaN material, R_s has greater impact on the port characteristics than using more traditional materials. R_s is set to a fixed value for the first optimization step, while further tuning of R_s at each bias point is necessary to permit good matching between simulated results and measured data.

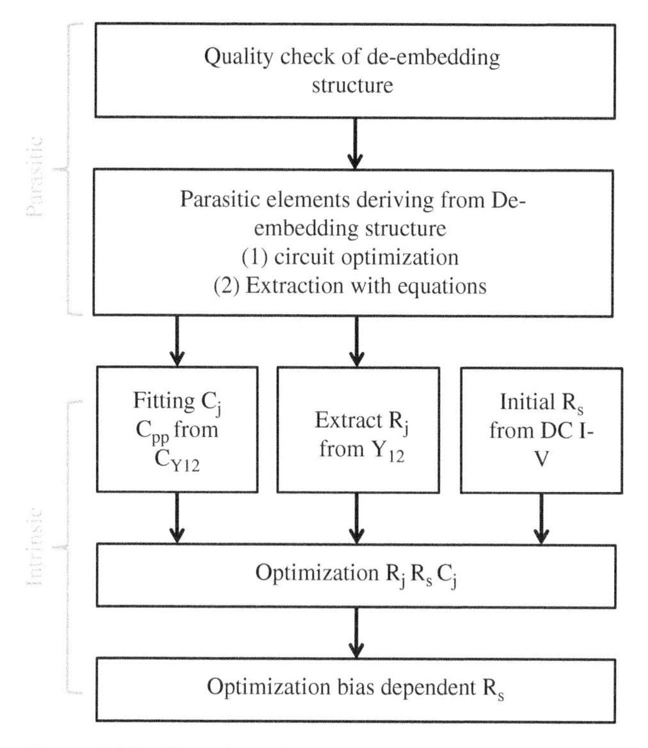

Figure 10.33 Flow chart of EC modeling based on small-signal S-parameters.

Table 10.7 EC parameters of 4 μm diode.

	Parasitics		
R_{P_1}	5 Ω	R_{P_2}	5 Ω
C_{P_1}	10.74 fF	C_{P_2}	10.57 fF
CPP	2 fF		
R_{F_1}	0.5 Ω	R_{F_2}	0.5 Ω
L_{F_1}	45 pH	L_{F_2}	45 pH
Intrinsic			
R_j	10 kΩ	C_j	5.5∼27.9 fF
R_s	30∼50 Ω		

The realized modeling procedure proved to be highly efficient and accurate.

The derived EC elements of this 4 μm diode are summarized in Table 10.7. Diodes with other anode diameters have the same pad dimensions but differ in the dimensions of air-bridge width. Thus the pad parasitics remain the same as the diode analyzed here, while the finger parasitics change slightly.

10.4.3 Results

The main objective of modeling is to find the values of intrinsic elements, which represent the core part of the Schottky diode. This objective was reached by the described modeling method, which allowed to evaluate the bias and diameter dependence of the C_j and R_S as shown in Figure 10.34. The evaluated C_j and R_s for different diameters provided a good reference for circuit design.

The diode cutoff frequency (f_c), defined in Eq. (10.5) was calculated from C_j to R_s and is shown in Figure 10.35. Higher operation frequency is shown to be possible for smaller anode diameter diodes. For example, a cutoff frequency of 1.2 THz is expected from 2.5 μ m size anodes, while 400 GHz are expected from 7.5 μm size anodes.

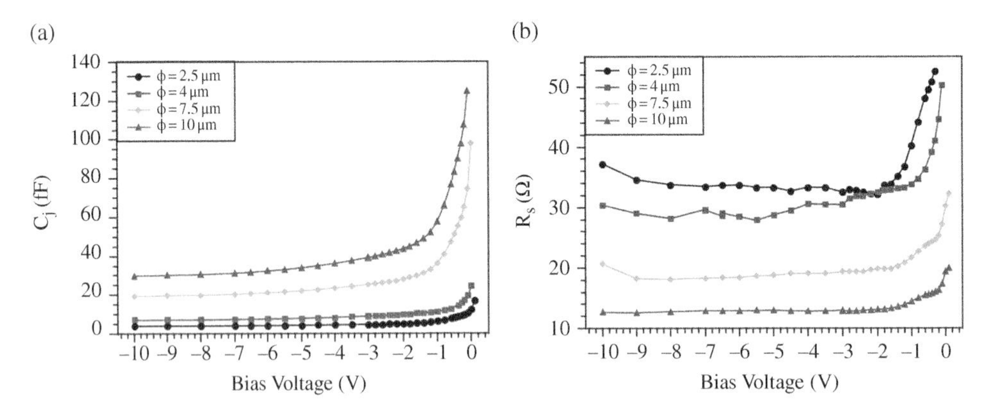

Figure 10.34 Extracted intrinsic parameters of diodes diameter from 2.5 to 10 μm.

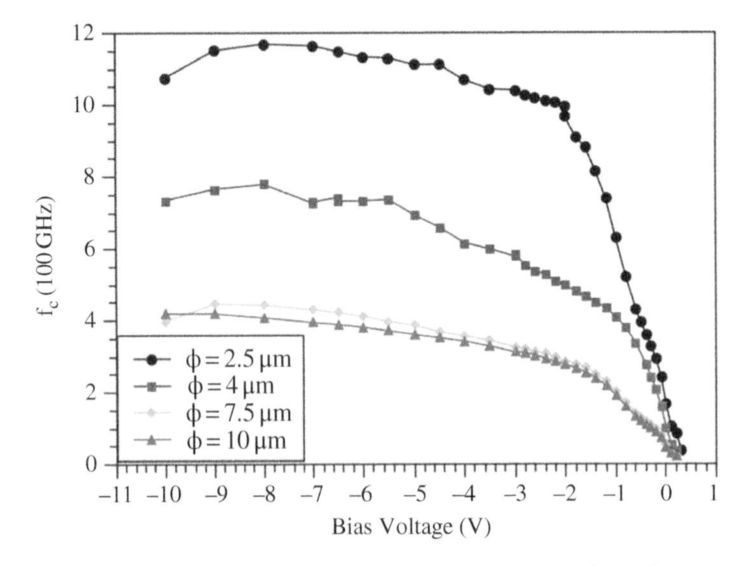

Figure 10.35 Calculated cutoff frequency from extracted C_j and R_S.

The diode cutoff frequency is defined by the inverse of the RC time-constant, where R and C correspond to the diode's R_s and C_j. This implies an intrinsic physical limitation on diode operation frequency as determined by C_j and R_s. f_c is correlated to harmonic conversion efficiency as shown in Figure 10.3. f_c becomes higher for either a smaller C_j or R_s. R_s is in practice more important since a smaller C_j is corresponding to lower power handling capability. The obtained R_s in this work is high compared with that of GaAs-based diodes, leaving therefore room for improvement.

10.4.4 Conclusion

The characteristics of fabricated GaN Schottky diodes were presented. High-frequency small-signal measurements were performed for diodes with variable diameter and bias voltage values. The obtained S-parameters were analyzed by EC modeling. This provided information on the Schottky junction and the way it is described by intrinsic elements of the EC model. The difficulties arising from periphery effects, dislocation assisted reverse current, as well as the way intrinsic elements interact with each other were explained. A modeling procedure was established and explained in detail using a 4 μm device as an example. The obtained intrinsic elements were shown for different diode diameters and the diode cutoff frequency was derived.

10.5 Large-signal On-wafer Characterization

The complexity of multiplier designs arises from the nonlinear nature of the device/circuit, which prevents one from an easy, direct evaluation of the multiplication characteristics. Prediction of the multiplication performance is presented here using a Large-Signal Network Analyzer (LSNA) as well as harmonic load-pull and an on-wafer measurement approach instead of a waveguide-based approach.

10.5.1 Characterization Approach

Methods for nonlinear measurements [52] include the use of a LSNA and nonlinear vector network analyzer (NVNA). These two methods are differentiated by the way of test signal down-conversion, LSNA being sampler-based, while NVNA being mixer-based. In order to capture accurately time-domain waveforms, the calibration of signal power and phase is necessary.

The approach described here utilizes a Maury MT4463A, whose block schematic is shown in Figure 10.36. The incident and scattered voltage waves are firstly sensed by couplers at both device under test (DUT) ports. After attenuation, the sensed signals are sent to an RF–IF converter, which converts all the spectral components coherently to a lower frequency "copy." The resulting "copied" low-frequency signals are further digitized by analog-to-digital converters (ADCs) and collected by a control PC for data processing. The RF–IF converters define the input bandwidth of the LSNA, which is 600 MHz to 50 GHz for MT4463A. A detailed introduction of LSNA operation principle and calibration can be found in [53].

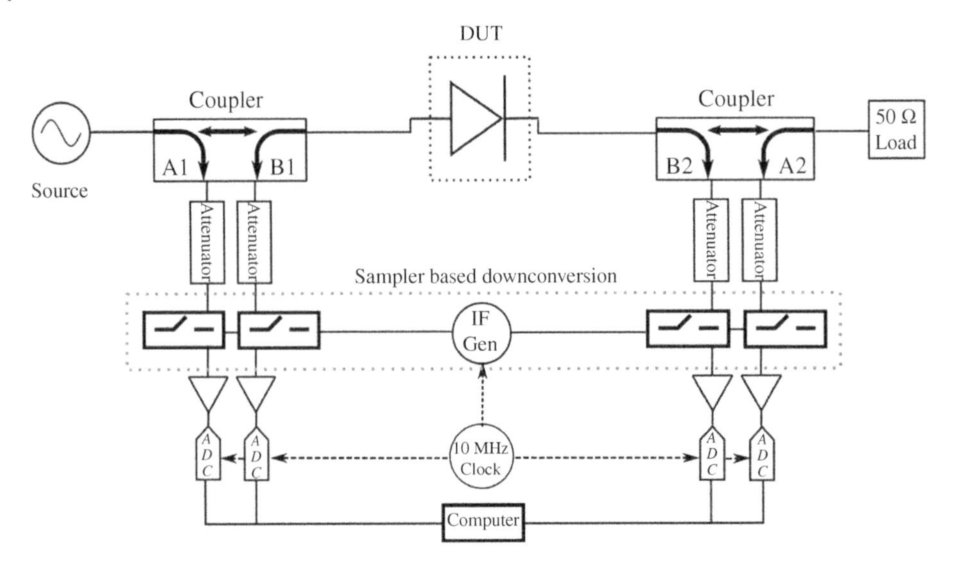

Figure 10.36 Simplified block schematic of sampler-based LSNA.

The results provided by LSNA can be in the format of voltage/current (V/I), incident/reflected voltage waves (A/B), time domain, frequency domain, or envelop domain. Both A/B and V/I formats were used in the measurements reported here.

Load-pull was used for characterization and evaluating the best load impedance for maximum output power or maximum efficiency. As conventional passive tuners provide a limited reflection factor (Γ), usually not exceeding 0.8 [54], active load-pull can be employed, where a signal is injected into the DUT to analogue the reflected signal from load. This turns out to be necessary for diodes since their S_{22} are often larger than 0.9.

LSNA has been applied to various types of devices and applications. Characterization may be performed under the same conditions (Bias voltage, incident power, input/output impedance, etc.) as those applied to component use in a circuit or subsystem. Measurement results can, therefore, represent the component features under real operation conditions. By way of an example, efficiency enhancement of a power amplifier was demonstrated by time-domain waveform optimization [55]. By carefully tuning the load impedance at harmonics, the voltage and current waveforms may be monitored to obtain minimum overlap of the voltage and current. Power added efficiency (PAE) as high as 84% at 1.8 GHz was achieved in this way. LSNA is also used as diagnostic tool for devices and circuits. The nonideal characteristics of a mixer were identified by LSNA measurements with a special odd multi-sine IF signal [56]. Models accounting for nonlinear effects benefit enormously from LSNA techniques where a circuit can be optimized by considering the waveforms provided by LSNA [57].

10.5.2 Large Signal Measurements of GaN Schottky Diodes

10.5.2.1 LSNA With 50 Ω Load

GaN Schottky diodes were characterized first with the default load impedance of 50 Ω. The measurement configuration is shown in Figure 10.37. The time-domain voltage and current

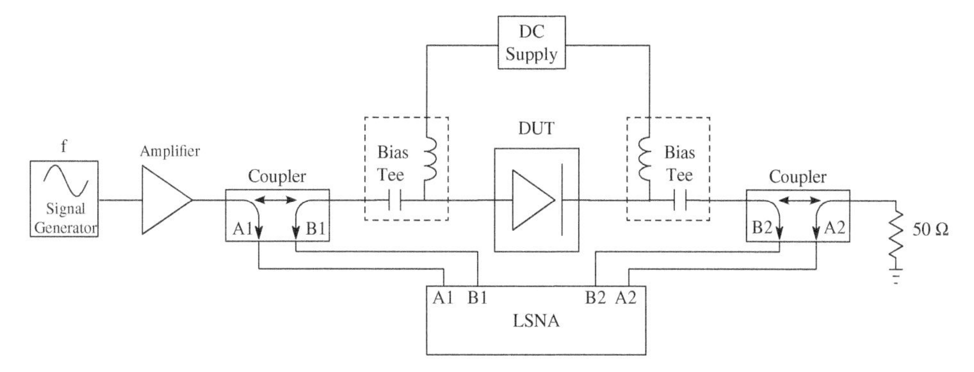

Figure 10.37 Configuration of measurement with 50 Ω.

waveforms at both anode and cathode are monitored by LSNA. Either power or bias voltage can be swept in order to evaluate the diode nonlinear properties. The high reactance impedance presented by the diode led in considerable reflection of the incident power and limited the RF voltage swing across the diode terminals to a rather small value, which was below the required one for nonlinear operation even for the highest power available from the setup. Thus, bias voltage sweeping was used combined with a small amplitude RF excitation to drive the diode into nonlinear operation.

The results obtained by LSNA are in phasor form for either V/I or A/B, which contain naturally all harmonics from DC to the highest specification frequency of the setup. Given the maximum frequency of 50 GHz for LSNA, the fundamental frequency was chosen to be 14 GHz. Thus, the output contained four frequency components, DC, 14, 28, and 42 GHz, by which the time-domain waveforms can be accurately represented.

10.5.2.2 Time Domain Waveforms

Figure 10.38a shows typical time-domain waveforms obtained from diode characterization. The current follows the RF voltage signal periodically thus forming the circle locus curves

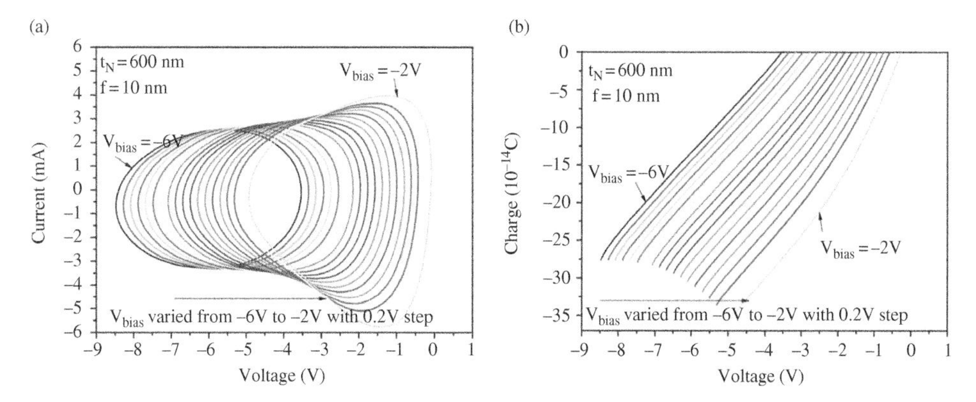

Figure 10.38 (a) Typical time-domain waveform in I–V plot; (b) Q–V relation obtained by integration of current over time. The device is 10 μm on 600 nm N^- GaN layer, measured at 15 dBm source power and bias sweeping.

on the IV plane. The space between each I–V circle represents the bias sweep step. Most of the spaces are uniform and equal to the value used here, i.e. 0.2 V, except when the RF voltage swing reaches 0 V, which indicates a pronounced nonlinearity.

Comparing the I–V locus plots obtained by bias sweeping, the following could be concluded regarding the diode operation.

1) Each I–V locus curve is the instant current and voltage amplitude presented at the diode in one signal cycle. Its area stands for the power exchange between the diode and source. The more irregular is its shape the greater is its nonlinearity.
2) *Bias dependence*: The instantaneous current is greater due to a larger displacement current required for charging/discharging the junction capacitance under bias conditions corresponding to less negative values. Correspondingly, the power exchange is higher leading to a more deformed I–V locus curves, i.e. nonlinearity enhancement.
3) *Diameter and frequency dependence*: The same principle applies when the diode diameter increases or the working frequency increases, i.e. the displacement current and thus the power will be higher.

10.5.2.3 Instant C–V Under Large-signal Driven Conditions

Figure 10.38b shows how the total amount of charge increases over a half cycle, i.e. by a monotonous increase from the lowest voltage peak to highest voltage peak. The instantaneous capacitance under large-signal condition can be obtained by the differential of the charge over the RF voltage swing. Since the charge is instantaneously following the RF voltage swing, the capacitance calculated in this way corresponds better to the conditions encountered by the diode under real operation than under quasi-static and small-signal capacitance measurement conditions. Figure 10.39 shows the instantaneous capacitance results versus instantaneous voltage. The observed trend resembles that extracted from

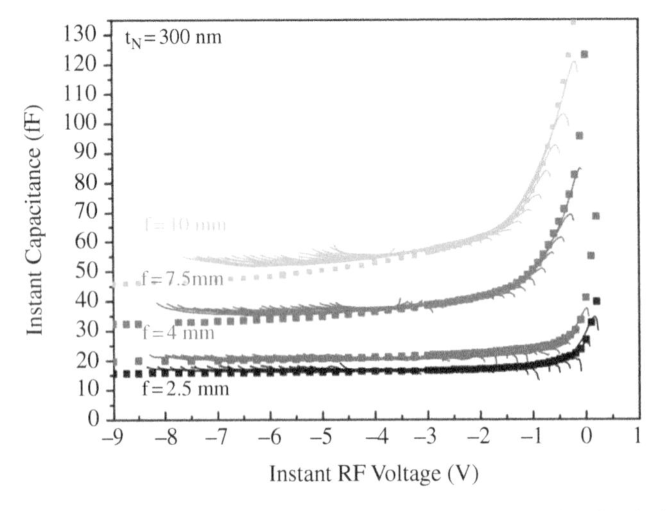

Figure 10.39 Derived instant Capacitance–Voltage relationship (solid lines), in comparison with extracted Capacitance from Small Signal condition (scatter dots).

small-signal measurements, but the minimum capacitance is larger than the small-signal one, while the maximum capacitance is found to be smaller. Thus, the C_{max}/C_{min} ratio evaluated by LSNA is smaller. This needs to be accounted for in large-signal modeling.

A source signal of 18 and 23 dBm applied to devices with 300 and 600 nm thick N^- layer, respectively. The instantaneous minimum capacitance is smaller for the 600 nm diode. However, due to the overall larger charge stored in the 600 nm diode, the overall capacitance is larger when the voltage is not negative enough. The larger maximum and smaller minimum capacitance are, therefore, expected to offer higher power handling capability.

10.5.2.4 Power Handling Characteristics

Measurements for the evaluation of power handling capability were performed with fixed bias voltage and swept-source power. Then a source power level was chosen under certain bias voltage to allow a voltage swing between 0 V and double the bias voltage. A comparison was made based on the same large-signal voltage waveform of each diode. Diodes with diameters ranging from 2.5 to 15 μm were characterized with a bias of −4 and −6 V and the results are shown in Figure 10.40.

The larger capacitance of larger diameter diodes leads in higher handling power and a value of about 100 mW was found for a 15 μm diode biased at −6 V. For the −4 V biased condition, this value is about 40 mW. Despite the thickness difference, the absorbed power is quite similar for diodes with the same diameter and bias voltage. However, the reactive power for the diodes on 600 nm N^- layer wafer is lower than the ones on 300 nm N^- layer wafer. This is interpreted by a larger resistance arising from the thicker N^- layer, as already seen in the simulation results.

10.5.3 LSNA With Harmonic Load-pull

Due to the impedance mismatch between the diode output and the 50 Ω load considered so far, the harmonic power at the load is weak. In order to explore the nonlinearity of the diodes by considering the harmonic output of all orders, harmonic load-pull was employed to match the diode's output and allow maximum power of second harmonic transfer to load. The measurement setup had the 50 Ω load been replaced by an active load realized with an

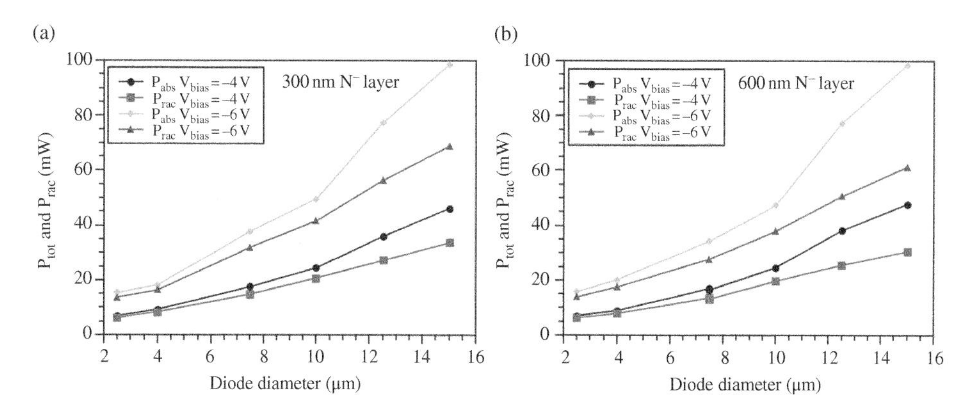

Figure 10.40 Power handling capability for (a) 300 nm thick N^- layer (b) 600 nm thick N^- layer.

additional signal source operating at the second harmonic. The fundamental frequency was chosen to be 14 GHz. Thus the maximum frequency of 50 GHz for LSNA allows to obtain harmonics up to third order.

Time-domain waveforms were obtained and unlike the circular shape I–V locus observed in the 50 Ω load case, the cathode I–V locus was crosslinked, which indicates enhanced harmonic amplitude. This was also evidenced by a remarkably increased amplitude of the second harmonic Enhancement of harmonic generation has been found to arise from better matching of the load impedance at the second harmonic. Considering the high loss and low reactance power offered by 2.5 μm diodes, larger diodes are expected to provide higher power as well as second harmonic power.

Overall, LSNA measurements provided basic information on the large-signal performance of the GaN Schottky diodes.

10.5.4 Conclusion

Diode characterization under large-signal conditions was performed using LSNA and on-wafer tests. This novel characterization method allowed access of the time-domain large-signal waveforms of GaN Schottky diodes. The characterization demonstrated allowed parameters such as instant capacitance, total power, reactance power to be derived under large-signal conditions and opened the way to the understanding of the large-signal diode operation. The described LSNA method can be standardized to provide rapid evaluation of diodes. This is important for device development as the case of Schottky diodes based multipliers, where the performance cannot be easily predicted until a circuit is realized.

10.6 GaN Diode Implementation for Signal Generation

The use of GaN Schottky diode in frequency multipliers for signal generation will be discussed next. Circuit simulation requires the development of a large-signal EC model for the diodes was developed. A harmonic load-pull approach can assist in circuit optimization for maximum output power.

10.6.1 Large-signal Modeling of GaN Schottky Diodes

Parameter extraction using small signal S-parameter measurements and optimization/verification combined with large-signal measurements is the recommended approach for diode modeling and multiplier circuit implementation. The large-signal model replaces the bias-dependent components of the small-signal model by a mathematic description, which allows simulation under nonstatic operation conditions. In the case of diodes, the bias-dependent components correspond to R_j, R_s, and C_j. The developed large-signal modeling and optimization procedure resemble a "LEGO®" piecewise method as described in [58], where Symbolically Defined Devices (SDDs) in ADS (Advanced Design System) by Keysight (former Agilent) have been employed [59].

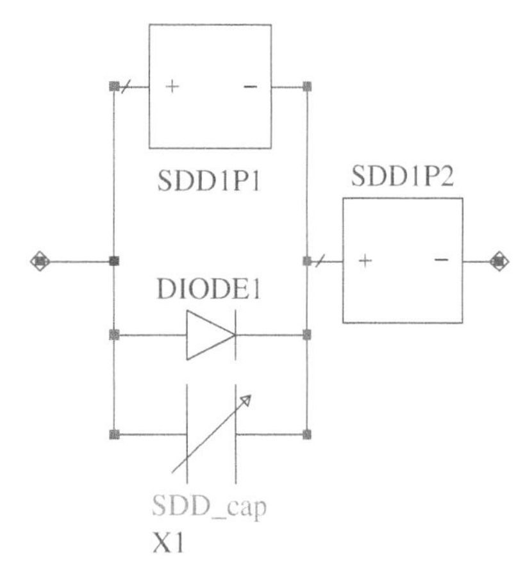

Figure 10.41 Large-signal diode model schematic for GaN Schottky diodes.

The realized model is composed of four building blocks as shown in Figure 10.41, where "DIODE1" was used to represent the forward current and breakdown voltage, "SDD1P1" stands for R_j for reverse DC current, "SDD1P2" for R_s and "SDD_cap" for the C_j.

The diode capacitance C_j is the most critical component in device and circuit modeling. As discussed earlier, the C_j derived by LSNA differs from the values derived from small signal S-parameters. C_j has, therefore, to be adjusted in order to obtain a model that matches well the measured characteristics. Figure 10.42 shows the difference in results obtained by simulation based on small-signal vector network analyzer results (S_VNA) versus those

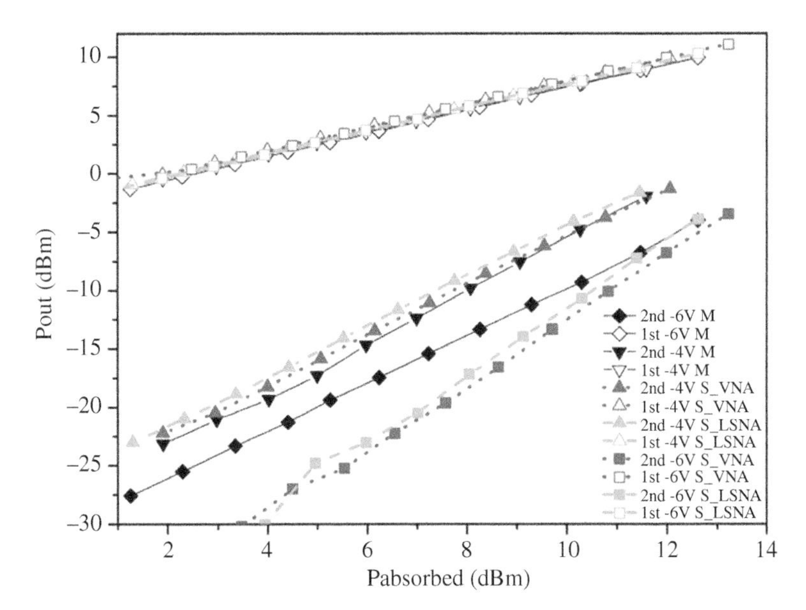

Figure 10.42 P_{out} (dBm) versus $P_{absorbed}$ (dBm). Symbols with solid lines: measured results (abbreviation as M); symbols with dotted lines: simulation results with model extracted from small-signal S-parameter (abbreviation as S_VNA); symbols with dashed lines: simulation results with model tuned toward LSNA (abbreviation as S_LSNA); Solid symbols: second harmonic; Hollow symbols: fundamental harmonic.

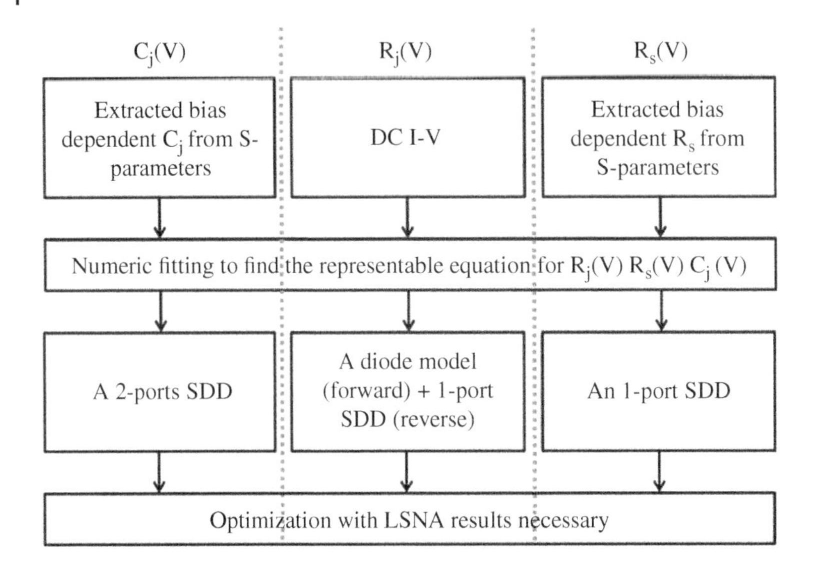

Figure 10.43 Flow chart of large-signal modeling for GaN Schottky diodes.

obtained by simulation based on large-signal network analyzer results (S_LSNA). Since the power absorbed and stored by a capacitor is directly proportional to the capacitance value, the higher absorbed power suggested by S_VNA indicates an over-estimated junction capacitance C_j. By tuning the junction capacitance C_j, a new S_LSNA model was determined and found to better match the large-signal measurement results. As shown in Figure 10.42 by the symbols with dashed lines, the results from this model match the measurement results for both the absorbed and output power under large absorbed power conditions. For low absorbed power levels, such as for example at −6 V bias, the simulation results from the S_LSNA model lead to underestimated characteristics of output power.

The modeling procedure using the large-signal EC model is reviewed in Figure 10.43. As can be seen, a key step is to find the right equation describing the bias dependence of each component; this equation has then to be included in the circuit simulator. DC and Small-signal measurements provide all necessary information for the large-signal modeling. However, the model built in such a way relies very much on the bias range of small-signal measurements. In principle, the resulting large-signal model cannot be used for predicting performance outside the measured bias range. For example, the exponentially fitted R_S were valid for negative bias values, but not in the forward region, where it predicts values much higher than in practice. Overall, this procedure allowed to obtain a large-signal diode model for circuit design and simulation.

10.6.2 Frequency Doubler

The characteristics of a 94 GHz frequency doubler will be discussed next using large-signal models for a GaN 10 μm diameter diode.

The diagram in Figure 10.44 shows the configuration used for frequency doublers. The input (fundamental) and output (second harmonic) signals are separated by filters. Matching networks at both input and output are also required for delivering maximum power to the load.

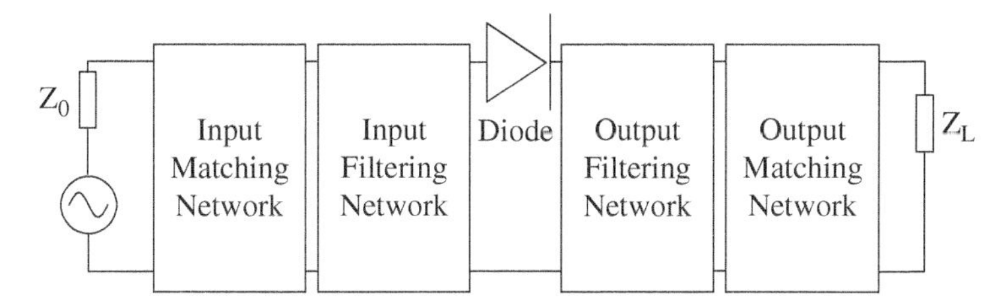

Figure 10.44 Block diagram of a single diode frequency doubler.

A schematic of the diode model is shown in Figure 10.45, with all the parasitic and intrinsic elements included and all parameters listed in separate "VAR"(= variable) components.

The filters can be realized by $\lambda/4$ stubs [27], as shown in Figure 10.47. The input short stub "TL2" had a length $\lambda/4$ at 47 GHz, which is equivalent to $\lambda/2$ at 94 GHz. It creates, therefore, reflective short for currents at 94 GHz without affecting currents at 47 GHz. The open stub "TL1" had also a length $\lambda/4$ at 47 GHz, behaving as reflective short for currents at 47 GHz. The stub acts as a $\lambda/2$ open section at 94 GHz and does not consequently perturb the output signal. The resulting S_{21} characteristics of the two stubs are shown in Figure 10.46. In the absence of stubs, the transmission characteristics resemble those of a high-pass filter that allows both the fundamental and second harmonic to travel through all parts of the circuit. The combination of the two stubs results in the solid curve that illustrates the possibility of isolating ($S_{21} = 0$) the fundamental and harmonic frequency. A study of the impact of substrate on the stub characteristics showed that an increase in substrate thickness leads in sharper transition from stop to pass frequency.

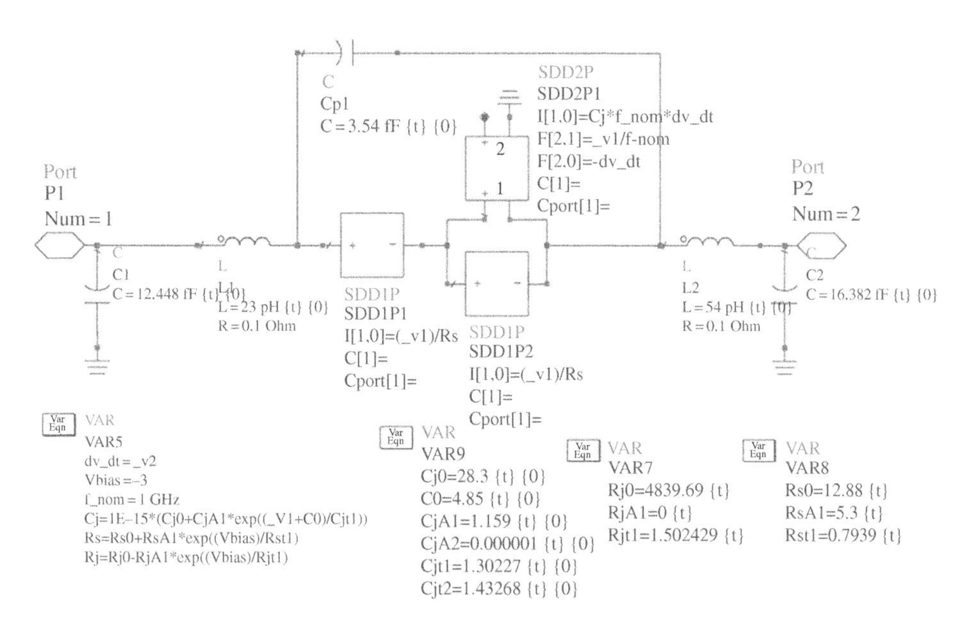

Figure 10.45 Schematic of large-signal model of the 10 μm diode.

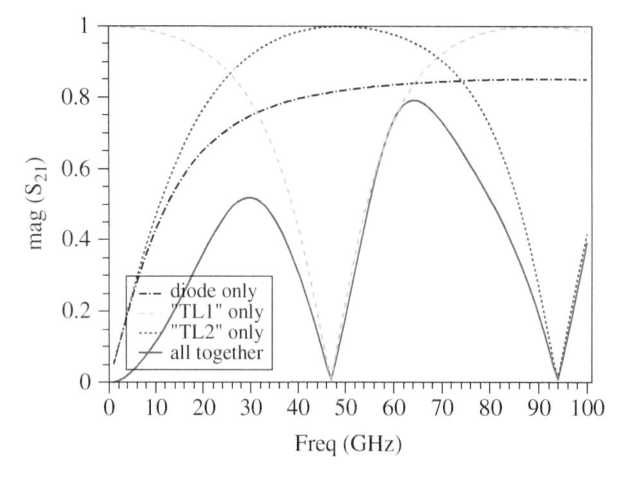

Figure 10.46 S_{21} of the diode and with the presence of stubs.

The doubler circuit can be considered as consisting of two separate ECs. One included the diode and the output filter, and the other included the input filter and diode. Input matching was established by considering the former network, while output matching was calculated based on the latter. Such a configuration allowed the load and source impedance not to interact with each other during the design of the matching network.

Harmonic balance (HB) and load-pull techniques were used to find the optimum source and load impedance providing maximum output power. Since maximum output power is sought at the second harmonic, load-pull was performed at this frequency. With both the input and output matched, the ratio of second harmonic output power over fundamental source power is defined to be the conversion efficiency. This was indexed by S_{2121} under the large-signal S-parameter definition [60], which is shown to be useful for the optimization of frequency doublers.

A schematic of the doubler simulation approach used in this study is shown in Figure 10.47. The HB simulation was set to include up to fifth order harmonic. The source and load impedance were also set to respond up to fifth harmonic, while adjustment of their values was possible at each harmonic independently. Load-pull mapping in the impedance plane allowed to find the maximum output power and corresponding load impedance. Moreover, the input impedance (reflection coefficient) at the fundamental frequency could also be obtained for this specific load impedance.

The results obtained are shown in Figure 10.48, where the diode was biased at −10 V and the input power was swept up to a value allowing a maximum operation range between −20 and 0 V. The output power increased with input power monotonously and the conversion loss indicated for the diode was found to approach saturation. This 10 μm diode biased at −10 V could absorb an input power of 22.2 dBm (166 mW). An optimum load impedance of $14.46 - j8.57$ was found from the load-pull impedance mapping contour, to allow a maximum output power of 13.05 dBm (20.2 mW). The source impedance of $11.4 + j56.79$ assured delivery of 22.2 dBm source power to the diode. The corresponding conversion efficiency was 12.2%. The time-domain waveforms were monitored to assure the diode operates well

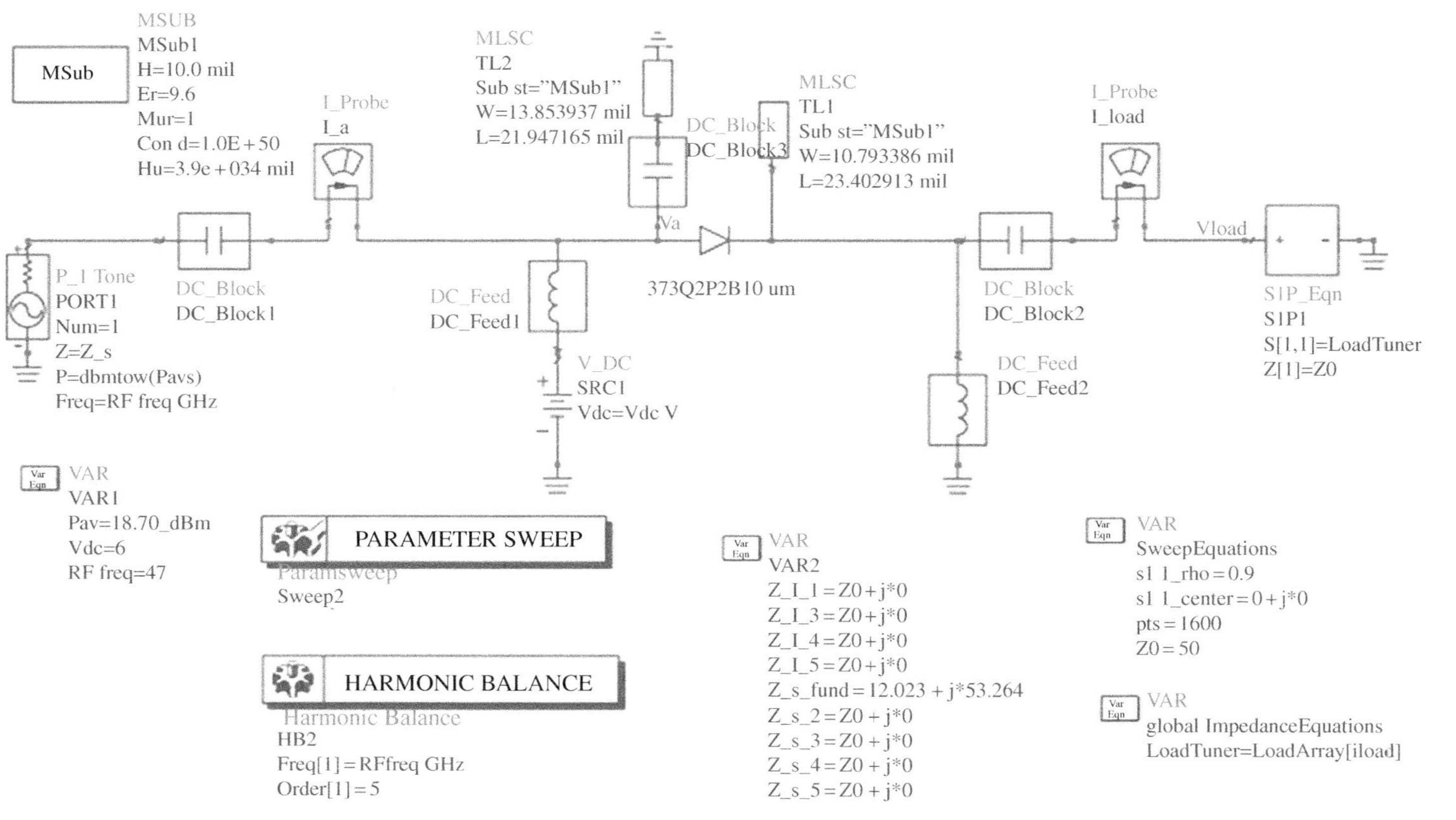

Figure 10.47 Schematic of circuit setup used for combining harmonic balance and harmonic load-pull for doubler design.

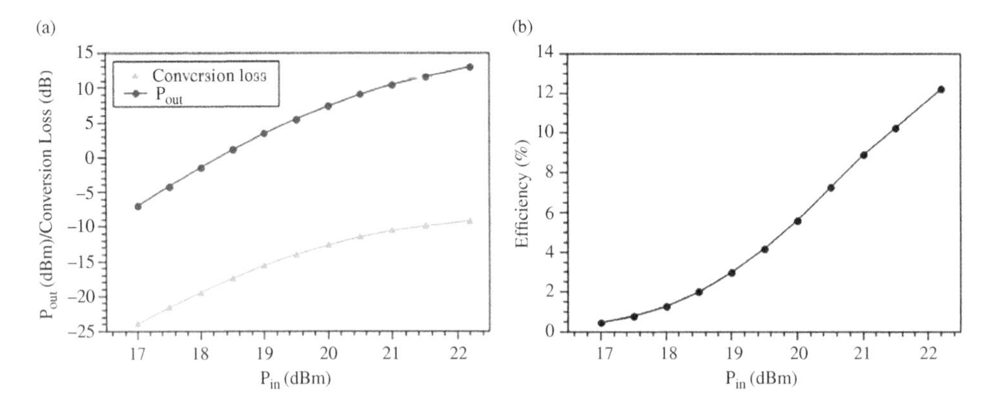

Figure 10.48 Simulated results of the 10 μm diode biased at −10 V, input power swept (a) Pout and conversion loss (b) Efficiency.

in the determined range. Both voltage and current waves at the cathode side showed frequency doubling compared with waves at the anode side.

A 16 μm GaAs diode with N^- layer thickness of 650 nm and 7×10^{16} cm^{-3} doping has been found to convert an input power of 330 mW into 65 mW output with 19.7% efficiency [27]. Taking the diameter difference into account, the 10 μm GaN diode understudy had shown better power handling capability per unit anode area than the GaAs diode. It's worth noting that an extension of the operation range into the forward region would further enhance the doubler performance.

10.7 Multiplier Considerations for Optimum Performance

This section addresses various multiplier device types, designs, and fabrication approaches that are important to consider when optimizing the performance characteristics of frequency multipliers. This includes GaN-based vertical devices on sapphire and bulk substrates, heterojunction designs, i.e. InN/GaN, transistor-based approaches, as well as, efficiency and noise considerations using Mote–Carlo techniques.

Early work on MOCVD-grown $Al_{0.4}Ga_{0.6}N$/GaN heterostructure barrier varactor (HBV) diodes showed that piezoelectric and spontaneous polarization effects lead in asymmetric electrical characteristics [61]. HBV diodes are of interest since they suppress even harmonics and allow, therefore, simplicity in the realization of multipliers for higher-order signal generation. Diodes of this type have been explored for both GaAs and InP-based heterostructure systems and led to more than 10% efficiency. The current–voltage and capacitance–voltage characteristics of the AlGaN/GaN HBVs shifted asymmetrically due to piezoelectric stress-induced and spontaneous polarization in the barriers. The C–V characteristics of HBVs are ideally symmetric at given bias point. However, this turned out not to be the case for the reported diodes and a polarization-dependent shift was observed in the position of the maximum capacitance. The maximum value of the capacitance (C_{max}) shifted occurred at −0.5 V rather than 0 V while the minimum capacitance (C_{min}) at increased negative biases was higher than at positive biases. Altogether, this led in capacitance modulation ratios C_{max}/C_{min} that differed depending on bias condition.

InN/GaN HBV multipliers have also been explored for signal generation [62]. Figure 10.49 shows the predicted harmonic amplitude characteristics for such diodes together with those of in $In_{0.53}Ga_{0.47}As/AlAs$ for comparison.

Diodes of this type are expected to operate as efficient triplers up to 1 THz. Their good performance is due to the following factors: (i) InN optical phonon emission rate of $2.5 \times 10^{13}\ s^{-1}$ and thus higher by about four times than for InGaAs or GaAs. The presence of strong electronphonon coupling results in electron heating in the InN modulation layers that occurs at much higher frequencies; (ii) the peak steady-state electron drift velocity in InN is 5×10^{7} cm/s and thus by a factor of about two higher than that of GaAs and in InGaAs; (iii) The InN/GaN conduction band offset is large and blocks efficiently the conduction current over the barriers. As a result, the inertia of electron redistribution during high-frequency operation is reduced in InN cladding layers, while the electron response and therefore high-frequency performance benefits from the faster transient effects in InN which presents higher velocity overshoot than GaAs or InGaAs.

Other important features shown by the results of Figure 10.49 include a near-linear dependence on excitation frequency beyond 200 GHz for the third harmonic amplitude in InN/GaN versus a decreased with frequency amplitude for InGaAs/AlAs and a much higher third to first harmonic ration for InN/GaN.

HEMTs can also be used for frequency multiplication and present an advantage compared with passive approaches based on diodes since they can lead to gain rather than losses. The main limitation in transistor-based multipliers is the cutoff frequency of operation of the employed transistor. Most reports on the use of wide bandgap semiconductors for multipliers are at relatively low frequencies. A wideband high power active doubler has for example demonstrated at $f_0 = 3.33$ GHz with a maximum $2f_0$ output power of 30.2 dBm (~1 W), bandwidth of 13.2%, a maximum conversion gain of 0.2 dB, and a maximum drain efficiency of 4.8% [63]. The use of III-nitride-based HEMTs allows one to achieve high power and deliver it to arrays consisting of many subsystems. Although the employed

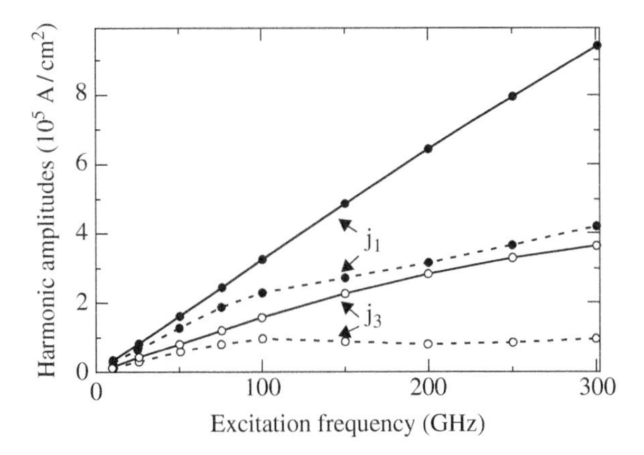

Figure 10.49 First (j_1) and third (j_3) harmonic amplitudes of the current oscillations versus excitation frequency in single-barrier varactor diodes (solid curves: InN/GaN; dashed curves: $In_{0.53}Ga_{0.47}As/AlAs$). An input sinusoidal excitation of 5 V and a DC bias voltage of 0.6 V are assumed for operation.

HEMT delivered an adequate amount of power (6 W), its frequency of operation was limited to 6 GHz. GaN HEMTs with higher frequency of operation are, therefore, necessary for use at THz frequencies. This is nowadays feasible as demonstrated by InAlGaN/AlN/GaN devices with a 50-nm-long gate which exhibited a maximum current density of 2.22 A/mm, a high current-gain cutoff frequency (f_T) of 182 GHz and a high power-gain cutoff frequency (f_{max}) of 402 GHz simultaneously [64]. Devices of this type open, therefore, the possibility of III-N-based HEMT amplifiers at THz.

The high-frequency potential of HEMTs has been explored for multipliers [65] and oscillators [66]. Among the resistive and capacitive nonlinearities of HEMTs, the transconductance (g_m) and the output conductance (g_d) are the most important ones, causing drain current clipping and higher harmonics generation. The capacitances (C_{gs} and C_{gd}) can be considered as parasitic elements which degrade the device gain and isolation at high frequencies. Large-signal analysis showed that the efficiency of generating higher harmonics is better with g_m than g_d. The bias of operation is also important and operation in saturation and pinch-off is attractive since it presents a low C_{gs}, small power consumption and is free of forward conduction current at the gate. It is also preferred for high-frequency operation due to the lower leakage current through C_{gs} since leakage current through this capacitance degrade the device gain and conversion gain.

In terms of direct signal generation using III-N-based HEMTs, large-signal analysis [66] indicates that optimum output power is obtained when the load impedance is of the order of 1/2 to 1/4 of the small signal negative resistance of the oscillator circuit at frequencies not too close to maximum frequency of oscillation (f_{max}). Generation of reasonable power levels appears to be feasible up to a frequency of 2/3 of f_{max} which would be more than 250 GHz for the devices of [64].

The relatively large bandwidth demonstrated in [63] compared with most commonly achieved bandwidths of ~5% is due to the special design of the network placed at the transistor output. Increased harmonic power output can be obtained by reflecting the FO signal at the HEMT output back to the drain at the correct phase angle to allow proper mixing by the nonlinear output conductance. By selecting an appropriate output reflector network, i.e. a 3-pole maximally flat 33.3% bandpass filter (BPF), centered at $2f_0$ [63], it was possible to ensure an almost unity reflection coefficient over a large bandwidth.

A cutoff frequency of 900 GHz was demonstrated [67] using a 4 μm diameter GaN-based air-bridge varactor diode with low-leakage current and breakdown voltage of 21 V. A power of 7.2 dBm with an efficiency of 16.6% was estimated for 47 to 94 GHz doubling with it. To reduce the parasitics, diodes of similar type were fabricated using dry etching to produce an isolation slot by reaching the sapphire substrate [68]. Air-bridge technology was also used in this case, but the substrate was thinned down to 50 μm and discrete diodes were obtained by laser wafer scribing. The diode total capacitance was 12 fF and the series resistance was 1.6 Ω and their cutoff frequency was 796 GHz.

The performance of GaN Schottky diodes when used for frequency multiplication was studied using a Monte Carlo approach coupled to a HB technique [69]. Both the electrical performances, i.e. conversion efficiency and noise, were calculated and a comparison was made between GaN and GaAs diodes; the spectral density of the current fluctuations can be found directly from a multi-particle history simulated during a sufficiently long time interval.

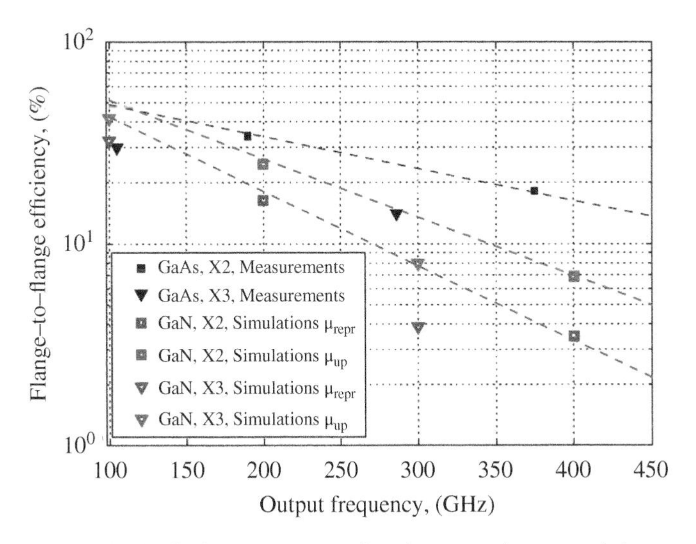

Figure 10.50 Performance comparison between the state-of-the-art measured efficiencies of GaAs multipliers and the efficiencies of the GaN multipliers simulated by Monte Carlo [69]. *Source*: Pardo and Grajal [69]; © 2015 IOP Publishing Ltd.

The results of Figure 10.50 show that the GaN multiplier efficiency is in general lower than GaAs-based components. Moreover, as the input frequency is increased, a significant degradation is observed in the efficiency of the GaN multipliers. These characteristics are due to the higher series resistance of GaN diodes related to the lower mobility of GaN compared with GaAs. The maximum conversion efficiency is expected to be ∼30–40% for a 100 GHz GaN tripler and ∼16–25% for a 200 GHz GaN doubler. Although lower than for GaAs, they are still attractive for millimeter-wave applications. Considering the higher power handling capability per unit anode area of GaN diodes, it appears, therefore, that GaN diodes may be more attractive for millimeter-wave rather than THz source signal multiplication, since efficiencies are there still comparable to GaAs and a major advantage exists in terms of power handling and thus small diode size; the latter leads to a multiplier design simplification at the first stages of a multiplier chain and a smaller overall number of diodes.

The current noise spectral density SI(f) of one diode used in a GaN and GaAs 200 GHz doubler are shown in Figure 10.51 under conditions of input power necessary for maximum diode efficiency [69]. The low-frequency region of the noise spectra is governed by shot and thermal noise, whose contribution to the noise increases at higher input power levels. At higher frequencies, one distinguishes two peaks: (i) the returning carrier (at ∼200 GHz for GaN and ∼2 THz for GaAs) and (ii) the hybrid plasma (at ∼2.5 THz for GaN and ∼4.6 THz for GaAs) peaks. For the input power levels used, the above noise types are the key contributors to the noise content of the 200 GHz output signal. Although intrinsic noise is not an important consideration for multipliers at millimeter-wave frequencies, one should note that the for equivalent power performance, one will require the use of a larger number of GaAs than GaN diodes. Noise risks, therefore, to be higher for GaAs than GaN multipliers.

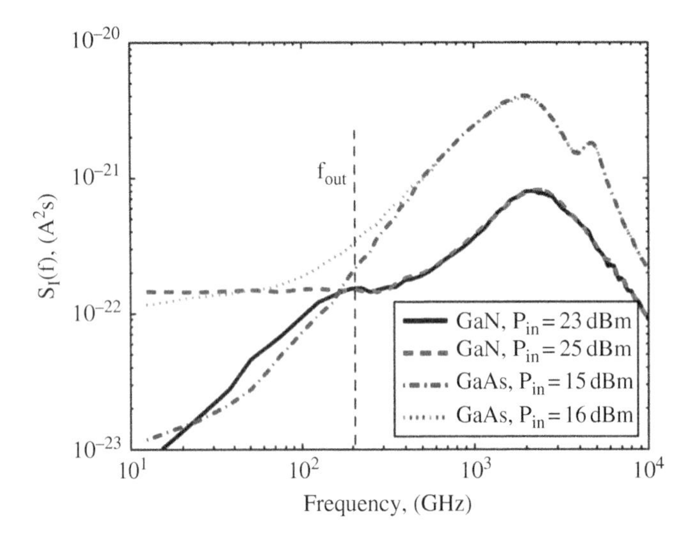

Figure 10.51 Current noise spectral density of single diodes in GaN- and GaAs-based doublers operating at 200 GHz. The input powers used are selected for maximum efficiency [69]. *Source*: Pardo and Grajal [69]; © 2015 IOP Publishing Ltd.

The performance reported for vertical devices with Schottky contacts directly deposited on GaN suggests that an improvement can be expected using lower dislocation density material. Using low defect density material with a dislocation density of, for example 10^4 cm^{-2}, which is by three to four orders of magnitude lower than traditionally used GaN material grown on sapphire should, therefore, present an advantage [70]. Of major importance in terms of expected improvement is the reduction of leakage current which limits the power handling capability. Scanning current–voltage microscopy performed on GaN samples, having similar total dislocation density (N_{DD}) but different ratio of screw dislocation, showed that the density of reverse leakage spots correlates well with the pure screw dislocation density [71]. Schottky contacts on low dislocation density appear, therefore, to be of interest for improving the diode characteristics. The dislocation density of homo-epitaxial GaN films can be less than 10^6 cm^{-2}, while the value for epitaxial films grown on other substrates is normally in the order of 10^8 to 10^9 cm^{-2} [72]. Bulk GaN substrates of 250 μm thickness were used for the diodes described below and had a doping of 1.3×10^{19} cm^{-3}. They were produced by the ammonothermal method, which involves treatment in supercritical ammonia in one zone and crystallization in another. The epitaxial structure was grown by MOCVD and consisted of a 0.2 μm thick N^{++} (1.3×10^{19} cm^{-3}) "buffer" layer grown first and followed by a 0.35 μm thick n-layer of 1×10^{17} cm^{-3} doping.

The current of the Schottky diode was theoretically estimated [48], by considering tunneling field emission through dislocations being the dominant mechanism responsible for leakage current. The current (I) was estimated by

$$J = \frac{A^*T}{k}\int_{-q(V_r-V_b)}^{q\varphi_D} \frac{1}{1+\exp\left(\frac{E}{kT}\right)} \times exp\left[-\frac{4}{3}\frac{\sqrt{2m^*}}{\hbar}\frac{(q\varphi_D-E)^{\frac{3}{2}}}{q\xi}\right] dE \qquad (10.19)$$

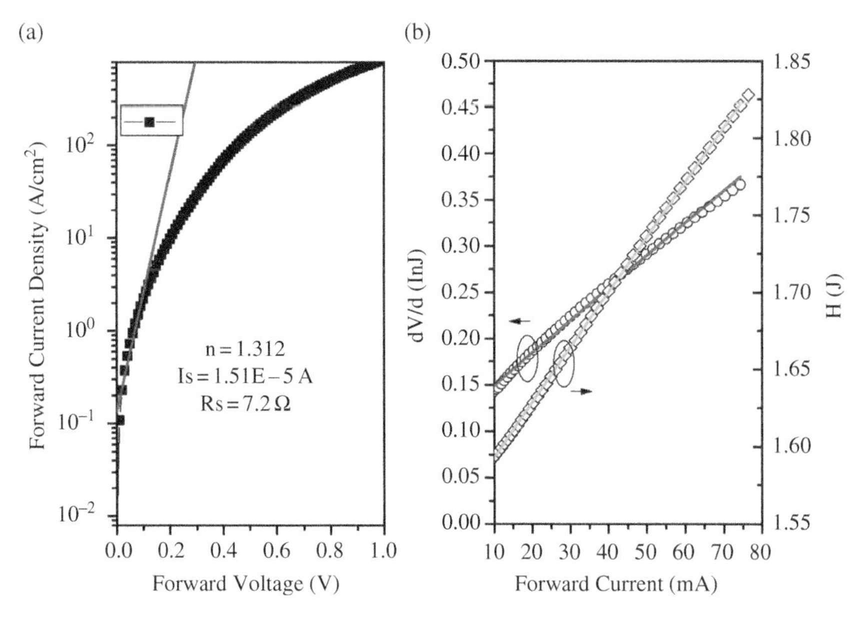

Figure 10.52 (a) Forward I–V and fitting with TE in the low voltage range; (b) fitting of forward region by Cheung's method.

The effective area of the dislocation AD is given by

$$A_D = A \times N_{DD} \times \pi \times {\lambda_D}^2 \tag{10.20}$$

where λ_D is the effective radius of dislocation. The voltage dependence of the reverse leakage current can then be calculated assuming an effective radius of dislocation leakage area AD of 13.1 nm and room temperature Schottky barrier φ_D (0.6 eV) formed on dislocations as in [48]. For Schottky diodes made on substrates with N_{DD} of 5×10^4 cm^{-2} it is expected to be 10^{-3} A/cm^2 at 20 V.

Direct current–voltage measurements were performed on large diameter (120 μm) devices. The TE model was used to fit the lower range of the forward I–V characteristics and allowed to estimate an ideality factor of 1.3 and a reverse saturation current of 1.5×10^{-5} A. The voltage drop on R_s (7.2 Ω) was relatively high due to the high current through the large contact area. Cheung's method [73] taking the high R_s value into consideration, was therefore used to extract the barrier height and good fitting was obtained as shown in Figure 10.52.

A barrier height of 0.354 eV was obtained in this way and the reverse leakage current was found to be in the order of 10^{-2} A/cm^2. The measured reverse current of 10^{-2} A/cm^2 for diodes made on GaN films that were epitaxially grown on the low dislocation density bulk GaN substrates was found to be close to the current density of 10^{-3} A/cm^2 predicted for this material. This agrees with the expectation of epitaxial layer properties with about the same characteristics as the bulk material on which they are grown and suggests the possibility of improved GaN Schottky diode characteristics as necessary for large-signal operation.

Exercises

E10.1 Consider W_m and W_s to be the work function of the Schottky metal and semiconductor with χ being the affinity.

Draw the band diagram before and after contact and indicate in it the barrier height φ_B and built-in voltage V_{bi}.

Assume an applied voltage V and kT/q being a voltage correction factor for the carrier distribution tail. Derive the space charge Q_{sc} and depletion capacitance CD of the diode for the case of constant doping.

Derive an expression for the diode junction capacitance CD as a function of the junction capacitance C_{j_0} at 0 V.

A hyperabrupt diode can be realized using an epitaxial or ion-implanted process. Sketch the doping profile in such a case and comment on how the derived CD vs. C_{j_0} relation will have to be modified to account for a hyperabrupt profile.

Consider a diode with $\varphi = 0.5$ V and a breakdown voltage of 40 V. Various electronic applications can take advantage of a large C_{max}/C_{min} ratio. Consider a variation of the applied voltage between φ and the maximum voltage and find C_{max}/C_{min} for the case of a constant doping and hyperabrupt profile. Discuss the obtained results.

E10.2 Consider a GaN Varactor diode with a capacitance (C_{j0}) of 40 fF at 0 V and a barrier height (φ) of 0.80 V. Assuming a breakdown voltage of 40 V.

Define the C–V relation and plot the C–V characteristics of the diode.

What would be the maximum DC bias voltage that can be applied to the diode for a use as multiplier?

Define the elastance modulation factor (S) and the equation providing its nonlinearity consider harmonic presence that does not exceed second order.

Assume that the access series resistance (R_s) of the diode is 2 Ω and find the diode's cutoff frequency. How should the operation frequency of the doubler compare with the diode's cutoff frequency for proper operation of the diode?

Consider the elastance modulation factors for the fundamental and second harmonic are $|m_1| = 0{:}502$ and $|m_2| = 0{:}166$ respectively and further harmonics are considered. What are the assumptions made in the event of fully driven operation?

How does the power handling capability of a diode depend on its breakdown? Provide an equation and discuss.

How does the power absorbed by the diode in a doubler depend on φ and C_{j_0}? How will an increase of φ and C_{j_0} impact loss in the diode and why? Will this have an impact on conversion efficiency and how is it related to cutoff frequency?

E10.3 The quality of GaN plays a major role in assuring good electrical performance such as low reverse leakage for GaN Schottky diodes. A lowe dislocation density material is desired for this purpose as for example obtained with bulk growth. Where values as low as 10^4 cm^{-2} can be obtained. Read more about reverse leakage and dislocation density effects in GaN in the paper "Reverse leakage mechanism of Schottky barrier diode fabricated on homoepitaxial GaN" vy Lei, Yong, Hai Lu, Dongsheng Cao, Dunjun Chen, Rong Zhang, and Youdou Zheng; Solid-State Electronics 82 (2013): 63–66.

Consider JTU to be the tunneling current density from metal to semiconductor within the defective regions.

Provide the equation for tunneling current density JTU.

Calculate the effective area of the dislocation A_D by considering the effective radius of dislocation λ_D, the dislocation density NDD, and the Schottky contact area A.

Consider an effective radius of dislocation leakage area A_D of 13.1 nm and room temperature Schottky barrier φ_D of 0.6 eV. Calculate and plot the leakage current density of a GaN diode for N_{DD} of 5×10^4 cm^{-2} and $N_{DD} = 0$ with φ_D of 0.95 eV. Comment on the obtained results.

E10.4 The series resistance R_s of a diode is one of the factors dictating its performance. To evaluate the process technology used for diode realization, the TLM is used.

- Discuss the TLM method, define the key equations and parameters involved.
- What information is provided by the slope of the obtained graphs and their intercept points on the on the resistance-axis distance-axis.
- Consider the parameters below obtained from the characterization of GaN diodes

```
Rc=0.18 Ω·mm Rsh=113 Ω/sq Lt=1.55 μm ρc=2.72e-6 Ω·cm2
Rc=0.19 Ω·mm Rsh=108 Ω/sq Lt=1.73 μm ρc=3.23e-6 Ω·cm2
Rc=0.17 Ω·mm Rsh=98.8 Ω/sq Lt=1.7 μm ρc=2.84e-6 Ω·cm2
Rc=0.2 Ω·mm Rsh=110 Ω/sq Lt=1.81 μm ρc=3.59e-6 Ω·cm2
```

and plot the TLM graphs corresponding to them. The ohmic contact of the fabricated diodes are required due to design requirements to be smaller than 1 μm in size. Will this have an impact on the diode characteristics?

E10.5 Sketch the layers necessary for realizing a GaN diode and discuss the fabrication process involved by commenting on the purpose of each step.

E10.6 Compare GaN-based to GaAs-based diodes in terms of

- Power handling capability
- Loss
- Nonlinearity
- Eventual changes of nonlinearity with frequency of operation
- Robustness

E10.7 Consider the design of a doubler using microstrip lines.

- What will be the advantage of using a series-connected two-diode approach?
- Generalize your answer to the above point by considering n series-connected diodes and comment on the resulting change of power capability.
- How should the capacitance of the individual diodes be adjusted to maintain the overall capacitance and resistance and thus cutoff frequency of a n-series connected diode scheme?
- Sketch a simple multiplier circuit using microstrip lines. What type of filers needs to be implemented to ensure proper operation?

References

1 Chattopadhyay, G. (2011). Technology, capabilities, and performance of low power terahertz sources. *IEEE Transactions on Terahertz Science and Technology* **1** (1): 33–53.

2 Saini, K.S. (2003). Development of frequency multiplier technology for ALMA. http://www.alma.nrao.edu/memos/html-memos/alma337/memo337.pdf (accessed 20 September 2020).

3 Maestrini, A., Mehdi, I., Siles, J.V. et al. (2012). Design and characterization of a room temperature all-solid-state electronic source tunable from 2.48 to 2.75 Thz. *IEEE Transactions on Terahertz Science and Technology* **2** (2): 177–185.

4 Martin, S., Nakamura, B., Fung, A. et al. (2001). Fabrication of 200 to 2700 ghz multiplier devices using gaas and metal membranes. Microwave Symposium Digest, 2001 IEEE MTT-S International, IEEE, vol. 3, 1641–1644.

5 Maestrini, A., Ward, J., Chattopadhyay, G. et al. (2008). Terahertz sources based on frequency multiplication and their applications. *Frequenz* **62** (5–6): 118–122.

6 Siles, J.V., Maestrini, A., Alderman, B. et al. (2011). A single-waveguide in-phase power-combined frequency doubler at 190 Ghz. *IEEE Microwave and Wireless Components Letters* **21** (6): 332–334.

7 Tang, A.Y., Schlecht, E., Lin, R. et al. (2012). Electro-thermal model for multi-anode Schottky diode multipliers. *IEEE Transactions on Terahertz Science and Technology* **2** (3): 290–298.

8 Lee, C., Ward, J., Lin, R. et al. (2009). A wafer-level diamond bonding process to improve power handling capability of submillimeter-wave Schottky diode frequency multipliers. 2009. MTT'09. IEEE MTTS International Microwave Symposium Digest, IEEE, 957–960.

9 Maestrini, A., Mehdi, I., Lin, R. et al. (2011). A 2.5–2.7 Thz room temperature electronic source. Proceedings of the 22nd International Symposium on Space Terahertz Technology.

10 Maestrini, A., Ward, J.S., Gill, J.J. et al. (2010). A frequency-multiplied source with more than 1 mw of power across the 840–900-Ghz band. *IEEE Transactions on Microwave Theory and Techniques* **58** (7): 1925–1932.

11 Siles, J.V. and Grajal, J. (2008). Capabilities of GaN Schottky multipliers for lo power generation at millimeter-wave bands. Proceedings of the 19th International Symposium on Space Terahertz Technology, 504–507.

12 Shur, M.S., Gaska, R., and Bykhovski, A. (1999). Gan-based electronic devices. *Solid-State Electronics* **43** (8): 1451–1458.

13 IOFFE Gallium nitride band structure and carrier concentration. http://www.ioffe.ru/SVA/NSM/Semicond/GaN/bandstr.html (accessed 20 September 2020).

14 Kolberg, E.L., Tolmunen, T.J., Frerking, M.A., and East, J.R. (1992). Current saturation in submillimeter-wave varactors. *IEEE Transactions on Microwave Theory and Techniques* **40** (5): 831–838.

15 Mion, C., Muth, J.F., Preble, E.A., and Hanser, D. (2006). Accurate dependence of gallium nitride thermal conductivity on dislocation density. *Applied Physics Letters* **89** (9): 092123.

16 Yue, Y., Hu, Z., Guo, J. et al. (2012). InALN/ALN/GaN HEMTs with regrown ohmic contacts and f_T of 370 Ghz. *IEEE Electron Device Letters* **33** (7): 988–990.

17 Brown, D.F., Williams, A., Shinohara, K. et al. (2011). W-band power performance of AlGaN/GaN DHFETs with regrown n+ GaN ohmic contacts by MBE. 2011 IEEE International Electron Devices Meeting (IEDM), 19.3.1–19.3.4.

18 Micovic, M., Kurdoghlian, A., Moyer, H.P. et al. (2008). GaN MMIC pas for e-band (71 Ghz – 95 Ghz) radio. *Compound Semiconductor Integrated Circuits Symposium*, 2008. IEEE CSIC, 1–4.
19 Fung, A., Ward, J., Chattopadhyay, G. et al. (2011). Power combined gallium nitride amplifier with 3 watt output power at 87 Ghz. Proceedings of the International Symposium on Space THz Technology.
20 Schellenberg, J., Watkins, E., Micovic, M. et al. (2010). W-band, 5W solid-state power amplifier/combiner. 2010 IEEE MTT-S International Microwave Symposium Digest (MTT), 240–243.
21 Panda, A.K., Dash, G.N., Agrawal, N.C., and Parida, R.K (2009). Studies on the characteristics of GaN-based Gunn diode for Thz signal generation. APMC 2009. Asia Pacific Microwave Conference, 1565–1568.
22 Yilmazoglu, O., Mutamba, K., Pavlidis, D., and Karaduman, T. (2008). First observation of bias oscillations in GaN Gunn diodes on GaN substrate. *IEEE Transactions on Electron Devices* **55** (6): 1563–1567.
23 Manley, J.M. and Rowe, H.E. (1956). Some general properties of nonlinear elements–Part I. General energy relations. *Proceedings of the IRE* **44** (7): 904–913.
24 Pantell, R.H. (1958). General power relationships for positive and negative nonlinear resistive elements. *Proceedings of the IRE* **46** (12): 1910–1913.
25 Penfield, P. and Rafuse, R.P. (1962). *Varactor Applications*. Cambridge: MIT Press.
26 Dunaevskii, G.E. and Perfil'ev, V.I. (2009). Analysis and synthesis of varactor frequency doublers. Part 1. Basic relationships. *Russian Physics Journal* **52** (10): 1052–1059.
27 Chen, S.-W., Ho, T.C., Pande, K., and Rice, P.D. (1993). Rigorous analysis and design of a high-performance 94 GHz MMIC doubler. *IEEE Transactions on Microwave Theory and Techniques* **41** (12): 2317–2322.
28 Calviello, J.A. (1979). Advanced devices and components for the millimeter and submillimeter systems. *IEEE Transactions on Electron Devices* **26** (9): 1273–1281.
29 Farahmand, M., Garetto, C., Bellotti, E. et al. (2001). Monte Carlo simulation of electron transport in the III-nitride wurtzite phase materials system: binaries and ternaries. *IEEE Transactions on Electron Devices* **48** (3): 535–542.
30 ATLAS (2012). *User's Manual*. Santa Clara, CA: Silvaco International.
31 Ieong, M., Solomon, P.M., Laux, S.E. et al. (1998). Comparison of raised and Schottky source/drain MOSFETs using a novel tunneling contact model. IEDM '98. Technical Digest, International Electron Devices Meeting, 733–736.
32 Hong, K., Marsh, P.F., Ng, G.-I. et al. (1994). Optimization of MOVPE grown $in_xal_{1-x}as$/$in_{0.53}ga_{0.47}as$ planar heteroepitaxial Schottky diodes for terahertz applications. *IEEE Transactions on Electron Devices* **41** (9): 1489–1497.
33 Pearton, S.J., Zolper, J.C., Shul, R.J., and Ren, F. (1999). GaN: processing, defects, and devices. *Journal of Applied Physics* **86** (1): 1–78.
34 Kwak, J.S., Mohney, S.E., Lin, J.-Y., and Kern, R.S. (2000). Low resistance al/ti/n-GaN ohmic contacts with improved surface morphology and thermal stability. *Semiconductor Science and Technology* **15** (7): 756.
35 Pelto, C.M., Chang, Y.A., Chen, Y., and Williams, R.S. (2001). Issues concerning the preparation of ohmic contacts to n-GaN. *Solid-State Electronics* **45** (9): 1597–1605.

36 Van Daele, B., Van Tendeloo, G., Ruythooren, W. et al. (2005). The role of al on ohmic contact formation on n-type GaN and AlGaN/GaN. *Applied Physics Letters* **87** (6): 061905.

37 Miura, N., Oishi, T., Nanjo, T. et al. (2004). Effects of interfacial thin metal layer for high-performance pt-au-based Schottky contacts to AlGaN-GaN. *IEEE Transactions on Electron Devices* **51** (3): 297–303.

38 Jin-Ping, A., Kikuta, D., Kubota, N. et al. (2003). Copper gate AlGaN/GaN hemt with low gate leakage current. *IEEE Electron Device Letters* **24** (8): 500–502.

39 Lee, N.-H., Lee, M., Choi, W. et al. (2014). Effects of various surface treatments on gate leakage, subthreshold slope, and current collapse in AlGaN/GaN high-electron-mobility transistors. *Japanese Journal of Applied Physics* **53** (4S): 04EF10.

40 Kordoš, P., Bernàt, J., Gregušovà, D. et al. (2006). Impact of surface treatment under the gate on the current collapse of unpassivated AlGaN/GaN heterostructure field-effect transistors. *Semiconductor Science and Technology* **21** (1): 67.

41 Suzue, K., Mohammad, S.N., Fan, Z.F. et al. (1996). Electrical conduction in platinum–gallium nitride Schottky diodes. *Journal of Applied Physics* **80** (8): 4467–4478.

42 Hacke, P., Detchprohm, T., Hiramatsu, K., and Sawaki, N. (1993). Schottky barrier on n-type GaN grown by hydride vapor phase epitaxy. *Applied Physics Letters* **63** (19): 2676–2678.

43 Huh, C., Kim, S.-W., Kim, H.-S. et al. (2000). Effective sulfur passivation of an n-type GaN surface by an alcohol-based sulfide solution. *Journal of Applied Physics* **87** (9): 4591–4593.

44 Rongming, C., Shen, L., Fichtenbaum, N. et al. (2008). Plasma treatment for leakage reduction in AlGaN/GaN and GaN Schottky contacts. *IEEE Electron Device Letters* **29** (4): 297–299.

45 Sun, Y., Shen, X.M., Wang, J. et al. (2002). Thermal annealing behaviour of ni/au on n-GaN Schottky contacts. *Journal of Physics D: Applied Physics* **35** (20): 2648.

46 Guo, J.D., Pan, F.M., Feng, M.S. et al. (1996). Schottky contact and the thermal stability of ni on n-type GaN. *Journal of Applied Physics* **80** (3): 1623–1627.

47 Wang, J., Zhao, D.G., Sun, Y.P. et al. (2003). Thermal annealing behaviour of pt on n-GaN Schottky contacts. *Journal of Physics D: Applied Physics* **36** (8): 1018.

48 Lei, Y., Lu, H., Cao, D. et al. (2013). Reverse leakage mechanism of Schottky barrier diode fabricated on homoepitaxial GaN. *Solid-State Electronics* **82**: 63–66.

49 Gao, J. (2010). *RF and Microwave Modeling and Measurement Techniques for Field Effect Transistors*. SciTec.

50 Tang, A.Y., Drakinskiy, V., Yhland, K. et al. (2013). Analytical extraction of a Schottky diode model from broadband s-parameters. *IEEE Transactions on Microwave Theory and Techniques* **61** (5): 1870–1878.

51 Louhi, J.T. and Raisanen, A.V. (1995). On the modeling and optimization of Schottky varactor frequency multipliers at submillimeter wavelengths. *IEEE Transactions on Microwave Theory and Techniques* **43** (4): 922–926.

52 Van Moer, W. and Gomme, L. (2010). NVNA versus LSNA: enemies or friends? *IEEE Microwave Magazine* **11** (1): 97–103.

53 Verspecht, J. and Root, D.E. (2006). Polyharmonic distortion modeling. *IEEE Microwave Magazine* **7** (3): 44–57.

54 Ghannouchi, F.M. and Hashmi, M.S. (2012). *Load-Pull Techniques With Applications to Power Amplifier Design*, vol. **32**. Springer.

55 Barataud, D., Blache, F., Mallet, A. et al. (1999). Measurement and control of current/voltage waveforms of microwave transistors using a harmonic load-pull system for the optimum design of high efficiency power amplifiers. *IEEE Transactions on Instrumentation and Measurement* **48** (4): 835–842.

56 Vandermot, K., Van Moer, W., Schoukens, J., and Rolain, Y. (2006). Understanding the nonlinearity of a mixer using multisine excitations. IMTC 2006. Instrumentation and Measurement Technology Conference, Proceedings of the IEEE, 1205–1209.

57 Verspecht, J., Verbeyst, F., and Bossche, M.V. (2003). Large-signal network analysis: going beyond s-parameters. 62nd ARFTG Conference Digest.

58 Cojocaru, V. and Sischka, F. (2003). Non-linear modelling of microwave pin diode switches for harmonic and intermodulation distortion simulation. 2003 IEEE MTT-S International Microwave Symposium Digest, vol. 2, 655–658.

59 Agilent Technologies (2009). Symbolically defined devices (sdd) in ads. https://edadocs.software.keysight.com/display/ads2009/User-Defined+Models+Print+View#User-DefinedModelsPrintView-CustomModelingwithSymbolically-DefinedDevicesĐ'd (accessed 20 September 2020).

60 Jargon, J.A., DeGroot, D.C., and Gupta, K.C. (2004). Frequency-domain models for nonlinear microwave devices based on large-signal measurements. *Journal of Research National Institute of Standards and Technology* **109** (4): 407.

61 Saglam, M., Mutamba, K., Megej, A. et al. (2003). Influence of polarization charges in Al 0.4 Ga 0.6 n/GaN barrier varactors. *Applied Physics Letters* **82** (2): 227–229.

62 Reklaitis, A. (2008). Terahertz-frequency inn/GaN heterostructure-barrier varactor diodes. *Journal of Physics: Condensed Matter* **20** (38): 384202.

63 Wong, C., Yuk, K., Branner, G.R., and Bahadur, S.R. (2011). High power, wideband frequency doubler design using AlGan/Gan HEMTs and filtering. *2011 41st European Microwave Conference (EuMC)*, IEEE, 587–590.

64 Zhu, G., Zhang, K., Kong, Y. et al. (2017). Quaternary InAlGan barrier high-electron-mobility transistors with f max> 400 Ghz. *Applied Physics Express* **10** (11): 114101.

65 Kwon, Y., Pavlidis, D., Marsh, P. et al. (1991). 180 Ghz InAlAs/InGaAs HEMT monolithic integrated frequency doubler. *Technical Digest* 1991, *13th Annual Gallium Arsenide Integrated Circuit (GaAs IC) Symposium*, IEEE, 165–168.

66 Kwon, Y., Pavlidis, D., and Tutt, M.N. (1991). An evaluation of HEMT potential for millimeter-wave signal sources using interpolation and harmonic balance techniques. *IEEE Microwave and Guided Wave Letters* **1** (12): 365–367.

67 Jin, C., Pavlidis, D., and Considine, L. (2012). DC and high-frequency characteristics of GaN Schottky varactors for frequency multiplication. *IEICE Transactions on Electronics* **95** (8): 1348–1353.

68 Liang, S., Fang, Y., Xing, D. et al. (2014). Realization of GaN-based high frequency planar Schottky barrier diodes through air-bridge technology. 2014 12th IEEE International Conference on Solid-State and Integrated Circuit Technology (ICSICT), 1–3.

69 Pardo, D. and Grajal, J. (2015). Analysis and modelling of GaN Schottky-based circuits at millimeter wavelengths. *Semiconductor Science and Technology* **30** (11): 115016.

70 Jin, C., Zaknoune, M., and Pavlidis, D. (2014). Low-dislocation density bulk GaN Schottky diodes. Proceedings of the 38th Workshop on Compound Semiconductor Devices and Integrated Circuits held in Europe (WOCSDICE 2014).

71 Hsu, J.W.P., Manfra, M.J., Molnar, R.J. et al. (2002). Direct imaging of reverse-bias leakage through pure screw dislocations in GaN films grown by molecular beam epitaxy on GaN templates. *Applied Physics Letters* **81** (1): 79–81.

72 Paskova, T., Hanser, D.A., and Evans, K.R. (2010). Gan substrates for iii-nitride devices. *Proceedings of the IEEE* **98** (7): 1324–1338.

73 Cheung, S.K. and Cheung, N.W. (1986). Extraction of Schottky diode parameters from forward current-voltage characteristics. *Applied Physics Letters* **49** (2): 85–87.

11

THz Resonant Tunneling Devices

Masahiro Asada[1] *and Safumi Suzuki*[2]

[1] *Institute of Innovative Research, Tokyo Institute of Technology, Tokyo, Japan*
[2] *Department of Electrical and Electronic Engineering, Tokyo Institute of Technology, Tokyo, Japan*

11.1 Introduction

The terahertz (THz) frequency range located approximately between 0.1 and 1 THz has been receiving considerable attention recently because of its many applications, such as ultrahigh-speed wireless communications, spectroscopy, and imaging [1, 2]. For these applications, compact and coherent solid-state sources are important components. Because the THz range is located between lightwaves and millimeter waves, both optical and electronic devices are being investigated for THz sources.

Figure 11.1 shows the current status of semiconductor single oscillators in the THz range. In the optical device side, p-type germanium lasers [3] and quantum cascade lasers (QCLs) are studied [4–7]. Electron devices are also intensively studied from the millimeter-wave side. In particular, transistors including heterojunction bipolar transistors (HBTs), high electron mobility transistors (HEMTs), and silicon complementary metal–oxide–semiconductor (Si–CMOS) transistors are making significant progress in operation frequency, recently [8–12]. In addition to oscillators, amplifiers operating at 1 THz with HEMTs have been achieved [13]. Oscillators with tunneling transit-time (TUNNETT) diodes, impact ionization avalanche transit-time (IMPATT) diodes, and Gunn diodes have also been reported [14–16].

Among the electron devices, resonant tunneling diodes (RTDs) have also been considered as one of the candidates for THz oscillators at room temperature. Research of RTDs started with the theoretical prediction by Tsu and Esaki in 1973 [17], and their behavior of negative differential resistance was experimentally demonstrated at liquid nitrogen temperature in 1974 [18] and at room temperature in 1985 [19, 20]. Oscillation in the microwave range was demonstrated at a low temperature in 1984 [21]. The oscillation frequency was then updated many times to several hundred GHz [22], and room-temperature fundamental

Fundamentals of Terahertz Devices and Applications, First Edition. Edited by Dimitris Pavlidis.

Companion website: www.wiley.com/go/Pavlidis/FundamentalsofTHz

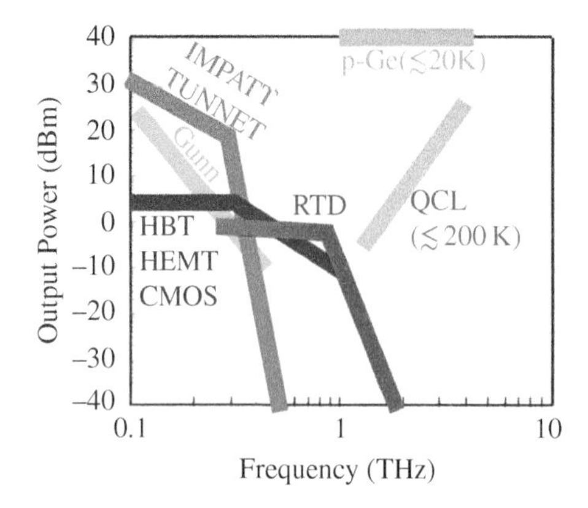

Figure 11.1 Current status of semiconductor single sources in THz range; output power as a function of oscillation frequency (as of the end of 2018). Operation temperatures are indicated except for room-temperature devices.

oscillation at 712 GHz was reported in 1991 [23]. RTDs with waveguides were mainly used in these oscillators. RTDs with planar antennas and their oscillation up to 650 GHz were also reported in 1997 [24]. Although the increase in oscillation frequency has stopped after achievement of 712 GHz until recently, oscillation at 831 GHz was reported in 2009 [25], and then, oscillation above 1 THz was achieved in 2010 [26]. The oscillation frequency was further increased by improvement of the structure of RTD mainly for short electron delay time [27–29] and the antennas for low conduction loss [30–32] and was now updated to 1.98 THz [32].

Several material systems have been used for RTDs [22]. These are combinations of materials with lattice constants close to each other for heterostructures. GaAs/GaAlAs heterostructures were used at first. InAs/AlSb system was also used [23], which has an advantage of low contact resistance. RTDs with InGaAs/AlAs system on InP substrate were shown to have high peak-to-valley current ratios [33, 34]. High current density, which is required for high-frequency oscillation as shown later, has also been reported in GaInAs/AlAs system [26, 35, 36]. The experimental results of RTD oscillators with this material system are described in this chapter. GaN-based system is also being investigated, as described in Chapter 1. Although RTDs with this system are presently in an early stage of research [37–40], oscillators with high output power may be possible, because a large negative-differential-conductance region in voltage can be designed owing to high breakdown voltage.

The output power of RTD oscillators was also increased by the improvements of RTDs and antennas [41, 42]. In addition to the improved structures for frequency and output power, various types of resonator and radiator structures were reported [43–53]. A frequency-tunable RTD oscillator was also proposed and fabricated [54, 55]. Recently, some preliminary applications of RTD oscillators were reported as compact THz sources to wireless data transmission [45, 56–60], spectroscopy [61], and imaging [62]. A THz sensing system using RTD oscillators was also investigated [63]. Application of RTDs to the THz detector was also reported [64]. In this chapter, we describe the current status of RTD oscillators, including the device structure, oscillation principle and characteristics, structures for high-frequency and high-output power operations, control of oscillation spectra and frequency, some applications, and expected future development.

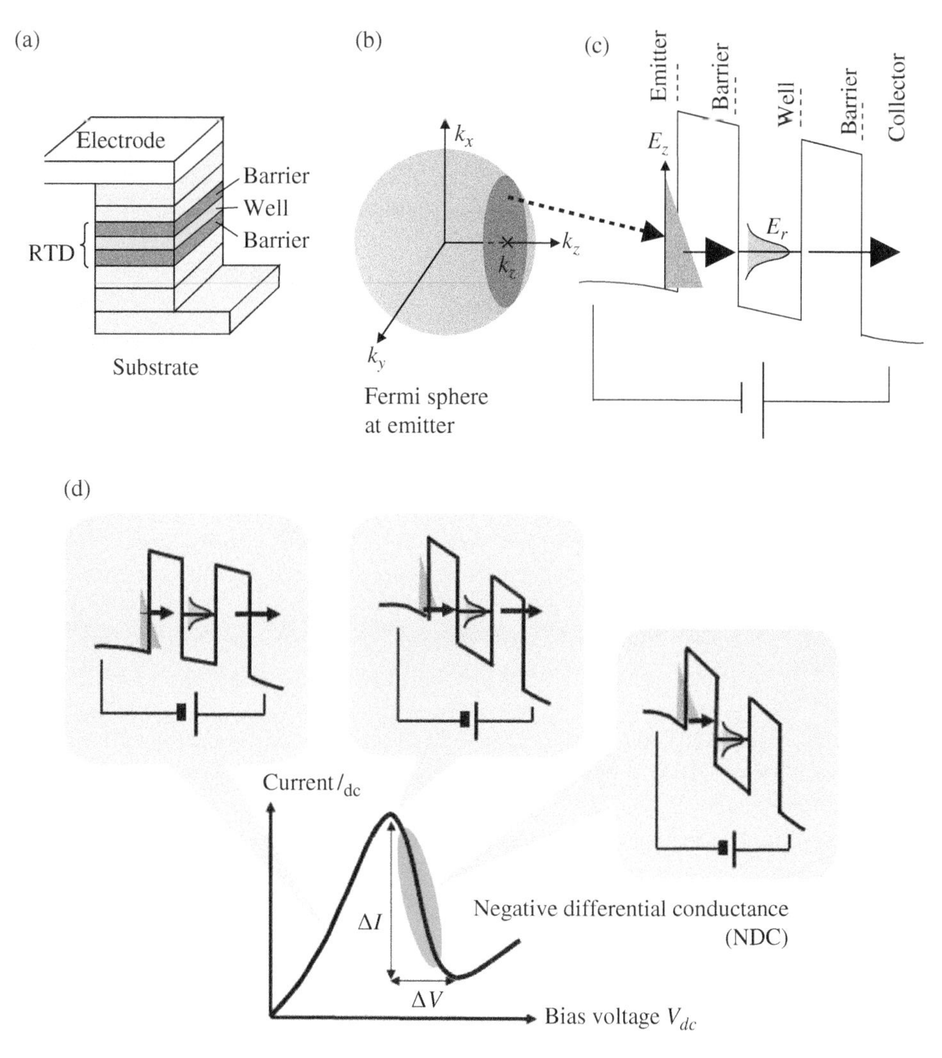

Figure 11.2 Operation principle of RTD. (a) Layer structure, (b) Fermi sphere at the emitter, (c) potential profile with electron energy distribution, and (d) I–V curve together with potential change with bias voltage.

11.2 Principle of RTD Oscillators

11.2.1 Basic Operation of RTD

The main part of a typical RTD is composed of double barriers and a quantum well made of a semiconductor heterostructure, as shown in Figure 11.2a. An example of detailed layer structures is shown later (Figure 11.6). As shown in Figure 11.2b, electrons travel by the resonant tunneling from the emitter to the collector through the resonance level E_r in the quantum well.

Electrons in the emitter have various directions of motion presented by the wavenumber components k_x, k_y, and k_z for the x, y, and z-directions, respectively, where the z-direction is perpendicular to the heterointerface. The electron kinetic energy E_i in the i direction ($i = x$, y, and z) is related to the wavenumber components as $E_i(k_i) = \hbar^2 k_i^2/(2m_e)$, where m_e is the electron effective mass assumed to be isotropic. As shown in Figure 11.2b, most of the electrons exist inside the Fermi sphere in the k space expressed by $E_x(k_x) + E_y(k_y) + E_z(k_z) \leq E_F$, where E_F is the Fermi energy. The disk in the Fermi sphere in Figure 11.2b indicates the assembly of electrons having a specified $E_z(k_z)$ and any values of $E_x(k_x)$ and $E_y(k_y)$. Because the tunneling is independent of motion in the x- and y-directions, these electrons are discussed here.

The gray triangle at the emitter in Figure 11.2c indicates the density of electrons incident into unit area of the barrier interface from the emitter per unit time. This density is a function of E_z given by the electron density at the disk in the Fermi sphere multiplied by the velocity in the z-direction. The tunneling current density is obtained by the integration over E_z of the product between the density of incident electrons, the resonant tunneling probability, and the electron charge. The shape of the resonant tunneling probability is shown in the quantum well in Figure 11.2c. The detailed equations on the tunneling current density are described, e.g., in [22].

The current–voltage (I–V) curve is shown in Figure 11.2d. At low bias voltage, as shown in the left-hand inset of the figure, the density of incident electrons at k_r is small, and the current is small. The current increases with increasing bias voltage. When the bias voltage is set so that the conduction band edge is near E_r, as shown in the middle inset of Figure 11.2d, the density of incident electrons is maximized and the current shows the peak. With further increase in bias voltage, the conduction band edge becomes higher than E_r, as shown in the right-hand inset of Figure 11.2d. Around this bias voltage, the current decreases with increasing bias voltage and shows the negative differential conductance (NDC) region. At higher bias voltage, the conduction band edge approaches the second resonance level if it exists, and the current increases again, indicating the valley in the I–V curve. A leakage current may also cause the current increase. The NDC region is used in the RTD THz oscillators described in this chapter. Details of the layer structure and observed I–V curves are shown and discussed in Section 11.3.

For the analysis of oscillation characteristics discussed later, the bias current I_{dc} of RTD as a function of bias voltage V_{dc} is approximated by the following expansion form around the center of the NDC region:

$$I_{dc}(V_{dc}) \simeq \left(\frac{\partial I_{dc}}{\partial V_{dc}}\right) V_{dc} + \frac{1}{3!}\left(\frac{\partial^3 I_{dc}}{\partial V_{dc}^3}\right) V_{dc}^3 \equiv -aV_{dc} + bV_{dc}^3, \tag{11.1}$$

where the origin of V_{dc} is at the center of the NDC region, $a = -\partial I_{dc}/\partial V_{dc}$, and $b = (1/6)\partial^3 I_{dc}/\partial V_{dc}^3$ at $V_{dc} = 0$. The term for the second derivative is neglected, assuming that the I–V curve is antisymmetric with respect to the center of the NDC region. If the I–V curve is well approximated by Eqs. (11.1) in the entire region of the NDC, $a \simeq (3/2)$ $(\Delta I/\Delta V)$ and $b \simeq 2\Delta I/\Delta V^3 \simeq (4/3)(a/\Delta V^2)$, where ΔI and ΔV are the widths of current and voltage of the NDC region, respectively, as shown in Figure 11.2d. From Eqs. (11.1), the NDC against the voltage $v_{ac}(t)$ oscillating around the center of the NDC region is given by $-G_{RTD} = \partial I_{dc}(v_{ac})/\partial v_{ac} = -a + 3bv_{ac}^2$. For high-frequency oscillation, a and b change with frequency from the values at DC, as discussed in Section 11.2.3.

11.2.2 Principle of Oscillation

Figure 11.3a shows the fundamental structure of the RTD oscillator. The RTD is located in a slot (or cavity) in a metal film (or block). This slot forms a standing wave of the electromagnetic field as a resonator and also acts as an antenna by radiating output power. The equivalent circuit for this structure is illustrated in Figure 11.3b, where $-G_{RTD}$ is the NDC of RTD as shown above, G_{ANT} is the conductance of the slot due to radiation and conduction loss, L is the inductance of the slot, and C is the capacitance composed of those of RTD and the slot. The parasitic elements such as the contact resistance are discussed later and not shown in Figure 11.3b.

Without RTD, the standing wave generated in the slot decays because of the power consumption at G_{ANT}. By putting $-G_{RTD}$ of RTD to compensate for the power consumption, the standing wave is sustained to oscillate. The oscillation frequency is determined by the parallel resonance of L and C in Figure 11.3b, which corresponds to the frequency of the standing wave in the slot.

We discuss this situation in some more detail with equations. The voltage $v_{ac}(t)$ in Figure 11.3b is calculated by the following van der Pol equation:

$$C\frac{d^2v_{ac}}{dt^2} - \left(a - 3bv_{ac}^2 - G_{ANT}\right)\frac{dv_{ac}}{dt} + \frac{1}{L}v_{ac} = 0. \tag{11.2}$$

Equation (11.2) has an approximated solution $v_{ac}(t) = V_{ac}\cos\omega t$. Substituting this formula in Eq. (11.2), and neglecting the third-harmonic term oscillating with 3ω, we obtain

$$\omega = \omega_{osc} \equiv \frac{1}{\sqrt{LC}} \quad \text{and} \quad V_{ac} = \left[\frac{4}{3b}(a - G_{ANT})\right]^{1/2} \tag{11.3}$$

if $a \geq G_{ANT}$, and $V_{ac} = 0$ if $a < G_{ANT}$. Thus, the oscillation frequency f_{osc} is $(1/2\pi)/\sqrt{LC}$, and the necessary condition for the oscillation is $a \geq G_{ANT}$. The second formula in Eq. (11.3) gives the output power, as shown later. In the calculation process obtaining the second formula in

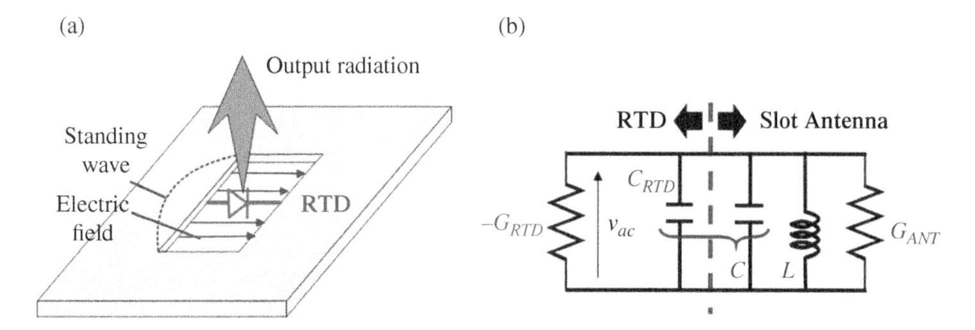

Figure 11.3 Fundamental structure of RTD oscillator. (a) RTD with slot resonator. *Source*: Copyright (2008) The Japan Society of Applied Physics [17]. (b) Basic equivalent circuit of (a). *Source*: Reprinted, with permission, from Ref. [65]. Copyright (2016) Springer Nature.

Eq. (11.3), we obtain $a - (3/4)bV_{ac}^2 = G_{ANT}$. This implies that, after the oscillation starts, G_{ANT} is compensated by the effective NDC given by [66]

$$\overline{G}_{RTD} = a - \frac{3}{4}bV_{ac}^2. \tag{11.4}$$

For the beginning of the oscillation, a has to compensate for G_{ANT}, and for the stationary state after the oscillation grows, $\overline{G}_{RTD}$ has to compensate for G_{ANT}. Figure 11.3b is the basic equivalent circuit without parasitic elements. An actual equivalent circuit including parasitic elements around the intrinsic RTD is discussed in Section 11.3.2.

In the exact solution of Eq. (11.2), the harmonic components are not negligible, if $a - G_{ANT}$ is comparable to or larger than $\omega_{osc}C$. Equation (11.2) has also been applied to analyses of the harmonic generation, injection locking between oscillators, and noise properties [67–69].

11.2.3 Effect of Electron Delay Time

11.2.3.1 Degradation of NDC at High Frequency

The effective NDC in Eq. (11.4) changes with frequency because of the electron delay time composed of the dwell time τ_{RTD}in the region from the emitter barrier to the collector barrier and the transit time τ_{dep} at the collector depletion layer, as shown in Figure 11.4. The effect of the electron delay times is analyzed here.

By the injection of electrons from the emitter into the barrier, positive charges are left at the emitter. As the electrons move from the emitter to the collector, the positive charges also move through the external circuit to the collector, resulting in an induction current in the external circuit. It is this current that acts as the usual RTD current, and it has a delay related to the residence of electrons in RTD. This current is given by the following formula [70–72] for the voltage $v_{ac}(t)$ applied to RTD together with the DC voltage V_{dc} to set the bias at the center of the NDC region:

$$i(t) = \frac{1}{d_{dep}}\int_0^{d_{dep}} I_{dc}\left(v_{ac}\left(t - \tau_{RTD} - \frac{z}{v_s}\right)\right)dz + \frac{\varepsilon S}{d}\frac{dv_{ac}}{dt}, \tag{11.5}$$

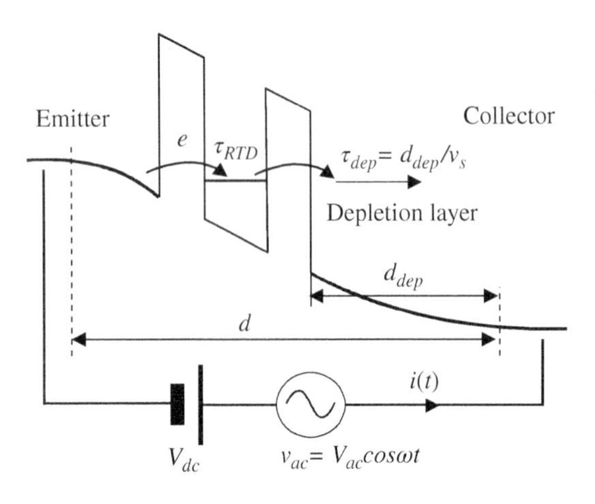

Figure 11.4 Electron movement with delay time in RTD. *Source*: Copyright (2008) The Japan Society of Applied Physics [17].

where I_{dc} is given by Eq. (11.1), $d_{dep}$is the thickness of the collector depletion layer; d is the total thickness including the emitter accumulation layer, barriers and well, and d_{dep}, as shown in Figure 11.4; ε is the dielectric constant; and S is the area of the RTD. The electrons are assumed to be resident for a time of τ_{RTD} between the barriers, where the movement is neglected, and then to move to the collector through the depletion layer with saturation velocity v_s. The change of d_{dep} with $v_{ac}(t)$ is neglected, because d_{dep} is determined by the thickness of the undoped collector spacer layer between the collector-side barrier and the heavily doped collector as mentioned in Section 11.3.1, and also because of the change in the depletion thickness of the heavily doped collector is negligibly small.

Substituting $v_{ac}(t) = V_{ac} \cos \omega t$ in Eq. (11.5), the effective NDC at ω is calculated from the current component varying as cosωt included in $i(t)$ as

$$\overline{G}_{RTD}(\omega) = -\frac{1}{V_{ac}}\left(\frac{\omega}{\pi}\int_0^{\frac{2\pi}{\omega}} i(t)\ \cos\omega t\ dt\right)$$

$$= \overline{G}_{RTD}(0)\cos\omega\tau \frac{\sin\left(\omega\tau_{dep}/2\right)}{\omega\tau_{dep}/2}, \tag{11.6}$$

where $\tau_{dep} = d_{dep}/v_s$ is the transit time described above, $\tau = \tau_{RTD} + \tau_{dep}/2$ is the total delay time, and $\overline{G}_{RTD}(0) = \overline{G}_{RTD}$ given by Eq. (11.4). From Eq. (11.6), $\overline{G}_{RTD}(\omega)$ is positive, i.e. RTD serves as an NDC, up to the frequency $f_c = 1/(4\tau)$. For $f > f_c$, $\overline{G}_{RTD}(\omega)$ repeatedly takes positive and negative values with increasing f, although the peak values of $\left|\overline{G}_{RTD}(\omega)\right|$ at $f > f_c$ are smaller than $\overline{G}_{RTD}(0)$. This property of $\overline{G}_{RTD}(\omega)$ is equivalent to the transit-time effect utilized in IMPATTs and TUNNETTs. The last factor in the right-hand side of Eq. (11.6) is negligible at $f < f_c$, and $\overline{G}_{RTD}(\omega) \simeq \overline{G}_{RTD}(0)\cos\omega\tau$. Because the frequency dependence of $\overline{G}_{RTD}$ is originated from those of a and b in Eq. (11.4), these factors also vary with frequency as $a(\omega) \simeq a(0)\cos\omega\tau$ and $b(\omega) \simeq b(0)\cos\omega\tau$.

In the actual structures described in Section 11.3, τ is estimated from the comparison between the measurement and calculation to be roughly 35–70 fs [73]. f_c is thus estimated to be 4–7 THz. Because the treatment of τ_{RTD} in the above analysis on τ is simplified, a more detailed discussion is a future subject for the study of high frequency response of RTD. The frequency response of the NDC in RTD without the depletion region has been discussed in the view points of the coherent tunneling [22, 74–76] and the sequential tunneling [77, 78]. The effect of the photon-assisted tunneling has been analyzed in the coherent-tunneling model, and the effect of the Coulomb interaction and displacement current, which is equivalent to the response of NDC with charge in the capacitances of the barriers, has been taken into account in the sequential-tunneling model.

11.2.3.2 Generation of Reactance at High Frequency

The current $i(t)$ in Eq. (11.5) has a component varying as sinωt for the applied voltage-$v_{ac}(t) = V_{ac}\cos\omega t$, which is equivalently expressed with the susceptance connected in parallel with RTD given by

$$\frac{1}{V_{ac}}\left(\frac{\omega}{\pi}\int_0^{\frac{2\pi}{\omega}} i(t)\ \sin\omega t\ dt\right) = \omega C_{RTD} \tag{11.7}$$

with

$$C_{RTD} = C_0 + \tau \overline{G}_{RTD}(0) \frac{\sin \omega\tau}{\omega\tau} \frac{\sin(\omega\tau_{dep}/2)}{\omega\tau_{dep}/2}, \tag{11.8}$$

where $C_0 = \varepsilon S/d$ is the capacitance constructed between the electrodes of RTD, and the second term is the additional capacitance due to the electron delay time which is approximated by $\tau\overline{G}_{RTD}(0)$ at low frequency. The last factor in the second term of Eq. (11.8) is negligible at $f < f_c$ similarly to $\overline{G}_{RTD}(\omega)$ in Eq. (11.6). Thus, a parallel capacitance was added due to the delay time effect as the second term in Eq. (11.8). The delay time effect has also been expressed as a series negative inductance [22], which is equivalent to the positive parallel capacitance in the present analysis. The capacitance due to the carrier accumulation in the quantum well region has been observed and theoretically analyzed [79–81], which is equivalent to the second term of Eq. (11.8). The effective admittance of RTD with the delay-time effect is written as

$$Y_{RTD}(\omega) = -\overline{G}_{RTD}(\omega) + j\omega C_{RTD} \simeq -\overline{G}_{RTD}(0)e^{-j\omega\tau} + j\omega C_0. \tag{11.9}$$

11.3 Structure and Oscillation Characteristics of Fabricated RTD Oscillators

11.3.1 Actual Structure of RTD Oscillators

For the actual structures of RTD oscillators, the waveguide type [22, 23] and planar type [24, 25] have been mainly reported. Figure 11.5a shows one of the fabricated planar types [31, 32]. A planar slot antenna is integrated with RTD, which works as a resonator and a radiator for THz wave. The lower electrode of the RTD is directly connected to the antenna electrode, and the upper electrode is connected to the top of MIM (metal–SiO_2–metal) capacitor via an air bridge. Because the MIM capacitor shunts a high-frequency field, the upper electrode of the RTD is connected to the slot antenna for the THz wave, while it is isolated for the DC

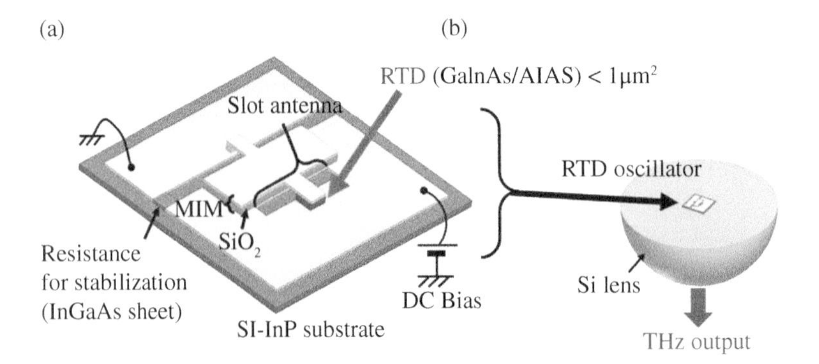

Figure 11.5 Structure of RTD oscillators. (a) Structure of fabricated planar RTD oscillator and (b) RTD oscillator on hemispherical Si lens. *Source*: Reprinted, with permission, from Ref. [65]. Copyright (2016) Springer Nature.

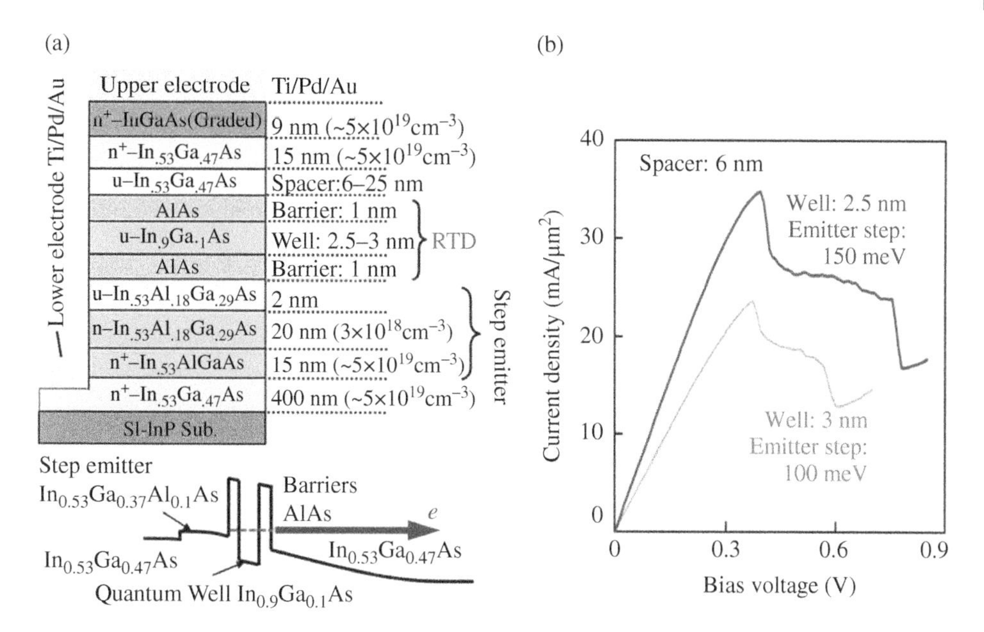

Figure 11.6 Structure and I–V characteristics of RTD. (a) Layer structure (upper) and potential profile (lower). *Source*: Reprinted, with permission, from Ref. [65]. Copyright (2016) Springer Nature. (b) I–V curves of RTDs with different well thickness and emitter step height.

bias. A sheet resistance with a heavily n-doped InGaAs layer is connected in parallel with the RTD to suppress low-frequency parasitic oscillation in external resonance circuits including the bias line. The fabrication process is described in [31]. The output power is extracted from the substrate side through a hemispherical silicon lens, as shown in Figure 11.4b, because approximately 98% of the output power, which is estimated with the relative dielectric constant $\varepsilon_r \cong 12$ of the substrate, is radiated into the substrate with high dielectric constant [82]. Although the silicon lens is convenient to extract the output power, it is bulky and the alignment of the oscillator position is needed. Moreover, the transmission loss at the lens surface reduces the output power. In particular, the transmission loss caused by the total reflection at the edge region of the spherical surface is significant in collimated hemispherical lens for a sharp output beam. More compact RTD planar oscillators without the silicon lens have also been reported [43–53].

The oscillation frequency was measured with a Fourier transform infrared spectrometer (FTIR), and the output power is measured with a calibrated power meter (Erickson PM5) or a calibrated Si-composite bolometer in [25–43].

Figure 11.6 shows an example of the structure and I–V characteristics of RTD [31]. As shown in the upper figure of Figure 11.6a, the main part of RTD is composed of AlAs barriers and a GaInAs quantum well. Thin barriers and a thin quantum well are used to obtain a high current density as well as a short dwell time of electrons in the resonant tunneling structure. A thin quantum well is also useful for a wide separation between the lowest and second lowest resonance levels, resulting in a low valley current density. However, a high bias voltage is required for NDC because of the elevated resonance levels in a thin quantum well. To reduce the bias voltage, a deep quantum well with In-rich GaInAs layer and a step emitter structure with AlGaInAs layer are used, as shown in upper and lower figures of Figure 11.6a. By these structures, the height of the resonance level relative to the emitter conduction-band edge is reduced. An undoped collector spacer layer is inserted between the collector-side barrier and the heavily doped collector to reduce the capacitance. An In-rich graded GaInAs, in which the In composition is gradually increased from the bottom to top, is also used for the top layer to reduce the contact resistance between the RTD and the upper electrode. The substrate is a semi-insulating InP.

As shown in Figure 11.6b, the peak current density increases with decreasing thickness of the quantum well. However, the peak voltage is kept unchanged by increasing the step emitter height. The observed NDC region is unstable because of the parasitic oscillation as described above. The peak current density is around 30 mA/μm^2, and the bias voltage at the center of NDC region is around 0.5 V. The peak-to-valley current ratio is ~2.

11.3.2 High-frequency Oscillation

For high-frequency oscillation, the following conditions are required: (i) NDC must be maintained at high frequency, (ii) losses of the antenna, such as the conduction loss, must be reduced, and (iii) parasitic elements around RTD, such as the series resistance and parallel capacitance, must be suppressed. The harmonic oscillation can further increase the frequency [69, 83].

The above condition (i) is investigated with the necessary condition for oscillation in the previous section, $a(\omega) \simeq a(0)\cos\omega\tau \geq G_{ANT}$. Because $a(\omega)$ decreases with frequency, the upper limit of the oscillation frequency exists as the frequency above which G_{ANT} cannot be compensated by $a(\omega)$, even if the oscillation frequency is designed according to the first formula in Eq. (11.3). This situation is illustrated in Figure 11.7. From this discussion, $a(0)$ (NDC in the I–V curve at DC) must be large and the delay time τ must be small for the condition (i). For large $a(0)$, a large value of $\Delta I/\Delta V$ is required in Figure 11.2, where ΔI and ΔV are the voltage and current widths of the NDC region of the I–V curve. A high peak current density with a high peak-to-valley current ratio is required for this purpose. A large RTD area also makes $a(0)$ large with the current density unchanged. In a large RTD, however, a small inductance is required in the antenna side in Figure 11.3b for high-frequency oscillation because of a large capacitance of RTD. This will be discussed later together with the condition (ii) for the antenna. Although a small ΔV, which is obtained by a thin collector spacer, also increases $a(0)$, the output power decreases as discussed in Section 11.3.3. The capacitance also increases with a thin collector spacer, as shown below.

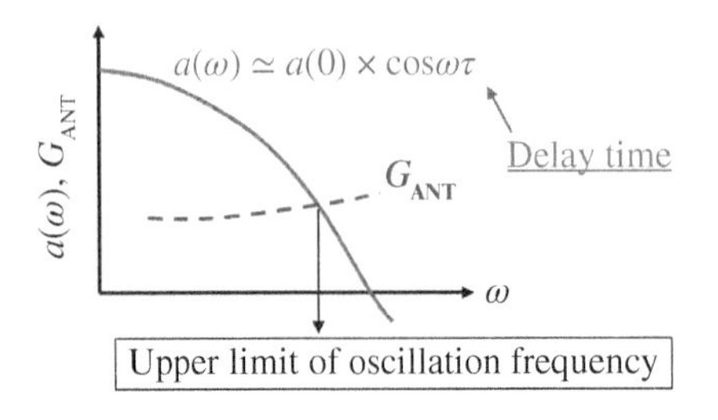

Figure 11.7 Schematic illustration of decrease of NDC with frequency indicating upper limit of oscillation condition. *Source*: Reprinted, with permission, from Ref. [65]. Copyright (2016) Springer Nature.

The layer structure of RTD in Figure 11.6 explained above has thin barriers and a thin quantum well for high peak current density. The electron delay time τ also decreases by these thin layers through the decrease in dwell time τ_{RTD} in the resonant tunneling structure. High-frequency oscillations is investigated with these thin layers, and the first oscillation above 1 THz [26] and the frequency increase to 1.3 THz [27] have been obtained. The collector transit time τ_{dep} , which is the other factor in the delay time τ, is reduced by thinning the collector depletion region. However, the spacer layer is inserted at the collector for the purpose of reducing the capacitance of RTD, as shown in Figure 11.6. Thus, the thickness of the collector spacer layer must be optimized because of the trade-off relation between a small capacitance and a short transit time.

Figure 11.8 shows the measured and calculated oscillation characteristics for RTDs with different thicknesses of the collector spacer [28, 73]; (i) oscillation frequency as a function of RTD mesa area, and (ii) output power, which will be discussed in Section 11.3.3, as a function of frequency. Details of the theoretical calculation including the parasitic elements are described in the last part of this section [72, 73]. The antenna parameters were calculated by a three-dimensional electromagnetic simulation (Ansys, HFSS). As shown in Figure 11.8a, the oscillation frequency increases with decreasing RTD mesa size because of the decrease in capacitance. At a fixed mesa size, the oscillation frequency increases with increasing spacer thickness also because of the decrease in capacitance. With decreasing mesa size, $a(\omega)$ decreases because of the two factors; the increase in oscillation frequency and the decrease in current. By these effects, the oscillation stops at the upper limit of oscillation frequency, as discussed above, which is indicated by the upper edges of theoretical curves in Figure 11.8a. The optimum spacer thickness, with which the upper limit is the highest, was 12 nm in this experiment, and an oscillation up to 1.42 THz was obtained at room temperature. The output power was 20–30 μW around 1 THz. The output power quickly goes down to zero at the upper limit of oscillation frequency, as shown in Figure 11.8b.

The electron delay time τ was estimated from the comparison between the measurements and calculations as a function of collector spacer thickness in Figure 11.8a. The dwell time

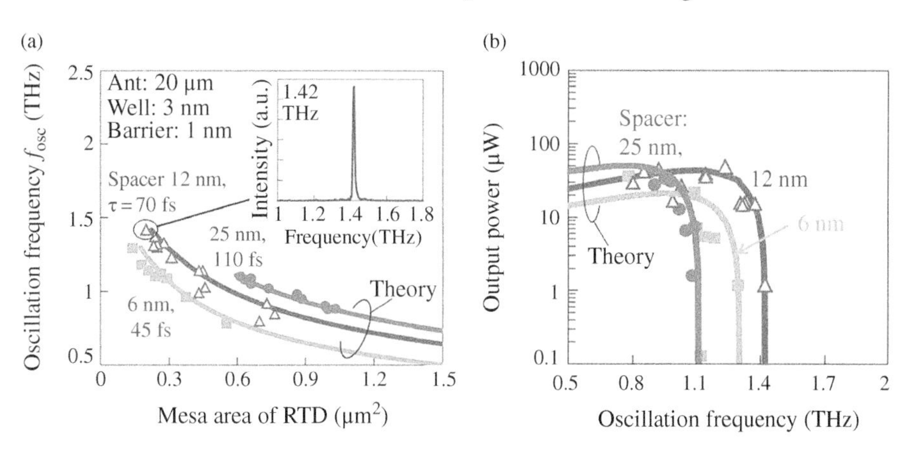

Figure 11.8 Oscillation characteristics of RTD oscillators with different thicknesses of the collector spacer. (a) Oscillation frequency as a function of RTD mesa size and (b) output power as a function of oscillation frequency. *Source*: Reprinted, with permission, from Ref. [28]. Copyright (2014) Springer Nature.

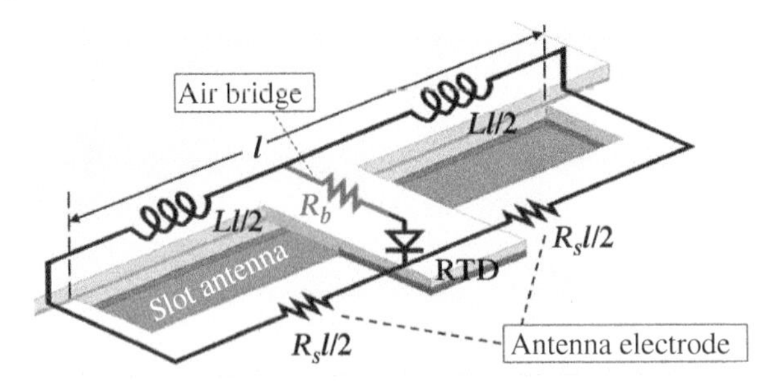

Figure 11.9 Schematic illustration of slot antenna and RTD with equivalent circuit of antenna electrode.

τ_{RTD} is estimated from τ in the limit of zero spacer thickness [73]. The decrease in τ_{RTD} with decreasing quantum-well thickness has also been obtained [73].

The above condition (ii) for high-frequency oscillation corresponds to the increase in the upper limit of oscillation frequency by the reduction in the antenna conductance G_{ANT} in Figure 11.7. G_{ANT} is approximately written as [31] $G_{ANT} \simeq G_{rad} + G_{loss}$, where G_{rad} and G_{loss} are the radiation conductance and nonradiative loss conductance, respectively. For a short antenna, G_{rad} is approximately proportional to square of the antenna length l [84]. G_{loss} is dominated by the conduction loss of the metal parts around RTD, which is given by Maekawa et al. [31] $G_{loss} \simeq (R_s l/4 + R_b)/(\omega L l/4)^2$, where R_s and L are the resistance and inductance per unit length along the slot antenna, and R_b is the resistance at the air bridge between the top contact of RTD and the slot antenna, as shown in Figure 11.9. From these relations, G_{rad} decreases while G_{loss} increases with decreasing antenna length l. The optimum antenna length thus exists in terms of the minimum G_{ANT} for the highest upper limit of oscillation frequency. By optimizing the antenna length, the oscillation frequency was increased to 1.55 THz [30].

Independent of antenna length, G_{loss} can be reduced by the reduction in R_b, as shown in the above equation. For this purpose, the air-bridge structure was improved [31]. In the structures described above, an n-doped GaInAs layer was covering the bottom surface of the metal air bridge connecting the RTD and antenna. From the three-dimensional electromagnetic simulation [31], the high-frequency current is found to flow in the GaInAs layer instead of the bottom surface of the metal air bridge due to the skin effect, resulting in high R_b. By improving the fabrication process, the GaInAs layer was removed, and the oscillation frequency was further increased to 1.92 THz [31], as shown in Figure 11.10. In this experiment, the highest frequency was obtained for the antenna length of 12 μm. This antenna length is not the optimized one, and a higher frequency is expected by 9 μm-long antenna, as shown in the theoretical curve in Figure 11.10. The output power is small at present (0.4 μW at 1.92 THz) because the oscillation frequency is close to the upper limit. As the upper limit grows away to higher frequency by the improvement, the output power will increases. The improved antenna structures described in the next section will also increase the output power.

The resistance R_s included in G_{loss} is reduced with increasing thickness of the antenna electrode. However, the decrease of R_s almost saturates at the thickness comparable to

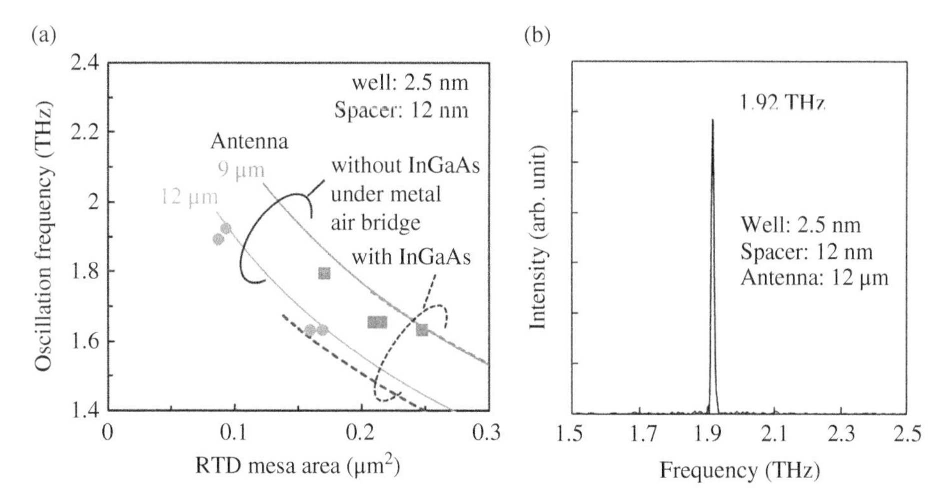

Figure 11.10 Oscillation characteristics of RTD oscillators with and without InGaAs layer under the metal air bridge. (a) Oscillation frequency as a function of RTD mesa size. *Source*: Copyright (2016) The Japan Society of Applied Physics [31]. (b) Oscillation spectrum for the sample with the highest frequency. *Source*: Reprinted, with permission, from Ref. [65]. Copyright (2016) Springer Nature.

double the skin depth (~150 nm for Au at 1 THz), because the high-frequency current penetrates the top and bottom surfaces of the electrode just by the skin depth.

With the increase in antenna thickness, the antenna inductance L also decreases. The decrease in L has the following two effects. One is the increase in G_{loss} which is inversely proportional to L^2 as mentioned above. Together with the change in R_s mentioned above, G_{loss} decreases with increasing antenna thickness when the antenna is thin, and then, increases. The other effect of the small L is that a large-area RTD having a large capacitance can be used even for high frequency. As a result, a large $a(0)$ can be used.

With these changes in G_{loss} and $a(0)$, the calculated upper limit of oscillation frequency increases with increasing antenna thickness up to roughly 1 µm, and then, almost saturated and slightly decreased with further increase in antenna thickness [32]. The experimental result was consistent with this calculation, and an oscillation up to 1.98 THz was obtained [32]. This is the highest oscillation frequency of room temperature electronic single oscillators to date. An oscillation above 2 THz may be expected by the combination of the optimized antenna length and thickness together with the optimized layer structure of RTD. Although it is difficult at present to universally estimate the ultimate limit of oscillation frequency, resonator structures other than the slot antenna, in which R_s and R_b can be largely reduced, must also be discussed for further increase in oscillation frequency.

For the condition (iii) mentioned above for high-frequency oscillation, the parasitic elements around the intrinsic RTD are shown in Figure 11.11. C_{RTD} is given by Eq. (11.8) including the delay-time effect, and R_{series} is the series resistance composed of the mesa-body resistance R_{mesa} and the spreading resistance R_{spread} between the foot of the RTD mesa and the lower electrode. R_c and C_c are the contact resistance and capacitance between the mesa top and the upper electrode, respectively. The mesa-body inductance and the kinetic inductance caused by the electron inertia in the region of R_{series} are negligibly small.

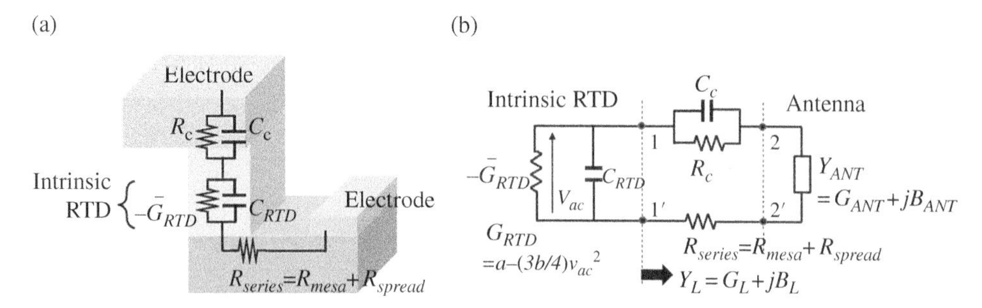

Figure 11.11 Parasitic elements around RTD. (a) Parasitic elements with schematic illustration of RTD mesa and electrodes, and (b) equivalent circuit including antenna.

The NDC viewed from the upper and lower electrodes (the real part of the admittance in the left-hand side of 2-2′ in Figure 11.11b) turns to positive at the frequency $f_1 \simeq 1/(2\pi C_{RTD})\sqrt{a/R_{series}}$, where a is $a(\omega)$ at f_1. The dominant parasitic elements in $f_1$are C_{RTD} and R_{series}. The effect of R_c on f_1 is small in case of high f_1 owing to the parallel capacitance C_c. The values of f_1 are estimated to be about 3.5 and 4 THz for the mesa areas S of 1 and 0.1 μ m^2, respectively, using the following parameter values: $C_{RTD} = 10$ fF/μm^2, $a = 60{\cdot}\cos\omega\tau$ mS/μm^2 with $\tau = 35$ fs, $R_{mesa} = \rho_{mesa} d_{mesa} = 0.4$ Ωμm^2 with the mesa resistivity $\rho_{mesa} = 10$ μΩm and mesa height d_{mesa}=40 nm, and $R_{spread} = \rho_{spread}/(\pi\, d)\ln(r/r_{mesa})$ where $\rho_{spread} = 2$ μΩm and $d = 0.3$ μm are the resistivity and thickness of the bottom n$^+$-GaInAs layer, respectively, r_{mesa} is the equivalent mesa radius given by $(S/\pi)^{1/2}$, and $r = r_{mesa} + 0.16$ μm is the distance between the mesa center and the edge of the lower electrode. The skin effect is neglected, because r_{mesa} and d are smaller than the skin depth.

f_1 is higher for smaller S, because a and C_{RTD} are proportional to S while R_{series} increases more slowly with S. f_1 is higher than the upper limit of oscillation frequency determined by the other factors mentioned above, and thus, the effect of the parasitic elements on the upper limit appears small at present. With the increase in oscillation frequency by future improvements, the effect of the parasitic elements may become non-negligible.

In the actual calculations of oscillation characteristics [72, 73], the parasitic elements are taken into account in the admittance of the right-hand side of 1-1′ in Figure 11.11b, which is the load connected to the intrinsic RTD. In this case, the oscillation condition is given by $\bar{G}_{RTD} = G_L$, where G_L is the real part of the admittance Y_L of the right-hand side of 1-1′ in Figure 11.11b, and the oscillation frequency is obtained by solving the equation $\omega C_{RTD} + B_L = 0$ with respect to frequency, where B_L is the imaginary part of Y_L. If the parasitic elements between 1-1′ and 2-2′ are negligible, the oscillation condition and the oscillation frequency reduce to Eq. (11.3). Theoretical calculations in Figure 11.8 were obtained by these methods together with the output power discussed in the next section.

11.3.3 High-output Power Oscillation

The output power of RTD oscillators is written using Eq. (11.3) as

$$P_{out} = \frac{1}{2} G_{rad} V_{ac}^2 = \frac{2}{3b} G_{rad}(a - G_{ANT}) \simeq \frac{1}{2}\frac{G_{rad}}{a}(a - G_{loss} - G_{rad})\Delta V^2, \tag{11.10}$$

where the approximation $b \simeq (4/3)(a/\Delta V^2)$ in Section 11.2.1 and $G_{ANT} = G_{rad} + G_{loss}$ with the radiation conductance G_{rad} and the nonradiative loss conductance G_{loss} are used in the last formula. P_{out} is changed with G_{rad}, and maximized at $G_{rad} = (a - G_{loss})/2$ as

$$P_{max} = \frac{1}{8}a\left(1 - \frac{G_{loss}}{a}\right)^2 \Delta V^2 \simeq \frac{3}{16}\Delta I \Delta V\left(1 - \frac{G_{loss}}{a(\omega)}\right)^2 \cos\omega\tau, \tag{11.11}$$

where the approximations $a = a(\omega) \simeq a(0)\cos\omega\tau$ in Section 11.2.3 and $a(0) \simeq (3/2)(\Delta I/\Delta V)$ in Section 11.2.1 are used in the last formula. For the precise analysis including the existence of the parasitic elements in Figure 11.11, the real part of the admittance $G_L$of the left-side of 1-1′ must be used instead of G_{ANT} in the middle of Eq. (11.10). $G_L$is approximated as $G_L \simeq G_{rad} + G'_{loss}$, where G'_{loss}includes the effect of the parasitic elements in addition to G_{loss}. G'_{loss} is independent of G_{rad} and slightly larger than G_{loss}, if the parasitic resistances are small compared to $1/|Y_{ANT}|$ and $G_{ANT} \ll B_{ANT}$. Thus, Eqs. (11.10) and (11.11) can be used even under the existence of the parasitic elements by replacing G_{loss} to G'_{loss}.

If $G_{loss} \ll a(\omega)$ and $\omega\tau \ll 1$ in Eq. (11.11), the maximum output power is given by $(3/16)\Delta I\Delta V$, which is proportional to the NDC area with the peak and valley points in the I–V curve as the diagonal positions in Figure 11.2d. As an example, this value is approximately 0.4–2 mW for the RTD in Figure 11.6 with the mesa area of 0.3–1.5 μm^2. However, experimental output powers are usually low (~10 μW). This is mainly because G_{rad} is much smaller than $a(\omega)$ and far away from the condition for the maximum output power, even if $G_{loss} \ll a(\omega)$ and $\omega\tau \ll 1$.

In order to make G_{rad} large and to fulfill the condition for the maximum output power, we proposed the offset slot antenna, in which the position of RTD is shifted from the center of the slot antenna, as shown in Figure 11.12a [85]. Using the offset slot antenna, we reported an RTD oscillator with the output power of 0.4 mW at 0.55 THz in a preliminary experiment [41]. The DC-to-terahertz conversion efficiency was 1.45% except for the power consumption at the resistance for stabilization shown in Figure 11.5. Figure 11.12b shows theoretical output power as a function of slot antenna width for different values of the current-density

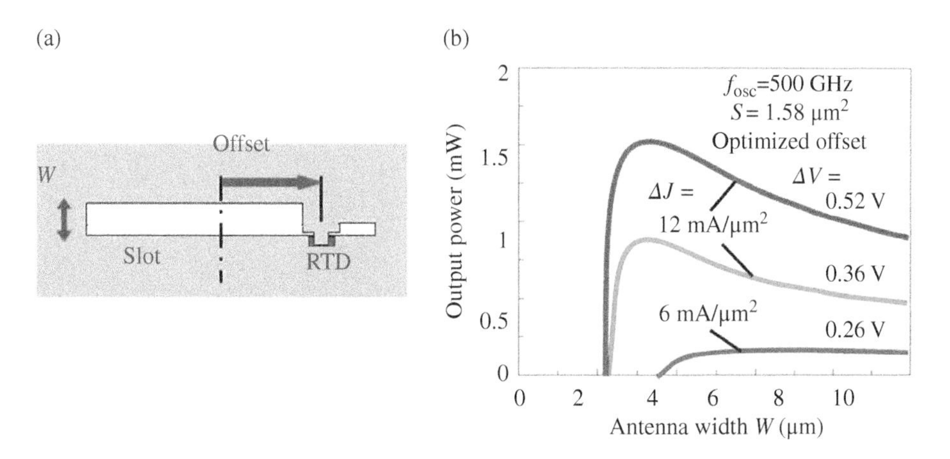

Figure 11.12 (a) Offset slot antenna for high output power and (b) theoretical output power as a function of slot antenna width for different values of current-density and voltage widths of NDC region in I–V curve of RTD. *Source*: Reprinted, with permission, from Ref. [65]. Copyright (2016) Springer Nature.

and voltage widths (ΔJ and ΔV) of the NDC region in the I–V curve of RTD [42, 86]. The offset position is optimized to maximize the peak of each curve. The output power of more than 1 mW is expected with an RTD having large ΔJ and ΔV and an optimized antenna structure. Thin barriers and a thin quantum well are effective for a large ΔJ as mentioned in Section 11.3.1, and a large ΔV is obtained by a thick collector spacer layer [73, 79], although the electron delay time increases.

In the offset slot antenna, the oscillation frequency is mainly determined by the resonance of electromagnetic waves at the short side (right-hand side from RTD in Figure 11.12a), while the output power is mainly determined by the radiation conductance of the long side (the left-hand side from RTD in Figure 11.12a). These separated roles of the both sides are the advantage of the offset slot antenna. However, the interaction between the both sides causes a small split of the resonance frequency of the short side. This results in that two frequencies with a small difference can oscillate, and a careful design of the structure is required for a stable oscillation at a desired frequency with the maximum output power. An alternative structure for high output power, in which the radiator and resonator are separated, was proposed [50, 51].

The array configuration for power combining is also effective for high output power [42, 48, 87–92]. Using the offset slot antenna and two-element array, we obtained the output power of 0.61 mW at 0.62 THz in a preliminary experiment [42], as shown in Figure 11.13. Two oscillator elements are coupled with each other through a MIM (metal–insulator–metal) stub structure in Figure 11.13a. The oscillation spectrum has a single peak as shown in Figure 11.13b. In a coupled array, the frequency and phase of each oscillator element coincide with each other because of the mutual injection locking, if the dispersion in oscillation frequencies between the elements is small in the individual free-running operation [67]. Because of this effect, the output from each element becomes coherent with each other, resulting in that the front power density and directivity increase in proportion with square of the element number. The combined total output power increases in proportion with element number. The dispersion in free-running frequencies allowed for the mutual locking is smaller for larger element number and for weaker coupling between the elements [67]. It is necessary to suppress the dispersion of the RTD sizes between the

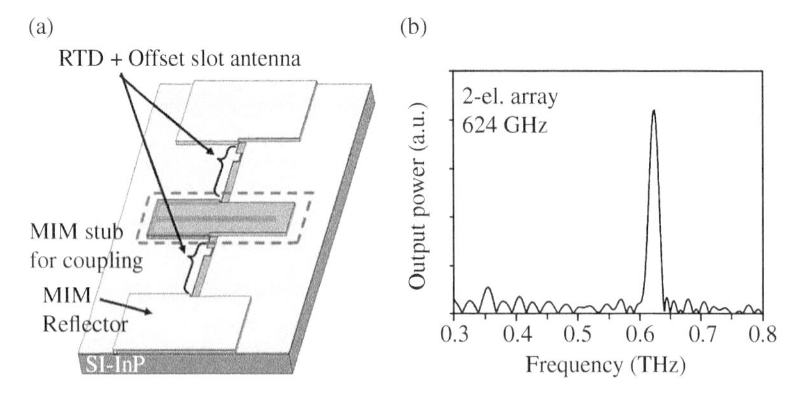

Figure 11.13 Array configuration of RTD oscillators for high output power. (a) Structure of two-element coupled array and (b) oscillation spectrum. *Source*: Copyright (2013) IEEE. Reprinted, with permission, from Ref. [42].

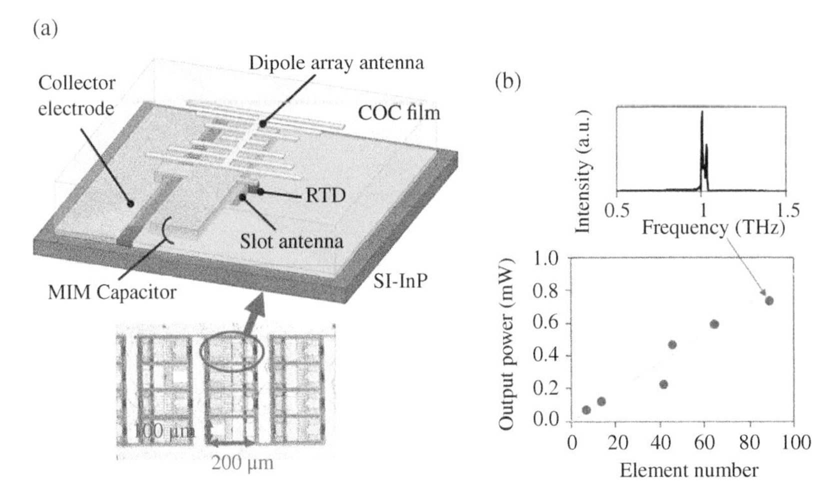

Figure 11.14 Large-scale array and output power. (a) Oscillator element of large-scale array constructed with dipole array for upward radiation and low-dielectric COC (cyclic olefin copolymer) film on RTD and slot antenna, and (b) output power as a function of element number with oscillation spectrum for maximum element number. *Source*: Reprinted from Ref. [48], with the permission of AIP Publishing.

elements, in order to obtain uniformity of the free-running oscillation frequencies. This condition may be satisfied by a precise fabrication process.

If the dispersion in free-running oscillation frequencies of the array elements is large, the mutual locking does not occur, and the output is incoherent. Even in this case, the total output power is proportional to element number. In some applications, such as imaging, the coherence of the source is not necessarily required. An incoherent high output power has been obtained by a large-scale array without taking care of the sufficient coupling between the elements (0.73 mW at 1 THz with 89-element array) [48], as shown in Figure 11.14. The average power per element is small in this preliminary experiment because of the decrease in radiation conductance, which is caused by the interaction of the dipole array between the elements. Higher output power is expected by an optimum arrangement of the array elements.

11.4 Control of Oscillation Spectrum and Frequency

11.4.1 Oscillation Spectrum and Phase-Locked Loop

The oscillation spectrum of RTD oscillators broadens due to the phase noise included in the output THz wave. The linewidth (full width at half maximum, FWHM) of the oscillation spectrum is given by [68]

$$\Delta f = \frac{1}{8\pi}\left(\frac{1}{CV_{ac}}\right)^2 \langle i_n^2 \rangle = \frac{\pi}{4}\frac{p_n \Delta f_r^2}{P_{out}}, \tag{11.12}$$

where C and V_{ac} are the capacitance and voltage amplitude across the RTD in the equivalent circuit shown in Figure 11.3b, $\langle i_n^2 \rangle$ is the average squire of the noise current per frequency, which is the sum of those generated at the RTD and G_{ANT}, P_{out} is the output power given by Eq. (11.10), $p_n = \eta \langle i_n^2 \rangle / G_{ANT}$ with $\eta = G_{rad}/G_{ANT}$, and $\Delta f_r = f/Q$ is the FWHM of the linewidth of the resonator without NDC (i.e. cold resonator), where f is the oscillation frequency, and $Q = 2\pi fC/G_{ANT}$ is the Q factor of the cold resonator.

The spectral linewidth of the RTD oscillator was measured by a harmonic heterodyne detection using an InP Schottky-barrier-diode detector [93]. Although the center of the spectral line was slowly fluctuating, the measured FWHM of the linewidth was less than 10 MHz, which was in reasonable agreement with the theoretical estimation (~6 MHz) obtained by Eq. (11.12) with the shot noise in the bias current of the RTD [93]. The measured linewidth was broad because of the low output power (~1 μW) in this experiment. The linewidth with the same order was obtained in the measurement using the beat between the outputs of two RTD oscillators [94].

The oscillation frequency slightly changes (1–5%) with bias voltage [93]. This property is originated from the change in dwell time in the resonant tunneling region discussed in Section 11.2.3.1. The capacitance changes according to the discussion in Section 11.2.3.2, and thus, the resonance frequency changes with bias voltage. The change in capacitance at the collector depletion layer with bias voltage is small, because the thickness of the depletion layer is almost determined by the spacer layer. Theoretical results were in reasonable agreement with measurements [95]. Using the bias-dependent capacitance change, the phase-locked loop (PLL) system was constructed, as shown in Figure 11.15 [96]. The output of an RTD oscillator is down-converted to 400 MHz by the heterodyne detection at first. The phase noise in the output of the RTD oscillator is then transformed to the amplitude noise by the balanced mixer and fed back into the bias voltage. Details of the operation of this PLL system are described in [97, 98]. The linewidth as narrow as 1 Hz was achieved by this system. The spectral narrowing by the PLL system of the frequency-tunable oscillators described in the next section was also demonstrated [97, 98].

The effect of the feedback by external reflection on oscillation characteristics is discussed [99]. Because the oscillation frequency, output power, and device current are influenced by this feedback, external reflections must be suppressed carefully.

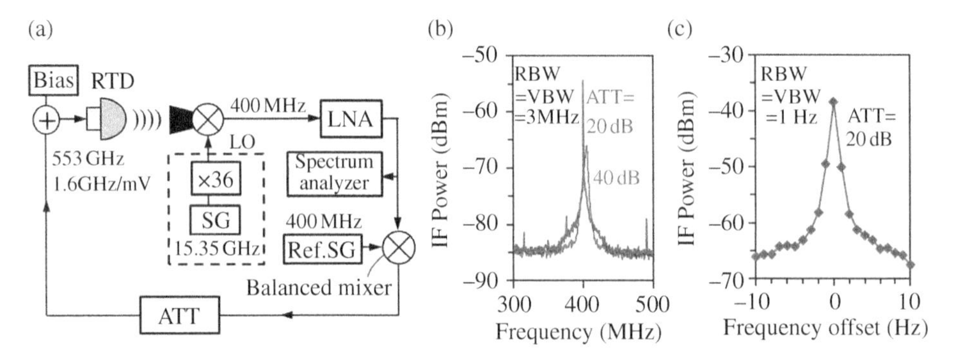

Figure 11.15 PLL system of an RTD THz oscillator. (a) Block diagram, (b) down-converted oscillation spectra with different feedback intensities, and (c) expanded spectrum under PLL.

11.4.2 Frequency-tunable Oscillators

Frequency sweep is a useful function of THz sources for various applications related to spectroscopy. However, the oscillation frequency of the RTD oscillators described above is fixed, once the antenna length and RTD mesa size are determined, except for a slight change with bias voltage mentioned in the previous section. To introduce a wide tunability of oscillation frequency into the RTD oscillators, we integrated a varactor diode in the slot antenna [54, 55], as shown in Figure 11.16a. The cross section of the varactor-integrated RTD oscillator is shown in Figure 11.16b along the RTD, slot antenna, and varactor. The varactor is composed of GaInAs layers with a p–n junction. The capacitance of the varactor is changed with reverse bias voltage through the change in thickness of the depletion layer at the p–n junction (mainly in the n-InGaAs layer side), resulting in the ability of electrical frequency tuning. The layer structure in Figure 11.16b is made with one-time epitaxial growth. The RTD layers under the varactor are negligible, because the voltage drop across this part is much smaller than that across the varactor.

The oscillation frequency increases with decreasing bias voltage (i.e. increasing negative bias voltage). The upper limit of the frequency tuning range is determined by the minimum capacitance of the varactor which is limited by the breakdown voltage. The lower limit of the tuning range is determined by the stop of oscillation with the increase in loss due to series resistance, which increases with increasing bias voltage because of the increase in thickness of the undepleted layer in the n-InGaAs layer in the varactor. Therefore, the doping

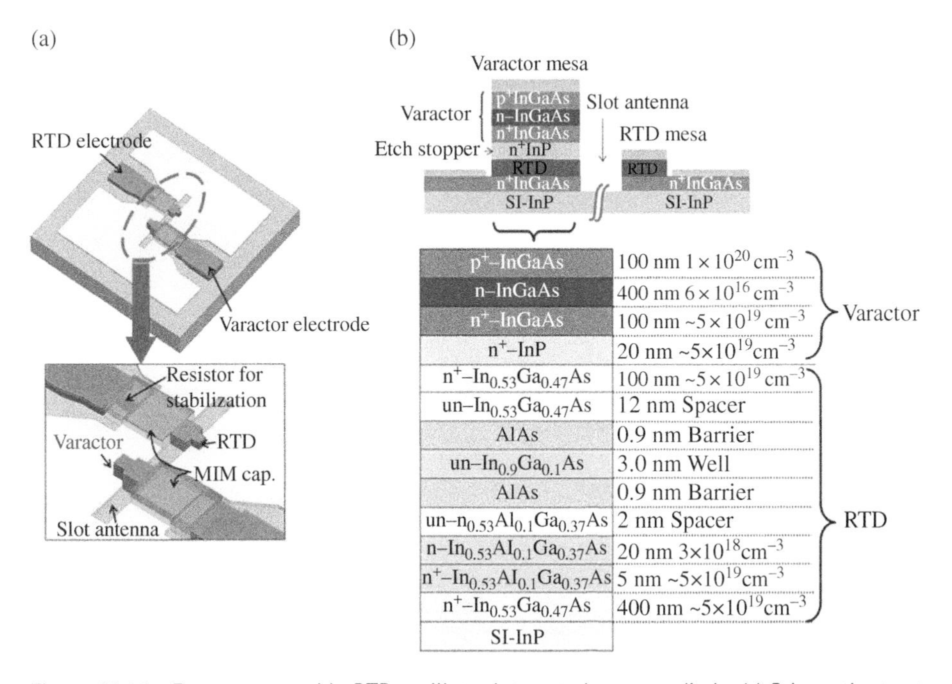

Figure 11.16 Frequency-tunable RTD oscillator integrated varactor diode. (a) Schematic structure and (b) cross section along the RTD, slot antenna, and varactor. *Source*: Reprinted, with permission, from Ref. [65]. Copyright (2016) Springer Nature.

concentration of the varactor must be optimized for wide tuning range. The ratio between the RTD and varactor areas must also be optimized for efficient use of the capacitance change of the varactor. By taking into account, these conditions in the device design, a wide tuning range of 580–700 GHz was obtained with the change in bias voltage of the varactor from −4 to +0.2 V [55]. Theoretical calculation agreed well with the measurement. The output power was typically 1–10 μW [54], which decreased with decreasing frequency because of the increase in series resistance of the undepleted layer in the varactor. The offset slot antenna may be useful for high output power.

To extend the tuning range, we fabricated arrays of the varactor-integrated RTD oscillators [55, 61]. Each oscillator element covers different frequency range. Figure 11.17 shows an example of array in which the frequency change of 410–970 GHz was obtained with sequential operation [61]. A wider range is possible by the structure optimization of each element as well as with more element number. Because the tuning range of a single device is typically 50–100 GHz, a wide range of 1 THz is expected with 10–20 element array.

The frequency tuning was also obtained by applying forward bias to the varactor [100]. With increasing forward bias voltage from 0.4 to 0.8 V, a steep increase of oscillation frequency from 600 to 730 GHz was observed. This behavior is not attributed to the capacitance change discussed in the reverse bias case, but to the inductance of the air bridge connected to the varactor. The inductance of the air bridge is connected in parallel to the slot antenna through the forward-biased varactor. The magnitude of the inductance viewed from the slot antenna steeply but continuously changes because of the continuous change in the series resistance of the varactor with the forward bias voltage. Thus, the total inductance of the resonance circuit decreases, and the oscillation frequency increases. Theoretical analysis of the frequency change agreed well with the measurement, and a wider tuning range is expected by optimizing the layer structure of the varactor [100].

The application of the frequency-tunable RTD oscillators to spectroscopy is shown in the next section. Spectral narrowing with the PLL system, in which the phase noise in the output of the RTD is fed back into the varactor, is also demonstrated [97, 98]. Wide tuning of a coherent THz wave with narrow-spectrum is possible with this system.

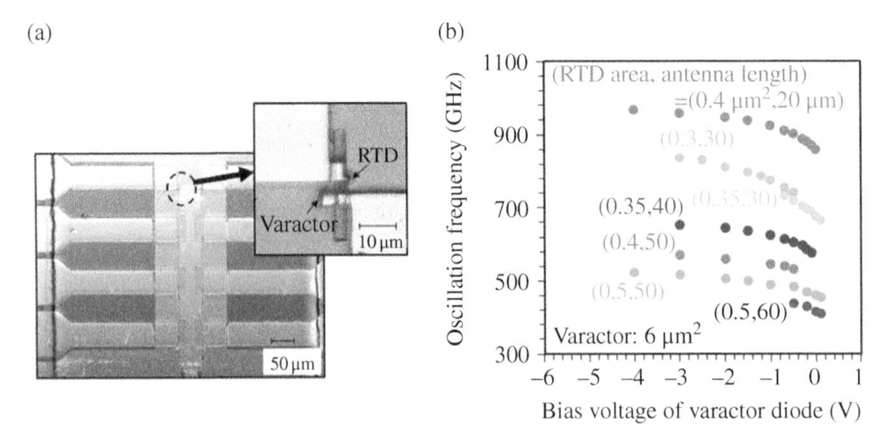

Figure 11.17 Array configuration of frequency-tunable oscillators for wide tuning range. (a) Microphotograph of array and (b) oscillation frequency as a function of bias voltage applied to varactor diodes. *Source*: Copyright (2017) The Japan Society of Applied Physics [61].

11.5 Targeted Applications

11.5.1 High-speed Wireless Communications

High-speed wireless communication is an important application of the THz frequency [101, 102]. The output power of RTD oscillators can be directly modulated with bias voltage for signal transmission. Because of this property, the RTD oscillator can be used as a compact and simple source. THz wireless communications with directly modulated RTDs have been demonstrated [45, 56–60].

The capacitance at the MIM structure in RTD oscillator in Figure 11.5a works a low-pass filter for external modulation signal and limits the modulation frequency. We fabricated an RTD oscillator with reduced MIM capacitance for high-frequency modulation, and obtained the cutoff frequency of 30 GHz [65, 103], as shown in Figure 11.18. Higher modulation frequency is possible with smaller MIM capacitance.

Wireless data transmission was demonstrated with the direct modulation of RTD and detection of Schottky barrier diode (SBD), as shown in Figure 11.19. An error-free transmission with the data rate of 25 Gbps and a transmission with 44 Gbps and the bit error rate (BER) of 5×10^{-4}, which is below the forward error correction (FEC) limit (BER $= 2 \times 10^{-3}$), were obtained. These are updated from the results in [59]. As shown in Figure 11.19b, BER rapidly increased with increasing data rate above 25 Gbps, because the modulation amplitude of the output power degrades by the limit of modulation frequency due to the external circuits, the dominant element of which is the inductance of the bonding wires. The signal-to-noise power ratio (SNR) decreases by the degradation of the modulation amplitude, resulting in the increase in BER. By improving the external circuits for modulation, higher data rates with lower BERs are expected.

The SNR at the receiver is determined by the antenna gain and modulation output power. In a theoretical calculation, a 50-Gbps simple amplitude-shift-keying (ASK) transmission at the carrier frequency of 800 GHz with BER $< 10^{-12}$ (SNR > 23 dB) requires the antenna gain

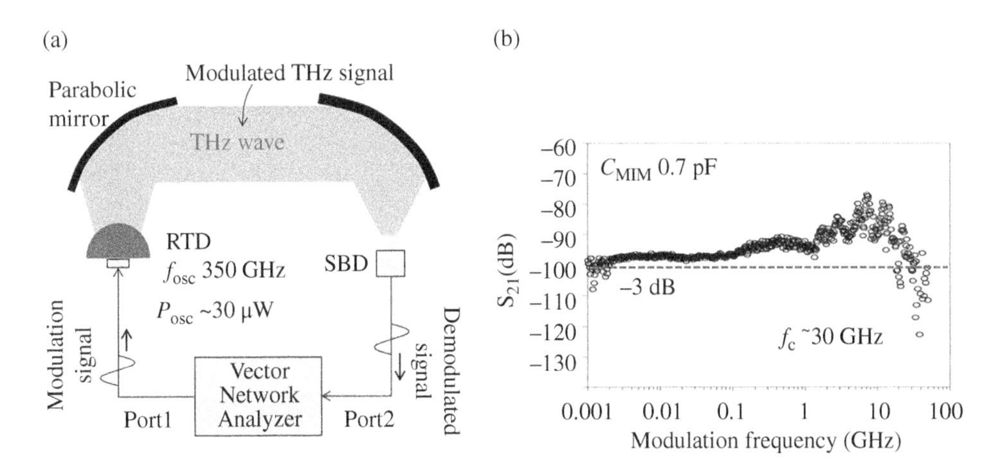

Figure 11.18 Direct modulation of RTD oscillator. (a) Measurement setup and (b) frequency response of direct modulation using a network analyzer as the modulation source. *Source*: Reprinted, with permission, from Ref. [65]. Copyright (2016) Springer Nature.

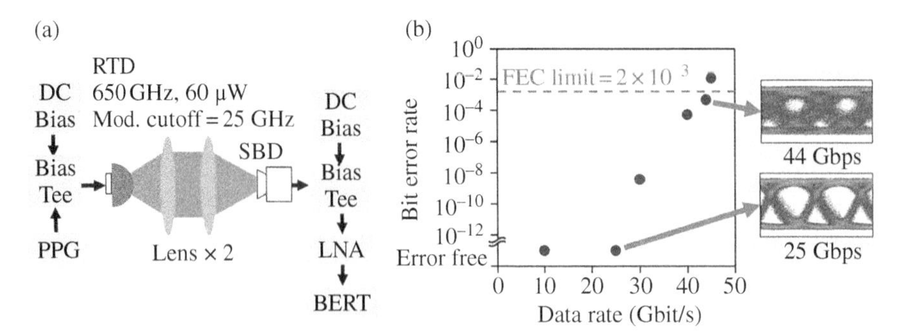

Figure 11.19 Wireless data transmission using direct modulation of RTD oscillator. (a) Measurement setup and (b) bit error rate as a function of data rate.

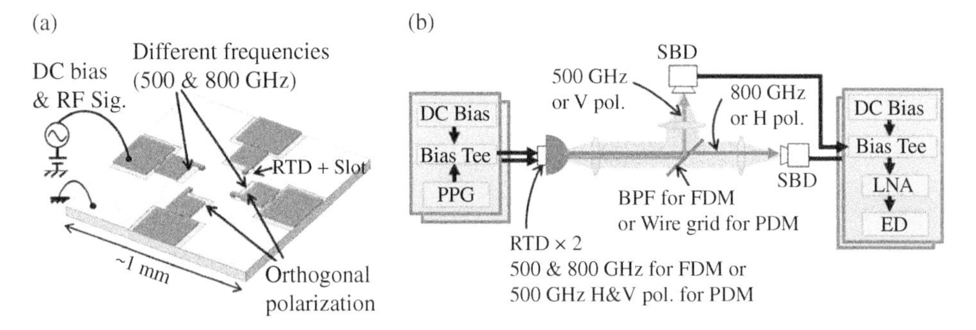

Figure 11.20 Wireless data transmission with frequency or polarization division multiplexing. (a) Integrated RTD oscillators for transmitter, and (b) measurement setup. *Source*: Copyright (2017) IEEE. Reprinted, with permission, from Ref. [60].

of 25 dBi for both of the transmitter and receiver and the output power of 100 μW for the transmission distance of 0.5 m [60]. This kind of short-distance high-capacity transmissions may be useful for inter-chip, inter-board, and kiosk-download uses. On the other hand, the antenna gain of 40 dBi and the output power of 1 mW are required for the same data rate with a relatively long distance of 50 m, which may be useful for indoor and inter-building transmissions. These applications can be targeted by compact transmitters using RTD oscillators.

In order to increase the transmission capacity, we fabricated integrated RTD arrays for multi-channel transmissions, as shown in Figure 11.20 [60]. Four RTD oscillators with two different frequencies (500 and 800 GHz) and two orthogonal polarization are integrated into one chip of this array. In a preliminary experiment, two-channel transmissions with frequency division multiplexing (FDM) or polarization division multiplexing (PDM) were separately investigated. In the experimental setup shown in Figure 11.20b, two outputs from the RTD oscillators are separated by a band-pass filter (BPF) in FDM (or by a wire-grid polarizer in PDM) and detected by two Schottky barrier diodes. Although the detectors are discrete, a compact integrated detector similar to the transmitter is possible. Two-channel transmission with the data rates of 28 Gbps × 2 was obtained below the FEC limit both in FDM and PDM [60]. Although the experiment was limited in two-channel transmissions because of the limit of the apparatus, four-channel transmission with FDM and PDM is possible, and the data rate >100 Gbps is expected. Higher data rates are further possible with more frequency channels as well as with multi-level modulations. As an example, 1.6 Tbps is estimated for the transmission with eight frequencies × two polarizations × four-level pulse amplitude modulation (PAM4). Combinations with frequency- or phase-shift keying (FSK or PSK) may also be possible with the frequency-tunable oscillators described in the previous section. Possibility of frequency division in the THz range for multi-channel short-distance communications has been discussed taking account of the atmospheric attenuation [60].

RTD oscillators with circular-polarized output have been reported [49]. The circular polarization is convenient for various communications, because the alignment of polarization axis is not necessary. In addition, the circular polarization is robust for the feedback by external reflection which disturbs communications through the changes in oscillation frequency and output power [99], because the polarization of the reflected wave rotates in reverse direction.

In the transmission systems mentioned above, direct detections at the receivers are used. A high sensitivity is expected with the heterodyne receivers. RTD oscillators and their integrated circuits are useful for the local oscillators in these systems.

11.5.2 Spectroscopy

The frequency-tunable RTD oscillators described in Section 11.4.2 can be applied to spectroscopy. Figure 11.21 shows the absorbance measurement of allopurinol in the THz range [61]. The absorbance is defined as $-\log_{10}(I/I_0)$, where I and I_0 are the intensities of the detected signals with and without specimen. Seven oscillators covering 410–970 THz in total were used for the source in this measurement. Although a Si-composite bolometer was used for the detector in this measurement, a compact Schottky barrier diode can also be used. As

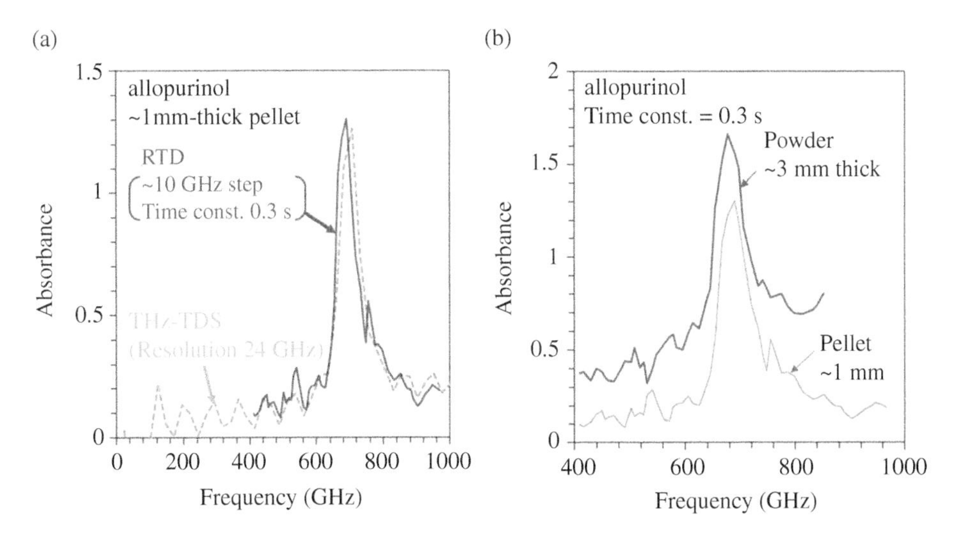

Figure 11.21 Absorbance of allopurinol measured by frequency-tunable RTD oscillators. (a) Comparison with TDS measurement and (b) comparison between powder and pellet samples. *Source*: Copyright (2017) The Japan Society of Applied Physics [61].

shown in Figure 11.21a, the measured result was in good agreement with the TDS measurement within resolutions of these measurements. By reducing the frequency step in the measurement, high resolution with short measurement time may be possible compared with the TDS measurement. The results were similar between powder and pellet samples, as shown in Figure 11.21b, although scatterings of THz wave may differently work in these sample configurations.

Compactness is the advantage of the RTD oscillators also in the application to spectroscopy. By the integration of the frequency-tunable RTD oscillator with a detector via a transmission line and a space for a drop of specimen, a microchip for spectroscopy is expected. Fast measurement may be possible even with a small output power of the oscillator because of the short distance between the source and detector. The output power required for the oscillator is estimated to be $P = \mathrm{SNR} \cdot \mathrm{NEP}/\sqrt{2\pi T}$ ~0.5–1.1 μW with the signal-to-noise power ratio (dynamic range) SNR = 30 dB at the detector, the noise equivalent power NEP = 20–50 pW/Hz$^{1/2}$ (typical for SBDs), and the time constant of the measurement T = 0.1 ms, neglecting the transmission loss between the source and detector. Although the spectral linewidth is as wide as 10 MHz, narrow linewidth for accurate measurement is possible by introducing the PLL, as mentioned in Section 11.4.1.

11.5.3 Other Applications and Expected Future Development

Imaging and sensing are also important applications of RTD THz oscillators [62, 63]. Two-dimensional transmitted and reflected images have been obtained by mechanically moving the object stage with the resolution of approximately 1 mm corresponding to the wavelength of the used frequency 300 GHz [62]. Interferometric reflection imaging for depth information has also been demonstrated.

For short measurement time, the line or area sensing with detector array is effective. Assuming a single source is used, the required output power is estimated as $P = N_{pixel} \cdot \mathrm{SNR} \cdot \mathrm{NEP}/\sqrt{2\pi T}$, where N_{pixel} is the pixel number of the detector array, SNR and NEP are the SNR and the noise equivalent power of the detector element, respectively, and T is the time constant of the measurement (~ reciprocal of the frame rate in the area sensing). As an example, P is approximately 0.5–1.3 mW for $N_{pixel} = 160 \times 120$, SNR = 30 dB, NEP = 20–50 pW/Hz$^{1/2}$, and T = 100 ms. Higher output power is required for more pixel number (higher resolution) and higher frame rate.

The coherence of the source is sometimes a drawback in a simple imaging, because an interference pattern disturbs a clear imaging. An array without synchronization between the source elements as shown in Figure 11.14 as well as the spectral broadening with direct modulation may be useful. The change in bias current due to the feedback with external reflection [99] can be utilized in the reflection imaging system. Three-dimensional imaging may be expected with various methods including frequency-modulated continuous wave and other time-of-flight measurement techniques.

The compactness and low-power consumption are the features of the RTD oscillators, and applications utilizing them are expected to increase. An ability of constructing integrated circuits is also an advantage of RTD oscillators, which is convenient for high functionalities. Circuit design in the THz range is important to utilize this property [104]. As regards the oscillation characteristics, further studies are required on higher-frequency oscillation, higher output power, and various functionalities such as the beam steering. As the oscillation frequency increases, difference of the oscillation mechanism from that of QCLs using the electron transition appears to become ambiguous [105]. Consideration on the electron transition in RTD oscillators may be useful for further improvement of the device characteristics, such as the output power.

Exercises

E11.1 a) Express the coefficients a and b in the right-hand side of Eq. (11.1) with the current and voltage widths of the NDC region (ΔI and ΔV in Figure 11.2d).
b) Derive and depict the NDC (dI_{dc}/dV_{dc}) as a function of V_{dc} using Eq. (11.1).

E11.2 a) Derive Eq. (11.2) for v_{ac} in the circuit shown in Figure 11.3b.
b) Derive Eq. (11.3) by substituting the approximate solution $v_{ac}(t) = V_{ac} \cos \omega t$ in Eq. (11.2) and neglecting the third harmonic terms.
c) V_{ac} in Eq. (11.3) is the result for the case of $V_{dc} = 0$ (center of NDC region). Discuss how V_{ac} changes with V_{dc} using the result of Problem 11.2b.

E11.3 a) Explain the reason why the parallel resistance for stabilization is necessary for the oscillator in Figure 11.5 (a), assuming the DC-bias source has a parallel resonance circuit with a low resonance frequency viewed from RTD.
b) Depict schematically the current-voltage curve of the oscillator including the resistor for stabilization.

E11.4 Discuss on the structures of RTD and antenna to obtain high-frequency oscillation.

E11.5 a) Derive Eq. (11.10) and depict the output power as a function of G_{rad}.
b) Equation (11.10) is the result of P_{out} for the case of $V_{dc} = 0$ (center of NDC region). Discuss how P_{out} changes with V_{dc} using the result of Problem 11.3c.
c) Derive Eq. (11.11) using the result of (a).
d) Calculate approximate values of P_{max} in Eq. (11.11) for the samples in Figure 11.6b with the RTD area of 0.3–1 μm^2.

E11.6 Suppose an antenna array with N elements shown in Figure 11.22. Each element has its own oscillator, and the magnitude of electric field for the radiation of one

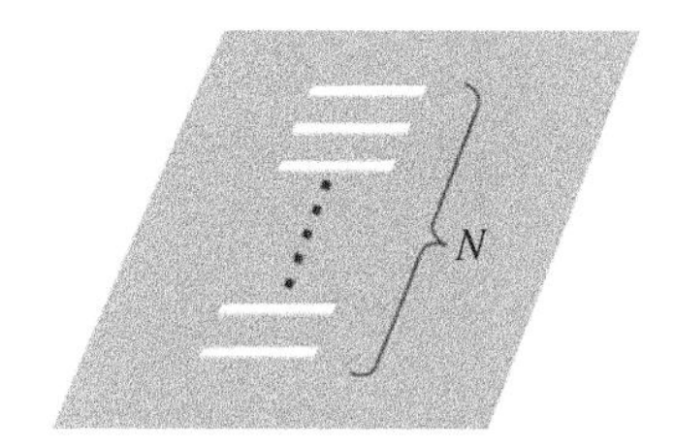

Figure 11.22 *N*-element antenna array. Each element is assumed to have its own oscillator, and all the oscillators are assumed to be synchronized.

element is E_0 at a point far enough in the direction just above the antenna. If the whole array is synchronized, the total magnitude for the radiation of all the elements is NE_0 at the same point. Because the power density is proportional to the square of the magnitude of the electric field, the power density at this point is proportional to N^2.

Does this mean that the radiated power of N-element array is proportional to N^2?

E11.7 Discuss on the structures of RTD and antenna to obtain high-output power oscillation.

E11.8 Discuss on the upper limit of modulation frequency for the oscillator in Figure 11.5 (a) in the direct modulation of output power by the superposition of modulation signal on bias voltage.

References

1 Tonouchi, M. (2007). Cutting-edge terahertz technology. *Nat. Photonics* **1**: 97–105.

2 Hangyo, M. (2015). Development and future prospects of terahertz technology. *Jpn. J. Appl. Phys.* **54**: 120101.

3 Komiyama, S. (1982). Far-infrared emission from population-inverted hot-carrier system in p-Ge. *Phys. Rev. Lett.* **48**: 271–274.

4 Köhler, R., Tredicucci, A., Beltram, F. et al. (2002). Terahertz semiconductor heterostructure laser. *Nature* **417**: 156–159.

5 Williams, B.S. (2007). Terahertz quantum-cascade lasers. *Nat. Photonics* **1**: 517–525.

6 Fathololoumi, S., Dupont, E., Chan, C.W.I. et al. (2012). Terahertz quantum cascade lasers operating up to ~200 K with optimized oscillator strength and improved injection tunneling. *Opt. Express* **20** (4): 3866–3876.

7 Liang, G., Liu, T., and Wang, Q.J. (2017). Recent developments of terahertz quantum cascade lasers. *IEEE J. Select. Top. Quant. Electr.* **23** (4): 1200118, July/August.

8 Urtega, M., Griffith, Z., Seo, M. et al. (2017). InP HBT technologies for THz integrated circuits. *Proc. IEEE* **105** (6): 1051–1067.

9 Tousi, Y.M., Momeni, O., and Afshari, E. (2012). A novel CMOS high-power terahertz VCO based on coupled oscillators: theory and implementation. *IEEE J. Solid-State Circuits* **47** (12): 3032–3042.

10 Pfeiffer, U.R., Zao, Y., Grzyb, J. et al. (2014). A 0.53 THz reconfigurable source module with up to 1 mW radiated power for diffuse illumination in terahertz imaging applications. *IEEE J. Solid-State Circuits* **49** (12): 2938–2950.

11 Steyaert, W. and Reynaert, P. (2014). A 0.54 THz signal generator in 40 nm bulk CMOS with 22 GHz tuning range and integrated planar antenna. *IEEE J. Solid-State Circuits* **49** (7): 1617–1625.

12 Yun, J., Yoon, D., Kim, H., and Reih, J.-S. (2014). 300-GHz InP HBT oscillators based on common-based cross-coupled topology. *IEEE Trans. Microwave Theor. Tech.* **62** (12): 3053–3064.

13 Mei, X., Yoshida, W., Lange, M. et al. (2015). First demonstration of amplification at 1 THz using 25-nm InP high electron mobility transistor process. *IEEE Electron. Dev. Lett.* **36** (4): 327–329.

14 Nishizawa, J., Plotka, P., Kurabayashi, T., and Makabe, H. (2008). 706-GHz GaAs CW fundamental-mode TUNNETT diodes fabricated with molecular layer epitaxy. *Phys. Stat. Sol. (C)* **5** (9): 2802–2804.

15 Eisele, H. and Kamoua, R. (2004). Submillimeter-wave InP gunn devices. *IEEE Trans. Microwave Theor. Tech.* **52** (10): 2371–2378.

16 Eisele, H. (2010). 480 GHz oscillator with an InP Gunn device. *Electr. Lett.* **46** (6): 422–423.

17 Tsu, R. and Esaki, L. (1973). Tunneling in a finite superlattice. *Appl. Phys. Lett.* **22** (11): 562–564.

18 Chang, L.L., Esaki, L., and Tsu, R. (1974). Resonant tunneling in semiconductor double barriers. *Appl. Phys. Lett.* **24** (12): 593–595.

19 Tsuchiya, M., Sakaki, H., and Yoshino, J. (1985). Room temperature observation of differential negative resistance in an AlAs/GaAs/AlAs resonant tunneling diode. *Jpn. J. Appl. Phys.* **24** (6): L466–L468.

20 Shewchuk, T.J., Chapin, P.C., Colemen, P.D. et al. (1985). Resonant tunneling oscillations in a GaAs–AlxGa1–xAs heterostructure at room temperature. *Appl. Phys. Lett.* **46** (5): 508–510.

21 Sollner, T.C.L.G., Tannenwald, P.E., Peck, D.D., and Goodhue, W.D. (1984). Quantum well oscillators. *Appl. Phys. Lett.* **45**: 1319–1321.

22 Liu, H.C. and Sollner, T.C.L.G. (1994). *High-Speed Heterostructure Device*", Chapter 6 (eds. R.A. Kiehl and T.C.L.G. Sollner). Academic, and references therein.

23 Brown, E.R., Sönderström, J.R., Parker, C.D. et al. (1991). Oscillations up to 712 GHz in InAs/AISb resonant-tunneling diodes. *Appl. Phys. Lett.* **58**: 2291–2293.

24 Reddy, M., Martin, S.C., Molnar, A.C. et al. (1997). Monolithic Schottky-Collector resonant tunnel diode oscillator arrays to 650 GHz. *IEEE Electr. Dev. Lett.* **18**: 218–221.

25 Suzuki, S., Teranishi, A., Hinata, K. et al. (2009). Fundamental oscillation of up to 831 GHz in GaInAs/AlAs resonant tunneling diode. *Appl. Phys. Exp.* **2**: 054501.

26 Suzuki, S., Asada, M., Teranishi, A. et al. (2010). Fundamental oscillation of resonant tunneling diodes above 1 THz at room temperature. *Appl. Phys. Lett.* **97**: 242102.

27 Kanaya, H., Shibayama, H., Sogabe, R. et al. (2012). Fundamental oscillation up to 1.31 THz in resonant tunneling diodes with thin well and barriers. *Appl. Phys. Exp.* **5**: 124101.

28 Kanaya, H., Sogabe, R., Maekawa, T. et al. (2014). Fundamental oscillation up to 1.42 THz in resonant tunneling diodes by optimized collector spacer thickness. *J. Infrared, Millimeter, THz Waves* **35**: 425–431.

29 Feiginov, M., Kanaya, H., Suzuki, S., and Asada, M. (2014). Operation of resonant-tunneling diodes with strong back injection from the collector at frequencies up to 1.46 THz. *Appl. Phys. Lett.* **104**: 243509.

30 Maekawa, T., Kanaya, H., Suzki, S., and Asada, M. (2014). Frequency increase in terahertz oscillation of resonant tunnelling diode up to 1.55 THz by reduced slot-antenna length. *Electr. Lett.* **50** (17): 1214–1216.

31 Maekawa, T., Kanaya, H., Suzuki, S., and Asada, M. (2016). Oscillation up to 1.92 THz in resonant tunneling diode by reduced conduction loss. *Appl. Phys. Express* **9**: 024101.

32 Izumi, R., Suzuki, S. and Asada, M. (2017). 1.98 THz resonant-tunneling-diode oscillator with reduced conduction loss by thick antenna electrode. Int. Conf. Infrared, Millimeter, and THz Waves, MA3.1, Cancun, August.

33 Inata, T., Muto, S., Nakata, Y. et al. (1987). A pseudomorphic $In_{0.53}Ga_{0.47}As/AlAs$ resonant tunneling barrier with a peak-to-valley current ratio of 14 at room temperature. *Jpn. J. Appl. Phys.* **26** (8): L1332–L1334.

34 Broekaert, T.P.E., Lee, W., and Fonstad, C.G. (1988). Pseudomorphic $In_{0.53}Ga_{0.47}As/AlAs/InAs$ resonant tunneling diodes with peak-to-valley current ratios of 30 at room temperature. *Appl. Phys. Lett.* **53** (16): 1545–1547.

35 Broekaert, T.P.E. and Fonstad, C.G. $In_{0.53}Ga_{0.47}As/AlAs$ resonant tunneling diodes with peak current densities in excess of 450 kA/cm^2. *J. Appl. Phys.* **68** (8): 4310–4312.

36 Shimizu, N., Nagatsuma, T., Waho, T. et al. (1995). $In_{0.53}Ga_{0.47}As/AlAs$ resonant tunneling diodes with switching timeof 1.5 ps. *Electr. Lett.* **31** (19): 1695–1696.

37 Kikuchi, A., Bannai, R., Kishino, K. et al. (2002). AlN/GaN double-barrier resonant tunneling diodes grown by rf-plasma-assisted molecular-beam epitaxy. *Appl. Phys. Lett.* **81** (9): 1729–1731.

38 Bayram, C., Vashaei, Z., and Razeghi, M. (2010). AlN/GaN double-barrier resonant tunneling diodes grown by metal-organic chemical vapor deposition. *Appl. Phys. Lett.* **96**: 042103.

39 Growden, T.A., Zhang, W., Brown, E.R. et al. (2018). 431 kA/cm^2 peak tunneling current density in GaN/AlN resonant tunneling diodes. *Appl. Phys. Lett.* **112**: 033508.

40 Encomendero, J., Yan, R., Verma, A. et al. (2018). Room temperature microwave oscillations in GaN/AlN resonant tunneling diodes with peak current densities up to 220 kA/cm^2. *Appl. Phys. Lett.* **112**: 103101.

41 Shiraishi, M., Shibayama, H., Ishigaki, K. et al. (2011). High output power (~400 μW) oscillators at around 550 GHz using resonant tunneling diodes with graded emitters and thin barriers. *Appl. Phys. Exp.* **4**: 064101.

42 Suzuki, S., Shiraishi, M., Shibayama, H., and Asada, M. (2013). High-power operation of terahertz oscillators with resonant tunneling diode using impedance-matched antennas and array configuration. *IEEE J. Selected Top. Quant. Electr.* **19** (1): 8500108.

43 Urayama, K., Aoki, S., Suzuki, S. et al. (2009). Sub-terahertz resonant tunneling diode oscillators integrated with tapered slot antennas for horizontal radiation. *Appl. Phys. Exp.* **2**: 044501.

44 Feiginov, M., Sydlo, C., Cojocari, O., and Meissner, P. (2011). Resonant-tunnelling-diode oscillators operating at frequencies above 1.1 THz. *Appl. Phys. Lett.* **99**: 233506.

45 Mukai, T., Kawamura, M., Takada, T., and Nagatsuma, T. (2011). 1.5-Gbps wireless transmission using resonant tunneling diodes at 300 GHz. Int. Conf. Opt. Terahertz Sci. and Technol., MF42, Santa Barbara, March.

46 Koyama, Y., Sekiguchi, R., and Ouchi, T. (2013). Oscillations up to 1.40 THz from resonant-tunneling-diode-based oscillators with integrated patch antennas. *Appl. Phys. Express* **6**: 064102.

47 Okada, K., Kasagi, K., Oshima, N. et al. (2015). Resonant-tunneling-diode terahertz oscillator using patch antenna integrated on slot resonator for power radiation. *IEEE Trans. THz Sci. Tech.* **5** (4): 613–618.

48 Kasagi, K., Suzuki, S., and Asada, M. (2019). Large-scale array of resonant-tunneling-diode terahertz oscillator for high output power at 1 THz. *J. Appl. Phys.* **125**: 151601.

49 Horikawa, D., Suzuki, S., and Asada, M. (2016). Resonant-tunneling-diode terahertz oscillator integrated with radial line slot antenna for circularly polarized wave radiation. Int.

Conf. Infrared and Millimeter Wave and Terahertz (IRMMW-THz), W5P-08-38, Copenhagen, September.

50 Alharbi, K.H., Khalid, A., Ofiare, A. et al. (2017). Diced and grounded broadband bow-tie antenna with tuning stub for resonant tunneling diode terahertz oscillators. *IET Microwave Antennas Prop.* **11** (3): 310–316.

51 Sasaki, S., Kawamura, S., Suzuki, S., and Asada, M. (2017). Resonant-tunneling-diode terahertz oscillators integrated with broadband bow-tie antenna. Compound Semiconductor Week, P2.31, Berlin, May.

52 Tateishi, Y., Koyama, Y., Sekiguchi, R., and Ouchi, T. (2013). Room temperature operation of RTD based stripe oscillators using double-sided metal waveguide. Optical Terahertz Science and Technology Conf. (OTST), Tu4-48, Kyoto, April.

53 Feiginov, M. (2015). Sub-terahertz and terahertz microstrip resonant-tunneling-diode oscillators. *Appl. Phys. Lett.* **107**: 123504.

54 Kitagawa, S., Suzuki, S., and Asada, M. (2014). Wide range varactor-integrated terahertz oscillator using resonant tunneling diode. *J. Infrared, Millimeter, THz Waves* **35**: 445–450.

55 Kitagawa, S., Suzuki, S., and Asada, M. (2016). Wide frequency-tunable resonant tunneling diode terahertz oscillators using varactor diodes. *Electron. Lett.* **52** (6): 479–481.

56 Wang, J., Al-Khalidi, A., Zhang, C. et al. (2017). Resonant tunneling diode as high speed optical/electronic transmitter. UK–Europe–China Workshop on Millimetre Waves and Terahertz Technologies (UCMMT), #63, Liverpool, September.

57 Diebold, S., Nishio, K., Nishida, Y. et al. (2016). High-speed error-free wireless data transmission using a terahertz resonant tunneling diode transmitter and receiver. *Electr. Lett.* **52** (24): 1999–2001.

58 Ishigaki, K., Shiraishi, M., Suzuki, S. et al. (2012). Direct intensity modulation and wireless data transmission characteristics of terahertz-oscillating resonant tunneling diodes. *Electr. Lett.* **48** (10): 582–583.

59 Oshima, N., Hashimoto, K., Suzuki, S., and Asada, M. (2016). Wireless data transmission of 34 Gbit/s at a 500-GHz range using resonant-tunnelling-diode terahertz oscillator. *Electr. Lett.* **52** (22): 1897–1898.

60 Oshima, N., Hashimoto, K., Suzuki, S., and Asada, M. (2017). Terahertz wireless data transmission with frequency and polarization division multiplexing using resonant-tunneling-diode oscillators. *IEEE Trans. THz Sci. Technol.* **7** (5): 593–598.

61 Kitagawa, S., Mizuno, M., Saito, S. et al. (2017). Frequency-tunable resonant-tunneling-diode oscillators applied to absorbance measurement. *Jpn. J. Appl. Phys.* **56**: 058002.

62 Miyamoto, T., Yamaguchi, A., and Mukai, T. (2016). Terahertz imaging system with resonant tunneling diodes. *Jpn. J. Appl. Phys.* **55**: 032201.

63 Okamoto, K., Tsuruda, K., Diebold, S. et al. (2017). Terahertz sensor using photonic crystal cavity and resonant tunneling diodes. *J. Infrared Millimeter THz Waves* **38**: 1085–1097.

64 Shiode, T., Mukai, T., Kawamura, M., and Nagatsuma, T. (2011). Giga-bit wireless communication at 300 GHz using resonant tunneling diode detector. *Proc. Asia Pacific Microwave Conf.*: 1122–1125.

65 Asada, M. and Suzuki, S. (2016). Room-temperature oscillation of resonant tunneling diode close to 2 THz and their functions for various application. *J. Infrared, Millimeter, THz Waves* **37**: 1185–1198.

66 Kim, C.S. and Brändli, A. (1961). High-frequency high-power operation of tunnel diodes. *IRE Trans. Circuit Theor.* **8**: 416–425.

67 Asada, M. and Suzuki, S. (2008). Theoretical analysis of coupled oscillator array using resonant tunneling diodes in subterahertz and terahertz range. *J. Appl. Phys.* **103**: 124514.

68 Asada, M. (2010). Theoretical analysis of spectral linewidth of terahertz oscillators using resonant tunneling diodes and their coupled arrays. *J. Appl. Phys.* **108**: 034504.

69 Orihashi, N., Suzki, S., and Asada, M. (2005). One THz harmonic oscillation of resonant tunneling diodes. *Appl. Phys. Lett.* **87**: 233502.

70 Shockley, W. (1938). Currents to conductors induced by a moving point charge. *J. Appl. Phys.* **9**: 635–636.

71 Ramo, S. (1939). Currents induced by electron motion. *Proc. IRE* **27**: 584–585.

72 Asada, M., Suzuki, S., and Kishimoto, N. (2008). Resonant tunneling diodes for sub-terahertz and terahertz oscillators. *Jpn. J. Appl. Phys.* **47** (6): 4375–4384.

73 Kanaya, H., Maekawa, T., Suzuki, S., and Asada, M. (2015). Structure dependence of oscillation characteristics of resonant-tunneling-diode oscillators associated with intrinsic and extrinsic delay times. *Jpn. J. Appl. Phys.* **54**: 094103.

74 Liu, H.C. (1991). Analytical model of high-frequency resonant tunneling: the first-order ac current response. *Phys. Rev. B* **43** (15): 12538–12548; (1993) Erratum, vol. 48, p. 4877.

75 Sugimura, A. (1994). Resonant enhancement of terahertz dynamics in double barrier resonant tunneling diodes. *Semicond. Sci. Technol.* **9**: 512–514.

76 Asada, M. (2001). Density-matrix modeling of terahertz photon-assisted tunneling and optical gain in resonant tunneling diodes. *Jpn. J. Appl. Phys.* **40** (9A): 5251–5256.

77 Feiginov, M.N. (2000). Effect of the Coulomb interaction on the response time and impedance of the resonant tunneling diodes. *Appl. Phys. Lett.* **76** (20): 2904–2906.

78 Feiginov, M.N. (2001). Displacement currents and the real part of high-frequency conductance of the resonant tunneling diode. *Appl. Phys. Lett.* **78** (21): 3301–3303.

79 Shimizu, N., Waho, T., and Ishibashi, T. (1997). Capacitance anomaly in the negative differential resistance region of resonant tunneling diodes. *Jpn. J. Appl. Phys.* **36** (3B): L330–L333.

80 Wei, T., Stapleton, S., and Berolo, E. (1993). Equivalent circuit and capacitance of double barrier resonant tunneling diode. *J. Appl. Phys.* **73** (2): 829–834.

81 Lake, R. and Yang, J. (2003). A physics based model for the RTD quantum capacitance. *IEEE Trans. Electr. Dev.* **50** (3): 785–789.

82 Rutledge, D.B., Neikirk, D.P., and Kasilingam, D.P. (1983). *Infrared and Millimeter Waves*, vol. **10**, Chapter 1 (ed. K.J. Button). Academic Press.

83 Lee, J., Kim, M., and Yang, K. (2016). A 1.52 THz RTD triple-push oscillator with a μW-level output power. *IEEE Trans. THz Sci. Tech.* **6** (2): 336–340.

84 Baranis, C.A. (2005). *Antenna Theory*", Sections 4.2.2 and 12.8. Wiley.

85 Suzuki, S. and Asada, M. (2007). Proposal of resonant tunneling diode oscillators with offset-fed slot antennas in terahertz and sub-terahertz range. *Jpn. J. Appl. Phys.* **46** (1): 119–121.

86 Shibayama, H., Shiraishi, M., Suzuki, S., and Asada, M. (2012). Dependence of output power on slot antenna width in terahertz oscillating resonant tunneling diodes. *J. Infrared, Millimeter, Terahertz Waves* **33**: 475–478.

87 Stephan, K.D., Wang, S.-C., Brown, E.R. et al. (1992). 5 mW parallel-connected resonant-tunneling diode oscillator. *Electr. Lett.* **28** (15): 1411–1412.

88 Steenson, D.P., Miles, R.E., Pollard, R.D. et al. (1994). Demonstration of power combining at W-band from GaAs/AlAs resonant tunneling diodes. *Int. Symp. Space Terahertz Technol.*: 756–767.

89 DeLisio, M.P., Davis, J.F., Li, S.-J. et al. (1995). A 16-element tunnel diode grid oscillator. *IEEE AP-S, Int. Symp.*: 1284–1287.

90 Fujii, T., Mazaki, H., Takei, F. (1996). Coherent power combining of millimeter wave resonant tunneling diodes in a quasi-optical resonator. IEEE MTT-S, Int. Microwave Symp., WE3F-28, June, San Francisco, 919–922.

91 Cantu, H.I. and Trunscott, W.S. (2001). Injection locking and power combining with double barrier resonant tunneling diodes. *Electr. Lett.* **37** (20): 1264–1265.

92 Kasagi, K., Oshima, N., Suzuki, S., and Asada, M. (2015). Power combination in 1 THz resonant-tunneling-diode oscillators integrated with patch antennas. *IEICE Trans. Electr.* **E98-C** (12): 1131–1133.

93 Karashima, K., Yokoyama, R., Shiraishi, M. et al. (2010). Measurement of oscillation frequency and spectral linewidth of sub-terahertz InP-based resonant tunneling diode oscillators using Ni–InP Schottky barrier diode. *Jpn. J. Appl. Phys.* **49**: 020208.

94 Suzuki, S., Karashima, K., Ishigaki, K., and Asada, M. (2011). Heterodyne mixing of sub-terahertz output power from two resonant tunneling diodes using InP Schottky barrier diode. *Jpn. J. Appl. Phys.* **50**: 080211.

95 Asada, M., Orihashi, N., and Suzuki, S. (2006). Experiment and theoretical analysis of voltage-controlled sub-THz oscillation of resonant tunneling diodes. *IEICE Jpn. Trans. Electr.* **E89-C** (7): 965–972.

96 Ogino, K., Suzuki, S., and Asada, M. (2017). Phase locking of resonant-tunneling-diode terahertz oscillator using bias-dependent oscillation frequency. Int. Commission for Optics, Tu2G-04, Tokyo, August.

97 Ogino, K., Suzuki, S., and Asada, M. (2017). Spectral narrowing of a varactor-integrated resonant-tunneling-diode terahertz oscillator by phase-locked loop. *J. Infrared, Millimeter, THz Waves* **38**: 1477–1486.

98 Ogino, K., Suzuki, S., and Asada, M. (2018). Phase locking and frequency tuning of resonant-tunneling-diode terahertz oscillators. *IEICE Brief Paper* **E101-C** (3): 183–185.

99 Asada, M. and Suzuki, S. (2015). Theoretical analysis of external feedback effect on oscillation characteristics of resonant-tunneling-diode terahertz oscillators. *Jpn. J. Appl. Phys.* **54**: 070309.

100 Kitagawa, S., Ogino, K., Suzuki, S., and Asada, M. (2017). Wide frequency tuning in resonant-tunneling-diode terahertz oscillator using forward-biased varactor diode. *Jpn. J. Appl. Phys.* **56**: 040301.

101 Song, H.-J. and Nagatsuma, T. (2011). Present and future of terahertz communications. *IEEE Trans. THz Sci. Tech.* **1** (1): 256–263.

102 Kleine-Ostmann, T. and Nagatsuma, T. (2011). A review on terahertz communications research. *J. Infrared, Millimeter, Terahertz Waves* **32**: 143–171.

103 Ikeda, Y., Kitagawa, S., Okada, K. et al. (2015). Direct intensity modulation of resonant-tunneling-diode terahertz oscillator up to ~30GHz. *IEICE Electr. Exp.* **12** (3): 2014116.

104 Diebold, S., Nakai, S., Nishio, K. et al. (2016). Modeling and simulation of Terahertz resonant tunneling diode-based circuits. *IEEE Trans. THz Sci. Technol.* **6** (5): 716–723.

105 Faist, J. and Scalari, G. (2010). Unified description of resonant tunneling diodes and terahertz quantum cascade lasers. *Electr. Lett.* **46** (26): 46–49.

12

Wireless Communications in the THz Range

Guillaume Ducournau[1] *and Tadao Nagatsuma*[2]

[1] *Univ. Lille, CNRS, Centrale Lille, Univ. Polytechnique Hauts-de-France, UMR 8520; Institute of Electronics, Microelectronics and Nanotechnology (IEMN), Villeneuve d'Ascq Cedex, France*
[2] *Graduate School of Engineering Science, Osaka University, Toyonaka, Osaka, Japan*

12.1 Introduction

Connecting ultra-fast mobile users is one of the great challenges of the twenty-first century. Wireless data networks are under tremendous pressure on the amount of data to be delivered to citizens, and conventional approaches are no longer sufficient. In a world that has become hyper-connected, digital information densities of several Terabits per second/km^2 are predicted. These new uses, such as video streaming in high definition (4/8 K), gaming, augmented reality, soon autonomous vehicles, real-time remote surgery, require the routing of masses of data, with a low latency time. Technologies based on "terahertz (THz)" or "T" waves can provide solutions with this need for wireless connectivity. Since the early days of radiocommunications, the capacity of transmission links has steadily increased but nowadays the electromagnetic spectrum is saturated on most frequencies already allocated. New frequency resources are being explored above 100 GHz, and more particularly in the 200–320 GHz range have great potential for very high-speed wireless transmissions. Since the first demonstrations of first wireless terahertz communication links in the laboratory, the technologies have evolved a lot, based on electronic and photonic building blocks. An overview of the systems already demonstrated in the laboratory or in an operational situation will be presented, giving the potential future trends.

12.2 Evolution of Telecoms Toward THz

12.2.1 Brief Historic

The use of mm and sub-mm (>300 GHz) waves for communications was considered very early, probably during the Second World War. At the end of the 1940s, mm-wave sources

Fundamentals of Terahertz Devices and Applications, First Edition. Edited by Dimitris Pavlidis.

Companion website: www.wiley.com/go/Pavlidis/FundamentalsofTHz

were beginning to be used for applications and spectroscopic studies showed that atmospheric absorption bands existed, separated by bands of good transparency. The variability of weather-dependent transmissions has prompted Bell Telephone Labs (BTL) engineers to develop long-distance, high-bandwidth communications in low-loss waveguides at 50 GHz. However, the discovery of the laser in 1960s and the gradual reduction of losses in silica fibers led to the discontinuation of this research on guided communications in the 1970s. However, they were continued for wireless applications, and today frequencies around 38, 50, and 70–80 GHz are used for point-to-point links and frequencies around 60 GHz are used for high-speed and short-distance communications (the atmosphere has a specific attenuation around 60 GHz). However, the saturation of the frequency bands used today and the very significant increase in requirements has prompted researchers to examine the potential use of THz waves for communications since the early 2000s.

12.2.2 Data Rate Evolution

The need to communicate has always guided the development of information transport technologies. The revolution in optical communications leads to the fact that the capacity of copper cables or wireless is far exceeded by fiber-guided optical technologies. This applies for both long-haul (submarine, metropolitan) networks but also on a smaller scale with free-space optical links to replace data buses in high-speed computers, these technologies being increasingly developed in "Photonic on silicon" technology, with a high level of integration. It is likely that the twenty-first century will partly see the advent of technologies for the massive development of "wireless." With 11.5 exabytes (11.5 EB (ExaByte)) of mobile data per month in 2017, more than 275 TB of data are exchanged worldwide every ... minutes! By 2022, telephones will exceed 90% of mobile data traffic [1] and 5G will account for more than 10% (12%) of total mobile traffic. Thus, while fiber-optic communications have reached very fast transmission speeds, the wireless connection speed, which is doubled every 18 months according to Edholm's law [2], remains the bottleneck of data exchanges.

Since the beginning of the radio with Marconi's experiments, the "carrier" frequency has never stopped to increase. Thus, in the perspective of the future communication networks, the wireless transmission systems will be required to use the beginning of the terahertz frequency range by 2025–2030, to increase the capacity of the network between each base station.

12.2.3 THz Waves: Propagation, Advantages, and Disadvantages

The main advantage of the THz frequencies compared to the lower frequencies relates to the data rate: with carrier frequencies of several hundred of GHz, it is theoretically possible to support 100 Gbit/s and beyond. The reduction of the wavelength makes it also possible to get much more directional beams for comparable antenna dimensions. The concentration of the beams also makes it possible to consider the reuse of the frequencies locally, which is very difficult at usual microwave frequencies.

The atmosphere has absorption bands mainly due to two components: water vapor (H_2O) and oxygen (O_2). This is strongly affecting the losses at particular frequencies (60, 119, 183, 335, 443 GHz, etc.). Thus, to limit the effects of distortions/losses in future THz radiocommunication systems on these frequency range, the future channels will have to use a

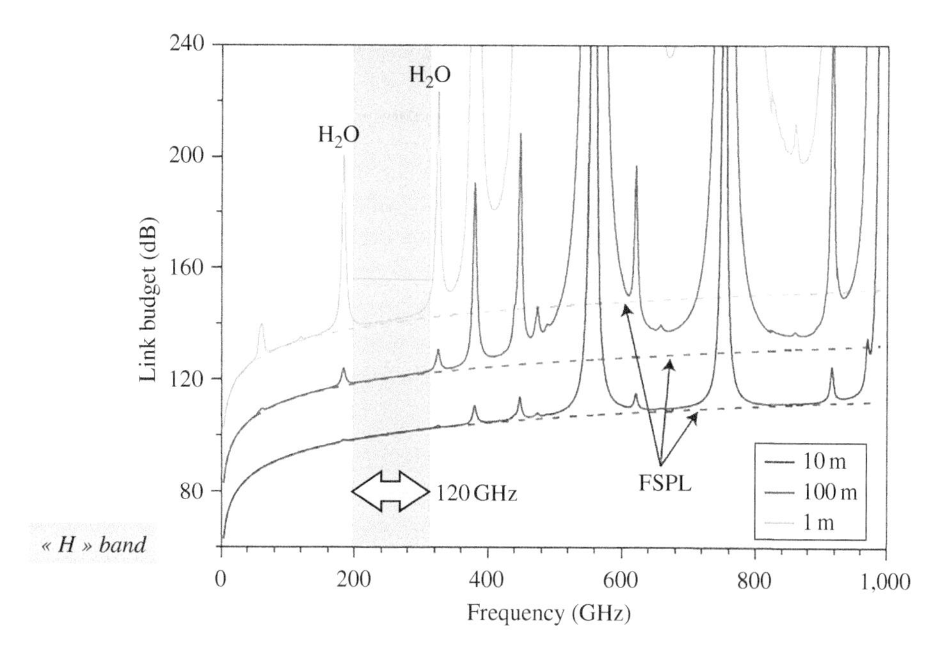

Figure 12.1 Link budget (L_{dB}) and FSPL up to 1 THz [3], with antenna gains of 0 dBi (isotropic antennas). Around 300 GHz, the atmospheric contribution is limited for zero or low precipitation (<5 mm/h). Higher frequencies are reserved for short-distance applications.

bandwidth free of strong absorption lines, and in this perspective, the 200–320 GHz band looks promising, and moreover, the frequencies beyond 275 GHz are still free (discussions are on the way at the end of 2019). Inside the transmission windows (300, 340, 410, and 480 GHz) and very dry atmosphere attenuations of 1–6 dB/km are observed. However, in standard conditions, attenuations of 15–50 dB/km are present and are comparable to the 60 GHz attenuation. THz communications (>300 GHz) are, in principle, limited at sea level at short distances (from a few cm to a few 100 m), or even km-range around 300 GHz. Above the sea level, greater distances can be envisaged and communications between aircraft or between an aircraft and a satellite or between satellites are possible. The path attenuation in the absence of particular phenomena (see below) can be calculated by the following formula:

$$L_{dB} = 92.4 - G_e - G_r + 20\log f + 20\log d + Ad = \mathrm{FSPL} - G_e - G_r + Ad \tag{12.1}$$

G_e and G_r are the gain in dB of the transmit and receive antennas, f is the frequency in GHz, d is the distance in km, and A is the atmospheric absorption in dB/km. The free space path loss (FSPL) term corresponds to the loss associated with the spherical scattering of waves according to the distance, frequency, and gain of the antenna systems used.

Looking at Figure 12.1, the THz link budget remains dominated by the FPSL up to 300 GHz within communication windows (where no absorption line occurs), which means that by increasing the antenna gains, the link budget can be managed. For example, the use of two 40 dBi antennas over a 100-m link at 300 GHz leads to a link budget of 120–40–40 = 40 dB. Beyond the km, the attenuation increases starts to increase very quickly with the distance and is no longer isotropic (link budget = FSPL).

The diffusion of the waves by the liquid or solid particles present in the air also induces attenuation. This is more important in the THz range than in the microwave range, however, for sub-millimeter-sized particles (fog, dust, etc.), the diffusion in the THz range remains much lower than in the visible or infrared range. In addition, the fluctuations in the refractive index of the atmosphere due to fluctuations in air density (well known in optics) produce scintillation effect. It is much less troublesome in the THz range and rare in the microwave range. Experiments have already been conducted to confirm these effects experimentally [4, 5].

12.2.4 Frequency Bands

Figure 12.2 specifies atmospheric attenuation before 450 GHz [6]. Recently, the 71–76 GHz and 81–86 GHz communication bands have been opened ("E" band) and commercial devices now exist in these frequency ranges [7, 8]. Beyond the E band comes the "W" band up to 110 GHz then the "D" band up to 170 GHz. However, all these bands are already allocated internationally [6]. Beyond the D band is the "H" band or "300 GHz" band, and there is currently no frequency allocation beyond 275 GHz. In a way related to the new opportunities created by THz frequencies for communication applications, the international organization IEEE [9] recently proposed (2017) a first standard, the IEEE 802.15.3d (an 802.11 equivalent for WiFi but using the THz range) considering these new potentialities. Thus, even though these standards are still under discussion, there seems to be a clearly identified target to provide a framework for these new wireless communications in the 300 GHz band in the under-explored ranges, for the 100 Gbit/s. For example, the use of frequencies beyond 275 GHz was discussed at the World Radiocommunication Conference, WRC 2019 [10].

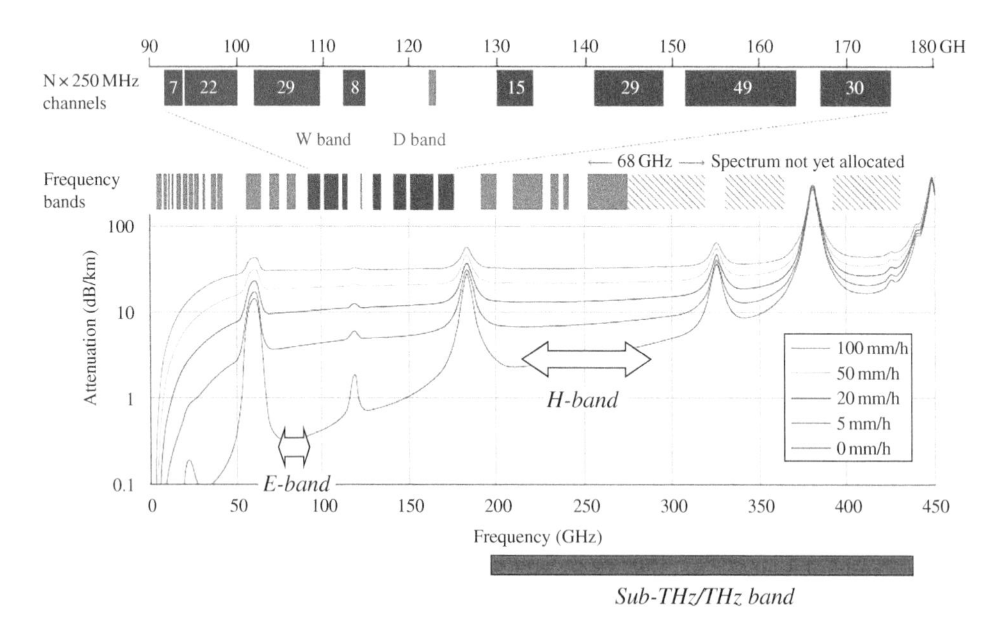

Figure 12.2 Frequency bands and atmospheric attenuation. The frequency range between 200 and 320 GHz has interesting features for THz coms. *Source*: Adapted from [6].

Thus, between the first laboratory demonstrations in this frequency range and the first standard, about 15 years have elapsed.

Shannon's theorem [11], used to evaluate the capacity of transmission systems ($C = B \cdot \log_2[1 + \mathrm{SNR}]$, where C is the capacity in bit/s, B the bandwidth in Hz, and SNR the signal-to-noise ratio), reminds us that the increase in capacity will pass through an increase in bandwidth B. With an available band $B = 120$ GHz in the window of 300 GHz, the potential bit rates would far exceed 100 Gbit/s. However, due to the technological challenges in the THz range, for the moment the first developed systems tend to take advantage of the bandwidth because the SNR remains limited. Several studies and technology developments are now focusing on power generation in the THz range to tackle these limitations.

12.2.5 Potential Scenarios

From the early developments of THz links, one could summarize the possible scenarios/use cases hereafter:

- Very high-speed communications over very short or short distances: From an electronic card to an electronic card (cm), from a download terminal to a moving object (several 10 cm), from a computer bay to a computer bay (couple of meters) or a THz hotspot to a moving object, which is more challenging and have not been yet demonstrated. These applications include wireless local area networks (WLAN, Wireless LAN), access networks in urban networks, or for the realization of fast kiosk downloading. One of the technological difficulties associated with this class of applications is related to the antennas used that must be reconfigurable (manually or automatically) to adapt to the constantly changing link budget. This is still a big challenge in the THz range. Last, one big advantage of these applications, in connection with short-range propagation, lies in the possibility of reusing frequencies on a small scale.
- Very high-speed broadband communications: Fixed wireless links, ranging in coverage from 100 m to several miles, can be used as backhaul for wireless networks. This application, being mainly an "operator" technology, represents the change in bit rate of the usual radio-relay links up to the Ka band (17–31 GHz). Beyond 100 GHz experiments have been carried out: point-to-point communications between an access point and a high-definition TV camera (demonstration of NTT over 1 km at the Beijing Olympics in 2008 at 125 GHz carrier frequency [12], demonstration of Dunkirk (France) at 300 GHz in 2017 [13] (Figure 12.3).
- Broadband communications over long distances: between an aircraft and the ground (couple of km), between an aircraft and a satellite (100 km), between the ground and a satellite and also potentially between satellites. For geostationary, terrestrial-to-satellite transmission links up to Tbit/s the ground-based radio-astronomy stations already established can be used. Leveraging on very large antenna gains, devices power limitations can be overcome [14] in this frequency range. These applications will, however, be directly dependent on the availability of solid state electronic amplification technologies (transistors) or vacuum electronics (traveling wave tubes).

The benefits of THz for data links applications are established; however, commercial developments will only occur if the costs associated with this emerging technology are low enough compared to solutions provided by the other frequency bands (microwave

Figure 12.3 (a) Outdoor THz experiments: left: communication at 125 GHz over several km (2008): NTT-Japan [12]. (b) 300 GHz km-range link: France University of Lille-CNRS-IEMN (2017) [13].

and optics) leveraging well-established technologies. As presented later on, many technologies are considered, and the use of a particular technology depends directly on the level of maturity, performance, compactness/consumption, and associated costs.

12.2.6 Comparison Between FSO and THz

One of the solutions currently envisaged as a wireless broadband communication solution is to use free-space optics (FSO), but one of the main limitations of these systems is directly related to the wavelength itself. In fact, at 1 or 1.55 μm, the IR optical signals are strongly affected by the diffusion, and, if the attenuation in dry air remains limited to less than 1 dB per km, this one goes up at more than 100 dB/km in the presence of fog [15], which makes it impossible to maintain a point-to-point (fixed) link in an operational scenario. In the presence of rain (25 mm/h), the attenuation is around 10 dB/km, which is in the same order of magnitude as in the THz range with a high degree of humidity. As mentioned above, recent work [8, 9] has been able to verify, by emulating the climatic conditions of propagation, that a THz transmission in system configuration was much more robust than their near infra-red counterpart.

In addition, the other practical issue in FSO links is a precise alignment or positioning of infrared light beams between the transmitter and receiver. A heavy-weight mechanical beam alignment apparatus is usually equipped with FSO links. We can quantitatively compare the ease of alignment between THz and FSO links by using the following equation:

$$\frac{\lambda_{THz}}{\lambda_{FSO}} \cong \frac{w_{THz}}{w_{FSO}} \cdot \frac{\alpha_{THz}}{\alpha_{FSO}},$$

where λ_{THz} and λ_{FSO} are wavelengths of THz and FSO links, respectively, and $\omega_{THz}/\omega_{FSO}$ and $\alpha_{THz}/\alpha_{FSO}$ correspond to the ratio of alignment margin for THz and FSO links with

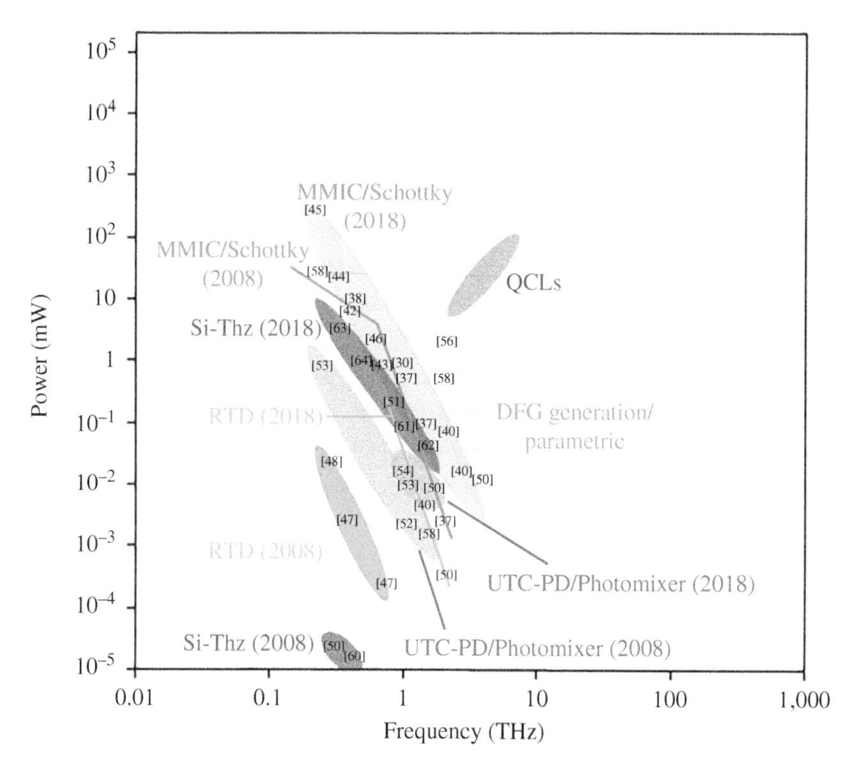

Figure 12.4 State of the art of THz sources. *Source*: Adapted from [16] (references in the graph from [16]).

respect to position (axis) and angle, respectively. For example, $\lambda_{THz}/\lambda_{FSO}$ = 1 mm (300 GHz)/1.55 μm ≅650.

12.3 THz Technologies: Transmitters, Receivers, and Basic Architecture

12.3.1 THz Sources

The active components for generating or detecting THz waves play a central role in the emergence of these technologies. Figure 12.4 gives an overview of the current performance of THz sources [16]. For electronic components (transistors, diodes), there is a significant drop in power with frequency (about $1/f^3$). In the sub-THz range of 0.1–1, THz powers of the order of 1 to >100 mW (even above for 100 GHz) are now possible in the solid state and at room temperature which allows to consider communication systems in this range. For quantum cascade lasers (QCLs), the opposite trend is observed. Nevertheless, the latter must be cooled to cryogenic temperatures and, taking into account the atmospheric absorption and the powers generated, it seems unlikely that communications at frequencies above 1 THz will emerge in the near future. The resonant tunneling diodes (RTDs) also have an interesting compactness/performance ratio at room temperature. Last, the optoelectronic components; uni-traveling-carrier photodiodes (UTC-PDs) or photoconductor-type photomixers have a fairly wide operating range, allowing an optical-to-THz conversion interesting for spectroscopy or communications.

Beyond the power aspect, looking toward communication system perspective, it is also necessary to be able to modulate the THz source with good efficiency to enable high data rates. The most direct solution is the modulation of the THz carrier. However, to date, there is not yet an effective and fast amplitude and/or phase modulator in the sub-THz or THz frequency range. Thus, the modulation of the THz signal is often performed in the usual electronic microwave domains (using mixers followed by frequency change stages) or by optical means, before down-conversion in the THz range.

12.3.2 THz Receivers

Two types of detection schemes can be envisaged for THz communications: direct detection and heterodyne detection:

- In direct detection, it is necessary to have a nonlinear element, such as a diode, capable of operating in the THz range, the nonlinearity being provided by the signal to be detected itself. This detection scheme is sensitive only to the envelope of the signals, thus handling vectorial signals is not possible (the information on the phase being lost). However, this type of detection is very simple to implement and enable straightforward bit error rate (BER) measurements in real-time. The commercially available THz detectors, based on Schottky diodes [17–19] in gallium arsenide (GaAs), make it possible to reach sensitivities of several V/mW and thus make it possible to work with detected powers between 10 and 100 μW usually.
- Heterodyne detection, or "mixing," in which a mixing of two signals (the local oscillator (LO) and the signal to be detected (RF)) is done. In this case, the nonlinearity comes from the LO signal, and the detection is linear on the RF channel as long as the signal remains of limited amplitude (less than the LO). This solution, much more sensitive than the direct detection, requires having a LO power usually around the mW and having a good spectral purity. In heterodyne detection, the orders of magnitude of the conversion losses of the subharmonic mixers are of the order of 8–10 dB at 300 GHz [17, 18] which makes it possible to process signals of smaller power than in direct detection. In this scheme, transmitter and receiver have generally to be synchronized to support a real-time operation of the THz link.

According to the specifications of these detection schemes, it is thus interesting to enable a direct detection with a good sensitivity. Thus, low barrier detection diodes are very interesting for sensitive direct-detected THz links [20, 21] (Figure 12.5).

12.3.3 Basic Architecture of the Transmission System

The basic architecture of an RF transmission chain is shown in Figure 12.6. Key elements are integrated into the "front-end": the mixer, the LO, and a possible amplifier. At the output of the front-end, a power amplifier (PA) and low noise in reception (LNA) can also be added. In the case of an electronic mixing, the mixer operates in a nonlinear range on the

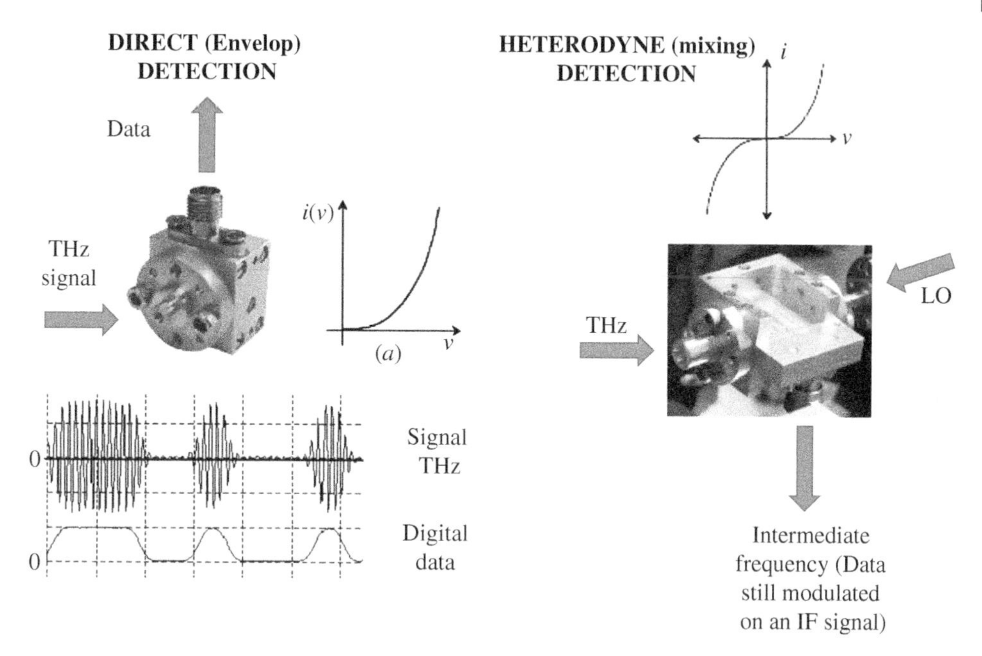

Figure 12.5 GaAs Schottky diode-based direct detection embedded in hollow waveguides associated with metallic horn antennas (a). Heterodyne detection in which a mixing is made between a local oscillator (LO) and the signal to be detected.

LO channel and in a linear range on the IF channel, enabling the phase detection of input THz signal. This RF front-end can also be replaced by a photomixer system using THz generation from optical signals (photomixing technique).

In the THz range, the advanced electronic components enabling the "front-end," and in particular, the active functions (PA and LNA) are derived from micro/nano-electronics technologies, today mainly in III/V (HBTs or HEMTs), these being pushed to their limits. Silicon technologies, while intrinsically less dedicated to this frequency range by cutoff frequencies, are also actively developed below 300 GHz for shorter range communication applications (SiGe and BiCMOS technologies). Nevertheless, demonstrators of THz transmission systems might use these technologies at some point as soon as reduction of the overall cost of the system is the limiting factor or direct integration with other functions outside the RF front end.

At the reception level, down-conversion is achieved by means of mixers. The performance of the mixers is fixing the sensitivity of the receivers, but the emergence of amplifiers would make it possible to gain in performances. On the other hand, these receivers require a relatively powerful LO which can generate a significant energy consumption of the receiver. Direct detection by a nonlinear element (diode or transistor) is an interesting alternative to simplify and reduce the overall consumption of the receiver with however lower sensitivity. This approach is particularly interesting for very high-speed and short-distance transmissions.

Several types of transmitters and receivers are possible to enable THz links. For Tx, electronic or photonic solutions are possible The last has the major advantage of being directly

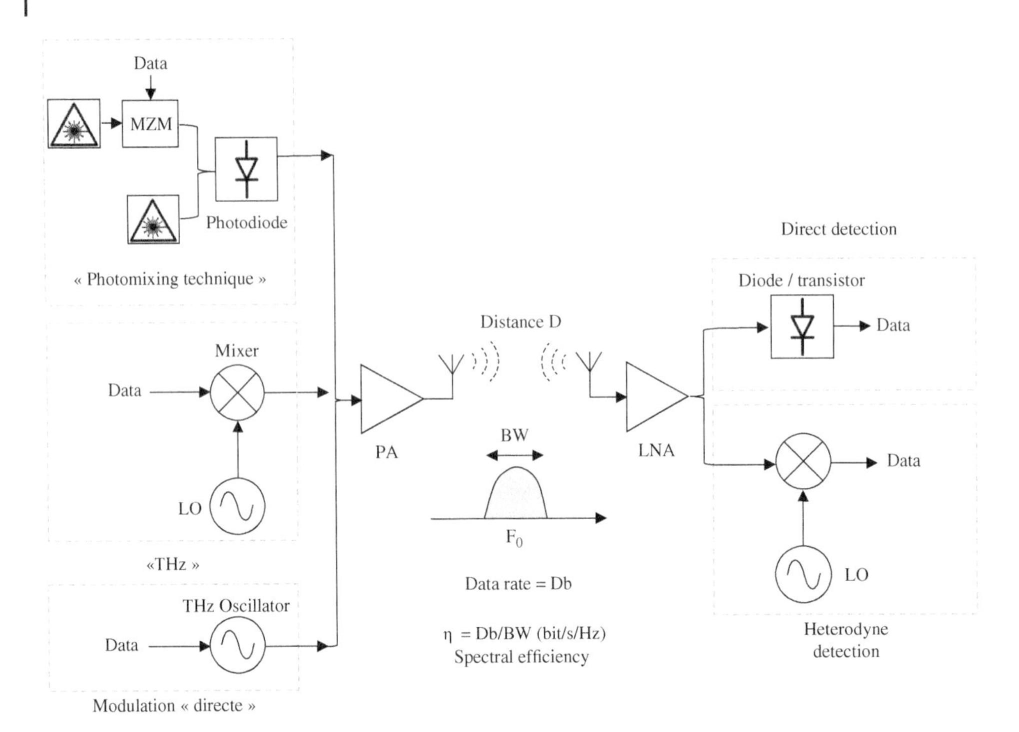

Figure 12.6 Architectures of the first THz wireless links. Several approaches have been tested, and combinations in emission and in detection. The data corresponding to the digital signals to be transmitted are either in "baseband" (unmodulated on a high frequency carrier) or using signals already modulated on a carrier, in this case, THz system realizes the up and down conversion. The "direct" modulation groups the multiplication chains, modulated at their input and also THz oscillators (RTDs, integrated oscillators). In the case of integrated approaches to transceivers, the scheme is more complex and depends directly on the architecture of the circuit.

fed by existing optical networks and can be easily agile in frequency. Electronic solutions, quite compact in general, potentially benefit from an advanced level of integration well beyond the front end. The overall performance of the system will depend on many parameters: the power available at transmitter, the modulation bandwidth and linearity of the whole chain, antennas (gain, bandwidth), the linearity of the receiver, phase noise (transmit/receive), amplitude, and phase distortions.

12.4 Devices/Function Examples for T-Ray CMOS

12.4.1 Photomixing Techniques for THz CMOS

Photomixing is a technique for generating mm/THz waves using a fast photodetector (PD). When shining this PD with two continuous waves optical signals, a signal at the difference frequency will be created (Figure 12.7). In the THz range, the photomixing components are photoconductors and photodiodes, and efficient devices are photoconductors in low temperature GaAs (LT-GaAs, photosensitive at 0.8 μm) and the UTC-PD [22], operating at 1.55 μm.

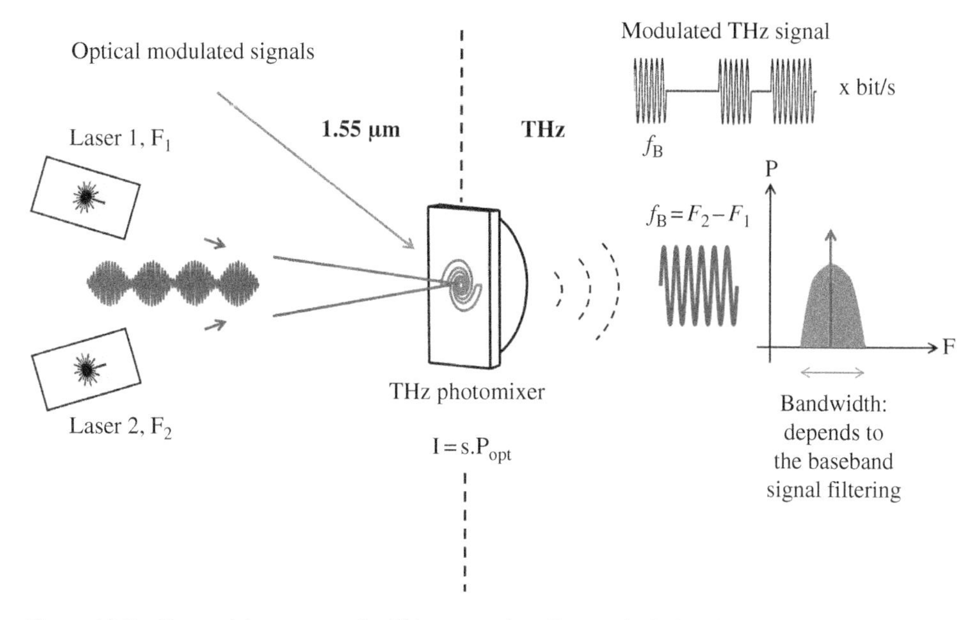

Figure 12.7 Photomixing process for THz generation. Two optical signals are coupled in a photodiode or an ultra-fast photoconductor (photomixer). This technique makes it possible to generate continuous THz signals if the optical sources are continuous, or modulated if one of the two waves in the optical domain is modulated. This modulation can be performed in amplitude, in phase, or both at the same time for vectorial modulations, in which the amplitude and the phase are modulated for the encoding of the signal.

The THz signal, once generated by photomixing, must be radiated in the free space, using an antenna. This antenna can take several forms: integrated planar antennas (bow-tie, spiral) [23], that have to be coupled with a silicon lens to extract the signal. Other integration types are using metallic waveguides associated with metal antennas horn type, horn-lens, or horn+lens/reflector, to further increase the gain. We will see in the state of the art of THz communications that the photomixing technique has been a technology driver of the first links. These numerous demonstrations of THz communications systems have taken advantage of the ability of this technique to transfer any modulation type from the optical domain to the THz domain.

Using such techniques, the current state of the art for signal generation in the 300 GHz range is around mW [24], either using a single PD or combination approach [25] (Figure 12.8). This allowed early practical applications, such as short or medium range THz communications. However, one of today's major challenges is to increase power in the devices: then, a coupling between photonic devices (photomixers) and transistor-based active devices is an appealing approach, leveraging on photonics linearity, and power level of active devices (Figure 12.9).

12.4.2 THz Modulated Signals Enabled by Photomixing

As shown in Figure 12.10, the main interest in the use of photomixers for THz communications lies in the fact that the signal modulation part (in amplitude or in phase) can be

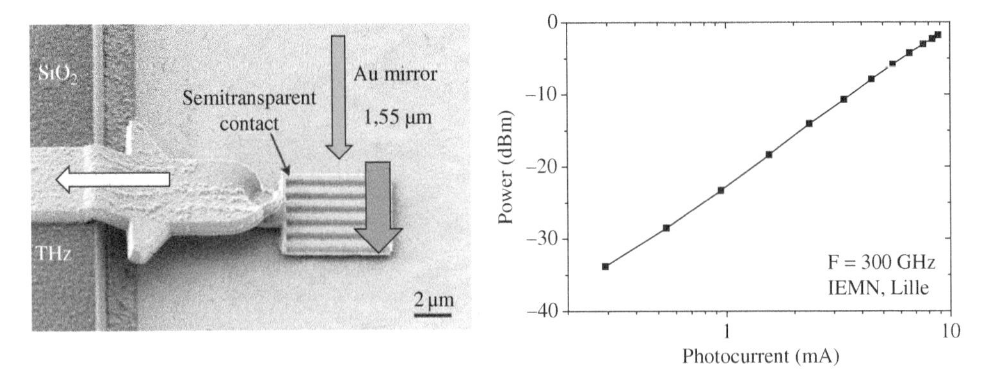

Figure 12.8 (a) SEM view of a UTC-PD photomixer [24]. (b) Power level generated at 300 GHz using UTC-PD by the photomelange technique. *Source*: Courtesy of J. Kooi, JPL. Source: Courtesy of G. Ducournau, IEMN Lille, France.

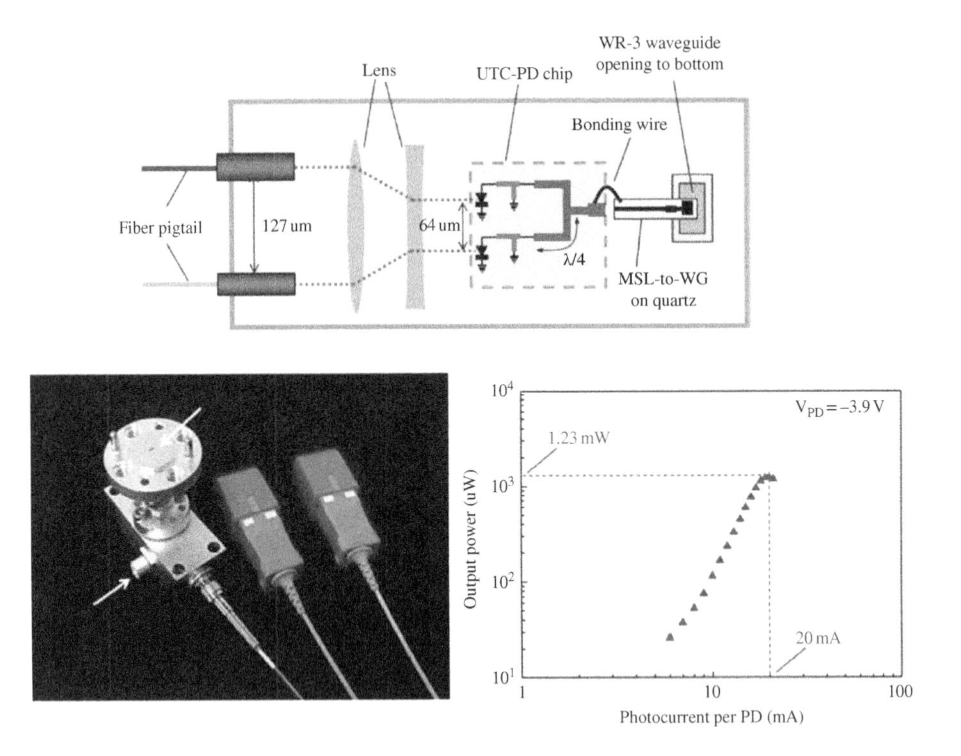

Figure 12.9 UTC-PD photomixer combination [25], reaching 1.2 mW at 300 GHz. *Source*: Courtesy of NTT, Japan.

realized in the optical domain before to be transposed into the THz range by the photomixer. In this case, the optical signal consists of a wavelength modulated with the desired format (QPSK, QAM-16, etc.) and a second unmodulated wavelength ("pilot" carrier). Figure 12.10 gives an example of the generation of a signal by photomixing of two optical signals [26]. In this example, two four-level baseband (I and Q) digital signals ("PAM4" type

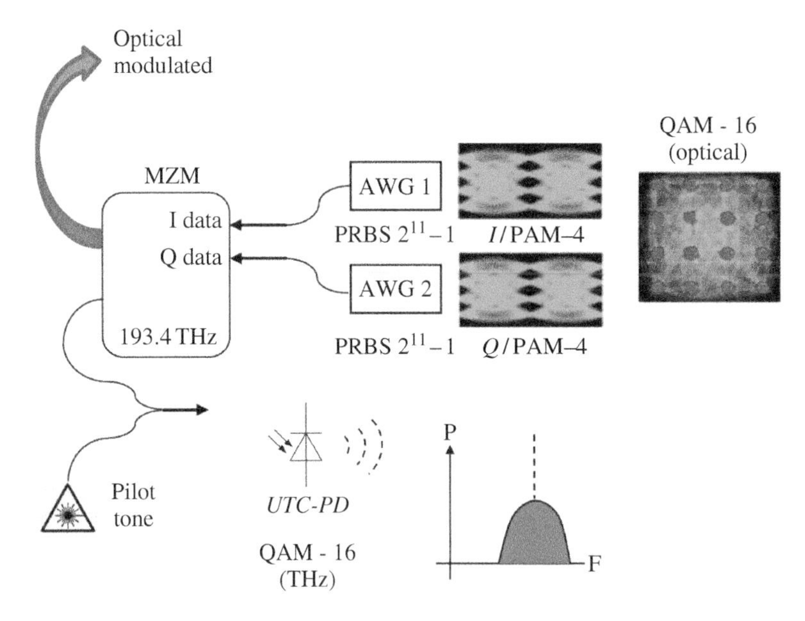

Figure 12.10 Complex THz modulation using photomixing technique.

signal) are used to modulate an optical signal at "telecom" wavelengths, i.e. 1.55 μm (frequency 193.4 THz). The modulated optical signal is then coupled to a "pilot" signal whose frequency is shifted by about 0.3 THz (2.4 nm in the optical range). This enables to generate the signal at 300 GHz at photodiode output.

One great advantage of the photomixing technique is that the principle can be easily extended to several wavelengths to generate several THz carriers by a single device, while keeping the photodiode in its linear region. Thus, THz photomixers can be used as "THz hotspot" to downconvert the guided optical signals in the fibers to millimeter/submillimeter radio. This frequency multiplexing concept is illustrated in Figure 12.11.

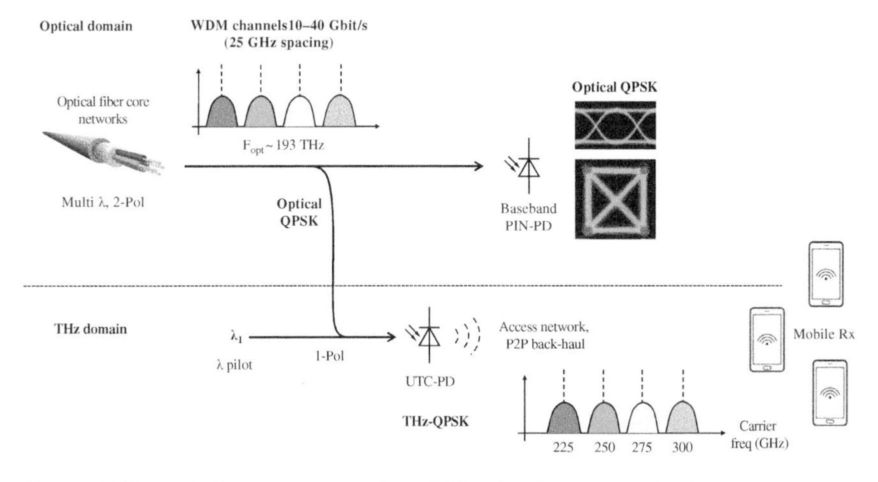

Figure 12.11 Multifrequency generation of THz signals using photomixing.

12.4.3 Other Techniques for the Generation of Modulated THz Signals

Beyond the photomixing techniques, several other techniques for generating modulated THz signals can be used, such as frequency multiplication [27], RTDs [28], or up/down-conversion chains [29–31], using many III–V semiconductors. Of course, even if the basic concepts could be demonstrated in millimeter or THz range with nonintegrated devices (on the shelf), more and more approaches of integrated transmit/receive circuits using CMOS and SiGe HBT have now entered the race.

Each technique has its advantages and limitations. For example, the frequency multiplication uses the nonlinearity of the diodes for the generation of harmonics [27]. Thus, it is possible to use this type of source to generate modulated signals, if the multiplier input is also modulated. However, the accessible bandwidth is generally limited by the architecture of this type of source but this works well at a few Gbit/s, having interesting power levels beyond 10 mW. This has made it easy to demonstrate high-frequency links [32] and also propagation studies [33, 34]. This solution is simple but in principle limited to amplitude modulations, because of the nonlinearity of the chain. It is technically possible, however, to generate PSK signals (constant power constellations) by adjusting the waveform at the input of the multipliers [35]; however, this is very difficult to achieve for more complex modulations usually used in wireless communications (multistate quadrature amplitude modulation (QAM)).

In order to reach more complex signals (QAM), it is necessary to have an end-to-end linear transmission chain, which is made possible by mixing stages or frequency transposition. This is called heterodyne architecture, or super-heterodyne if several mixing stages are used. This possibility opens the way for the use of THz bands by parallelization of already existing signals in the lower bands, for example in V and E (60 and 70–80 GHz) bands. This makes it possible to increase the data rate while being compatible with the systems already in place. Finally, for shorter-range communications, RTD oscillators that offer interesting power levels of a few hundred GHz [36] and also beyond THz [37] can also be compact solutions for THz communications.

It is also worth to mention that QCLs have also been used for the transmission of signals in the THz range [38]. However, as telecom is generally a volume market, the high-cost pressures will tend to favor integrated silicon approaches for volume applications. On specialized links or very high performance, III–V devices, at the state of the art, might play a role. However, at this moment, even if the progress of silicon is impressive, to date beyond 300 GHz the overall performance of III–V circuits and devices is still competitive. Eventually, the situation will evolve with regard to the progress made by silicon technologies and the scenarios/use cases that will set volumes and markets.

12.4.4 Integration, Interconnections, and Antennas

12.4.4.1 Integration

The integration and coupling of devices in the THz range are coming with both physical and technological difficulties from the significant absorption by free charge carriers, the materials often having losses, but also by the wavelength itself concerning the dimensionality of objects and subassemblies for packaging. In the THz range, many approaches are possible [39], coming from the optical domain or microwave or millimeter-wave engineering. The packaging part in THz range representing a complete field of engineering, we only mention the main approaches here. Traditional waveguide machining approaches work very well up to several hundred of GHz to make amplifiers [40] detectors, and mixers. However, the cost and compactness of this approach are not optimal, and at this moment, limits the diffusion of THz technologies to specialized environments. Several other routes have thus been initiated [39], as keeping a waveguide approach at lower costs and new manufacturing techniques, such as the use of LTCC (low-temperature co-fired ceramic) or 3D polymer printing metalized or 3D metal printing techniques. We can also mention, in order to improve compactness while guaranteeing the possibilities of scaling beyond the THz, deep RIE techniques [41] are a very advanced technological solution, in the perspective for example of "CubeSats" [42].

12.4.4.2 Antennas

For free-space (wireless) communication applications, several types of antenna approaches are used [43]. First, based on the metallic waveguide technologies, horn type antennas have the best performance at this moment and are the most conservative solution. Moreover, the use of reflective systems (Cassegrain or parabolic type) or lenses makes it possible to obtain considerable gains at 300 GHz as soon as the size of the reflector/lens reaches a few tens of cm in diameter [44]. Planar antenna systems on substrate or membrane can also be used, having the advantage of monolithic integration with the THz component. However, limitations in both radiation efficiency and dispersion in the bandwidth often occur, due to the presence of a silicon lens used for the extraction of the THz signal or resonant design. In order to reduce these limitations, planar antennas on low permittivity substrates can be a low-cost solution [45], as well as suspended dielectric antenna approaches [46].

To date, all the first demonstrations of communication in the THz range use antennas, directives in general, in point-to-point mode. In the long term, one of the challenges in this frequency range will be the experimental realization of reconfigurable antennas, in order to find a compromise between point-to-point links (imposed by the link budget) and the possibility of addressing mobile users. Beamforming/steering is also an important topic for the development of future THz systems, and the reference [47] provides good coverage of these aspects.

12.5 THz Links

12.5.1 Modulations and Key Indicators of a THz Communication Link

In the THz range, different approaches have been followed, coming from microwave domains, long-distance optical communications, coherent optical communications, and even more conventional radio domains. As a result, amplitude and phase/amplitude vector modulations (QPSK, QAM-16, QAM-32, and beyond) are used.

The diversity of scientific communities in the THz research has induced the use of numerous performance indicators, sometimes difficult to compare between different systems. The main indicators are summarized in the following:

- The spectral efficiency of a modulation, defined by the ratio data-rate/occupied bandwidth (bit/s/Hz), is the parameter to be optimized in order to maximize the amount of information transmitted for a given bandwidth. Since bandwidth constraint is the most critical in wireless networks, this parameter is fundamental for this type of application. Of course, the higher the spectral efficiency, the higher the SNR is required for an acceptable error rate.
- The energy per bit, usually expressed in pJ/bit, also qualifies the energy performance of a transmission system, including all the elements of the system, both the RF front-end and the signal generation/detection elements and baseband components. This figure has to take into account the whole transmission chain to enable fair comparison between systems.
- BER: Since the communication link has the role of ensuring end-to-end transmission of the initial signal (usually in digital baseband form), the main performance indicator must be the BER, which is evaluated by comparing the transmitted and received signals. This should be done in real-time, using bit error rate testers (BERTs).

For systems operating in amplitude modulation and direct detection, the output signal of the detector is found in the baseband, and an error rate measurement is directly possible, without the need to process the signal. One should keep in mind that if a noisy signal is obtained in reception (BER degraded), data can be corrected up to BERs of the order of 2×10^{-3} [48]. Indeed, error correction techniques (FEC, forward error correction) which are widely used in long-distance optical transmissions can also be used for THz communications.

Nevertheless, these techniques require (i) the addition of correction data, which reduces the useful information, rate (part of the bit rate being allocated to the FEC) and (ii) a system working in real-time.

- For vector modulations, the EVM (error vector magnitude) parameter is usually used to quantify the quality of the received modulated signal. Thus, performance estimation is performed on an intermediate signal (the EVM of the constellation) without the baseband signal being physically available. Thus, the measurement of the BER on the physical signal analysis in real time is not possible. It is, therefore, necessary to distinguish the systems in which BER measurement is performed in real time and those where system performance is evaluated in non-real time, "off-line."

Finally, as any communication system's main target is to transmit data over a fixed distance, the speed-distance product, expressed in (Gbit/s) km can be used as an indicator.

It is thus not always easy to compare the performance of transmission systems in the THz range because often all the indicators are not measured with the same procedures. Thus, the metrology and methods to measure the system performance is an important point to consider for the development of these systems.

12.5.2 State-of-the-Art of THz Links

12.5.2.1 First Systems

The first age of communications in the THz range began about 15 years ago, in a few institutes in the world. It has to be mentioned, however, that the NTT (Japan) was the pioneer in the field: as early as the 2000s, the first tests were done at 120 GHz [49, 50], even if this band is at the limit of the beginning of the "THz" range. Even if many types of sources and detectors have been tested, we can notice that, to date, the diversity of transmitters is much greater than that of transmitters. Thus, on the receiver side, the Schottky GaAs diode devices, in direct detection or subharmonic mixing (with or without LNA), have been by far the most used in the laboratories, even if several demonstrations are carried out with specific receptors (InP and more recently silicon). Nevertheless, the development of these technologies is now diversifying in anticipation of future systems, and it is now accepted that the THz frequencies will be a constituent element for the 6G (or at least for the "beyond 5G") [51].

As in the microwave range, the coding of the signal can be carried out simply binary using two levels (NRZ) to obtain a digital amplitude modulation (OOK or ASK) on carrier THz. This solution is well adapted to frequency-multiplication systems (so-called "multiplied" sources) and makes possible to produce the first laboratory systems at a few Gbit/s [32] by reusing in certain cases advantageously the duobinary technique (well-known in optics, featuring reduction the size of the spectrum emitted) to tackle the available bandwidth limitations. We can also mention the demonstration of transmission of a THz carrier audio signal [38] with a QCL, even if the bit rate of this type of solution is limited.

12.5.2.2 Photonics-Based Demos

Many systems have been developed on the basis of photomixing type emitters. Most of the time, single components have been developed, and enable "laboratory type" demonstrations where the integration of the devices is not blocking. As discussed above, the main advantage of photonics is to obtain a very high modulation index associated with a large modulation bandwidth (because of the high frequency of the optical carrier), using the fiber technologies [52]. This has made possible several important demonstrations since 2010, where fiber networks data-rates (per single polarization/wavelength) were reached in the THz radio range. For example, 50 Gbit/s at 300 GHz in real time [53], 60 Gbit/s at 200 GHz [54], 46 Gbit/s at 400 GHz [55]. The 100 Gbit/s threshold was also reached in several ways, and for the first time with photonic-based THz transmitters again: in 2013, the combination of a UTC photodiode with an active electronic receiver at 237.5 GHz enabled a 100 Gbit/s multi-channel system [56], using signal processing at the receiver.

In 2016, the first (real-time) 100 Gbit/s transmission (QPSK) is obtained [57]. In 2018, the first 100 Gbit/s transmission on a single THz carrier is realized in QAM-16 modulation [26], in line with the new IEEE 802.15.3d standard which will be mentioned in the last part. This type of demonstration was also performed using frequency comb techniques to facilitate detection due to a carrier with lower phase noise [58]. Although the InP III–V photonic technologies for THz are at state-of-the-art, photonic approaches can also benefit from the integration level of silicon photonics due to their compatibility with industrial foundry. For example, the use of Ge photodiodes has been considered since 2014 for generation at 180 GHz [59] and the use of ST Microelectronics (France) PIC25 germanium photodiodes has shown the first link in the 300 GHz range using an industrial level photonic emission both in amplitude modulation [60] and in vectorial modulation [61]. All of these photonics-enabled links used direct-detection with GaAs Schottky diode electronic receivers [62] or harmonic mixing [63], the latter operating in real-time thanks to synchronization between transmission and detection. Transmitters based on UTC photodiode have also been used in the band beyond 600 GHz. For example, HD-TV signal transmissions (some Gbit/s) were obtained at 600 GHz [64] and more recently at 720 GHz [65] with a bit rate as high as 12.5 Gbit/s.

To date, the highest rates have, therefore, been obtained with THz transmitters of the "photonic" type in combination with mixer-based receivers. However, the transmission power of these devices being so far limited, the range of these links is still limited. A combination of technologies is also envisaged in order to take advantage of the advantages of different approaches and performance [66].

12.5.2.3 Electronic-Based Demos

The first outdoor demonstration over large transmission distances (5.8 km) was conducted in Japan by NTT [67] at 120 GHz (at the beginning of the THz band). On the "all-electronic" systems side (beyond the use of electronic receivers combined with the aforementioned photonic devices), first complete electronic systems were realized using standard waveguide devices originally developed in GaAs technology for radio-astronomy (for example ALMA). Through the use of these components; multipliers, mixers, sources, and detectors (diodes) available in a laboratory environment, researchers have developed THz links in many configurations. The first use is direct amplitude modulation [32], which is a very simple technique, but limited bandwidth and not suitable for phase coding. In this approach, there are several limitations: (i) the nonlinear behavior of the multiplication chains which limits amplitude modulation and (ii) the relatively high impedance (kΩ) of Schottky barrier diode-based direct detectors. The researches on low barrier-height detectors are also an attractive route [20, 68] to overcome these limitations. By using harmonic mixers in "up-conversion" mode, a linear behavior on the IF-to-RF conversion can be obtained and complex first signals were transmitted as early as 2010 [69].

In parallel with these first systems, research was carried out on the realization of integrated THz circuits, or assembled subsystems, and many realizations have emerged, firstly in III/V due to the performance of the active unitary components (transistors). However, the evolution of silicon (SiGe HBT, CMOS) will certainly bring complementary solutions to larger production volumes. For example, on the receivers, the demonstration of a fully integrated 300 GHz ASK receiver on InP [70] in 2014. A complete system (transmitter/receiver) using InP HEMT technology has also been validated for 50 Gbit/s QPSK modems [71]. Also, in GaAs HEMT, complete circuits have been validated at 300 GHz for QPSK modulations at 64 Gbit/s [72]. In the 2013 InP HEMT technology, examples of complete transmission/reception chain were validated in G-band (190 GHz), with 200 Mbit/s combined to advanced modulations (QAM-64) [73]. Another example is the 2018 demonstration of a HEMT-InP transmitter/receiver that also validates the 100 Gbit/s data-rate at 300 GHz. Finally, on DHBT InP technology, fully integrated modulation and demodulation circuits were also realized in the "D" band, 110–170 GHz [74].

Until 2015, for communication distances beyond a few meters, at frequencies above 100 GHz, the GaAs and InP have been the main technological solutions in research on fully electronic THz communications. This was mainly due to the high cutoff and maximum frequencies of transistors (for example, f_t and f_{max} of the order of 400–500 GHz/1000 GHz with InP DHBT [75] and 660 GHz/>1 THz with 20 nm mHEMT [76]. In 2015, an outdoor demonstration on 850 m was carried out at 240 GHz [77], then at 300 GHz in the laboratory [72]. However, other technologies are also expected to pave the way for THz communications, and a perspective of THz chipsets compatible with mass production will push these developments. For example, in 2015, SiGe HBTs reached values of 700 GHz in cut-off frequency [78], and the first chipsets have already been achieved. At 160 GHz, 10 mW of output power (1-W total consumption) was obtained and for 240 GHz output powers above 100 μW as well as 20 GHz bandwidth for Tx and 10 dB receptor conversion gain [79]. On silicon technologies, experiments in free space using planar antennas and silicon lenses, indoors for the moment (0.3 m) [78]. CMOS electronics also showed early results at the system level: for example, a full 130 GHz/11 Gbit/s, real-time demonstration over 3 m was shown [80].

It is not yet certain that the maximum operating frequencies of Si-CMOS devices will reach the THz, opening the way for viable integrated circuits at 300 GHz but could offer in some cases of use at a short-distance interesting solutions for compact transceiver chipsets. However, to date, the main limitation of the THz transmission systems being the available output power, the GaN and InP technologies, with high breakdown voltages, are particularly envisaged for the realization of the power stages of these future systems. In this case, the co-integration of different technologies becomes a crucial point. Another approach may be to use the output power combinations of the THz transmitters: For example, we can cite the use of integrated antenna arrays in Si-CMOS transmitters [81, 82], with RTD [83]. On this last point, antenna array technologies can also provide solutions.

On silicon technologies side, two approaches are currently being pursued; using for one the SiGe HBT and the other a CMOS approach. It was thus shown in 2017 a 190 GHz transceiver featuring 50 Gbit/s a very small (<cm) distance [84]. In 2018, a 25 Gbit/s system at 240 GHz has been demonstrated [85] and then a 90 Gbit/s system in the 220–260 GHz band [86]. In 2019, the first demonstration of QAM-16100 Gbit/s has been performed [87], although EVM values are still a little bit high. Last, the CMOS (40 nm) technology has also demonstrated a QAM16 receiver with 32 Gbit/s in 2017 [88] for the 300 GHz band. Also, in 2018, a 120 Gbit/s system in QAM16, at a lower frequency (E band, 70 GHz) was achieved [71]. In 2019, a 100 Gbit/s 300 GHz transmitter/receiver has been demonstrated [89] with frequency channels compliant with the new IEEE 802.15.3d standard. However, because of the limitation of F_{max} for 40 nm CMOS technology node, a "PA-LNA less" architecture (without PA or LNA) is only possible for the moment, making the transmission distance limited to a few cm. This, therefore, validates the modulation/demodulation parts but, to take advantage of these integrated circuits, again an association with amplifiers is mandatory.

For short-distance links, the RTD-based devices have also been used for the realization of THz data transmission systems. Due to the modulation of the device, the amplitude formats are only possible for the moment. For example, speeds of 22 Gbit/s were obtained in 2017 [90]. We can also mention a mixed link with photomixer transmitter and RTD receiver, featuring 27 Gbit/s [91] and a 30-Gbit/s error-free link with RTD in both transmission and detection [92].

12.5.2.4 Beyond 100 GHz High Power Amplification

Beyond the solid-state electronics technologies, traveling wave tube amplifiers (TWTA) also have a high potential for power amplification at higher frequencies. In W band (around 100 GHz), several achievements have been made. We will limit here to the systems used in

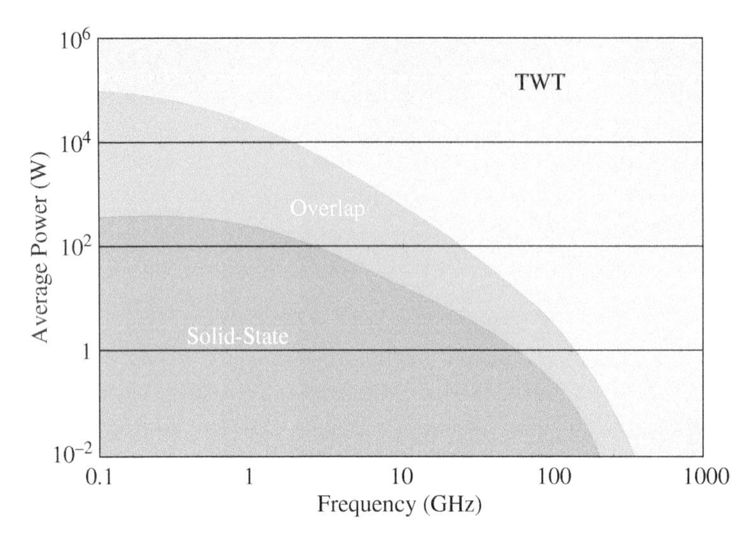

Figure 12.12 Solid-state and TWT power-levels as function of carrier frequency. Beyond 300 GHz, the TWT has leading performances.

communication applications. The realization of these amplifiers is very difficult at frequencies beyond 150–300 GHz, so far only a few research institutes have been able to demonstrate functional systems. Therefore, there is, up to now, very few reported works on system-level demonstrations.

For example, a 140 GHz TWT with an output power of around 500 mW resulted in a 21 km transmission [93]. Also, at 300 GHz, an output power of 1 watt was obtained in 2019 [75], which is considerable. However, the bandwidth is currently limited to a few GHz. Finally, the highest frequency and bit rate demonstration (9.5 Gbit/s) was reported by Northrop Grumman in 2019, with a TWT at 640–650 GHz, with an output power of 150 mW [94]. It is yet difficult to predict which technology between GaN and vacuum electronics (TWT) will provide the best cost/performance/reliability compromise. GaN is developing fast beyond 100 GHz, but in 2019, an active 190 GHz circuit has been realized [95]. Figure 12.12 gives an idea (https://www.microwavejournal.com/articles/32759-twtas-still-dominate-high-power-and-mmwave-applications) of the power levels accessible between solid state dies (all together) and tube amplifiers.

12.5.2.5 Table of Reported Systems

The following table summarizes together different technological solutions or demos for THz or sub-THz communications. Real-time performance measured with an error rate tester is indicated in blue (Figure 12.13, Table 12.1).

12.6 Toward Normalization of 100G Links in the THz Range

The IEEE 802 Committee (LAN/MAN Standard Committee), which has a significant influence on the interconnection specification of wireless communications, set up a working

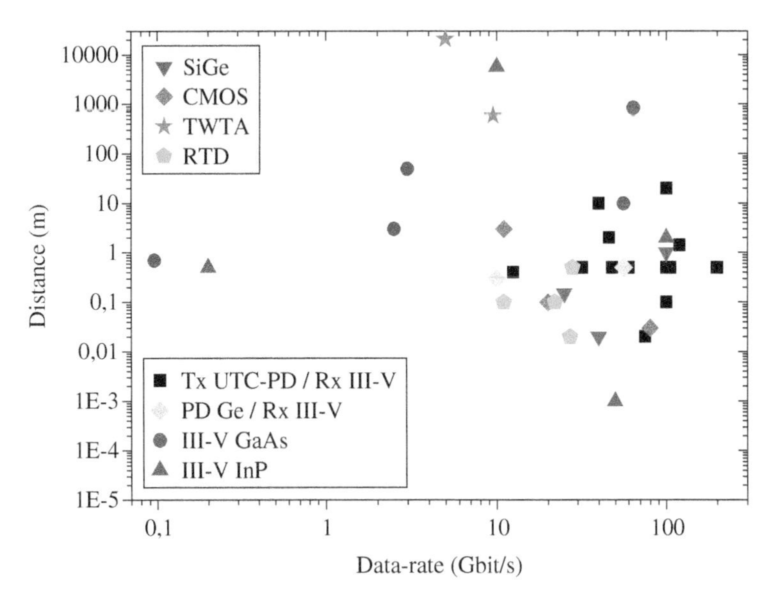

Figure 12.13 Representation of the current performance of THz communication links with different transmitter/receiver technologies. It is clear that photonic systems have, to date, shown the highest data-rates. It also appears that few systems (only in III–V electronics) have been able to show long-range (km) performance, in relation to the power performances of the devices used. SiGe and CMOS channels start to take a place in this flow race but not yet in transmission distance. The links made with TWTs (only in emission), although the data-rates are limited, present outstanding performances in transmission distances.

group on THz communications (http://www.ieee802.org/15/pub/TG3d/index_IGthz.html) as early as 2008. In 2013, the group focused on a 100-Gbit/s radio systems perspective. Finally, the task group IEEE802.15.TG3d, which deals with the standardization of 100-Gbit/s point-to-point wireless links using the 300-GHz band, was founded in May 2014. This group completed a standardization of the PHY layer and the MAC layer of wireless communication systems using frequencies from 252 to 325 GHz and released the standard IEEE Std 802.15.3d in October 2017 [9]. This standard assumes several use cases such as kiosk downloading, inter-chip/inter-board communications, data center wireless communications, and mobile fronthaul/backhaul. Figure 12.14 gives an example of the channels provided in this IEEE 802.15.3d standard for THz communication systems.

International telecommunication union radio sector (ITU-R) holds the world radio conference (WRC) once every 3 or 4 years, discussing revision of radio regulations (RR). Based on the discussion in the WRC-15 in 2015, they resolved to recommend that the WRC-19 held in 2019 discusses about the agenda item 1.15, "to consider identification of frequency bands for use by administrations for the land mobile and fixed services applications operating in the frequency range 275–450 GHz, in accordance with Resolution 767 (WRC-15)." In the frequency range from 275 to 1000 GHz, the frequency has already been identified for passive services, and the WRC-19 agenda item 1.15 was to consider the use of this range for active services. "Identification" differs from "allocation," and means pre-assertion before concrete allocation.

Table 12.1 Various THz communication systems and associated technologies.

Reference	Freq (GHz); channel number	*D* (m)	Data-rate (Gbit/s)	Technology Tx/Rx	Modulation	BER, EVM%	Year
[69]	300	0.7	0.096	GaAs	64QAM	EVM 6.89%	2010
[32]	625 (1)	3	2.5	GaAs–GaAs (diode)	ASK	10^{-10}[b]	2011
[67]	120 (1)	5800	10	InP/InP-HEMT[a]	ASK	10^{-9}[b]	2012
[96]	100 (2)	0.5	200	InP/GaAs (SHM)[a]	QPSK	10^{-3}	2013
[56]	237.5 (3)	20	100	InP/InGaAs (m-HEMT)[a]	16-QAM	2×10^{-3}	2013
[73]	195 (1)	0.5	0.2	SHM/SHM (InP-HEMT)	64-QAM	—	2013
[62]	300 (2)	0.5	48	InP/GaAs (diode)	ASK	10^{-9}	2013
[54]	200 (3)	0.02	75	InP/GaAs (SHM)	QPSK	10^{-5}	2014
[97]	340 (1)	50	3	SHM/SHM (GaAs)	16-QAM	10^{-10}	2014
[98]	300 (1)	B2B	40/50	InP-HEMT	QPSK	10^{-11}/n.a.	2014
[55]	400 (1)	2	46	InP/GaAs (SHM)	ASK	10^{-3}	2014
[80]	130 (1)	3	11	40-nm CMOS	ASK	10^{-9}	2015
[99]	385 (1)	0.5	32	InP/GaAs (SHM)	QPSK	10^{-5}	2015
[77]	240 (1)	850	64	MMIC m-HEMT InGaAs	QPSK	5×10^{-3}	2015
[74]	115–155 (–)	—	—	InP-DHBT	—	—	2015
[72]	300 (1)	1	64	MMIC m-HEMT InGaAs	QPSK	—	2015
[100]	300 (1)	10	40	InP/GaAs (SHM)	QPSK	10^{-4}	2015
[53]	300 (1)	20	30	InP/GaAs (SHM)	ASK	10^{-9}	2015
[101]	400 (4)	0.5	60	InP/GaAs (SHM)	QPSK	10^{-3}	2015
[57]	300 (1)	0.1	90/100	InP/GaAs (SHM)	QPSK	1.7×10^{-3}/–	2016

Table 12.1 (Continued)

Reference	Freq (GHz); channel number	*D* (m)	Data-rate (Gbit/s)	Technology Tx/Rx	Modulation	BER, EVM%	Year
[84]	190	0.02/ 0.006	40/50	SiGe HBT	BPSK	$10^{-9}/10^{-3}$	2017
[28]	490/790	0.5	28/28	RTD	ASK	$2.3\times10^{-4}/1.5\times10^{-3}$	2017
[93]	140	21 000	5	GaAs + TWTA	16QAM/ ASK	10^{-10}/n.a.	2017
[90]	490	0.1	22	RTD/GaAs (diode)	ASK	10^{-12}	2017
[60]	300	0.3	10	PD Ge/GaAs ZBD		10^{-11}	2018
[85]	240	0.15	25	SiGe: C BiCMOS (130 nm)	BPSK	2×10^{-4}	2018
[102]	300	0.1	20	CMOS	—	—	2018
[103]	300 (1)	2	100	InP-HEMT	—	—	2018
[58]	400 (1)	0.5	106	InP/GaAs (SHM)[a]	16QAM	10^{-2}	2018
[26]	300 (1)	0.5	100	InP/GaAs (SHM)[a]	16QAM	$10^{-4}/10^{-6}$	2018
[91]	322 (1)	0.02	27	InP/RTD	ASK	10^{-11}[b]	2018
[87]	230 (1)	1	100	SiGe HBT	16QAM	17% EVM	2019
[61]	280	0.5	56	PD Ge/GaAs (SHM)	16QAM	4.3×10^{-4}	2019
[74]	265 (–)	0.03	80	40 nm CMOS	16QAM	12% EVM	2019
[96]	300 (1)	10	56	InGaAs mHEMT (35 nm)	16QAM	12% EVM	2019
[92]	340 (1)	0.1	30	RTD/RTD	ASK	10^{-11}	2019
[104]	375–500 (6)	1.42	120	InP/GaAs (SHM)[a]	QPSK	10^{-3}	2019
[94]	666 (1)	9.5	590	InP-HEMT + TWTA	QPSK/ 16APSK	<5% EVM	2019

B2B: on-wafer performance (Tx and Rx directly connected, i.e. back-to-back).
a) Photonics-based transmitter.
b) Real-time performance.

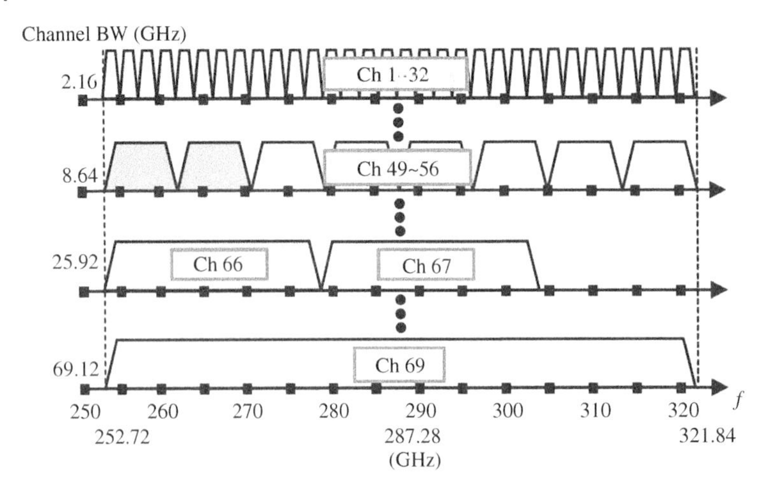

Figure 12.14 Transmission channels in IEEE 802.15.3d format. *Source*: From [89].

12.7 Conclusion

Last "free" region of the electromagnetic spectrum, the THz range has begun to show its potentiality for very high-speed wireless communication systems. The development of the first THz communications was initiated mainly with III–V GaAs and InP-based electronic/photonic systems due to the higher dynamics and power performance in the frequency ranges considered. This allowed to build the bases of the first systems, concepts, and techniques. These first systems, obtained in a laboratory environment, have already evolved considerably. However, to date, many technological developments are still to be made, especially on volume technologies, with use case and scenarios to be clarified. Currently, if the cost of the available advanced technologies could be absorbed on network cores, it would be harder for mass-market applications. Thus, the mass deployment of these technologies for consumer applications will have to go through the development of silicon-based circuits. Highest operating frequencies (i.e. beyond 300 GHz) or long -range links might leverage on specific III–V or TWTA channels. Compared to fiber-optic communications, with 50 years of development, THz communications are still in their infancy and the main part of the story is ahead.

12.8 Acronyms

ASK	amplitude shift keying
BER(T)	bit error rate (tester)
BTL	Bell Telephone Laboratories
DFG	difference frequency generation
GaAs	gallium arsenide
GaN	gallium nitride
HBT	heterojunction bipolar transistor
HEMT	high electron mobility transistor

ITU	international telecommunication union
InP	indium phosphide
MMIC	monolithic microwave integrated circuit
NTT	Nippon Telephone and Telegraph
OOK	oon-off keying
PIC	photonic-integrated circuit
PRBS	pseudo random binary sequence
QAM	quadrature amplitude modulation
QPSK	quadrature phase shift keying
QCL	quantum cascade laser
PAE	power added efficiency
RTD	resonant tunneling diode
Rx	receiver
SiGe	silicon germanium
SHM	sub-harmonic mixer
TMIC	terahertz monolithic-integrated circuit
Tx	transmitter
TWT(A)	travelling wave tube (amplifier)
UTC-PD	uni-travelling carrier-photodiode

E12.1 Link Budget of a THz Link

In this exercise, we consider a 300 GHz link, using a transmitter with a moderate power (4 mW), associated with a high gain antenna of gain $G_E = 50$ dBi. This is displayed in Figure 12.5.

1) Calculate the EIRP (effective isotropic radiated power) of the association Tx + emission antenna.

   The 300 GHz signal is modulated at 20 Gbit/s using a QPSK, and transmitted over the distance d, and the signal is detected using a high gain antenna of G_R. The receiver (Rx) required SNR (signal-to-noise ratio) is 20 dB to recover properly the data.

2) A colleague has drawn this schematic of the situation. Please comment on it.

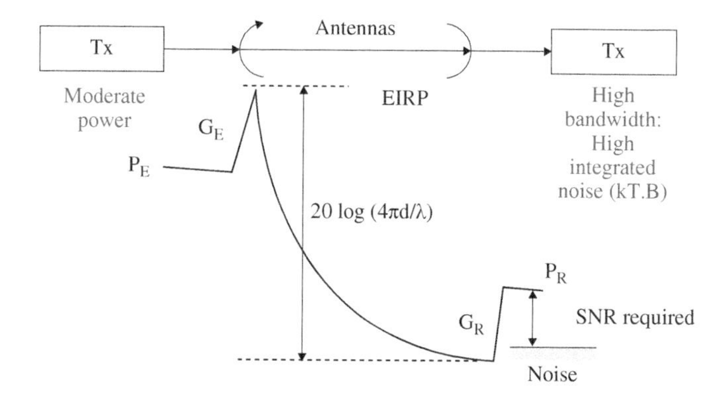

3) What is behind the integrated noise calculation when using k.T.B? Is this still valid at 300 GHz?
4) Assuming a 20 Gbit/s QPSK with no filtering, how can you estimate the size of the modulated spectrum? Do you expect any disturbance in the atmosphere for a 300 GHz carrier? Can we use a 325 GHz carrier?
5) Calculate the possible range of the link with a receiving antenna with 50 dBi, in an outdoor configuration with some rain, featuring a loss of 10 dB/km (Figure 12.2).
6) Does it make a difference to operate the 300 GHz link in space (considered as pure vacuum media) in terms of transmission distance?

Exercise contact: Prof G. Ducournau at guillaume.ducournau@univ-lille.fr

References

1 Cisco, On-Line Resource. (2020). Cisco Visual Networking Index: Global Mobile Data Traffic Forecast Update 2017–2022. https://www.cisco.com/c/en/us/solutions/collateral/service-provider/visual-networking-index-vni/white-paper-c11-738429.html (accessed 21 September 2020).

2 Cherry, S. (2004). Edholm's law of bandwidth. *IEEE Spectrum* **41**: 50.

3 Nagatsuma, T., Ducournau, G., and Renaud, C.C. (2016). Advances in terahertz communications accelerated by photonics. *Nature Photonics* **10**: 371–379. https://doi.org/10.1038/nphoton.2016.65.

4 Su, K., Moeller, L., Barat, R.B. et al. (2012). Experimental comparison of performance degradation from terahertz and infrared wireless links in fog. *Journal of the Optical Society of America A* **29**: 179–184.

5 Ma, J., Vorrius, F., Lamb, L. et al. (2015). Experimental comparison of terahertz and infrared signaling in laboratory-controlled rain. *Journal of Infrared, Millimeter, and Terahertz Waves* **36** (9): 856–865. https://doi.org/10.1007/s10762-015-0183-3.

6 Ericsson Technology Review, #2 (2017). Microwave Backhaul beyond 100 GHz. https://www.ericsson.com/en/ericsson-technology-review/archive/2017/microwave-backhaul-evolution-reaching-beyond-100ghz (accessed 21 September 2020).

7 ELVA (2020). http://elva-1.com/products/a40106 (accessed 21 September 2020).

8 SIKLU Communications (2020). https://www.siklu.com/ (accessed 21 September 2020).

9 IEEE Standard for High Data Rate Wireless Multi-Media Networks – Amendment 2 (2017). 100 Gb/s wireless switched point-to-point physical layer. IEEE Std 802.15.3d-2017 (Amendment to IEEE Std 802.15.3–2016 as Amended by IEEE Std 802.15.3e-2017), 1–55.

10 World Radiocommunication Conference 2019 (WRC-19), Sharm el-Sheikh, Egypt, 28 October to 22 2019 https://www.itu.int/en/ITU-R/conferences/wrc/2019/Pages/default.aspxe

11 Shannon, C.E. (1949). Communication in the presence of noise. *Proceedings of the Institute of Radio Engineers* **37**: 10–21.

12 Phileweb (2008). HD-TV link by NTT in real-time at 120 GHz, Beijing Olympic. https://www.phileweb.com/news/d-av/200811/25/22581.html (accessed 21 September 2020).

13 G. Ducournau (2017). HD-TV link in real-time at 300 GHz, Dunkirk, Dunkerque, France. https://www.youtube.com/watch?v=laE-K6Pj_Rk (in French) (accessed 21 September 2020).

14 Suen, J.Y., Fang, M.T., Denny, S.P. et al. (2015). Modeling of terabit geostationary terahertz satellite links from globally dry locations. *IEEE Transactions on Terahertz Science and Technology* **5** (2): 299–313. https://doi.org/10.1109/TTHZ.2015.2399694.

15 Miyauchi, K. (1983). Millimeter-wave communications. IRMMW – vol. 9 – Millimeter Components and Techniques, Part I, ISBN 0-12147 709-6.

16 Sengupta, K., Nagatsuma, T., and Mittleman, D.M. (2018). Terahertz integrated electronic and hybrid electronic–photonic systems. *Nature Electron* **1**: 622–635. https://doi.org/10.1038/s41928-018-0173-2.

17 Virginia DIODES Inc (2020). https://www.vadiodes.com/en/ (accessed 21 September 2020).

18 Radiometer Physics GmbH (2020). https://www.radiometer-physics.dpp.e/ (accessed 21 September 2020).

19 ACST Company (2020). https://acst.de/?page&view=schottky (accessed 21 September 2020).

20 Nadar, S., Zaknoune, M., Wallart, X. et al. (2017). High performance heterostructure low barrier diodes for sub-THz detection. *IEEE Transactions on Terahertz Science and Technology* **7** (6): 780–788. https://doi.org/10.1109/TTHZ.2017.2755503.

21 Ito, H. and Ishibashi, T. (2017). Low noise homodyne detection of terahertz waves by zero-biased InP/InGaAs Fermi-level managed barrier diode. *IEICE Electronic Express* https://doi.org/10.1587/elex.14.20170722.

22 Ishibashi, T., Muramoto, Y., Yoshimatsu, T. et al. (Nov.-Dec. 2014). Unitraveling-carrier photodiodes for terahertz applications. *IEEE Journal of Selected Topics in Quantum Electronics* **20** (6): 79–88. https://doi.org/10.1109/JSTQE.2014.2336537.

23 Bründermann, E., Hübers, H.-W., Kimmitt, M. FitzGerald, "Terahertz techniques," Springer, 2013, ISBN 978-3-642-02592-1.

24 Latzel, P., Pavanello, P., Billet, M. et al. (Nov. 2017). Generation of mW level in the 300-GHz band using resonant-cavity-enhanced unitraveling carrier photodiodes. *IEEE Transactions on Terahertz Science and Technology* **7** (6): 800–807. https://doi.org/10.1109/TTHZ.2017.2756059.

25 Song, H., Ajito, K., Muramoto, Y. et al. (2012). Uni-travelling-carrier photodiode module generating 300 GHz power greater than 1 mW. *IEEE Microwave and Wireless Components Letters* **22** (7): 363–365. https://doi.org/10.1109/LMWC.2012.2201460.

26 Chinni, V.K., Latzel, P., Zégaoui, M. et al. (2018). Single-channel 100 Gbit/s transmission using III–V UTC-PDs for future IEEE 802.15.3d wireless links in the 300 GHz band. *Electronics Letters* **54** (10): 638–640. https://doi.org/10.1049/el.2018.0905.

27 Maestrini, A., Ward, J., Chattopadhyay, G. et al. (2019). Terahertz sources based on frequency multiplication and their applications. *Frequenz* **62** (5–6): 118–122.

28 Oshima, N., Hashimoto, K., Suzuki, S. et al. (2017). Terahertz wireless data transmission with frequency and polarization division multiplexing using resonant-tunneling-diode oscillators. *IEEE Transactions on Terahertz Science and Technology* **7** (5): 593–598. https://doi.org/10.1109/TTHZ.2017.2720470.

29 Jastrow, C., Priebe, S., Spitschan, B. et al. (2010). Wireless digital data transmission at 300 GHz. *Electronics Letters* **46** (9): 661–663.

30 Ducournau, G., Szriftgiser, P., Bacquet, D. et al. (2014). Multi-carrier transmission of vectorial modulation schemes using electronic transceivers at 588 GHz. *Electronics Letters* **50** (23): 1715–1717. https://doi.org/10.1049/el.2014.3652.

31 Dan, I., Ducournau, G., Hisatake, S. et al. (2020). A terahertz wireless communication link using a superheterodyne approach. *IEEE Transactions on Terahertz Science and Technology* **10** (1): 32–43.

32 Moeller, L., Federici, J., and Su, K. (2011). 2.5 Gbit/s duobinary signaling with narrow bandwidth 0.625 terahertz source. *Electronics Letters* **47** (15): 856–858. https://doi.org/10.1049/el.2011.1451.

33 Ma, J., Shrestha, R., Moeller, L. et al. (2018). Channel performance for indoor and outdoor terahertz wireless links. *APL Photonics* **3**: 051601. https://doi.org/10.1063/1.5014037.

34 Ma, J., Shrestha, R., Zhang, W. et al. (2019). Terahertz wireless links using diffuse scattering from rough surfaces. *IEEE Transactions on Terahertz Science and Technology* **9** (5): 463–470. https://doi.org/10.1109/TTHZ.2019.2933166.

35 Nopchinda, D., He, Z., Granström, G. et al. (2018). 8-PSK upconverting transmitter using E-band frequency sextupler. *IEEE Microwave and Wireless Components Letters* **28** (2): 177–179. https://doi.org/10.1109/LMWC.2018.2792687.

36 Suzuki, S., Hinata K., Shiraishi M. et al. (2010). RTD oscillators at 430–460 GHz with high output power (~200 μW) using integrated offset slot antennas. 22nd International Conference on Indium Phosphide and Related Materials (IPRM2010), 1–4.

37 Izumi, R., Suzuki S., and Asada, M. (2017). 1.98 THz resonant-tunneling-diode oscillator with reduced conduction loss by thick antenna electrode. 42nd International Conference on Infrared, Millimeter, and Terahertz Waves (IRMMW-THz), Cancun, 1–2. https://doi.org/10.1109/IRMMW-THz.2017.8066877.

38 Grant, P.D., Laframboise, S.R., Dudek, R. et al. (2009). Terahertz free space communications demonstration with quantum cascade laser and quantum well photodetector. *Electronics Letters* **45** (18): 952–954. https://doi.org/10.1049/el.2009.1586.

39 Song, H. (2017). Packages for terahertz electronics. *Proceedings of the IEEE* **105** (6): 1121–1138. https://doi.org/10.1109/JPROC.2016.2633547.

40 Deal, W., Mei, X.B., Radisic, V. et al. (2010). Demonstration of a 0.48 THz amplifier module using InP HEMT transistors. *IEEE Microwave and Wireless Components Letters* **20** (5): 289–291.

41 Jung-Kubiak, C., Reck, T., Siles, J.V. et al. (2016). A Multistep DRIE process for complex terahertz waveguide components. *IEEE Transactions on Terahertz Science and Technology* **6** (5): 690–695. https://doi.org/10.1109/TTHZ.2016.2593793.

42 Chahat, N., Decrossas, E., Gonzalez-Ovejero, D. et al. (2019). Advanced cubesat antennas for deep space and earth science missions: a review. *IEEE Antennas and Propagation Magazine* **61** (5): 37–46. https://doi.org/10.1109/MAP.2019.2932608.

43 Nagatsuma, T. (2018). Antenna technologies for terahertz communications. International Symposium on Antennas and Propagation (ISAP), Busan, Korea (South), 1–2.

44 Nagatsuma, T., Oogimoto K., Yasuda, Y. et al. (2016). 300-GHz-band wireless transmission at 50 Gbit/s over 100 meters. 41st International Conference on Infrared, Millimeter and Terahertz Waves (IRMMW-THz), Copenhagen, 1–2. https://doi.org/10.1109/IRMMW-THz.2016.7758356.

45 Inoue, M., Hodono, M., Horiguchi, S. et al. (2014). Ultra-broadband terahertz receivers using polymer substrate. *IEEE Transactions on Terahertz Science and Technology* **4** (2): 225–231. https://doi.org/10.1109/TTHZ.2013.2296994.

46 Withayachumnankul, W., Yamada, R., Fumeaux, C. et al. (2017). All-dielectric integration of dielectric resonator antenna and photonic crystal waveguide. *Optics Express* **25**: 14706–14714.

47 Withayachumnankul, W., Yamada, R., Fujita, M. et al. (2018). All-dielectric rod antenna array for terahertz communications. *APL Photonics* **3** (5): 051707.

48 Chang, F., Onohara, K., and Mizuochi, T. (2010). Forward error correction for 100 G transport networks. *IEEE Communications Magazine* **48** (3): S48–S55. https://doi.org/10.1109/MCOM.2010.5434378.

49 Nagatsuma, T., Hirata A., Royter Y. et al. (2000). A 120-GHz integrated photonic transmitter. IEEE Topical Meeting on Microwave Photonics (MWP2000), 225–228.

50 Hirata, A., Kosuji, T., Takahashi, H. et al. (2006). 120-GHz-band millimeter-wave photonic wireless link for 10-Gb/s data transmission. *IEEE Transactions on Microwave Theory and Techniques* **54** (5): 1937–1944. https://doi.org/10.1109/TMTT.2006.872798.

51 6G Flagship (2020). https://www.oulu.fi/university/news/6g-white-paper (accessed 21 September 2020).

52 Richardson, D.J., Fini, J.M., and Nelson, L.E. (2013). Space-division multiplexing in optical fibres. *Nature Photonics* **7**: 354–362.

53 Nagatsuma, T. and Carpintero, G. (2015). Recent progress and future prospect of photonics-enabled terahertz communications research. *IEICE Transactions on Electronics* **E98-C**: 1060–1070.

54 Shams, H., Fice, M.J., Balakier, K. et al. (2014). Photonic generation for multichannel THz wireless communication. *Optics Express* **22**: 23465–23123472.

55 Ducournau, G., Szriftgiser, P., Beck, A. et al. (2014). Ultrawide bandwidth single channel 0.4 THz wireless link combining broadband quasi-optic photomixer and coherent detection. *IEEE Transactions on Terahertz Science and Technology* **4**: 328–337.

56 Koenig, S., Lopez-Diaz, D., Antes, J. et al. (2013). Wireless sub-THz communication system with high data rate. *Nature Photonics* **7**: 977–981.

57 Nagatsuma, T., Y. Fujita, Y. Yasuda et al. (2016). Real-time 100-Gbit/s QPSK transmission using photonics-based 300-GHz-band wireless link. IEEE International Topical Meeting on Microwave Photonics (MWP2016), Long Beach, CA, 27–30. https://doi.org/10.1109/MWP.2016.7791277.

58 Jia, S., Pang, X., Ozolins, O. et al. (2018). 0.4 THz photonic-wireless link with 106 Gbps single channel bitrate. *Journal of Lightwave Technology* **36** (2): 610–616.

59 Bowers, S.M., Abiri, B., Aflatouni, F., et al. (2014). A compact optically driven travelling-wave radiating source. Optical Fiber Communication Conference Exhibits (OFC2014), Tu2A.3.

60 Lacombe, E., Belem-Goncalves, C., Luxey, C. et al. (2018). 10-Gb/s indoor THz communications using industrial Si photonics technology. *IEEE Microwave and Wireless Components Letters* **28** (4): 362–364. https://doi.org/10.1109/LMWC.2018.2811242.

61 Belem-Goncalves, C., Lacombe, E., Gidel, V. et al. (2019). 300 GHz quadrature phase shift keying and QAM16 56 Gbps wireless data links using silicon photonics photodiodes. *Electronics Letters* **55** (14): 808–810. https://doi.org/10.1049/el.2019.1242.

62 Nagatsuma, T., Horiguchi, S., Minamikata, Y. et al. (2013). Terahertz wireless communications based on photonics technologies. *Optics Express* **21**: 477–487.

63 Ducournau, G., Yoshimizu, Y., Hisatake, S. et al. (2014). Coherent THz communication at 200 GHz using a frequency comb, UTC-PD and electronic detection. *Electronics Letters* **50**: 386–388.

64 Ducournau, G., Pavanello, F., Beck, A. et al. (2014). High-definition television transmission at 600 GHz combining THz photonics hotspot and high-sensitivity heterodyne receiver. *Electronics Letters* **50** (5): 413–415. https://doi.org/10.1049/el.2013.3796.

65 Nagatsuma, T., Sonoda, M., Higashimoto, T. et al. (2019). 12.5-Gbit/s wireless link at 720 GHz based on photonics. 44th International Conference on Infrared, Millimeter, and Terahertz Waves (IRMMW-THz), Paris, 1–2. https://doi.org/10.1109/IRMMW-THz.2019.8873870.

66 Ducournau, G. (2018). Silicon photonics targets terahertz region. *Nature Photonics* **12**: 574–575. https://doi.org/10.1038/s41566-018-0242-0.

67 Hirata, A., Kosuji, T., Takahashi, H. et al. (2012). 120-GHz-band wireless link technologies for outdoor 10-Gbit/s data transmission. *IEEE Transactions on Microwave Theory and Techniques* **60**: 881–895.

68 Nagatsuma, T., Sonoda, M., Higashimoto, T. et al. (2019). 300-GHz-band wireless communication using Fermi-level managed barrier diode receiver. IEEE MTT-S International Microwave Symposium (IMS2019), Boston, MA, USA, 762–765. https://doi.org/10.1109/MWSYM.2019.8700954

69 Jastrow, C., Priebe, S., Spitschan, B. et al. (2010). Wireless digital data transmission at 300 GHz. *Electronics Letters* **46** (9): 661–663. https://doi.org/10.1049/el.2010.3509.

70 Song, H.J., Kim, J.Y., Ajito, K. et al. (2013). Fully integrated ASK receiver MMIC for terahertz communications at 300 GHz. *IEEE Transactions on Terahertz Science and Technology* **3**: 445–452.

71 Tokgoz, K.K., Maki S., Pang J. et al. (2018). A 120Gb/s 16QAM CMOS millimeter-wave wireless transceiver. IEEE International Solid-State Circuits Conference (ISSCC2018) (February 2018), 168–170.

72 Kallfass, I., Dan, I., Rey, S. et al. (2015). Towards MMIC-based 300 GHz indoor wireless communication systems. *IEICE Transactions on Electronics* **E98-C**: 1081–1088.

73 Sarkozy, S., Vukovic, M., Padilla, J.G. et al. (2013). Demonstration of a G-band transceiver for future space crosslinks. *IEEE Transactions on Terahertz Science and Technology* **3** (5): 675–681. https://doi.org/10.1109/TTHZ.2013.2276118.

74 Carpenter, S., Abbasi, M., and Zirath, H. (2015). Fully integrated D-band direct carrier quadrature (I/Q) modulator and demodulator circuits in InP DHBT technology. *IEEE Transactions on Microwave Theory and Techniques* **63**: 1666–1675.

75 Hu, P., Lei, W., Jiang, Y. et al. (2019). Demonstration of a watt-level traveling wave tube amplifier operating above 0.3 THz. *IEEE Electron Device Letters* **40** (6): 973–976. https://doi.org/10.1109/LED.2019.2912579.

76 Schlechtweg, M. (2015). Multifunctional circuits and modules based on III/V mHEMT technology for (sub-)millimeter-wave applications in space, communication and sensing. IEEE International Microwave Symposium (IMS2015) Workshop WSI-6.

77 Kallfass, I., Boes, F., Messinger, T. et al. (2015). 64 Gbit/s transmission over 850 m fixed wireless link at 240 GHz carrier frequency. *Journal of Infrared, Millimeter, and Terahertz Waves* **36**: 221–233.

78 Pfeiffer, U. (2015). RF front-ends for mm-wave and THz application in SiGe/CMOS. IEEE International Microwave Symposium (IMS2015) Workshop WSI-5.

79 Zhao, Y., Ojefors, E., Aufinger, K. et al. (2012). A 160 GHz subharmonic transmitter and receiver chipset in an SiGe HBT technology. *IEEE Transactions on Microwave Theory and Techniques* **60**: 3286–3299.

80 Fujishima, M., Amakawa, S., Takano, K. et al. (2015). Terahertz CMOS design for low-power and high-speed wireless communication. *IEICE Transactions on Electronics* **E98-C**: 1091–1104.

81 Sengupta, K. and Hajimiri, A. (2012). A 0.28 THz power-generation and beam-steering array in CMOS based on distributed active radiators. *IEEE Journal of Solid-State Circuits* **47**: 3013–3031.

82 Han, R. and Afshari, E. (2013). A CMOS high-power broadband 260-GHz radiator array for spectroscopy. *IEEE Journal of Solid-State Circuits* **48**: 3090–3104.

83 Kasagi, K., Oshima, N., Suzuki, S. et al. (2015). Power combination in 1 THz resonant-tunneling-diode oscillators integrated with patch antennas. *IEICE Transactions on Electronics* **E98-C** (12): 1131–1133.

84 Fritsche, D., Stärke, P., Carta, C. et al. (2017). A low-power SiGe BiCMOS 190-GHz transceiver chipset with demonstrated data rates up to 50 Gbit/s using on-chip antennas. *IEEE Transactions on Microwave Theory and Techniques* **65** (9): 3312–3323.

85 Eissaet, M.H., Malignaggi, A., Wang, R. et al. (2018). Wideband 240-GHz transmitter and receiver in BiCMOS technology with 25-Gbit/s data rate. *IEEE Journal of Solid-State Circuits* **53** (9): 2532–2542.

86 Rodríguez-Vázquez, P., Grzyb, J., Heinemann, B., et al. (2018). Performance evaluation of a 32-QAM 1-m wireless link operating at 220–260 GHz with a data-rate of 90 Gbps. Proceedings of the Asia-Pacific Microwave Conference, 723–725.

87 Rodríguez-Vázquez, P., Grzyb, J., Heinemann, B. et al. (2019). A 16-QAM 100-Gb/s 1-m wireless link with an EVM of 17% at 230 GHz in an SiGe technology. *IEEE Microwave and Wireless Components Letters* **29** (4): 297–299. https://doi.org/10.1109/LMWC.2019.2899487.

88 Hara, S., Katayama K., Takano K. et al. (2017). A 32 Gbit/s 16QAM CMOS receiver in 300 GHz band. IEEE MTT-S International. Microwave Symposium (IMS2017) (June 2017), 1703–1706.

89 Lee, S., Hara, S., Yohsida, T. et al. (2019). An 80-Gb/s 300-GHz-band single-chip CMOS transceiver. *IEEE Journal of Solid-State Circuits* **54** (12): 3577–3588. https://doi.org/10.1109/JSSC.2019.2944855.

90 Suzuki, S. and Asada, M. (2017). Terahertz communications using resonant-tunneling-diode oscillators. 11th European Conference on Antennas and Propagation (EUCAP), Paris, 1615–1617. https://doi.org/10.23919/EuCAP.2017.7928431

91 Nishigami, N., Nishida, Y., Diebold, S. et al. (2018). Resonant tunneling diode receiver for coherent terahertz wireless communication. Asia-Pacific Microwave Conference (APMC2018), Kyoto, 726–728. https://doi.org/10.23919/APMC.2018.8617565.

92 Nishida, Y., Nishigami, N., Diebold, S. et al. (2019). Terahertz coherent receiver using a single resonant tunnelling diode. *Scientific Reports* **9** (18): 125.

93 Wu, Q., Lin C., Lu, B. et al. (2017). A 21 km 5 Gbps real time wireless communication system at 0.14 THz. 42nd International Conference on Infrared, Millimeter, and Terahertz Waves (IRMMW-THz), Cancun, 1–2. https://doi.org/10.1109/IRMMW-THz.2017.8066870.

94 Deal, W. R., Foster T., Wong M.B. et al. (2017). A 666 GHz demonstration crosslink with 9.5 Gbps data rate. IEEE MTT-S International Microwave Symposium (IMS2017), Honololu, 233–235. https://doi.org/10.1109/MWSYM.2017.8059083.

95 Ćwikliński, M., Bruckner P., Leone S. et al., "190-GHz G-band GaN amplifier MMICs with 40GHz of bandwidth," IEEE MTT-S International Microwave Symposium (IMS2019), Boston, USA, pp. 1257–1260, 2019. doi: https://doi.org/10.1109/MWSYM.2019.8700762

96 Li, X., Yu, J., Zhang, J. et al. (2013). A 400G optical wireless integration delivery system. *Optical Express* **21**: 187894–187 899.

97 Wang, C., Lu, B., Lin, C. et al. (2014). 0.34-THz wireless link based on high-order modulation for future wireless local area network applications. *IEEE Transactions on Terahertz Science and Technology* **4**: 75–85.

98 Song, H.J., Kim, J.Y., Ajito, K. et al. (2014). 50-Gb/s direct conversion QPSK modulator and demodulator MMICs for terahertz communications at 300 GHz. *IEEE Transactions on Microwave Theory and Techniques* **62**: 600–609.

99 Ducournau, G. et al. (2015). 32 Gbit/s QPSK transmission at 385 GHz using coherent fibre-optic technologies and THz double heterodyne detection. *Electronics Letters* **12**: 915–917.

100 Kanno, A., Kuri, T., Morohashi, I. et al. (2015). Coherent terahertz wireless signal transmission using advanced optical fiber communication technology. *Journal of Infrared, Millimeter, and Terahertz Waves* **36**: 180–197.

101 Yu, X., Asif R., Peils M. et al. (2015). 60 Gbit/s 400 GHz wireless transmission. International Conference on Photonics in Switching (PS2015), 4–6.

102 Hara, S., Takano, K., Katayama, K. et al. (2018). 300-GHz CMOS transceiver for terahertz wireless communication. Asia-Pacific Microwave Conference (APMC2018), 429–431.

103 Hamada, H., Fujimura, T., Abdo, I. et al. (2018) 300-GHz 100 Gbps InP-HEMT wireless transceiver using a 300-GHz fundamental mixer. IEEE/MTT-S International Microwave Symposium (IMS2018) (June 2018), 1480–1483.

104 Li, X., Yu, J., Wang, K. et al. (2019). 120 Gb/s wireless terahertz-wave signal delivery by 375 GHz–500 GHz multi-carrier in a 2×2 MIMO system. *Journal of Lightwave Technology* **37** (2): 606–611. https://doi.org/10.1109/JLT.2018.2862356.

105 https://www.microwavejournal.com/articles/32759-twtas-still-dominate-high-power-and-mmwave-applications

106 http://www.ieee802.org/15/pub/TG3d/index_IGthz.html

13

THz Applications: Devices to Space System

Imran Mehdi

Jet Propulsion Laboratory (JPL), Pasadena, CA, USA

13.1 Introduction

The THz or submillimeter-wave frequency range is loosely defined as 300 GHz to 10 THz. While research and applications in the millimeter-wave have been on-going for more than a century, the interest in the submillimeter-wave is relatively recent. A main reason for this can be attributed to the recent availability of robust components in this frequency range. The THz or submm-wave spectrum falls between well-established microwave technology at the low end and optical techniques and technologies at the higher end, thus giving rise to the term "THz gap." Several notable technical breakthroughs in the 1980s resulted in a surge of activity in this frequency range. For the first time, commercially available 3D electromagnetic simulators now made it possible for researchers to simulate designs with confidence instead of having to build and characterize labor-intensive lower-frequency models. Similarly, introduction of high precision multiple-axis milling machines made it possible to build waveguide blocks with the required precision. The machines were capable of directly importing 3-D solid models reducing delivery time and human errors while increasing throughput. However, arguably the most significant was the advent of planar Schottky diodes with cutoff frequencies in the THz, which made it possible to build and conceptualize components such as multipliers and mixers in this frequency range. This work was spearheaded by researchers at the University of Virginia and they first demonstrated the potential of these planar devices [1] compared with whisker contacted diodes. This was followed by a number of different groups around the world making planar Schottky devices and demonstrating excellent performance only a few years later [2–4]. And a dozen years later, a fully solid-state source with planar diodes at 2.7 THz with sufficient power to make spectroscopy possible had been demonstrated [5]. See [6] for an interesting historical perspective with regards to the early activity regarding planar diodes.

Further impetus for interest in this frequency range, at least as far as space science was concerned, came from the ground-breaking observations by the Infrared Astronomical

Fundamentals of Terahertz Devices and Applications, First Edition. Edited by Dimitris Pavlidis.

Companion website: www.wiley.com/go/Pavlidis/FundamentalsofTHz

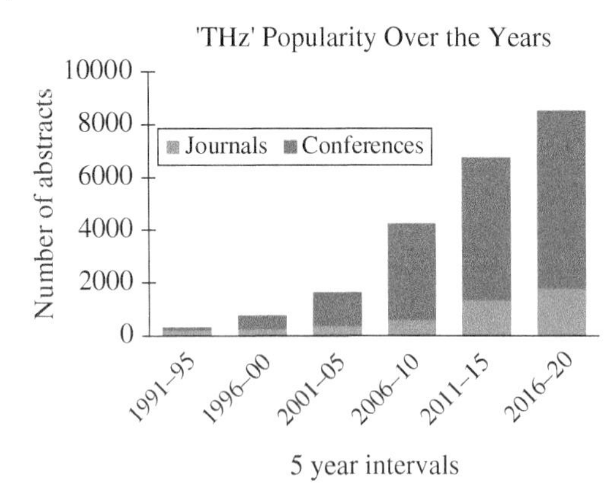

Figure 13.1 The increased interest in THz technology is apparent from recent research associated with THz science and technology. The data above is obtained only from IEEE journals and conferences and quantifies the use of "THz," or "Terahertz" or "Submillimeter" in the abstract of the published article or conference abstract.

Satellite (IRAS) [7]. By being in space, IRAS could penetrate through the dusty universe and was able to collect photons from over a half a million sources of infrared radiation. Besides several key discoveries, IRAS data showed that half of the energy released by galaxies is in the form of infrared emission, created when interstellar dust absorbs light from young stars and radiates it as heat. This discovery provided a fresh and unprecedented insight into solar system objects and the interstellar medium (ISM). This provided sound rationale for the various national space agencies to start investing in infrared technologies and space instruments.

Today, research and development of submillimeter-wave components and science is a worldwide phenomenon with IEEE having initiated a dedicated transaction to this subject starting in 2011 [8] and numerous international conferences and workshops devoted to this subject matter. Figure 13.1 shows the usage of the word "THz" in only IEEE journal and conference abstracts over the last three decades. While space applications continue to be a niche application (and this chapter is devoted to this), non-space applications continue to grow with impressive results and market penetration. Table 13.1 provides a summary of various applications that have been demonstrated in this frequency range. It is not possible for a single chapter to provide in-depth review of all these applications, instead, references are included for the reader who might be interested in further exploring any one of these applications. The present chapter will focus on the technology and engineering aspects of the heterodyne receiver which is the system of choice for conducting high-resolution spectroscopy for space applications.

Remainder of this section will discuss the fundamentals of THz spectroscopy and why it is an important tool for space scientists. Section 13.2 will describe the tool of choice for high-resolution spectroscopy, namely the heterodyne receiver, and a discussion of the current State-of-the-Art for its critical components such as mixers and local oscillators (LOs). Section 13.3 will discuss three distinct space science applications for THz instruments and how these applications are currently driving technology development.

13.1.1 Why Is THz Technology Important for Space Science?

Space scientists and technologists have long been interested in this regime as it provides measurement and probing capabilities that are hard to achieve in other parts of the

Table 13.1 Development of THz technology can be sorted based on a few general applications.

Application	Qualitative goals
High spectral resolution spectroscopy for space science	Near quantum-limited detection; broadband and robust systems; cryogenic and noncryogenic detectors; large sky maps; velocity resolved spectroscopy; planetary atmospheric chemistry; hunt for water; plume detection and phenomenology; wind velocity measurements, etc. References: Waters et al. [9], Melnick et al. [10], Gulkis et al. [11], De Graauw et al. [12], and Hartogh et al. [13].
Passive imaging (far field)	Imaging through clothes for contraband; real-time throughput; cm resolution; large FOV capability; imaging from large distances, i.e. > 10 m; imaging of crowded scenes References: Luukanen et al. [14] and Appleby and Anderton [15].
Active imaging	Imaging through clothes for contraband; real-time throughput; cm resolution; large FOV capability; imaging from large distance, imaging crowded scenes; 3-D imaging; being able to detect unique signatures from target; remote sensing of atmospheres; spectroscopic as well as dynamic characterization of plumes, etc. References: Cooper et al. [16], Robertson et al. [17], Grajal et al. [18], Robertson et al. [19], and Coulombe et al. [20].
Medical imaging	Identification of abnormal physiology; subcutaneous imaging; therapeutic usage; breast and skin cancer; burn wound, skin hydration, corneal hydration; melanoma/carcinoma identification References: Taylor et al. [21], Wang et al. [22], Taylor et al. [23], Woodward et al. [24], Wallace et al. [25], Brown et al. [26], Ashworth et al. [27], and Arbab et al. [28].
Nonmedical imaging	Nondestructive testing; agricultural usage; art and manufacturing References: Cosentino [29], Kawase et al. [30], and Afsah-Hejri et al. [31].
Communications	Secure local area networks; high bandwidth links; rural internet penetration References: Akyildiz et al. [32], Kurner et al. [33], Song and Nagatsuma [34], and Mumtaz et al. [35].

frequency spectrum. One of the main tools utilized by space scientists in this frequency range is high-spectral resolution spectroscopy. Every dipole molecule emits a unique spectroscopic signature that can be measured with high sensitivity detector systems. Thus, in a way, nature predicates the use of this frequency range as a number of molecules that provide physical insight to the workings of our universe have thermal absorption and emission lines in this frequency range.

The capabilities of microwave observations for retrieval of atmospheric parameters have been widely demonstrated for planets, in particular Earth and Mars. Among numerous examples, the Microwave Limb Sounder (MLS) on the Upper Atmosphere Research Satellite (UARS) and the Earth observing system (EOS) Aura spacecraft demonstrate the value of microwave limb sounding for trace constituent detection and wind on Earth [36]. Mars has been an intriguing target for exploration and investigation. For example, the Submillimeter Wave Astronomy Satellite used the rotational transition of H_2O (557 GHz) and 13CO (551 GHz) to derive the temperature structure and altitude distribution of water vapor on Mars [10]. Understanding Martian atmosphere and water cycle have been a major theme at NASA, European space agency (ESA), and JAXA. THz detectors have proven useful for measuring trace constituent abundances and physical properties under all climate conditions, including dust clouds. The strong submillimeter transitions of polar molecules permit detection of numerous trace species at parts per trillion to parts per billion sensitivity. As an emission measurement, observations are carried out continuously in a passive mode without the need for any time restrictive event such as a solar occultation. At these wavelengths, a moderate-sized antenna (30 cm effective diameter) can yield high-spatial resolution measurements, while the heterodyne technique permits ultrahigh spectral resolution ($>3 \times 10^6$) which provides clear line separation and well-defined line profiles.

A second reason, as mentioned before, for space scientist's interest in this frequency range is the fact that most of the electromagnetic radiation arriving from the local universe is generated as thermal radiation from dust and arrives in the far-infrared or submillimeter-wave regime. Almost half of the radiation from the Milky Way is in this band and the Ultra Luminous (most luminous galaxies in the local Universe) emit around 99% of their radiation in this band [7]. Similarly, early glow from protostars is in the submillimeter-wave range as the protostar's gravitational energy is converted to heat. The star-forming process itself needs instruments in the submillimeter to investigate, as the important cooling lines such as H_2O, SO_2, and CO and rotational fine structure lines CI, CII, and NII can only be detected in the far-infrared or submillimeter-wave range [37].

A third but significant reason for THz instruments in space is the presence of water vapor in our atmosphere. The water vapor and thus the absorption is dependent on location, altitude, and wavelength. A number of theoretical models can predict absorption through this band based on various conditions, but generally, it is impossible to carry out submillimeter astronomy from the earth at sea level altitudes. This is the reason why submm instruments are usually placed on mountain tops or balloons or aircraft. Obviously, the best spot for submm astronomy is from space, hence the interest of the space community in this frequency range.

The submillimeter-wave band provides unique capabilities when investigating planetary bodies [38]. Presence of water vapor in the Martian atmosphere is widely accepted; however, quantitative measurement is still lacking. 3-D model of a prototype THz instrument that can be used on Mars or other planetary bodies is shown in Figure 13.2. Main subsystems of the instrument are outlined. This particular rendition can provide one-dimensional scanning. Of course, a more sophisticated instrument can be designed if the science drivers require that. With such an instrument lines from several isotopologues of water can be used to provide the dynamic range needed to observe the water abundance under different atmospheric conditions and altitudes. Using the $H_2{}^{18}O$ transition at 547.676 GHz and assuming knowledge of the ${}^{18}O/{}^{16}O$ ratio (which is not expected to vary significantly), the $H_2{}^{18}O$ distribution can serve as a proxy for the distribution of the optically thick main isotopologue. The single-limb measurement precision (half scale height resolution) is <1%

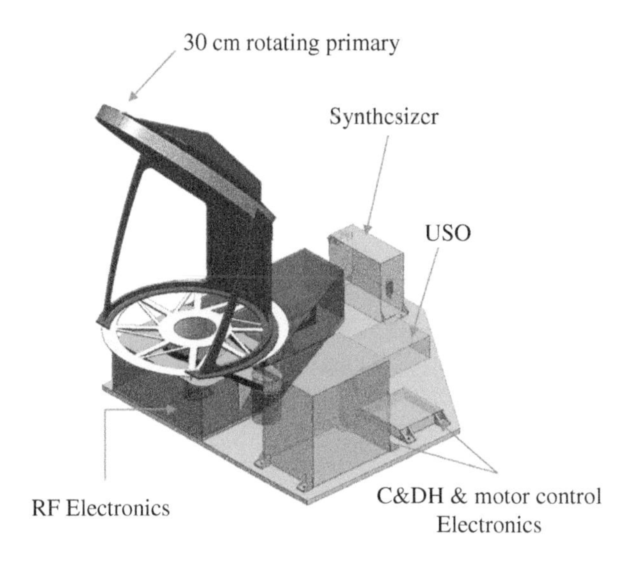

Figure 13.2 A prototype THz instrument for exploring the atmosphere and water dynamics on Mars. A 500–600 GHz receiver system and a 30 cm primary are sufficient to quantitatively measure water vapor on Mars.

below 30 km, increasing slowly to 20% at 60 km. Measurements to higher altitude can be obtained using the stronger water line at 556 GHz. In nadir viewing, measurements of H_2O (556 GHz) with atmospheric scale height (~8 km) vertical resolution are achievable in a single integration sample: precision of ~10% below 30 km, increasing to 50% at 50 km. Submm spectrometers with a 750 MHz bandwidth can allow the broad line to be measured, enabling, for example, detection of small variations in atmospheric water in the lowest scale height in a small, localized region providing high-resolution mapping. Table 13.2 lists several species of interest in the Martian atmosphere and their detectability from a 500 km orbit with a Schottky based room temperature detector. Even for a limited bandwidth of 530–585 GHz, several lines can be detected.

Submillimeter heterodyne techniques will also be important for studying the Jovian system. Voyager and Galileo have taken pictures of Europa that suggest the presence of liquid water on Europa. This geological evidence is tantalizing, but incomplete, as it suggests that liquid water could be present, but also allows for the possibility that the strange features we saw on Europa's surface could all have formed through the motion of soft ice. Models of Europa including gravitational field effects, radiogenic heating, and tidal bulging have all been inconclusive with regards to the presence of water on Europa. Also, the composition of the ice on Europa is not well understood. Most models assume that it is pure water ice, but a small amount of other material present in the ice, especially other volatiles such as ammonia or salts, could dramatically alter the rheology of Europa's ice. Most models also assume that the ice layer is solid, but the rheology could be changed if the ice layer is broken or fractured, or if the grain size is different than assumed in the models.

A passive submillimeter instrument has been selected on the payload of the JUpiter Icy moons Explorer (JUICE) mission [39]. The Submillimeter Wave Instrument (SWI) on JUICE will provide important measurements of both Jupiter and its satellites. As an

Table 13.2 A Submm spectrometer can measure a number of trace gases in the Martian atmosphere. 5000 K DSB receiver noise temperature and 750 MHz bandwidth is assumed for the receiver for these calculations. Calculations courtesy of Brian Drouln, JPL.

Trace gas	Frequency (GHz)	3 s limb measurement (pptv)
HCN	531.716	11
CH_3SH	534.698	1253
OCS	534.728	611
NO_2	566.981	5495
SO_2	571.553	511
NH_3	572.497	60
N_2O	577.578	3556
HO_2	578.196	182
H_2S	579.799	416
H_2CO	583.145	38
H_2CS	583.717	97
HCOOH	584.201	34 485

example, we briefly discuss here a few areas where such an instrument can impact our knowledge of the Galilean moon Ganymede. A prime objective of this mission at Ganymede is to characterize the extent of the ocean and its relationship to the ice crust on Ganymede. SWI will be able to characterize the thermal physical properties of Ganymede's surface as well as sensing its tenuous atmosphere. Both day and night side temperatures of Ganymede's surface can be measured without the need for a cooled receiver. These will permit the thermal physical properties of its surface to be measured. Global limb and nadir measurements of water will permit the detection of water from the ground up to 400 km with 5 km vertical resolution. The sensitivity will be sufficient to detect column abundances of 10^{11} molecules/cm^2. This is sufficient to characterize sputtering and sublimation. The atmosphere of Ganymede will be driven by high-speed winds and these will be detectable by measuring the Doppler widths of spectral lines of water. Finally, we mention the capability of measuring the D/H ratio on Ganymede using measurements of HDO and H_2O. Many other studies will be possible with this instrument on JUICE.

13.1.2 Fundamentals of THz Spectroscopy

Both passive and active radiometry can be performed at submillimeter wavelengths. These techniques provide identification and quantification of bulk and trace constituents of remote gases during fly-by and orbital mission phases. Passive radiometry can be accomplished through limb or nadir scanning as well as atmospheric occultation of blackbody sources. Active radiometry can be accomplished through radar techniques or via occultation with a beacon. So far, only passive methods have been employed in space, such as for Earth science (e.g. EOS–MLS); cometary science (MIRO); and astrophysics (Herschel Space Observatory).

The frequency and temperature-dependent integrated absorption or emission intensity of any molecular gas transition at any wavelength is calculated from the following equation after [10]:

$$I(\nu, T) = \Gamma(\nu)I(T)$$

where the frequency-dependent line shape $\Gamma(\nu)$ of the emission/absorption intensity is given as the Doppler profile and is calculated as:

$$\Gamma(\nu) = e\wedge - ((\nu - \nu o)/\Delta\nu D)\wedge 2$$

The temperature-dependent absorption intensity $I(T)$ contains the quantum mechanics parameters where h is Planck's constant, c is the speed of light, ν is the frequency, E'' and E' are the upper and lower state energies, respectively, k is Boltzmann's constant, T is the temperature, S is quantum line strength, μ is the dipole moment and Q is the partition function

$$I(T) = ((8\pi 3/3hc)\nu S\mu\wedge 2\,[\mathrm{e}\wedge(-E''/kt) - \mathrm{e}\wedge(-E'/kT)\,])/Q$$

In rotational spectroscopy, the energy exponents are typically $\mathrm{e}^{-E/kT} \sim 0.01$–$0.1$. Plugging in typical values for a generic species of say 100 atomic mass units and a 1 Debye dipole moment results in a peak absorption of $\Delta I/I = 10^{-4}\ \mathrm{cm}^{-1}\mathrm{Torr}^{-1}$, i.e. the rotational transitions (at 1 Torr pressure) will be optically thick (>10% absorption) in 1 m of absorption path length. Transition moments of infrared and electronic transitions are typically one to five orders of magnitude smaller than rotational transitions [40].

The intensity of a transition is distributed over the linewidth, $\Delta\nu$, the peak absorption ΔI is proportional to $I(\nu,T)\delta\nu/\Delta\nu$. The Doppler width ($\Delta\nu \propto \nu\sqrt{T/M}$) is typically 0.2 MHz in our frequency range. Pressure broadening (2–15 MHz/Torr) is independent of wavelength. Since all absorption and emission spectroscopy instruments detect peak contrast this is very favorable to rotational spectroscopy when pressure broadening is minimal. Submm spectrometry also allows for observations in the limb and nadir modes. Viewing molecular emission against the cold background sky toward the limb provides the highest sensitivity for detection and vertical profiling. Because thermal emission is linearly proportional to temperature, a few degrees of uncertainty in assumed or derived atmospheric temperature introduces only a small error (~1%) in retrieved abundance. For abundant species (e.g. H^2O, CO), nadir observations also are useful since one single nadir measurement provides abundance and temperature profiles with a horizontal resolution much better than at the limb and yields higher sensitivity nearest the surface. Nadir retrieval is possible when the gas emission/absorption is visible against a colder/warmer background—either a surface or an underlying optically thick gas layer (relatively typical conditions). This underlying physics makes submm-wave spectroscopy for space science an extremely important and powerful technique. It is an established technique for earth atmospheric studies, planetary exploration, astrophysics, as well as heliophysics.

13.1.3 THz Technology for Space Exploration

The development of technology specifically aimed at remote sensing of chemical species that thermally emit in the THz frequency range has been a major thrust area for national space agencies. Enormous progress has been made at JPL and elsewhere, in both heterodyne and direct detection THz sensor technology. Complete satellite instruments have now been deployed for the spectroscopic analysis of the Earth's stratosphere to further our understanding of ozone depletion chemistry and global warming (Upper Atmospheric Research Satellite and Aura's MLS). Similarly, heterodyne receivers have been built to measure water and carbon monoxide emissions from comet Churyumov–Gerasimenko and the surface temperature of the asteroids Steins and Lutetia, yielding new data on the structure and evolution of these bodies (Microwave Instrument for the Rosetta Orbiter-MIRO); and to explore the composition of interstellar and intragalactic space adding to our knowledge of cosmology, stellar evolution and galactic structure (ESA's Herschel Space Observatory: Heterodyne Instrument for the Far Infrared [HIFI]).

THz receivers allow the detection of THz radiation. Such receivers can be divided into two main classes, namely direct detectors and heterodyne detectors. Direct detectors are an important tool for space science, especially, astrophysics where their sensitivity is important for detecting very faint signals. Instruments that have used direct detectors and flown in space include Planck and HFI on Herschel and various other concepts being proposed for future missions [41, 42]. For direct detector systems, detectors such as microwave kinetic inductance detectors (MKIDs), transition edge sensor (TES), and quantum capacitance detector (QCD) have been proposed and are under investigation at a number of groups around the world. They are the detectors of choice for cosmic microwave background (CMB) science investigations that hold keys to our understanding of how our universe came to being. However, for high spectral resolution spectroscopy, the tool of choice by space scientists is the heterodyne receiver which will be discussed in more detail.

The long history of submm-wave space instruments has been chronicled by Siegel [43]. Table 13.3 builds on the table provide by Siegel and provides a list of heterodyne space instruments that have been launched or are under development since 2008. The only exception is the Origin Space Telescope which at the writing of this chapter has only been a proposal to the NASA decadal survey. These instruments all differ in their approach as well as in their sensitivity, dynamic range, and spectral coverage. Each offers unique advantages for meeting the science objectives of the mission. Many of the key spectral signatures of interest to atmospheric chemists, planetologists, and astrophysicists either peak in the THz frequency range or have their lowest order emission lines at these wavelengths. Emission profiles representing thermally excited vibrational and rotational motions contain key information on surrounding pressure, temperature, gas velocity, and molecular abundance. It is also worth pointing out that THz instruments are versatile as they can be used for earth science, astronomy, as well as planetary science.

13.2 THz Heterodyne Receivers

A nominal heterodyne receiver, shown in Figure 13.3, is comprised of three main subsystems; the nonlinear mixing device (often the IF amplifier is considered part of this); the locally generated oscillator signal, the LO; and the backend spectrometer. The nonlinear mixer produces a beat signal between the incoming signal and the LO signal which can be amplified and investigated further with the spectrometer. Such a receiver preserves both the amplitude as well as the phase of the incoming signal and is commonly used for

Table 13.3 A summary of space instruments since 2008.

THz instrument	Mission	Frequency range	Main science theme	Technology	status	References
Heterodyne instrument for far-infrared (HIFI)	On-board herschel space observatory	Multiple channels 480–1250 GHz and 1410–1910 GHz	Star formation; interstellar medium; D/H ratio on comets, etc	SIS mixer to ~1200 GHz, HEB mixers for higher bands. Schottky multiplier chains for LO	From May 2009 to April 2013	De Graauw et al. [12]
Superconducting submillimeter-wave limb-emission sounder, (SMILES)	Launched from Intl. space station	~650 GHz	Study the chemical makeup of the Earth's middle atmosphere	SIS mixers pumped by Schottky multipliers; cryocooler	Operated from September 2009 to April 2010	Ochiai et al. [44]
Stratospheric terahertz observatory II (STO-2)	Balloon flight from Antartica	1.4, 1.9, and 4.7 THz	Astrophysics; star formation	HEB mixers and Schottky diode LO for 1.4 and 1.9 THz; HEB mixer and QCL for 4.7 THz	December 2015	Walker et al. [45]
German receiver for astronomy at terahertz frequencies (GREAT & up-GREAT)	SOFIA (airborne)	1.4–1.9 THz; 2.5–2.7 THz; and 4.7 THz; 1.85–2.54 THz on Up-GREAT	Astrophysics; star formation; extra galactic observations	HEB mixers and Schottky diode LO technology for 1.4–1.9 THz channel; QCLs for higher channels	Flights since April 2011	Risacher et al. [46]
IceCube	CubeSat launched from ISS	883 GHz	Measure cloud ice in the middle-to-upper troposphere	Schottky diode	May 2017 to October 2018	Wu et al. [47]
Ice cloud imager (ICI)	On MetOp-SG satellite	183, 243, 325, 448, and 664 GHz	Remote sensing of high-altitude ice clouds	Schottky diode mixers and LO chains	Expected launch 2021	Thomas et al. [48]
Galactic/extragalactic ULDB spectroscopic terahertz observatory (GUSTO)	Balloon flight	1.4, 1.9, and 2.7 THz	Composition energetics and dynamics of the interstellar medium	HEB mixers and Schottky LO for lower bands; QCL for the 2.7 THz band	Expected launch from Antartica 2022	Walker et al. [49]

(Continued)

Table 13.3 (Continued)

THz instrument	Mission	Frequency range	Main science theme	Technology	status	References
Submillimeter-wave instrument (SWI)	JUICE (**JU**piter **IC**y moons **E**xplorer)	530–625 GHz and 1080–1275 GHz	Will investigate Jupiter's middle atmosphere and the Galilean moons	Schottky diode technology	Expected launch in 2022	Hartogh et al. [50]
ASTHROS	Balloon flight from Antartica	1.4–2.7 THz	High-resolution maps of ionized gas around star-forming regions	HEB mixers and Schottky LOs	Expected launch from Antartica in 2023	Cooper et al. [51]
Millimetron space observatory	Russian funded 10-m telescope	Instruments have not been selected yet	0.07–10 mm astronomy	TBD	Expected launch in 2029	Smirnov et al. [52]
Hetrodyne receiver for the origins space telescope (HERO)	Origins space telescope (OST)	468–4752 GHz (Concept 1); array receivers	Observe trail of water from the interstellar mdium to disks around protostars	SIS mixers to 1200 GHz; MgB2 based HEB detectors for high frequencies; Schottky multiplier chains for LO	Proposed to NASA Decadal Survey, 2019	Wiedner et al. [53] (see Table 13.4 for more details)

This table should be read along with Table 13.1 from [43] to get a complete picture of THz instrumentation in space.
Source: Based on Siegel [43]. Cooper et al. [51]. https://www.nasa.gov/feature/jpl/nasa-mission-will-study-the-cosmos-with-a-stratospheric-balloon

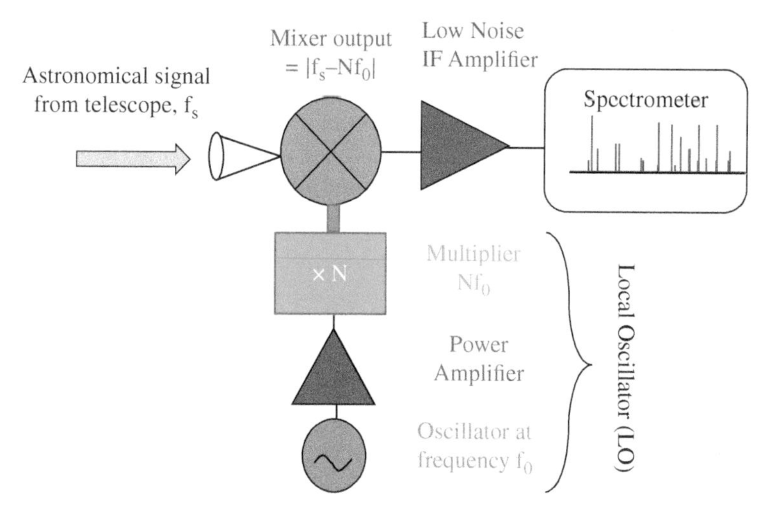

Figure 13.3 The availability of planar Schottky diodes has made it possible to build and fly THz heterodyne receivers, required for high-spectral resolution spectroscopy. Both multipliers as well as mixers use THz diodes for room temperature receivers.

molecular spectroscopy with high spectral resolution. The sensitivity of the receiver is usually characterized by measuring the minimum change in temperature that can be measured with the receiver, namely:

$$\Delta T = \frac{T\mathrm{sys}}{\sqrt{\Delta \nu \tau}}$$

Where tau is the integration time, delta ν is the bandwidth. T_{sys} is the system noise temperature and includes the noise power generated by the receiver itself, T_{rcvr}, referenced to the receiver input. All receivers generate noise, and any receiver can be represented by an equivalent circuit consisting of an ideal noiseless receiver whose input is a resistor of temperature T_{rcvr}.

T_{rcvr} is strongly dependent on the mixer device that is utilized along with operational temperature. Heterodyne instruments for astrophysics, are usually not background limited and thus rely on cryogenically cooled detectors for maximum sensitivity. Space instruments for planetary and Earth science, however, require room-temperature operation to reduce mass, volume, and complexity. The versatile THz Schottky diode plays a key role in such systems, working both as a detector and as a component of the LO. Similarly, for ground-based applications such as contraband scanning, near-field and far-field imaging, and for metrology, noncryogenic systems are preferred in order to effectively deploy them in the field.

13.2.1 Local Oscillators

The LO refers to the signal that is synthesized on-board the receiver and is used to mix (beat) with the signal obtained from the scene. It can be argued that lack of robust coherent sources in the THz frequency range has been a bottleneck for advancement in this area. A number of technologies can provide this capability and a summary plot of what can be achieved with different technologies as far as output power is concerned is shown in Figure 13.4. Note that this is a simplistic rendition as the plot does not address nuances such as beam quality or efficiency.

A number of important trends are worth discussing from Figure 13.4. From the optical domain, the quantum cascade lasers (QCLs) have made tremendous progress and are starting to provide pragmatic solutions in the THz range. QCLs are quickly progressing and offer attractive power levels but require cryogenic cooling [54, 55]. QCLs have disadvantages such as the need for frequency locking, and limited bandwidth. However, there is considerable effort in making QCLs compatible with space instruments and they might provide the ideal situation for certain architectures especially when it comes to array receivers [56]. More recent work with VECSEL based QCLs is very promising as it provides high power levels with controlled beam quality [57]. Most often the total output power is reported from QCLs, while due to poor beam quality, the useful output power is drastically reduced as reported in [58]. QCL output power numbers are also dependent on whether the laser is being used in pulse mode or CW and on the physical temperature. For these reasons, a range is presented for QCL output power in Figure 13.4.

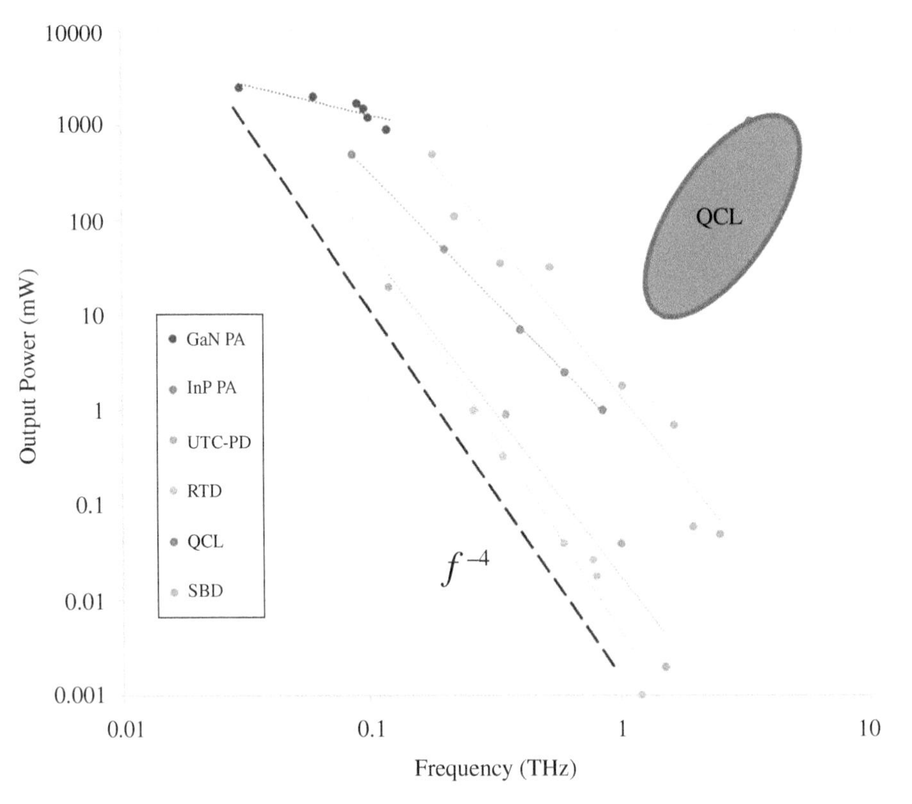

Figure 13.4 A number of different technologies can possibly be used as local oscillator sources for heterodyne receivers. However, for the THz range, the most robust, broadband, and easy to implement approach is the frequency multiplied sources. Even with recent success the output power rolls off significantly with increasing frequency with QCLs being the only exception.

Another rapidly developing technology is monolithic microwave integrated circuit (MMIC)-based power amplifiers which can be used in conjunction with low-power oscillators. Both GaN and InP-based transistor technology which allows for on-chip power-combining and massive integration to increase output power show encouraging results. GaN amplifiers to 122 GHz have been demonstrated and some representative data is shown in Figure 13.4 after [59]. InP-based PHEMT and heterojunction bipolar transistor (HBT) technology has both been demonstrated in the submillimeter. About 50 mW in the 200 GHz range Lai et al., 7 mW in the 400 GHz range, 2–3 mW in the 600 GHz [60] range and ~1 mW at 850 GHz [61] have been demonstrated using power combining techniques in pHEMT technology. Gain to 1 THz in this technology has also been demonstrated [62]. However, the achievable bandwidth is fairly restrictive and availability of this technology beyond 300 GHz is dependent on development of commercial markets.

Some of the other niche technologies that have demonstrated useful power levels in the THz range include photomixers [63, 64] and resonant-tunneling diodes (RTDs) [65–67]. These technologies are useful and provide advantages for certain applications. However, in the last two decades, the GaAs-based Schottky barrier diode (SBD) frequency multipliers have been the most popular due to compactness, relatively low dc power consumption, acceptable performance at room temperature (with possibility of improvement upon cooling), high-fractional bandwidth (15–20% typical), frequency agility and tunability, temperature stability, and spectral purity. A detailed review of this technology was recently presented by Mehdi et al. [68] and the chapter by Maestrini and Siles provides detailed information on this technology. Only a brief introduction is presented in the next sub-section.

13.2.1.1 Frequency Multiplied Chains

The LO chains developed for HIFI instrument in the late 1990s were able to provide only a few microwatts of power at 1.9 THz. Several factors over the last decade have now made it possible to provide considerably more power at this frequency. Availability of high power amplifiers, both at W-band as well as Ka-band (GaN), along with more nuanced thermal designs and power combining have enabled chips that can handle increased input power. Accurate device models, more precise control of waveguide dimensions during computerized numerically controlled (CNC) machining, and better circuits have increased efficiency and repeatability. The ability to design power combiners and dividers on-chip has provided a significant boost to output powers that can be expected in the THz range. Recent work by Siles et al. [69] confirms the significance of this breakthrough.

Figure 13.5 shows a summary plot of output power at room temperature from a number of institutions. A direct comparison between multiplier output power levels is not completely fair as it is possible that some circuits have been optimized for bandwidth. Nevertheless, this shows rapid decrease of output power as operating frequency is increased. This data is at room temperature (300 K). Nominally, a Schottky diode-based mixer needs around 1 mW of pump power, Superconducting-Insulator-Superconducting (SIS) junctions around 0.1 mW and hot-electron bolometric (HEB) close to 0.01 mW. Based on these criteria, multiplied chains can provide sufficient power to operate HEB receivers in the 2–3 THz range.

Another advantage that can be attributed to better quality of currently available diodes is the lack of excess noise and parametric instabilities in well-designed LO subsystems. While

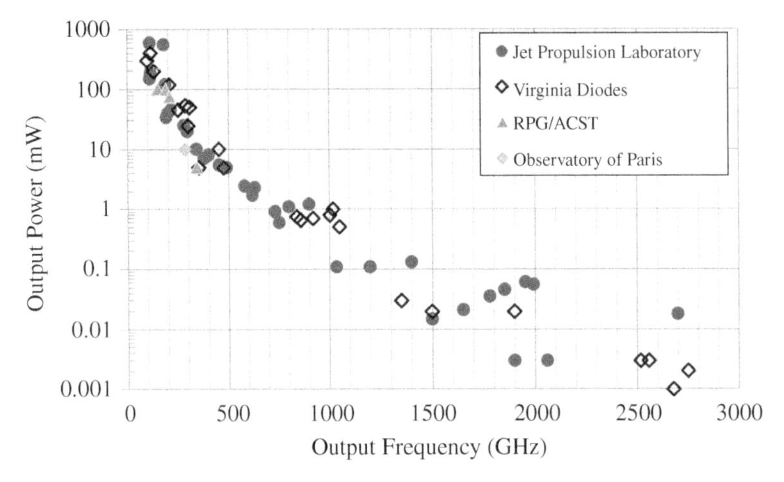

Figure 13.5 A compilation of measured output power from various multiplied sources demonstrates capability of this technology but also highlights the roll-off with frequency.

excess noise can still be a problem if the power amplifiers or other active components in the chain are not well conditioned, owning to better matching between diodes and improved circuit design, excess noise can be eliminated. While THz multiplier diode technology has made good progress in the last 20 years there are significant opportunities for improvement. Work on newer material systems such as GaN for the first stage multipliers is still desirable while packaging and fabrication of higher frequency diodes continues to be challenging. Another trend going forward is the development of array receivers, which will require the development of compact multi-pixel sources. This presents a significant challenge in terms of DC power requirement and ability to package compactly.

13.2.2 Mixers

13.2.2.1 Room Temperature Schottky Diode Mixers

GaAs Schottky diodes have been the workhorse of room temperature mixers. The high electron mobility of GaAs allows these mixers to work well in the THz range. Recent progress has been due to better designs along with tighter control of the dimensions both for the chip as well as waveguide blocks. Figure 13.6 shows a receiver front-end with Schottky receivers that was designed at 1200 GHz for a proposed mission to Europa [70]. This particular receiver architecture is based on a sub-harmonically pumped mixer. The great advantage with subharmonic topology is the ability to have the LO signal at half the frequency of the signal being measured, drastically reducing the difficulty of building the appropriate LO source.

While Schottky diode mixers work well at room temperature [71], their sensitivity can be further improved with cooling them. The mobility of GaAs peaks around 77 K and thus the mixers do not need cryogenic cooling but often passive radiators can supply enough cooling power to cool the mixers to 100–120 K. This modest cooling does provide significant enhancement in sensitivity [72].

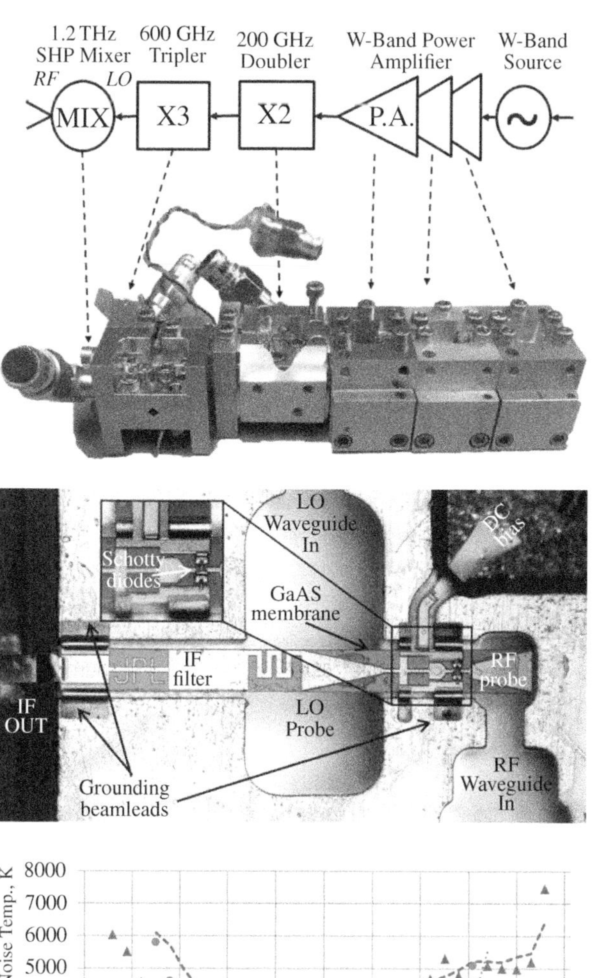

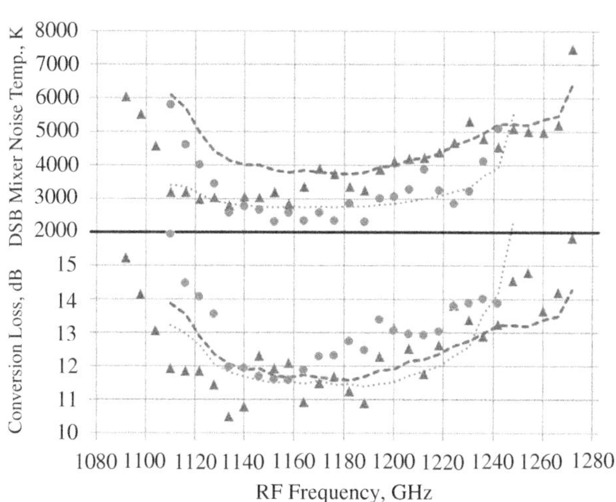

Figure 13.6 An all-solid-state receiver to 1200 GHz has been demonstrated. Block diagram and receiver front-end is shown on the top. The 1200 GHz sub-harmonically pumped mixer chip mounted in the waveguide block is shown in the bottom. Measured performance of the 1200 GHz receiver front-end. Top plot is the DSB mixer noise temperature at 300 K (triangles) and 120 K (circles). Bottom plot shows DSB conversion loss while dashed lines represent simulations.

13.2.2.2 SIS Mixer Technology

SIS-based receivers have been the most sensitive heterodyne detectors in the submm since they were first demonstrated in the 1980s in quasi-optical mounts [73, 74]. More recent progress has been reported by Uzawa et al. [75] and Karpov et al. [76] where excellent performance was achieved for NbTiN receivers to 1.1 THz. To tune out the parasitic capacitance of the SIS tunnel junction, a second junction is typically connected in parallel by means of an inductive microstrip line. In this way, the RC-L-RC equivalent lumped element circuits mimics a transmission line with a broad fractional tuning bandwidth [77]. Careful design makes it possible to tune the "twin" SIS junction configuration to cover the desired bandwidth.

A double side band (DSB) mixer utilizing a Nb "twin" SIS junction was recently demonstrated in the 500–600 GHz range. The device is fabricated on 6-micron Silicon-on-insulator (SOI) wafer. radio frequency (RF) decoupling occurs at the end of the RF choke structure by means of the two Au plated pads which form a 1 pF capacitance each to the block (ground). Contacting the pads (not shown) is a bias return wire bond to the block. The chip mounted in a waveguide block is shown in Figure 13.7.

The receiver was tested with multiplied chain as the LO and a low noise cryogenic amplifier. The receiver sensitivity as a function of frequency is shown in Figure 13.8. The blue curve is the measured data without any correction while the brown curve shows the expected performance in space. For reference, the DSB quantum noise limit ($h\nu/2kB$) is shown by the dashed green curve, indicating the obtained sensitivity is ~a few times the quantum noise limit.

Similar twin-junction tuning circuit mixers with Nb based junctions have demonstrated high sensitivity until about 700 GHz, the energy gap of Nb. For niobium, photons with energies greater than 700 GHz are able to break paired-electrons (Cooper pairs) in the superconductor, which results in significant absorption loss. The calculated loss for Nb vs NbTiN as a

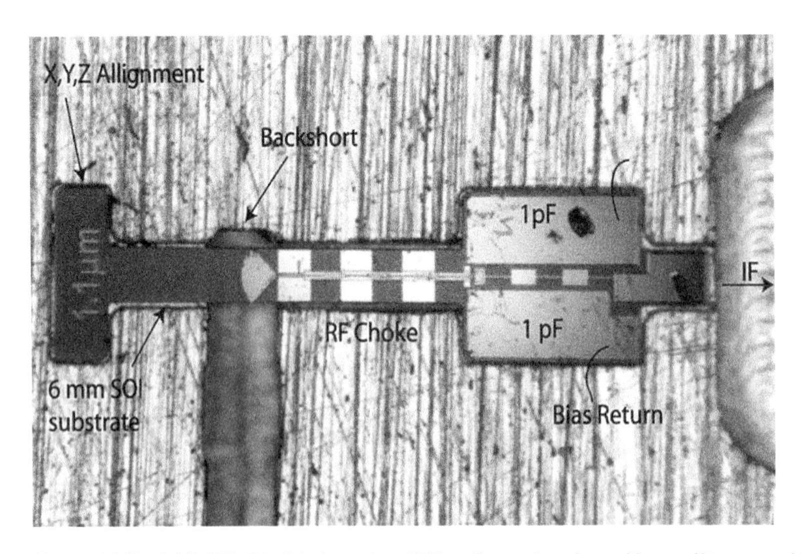

Figure 13.7 A Nb SIS chip fabricated on SOI wafer and packaged in a split waveguide block is shown. Circuit is designed for 500–600 GHz operation.

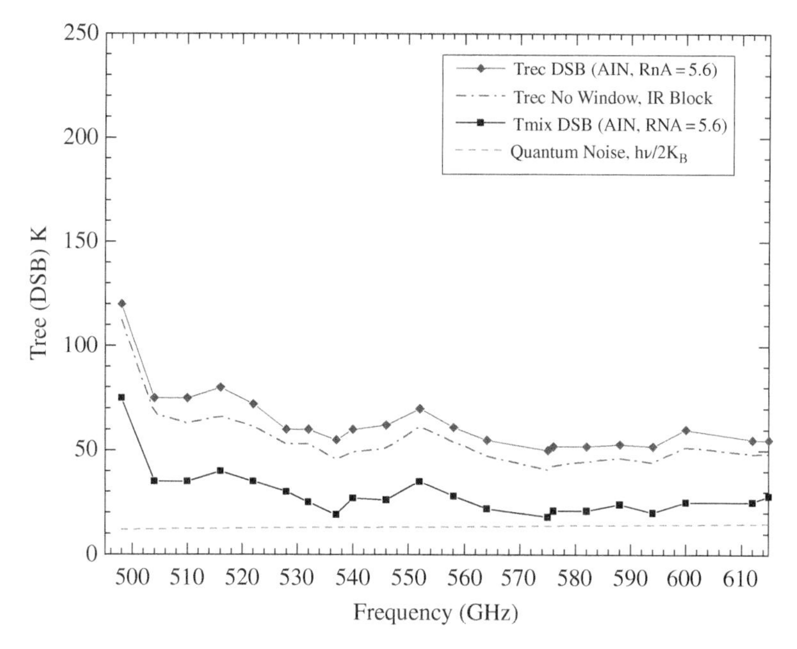

Figure 13.8 Recent measurements on high current density SIS receiver. The brown curve has the cryostat pressure window and IR block removed, actual flight scenario. *Source*: Courtesy of J. Kooi, JPL.

function of frequency is shown in Figure 13.9. Obviously, to go above 700 GHz, higher energy-gap superconductor film is needed. A number of different material systems have been proposed and tried to extend the frequency range beyond 700 GHz. JPL/Caltech championed the use of NbTiN which has been able to demonstrate performance but suffers from low yields. The Atacama Large mm/sub-mm Telescope (ALMA) has demonstrated Nb/AlN/NbTiN tunnel junction with a receiver noise temperature of 125 K DSB (just three times the quantum limit at 800 GHz). In this scheme, NbTiN has been used as the ground plane and base electrode and Nb to form the counter electrode, and aluminum (normal metal as the tuning inductor and top layer transmission line). We see that above ~1.3 THz superconducting films made of NbTiN also start to exhibit significant loss. It is for this reason that around this frequency superconducting detector technology switches to the use of Hot Electron Bolometers (discussed next).

13.2.2.3 Hot Electron Bolometric (HEB) Mixers

HEB superconducting mixers have clear applications for THz astrophysics where receiver sensitivity is critical and known SIS technology starts to fall apart. In the last few years, a great deal of progress has been made in developing this technology both for waveguide as well as quasi-optical systems. A recent review of this technology has been provided by Shurakov et al. [78]. As the HEB is a thermal device, the required LO power is quite reduced compared with Schottky or SIS receivers. Nominally, 200–400 nW of LO power is more than sufficient assuming 3–6 dB losses due to optical losses. Theoretically, the devices need only 50–100 nW. A number of research groups have demonstrated working receivers and these

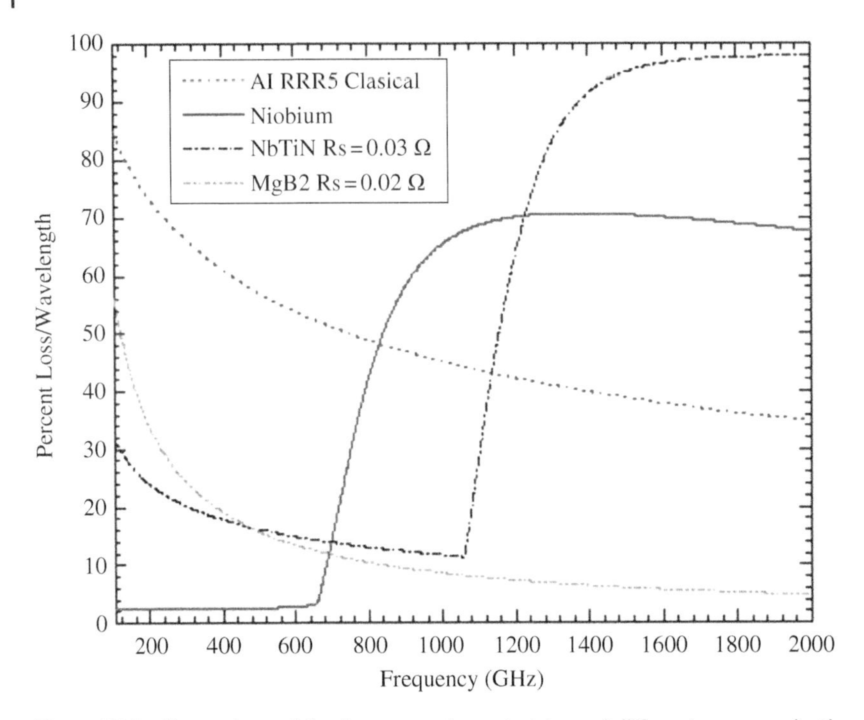

Figure 13.9 Comparison of the frequency-dependent loss of different superconducting materials. *Source*: Figure courtesy of J. Kooi, JPL.

receivers have flown on HIFI, STO-2, and Stratospheric Observatory for Far-Infrared Astronomy (SOFIA).

The physics of the device dictates that the IF bandwidth achievable is determined by the electron relaxation time. In order for an electron to relax, it has to transfer its energy to a phonon, which subsequently escapes from the NbN film into the substrate. This process is therefore highly dependent on the electron–phonon interaction time for a given material system. With an electron–phonon relaxation time of about 12 ps at 10 K for NbN, HEB mixers with a gain bandwidth of up to 3 GHz can be readily achieved. With further improvements in HEB mixer fabrication and design, a maximum gain bandwidth of up to 6.5 GHz has been recently realized. However, due to the low critical temperature Tc for ultrathin NbN films of nearly 10 K, such mixers require to be operated at temperatures close to 4.2 K. To investigate terahertz radiation within increasingly broader frequency ranges HEB mixers now require even higher IF bandwidths (up to 10 GHz) than those achievable with currently used materials, in order to capture Doppler broadened lines in interstellar molecular clouds.

A number of research groups have started investigating MgB2 as an alternative material for THz mixers. The electron–phonon interaction time in MgB2 has been estimated to be only a few picoseconds, suggesting the possibility of achieving IF bandwidths beyond those of NbN HEB mixers. The high Tc of 39 K is another characteristic which makes the use of MgB2 for HEB mixers extremely beneficial. Recent results from JPL and Chalmers

University have indicated that this is a promising material for THz mixers for larger IF bandwidths and IF bandwidths of 13 GHz have been demonstrated [79].

13.2.2.4 State-of-the-Art Receiver Sensitivities

Figure 13.10 shows a comparison of various detector technologies in the submillimeter-wave regime. While SIS mixers have unsurpassed sensitivity through 1 THz, there is a sizable degradation in sensitivity above ~700 GHz. Below this region, SIS mixers achieve about two to three times the quantum limit ($h\nu/k_B$), but the value quickly goes up to over 12 times the quantum limit at 1 THz. Based on these results, it is expected that SIS mixer technology should be able to achieve similar levels of sensitivities till around 1.2 THz. Above this frequency for nonbackground limited instruments, the choice of HEB mixers presents clear advantage. However, considerable effort is still required to demonstrate robust HEB devices in newer material systems where IF bandwidths of greater than 10 GHz can be obtained. As discussed previously, MgB_2 presents a compelling choice but perhaps there are other material systems that can be grown with better repeatability in the future. Schottky diode mixers continue to provide a niche capability, consistent with instruments that are constrained by mass and power requirements or unable to fly cryogenic detectors. It should be noted that Schottky mixers can achieve performance enhancement if they are cooled to noncryogenic temperatures of around 100–120 K which can be achieved with passive coolers in space instruments.

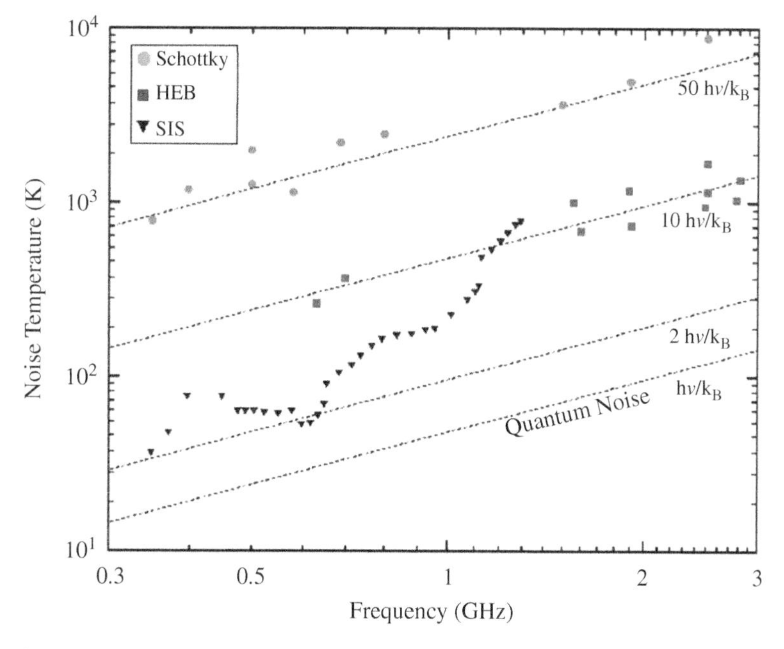

Figure 13.10 Measured sensitivity of SIS, HEB, and SBD is compared along with the theoretical quantum noise. Both SIS and HEB require very cold operating temperatures (2–4 K) and provide higher sensitivities. Notice that SIS technology approaches close to the quantum limit at the lower frequency end.

13.3 THz Space Applications

Table 13.2 lists the main submm-wave instruments that have flown or currently in development since 2008. The large and diverse list speaks to the potential of this technology for space exploration. Significant science data return from Astrophysics-centric instruments such as *Herschel's* HIFI and the German Receiver for Astronomy at Terahertz Frequencies (GREAT) on the SOFIA, is greatly impacting the future of this technology as origins space telescope (OST) is being proposed. Similarly, planetary instruments such as the MIRO have confirmed the usefulness of room temperature detector systems for planetary exploration. One of the common themes of these instruments is their ability to measure molecular abundances and production rates of a large body of different molecules (e.g. carbon monoxide, ammonia, water). Each space instrument is unique and is driven foremost by the specifications that are derived directly from the science investigation. For example, the choice of frequency range, IF bandwidth, and choice of mixer technology is driven by optimizing resources to achieve the science objectives. This section will present three distinct science themes and show how the science in each category is driving the technology development for the future.

13.3.1 Planetary Science: The Case for Miniaturization

Comets provide a unique window through time because they remain dormant until close enough to the sun to be thermally irradiated, vaporizing the surface, and exposing materials that remained virtually unprocessed since their formation. This fossil record of our initial solar nebula, captured in small bodies, provides direct measurement opportunities of the isotopic composition of water which is sensitive to the thermal history of the forming solar system (through D/H ratio) and the origin and composition of volatile reservoirs (through O isotopes). The volatile emissions have essentially the same composition as that around the birth time of our solar system at their forming location, and they can be used to trace the exact regions/components of our solar system which delivered volatiles such as water to the young planets. To understand if water on earth was brought by comets, one would measure the water D/H ratio on these comets and compare it with measurements on earth.

Measuring the D/H ratio requires determining the water abundance and the abundance of the HDO isotope. Water is particularly challenging due to *ortho* and *para* spins that result in effectively two distinct molecules with completely different excitation rates. Conversely, HDO has no spin symmetry and all states are connected. HDO has often been measured near 500 GHz and must assume the same excitation model as for normal water. Historically water content has been determined by probing only the ground state at 556 GHz and assuming that the *ortho*/*para* ratio has a thermally equilibrated value around 3.0, and using a model to deduce the excitation. However, a recent analysis of the *ortho* to *para* ratio (OPR) in several comets suggested that they span the range 1.8 to 3.0, with a mean value around 2.6 [80]. This departure from the expected thermal equilibrium value shows that assuming a specific OPR value when deriving water production rates from *ortho*- or *para*-lines only can yield a substantial uncertainty, on the order of 20%. This uncertainty would then naturally propagate to the D/H isotopic ratio. Thus, to provide precision on this measurement one must measure the *ortho*- and *para*- lines.

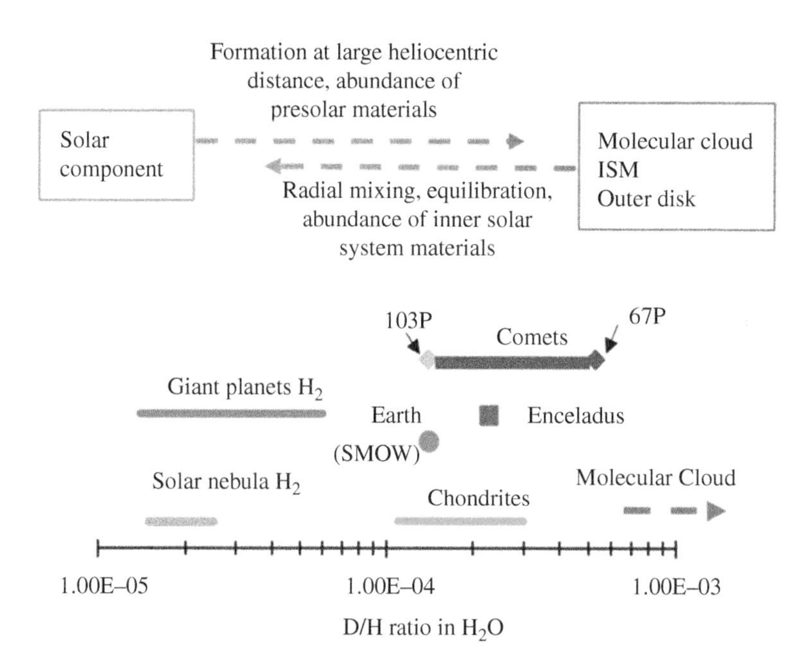

Figure 13.11 Measured D/H ratios indicate a large range for comets with only a handful of data points. Courtesy of Mathieu Choukroun, JPL.

HIFI observed the comet 103P/Hartley 2 in November 2010 from a distance of 0.21 A.U. and the D/H ratio was measured to be very similar to that of earth's oceans [81]. This is the first time ocean-like water was found on a comet. However, due to the large distance to comet requiring long integration times, HIFI did not measure the *ortho* and *para* molecules. More recently, ROSINA measured the D/H ratio of comet 67P/Churyumov–Gerasimenko (67P) and found a value about three times larger than that of Earth [82]. This complicates the known paradigm because both 103P and 67P are Jupiter Family Comets, yet span a larger range of D/H than Oort-Cloud Comets. This strongly calls for increasing the sample size for this measurement in order to better understand the water content in our solar system and its distribution early in its history. A factor of three variation exists in the very limited data set available to date for comets. Without being able to distinguish the *ortho* or *para* water molecule there is inherently a 30% uncertainty in the measurement. Thus far, all measurements of the water isotopes have been limited due to this fact. While measurements of D/H ratios in comets have provided some clues to the potential origins of Earth's water, due to limited samples and unknown redistribution of small bodies during solar system formation, it is currently impossible to draw definitive conclusions from the limited set of data (Figure 13.11).

To completely address this problem, a statistically larger sample, including *ortho* to *para* ratio measurements is needed. Such a data set would then ultimately provide clues to the origin of earth's water. A number of mission concepts can be envisioned to achieve this objective. Building a HIFI follow-on with coverage of only the water isotopes and HDO is a possibility. To increase the number of targets such a mission would need a very large

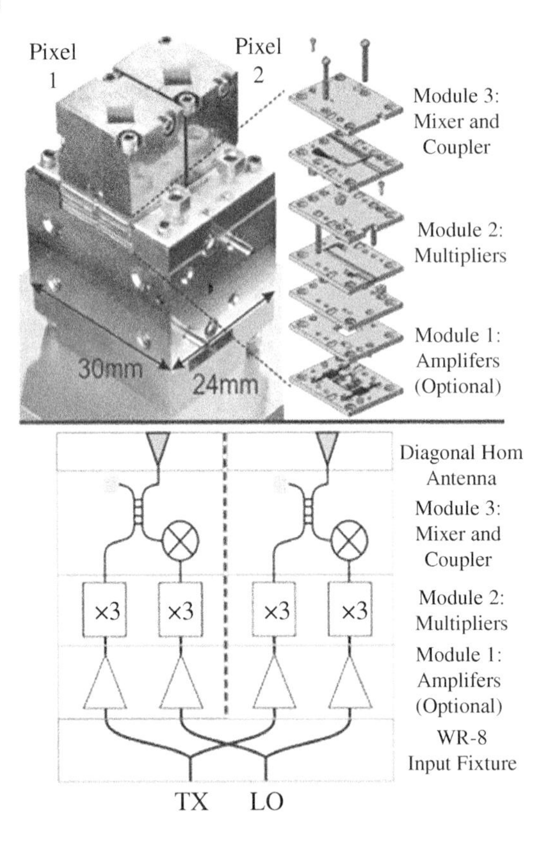

Figure 13.12 A two-pixel 340 GHz. (a) show the full assembly with WR-8 input fixture supporting the silicon micromachined stack. (b) shows an exploded view of the seven layers of each transceiver pixel. (c) is a system diagram of the unit cell.

collecting area and as best a sensitivity as possible. Such an instrument could be launched and parked at L2 and observe the various comets as they approach the Sun. The SIS receiver described in Section 13.2 would be an ideal detector choice as it provides exquisite sensitivity. Successful implementation of this concept requires SIS based receivers that cover up to 1200 GHz with near quantum-limited sensitivities.

A somewhat drastically different concept would be to fly small targeted instruments that rendezvous with the different comets. This would be similar to what the MIRO instrument on ROSETTA achieved. As these instruments will be traveling potentially large distances it would be helpful to make them as light as possible. The MIRO instrument was a single channel narrowband receiver on a nonarticulated primary. MIRO was 40 kg and required 72 W of average power. We are now developing broadband receiver systems that provide substantially better sensitivity and operate with much lower power and mass. Part of this is accomplished by utilizing silicon micromachining technology to package THz devices and provide capability of having multi-pixel receivers. A potential example is shown in Figure 13.12. An 8-pixel linear array, consisting of 4, 2-pixel unit cells at 340 GHz has been demonstrated using a stack of silicon wafers. Each of these cells supports two silicon micromachined pixels mounted between a metal waveguide block and two metal diagonal horns. The circuit consists of the three functionalities of the transceiver, namely, amplification, multiplication, and mixing (with coupling). The waveguide block, which supports the silicon stack, accepts the W-band power for the LO and transmitter (TX) signals on two sides, allowing the array to stacked along the other two faces. The LO and TX power is divided in the metal fixture between the two pixels with a waveguide Y-junction and then routed upward to the silicon stack with a step bend.

Pixel spacing, ease of assembly, and a desire for modularity guide the design of the array. All of these features are facilitated by a vertically stacked architecture [83]. This arrangement reduces the area of each pixel in the focal plane by separating the components of the system vertically into separate silicon modules, indicated in Figure 13.12. Each module consists of two or three 800 μm thick etched silicon wafers sputter coated in gold so that standard active components designed to couple into E-plane split waveguide can be housed. By separating each component in its own layer, assembly is simplified by reducing the number of active components mounted per layer and makes testing more manageable. By utilizing 3-D integration and wafer-level bonding it would be possible to make a THz array receiver. This technique becomes extremely useful when it can be combined with microlens coupling structures further miniaturizing the receivers. This work is still under progress as a number of technical issues need to be resolved before large pixel count arrays can be made. Mounting many components per module is a concern because each time the top-half of the module is mounted or removed, all active components can potentially be misaligned or damaged. Moreover, degradation of receiver sensitivity is not acceptable and a careful system thermal design will have to be carried out.

13.3.2 Astrophysics: The Case for THz Array Receivers

How stars are formed continues to be a perplexing question for astronomers. The details of the evolution of molecular clouds and the formation of the next generation of stars out of the dust, and gas in these dense regions remain mysterious. It is abundantly clear from a large number of Herschel photometric observations with photodetector array camera and spectrometer (PACS) and spectral and photometric imaging receiver (SPIRE) that the dust is filamentary in nature and that the filaments contain a large number of dense cores. Many of these are self-gravitating and likely to form stars, but evidence of star formation is far from universal. What cannot be determined from the photometry is the velocity structure of these filaments or their relationship to the various phases of the ISM. Photometry cannot trace the processes that lead to the origin of the filaments and the likelihood that stars will form in the cores that form within them. Velocity information is required to assess the role of turbulence and gravity in determining the kinematics of the ISM and its evolution into new stars. Not surprisingly, the best tracers of velocity are the strongest available lines: CII and OI, which are complementary in that they trace different regimes of extinction. The molecular tracer CH can be used as a surrogate for H_2, and thus traces the total gas column density. CH has a ground state that interacts strongly with magnetic fields through the Zeeman Effect making it a potential tracer of magnetic fields as well.

High-spectral resolution THz observations can facilitate determination of velocity along the line of sight allowing the relative motions and states of the ISM to be mapped and correlated with the observed dust structures and evolution of starless and pre-stellar cores. Simultaneous observation of combinations of lines correlated with previous photometric maps allows for the interactions between ISM phases and the connection of the gas with the dust to be directly observed. High-resolution is required to understand the turbulence and dynamics of this interaction as well as to de-convolve the contributions along the line of sight. Understanding how star formation proceeds in galaxies other than the Milky Way—with different metallicities and at different times in the evolution of the universe—is currently one of the hottest topics in astronomy, but understanding these sources will require advanced spectroscopy and better understanding of "prototypical" nearby sources in our own and in nearby galaxies. The 1.9 THz fine structure line of the CII gas is a tracer of the different phase of the ISM. The ISM contains ionized gas, atomic gas, molecular gas, and dense molecular gas that is directly associated with star formation. The [CII] line is the main cooling line in the ISM that is heated by newly formed stars and is therefore a tracer of star formation. This line was targeted by HIFI on Herschel, upGREAT on SOFIA, and on upcoming GUSTO. However, to be able to map this line for large areas one must have array receivers. A compelling argument for this has recently been put forward in [84].

Generation of LO power remains a critical challenge in enabling array receivers of the future. Again, with respect to various technology options, as depicted in Figure 13.4, a number of options are possible. QCLs with a Fourier grating have demonstrated capability of pumping multi-pixel receivers. Other solutions such as photomixers and RTDs also continue to attract interest. Schottky diode-based multiplier chains provide distinct advantages and are now being developed for array receivers. Availability of high-power amplifiers, both at W-band as well as Ka-band utilizing GaN technology have now made it possible to have compact modules that can provide power in the range of 1–3 W in the 90–110 GHz range. Advanced Schottky diode fabrication technology, along with better thermal design and utilization of on-chip power combining topologies, has enabled Schottky multiplier chips that can efficiently handle the increased input power. Two distinct approaches can be adopted for developing multi-pixel receivers and specifically the LO needed for such systems. In the first approach, a single-pixel high-power source is constructed that puts out enough power to pump the desired number of mixers. In such an approach a device such as a Fourier grating is used to multiplex the LO signal. The second approach is to build a multi-pixel LO source where there is a one-to-one correspondence between the mixer elements and the LO elements. This approach offers a number of advantages as it does not stress any single diode with high temperature and provides some cushion from single-point failures. Moreover, it allows for optimizing power for each pixel, which is critical for attaining optimum sensitivity.

Scheme for a compact 16-pixel source is shown in Figure 13.13. This is based on utilizing GaN amplifiers working in the 70–77 GHz range with >1 W of output power. Such amplifiers are now available and make it possible to design multi-pixel chains. The first stage tripler is a single chip that can divide the power and multiply it. Either a dual-chip or quad-chip topology can be utilized. A quad-chip can handle close to 3 W of input power and can produce around 500 mW at the desired frequency range [69].

The second stage tripler is based on the dual-chip architecture and can produce 20–25 mW of output power. This chain, and in particular the last stage that includes a built-in horn, was machined by developing a meticulous process for achieving mechanical tolerance of less than a few microns. Spacing between the horns is 2.5 mm but can be varied according to the optical design of the receiver. A compact 16-pixel LO chain based on the scheme of

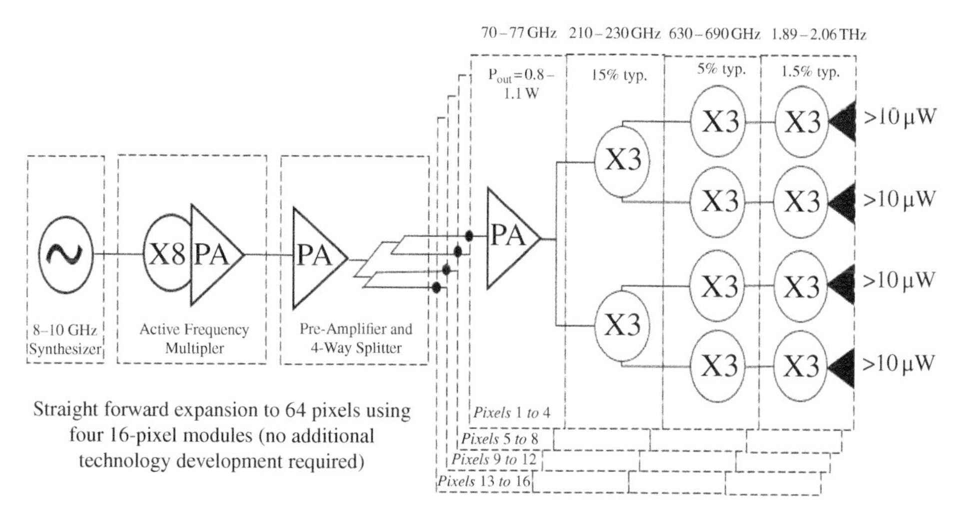

Figure 13.13 W-band GaN power amplifiers make it possible to design 1.9 THz sources with only three multiplier stages. The large input power from the amplifiers is effectively handled by designing on-chip power dividing and power combining topologies.

Figure 13.13 is shown in Figure 13.14 and typical measured results from a 1.9 THz source are shown in Figure 13.15, at room temperature.

The need for arrays is also stressed in upcoming missions such as OST. If approved, OST will have a heterodyne system and will have the capability described in Table 13.4 [53]. The architecture of the HERO instrument largely follows the classical design of heterodyne receivers used on such instruments as HIFI, however, what sets HERO for OST apart is the usage of large focal plane arrays and the continuous frequency coverage from 468 to 2700 GHz. As part of the engineering study for HERO, two concepts were developed. Table 13.4 refers to concept 1 which was more ambitious. Both of the concepts are detailed by [53].

13.3.3 Earth Science: The Case for Active THz Systems

High spectral resolution receivers provide compelling capabilities when it comes to monitoring and measuring Earth's atmosphere. From simple weather monitors to multiple satellite systems have been utilized. The MLS instrument launched as part of the EOS platform was a ground-breaking THz instrument that is still operational. Its main science goal was the investigation of the ozone chemistry and data is still being collected. Similarly, all other THz or submm-wave instruments for Earth science, such as IceCube, MetOp-SG, and other balloon-borne instruments are passive receivers. However, in the last few years, mostly due to the availability of high-power electronic sources in this frequency range, active receiver systems (radars) are starting to be feasible and they provide attractive and unique capabilities for atmospheric remote sensing applications.

While W-band and lower frequency radars have been used for atmospheric sensing as well as for planetary landing capability, higher frequency systems have not matured enough to be deployed in space yet. The underlying technology for a submm-wave radar is

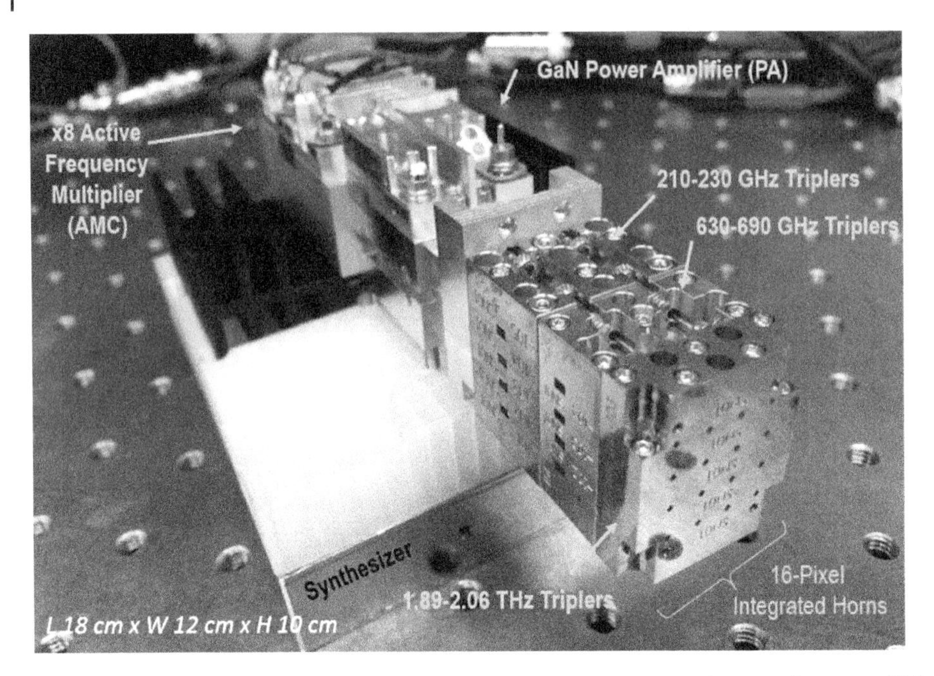

Figure 13.14 A compact 16-pixel LO chain has been built with diagonal horns at the output. This can directly couple power to the hot electron bolometer based mixers.

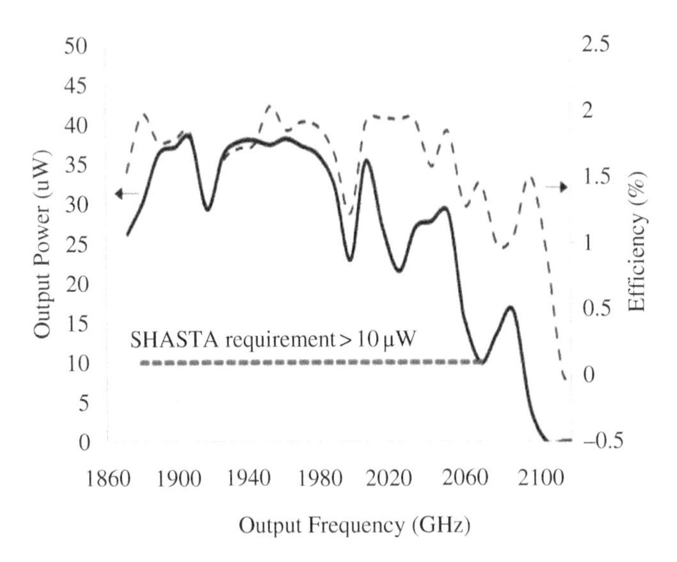

Figure 13.15 Nominal power requirement of 10 microwatts is demonstrated across a wide frequency range. This power level is sufficient to pump hot electron bolometer mixers.

Table 13.4 The HERO concept for OST is based on being able to build and characterizer several large array receivers in the submillimeter-wave range.

HERO design parameters (for OST Concept 1)							
Band	**F_{min} GHz**	**F_{max} GHz**	**Pixel**	**Trx K**	**Beam "**	**T_{rms}[a]**	**Line flux[b] Wm^{-2}**
1	468	648	2 × 16	40	14.6	2.0	2.1E−21
2	648	900	2 × 16	80	10.5	3.4	4.9E−21
3	900	1260	2 × 64	110	7.6	3.9	7.9E−21
4	1242	1836	2 × 64	200	5.3	6.0	1.7E−20
5	1836	2700	2 × 64	300	3.6	7.4	3.1E−20
6	4536	4752	2 × 64	500	1.8	8.6	7.5E−20

a) Receiver noise for one hour integration at 106 resolution (0.3 km/s) using one polarization.
b) Detectable line flux at 5 sigma, for one hour pointed integration (on+off source) in two polarization, with a 9.1 m primary mirror as designed for OST Concept 1.

essentially identical to the technologies that have been discussed thus far; the obvious difference being that unlike passive radiometers, radars need a TX. Radar at higher frequencies is also desirable due to the fact that in the Rayleigh limit the higher frequency signals are scattered strongly and are more sensitive to precipitation due to the λ^4 scaling law. Higher frequency systems also provide better spatial resolution for a given telescope size. Moreover, recent studies have suggested that the cloud precipitation dynamics and microphysical properties can only be inferred from Doppler and multi-frequency radar measurements at G-band or higher [85].

In spite of these advantages, the lack of robust radars in the submm-wave range is due mostly due to the unavailability of suitable components especially coherent sources. For earth observing systems another parameter to keep in mind is the atmospheric loss which can be significant if the appropriate frequency range is not selected carefully. For example, the 680 GHz band is usually considered as an atmospheric transmission window and a number of experiments have been conducted in this range. However, absorption in this range can be as high as 200 dB/km under humid conditions. This reason alone is sufficient to deploy such instruments in space where water vapor losses are not dominant.

Recently, a 167–174 GHz radar instrument named Vapor In-cloud Profiling Radar or VIPR has been developed and demonstrated at JPL [86]. This instrument has been specifically designed as a prototype for an airborne campaign with further enhancements for space applications. This instrument utilizes differential radar algorithms to measure cloud precipitation and water vapor inside the clouds. Figure 13.16 shows a generic block diagram for such an instrument. It is important to select the appropriate observing frequency in order to comply with regulatory agencies. Note that for this instrument a low-noise amplifiers (LNA) can be used in front of the subharmonic mixer but as the operating frequencies are increased this option is not always available or practical. Also, with available technology, even with the recent progress, output power drops sharply with increasing frequency. Detailed measurements and capability of VIPR are detailed in [86]. The instrument is capable of performing range-resolved absolute humidity profiling within volumes of clouds or

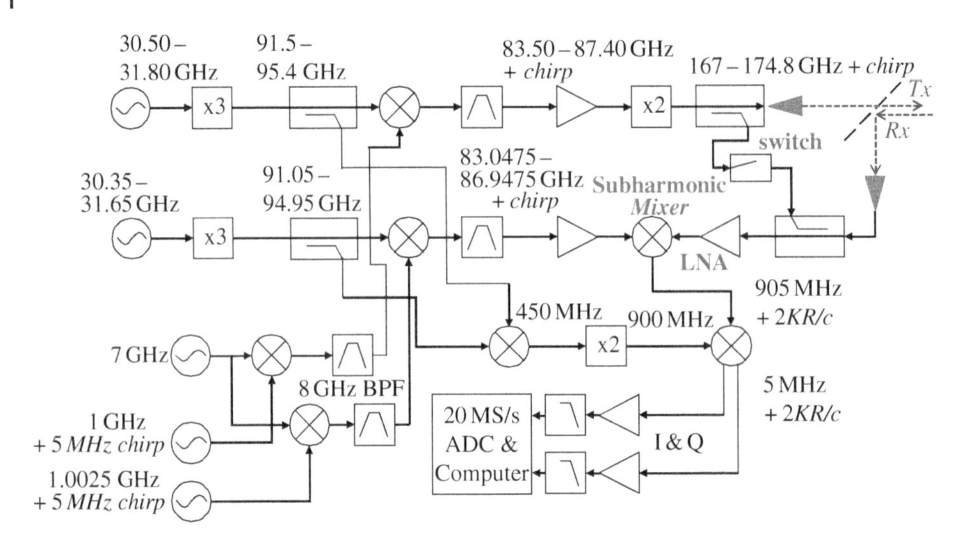

Figure 13.16 A nominal block diagram for an active instrument that can measure humidity and precipitation includes RF components as well as digital electronics. This instrument has been implemented at JPL for 167–174 GHz [51] and can be scaled to higher frequencies with development of the appropriate RF components. *Source:* Courtesy of Ken Cooper, JPL.

precipitation; total or partial column water vapor sounding in clear air between the radar and a distant target such as cloud cover; and calibrated, attenuated reflectivity detection from clouds and precipitation.

This instrument has now been successfully deployed on a twin-engine otter aircraft and several campaigns have been carried out. As part of calibration data obtained through this approach was compared directly with radiosondes and good agreement was shown [86], (Figure 13.12c of Cooper et al., 2020). Such instruments provide a key measurement capability and in future these instruments will be spaceborne and also advance in frequency range as RF components such as power amplifiers, LNA, switches, circulators, isolators, filters, mixers, couplers, and radomes become more readily available.

13.4 Summary and Future Trends

As can be seen from the various technologies discussed in this book, a number of promising technologies are being developed that will help in further expanding applications in the THz frequency range. Space science that is unique to this frequency range has a long history and continues to provide impetus for current technology development that can further enhance science return in this frequency range with unprecedented insight. This book provides details into the different technologies that can provide the building blocks for future applications while this chapter has focused on applications specifically relating to space science. One of the secular trends that ought to guide technology development is the need to build cost-effective heterodyne arrays in the submillimeter-wave range. This challenge is particularly compelling as compared to direct detector systems where kilo-pixel systems are readily available. Moreover, it is important to keep in mind that as we push for large pixel receiver systems, performance, or sensitivity cannot be compromised. If large arrays provide degraded sensitivity then they will be difficult to justify scientifically. The arrays also need to provide uniform performance which is another topic that needs concerted effort. Receiver systems also need to be low-power and low-mass. While this is desirable in general, it becomes extremely important for planetary science where instruments have to travel large distances and solar power becomes scarce. Finally, any technology such as detector devices or sources or LNA, that simplify system design, characterization, calibration, and implementation will ultimately provide the least risky way for future space instruments.

Acknowledgment

This work was carried out at the Jet Propulsion Laboratory, California Institute of Technology, Pasadena, CA, under a contract with National Aeronautics and Space Administration (NASA). Government sponsorship is acknowledged. The author is grateful for the opportunity to work with a talented group of scientists and technologists at JPL which has been key in developing THz technology for space applications.

Exercises

E13.1 You have been asked to design a submm-wave instrument for a prospective orbiter mission to Mars. The specified frequency range is 500–600 GHz. Nominal orbiter altitude is 400 km with a polar orbit.

(a) Identify the possible number of molecular lines that could possibly be detected?
(b) Determine the line strength for each of these molecules based on literature survey.
(c) Assuming that the specification asks for at least a horizontal spatial resolution of 5 km calculate the size of the main antenna to meet this requirement.
(d) Assuming SIS detector technology with DSB of 70 K across the band calculate the necessary integration time for each line.
(e) Assuming Schottky diode technology with DSB of 500 K across the band calculate the necessary integration time for each line.

E13.2 You have been tasked with designing a submm-wave instrument for measuring water plumes (existence as well as composition) on Europa. The mission necessitates room temperature detectors and a 30 cm primary.

(a) Which molecular line(s) would be your baseline target? What would be the desired RF bandwidth of your receiver system?
(b) What detector system would you select and what is the SOA detector sensitivity of the system for the frequency range of interest?
(c) From what distance will you be able to detect the plumes?

E13.3 The MIRO instrument on ROSETTA had a very limited bandwidth. Why was this? If you were asked to design an instrument with similar science goals as MIRO what technology choices would you make? Calculate total instrument mass and power and compare them with MIRO.

E13.4 Based on the block diagram of Figure 13.16, select appropriate parts from commercial vendors. Estimate total mass and power of such an instrument. How much extra mass and power is added by the transmitter?

References

1 Bishop, W.L., McKinney, K., Mattauch, R.J. et al. (1987). A novel whiskerless schottky diode for millimeter and submillimeter wave application. *IEEE MTT-S International Microwave Symposium Digest*, Palo Alto, CA, USA, 607.

2 Erickson, N.R. (1990). High efficiency submillimeter frequency multipliers. *Proceedigns of the International Microwave Symposium Digest*, Dallas, TX, USA (May 1990), 1301–1304.

3 Mehdi I., Siegel, P., and Mazed, M. (1993). Fabrication and characterization of planar integrated schottky devices for very high frequency mixers. Proceedings of the IEEE Cornell Conference on Advanced Concepts in High Speed Semiconductor Devices and Circuits (2–4 August 1993).

4 Siegel, P.H., Dengler, R.J., Mehdi, I. et al. (1993). Measurements on a 215 GHz subharmonically pumped waveguide mixer using planar back-to-back air-bridge schottky diodes. *IEEE Transactions on MTT* **41** (11): 1913.

5 Pearson, J.B., Drouin, B.J., Maestrini, A. et al. (2011). Demonstration of a room temperature 2.48-2.75 THz coherent spectroscopy source. *Review of Scientific Instruments* **82** (9): 093105.

6 Siegel, P. (2014). Terahertz pioneer: Robert J. Mattauch (two terminals will suffice). *IEEE Transactions On Terahertz and Science and Technology* **4** (6): 646–652.

7 Neugebauer, G., Beichman, C.A., Soifer, B.T. et al. (1984). Early results from the infrared astronomical satellite. *Science* **224** (4644): 14–21.

8 Siegel, P. (2011). Inaugural editorial. *IEEE Transactions On Terahertz Science and Technology* **1** (1): 1–4.

9 Waters, J.W., Froidevaux, L., Harwood, R.S. et al. (2006). The earth observing system microwave limb sounder (EOS MLS) on the aura satellite. *IEEE Transactions on Geoscience and Remote Sensing* **44** (5): 1075–1092.

10 Melnick, G.J., Stauffer, J.R., Ashby, M.L.N. et al. (2000). The submillimeter wave astronomy satellite: science objectives and instrument description. *The Astrophysical Journal* **539** (2): L77–L85.

11 Gulkis, S., Frerking, M., Crovisier, J. et al. (2007). MIRO: microwave instrument for Rosetta orbiter. *Space Science Reviews* **128** (1): 561–597.

12 De Graauw, T., Helmich, F.P., Phillips, T.G. et al. (2010). The herschel-heterodyne instrument for the far-infrared (HIFI). *Astronomy and Astrophysics* **518**: 7.

13 Hartogh, P., Jarchow, C., Lellouch, E. et al. (2010). Herschel/HIFI observations of mars: first detection of O_2 at submillimetre wavelengths and upper limits on HCl and H_2O_2. *Astronomy and Astrophysics* **521**: L49.

14 Luukanen, A., Helistö, P., Lappalainen, P. et al. (2009). Stand-off passive THz imaging at 8-meter stand-off distance: results from a 64-channel real-time imager. SPIE Proceedings 7309, Passive Millimeter-Wave Imaging Technology XII.

15 Appleby, R. and Anderton, R.N. (2007). Millimeter-wave and submillimeter-wave imaging for security and surveillance. *Proceedings of the IEEE* **95** (8): 1683–1690.

16 Cooper, K.B., Dengler, R.J., Llombart, N. et al. (2011). THz imaging radar for standoff personnel screening. *IEEE Transactions on Terahertz Science and Technology* **1** (1): 169–182.

17 Robertson, D.A., Macfarlane, D.G., Hunter, R.I. et al. (2018a). The CONSORTIS 16-channel 340-GHz security imaging radar. SPIE Proceedings on Passive and Active Millimeter-Wave Imaging XXI (May 2018).

18 Grajal, J., Badolato, A., Rubio-Cidre, G. et al. (2015). 3-D high-resolution imaging radar at 300 GHz with enhanced FoV. *IEEE Transactions on Microwave Theory and Techniques* **63** (3): 1097–1107.

19 Robertson, D.A., Macfarlane, D.G., Hunter, R.I. et al. (2018b). Real beam, volumetric radar imaging at 340 GHz for security applications. Proceedings of IEEE 2018 International symposium on Antennas and Propagation (July 2018).

20 Coulombe, M.J., Horgan, T., Waldman, J. et al. (1999). A 520 GHZ polarimetric CompactRange for scale model RCS measurements. Antenna Measurements and Techniques Association (AMTA) Proceedings (October 1999).

21 Taylor, Z.D., Garritano, J., Sung, S. et al. (2015). THz and mm-wave sensing of corneal tissue water content: electromagnetic modeling and analysis. *IEEE Transactions On Terahertz Science and Technology* **5** (2): 170–183.

22 Wang M., Yang, G., Li, W., and Wu, Q. (2013). An overview of cancer treatment by terahertz radiation. Microwave Workshop Series on RF and Wireless Technologies for Biomedical and Healthcare Applications (IMWS-BIO), IEEE MTT-S.

23 Taylor, Z.D., Singh, R.S., Bennett, D.B. et al. (2011). THz medical imaging: in vivo hydration sensing. *IEEE Transactions On Terahertz Science and Technology* **1** (1): 201–219.

24 Woodward, R.M., Cole, B.E., Wallace, V.P. et al. (2002). Terahertz pulse imaging in reflection geometry of human skin cancer and skin tissue. *Physics in Medicine and Biology* **47**: 3853.

25 Wallace, V.P., Fitzgerald, A.J., Shankar, S. et al. (2004). Terahertz pulsed imaging of basal cell carcinoma ex vivo and in vivo. *British Journal of Dermatology* **151**: 424–432.

26 Brown, E.R., Taylor, Z.D., Tewari, P. et al. (2012). THz imaging of skin tissue; exploiting the strong reflectivity of liquid water. *35th International Conference on Infrared, Millimeter, and Terahertz Waves (IRMMW-THz)* (5–10 September 2010), Rome, Italy, 1–2.

27 Ashworth, P.C., Pickwell-MacPherson, E., Provenzano, E. et al. (2009). Terahertz pulsed spectroscopy of freshly excised human breast cancer. *Optics Express* **17** (15): 12444–12454.

28 Arbab, M.H., Dickey, T.C., Winebrenner, D.P. et al. (2011). Characterization of burn injuries using terahertz time-domain spectroscopy. *SPIE Advanced Biomedical and Clinical Diagnostic Systems IX*. San Francisco, CA, 78900Q–7.

29 Cosentino, A. (2016). Terahertz and cultural heritage science: examination of art and archaeology. *MDPI Technologies* **4** (1): 1–6.

30 Kawase, K., Ogawa, Y., Watanabe, U., and Inoue, H. (2003). Non-destructive terahertz imaging of illicit drugs using spectral fingerprints. *Optics Express* **11** (20): 2549.

31 Afsah-Hejri, L., Hajeb, P., Ara, P., and Ehsani, R. (2019). A comprehensive review on food applications of terahertz spectroscopy and imaging. *Comprehensive Reviews in Food Science and Food Safety* **18** (5): 1563–1621.

32 Akyildiz, I.F., Jornet, J.M., and Han, C. (2014). Terahertz band: next frontier for wireless communications. *Physical Communication* **12**: 16–32.

33 Kurner, T., Britz, D., and Nagatsuma, T. (2013). *IEEE Journal of Communications and Networks*, special issue on THz communications **15** (6).

34 Song, H.J. and Nagatsuma, T. (2011). Present and future of terahertz communications. *IEEE Transactions on Terahertz Science and Technology* **1** (1): 256–263.

35 Mumtaz, S., Jornet, J.M., Aulin, J. et al. (2017). Terahertz communication for vehicular networks. *IEEE Transactions On Vehicular Technology* **66** (7): 5617–5625.

36 Waters, J.W., Froidevaux, L., Read, W.G. et al. (1993). Stratospheric ClO and ozone from the microwave limb sounder on the upper atmosphere research satellite. *Nature* **362**: 597–602. https://doi.org/10.1038/362597a0.

37 Langer, W.D., Veluswamy, T., Pineda, J.L. et al. (2010). C+ detection of warm dark gas in diffuse clouds. *Astronomy and Astrophysics* **521**: L17.

38 Lellouch, E., Vinatier, S., Moreno, R. et al. (2010). Sounding of Titan's atmosphere at submillimeter wavelengths from an orbiting spacecraft. *Planetary and Space Science* **58**: 1724–1739.

39 Grasset, O., Dougherty, M.K., Coustenis, A. et al. (2013). JUpiter ICy moons explorer (JUICE): an ESA mission to orbit Ganymede and to characterize the Jupiter system. *Planetary and Space Science* **78**: 1–21.

40 Gordy, W. and Cook, R.L. (1984). *Microwave Molecular Spectroscopy*. New York, NY: Wiley.

41 Bonetti, J.A., Turner, A.D., Kenyon, M. et al. (2011). Transition edge sensor focal plane arrays for the BICEP2, keck, and spider CMB polarimeters. *IEEE Transactions On Applied Superconductivity* **21** (3): 219–222.

42 Jones, G. (2014). LEKID-based instruments for cosmic microwave background polarimetry. 2014 United States National Committee of URSI National Radio Science Meeting.

43 Siegel, P. (2008). IRMMW-THz in space: the golden age. Proceedings of the 2008 IRMMW Conference.

44 Ochiai, S., Kikuchi, K., Nishibori, T. et al. (2013). Receiver performance of the superconducting submillimeter-wave limb-emission sounder (SMILES) on the international space station. *IEEE Transactions on Geoscience and Remote Sensing* **51** (7): 3791–3802. https://doi.org/10.1109/TGRS.2012.2227758.

45 Walker, C., Kulesa, C., Bernasconi, P. et al. (2008). The stratospheric terahertz observatory (STO). Proceedings of the 19th International Symposium Space Terahertz Technology (April 2008) W. Wild, Ed., 28–30.

46 Risacher, C., Güsten, R., Stutzki, J. et al. (2016). First supra-THz heterodyne array receivers for astronomy with the SOFIA observatory. *IEEE Transactions on THz Science and Technology* **6** (2): 199–211.

47 Wu, D., Piepmeier, J.R., Esper, J. et al. (2019). IceCube: spaceflight demonstration of 883-GHz cloud radiometer for future science. Proceedings of SPIE conference on CubeSat sans SmallSats for Remote Sensing III (August 2019).

48 Thomas, B., Brandt, M., Walber, A. et al. (2012). Submillimetre-wave receiver developments for ICI onboard MetOP-SG and ice cloud remote sensing instruments. Proceedings of the 2012 IEEE International Geoscience and Remote Sensing Symposium.

49 Walker C.K., Kulesa, C., Goldsmith, P. et al. (2018). GUSTO: Gal/Xgal U/LDB spectroscopic-stratospheric terahertz observatory. Proceedings of the American Astronomical Society Meeting (January 2018), Article no. 231.05.

50 Hartogh, P., Barabash, S., Beaudin, G. et al. (2013). The submillimeter wave instrument on JUICE. European Planetary Science Congress.

51 Cooper, K.B., Roy, R.J., Dengler, R. et al. (2020). G-band radar for humidity and cloud remote sensing. *IEEE Transactions on Geoscience and Remote Sensing*: 1–12. https://doi.org/10.1109/TGRS.2020.2995325.

52 Smirnov, A., Pilipenko, S., Golubev, E. et al. (2018). Current status of the millimetron space observatory. Proceedings of the 42nd COSPAR Scientific Assembly (July 2018).

53 Wiedner, M.C., Mehdi, I., Baryshev, A. et al. (2018). A proposed heterodyne receiver for the origins space telescope. *IEEE Transactions On Terahertz Science and Technology* **8** (6): 558–571.

54 van Marrewijk, N., Mirzaei, B., Hayton, D. et al. (2015). Frequency locking and monitoring based on bi-directional terahertz radiation of a 3rd-order distributed feedback quantum cascade laser. *Journal of Infrared, Millimeter, and Terahertz Waves* **36** (12): 1210–1220. https://doi.org/10.1007/s10762-015-0210-4.

55 Williams, B.S., Xu, L.Y., Curwen, C.A. et al. (2017). Terahertz quantum-cascade metasurface VECSELs. *SPIE OPTO* https://doi.org/10.1117/12.2251447.

56 Richter, H., Wienold, M., Schrottke, L. et al. (2015). 4.7-THz local oscillator for the GREAT heterodyne spectrometer on SOFIA. *IEEE Transactions on THz Science and Technology* **5**: 539.

57 Curwen, C.A., Reno, J.L., and Williams, B.S. (2018). Terahertz quantum cascade VECSEL with watt-level output power. *Applied Physics Letters* **113** (1): 011104.

58 Kloosterman, J., Hayton, D.J., Ren, Y. et al. (2013). Hot electron bolometer heterodyne receiver with a 4.7-THz quantum cascade laser as a local oscillator. *Applied Physics Letters* **102**: 011123.

59 Camargo, E., Schellenberg, J., Bui, L., and Estella, N. (2016). F-band, GaN power amplifiers. Proceedings of the 2018 IEEE International Microwave Symposium, 753–756.

60 Radisic, V.K.M.K., Leong, H., Mei, X.B. et al. (2012). Power amplification at 0.65 THz using InP HEMTs. *IEEE Transactions on Microwave Theory and Techniques* **60** (3): 724–729.

61 Leong, K.M.K.H., Mei, X., Yoshida, W.H. et al. (2017). 850 GHz receiver and transmitter front-ends using InP HEMT. *IEEE Transactions on Terahertz Science and Technology* **7** (4): 466–475.

62 Mei, X., Yoshida, W., Lange, M. et al. (2015). First demonstration of amplification at 1 THz using 25-nm InP high electron mobility transistor process. *IEEE Electron Device Letters* **36** (4): 327–329.

63 Nagatsuma, T., Ito, H., and Ishibashi, T. (2009). High-power RF photodiodes and their applications. *Laser and Photonics Reviews* **3** (1–2): 123–137.

64 Nagatsuma, T. and Ito, H. (2011). High-power RF uni-traveling-carrier photodiodes (UTC-PDs) and their applications. In: *Advances in Photodiodes*. Gian Franco Dalla Betta, IntechOpen, 291–314. doi: 10.5772/14800. https://www.intechopen.com/books/advances-in-photodiodes/high-power-rf-uni-traveling-carrier-photodiodes-utc-pds-and-their-applications.

65 Al-Khalidi, A., Alharbi, K.H., Wang, J. et al. (2020). Resonant tunnelling diode terahertz sources with up to 1 mW output power in the J-band. *IEEE Tranactions On Terahertz Science and Technology* **10** (1): 99.

66 Suzuki, S., Asada, M., Teranishi, A. et al. (2010). Fundamental oscillators of resonant tunneling diodes above 1 THz at room temperature. *Applied Physics Letters* **97** (24): 242102.

67 Suzuki, S., Shiraishi, M., Shibayama, H., and Asada, M. (2013). High-power operation of terahertz oscillators with resonant tunneling diodes using impedance-matched antennas and array configuration. *IEEE Journal of Selected Topics in Quantum Electronics* **19** (1): 8500108.

68 Mehdi, I., Siles, J., Lee, C., and Schlecht, E. (2017). THz diode technology: status, prospects, and applications. *Proceedings of the IEEE* **105**: 990–1007.

69 Siles, J., Cooper, K.B., Lee, C. et al. (2018). A new generation of room-temperature frequency-multiplied sources with up to 10×higher output power in the 160-GHz–1.6-THz range. *IEEE Transactions On Terahertz Science and Technology* **8** (6): 596–604.

70 Schlecht, E., Siles, J.V., Lee, C. et al. (2014). Schottky diode based 1.2 THz receivers operating at room-temperature and below for planetary atmospheric sounding. *IEEE Transactions on Terahertz Science and Technology* **4** (6): 661–669.

71 Treuttel, J., Gatilova, L., Maestrini, A. et al. (2016). A 520–620 GHz schottky receiver front-end for planetary science and remote sensing with 1070K–1500K DSB noise temperature at room temperature. *IEEE Transactions On THz Science and Technology* **6** (1): 148–155.

72 Thomas, B., Maestrini, A., Ward, J. et al. (2010). Terahertz cooled sub-harmonic Schottky mixers for planetary atmospheres. Proceedings of the 5th ESA Workshop on Millimetre Wave Technology and Applications & 31st ESA Antenna Workshop.

73 Phillips, T.G. and Jefferts, K.B. (1973). A low temperature bolometer heterodyne receiver for millimeter wave astronomy. *Review of Scientific Instruments* **44**: 1009–1014.

74 Richards, P.L., Shen, T.M., Harris, R.E., and Lloyd, F.L. (1979). Quasiparticle heterodyne mixing in SIS tunnel junctions. *Applied Physics Letters* **34** (5): 345–347.

75 Uzawa, Y., Fujii, Y., Gonzalez, A. et al. (2013). Development and testing of Band 10 receivers for the ALMA project. *Physica C Superconductivity* **494**: 189–194.

76 Karpov, A., Miller, D., Stern, J.A. et al. (2009). Development of low noise THz SIS mixer using an array of Nb/Al-AlN/NbTiN junctions. *IEEE Transactions on Applied Superconductivity* **19**: 305–308.

77 Zmuidzinas, J., Leduc, H.G., Stern, J.A., and Cypher, S.R. (1994). 2-Junction tuning circuits for submillimeter SIS mixers. *IEEE Transactions on Microwave Theory and Techniques* **42**: 698–706.

78 Shurakov, A., Lobanov, Y., and Goltsman, G. (2016). Superconducting hot-electron bolometer: from the discovery of hot-electron phenomena to practical applications. *Superconducting Science and Technology* **29** (2): 023001.

79 Acharya, N., Novoselov, E., and Cherednichenko, S. (2019). Analysis of the broad IF-band performance of MgB2 HEB mixers. *IEEE Transactions on Terahertz Science and Technology* **9** (6): 565–571. https://doi.org/10.1109/TTHZ.2019.2945203.

80 Bonev, B.P., Mumma, M.J., Villanueva, G.L. et al. (2007). A search for variation in the H2O ortho-para ratio and rotational temperature in the inner coma of Comet C/2004 Q2 (Machholz). *The Astrophysical Journal Letters* **661** (1): L97.

81 Altwegg, K., Balsiger, H., Bar-Nun, A. et al. (2015). 67P/Churyumov-Gerasimenko, a Jupiter family comet with a high D/H ratio. *Science* **347** (6220): 1261952.

82 Hartogh, P., Lis, D.C., Bockelee-Morvan, D. et al. (2011). Ocean-like water in the Jupiter-family comet 103P/Hartley 2. *Nature* **478**: 218–220.

83 Reck, T., Jung-Kubiak, C., Siles, J.V. et al. (2015). An eight pixel transceiver array for submillimeter-wave radar. *IEEE Transactions on Terahertz Science and Technology* **5** (2): 197–206.

84 Goldsmith, P. (2017). Sub-millimeter heterodyne focal-plane arrays for high-resolution astronomical spectroscopy. *The Radio Science Bulletin* **362** (53): 53–73.

85 Battaglia, A., Westbrook, C.D., Kneifel, S. et al. (2014). G band atmospheric radars: new frontiers in cloud physics. *Atmospheric Measurement Techniques* **7**: 1527–1546.

86 Cooper, K.B., Roy, R.J., Lebsock, M. et al. (2019). Validation measurements of humidity profiling in rain using a 170 GHz differential absorption radar. *Proceedings of the 2019 International Symposium on Space Terahertz Technology, ISSTT 2019 – 30th International Symposium on Space Terahertz Technology, Proceedings*, 227.

Index

Fundamentals of Terahertz Devices and Applications, First Edition. Edited by Dimitris Pavlidis.

Companion website: www.wiley.com/go/Pavlidis/FundamentalsofTHz

q

r